HANDBOOK of
**CULTURAL
GEOGRAPHY**

HANDBOOK of
CULTURAL
GEOGRAPHY

Edited by
KAY ANDERSON
MONA DOMOSH
STEVE PILE
and
NIGEL THRIFT

SAGE Publications
London • Thousand Oaks • New Delhi

First published 2003

SAGE Publications Ltd
6 Bonhill Street
London EC2A 4PU

SAGE Publications Inc.
2455 Teller Road
Thousand Oaks, California 91320

SAGE Publications India Pvt Ltd
32, M-Block Market
Greater Kailash - I
New Delhi 110 048

British Library Cataloguing in Publication data

A catalogue record for this book is available from
the British Library

ISBN 0 7619 6925 X

Library of Congress Control Number 2002102285

Typeset by C&M Digitals (P) Ltd., Chennai, India
Printed in Great Britain by The Cromwell Press, Trowbridge, Wiltshire

Contents

List of Figures

Chapter 26

Introduction to Section 9

Every effort has been made to trace all copyright holders, but if any have been overlooked, or if any additional information can be given, the publishers will make the necessary amendments at the first opportunity.

Notes on Contributors

John Agnew is Professor and Chair in the Department of Geography at UCLA, Los Angeles, USA. He is the author of *Place and Politics* (1987), *Geopolitics: Re-visioning World Politics* (1998) and *Place and Politics in Modern Italy* (2002). He is the co-editor of *The Power of Place* (1989), *American Space/American Place* (2002) and the *Companion to Political Geography* (2002).

Kay Anderson is Professor of Geography at Durham University where she teaches various threads of cultural geography, including colonial cultures of nature, race and identity politics. She has a long history of intellectual engagement with these issues in Australia and Canada, signalled by her book *Vancouver's Chinatown* (1991) and numerous other publications, as well as the genealogy of the culture concept and the subdiscipline *Cultural Geographies* (1999). She is also an editor of *Progress in Human Geography*.

Trevor J. Barnes is Professor of Geography at the University of British Columbia, Vancouver, Canada, where he has been since 1983. He writes primarily on economic geography, and most recently about its history. He is the author or editor of seven books including *Logics of Dislocation* (1996), *The New Industrial Geography* (with Meric Gertler, 1999) and *A Companion to Economic Geography* (with Eric Sheppard, 2000).

Liz Bondi is Professor of Social Geography at the University of Edinburgh, Scotland. She is currently conducting research on voluntary sector counselling services in Scotland, which develops out of her long-standing concern with gender dimensions of urban social change. She has published extensively in feminist geography and is founding editor of *Gender, Place and Culture*.

Alastair Bonnett is Reader in Social Geography in the Department of Geography at the University of Newcastle. He is the author of *Anti-racism* (2000), *White Identities* (2000) and *How to Argue* (2001) and is currently researching the construction of the idea of 'the west' within the Soviet Union.

Michael Brown is Associate Professor of Geography at the University of Washington, Seattle, USA. His interests include urban political and health geographies. His work has considered issues of sexuality through research on AIDS politics, the spatial metaphor of the closet and questions of care.

Noel Castree is Reader in Geography at Manchester University. Co-editor (with Bruce Braun) of *Remaking Reality* (1998) and *Social Nature* (2001), his interests are in the political economy of environmental change. His current research focuses on the 'dematerialization' of nature and the transnational exchange and accumulation of genetic information.

Daniel Clayton is Lecturer in Human Geography at the University of St Andrews, where he teaches courses on the geographies of colonialism and postcolonialism. He is the author of *Islands of Truth: The Imperial Fashioning of Vancouver Island* (2000), and is currently working on a book entitled *Colonialism's Geographies* and a funded research project entitled 'Tropicality in decolonisation: Pierre Gourou and French Indochina, 1926–1972'.

Denis Cosgrove is currently Alexander von Humboldt Professor of Geography at the University of California Los Angeles. His work focuses on the roles of vision and graphic images in the western geographical imagination and he has published widely on these issues. His books include *Social Formation and Symbolic Landscape* (1984; 1997), *The Palladian Landscape* (1993), *Mappings* (1999) and *Apollo's Eye: A Cartographic Genealogy of the Earth in the Western Imagination* (2001).

Tim Cresswell teaches Social and Cultural Geography at the University of Wales, Aberystwyth. He is the author of *In Place/Out of Place: Geography, Ideology and Transgression* (1996) and *The Tramp in America* (2001). He is co-editor (with Ginette Verstraete) of *The Politics of Place* (2002) and (with Deborah Dixon) *Engaging Film* (2002). His current research focuses on the politics of mobility from the body to the globe.

Simon Dalby is Professor of Geography and Political Economy at Carleton University in Ottawa. His research interests include political ecology, critical geopolitics, sustainability and environmental security. He is author of *Creating the Second Cold War* (1990) and *Environmental Security* (2002). He is also co-editor of *Rethinking Geopolitics* (1998) and *The Geopolitics Reader* (second edition, 2003).

Joyce Davidson is currently conducting postdoctoral research on experience and meanings of 'biophobias' at the Institute for Health Research, Lancaster University, UK. Her previous work on agoraphobia has been published in journals such as *Area* and *Sociology of Health and Illness*, and she has contributed chapters to *The Ethics of Place* and *Geographies of Women's Health*.

Mona Domosh is Professor of Geography at Dartmouth College. Her research lies at the intersection of cultural, feminist and historical geography. She is the author of *Invented Cities: The Creation of Landscape in Nineteenth-Century New York and Boston* (1996), and the co-author of *Putting Women in Place: Feminist Geographers Make Sense of the World* (1999), and *The Human Mosaic: A Thematic Introduction to Cultural Geography* (2001).

Isabel Dyck is Associate Professor in the School of Rehabilitation Sciences at the University of British Columbia. Her research and teaching focus on geographies of disability, feminist analyses of body and identity, resettlement and healthcare access issues for immigrant families, and qualitative methodology. Recent research explores the experiences of women with chronic illness; family and femininity in the lives of immigrant mothers and daughters; and the home as a site of long-term care.

Jody Emel is Associate Professor of Geography at Clark University. She teaches courses in resource geography, water resource management and materialist ecofeminism. Her research projects focus on the political economy of natural resource development and social activism, and corporate responsibility in the animal cloning industry. She and Jennifer Wolch are co-editors of *Animal Geographies: Place, Politics and Identity in the Nature–Culture Borderlands* (1998).

Meric S. Gertler is Professor of Geography, Goldring Chair in Canadian Studies and a member of the Centre for International Studies at the University of Toronto. He studies industrial practices, institutions and cultures in North America and Europe. Among his recent publications are *The New Industrial Geography: Regions, Regulation and Institutions* (1999, with Trevor Barnes) and *The Oxford Handbook of Economic Geography* (2000, with Gordon Clark and Maryann Feldman).

Nicky Gregson is Reader in Geography at Sheffield University. She has research interests in consumption and material culture and is the co-author of *Second Hand Worlds* (2002, with Louise Crewe) and *Servicing the Middle Classes* (1994, with Michelle Lowe) and was a member of the WGSG *Feminist Geographies* collective (1997).

Francis Harvey is Assistant Professor at the University of Minnesota and teaches Geography and GIScience at the University of Minnesota. His recent work emphasizes the political dimensions of local government bodies' geographic information practices. Working with grassroots groups, he is seeking ways to facilitate decision making processes in communities that engender learning from diverse forms of knowledge while respecting sensitive local knowledges. This work specifically addresses the political problems of knowledge representation.

Steve Hinchliffe works in the Faculty of Social Sciences at the Open University, England. He is co-editor of a number of texts including *The Natural and the Social* (2000), *Understanding Environmental Issues* (2003) and *Environmental Responses* (2003). He has published a range of papers on the place of nature in contemporary western society and is currently working on a project entitled 'Living Cities'.

Richard Howitt teaches Human Geography at Macquarie University, Sydney, Australia. He works closely with indigenous peoples throughout Australia on issues of the social and environmental impacts of mining, infrastructure projects and regional development. He is currently working with native title claimants in South Australia on negotiations with the state. In 1999, he was awarded the Australian Award for University Teaching (Social Science).

Peter Jackson is Professor of Human Geography at the University of Sheffield. His research focuses on the cultural politics of identity and the geographies of contemporary consumption. Recent projects include studies of men's 'lifestyle' magazines, domestic consumption and transnational commodity cultures. Publications include *Maps of Meaning* (reprinted 1992), *Constructions of Race, Place and Nation* (1993), *Shopping, Place and Identity* (1998), *Commercial Cultures* (2000) and *Making Sense of Men's Magazines* (2001).

Jane M. Jacobs is a Lecturer with the Department of Geography, University of Edinburgh. She has published widely in the area of cultural geography, especially in the cultural politics of cities, contested heritage and postcolonial spaces. She is the author of *Edge of Empire* (1996); co-author, with Ken Gelder, of *Uncanny Australia* (1988); and co-editor, with Ruth Fincher, of *Cities of Difference* (1998).

John Paul Jones III is Professor of Geography and Co-director of the Committee on Social Theory at the University of Kentucky. His theoretical research has examined spatial epistemology, whiteness, objectivity, anti-essentialist identity theory, theories of space and representation, and poststructuralist geography, among other topics. He also writes widely on geographic methodology. He served as editor of the *Annals of the Association of American Geographers* from 1997 to 2002.

Anthony D. King is Professor of Art History and of Sociology at the State University of New York at Binghamton. He has recently published essays in *Global Futures* (edited by J.N. Pietersee, 2000) and *China and Postmodernism* (edited by A. Dirlik and X. Zhang, 2000). With Tom Markus, he co-edits the Routledge Archi*Text* series on architecture and social/cultural theory, for which he is preparing *Spaces of Global Cultures*.

Larry Knopp is Professor and Head of Geography at the University of Minnesota, Duluth, USA. His interests cross many of geography's subfields but have tended to coalesce around questions of power as they relate to the spatiality of sexuality, gender and class.

Audrey Kobayashi is Professor of Geography and Women's Studies at Queen's University, Kingston, Ontario. Her research interests include human geography theories, racism, gender, immigration, geography and law, and employment equity. She has published widely on these themes in a range of journals and popular articles. She also works as an anti-racist activist, and as a consultant in the areas of employment equity and anti-racism.

Robyn Longhurst is Senior Lecturer in Geography at the University of Waikato, Hamilton, New Zealand. She is author of *Bodies: Exploring Fluid Boundaries* (2001) and a co-author of *Pleasure Zones: Bodies, Cities, Spaces* (2001). She is interested in feminist cultural geography and is currently researching spatial aspects of body size.

Linda McDowell is Professor of Economic Geography at University College London. She is the author of numerous papers and books about economic change, class and gender relations in the UK including *Capital Culture* (1997), *Gender, Identity and Place* (1999) and *Young Men Leaving School: White Working Class Masculinity* (2001). She is currently investigating the working lives of European migrant women in Britain in the late 1940s and 1950s using oral histories.

Cheryl McEwan is Lecturer in Human Geography at the University of Birmingham, UK. She is the author of *Gender, Geography and Empire* (2000) and numerous articles that draw on postcolonial and feminist theories in both contemporary and historical contexts. She is also co-editor (with Alison Blunt) of *Postcolonial Geographies* (2002).

David Matless is Reader in Cultural Geography at the University of Nottingham. He is the author of *Landscape and Englishness* (1998) and co-editor of *The Place of Music* (1998) and *Geographies of British Modernity* (2003). His current research focuses on cultures of nature in mid-twentieth-century England, relations of science and landscape, and the work of ecologist and artist Marietta Pallis.

Don Mitchell is Professor in the Department of Geography at Syracuse University. He is the author of *The Lie of the Land: Migrant Workers and the California Landscape* (1996) and *Cultural Geography: A Critical Introduction* (2000). He is the Director of the People's Geography Project. His research focuses on labour, landscape, homelessness and public space.

Katharyne Mitchell is Associate Professor in the Department of Geography at the University of Washington. She has published extensively in the area of migration, urban geography and transnational studies. Her co-edited book (with Gerard Toal and John Agnew) *A Companion Guide to Political Geography*, is to be published in 2002. Mitchell's current work centres on the impact of transnational migration on conceptions of education, with a particular focus on how children are educated to become citizens of a particular nation-state.

Pamela Moss is Professor in the Studies in Policy and Practice Program at the University of Victoria, Canada. Her research coalesces around themes of power and body emerging from women's experiences of changing environments – women with chronic illness, women's collective autobiographies, and organization of support services for women in crisis. She edited *Placing Autobiography in Geography* (2001) and *Feminist Geography in Practice* (2002). She is also active in feminist community politics around issues in women's housing.

Anoop Nayak is Lecturer in Social and Cultural Geography in the Department of Geography at the University of Newcastle. He is the editor of *Invisible Europeans* (1993, with Les Back) and has published papers on racialization and masculinity in *Body and Society*, *Gender and Education*, *International Journal of Sociology in Education* and many other journals.

Clare Newstead is a PhD candidate in the Department of Geography at the University of Washington. She is currently completing her dissertation on the political implications of regional economic intergration in the Caribbean. In particular she is interested in the struggles of regional social movements to negotiate a public political space at the supranational level.

Anssi Paasi is Professor of Geography at the University of Oulu in Finland. He has written numerous articles on the problems of region and territory building and the social and cultural construction of boundaries and spatial identities. His recent books include *Territories, Boundaries and Consciousness: The Changing Geographies of the Finnish–Russian Border* (1996) and *J.G. Granö: Pure Geography* (editor, with Olavi Granö, 1997).

Steve Pile is Senior Lecturer in Human Geography in the Faculty of Social Sciences, The Open University. He is author of *The Body and the City* (1996) and has co-edited a number of books, including *City A–Z* (2000, with Nigel Thrift) and *Social Change* (2002, with Tim Jordan). He is currently researching a book on affect and city life.

Elspeth Probyn is Associate Professor in the Department of Gender Studies at the University of Sydney. Her publications include *Carnal Appetites: FoodSexIdentity* (2000), *Outside Belongings* (1996), *Sexy Bodies: The Strange Carnalities of Feminism* (co-editor with Elizabeth Grosz, 1995), and *Sexing the Self: Gendered Positions in Cultural Studies* (1993). She is currently working on a book entitled *Dis/connect: Bodies Affect Writing* about forms of writing and theories of affect.

Carolina K. Reid is a PhD candidate in Geography at the University of Washington. She is interested in the interlinkages between welfare and urban housing policy in the United States. Her dissertation focuses on the home ownership experiences of low-income and immigrant families.

Jennifer Robinson is Lecturer in Geography at the Open University. Her book *The Power of Apartheid* (1996) explored the relations between space and power in the construction of apartheid cities, and in the emergence of a post-apartheid urban form. More recently, her work aims to develop a postcolonial critique of urban theory and urban development policy. Other interests include feminist political theory. Jenny is also joint editor of *Geoforum*.

Joanne P. Sharp is Senior Lecturer in Geography at the University of Glasgow. Her research interests are in political, cultural and feminist geography with a particular interest in popular geopolitics. She recently published a monograph on the role of the media in the construction of US political culture as *Condensing the Cold War: Reader's Digest and American Identity* (2000).

David Slater is Professor of Social and Political Geography at Loughborough University, England. He is author of *Territory and State Power in Latin America* (1989), editor of *Social Movements and Political Change in Latin America* (1994) and co-editor of *The American Century* (1999). He is also an editor of *Political Geography*.

Don Slater is Reader in Sociology at the London School of Economics. His main research concerns are consumption and consumer culture; the relation between culture and economy; and ethnographies of new media. Recent publications include *Consumer Culture and Modernity* (1997), *The Internet: An Ethnographic Approach* (with Daniel Miller, 2000) and *Market Society: Markets and Modern Social Theory* (with Fran Tonkiss, 2001).

Matthew Sparke is an Associate Professor at the University of Washington. He is the author of *Hyphen-Nation-States: Critical Geographies of Displacement and Disjuncture* (forthcoming), and is currently working on a National Science Foundation CAREER project to integrate his research on the transnationalization of civil society with a series of educational outreach initiatives to schools in minority neighbourhoods of Seattle.

Ulf Strohmayer is Professor of Geography at the National University of Ireland, Galway. He has studied and worked in his native Germany, Sweden, France, the United States and Wales before moving to the west of Ireland. In addition to being a passionate geographer, he is interested in social philosophies and architecture. Combining all three interests, he is currently finishing a book entitled *Modernity and the Urban Geography of Paris, 1550–2000*.

Sandra Suchet-Pearson is a Lecturer in Human Geography, Macquarie University, Sydney, Australia. She has worked closely with indigenous and local communities in Australia, Canada and southern Africa. Her recent doctoral work considered the nature and implications of indigenous involvement in wildlife management and involved fieldwork in several areas.

Nigel Thrift is Professor in the School of Geographical Sciences at the University of Bristol. His main interests are in social and cultural theory (especially the development of non-representational theories), the cultural impacts of science and technology, cultural economy and the joys of performance. His most recent publications include *Thinking Space* (2000, co-edited with Mike Crang), *TimeSpace* (2001, co-edited with Jon May) and *Cities* (2002, with Ash Amin).

Adam Tickell is Professor of Human Geography at the University of Bristol and has worked at Leeds, Manchester and Southampton. His main research interests lie in the cultural and political economies of finance and the seemingly endless restructuring of the state, and he is currently exploring the transformative capacities of neoliberalism.

Gerard Toal (Gearóid Ó Tuathail) is Director of the Masters of Public and International Affairs at Virginia Tech. He is the author of *Critical Geopolitics* (1996) and a co-editor of *A Companion to Political Geography* (2002) and *The Geopolitics Reader* (second edition, 2003) among other works. He is Associate Editor of the journal *Geopolitics*. His current research interests are in the critical geopolitics of world risk society, and US foreign policy towards the Balkans in the 1990s.

Michael Watts is Chancellor's Professor of Geography and Director of the Institute of International Studies at the University of California, Berkeley, where he has taught for over 20 years. He has published widely in development and geography and on Nigerian history and politics. He is currently working on a book on oil and politics in Nigeria.

Sarah Whatmore is Professor of Geography at the Open University. She has written widely on the labours of division that mark nature off from society, particularly in relation to the socio-material complications of agriculture and food, biodiversity and biotechnology. Her most recent book on these issues is *Hybrid Geographies* (2002).

Chris Wilbert is Lecturer in Geography at Anglia Polytechnic University, England. He is the co-editor (with Chris Philo) of *Animal Spaces, Beastly Places: New Geographies of Human–Animal Interactions*. Other publications focus on non-human animals and agency, leisure practices in virtual spaces, and environmental politics. He is also a member of the Animal Studies Group in England, which promotes cultural and social studies of animals.

Jennifer Wolch is Professor of Geography at the University of Southern California, where she co-directs the Sustainable Cities Program and conducts research on cultural diversity and attitudes toward animals, and the impacts of urbanization and urban design on animal life. With Jody Emel, she is co-editor of *Animal Geographies: Place, Politics and Identity in the Nature–Culture Borderlands* (1998).

Brenda S.A. Yeoh is Associate Professor in the Department of Geography at the National University of Singapore. She teaches social and historical geography and her research foci include the politics of space in colonial and postcolonial cities, and gender, migration and transnational communities. She has published a number of scholarly journal papers and books in these areas, including *Contesting Space: Power Relations and the Urban Built Environment in Colonial Singapore* (1996).

Preface

Kay Anderson, Mona Domosh, Steve Pile and Nigel Thrift

The *Handbook of Cultural Geography* is one of a series of handbooks published by Sage. For Sage these books represent the 'state of the art' in their specific fields, and they are pitched at an audience who have a degree of familiarity with the subject, but would like to know more about a specific topic or extend their understanding of the breadth of work in the area. For us – as editors – this has set some very interesting puzzles. It is quite a challenge to think about the state of the art in any one field, but for us the deeper problem was that we found it very hard to delineate 'our' field. Indeed, if there is one thing about cultural geography that we know for sure, it is that it is not a field! As we debated this 'border' problem, it became clear to us that the field of cultural geography was better marked both by its disruption of the usual academic boundaries and by its insatiable enthusiasm for engaging new issues and ideas – whatever their source. As we began to sketch out possible contents and structures for the book, we decided that we should try to make this particular *Handbook* reflect this interdisciplinarity and this passion for thinking more broadly about what counts as geography. To this end, we contacted a number of people and asked them to edit sections that reflected both the breadth of cultural geography's thematic interests, and also some of the most interesting areas of debate and research within these (let's call them) 'fields of engagement'. Of course, the result is far from comprehensive in either content or debate. However, our purpose was to reflect the varieties of cultural geographies being undertaken and suggest that there is far more to be done.

The *Handbook of Cultural Geography* is not a fixed map of a discipline that has a clearly identifiable boundary and a terrain permanently marked out for itself. Instead, this book contains leading representatives of the kinds of issues that have preoccupied cultural geographers (and, like as not, will do in the future) and of the kinds of debates that geographers are now engaged in. The *Handbook* is not, then, a signpost that the traveller will pass on their way somewhere else, but a resource for travellers along a journey. What we are suggesting here is that cultural geography is less of a fixed inventory of objects, and more a way of changing how we understand the objects of knowledge. What is distinctive about cultural geography – we'll return to the cultural part in a moment – is that it brings a geographical imagination to bear on these objects. This imagination is not simply attendant on the whereness of things (where on earth?!), as geography is often caricatured. Whereness now provokes a whole series of questions about the spatial relations that constitute things, about the movements and gatherings of things, and about the very constitution of space, place and nature. Geographers wonder why things are where they are, why they are represented in particular ways, how things move and settle, how they are brought together and kept apart, and – and this is crucial – how this came about. Cultural geography is, then, a style of thought – a way of expanding and illuminating geographies. Or, perhaps better, cultural geography is a style of thought that gathers to it a wide variety of questions, and ways of answering.

Cultural geography as a style of thought is not, to be sure, a singular worldview (not one way of thinking about the world; not a fixed question and answer session, as if in the latest game show), but a place from which to ask valid and urgent questions of the world; one in which the geographical is seen as constitutive of how the world is 'made up'. More than this, it is also a small 'p' politics of the object (of all possible objects of knowing and unknowing) and of geographical relationships. It intends to change our minds about how those geographies came about – and, thereby, about what possibilities there are for changing things in the present, and in the future. This may seem hopelessly naive, but it is a modest endeavour. The cultural has modified the geographical, making it possible to study more and more 'things', but also to bring more and more 'things' under critical scrutiny. In some small way, then, it is about democratizing understanding, about being able to look to the world for the different things that are going on there. And to learn lessons from it. It is therefore no accident that this book has sections that seem to belong to other books – on the economy, maybe, or on the social (and we even begin the book with these).

The *Handbook of Cultural Geography* ultimately is a slightly (and deliberately) unruly affair. In this, we hope to surprise readers, to help them appreciate not just what is out there, but what else can be done – achieved – with these ideas. At turns, this book might intrigue, annoy, frustrate or surprise – but this is exactly what cultural geography is about. There is really something to get to grips with here. To this extent, the *Handbook* is also an invitation. We invite you to engage with the ideas that are presented. We invite you to share the passions of this book. And, perhaps most importantly, we invite you to do cultural geography for yourself, to change it – to change the very ways in which we think the cultural and the geographical, and how it is that we can do them.

Finally, the editors have some debts of gratitude to acknowledge. In the first place, we would like to express our appreciation for the outstanding hard work of our editor at Sage, Robert Rojek. From the earliest meetings for this project, he would leave us by saying that this was going to be 'a fine thing'. His commitment to this book as a fine thing has been shown both in the ways he has made it possible for this ambitious work to actually happen, and also in his intellectual engagement with the project in its fine grain and its big picture. As we have said, from the outset we wished to make this book a collaboration amongst geographers. So we would like to thank all the section editors for their energy and enthusiasm for this project. For most, this has required patience and endurance as much as inspiration and passion. Ultimately, we hope, this has been as rewarding an experience for them as it has been for us. Last, but far from least, we would like to thank Michèle Marsh. The *Handbook* has involved a constant stream of draft chapters, which have had to be collated and distributed. This process would have been a logistical nightmare but for Michèle's secretarial efficiency and her cheerful enthusiasm.

A Rough Guide

Kay Anderson, Mona Domosh, Steve Pile and Nigel Thrift

GEOGRAPHY'S CULTURE

This is a *Handbook of Cultural Geography*. It is, therefore, reasonable to expect some working definition of what cultural geography *is*, or (really) was, and how it got here. Even working definitions can end up being tombstones …

Figure 1   *Here Lies Cultural Geography,*
Born 1925, Died 2002.
In Loving Memory

In this introduction, we are not going to provide a history – or life story – of cultural geography, as if it was a character in an academic drama. Partly, our reluctance to do this has to do with a refusal to institute a canon of work that is identifiably cultural geography. Partly, this reluctance is about installing (or not) certain figures as foundational to the discipline, but it is also about seeing cultural geography as being motivated differently in different places. We see cultural geography, therefore, as a contested terrain of debate – in fact, a far-flung set of debates. We do not wish, that is, to stitch these different debates together into a seamless story: birth, life … and ultimately death. Cultural geography is a living tradition of disagreements, passions, commitments and enthusiasms. It is something of this that we wish to evoke in this book.

For the purposes of this book, cultural geography is better thought of as a series of intellectual – and, at core, politicized – engagements with the world. It is a style of thought, fixed in neither time nor space. It is nevertheless possible to pull out certain strands that go to make up this style of thought. So, we would like to begin this introduction by setting out some of these strands in a little depth. These strands are multiform, and they are far from discrete, being instead knotted together in ways that bind various kinds of geographies in variable geometries and patterns. While this might offer a disturbingly flexible 'map' of cultural geography, we would like to make a particular case for cultural geography. In the next section, we would like to argue that – at this point in time – the most radical of these agendas has involved the injunction to think spatially about the world. Later, we will exemplify these issues through the use of four 'vignettes': each will show how various geographies can be read out of particular events, objects, situations and locations. In this way, we hope to introduce what it is that cultural geography *does*: not as a closed and bounded field of academic endeavour, but as an open and engaged style of thought.

Our working definition of cultural geography, then, is more related to an unfolding intellectual terrain than it is to a long path, littered with iconic texts that mark the discipline's sure-footed academic journey. As we begin to think about cultural geography, and especially as we read over the contributions to this book, we are better able to see that these unfolding arguments and debates are about specific 'geographical problems', rather than keeping within an identifiable terrain of thought. For us, we can see that five particular themes are evident across this book – and cultural geography as a whole. No doubt these are not the only ones, but they make sense in terms of the intellectual frontiers this *Handbook* opens up. They are:

- culture as **distribution of things**
- culture as **a way of life**
- culture as **meaning**
- culture as **doing**
- culture as **power.**

Let us think further about the various strands of thought that have gone to make up cultural geography. We have noted that these strands are bound together in different ways, and you will find various aspects of the thinking described in each of the chapters – in sometimes more and less apparent ways. These strands are long standing and represent aspects of cultural geography as both a style of thought and a substantive arena of research and debate. We have decided to stabilize these threads on aspects of culture, as a convenient heuristic.

Culture as distribution of things   All groups of people produce cultural artefacts, from the everyday personal items we see around us like furniture and clothing, to the larger-scale and more public artefacts such as buildings and roads. But how exactly do we understand the relationships between the patterning of those artefacts and the values, livelihoods, beliefs and identities of the cultures who have produced them? What really can the pattern of material artefacts tell us about the social, economic and political dynamics of cultures? These concerns are central to cultural geography.

In the first half of the twentieth century, cultural geographers concentrated on charting the movement and locations of material artefacts in the landscape. Some geographers of the Berkeley School of cultural geography spent considerable energy mapping the locations of certain key, and primarily vernacular, artefacts within the United States in order to delineate cultural regions, that is, regions that expressed a defined cultural homogeneity. Fred Kniffen, for example, used folk housing styles as a diagnostic for deciphering cultural regions. But as many cultural geographers have reminded us since, housing, and by implication material artefacts, can tell us much more than this about culture. Studying the distribution of cultural artefacts involves asking *whose* artefacts, *how* did they get put in place, and *for what reasons*. For example, housing style and decor are among the most public forms of identity expression; they can reveal the economic class of the inhabitants, their ethnicity, perhaps their attitudes toward nature, their sense of belonging in a community. Open the door to the interior, and its design can suggest the gendered relationships in the family, their social and economic aspirations, their work situations. Conversely, housing often reveals more about societal structures than about individual identity. The disinvestment in housing that is evident in many inner-city neighbourhoods reveals a common pattern of capital mobility by the large stakeholders in urban real estate. Charting the distribution of these neighbourhoods can tell us as much about the workings of post-Fordist capitalism as it does about the cultural attributes of its inhabitants. Deprived of formal expression in the design of artefacts, people may express their views through the graffiti that decorate the walls and buildings in these neighbourhoods – symbols and signs of community and resistance.

Cultural geographies of artefacts, then, are as much about the graffiti themselves as they are about the locations of the graffiti-marked buildings; as much about the idea of home as they are about the distribution of housing; and as much about the diversity within culture as they are about cultures *per se*. These geographies ask *why* and *how*, as much as *where* and *when*.

Culture as a way of life   Perhaps one of the most persistent of the democratizing moves marking out the field of cultural geography, at least in its Anglo/American/Australasian trajectories, has been that profoundly relativist appreciation of the diverse properties of people in place. From studies of the *genre de vie* of regionally based groups – typically rural and 'traditional' societies – through to more recent forays into the signifying systems of all social groups – 'us' as well as 'them', and back again to 'them' with the tools of postcolonial criticism – a consistent focus has been the assortment of practices that constitute people and place, life and landscape. The values, beliefs, languages, meanings and practices that make up people's 'ways of life', however mobile and mutable, have been the stock-in-trade of cultural geography for close on a century.

Of course much has been made, and quite rightly, of the epistemological and political differences distinguishing various conceptions of people's diverse 'lifeworlds'. The shift in cultural geography's focus from the exotic livelihoods and landscape imprints of 'other', usually non-western groups, to the habits of mind and practice of western social groups, has developed in tandem with poststructuralist critiques of knowledge in human geography more generally. It is hard to overestimate the impact of the critique of essentialism on cultural geography, and the subject's inherent engagement with the different lifeways that transform abstract space into lived worlds. Certainly, culture's constituency has been radically enlarged and diversified, and many scholars would now accept that a 'way of life' can be defined as much around an oppositional identity politics in an urban housing estate as around a system of farming the land in rural Minnesota. Refinements are ongoing too, to today, in how we might think about the 'actants' and agencies out of which 'world-building' takes place. But one detects, also, an intriguing continuity of ethical commitment in the efforts that venture to foreground the lived lives of geography's diverse beings, and all that fuses and fragments them.

Culture as meaning   Understanding the meanings of particular landscapes and places is no small matter: battles are waged every day over control of religious sites; city streets can be thronged with groups protesting the loss of public space; and feelings of grief and loss fuel political conflicts over the design and siting of war memorials. Yet understanding how and why landscapes become embedded with individual and cultural meaning and in turn create new meanings is fraught with complications. What exactly do we mean by *meaning*? Does it refer to individual emotions, experiences and memories, or to group values, attachments and ideals? How do we interpret meaning from place? And whose meanings are given precedence in those interpretations? Interpretation of 'ordinary' landscapes – places that we often take for granted in our everyday life, like our homes and towns – requires in-depth, often intimate, knowledge of local history, cultural values and economic structures. Interpretation of symbolic landscapes – places that are imbued with special meaning beyond the everyday – requires investigations at a different scale. A city's skyline often becomes the dominant icon for the city, and can resonate as a symbol of that city within the nation and beyond. Deciphering the meanings in these cases requires delving into issues of civic pride, national identity and global circulation – into, therefore, urban and national political agendas, constructions of local and national identity, and the global market of image circulation.

Finding an appropriate method for such an investigation is a vexing issue. Some geographers have adapted methods from literary criticism to provide a textual 'reading' of landscape, and with it the potential for a deconstructive interpretation; others have borrowed the technique of iconography from art history in order to interpret landscape as visual image.

But these approaches have been criticized as too reliant on the discursive, ignoring the very real material conditions of landscape production and thereby erasing the meanings of its producers. Conversely, some geographers call for studies of meaning that focus on the non-material aspects of landscape, that is, on people's intimate experiences of and performances in places. These methodological debates only underscore the importance of trying to understand the powerful, and often conflict-ridden, relationship between cultural meanings and the places and landscapes that embody, reflect and shape those meanings.

Culture as doing   For some, the idea that culture is 'done' is associated with ideas within Marxism, and particularly Marx's understanding of consciousness as practical. Following Epicurus, it seems that Marx thought of matter as too recalcitrant or unruly to be part of an order of natural necessity; matter therefore had a kind of vitalizing property. But, under the influence of Hegel, Marx went on to ally this recalcitrance and vitality with human self-consciousness, so losing touch with an appreciation of agency within nature and producing a division which still plagues so many accounts of culture. So on the one hand, there is a philosophical anthropology in which the 'swerve of the atom' – an idea taken from Epicurus to describe deviation from assigned paths – becomes man's capacity to resist physical form. On the other hand, there is a view in which the 'swerve' is much more deeply embedded in materiality, in which the division between nature and culture is challenged, and where agency is distributed across all kinds of hybrid actors.

Recent work in cultural geography has similarly moved away from philosophical anthropology towards more general notions of life that can reinstate the Epicurean notion of materiality (more and more often, under the general heading of performance), so attempting to reinstate the richness of the world which is so often rendered in conventional academic accounts as 'just' everyday life. Such attempts have taken up various traditions, as diverse as certain kinds of phenomenology, certain ethnomethodological attempts to follow outcomes of action and affect, and certain kinds of virtualism, reworking them through notions like 'habitus', 'actor network' and immanent 'becoming' to produce new modes of thinking and harrying space which, at one and the same time, create new spaces. Such cultural geography tends to be hyperactive, caught up especially in forms of life as varied as theatrical performance and various kinds of protest, forms of life which tend to refuse the directive cultural politics so beloved of academics for a quicksilver cultural politics that is not only more tenuous but also, in its emphasis on the mobility of circumstance, more alert to differences and so more able to follow some of those paths to freedom.

Culture as power   In some ways, this theme stands out from the other four that we have identified. This is because the analysis of 'power' is, at least, implicit in each of these themes. Meaning, for example, is contested. Geographers have asked questions about how artefacts get made, how they get from one place to another, and (sometimes) who benefits from all this trading and placing. Often, power itself is not the object of cultural geographical analysis. Nevertheless, critiques of power relations are usually uppermost in cultural geographical work, much of which is intended to be an active engagement with what we might call (following the work of people such as Stuart Hall) cultural politics. Early writings in cultural politics were interventions in debates within Marxism about the role of culture in the functioning of society. Terms such as 'Hegemony' and 'Ideology' were hotly debated within cultural studies and geographers were quick to see the implications of these debates.

Over time, understandings of power have shifted, away from models based on the power of one group over another, towards those involving the power to do things. This has suggested that power relations consist not only of domination, but also of seduction, influence, persuasion, capacity, ability, manipulation, consent, compromise, subversion, control and so on. More than this, class-based models of power have been supplemented (though, for some, this meant weakened) by analyses of power relations organized around politics, gender, lifestyle, nature, race, sexuality, nationality and so on. Cultural geographers continue to be interested in the analysis and critique of power because space is bound up in the constitution of injustice, inequality and oppression. In particular, geographers have pointed to the sites of oppression and resistance, the different scales through which power relations operate and how space is manipulated by the powerful and the weak. As a result, geographers have drawn on and contributed to debates within Marxism, feminism, development studies, international relations and queer and postcolonial studies. The problem remains, still, both how to identify and understand the ways in which space, place and nature are implicated in – and constitutive of – unjust, unequal and uneven power relations, and how to suggest ways of addressing and redressing these relations.

These themes weave in and out of the *Handbook*, and in many ways give this book a coherence of endeavour and perspective. Further, these themes are to be found in the four case studies that we will be presenting later. However, the next step in this introduction will be to think more carefully about the relationship between the geographical and the cultural, and how to understand the world through space offers particular ways of opening it up.

THINKING SPATIALLY ABOUT CULTURE, THINKING CULTURALLY ABOUT SPACE

Not surprisingly, thinking spatially about culture has had a long history in cultural geography, arising out of the general history of geography, its often protracted negotiations with other disciplines, and the changing spatial tapestry of historical events and discourses. Roughly speaking, we can identify a series of *styles* of thinking spatially about culture which, though not incompatible, cannot be just placed together since they often show up quite different things. One style of thought is what we might call 'building bloc'. Such a style attempts to identify large processes in order to produce large explanations. A good example is Harvey's highly influential work on 'time–space compression', a transmission mechanism by which something called 'economy' could be linked to something called 'culture', so allowing traffic to take place between the two blocs. Space then becomes both a central engine of change in the nature of capitalism and its expression. Notwithstanding recent attempts to nuance such accounts by appealing to different scales which allow for a similar set of spatial registers, cultural geographers have increasingly looked askance at these kinds of explanations, not only because they force pared-down narrative structures on to the geography of history which are just too easy but because they thereby offer an illusion of control over a perplexing world which is precisely what a spatial consciousness should be going against. Thus other styles of thought have arisen which emphasize spatial context and contingency to a much greater extent.

A second style of thought has therefore considered spaces of identity. Here the idea of a fixed identity unambiguously belonging to one group and unambiguously expressed in space has been replaced by notions of more fluid identities belonging to particular subject positions which can vary in intensity and can be combined in many different ways, so challenging homologous explanations. This emphasis on hybridity worked out in and through space is perhaps best expressed in the highly influential work of Gilroy (1993) which attempts to show how group identities were made through numerous circulations across the Atlantic. Gilroy's geographical imagination can be seen as an attempt to construct an imaginative geography of resistance which both excavates the past in new ways and can be used as a means of constructing more open and heterogeneous identities which are able to interconnect in new ways. It is no surprise that 'mobility' and 'hybridity' in numerous forms have become key tropes in recent work on identity, expressing new forms of identification in which aspiration can be both more freely expressed and more easily questioned.

This attack on the bounded spatial imagination is common to other styles of thinking culture in cultural geography. Thus a third style of thought has been concerned with the apprehensions of spatial surfaces. Arising originally out of a concern with 'landscape' and especially the structuring of the act of seeing the land, incarnated in particular 'scapes', this style of thought has now branched out to encompass many different means of sensing the land – not just vision but other senses (like sound, as in the growing body of work in cultural geography on music), not just landscapes but city and even earthscapes (Cosgrove, 2001). In part, this is simply an attempt to understand how we make expressive places, so allowing the memory of the senses to gain a foothold. In part, it is an attempt to destabilize surfaces so that, like in many new architectures, they can bend and curve and thereby point up the temporary nature of so much of what we regard as solid (for example, the trajectory of objects which, like habitual practice, are so often apprehended without recourse to discourse or representation). And, in part, it can be part of a grander attempt to ground in new ways and so give the earth voice. 'Our houses are tumuli erected over the slaughtered body of the giant ground; only our nervous decoration, our attention to monumental detail, or preoccupation with property, gives us away' (Carter, 1996: 2).

A further and related style of thought is concerned with dwelling. In its original phenomenological guise such work tended to be caught up with the human experience of place which traded heavily on notions of authenticity within 'nature' or culture. However, more recently these essentialist leanings towards fixed ways of life and definite embodiments have melted into more general ethological misgivings concerning how different actors – many of them non-human – can be related to one another through 'political wills' made in and from space. Such 'inhuman' thinking about place, informed by developments in the sociology of science, actor network and non-representational theory, has been seen by some commentators as simply a kind of foraging postmodernity, but this is surely ill-conceived. Rather, it is an attempt to map out interconnections not just as incidental to how places are constituted but as central and thereby articulate a 'politics of coexistence' (Latour, 1999). So, for example, as Whatmore and Thorne (1998) map out how animals were drawn into various projects which circulated them around the world, they are also trying to show how such shifting geographies of power can be used to question our understanding of the natural, so providing the bare bones of a new 'natural contract' in which 'habitus', the 'unconscious' and other such sedimentations can be dreamt in different, more active ways.

So we come to the final style of thought which we might call experimental. This kind of thinking attempts to culture a kind of delight in the intricacies of space, not from a dilettantish need to add more and more twists and turns, but rather as a means of finding new interconnections through which new kinds of humanity can be realized, so expanding out what can be thought, felt and done – into the silence. The increasing appeal to performance as a guiding metaphor goes hand in hand with attempts to produce new visions of place or, more accurately, space–time, which are able to operate on preconceived notions by articulating actors, most especially, in the collective register. Thus, work informed by experiments in performance studies can be laid alongside work that attempts to name new fluid forms of space which do not have shape or consistency but do still situate, in that both are attempts – using bodies, contexts and all manner of other actors – to defy conventional cognitive coordinates, by getting at something different, something that exceeds, something that can unlock the power of virtuality with its attachment to life. So, against the background of a weak ontology, a new kind of spatial ethics is being forged. This is what Bennett calls an 'ethical energetics', which can produce new stances to the world, 'fundamentally more capacious, generous and "unthreatened" becomings of the self' (2001: 93). There is all to play for.

UNFOLDING GEOGRAPHIES OF CULTURE

The first two sections to this introduction have been fairly general. In them, we have suggested that cultural geography is characterized by particular ways of knowing, or of questioning the world. Perhaps it is better thought of, in fact, as a way of unknowing – a way of making the familiar strange. By using a geographical imagination and through an understanding of culture that recognizes that we are bound up in our own apprehensions and appreciations, cultural geographers seek new ways of addressing their subjects, and of telling new stories about the world, and subverting others. In this, we hope to grasp something of that world, but perhaps it is better to say something with conviction, in the hope that these passions for our subject will make a difference.

Let us take this idea of passion for the subject one step further. In some ways, what we are presenting in this section is a series of 'case studies'. Geography, as an academic discipline, has a strong heritage in exploration and what contentiously became called 'discovery'. We would argue that one positive consequence of this legacy is a strong sympathy with 'the empirical', with empirical research and with fieldwork across diverse sites. For those outside geography, this has often led them to see the discipline as atheoretical, descriptive – though their evaluations of this as good or bad have differed markedly. Wherever one starts in this engagement, cultural geography is rightly seen as 'empirical' in some sense, but this is no naive stance. In many ways, geographers have attempted to describe, to understand, to explain, and to intervene in the world. 'Case studies' have performed a very particular role: part evidence, part validation, part intervention.

We would like to add to this, part evocation. Contrary to the lingering legacy of positivist thinking with which 'case studies' grew up, we do not conceive of them as simple descriptions of the world as it is. For cultural geographers, the case study is not a local application of an abstract model, or a 'micro' statement of a 'macro' series of events. Rather, case studies are passionate evocations of the world and an engagement in it. In what follows, we present accounts of different worlds: worlds through the nose; the worlds of Victorian literature; worlds gathered into a display of nature's resources; and worlds that collide in protest. In the case studies, we do not attempt to unfold all the worlds we can find there, but seek to drive a certain purpose through the evocation of very particular worlds. In that sense, this part of the introducion takes its shape less from an explication *about* cultural geography (something that might not have been possible but for the introductions to each section by the section editors, and for the work of the authors themselves) than from an immersion *in* it.

In the four case studies or vignettes that follow, we seek to show how different lines of argument and thought can be played out under the umbrella of cultural geography. In doing this, we seek to show that how cultural geographers go about telling their stories differs from person to person, and that they use different voices and evidence to do so, with different purposes in mind. We are, nevertheless, all trying to say something about the world, and to expand the possibilities for unfolding and engaging that world differently. To be sure, the four case studies do not provide four corners of a map within which cultural geography is to be found. Instead, these vignettes are intended to be evocative of the diversity of possible cultural geographies. Each case study offers an analysis of different kinds of 'focus of attention' or 'object': from smell, to a novel, to an agricultural show, to a site of protest. From these 'objects', different geographies are unfolded to provide arguments about the world. In this sense, these case studies are more than examples: they are invitations to explore the world, to focus attention differently, to open the world out.

'All nose'
Nigel Thrift

The world is populated by what Bruno Latour calls 'mediaries' – active means of crossing and linking, means of making new 'wes' and its', which share in so much of the agency we so often ascribe only to ourselves. Mediaries can make unexpected connections as they produce new means of holding things together, connections which not only code but also produce wonder and enchantment. Here I want to write about a set of mediaries which we constantly construct and inhabit but which are rarely written about, mediaries that force us to think in new ways about how the world is: aromas, smells, scents – the entities that impact our 6–10 million nasal neurons and set off all kinds of bodily reactions, some explicit, some veiled.

Aromas inhabit our world like a second skin. They are so familiar that, much of the time, we seem to hardly register them at all. Yet just like skin, they can also stimulate, and transmit love and distress. Aromas can create an ambience of wellbeing, they can evoke past situations, and they can produce reactions of disgust and shame. It is no surprise that aromas have been central to human practices for many thousands of years, as boosts to sensual desire, as gifts, even as an element in the intercession between God and human.

In current cultures, aromas surround us, from the body scents we secrete and wear, to the cooking smells purposely introduced into the air of supermarkets in order to stimulate purchases. Indeed it is possible to argue that the contemporary world plays host to a spectrum of smells even greater than those of Ancient Egypt (renowned for its smell culture) or Ancient China (which went as far as inventing incense clocks), the result of the chemical engineering of aroma that has been characteristic of the last 150 years.

Certainly even though contemporary cultures may still privilege the visual register, they cannot imagine worlds that exist without smells; just as most religious paradises were imagined as being sweet-smelling, and even some utopias (like that of Fourier) are conjured up complete with complex registers of smells (Classen, 1998), so an essential part of today's consumer paradises is the articulation of smell.

Aroma then is central to human activity, and yet somehow it is off-centre to how we write that activity. Its quiet intensity dislocates our abilities to describe it. Why this disjuncture? Well, to begin with, it is because aroma automatically conjures up the kind of sensuality which many academics are exactly trying to dispel. Written argument, it seems, has to be divorced from the whole body. Then again aromas, though readily detected and differentiated, are not easily described in language. They are difficult to 'read', though so frequently displayed. They lend themselves not to semantic reductions but rather to somatic reactions. Aromas are not easily made specific in ways which lend themselves to written categorization. And, finally, aromas seem to escape our cognitive consciousness. They belong to a realm of 'peripheral' psychomotorial actions, an insistent substrate of incessant movement that makes up so much of what we are, but which we so often choose not to register as thought, even though the stamp of the impressions of this movement constantly influences us. They are a part of the landscape of the body which we have so often tried to suppress (Dagognet, 1992).

So how can we register aroma? And, more to the point, what has this got to do with cultural geography now? To answer these two questions, let me move to a brief history of the economy of smell which shows its centrality, and then draw some 'sensational' conclusions.

Smell has always had economic value. In the ancient, classical and medieval worlds, that value was found in the trade in 'spices', a category that includes all kinds of aromatic items that now might well be traded separately: drugs, condiments, perfumes, incense and even dyestuffs (which might well be aromatic) were never clearly distinguished from each other. However categorized, these aromatics provided one of the economic pivots of the world economy, being used in religious ceremony, in perfumery, in diet and increasingly in medicine. 'In the medieval west, different kinds of aromata were purchased from apothecaries, spicers, pepperers, perfumers (aromarii), grocers and pigmentarii' (Donkin 1999: 2). The travels of aromatics were one of the keys to world trade, as they were moved from their source areas through major entrepôts like Constantinople (which by the seventeenth century had 600 apothecaries in 500 shops, 115 perfumers in 80 shops, 3000 grocers stocking items like juniper, cloves, pepper and cinnamon, and numerous 'merchants of rosewater') to their final landing places on the tables, in the ceremonies or on the bodies of the rich and middling sorts. They brought China, Central Asia, Russia, Iran and Iraq, Indonesia, Malaysia, India, Sri Lanka, Madagascar, Arabia, Egypt, West Africa and many European countries into conjunction with each other. The explosion of new spice routes post-Columbus brought even more regions into the story – Florida, Mexico, the Caribbean, Ecuador, Brazil and Peru (Dalby, 2000).

The effects of the spice trade on the landscapes of sensation that were available to the average European were considerable. For example, in seventeenth-century England whole new orders of worth (Thévenot, 2001) were established based on aroma. Take just gum benzoin, one of the products of the Malayan archipelago, and a favourite of the Chinese, which had become an important ingredient for perfumers by the time of Ben Jonson's play *Cynthia's Revels*, written in 1600.

> *Amorphus*: Is the perfume rich in this …?
> *Perfumer*: Taste! Smell! I assure you, Sir, pure beniamin, the only spirited scent that ever awakened a Neapolitan nostril. You would wish yourself all nose for the love on't. I frotted a jerkin, for a new-revered gentleman, yielded me threesome crowns but this morning, and the same titillation.

The phrase 'pure beniamin' here does not mean that gum benzoin was the only ingredient present in the perfume, for

> There follows a discussion and a list of them all – musa, civet, amber, turmeric along with thirteen others that came out of Ben Jonson's Latin books – but, as the perfumer says, 'it is the sorting, dividing, and the mixing and the tempering, and the searching, and the decocting' that make for success. The result is praised by Amorphus and his friends, as 'most worthy of a true voluptuary'. For courtship is in view, and in the England of 1600 men were still ready enough to discuss with one another the perfumes they wore on such occasion. (Dalby, 2000: 61)

The skills of sorting, dividing, mixing, searching, tempering and decocting smells still exist, of course, but now transformed into the modern global fragrance industry, worth, at current estimates, some $20 billion per annum. The industry manufactures smell from a vast array of sources: the 200 or so plants raised commercially for their perfume (plants as different as roses, jasmine, lavender, iris, ginger, laurel, geranium, orange, lemon, grapefruit, balsam, olibanium (better known as frankincense), galbanium), animal products and, increasingly, synthetics. In turn, development, packaging and marketing add massive value to products which are sometimes worth only a few dollars at source.

Any one fragrance will be complex, involving at least 60 to 100 ingredients; some fragrances can have more than 300, and one scent reportedly has 700. So producing aroma is complex and its applications are wide-ranging. After all, most smells are produced not for perfumery but for much more mundane uses: for soaps, detergents, air fresheners, fabric softeners, cat litter, shaving cream, baby powder, nappies – the list goes on almost endlessly. Wander down the aisle of a supermarket and note just how many products are scented.

But how can this ocean of smell be understood and worked with? There are three ways, each of which has its geography of cultural 'representation'. One is through language. But one of the most problematic aspects of the process of producing smell is precisely language.

> Speaking about fragrance can be like trying to get a toe-hold on a cloud … There's even a term for it – the olfactory verbal gap – according to Dr Harry Lawless, a psychologist and professor of food sciences at

Cornell University. Lawless also coined the term 'tip of the nose phenomenon'. Take away the rose, he says, and 25 to 50 per cent of those smelling its scent might not be able to identify it as a rose.

Tongue-tied, we compensate and translate description of fragrance into the language of other sensory experience. The language of fragrance employs color, for example, using the term 'a green note' to denote grass-like scents that derive from leafs, and shrubs. Another means is using the imagery of music, 'a top note' refers to substances that evaporate off the skin and hit the nose first, like citrus oils do. (Newman, 1998: 12)

A second means is through the vast network of science as it is laid out in spaces like laboratories. After all, all aromas can be represented as chemical formulae. Most particularly, gas chromatography mass spectrometers measure chemical traces in parts per million via spiked graphs. Some spikes will be irrelevant: they contribute nothing to smell. But others are easily identifiable aromas. Using such a method, 90–95 per cent of a scent's components can be captured (though significantly, the last 5 per cent or so still need to be picked out by a nose).

A third means is habituated experience. So, perfumers are trained very carefully over many years and they are perhaps best described as artists of smell – like musicians or painters or wine tasters. Over many years they come to be able to identify and mix smells ('notes') in new combinations. The 400 or so expert perfumers in the world build up their expertise scent by scent, usually associating each scent with an event, a memory, a picture. And for the best perfumers it is

Imagination – Fantasy. It's the difference between a chemist and a perfumer. You dream your perfume before you write the formula. It's not just chance. It's not just exact science. There will always be things that won't work. You begin your fragrance as a composer, putting elements together. You finish your fragrance as a sculptor, shaping and paring down. (1998: 49)

Some perfumers still use 'perfume organs' – vast arrays of different fragrances which allow different combinations to be built up note by note.

This economy of smell lays down a series of challenges to our understanding of the spaces of the world. Quite clearly smell is a powerful force in human life. It can become the subject of economic empires. It can produce a symphony of sensibility. It can conjure up particular conducts and stimulate memories that are often peculiarly evocative. And yet we seem unable to say much about it that isn't either trite or obvious. And, of course, smell is hardly the only sense to which this complaint can be applied: we have similar problems with taste, or touch, and even some aspects of vision and hearing (Marks, 2000). Why might this be? Almost certainly, it is because our 'vocabulary' (a word I will come back to) is too restricted to encompass anything other than certain dogmatic ways of thinking which arise from a scholastic way of life. Think of the typical model of the scholar. She is static. She usually inhabits a quiet space. She sits and contemplates the world through processes of 'internal' thought. She writes or taps a keyboard that both embodies and twists that thought. This way of thinking is culturally and historically specific. It has to be learnt at school and then as an adult. It comes generally from the time when the practices of reading first became internalized and reading aloud came to be considered as a childish thing (Johns, 2000).

Of course, this way of thinking is very powerful. Its mode of abstraction (quite literally) from the world produces a particular highly absorbed grip in which coherence and signification are all. It therefore values the emblems of its way of life: considered writing, the interpretation of images and signs, long periods of gestation, in ways which can too easily come to be seen as the only means of thought (Bourdieu, 1999). But it is not the way most people live most of the world. They live absorbed in the cares of the moment, reacting to embodied presentations not disembodied representations, to *events* not historical structures. And these events are an emotional wash in which tears and smiles are just as much of an intellectual currency as ideas and concepts. In other words, they live in a world in which space and time are means of inhabitation, not just metrics (Thrift, 1996; 2000; 2002).

What does this world look like? Fortunately, since the mid nineteenth century many writers have tried to describe its key features, using traditions as diverse as phenomenology, various forms of micro-sociology, and anthropological studies of the everyday. So we know that this world thinks through the body's 'non-representational' capacities for reaction to sensation as much as its capacities for cognition. In particular, it is profoundly affective. And we know that it thinks through objects which are not separated beings but are a part of general ethologies which think the body as much as the body thinks them, questioning what we regard as life by expanding what can count as the nerve centres of the world. And we know that the world thinks through an ethos of engagement. It works not to abstract moral rules but to ethical modulations which vary according to circumstance.

In turn, such a view of thinking has some consequence. First, the world is no longer conceived as full of large or small things. The world does not consist of the clash of 'big' forces, surrounded by mere detail. Rather, as Gabriel Tarde noted so long ago, the devil is precisely in the detail. Second, the world is patiently constructed by fixed mechanisms of regimentation which exactly play on detailing the public flesh. But it also consists of numerous spaces, very often those which are born out of unruly senses like smell, in which new elements of the world can live and work, spaces we are only now learning how to map. Third, it follows that the world is prodigious: it constantly overruns static categories of thought, because it is virtual, tending towards actualization without producing a fixed resolution. The world can produce solutions not previously contained in their formulation. The world is artful.

Some writers go further still. They conjure up new kinds of inhuman landscape, which consist of sweeping planes of sensuality which mobilize both flesh and stone, or 'universes' of becoming which coalesce (or 'concresce') in unpredictable ways and routinely bleed into each other (rather like smell), and interactions which may well exist outside the realm of subjectivity. They need to produce a vocabulary of movement and emergence which can form, couple and break apart just like the world itself, and can actually intercede by producing new expressive resonances, including new kinds of people which can make new senses (Bennett, 2001; Law and Mol, 2001).

Interestingly, such a dynamic vocabulary has an analogue in the vast archive of work on performance which consists primarily of knowledges which attempt to conjoin and articulate the unlike in order to produce new effects which function at not just an intellectual but also an affective level. So performance has produced a whole repertoire of knowledges of engagement of the body and other hybrids which understand how little can rapidly become large, how space can rapidly become time, and how sensation can rapidly spread its wings, touching so many as it constructs its refrains: singular and plural in one.

> The refrain is a prism ... it acts upon that which surrounds it ... extracting from it various vibrations, or decompositions, projections, or frustrations. The refrain also has a catalytic function, not only to increase the speed of the exchanges and reactions in that which surrounds it, but also to assure indirect interactions between elements devoid of so-called natural affinity. (Deleuze and Guattari, 1988: 349)

And so we come back to smell. For smell is previously a means of synthesizing affective engagement with the world which requires this kind of performative knowledge to understand, in that it aids or stimulates artful shifts from one pragmatic orientation to another using a sensual resource which operates as a foundation for conduct while remaining 'outside the foreground of our self-awareness' (Katz, 1999: 7).

In other words, smell is an affective shape-shifter which adds to our experience of the world, by producing new means of engaging the moment, and so new kinds of eventfulness (Foucault, 2000). And it does so through geographies of emotional labour that we are only now beginning to get to grips with as we begin to derive a new kind of spatial vocabulary. So cultural geography takes another turn, towards the world of affect.

Anxious geographies
Mona Domosh

Few evocations of the cultural geography of late nineteenth-century New York can rival the intricate and multilayered portrayal found in Edith Wharton's *The House of Mirth*. Published in 1905, the novel portrays the 'gilded age' city of the 1880s through the eyes of Lily Bart, an ingenue of precarious social standing whose economic stability resides solely in finding the right husband, a search that requires an intimate knowledge of the social and geographic codes of the city. As a woman, and particularly one of dubious background, she knows that appearance means everything, and she understands that appearance comprises not only her body but the spaces around it – the rooms, homes, streets and settings in which she is seen. As such, the city's landscape is more than a background for her performances – it is an integral part of that performance. She must be 'seen' in particular locations in the city and not in others to maintain her moral standing, 'framed' by elegant surroundings to highlight her beauty, and portrayed within a landscape that reflects that beauty. Knowing and using the correct codes of appearance, whether that pertains to her dress, hair and makeup, her actions or performances, or the stages on which those performances occur, are critical to Lily's survival, and are her one source of power. The one thing that she cannot control is how people interpret her actions, and this eventually leads to her demise. In scenarios repeated throughout the novel, Lily's actions and appearances are misread by others who assume the worst of her. Appearances, she learns too late, can deceive.

The House of Mirth is a powerful commentary on what it was like to be a middle-class white woman in turn-of-the-century New York: as visual objects, women had to play by the prescribed rules of appearance and performance; yet without a forum for expressing their subjectivity, this superficiality left them vulnerable to misinterpretation. It is also a powerful commentary on the geographies of late nineteenth-century New York, a city where surface appearances were all that mattered, and where, as Mr Selden (a main character in the novel) remarks about the drawing room of a member of the *nouveaux riches*, 'one had to touch the marble columns to learn they were not of cardboard, to seat one's self in one of the damask-and-gold arm-chairs to be sure it was not painted against the wall'. It was an ersatz landscape, indicating an ersatz society – both, we are to assume, characteristic of late nineteenth-century New York, and both projected onto and seen from within the tragic life of Lily Bart.

This intricate connection between landscape and society was not simply a literary convention. Edith Wharton, like many other people at the turn of the century, believed in a form of domestic environmentalism. In other words, she believed in an active relationship between the built environment and society – that the architecture and interior design of buildings reflect and actively shape people's character and behaviour. In *The House of Mirth*, the drawing room is singled out as a reflection of the woman of the house. Lily describes her aunt, with whom she is living, as a woman whose 'imagination is shrouded, like the drawing room furniture'. She daydreams about redecorating the room: 'If only I could do over my aunt's drawing room, I know I should be a better person.' Lily never fulfils this dream, and in fact finds herself in less and less commodious surroundings throughout the novel, reduced to working in a millinery shop and living in an unpleasant boarding house. Lily's identity and one source of power was her appearance; 'framed' within a landscape of the drab browns of factories, she is doomed.

Although gender relationships today may allow for more independent subjectivities on the part of women, we still do live in a world where women's identities are 'read' from their bodily appearances, and from the appearances of the worlds that surround them. And while we may not theorize relationships between built environments and society in such forceful and uncontested terms as the late nineteenth century, we do try to 'read' powerful societal relationships such as gender, class, ethnicity and sexuality from the visual world. Yet Wharton's depiction of Lily Bart's plight alerts us to the incredible vulnerability of an account based solely on the visual. How does one accurately, authentically, interpret the visual? What really is the relationship between the material/visual world and the discursive; between surface and depth; between outer appearance and inner meaning? I plan to use Wharton's depiction of Lily Bart as a woman caught between her inner self and outer appearance to highlight the anxieties caused by the new cultural geographies of late nineteenth-century New York, and conversely, to suggest that an examination of those geographies can help us make sense of gendered identities. In other words, by providing one 'story' of the often anxious relationship between landscape, gender and identity, I hope to suggest the usefulness of exploring geographies of cultural anxiety. The tension between surface and depth, appearance and meaning, came into full light within the relatively new and flourishing consumer culture of late nineteenth-century New York, and those tensions were played out upon and were displaced onto the women of the city.

Late nineteenth-century New York City was indeed a place characterized by superficiality. It was not quite a 'shock' city in the sense of industrial Manchester, whose factories, pollution and poverty elicited extreme reactions, but the landscape and society of New York City nonetheless struck commentators as something unusual and new, akin to what architectural historian Reynar Banham (1971) wrote of Los Angeles in the 1960s – it was instant architecture in an instant landscape. Growing northward at a rate of approximately two city blocks a year in the first half of the nineteenth century, the city seemed forever in movement. Slowed only by episodic economic downturns, New York's ebullient real estate and building industries took full advantage of the wealth flowing into and out of Wall Street, and the vanities of aspiring socialites and other members of the *nouveaux riches*, to construct appropriate settings to display that wealth (and along the way, of course, to create more venues for wealth creation). Standing out from the rows of speculative brownstones and small shops and factories were homes built to resemble European castles, department stores disguised as Venetian palazzos, and office buildings draped in Renaissance cloaks. By the 1860s, whole sections of the urban landscape were dedicated to consumer-related and leisure activities – to shopping, eating and drinking, strolling, carriage riding. Hotels, bars, restaurants and theatres began to line the streets around Union Square, elaborate department stores and boutiques clustered on Sixth Avenue, men's clubs dotted Fifth Avenue reaching up to the Park, where lovers strolled, men raced their carriages, and women and children socialized. This was a landscape that spoke of fast living and fast money; a place inhabited by, in Karen Haltunnen's (1982) words, confidence men and painted women; a landscape, in other words, of insincerity. Contemporary commentators questioned who was really behind the yards of silk and lace, beyond the façade of Corinthian columns and gilded bas-relief. What source of wealth was able to fund those mansions and carriages and clubs?

Such a decidedly 'phoney' landscape and society was subject to various forms of criticism. *The House of Mirth* can be read in this light: Wharton offers a harsh condemnation of a society governed by appearance. It is this phoney landscape and society that eventually destroys Lily Bart. Throughout much of *The House of Mirth*, Lily Bart's identity seems to be forever changing, completely reliant on her surroundings. Yet at the end of the book she realizes that she has an inner, 'real' self, removed from those contingencies – an identity that she cannot deny. She recognizes the importance of a unitary self, and sets it apart from most people's subjectivity that she knows, including her parents: 'Her parents too had been rootless, blown hither and thither on every wind of fashion, without any personal existence to shelter them from its shifting gusts. She herself had grown up without any one spot of earth being dearer to her than another: there was no centre of early pieties, of grave endearing traditions, to which her heart could revert and from which it could draw strength for itself and tenderness for others' (1986: 307). Part of Lily's tragedy is that she realizes her inner, 'true' self too late to act upon its directives.

Wharton seems to be issuing a warning about the importance of maintaining an identity rooted to place and a core set of values – an identity that is not dependent on context and that does not change with the whims of fashion. She is, of course, giving voice to a long-held view of subjectivity – one that scholars have argued is rooted in early modernism. Lily's demise is predicated upon an identity completely tethered to the material world, completely dependent on visual context, and therefore ever changeable. According to Wharton, this is an unsustainable form of subjectivity, a condition similar to what contemporary scholars argue is at the root of agoraphobia. According to Joyce Davidson (2000), agoraphobics experience an acute anxiety over the slippery nature of the self and how it is constituted in relationship to that which is outside the self. Agoraphobics experience a lack of separation between themselves and the places they inhabit, and since they cannot control that which is outside them, they feel out of control of themselves as well. As Davidson explains: 'agoraphobic anxiety seemingly threatens the dissolution of the self … when it surfaces it erodes the boundary between "inside" and "outside"' (2000: 218). With the world caving in on you, threatening to destroy you, the only recourse is to retreat hurriedly to the space you can control – that of the home.

Most agoraphobics are women: they constitute about 89 per cent of the sufferers (Davidson, 2000). Why? Davidson suggests that it is because of women's socialized tendencies to be ever aware of themselves as potential objects of visual consumption. In this way, women are always sensitive to how they and their surroundings appear before others; sensitive, therefore, always to the visual. Some women may use this sensitivity to how they are being seen by others for their own empowerment. In the first half of *The House of Mirth*, Lily Bart is presented in this light. In one of the most compelling scenes in the book, Lily participates in a tableau vivant – a pageant of sorts where women socialites dress up and perform as works of art. Appearing as Reynolds' *Mrs Lloyd*, she is a huge success since her choice allows her to present her beauty at its best, and because she appears almost without artifice. In fact, as Mr Selden and her friend Gerty Farish remark, she is more herself in this performance than otherwise: 'She had shown her artistic intelligence in selecting a type so like her own that she could embody the person represented without ceasing to be herself … It makes her look like the real Lily' (1986: 129–30). When Lily can control her setting, as she is able to do in the hyper-performance of the tableau vivant, she can be herself; in the everyday world of the drawing rooms of the *nouveaux riches*, a world she does not control, she does not know who she is. But because, as Selden reflects, 'this was the world she lived in, these were the standards by which she was fated to be measured!' (1986: 130), Lily has no recourse. If not quite an agoraphobic at the end of the novel, Lily nonetheless suffers greatly from a society that judges only by what it sees. Wharton passes her own judgement on that society – a time and a place where women bear the social *and* psychic costs of artifice.

Yet most nineteenth-century commentators on the New York scene bore other judgements. They condemned not society at large, but the women themselves. Indeed, many depicted the fashionable woman of the city as the embodiment of urban immorality. A common motif of late nineteenth-century urban writings was an exposé of the city's fashionable women – a style of writing part gossip column, part tabloid. In one of the most virulent of these chronicles, *The Women of New York, or The Underworld of the Great City,* published in 1869, the author sets out, in over 600 pages, to '*in*form the public and *re*form society' (emphasis in original). The goal of the book is stated clearly in the preface: 'the women of the Metropolis are boldly and truthfully unveiled, and every phase of society is thoroughly ventilated. Where sin and immorality have tainted women in high life, and where fashionable wives and daughters have yielded to the enticer's arts, it tears the fictitious robes from their forms and reveals their habits of life, their follies and frailties.' Notice the choice of words used here: unveiled, ventilated, tainted, fictitious, reveals – all words that point to the deceptiveness of the world of fashionable women, behind which is hidden a more real, moral and truthful world. Through hundreds of 'real-life' examples, the book provides an exegesis of the city's woes, uncovering its evils in the bodies and hearts of its fashionable women.

But why was the focus on women? As Karen Halttunen reminds us, late nineteenth-century anxieties about 'reading' character from outer appearance were expressed through the figure of the confidence man as well as the painted woman. These men were gamblers, petty thieves, dandies who, through tokens of professed sincerity, won the confidence, and then took advantage of naives in the city. But the focus of moral outrage was on the fashionable woman. It is not difficult to imagine why. In late nineteenth-century New York City, middle- and upper-class women were the most visible symbols of the excesses of consumer culture, and they inhabited its landscape – as they shopped for and wore fashionable clothing and jewellery, attended to social events, promenaded along Broadway. But perhaps even more importantly, because women in nineteenth-century society were seen as embodying its moral centre, any deviance from that position caused great anxiety. The ideology of separate spheres positioned women as its saints, that is, as the providers of moral guidance. To see women corrupted by fashion threatened therefore the very basis of a controlled society. As Halttunen argues, during this period many people 'feared above all fashion's power to transform women into hypocrites and thus undermine their moral influence over the larger society'. As a result, fashionable women, and the landscapes of consumption that they inhabited, were scorned and feared. A general societal anxiety over the rise of consumer culture was displaced onto the figure of the fashionable woman, and into the landscapes she inhabited – her home and the public consumer spaces of the city.

In the late nineteenth-century, then, women's morality was meant to moderate the tension between artifice and authenticity – between the look of spaces and things, and their meanings. This gave women some authority, but also positioned them to take the blame when artifice went wrong. In the twentieth century, bureaucratic institutions, corporations and experts wrested control over spaces and their designated meanings from women, but left them to bear the weight of a potentially malignant artifice. Edith Wharton, sensitive to the dangers of bad design and what it could say about the moral character of women, became an 'expert' herself, cowriting a book in 1897 called *The Decoration of Houses* that gave direct aesthetic advice about how best to design and decorate spaces so as to shape and reflect correct gender roles and a family's position in society. If *The House of Mirth* can be read as a warning of what can happen where there is a slippage between artifice and authenticity, then, the ideas expressed in *The Decoration of Houses* is meant as a corrective.

That slippage was readily apparent in late nineteenth-century New York City, with its displays of consumer culture set loose for all to see on the streets and in the shops of the city. For many, the fashionable women of the city, those not abiding by their proper roles as moderators of morality, were to blame for the excesses and potential moral dangers of overconsumption. For others, like Wharton, the blame lay elsewhere – in a time and a place that valued appearance over character. In either case, women bore the costs – both social and psychic. As *The House of Mirth* reminds us, understanding the anxiety-inducing relationship between landscape and identity is far from a trifling matter. For Lily Bart, what was at stake was nothing less than her life. Tracing the gendered (and, in other cases, the racialized and sexualized) geographies of cultural anxiety seems a task well worth pursuing.

Post-humanist geographies
Kay Anderson

Few events perform so ritualistically the triumphal narrative of human ingenuity and agency over the natural world as those mundane displays of produce and machines known as 'agricultural shows'. From Boston to Sydney, to Auckland and Toronto, to imperial London as well – those liminal forms of Herefords, harvesters, honey and the like, that spill over the cate-gories we think of as culture and nature, have been assembled since colonial times. Such events, typically staged in cities, engage the interest of the cultural geographer for many reasons. In the artefacts they display, their orderings and codings, urban agricultural shows speak to well-known themes in contemporary cultural geography: of power and performativity, bodies and objects, colonialism and nation-building, knowledge and discourse, cities and citizenship. They also afford a lens through which to bring the 'new vitalism' signalled by the study of non-human geographies into contact with cultural geography's most referential category, that of culture. In that sense, the focus in what follows on one such spectacle offers a point of insertion into some defining narratives of the subject to which this *Handbook* is devoted.

The Royal Easter Show held in Sydney (Australia) is a case in point of the genre of the metropolitan agricultural show. Since the 1820s, agricultural displays have been made to stand as testimony to 'civilization's' arrival and development in the colony of New South Wales (NSW). Shorthorns, stallions, shearing machines and other artefacts that sit in that borderland space between the human and the non-human have been annually assembled. This proud performance of the agricultural ideal through the props of cultivation – of livestock and produce, farm implements and machines – reiterates the tropes of Australia's colonial mapping. These include pastoralism, agriculture, property and the spatiality of 'settlement' in a landscape which, as *terra nullius*, had apparently been devoid of these 'improvements'. By today, the volatile mix of Aboriginal and environmental critiques of such land uses highlights the tensions within the massive binary structures of culture/nature, city/country and settlement/wilderness that have been the stock-in-trade framings of much of cultural geography's subject matter.

Conceived within these broad framings, the genre of the agricultural show speaks to problems at the heart of cultural geography's disciplinary identity. Within the long tradition of geographic learning and labour referred to as 'human–environment' or 'man–land', agricultural activities and their associated patterns of human settlement have figured in regionally based historical geographies. In the United Kingdom, a long tradition of historical geography continues to map the changing configurations of agriculture, industry and settlement that have been implicated in the 'making of the English landscape' (Hoskins, 1955). In the United States, the study of regional landscape features has been framed by a model of culture that continues to centre humans as key agents of landscape change (Rubenstein, 1999).

Armed with a mix of analytic tools, an alternative perspective on the agricultural show is possible. The detail of such a perspective is of less interest here than the strategy of using vignettes (of events, objects, anecdotes) to evoke new ways of *thinking spatially* – ways that articulate links rather than fissures, that signal geographies of connection and circulation rather than segregation and stratification. In this case, a focus on one city's agricultural show brings into view the complex entanglements of culture/nature categories that have stood on opposed sides of the conventional faultline of much geographic endeavour. That the chosen event occupies a site in the so-called 'New World' is also of no incidental significance. Rather, the choice is central to the tactic of disrupting oppositional talk – of deranging the conventional relation in colonial studies of a centred imperial subject impacting a distant margin 'out there'.

After all, and historically speaking, the technologies of agriculture and pastoralism were not coincident with Europe's Age of Discovery from the sixteenth century. They belong within a heritage of European-derived stories that both predate and outlive Europe's global extension. These are stories of 'civilization' conceived in a quite specific way as humanity's historical ascent *out* of nature. Indeed the earliest *and* most persistent notion of 'civilization' in the classical anthropology handed down by humanism is the story of the temporal process through which humanity strives to raise itself out of an animal state. Civilization in this sense is the realization of humanity as 'truly human', conceived as a movement away from mere living things that live (Glendinning, 2000). Still today, world historians see agriculture as the developmental threshold that propelled this movement, leading to the emergence of the great regional traditions of human civilization (for example, Smith, 1995). Nature's cultivation is construed as the turning point that launched humanity out of a raw existence and set it on its diverse 'civilizing' paths.

For Carl Sauer, writing from the 1920s in the American context, 'culture' was the name given to this evolutionary force behind the imprints that people left on the earth. The richness of Sauer's work lay in his attention to the variable expressions of its influence in regional contexts across the Americas and beyond. More generally, he wrote: 'Man alone ate of the fruit of the tree of Knowledge … and thereby began to acquire and transmit learning, or Culture' (1956: 2). His overriding point – taken up by numerous followers in the Berkeley School of cultural geography and beyond – was the instrumental role of culture as a 'universal capacity' of 'even the most primitive people, including the obtuse Tasmanians' (1952: 11) to turn nature into culture. Even they, it was granted by Sauer, were uniquely human.

Buried within Sauer's brand of cultural relativism lies some residual primitivism that continues to fuel that pervasive hierarchy of the premodern and the modern. Indeed the example of these 'obtuse' Australians, who neither domesticated plants and animals nor settled in nodes around them, affords an opportunity to augment the critical race theory of anti-colonial studies with a critique of an equally powerful civilizational discourse. For the linking of culture, cultivation and human potentiality turns a spotlight on a peculiar *humanist politics* – one that bears more scrutiny from cultural geographers. This applies in at least two senses, the excavation of which brings cultural geography's long-standing interest in the culture/nature interface into contact with themes of race and empire.

First, the coupling of nature's cultivation and 'civilization' reveals a representational conceit within many of the world's cultural traditions: that the hand of a universalized human entails a supersession of (animal) nature. By now, such a view sits uncomfortably with the critiques of the dualism of humanity/animality on which that and other premises of philosophical humanism depend (Glendinning, 1996; Pearson, 1997). Within cultural geography, one finds that evolutionary as well as more recent symbolic conceptions locate culture in a sealed, species-specific sphere of humanity. Both conceptions rely on a model of humanity as an essentialized condition and status opposed to animality. Crucially this includes *human animality*, which has long since been conceived as the substrate of 'biology' on which 'culture' is thought to sit (see Ingold, 1995). The capacities of consciousness, sociability and intentionality which make up that superstructure of 'culture', and which are thought to uniquely equip humans with the ability to transcend nature, are taken to be defining measures of the human. Not only does this deep-seated view overlook the evidence for such capacities in beings additional to the human – as witness the emerging field of 'animal geography' (see Section 3 in this book) – it also privileges a restrictive figuration of 'the human' and human history.

Second, the image in western tradition of the earth as garden assumes a teleological course of agrarian land use and livelihood that carries not only a heavy humanist baggage, but an ethnocentric one as well. In this tale, the movement of history is the march of humanity which has its ultimate manifestation in the secure enclosure of the bounded city. There the expanse of nature ends or is 'brought in' in proudly domesticated forms (Cosgrove, 1993). This was a key narrative projection that underpinned the diverse European extensions into the 'New World' from the seventeenth century. More specifically, the constitutive spatialities within that progression – of improvement, property and settlement – were ones through which Aboriginal non-cultivators on the Australian continent were dispossessed from parts which, as 'wastelands', were not in production or even on the path to humanity's proper dwelling space in the city.

There exists within these conceits an opportunity to engage in long-overdue analytical bridgework across the spheres of the human and non-human worlds. Crucially, this is a move that requires more than an excavation into the history of *ideas* – in this case, Eurocentric representations of humanity and animality, as well as different people and places. For in this case, questioning the figuration of the 'civilized' human in western discourse forces not only a narrative but also an ontological recognition of the copresence of all other living things to which the generic human has long since been discursively opposed. It compels us to take seriously the vitalism and materiality of non-human entities (Whatmore, 1999), not least those of the long devalued non-human animal.

The approach through which can be conceived a select agricultural show points well beyond itself, then, to interests at the heart of cultural geography. In vivid contrast to studies of 'human impact' that either commemorate or lament human change to 'the environment', an alternative writing tactic uses the hybrid materials of culture/nature to *think across* the divides so proudly inscribed by the agricultural show. In so doing it is possible to unseat the figure of the human who has pinned so much of his (*sic*) defining status on the capacity to turn nature into culture. And far from being unmarked, the historically situated bodies of this idealized figure were profoundly raced (and gendered), as can be seen through the following window on 'white natures' in colonial Sydney.

The show beyond its text Like agricultural societies elsewhere, the Agricultural Society of New South Wales undertook many functions from its establishment in the 1820s, including registering breeding societies, checking pedigrees, and compiling stud and herd books. A key promotional strategy of the society was the holding of competitive shows. By the end of the 1800s, Sydney was the stage to which regional societies throughout the colony and the government experimental farms forwarded their best products. In the words of Society president Sir John See, when opening the 1898 show and as reported in a local newspaper: 'The exhibition, in focussing local competition, is an object lesson which enables the people of Sydney to see for themselves the true capabilities of NSW. These shows are a marvellous incentive to producers in the country who are brought into contact with … all concerned with the material development of the colony.' As was the case in other emblematic spaces of civilization, including imperial London with its fancier's clubs and botanical and zoological gardens, there was an enthusiasm in colonial Sydney for domesticated natures – both the enclosures of wild nature, and activities that celebrated breeding achievements.

Competition was not only designed to increase resource production for the colony and British market. It was also thought to foster a basic building block of humanity's development. In 1909 the Royal Agricultural Society annual stated: 'The Show shows us by ocular demonstration what great things the pastoralist and farmer are doing for the advancement of the entire human race' (p. 9). In that sense, perfecting the raw material of nature was the key to an apparently loftier destiny still, of propelling the human race in teleological terms from potentiality to actuality. Failure to fulfil this destiny was to risk a degenerative lapse – as if, in the words of Deleuze and Guattari's critique of the anthropocentrism of origins, 'beneath civilization we would rediscover, in terms of resemblance, the persistence of a bestial and primitive humanity' (cited in Pearson, 1999: 180).

According to this humanist model of 'origins', inscribed in Darwinian anxieties about the natural history of man, certain humans on the Australian continent embodied the primitive potential within humanity's (interiorized) animality. As figurations of the bestial condition that it was humanity's distinction to transcend, the Aboriginal savage stood in for 'early man'. Thus, a Society publication noted in 1898 that those indigenous people who had become involved in pastoral pursuits on cattle and sheep stations in New South Wales would spare themselves their race's destiny of extinction. In point of fact, non-farmers had transformed the continent according to their own knowledges and technologies for centuries prior to British colonization, throwing into relief not only the politics of colonial land use imposition, but also the humanist ideal attached to nature's cultivation – in Britain as well as its colonies.

Display and judging were the primary purposes of the Sydney show. Regarding livestock, the cattle show attracted immense interest. Cattle and horses travelled to Central Station where they were escorted through the city to the showground. Shorthorn, Hereford, Devon, Jersey and Holstein were some of the major categories of awards for dairy and beef cattle. In the annual spectacle called the Grand Parade, commencing in 1907, champions from each 'class' of horse, cattle and goat circled the main arena in an intricate choreography. Controlled by the ringmaster, the procession of winners dramatized the perceived triumph of humanity's experimental elaboration of the non-human world.

There were other sculptured forms. Stallions were brought to the showground, put to the test of the obstacle course, and judged according to speed, size and symmetry. Regarding displays of produce, an initiative was proposed in 1899 to hold regional exhibits of farm, dairy and floral products at the show. A 'district exhibition court' was opened to enable comparative examination of the products of the soil. These were judged for their visual appeal, as well as the taste, smell and texture of sample pieces made available to show-goers and judges. These assemblages, which constitute the very ground 'in between' culture and nature, evoke less purified conceptions of humans and 'things' than those which obtain in the triumphal discourses of civilization. Highlighting the material 'impurity' of such artefacts in this colonial setting is to derange the object/subject relation of nature and culture which has long been the humanist geographer's referential axis.

Displays of farm machinery were principal features of the show. If one of the defining measures of Australian nation-building after federation of the colonies in 1901 was 'settlement' of the land, then the reapers, binders and threshing machines that had pride of place on the show's stages were some of its key instruments. In the case of the showground's 'machinery avenue', farmers were introduced to machines 'to conquer the worst wilderness in Australia, and reduce the most unyielding soil to a state of tractable fruitfulness' (*The Daily Telegraph*, 12 April 1900).

Modernity's technologies were obviously intimate partners in Britain's colonization of the Australian landscape. But there was more at stake here than a geopolitics of racialized power and possession, about which anti-colonial scholarship has already told us so much. For if we foreground the normative divide that had been inscribed under European modernity between, on the one hand, *certain* technologies (for example tractors and *not* boomerangs) and, on the other, a world of nature on which machines were thought to do their work, a critical spotlight is turned on another vector of power (see also Latour, 1993; Strathern, 1999). This has as its conceit the premise of human 'emancipation' from a passive world of objects in nature, according to which the likes of boomerangs could never match the instruments of nature's 'improvement'. In this sense the civilizational discourse handed down by classical humanism provided the connecting tissue between the regimes of modernity and colonialism. It directs our critical imagination to the *linked* oppressions of the non-human and human worlds. This civilizational discourse, of 'culture' as the supersession of nature, has served to inferiorize non-human animals, while simultaneously privileging those humans who have affected the greatest distance (alienation) from the non-human world.

In a recent review of anthropological epistemologies, Thomas (1997) claims that presences as well as representations, objects as well as texts, substances as well as signifiers, doings as well as meanings, should be central to contemporary cultural analysis. In the case of this story of an agricultural show, attending to the materiality as well as the iconic power of the exhibits helps advance the preoccupation of constructivist strands of critical cultural geography with discourse and representation. The Sydney showground was not only one of colonialism's quintessential representational spaces whose meanings we might read as 'texts'. By also evoking the impurity of the artefacts themselves – in ways that activate our imagining beyond stories of human agency and invention – it is possible to chip away at a much older and persistent story that western society tells itself. This is the tale of the human, or 'lord of creation' in Sauer's (1952: 104) words, as the being who transcends the merely natural. Challenging this cherished self-image forces an exciting revision of concepts at the heart of the subject to which this volume is devoted.

Struggles over geography
Steve Pile

This vignette is about geographies of protest: about a particular (urban) site of protest, Trafalgar Square in London; about the geographies that are folded up in moments. For a decade or so, Labour Day – 1 May – has been marked in London by a gathering of various leftist groups at Trafalgar Square, at the heart of the capital. Though many groups were represented, the largest single groups seemed to belong to various Turkish communist parties. At least, they had the biggest, reddest banners. However, things have changed recently, within the last three years.

In the wake of a series of three bombings, 1 May 1999 was marked by the confluence of divergent marchers, streaming down from different symbolic sites in London: from Brixton, from Brick Lane, from Soho. First a little reminder of the events leading up to the protests. On 17 April 1999, two stall-holders at Brixton's Electric Avenue market began to examine a plastic bag that someone had left behind. Some thought it might contain a bomb, so one of the men even took out a container from the bag. People were laughing, until some said 'it's ticking'. At 5.30 p.m., while police were moving people away, the bomb exploded. Ten pounds of metal nails were blown into the street, injuring 50 people. Immediately, there was speculation that the bomb had been aimed at Brixton's black community, but as yet no-one had any idea who had set the bomb, or why. A week later, on 24 April 1999, a second bomb exploded in Brick Lane, at the centre of one of London's most prominent Asian communities. This time, 13 people were injured. Fears that these attacks were motivated by racist hatred seemed confirmed. Six days later, in the early evening of 30 April 1999, a third bomb exploded in the Admiral Duncan pub in Soho. The holdall with the bomb was placed at the centre of the pub, so the blast was concentrated within the building. This time, three people were killed and 76 injured. The Admiral Duncan was (and is) a well-known 'gay' hangout, and speculation centred now on the possibility that a far right racist and homophobic organization was behind the bombing.

It seemed clear to most, probably all, commentators that a small cell of highly active 'fascists' was at fault. Indeed, soon after the first attack, a number of little groups – such as Combat 18, the White Wolves, the English National Party and the English Liberation Army – had claimed responsibility for planting the bombs. Preliminary research by police and journalists highlighted that groups such as these were keen to detonate a race war in Britain. In fact, the bombings had the exact opposite effect. Different groups with different political agendas, across the width of the political spectrum, began to mobilize against the 'Nazi' threat. Thus 1 May 1999 was going to have a very different feel to previous years.

As the usual heterogeneous groups of Marxists began to gather at Trafalgar Square, they were joined by anti-racist marchers from the south, having walked from Brixton, and from the east, from Brick Lane. From the north, gay and lesbian activists had occupied Soho and begun to drift into the Square. At the bottom of Nelson's Column, a remarkable array of different kinds of protests and protesters began to mingle together, as the different groups walked through one another to get to the podium, where speakers from various groups were set to speak. Though different protests, and different causes, were represented and being expressed, at that moment these turned into one crowd, one voice. What they did not know was that the police already had the name of a suspect and were tracking him down. On 30 June 2000, David Copeland was given six life sentences for the bombings. Chillingly, he showed no emotion, either during his police interviews or during his sentencing. Nor has he shown remorse for his actions. And, as you might expect, police found evidence of Copeland's membership of far right organizations. However, they could not find any evidence linking the bomb attacks to anyone in those groups. He had, it seemed, acted alone.

Of course, Trafalgar Square – and central London – had been associated with popular protest long before 1 May 1999. In London, on 1 May 1517, there had been attacks against buildings associated with rich foreign merchants and craftsmen. London, before and after, witnessed other riots – such as, to name but a few, those in 1660 (by Charles II loyalists), 1736 (against Irish workers living in the area where Brick Lane is today), 1780 (by 'conservative' elements led by Lord Gordon), 1866 (for political reform) and 1886 (by political radicals) – though for markedly different (conservative/radical) reasons. (It has to be remembered that the so-called left do not have a monopoly on either rioting or resistance.) Most recently, demonstrations in 1990 against the poll tax turned sour and fighting broke out between protesters and police. It was said, at the time, that a pall of smoke hung over Nelson's Column. It was lucky it was not beaten up and arrested.

By 1 May 2000, police were on their guard (Pile, forthcoming). Not this time against anti-racist or anti-homophobic demonstrators, but against the threat of 'anti-capitalist' demonstrations. Having been taken by surprise at the volume and ferocity of the demonstrations in the financial district on 18 June the previous year, the police were in no mood to underestimate the potential for trouble. However, they did. Within hours of the initial protesters arriving, attacks had been carried out on one of the great symbols of the evils of global capitalism: a McDonald's. With typical insensitivity and with a desire to overwhelm with force, there was a massive mobilization of police fully decked out in riot gear. Though the protest was widespread up and down a main thoroughfare connecting Parliament Square to Trafalgar Square, the police began to shunt all the protesters, peaceful or not, into Trafalgar Square. There, blocked in, protesters began to fight back in running skirmishes with the police. It was not until late in the day that the police eventually, in their words, 'retook' Trafalgar Square – as if it were some military objective, once lost to a powerful enemy. The parallel is interesting. Once again, Trafalgar Square echoed – not to the sound of a famous imperial victory, but to the cries of demonstrators and the violence of political oppression.

The demonstrations of 1 May 2000 left behind them the scars of various attacks on property, not just the McDonald's, but also various national monuments. Although venting anger on property and monuments is age-old – think back to 1517 and before – politicians of all hues were outraged, and the police were put on the back foot for allowing such 'deplorable' activities to take place. On 1 May 2001, the police were in no mood to give any possible protesters the freedom to create a carnival of violence. However, the anti-capitalist demonstrators had plans for protests – note: peaceful protests – that were to give the police severe headaches.

Figure 2 *Anti-capitalist protesters in Oxford Circus, 1 May 2001 (Photo: Steve Pile)*

Instead of focusing anti-capitalism on one site of protest, the diverse and heterogeneous groups used the allegory of the popular board game Monopoly to organize their demonstrations. Monopoly (invented by Charles Darrow in 1933) is based on various locations in Atlantic City. Many cities now have their own version of the game. In the game, players attempt to capitalize on those locations by charging larger and larger rents. It is model property capitalism. It was now being used as model anti-capitalism. The earliest protests began at Marylebone Station. A fairly small group of cyclists began a slow journey to King's Cross

Figure 3   *Riot police guard Laura Ashley near Oxford Circus, I May 2001 (Photo: Steve Pile)*

station, via Euston. (All sites on the London Monopoly board. You get the idea.) At King's Cross, the bikers handed out free food, breakfast, even to some police.

In a taste of things to come, the police reaction was extraordinary: transit vans full of police followed the cyclists, and roads around King's Cross had been blocked off. By mid day, small protests were taking place across London's 'real-life' Monopoly board – from the Elephant and Castle to Mayfair, from the Old Kent Road to Hyde Park Corner. Most events were relatively small, involving handfuls of protesters. Almost everywhere, there were twice as many police as demonstrators. Around Trafalgar Square itself, the police had thrown a cordon to prevent people from moving down Whitehall to Parliament. The protesters didn't seem that bothered: small groups angrily/joyously chanted and sang wherever they could get to. For example, a small group gathered at the top of Whitehall and began declaring, very loudly, the injustices befalling women outside the west. In the Square itself, the protest amounted to about half a dozen people dressed as Mary Poppins and feeding the pigeons (in protest against remarks by the supposedly radical left Mayor, Ken Livingstone) – and a ring of police around the base of Nelson's Column.

At each location of Monopoly London, the symbolic meaning of the place was being drawn on by different groups to protest different aspects of capitalism. In its diversity and heterogeneity, it marked a full-blown critique of the myriad and shifting injustices and inequities of lives led under capitalism. No. More than this, the variety of protests was show-ing exactly that there were as many capitalisms as forms of protest. And more besides. To bring all these strands of critique together, all the protests were due to converge on a central location, Oxford Circus. Although people were not meant to get there until 4 p.m., Oxford Circus had been occupied from early afternoon. The site is the symbolic centre of one of London's most famous shopping streets: significantly, NikeTown faces into the Circus. In preparation, as you might expect, precautions had been taken, and the shops along Oxford

Street had been boarded up. By about 2 p.m., thousands of people were converging on Oxford Circus – a vast panoply of anti-this and anti-that, but always anti-capitalism.

Police strategy was to run rings around the protesters. A small group of protesters was blocked into the Circus itself. Another ring of police stopped further protesters from joining the inner circle. From there, the police attempted to gain control of the situation by forcing people back along the main roads leading into the Circus, and by blocking off side streets to prevent escape and evasion. With twice as many police as protesters, and many donning riot gear prior to any kind of trouble, the overwhelming sense of intimidation mounted. There was, as a result, trouble. Shops were attacked; fires were lit. Within the Circus itself, some people attempted to burn down NikeTown. Police response was swift, and violent. Snatch squads ran into the crowds, hitting people indiscriminately on the way in, dragging individuals away on the way out. A tense stand-off began to develop.

Figure 4   *NikeTown under siege, Oxford Circus, 1 May 2001 (Photo: Steve Pile)*

After eight hours, thousands of protesters were still being 'detained' (illegally, it was said) at the Circus. For many of the demonstrators, though, there was a sense of salt being rubbed into the wound. *The Guardian* newspaper quoted Ross Bateman, aged 20, the following day:

We're being treated like terrorists but all we're doing is showing that we're a bit pissed off with the way things are. I am just here to make a stand because people don't seem to have a voice anymore but even when we're standing in the street they just section us off so we can't mingle with ordinary people. (p. 4)

While agreeing with this, it might be observed that the protesters were actually 'ordinary' people. The day was not organized by one overarching group or political agenda, but was disorganized by diverse people with different aims – and this is very ordinary. Of course, dressing up like Mary Poppins is not an ordinary act, but the wit of such acts draws on a carnival spirit that is ordinary. The spirit on the day drew on this ordinary spirit of frustration, injustice and playfulness. Within the Circus itself, after hours of detention, a small group playing drums began to sing various songs. Once they had been through a medley of protest songs, they began to strike up other popular songs – perhaps the loudest of which was 'We All Live in a Yellow Submarine'.

At this point – the profound idea that we do live in a yellow submarine – I'd like to step back from these stories and locate them within various geographies. Some of these will have already occurred to you, but the basic point I am trying to make is that these events provide a point of contact for a variety of overlapping geographies. It is not that these geographies are simply there, but that these events are formed out of the overlap of these geographies. In other words, reducing the above events to a single geography is to misunderstand the spatialities at play in their formation, and their motivation. There is one further idea. Geography is not neutral politically. The stories I have given above do not suggest that one geography is radical and another not: instead, we should say that geography is something to struggle over as well as struggle through.

Entangled in these stories are geographies of power (following Sharp et al., 2000). It is possible to track these power relations through their different manifestations, some of which are fairly hideous. To begin with, there are geographies of 'otherness' here. London is no exception in drawing in people from well beyond its administrative borders. There is no city in the world that contains a homogeneous population: they all gather in peoples, though through specific lines of migration. However, settlement within cities is often strongly marked by segregation, often most visibly along divisions of 'race' and class (see Pile et al., 1999), but also by sexuality, gender and other kinds of difference. There is clearly a link between social relations between divergent groups and the spatial interactions (or social distances) between them. Now, it may at first glance appear that the separation between groups socially and spatially implies that there is no interaction – or no good interaction – between people of different backgrounds. Indeed, it has been speculated that David Copeland's background in predominantly white suburban areas contributed to his hatred of London – and all the diverse groups that London accommodates. Nevertheless, London also mixes up people from different backgrounds. So, when Copeland bombed Brixton to maim and kill black people or the Admiral Duncan, his attacks fell indiscriminately on all different kinds of people. Indeed, two of the three who were killed in the Admiral Duncan were a couple celebrating news of her pregnancy, the third victim being a friends of theirs. Not that Copeland cared: anyone in the pub, or in Brixton or Brick Lane, was guilty simply by association. His disgust knew no bounds: London was simply guilty, all Londoners guilty.

So, from these events, we can witness not just geographies of separation, geographies of interaction and mixing, but also geographies of hatred and disgust. These geographies are incendiary, and not just in London. Race wars have geographies all over the world and they hardly need listing, they are so familiar. The point about these geographies is that they are about the ways in which people circulate – or are forced to circulate – around the world, about how they settle in – or are uprooted from – places, and about how they interact – or don't – with people as they live in the world. These questions do not have straightforward answers: segregation is no more a universal solution than is the compulsory mixing of people. Contained in the events of London 1999 are the very paradoxes of how people come to live with one another or apart from one another, and how they come to form hatreds and affiliations. More than this, though, we should also recognize that the protests on 1 May 1999 also saw the intersection – and joining – of divergent protests against racism, homophobia and the injuries of class.

Although 1 May 1999 was marked by the intersection of different protests, they were focused on the single issue of the bombings. Protests on 1 May 2000 and 1 May 2001 reveal a developing trajectory around so-called 'anti-capitalism'. There is something very remarkable in this. On 1 May 1999, the 'anti-capitalist' elements were Turkish communists and a small group of Socialist Workers. Little attention was paid to them. However, events in Seattle in late November and early December 1999 had become part of the folklore of popular radical politics. London had staged its own disorganized protests by 18 June the following year. Almost unnoticed, a new form of radical politics had sprung up – unnoticed, perhaps, because it had no overbearing or singular system of ideas underpinning it. Some complained of the lack of a positive political agenda, i.e. the lack of a formal, institutionalized, politics. However, it is more accurate to say that many positive agendas were now jostling to be heard, and to make themselves heard.

The noise of these different agendas can now be heard across cities in the west, wherever the leaders of self-styled democracies gather to talk 'power'. In many ways, the 'big' geographies of corporations and states were being contested by the 'small' geographies of protest groups. In fact, these groups are hard to map and track, and this is their strength. They spring up from all kinds of ideas and places: they form connections that are wide and unpredictable. They are, almost literally, uncontainable. But we should be careful not to romanticize the plurality and disorganization of these resistances. One irony of the 1 May 2001 protest is that, having been trapped for hours in Oxford Circus, it was the highly organized Socialist Workers' Party that had the capacity – as the police wished – to negotiate the release of the protesters from the Circus. More than this, they also had legal observers monitoring the police for illegal action – including the detention of people at the Circus.

If there are diverse geographies of resistance in these stories, then there are multifarious geographies of capitalism too. NikeTown takes us down other paths, from London, around the world, to its factories in the 'developing' world. Following these chains of commodities can lead to some fairly predictable tales of exploitation. These tales, of course, are still worth telling, lest we forget. However, they also show that there is no singular capitalist relation. Or, put another way, they show that there are multiform capitalist social relations, and they do not come in a neat 'capitalist versus worker' package (except in those circumstances where they do, of course!). The papers were quick to point out that one of those trying to burn down NikeTown was wearing Nike trainers. More than one of the protesters, of course, was wearing designer gear. A small fight even broke out amongst the protesters when one group decided to burn designer gear, and attempted to take such items off the backs of others by force. Perhaps paradoxically, the main sources of flammable material used to burn Nike were copies of the Socialist Worker newspaper.

Of course, another geography concerns the fabric of the city itself. In each of the stories above, we can see that Trafalgar Square organizes the spaces of repression and expression of the various groups. It acts as a focal point, drawing protest in. Of course, this might seem paradoxical. Not only does the monument commemorate Lord Nelson and his victory at Trafalgar; the Square is also a memorial ground to other great imperial adventures. On three of its four plinths, Victorian generals stand. Of course, the fourth plinth, in its emptiness, is almost a statue, in absence, to the fact of London's post-imperial present. Well, maybe. Trafalgar Square, built of empire, still marks imperial pretensions on behalf of the city. Its cosmopolitanism may mean empire is coming home, but it also means that London's connections to the wider world are strengthening rather than weakening. This is a postcolonial city, still walking imperial paths. The monument, then, represents the dead pasts of the city that still haunt its present endeavours and sense of itself. But, like all monuments, it gathers in alternative histories too. Geographers have been keen to map out the alternative stories to monuments, perhaps to challenge the feeling that monumental space deadens the idea that history can move forward (Atkinson and Cosgrove, 1998; Harvey, 1979; Johnson, 1994; 1995; Till, 1999).

I won't continue with these geographies, since themes touched on here will be picked up in various sections of this book, for example subjectivity (Section 5) and geopolitics (Section 8); lines of engagement could also be drawn to issues of postcolonialism (in Sections 6 and 7) and even social and economic relations (in Sections 1 and 2). I guess one question mark concerns the role of culture in all of this. In some ways, of course, this is obvious. We can talk of cultures of more or less anything, from protest to capitalism. However, these geographies are struggles over meaning. This is not meaning in some abstract space that lies as a thin film over the surface of the world, or that is the fixed outcome of some logic of language. These meanings are about interactions and communications between people: people situated in specific places and times, dealing with the multiple worlds they inhabit and as these worlds pass through them. Meaning, in this sense, is forged in these struggles over geography, and struggles for geography; it is not brought in, as it were, from the outside. More than this, meaning and significance are not to be reduced to the textual or the symbolic: they are thoroughly embodied, in gestures, dance, song, voice, noise, the rhythms of life. In these senses, geography makes culture as much as culture makes geography.

STRUCTURE OF THE *HANDBOOK*

We have described the broad context and purpose of the *Handbook* in the Preface. In the first three sections of this introduction, we hope to have conveyed a sense of the unfolding intellectual terrains that constitute cultural geography. Finally, now, we would like to take a little time just to explain the structure of the book and the intentions built into it.

As you will see, there are nine sections. There could have been more, and of course less. In some ways, our choice of sections may seem a little odd. Why, for example, does a book on cultural geography have a section on the economy? It is our intention to show that thinking culturally about space has impacted on the full range of human geography, and not just on an identifiable subdiscipline labelled as cultural geography because of the 'objects' it chooses to study. Our suggestion that cultural geography is a set of engagements with the world carries implications, then, beyond the recognizable subdisciplinary boundaries within human geography. More than this, it carries implications for thinking across the blurred boundaries between geography and academic disciplines with related (spatial, cultural) interests.

We begin the *Handbook* with sections on the social and the economic precisely because these have often been seen as incompatible with, or antagonistic to, the subject of cultural geography. Indeed, we include authors who remain sceptical about the value of thinking culturally, precisely to show where core debates are taking place in thinking culturally about social and economic relationships. Following this logic, nature has been seen as the exclusion of that which is cultural and therefore provides a next step in this exploration of cultural thinking. In fact, culture has been heavily associated in geography with the analysis of landscape, our fourth section. The sections on nature and landscape have been placed side by side in order to give a sense of both the boundaries and the connections between them. Cultural geography has, in many ways, been born out of an engagement with issues around cultural politics (see Jackson, 1989; Mitchell, 2000; Philo, 1991). Geographers have been keen to explore the ways in which culture and geography intersect in the production of identity, and latterly subjectivity. In particular, geography has been sensitive to difference in different places, through a sense both of the legacies of their pasts, and the possibilities for alternative trajectories. These form the core of the next four sections, on subjectivity, postcolonialism, post-developmentalism and geopolitics. Finally, cultural geography has been continually reflexive about its own practices and knowledges. This forms part of the core work of the discipline, so our final section looks at geographies of knowledge.

We would naturally expect readers to dip in and out of the book, selecting for themselves those chapters that look most relevant to their own work. Nevertheless, the section editors provide invaluable contextualizations for the chapters. Not only do they situate chapters within wider debates, they also provide small maps of the debates in these fields of endeavour. Readers will therefore find these essays invaluable prior to reading specific chapters.

At the outset to this introduction, we declared that cultural geography is a living tradition, characterized by debate, enthusiasm, passion and commitment. In reading this book, we hope you will find that – more than this – there is a commonwealth of different intellectual cultures: to be drawn on, to be contested, to be changed, to do (differently) yourself.

FURTHER READING

In producing the 'case studies' we presented earlier, we realized that we were drawing on a large stock of shared readings. Instead of relentlessly citing these in the vignettes, we decided to collate them into one list. This is not meant to be a comprehensive list of all the key works in cultural geography, but is rather a rough guide to help you explore the sheer variety of cultural geographies around. We have included some monographs, as examples of the kinds of research undertaken in cultural geography, but more in the way of anthologies gathering together work in specific areas of research.

Adams, P., Hoelscher, S. and Till, K. (eds) (2001) *Textures of Place: Exploring Humanist Geographies*. Minneapolis: University of Minnesota Press.

Agnew, J. and Corbridge, S. (1995) *Mastering Space: Hegemony, Territory and International Politics*. London: Routledge.

Anderson, K. (1991) *Vancouver's Chinatown: Racial Discourse in Canada, 1875–1980*. Montreal: McGill–Queen's University Press.

Anderson, K. and Gale, F. (eds) (1999) *Cultural Geographies*, 2nd edn. London: Longman.

Barnes, T. and Duncan, J.S. (eds) (1992) *Writing Worlds: Discourse, Text and Metaphor in the Representation of Language*. London: Routledge.

Bell, D. and Valentine, G. (eds) (1995) *Mapping Desire: Geographies of Sexuality*. London: Routledge.

Blunt, A. and Wills, J. (2000) *Dissident Geographies: An Introduction to Radical Ideas and Practice*. London: Prentice Hall.

Brown, M.P. (2000) *Closet Space: Geographies of Metaphor from the Body to the Globe*. London: Routledge.

Butler, R. and Parr, H. (eds) (1999) *Mind and Body Spaces: Geographies of Illness, Impairment and Disability*. London: Routledge.

Cook, I., Crouch, D., Naylor, S. and Ryan, J. (eds) (2000) *Cultural Turns/Geographical Turns: Perspectives on Cultural Geography*. London: Prentice Hall.

Cosgrove, D. (1998) *Social Formation and Symbolic Landscape (1984)*. Madison: University of Wisconsin Press.

Cosgrove, D. (2001) *Apollo's Eye*. Baltimore: Johns Hopkins University Press.

Crang, M. and Thrift, N. (eds) (2000) *Thinking Space*. London: Routledge.

Cresswell, T. (1996) *In Place/Out of Place: Geography, Ideology, and Transgression*. Minneapolis: University of Minnesota Press.

Daniels, S. (1993) *Fields of Vision: Landscape Imagery and National Identity in England and the United States*. Cambridge: Polity.

Daniels, S. (1999) *Humphrey Repton*. New Haven: Yale University Press.

Doel, M. (1999) *Poststructuralist Geographies: The Diabolical Art of Spatial Science*. Edinburgh: Edinburgh University Press.

Domosh, M. (1996) *Invented Cities: The Creation of Landscape in Nineteenth-Century New York and Boston*. New Haven: Yale University Press.

Driver, F. (2001) *Geography Militant: Cultures of Exploration and Empire*. Oxford: Blackwell.

Duncan, J. and Gregory, D. (eds) (1997) *Writes of Passage: Travel Writing, Place and Ambiguity*. London: Routledge.

Duncan, J. and Ley, D. (eds) (1993) *Place/Culture/Representation*. London: Routledge.

Duncan, N. (ed.) (1996) *BodySpace: Destabilizing Geographies of Gender and Sexuality*. London: Routledge.

Gould, P. and Olsson, G. (eds) (1982) *A Search for Common Ground*. London: Pion.

Gregory, D. (1994) *Geographical Imaginations*. Oxford: Blackwell.

Harvey, D. (1989) *The Condition of Postmodernity*. Oxford: Blackwell.

Harvey, D. (1996) *Justice, Nature and the Geography of Difference*. Oxford: Blackwell.

Jackson, P. (1989) *Maps of Meaning: An Introduction to Cultural Geography*. London: Unwin Hyman.

Jacobs, J. (1996) *Edge of Empire: Postcolonialism and the City*. London: Routledge.

Jones, J.P., Nast, H. and Roberts, S. (eds) (1997) *Thresholds in Feminist Geography: Difference, Methodology, Representation*. Lanham: Rowman and Littlefield.

Keith, M. and Pile, S. (eds) (1993) *Place and the Politics of Identity*. London: Routledge.

Kirby, K. (1996) *Indifferent Boundaries: Exploring the Space of the Subject*. New York: Guilford.

Lefebvre, H. (1991) *The Production of Space (1974)*. Oxford: Blackwell.

Ley, D. and Samuels, M.S. (eds) (1978) *Humanistic Geography: Prospects and Problems*. London: Croom Helm.

Livingstone, D. (1992) *The Geographical Tradition: Episodes in the History of a Contested Enterprise*. Oxford: Blackwell.

Massey, D. and Allen, J. (eds) (1984) *Geography Matters! A Reader*. Cambridge: Cambridge University Press.

Matless, D. (1998) *Landscape and Englishness*. London: Reaktion.

May, J. and Thrift, N. (eds) (2001) *Timespace*. London: Routledge.

McDowell, L. (1997) *Capital Culture: Gender at Work in the City*. Oxford: Blackwell.

McDowell, L. (1999) *Gender, Identity and Place: Understanding Feminist Geographies*. Cambridge: Polity.

Mitchell, D. (2000) *Cultural Geography: A Critical Introduction*. Oxford: Blackwell.

Nast, H. and Pile, S. (eds) (1998) *Places through the Body*. London: Routledge.

Ogborn, M. (1998) *Spaces of Modernity: London's Geographies, 1680–1780*. New York: Guilford.

Ó Tuathail, G. (1996) *Critical Geopolitics*. Minneapolis: University of Minnesota Press.

Philo, C. (compiler) (1991) *New Words, New Worlds: Reconceptualising Social and Cultural Geography*. Department of Geography, St David's University College, Lampeter.

Philo, C. and Wilbert, C. (eds) (2000) *Animal Spaces, Beastly Places*. London: Routledge.

Pile, S. (1996) *The Body and the City: Psychoanalysis, Subjectivity and Space*. London: Routledge.

Pile, S. and Thrift, N. (eds) (1995) *Mapping the Subject: Geographies of Cultural Transformation*. London: Routledge.

Pred, A. (1995) *Recognizing European Modernities: A Montage of the Present*. London: Routledge.

Pred, A. (2000) *Even in Sweden*. Berkeley: University of California Press.

Rose, G. (1993) *Feminism and Geography: The Limits of Geographical Knowledge*. Cambridge: Polity.

Ruddick, S. (1996) *Young and Homeless in Hollywood: Mapping Social Identities*. New York: Routledge.

Sharp, J.P., Routledge, P., Philo, C. and Paddison, R. (eds) (2000) *Entanglements of Power: Geographies of Domination/Resistance*. London: Routledge.

Shurmer-Smith, P. (ed.) (2001) *Doing Cultural Geography*. London: Sage.

Sibley, D. (1995) *Geographies of Exclusion: Society and Difference in the West*. London: Routledge.

Skelton, T. and Valentine, G. (eds) (1997) *Cool Places: Geographies of Youth Culture*. London: Routledge.

Soja, E. (1989) *Postmodern Geographies: The Reassertion of Space in Critical Social Theory*. London: Verso.

Teather, E.K. (ed.) (1999) *Embodied Geographies: Space, Bodies and Rites of Passage*. London: Routledge.

Thrift, N. (1996) *Spatial Formations*. London: Sage.

Tuan, Y.-F. (1974) *Topophilia: A Study of Environmental Perception, Attitudes and Values*. Englewood Cliffs: Prentice Hall.

Tuan, Y.-F. (1977) *Space and Place: The Perspective of Experience*. London: Arnold.

Unwin, T. (1993) *The Place of Geography*. Harlow: Longman.

Whatmore, S. (2002) *Hybrid Geographies*. London: Sage.

REFERENCES

Atkinson, D. and Cosgrove, D. (1998) 'Urban rhetoric and embodied identities: city, nation, and empire at the Vittorio Emanuale II Monument in Rome, 1870–1945', *Annals of the Association of American Geographers* 88 (1): 28–49.

Banham, R. (1971) *Los Angeles: The Architecture of Four Ecologies*. New York: Harper and Row.

Bennett, J. (2001) *The Enchantment of Modern Life. Attachments, Crossings, and Ethics*. Princeton: Princeton University Press.

Bourdieu, P. (1999) *Pascalian Meditations*. Cambridge: Polity.

Carter, P. (1996) *The Lie of the Land*. London: Faber and Faber.

Classen, C. (1998) *The Colour of Angels*. New York: Routledge.

Cosgrove, D. (1993) 'Landscapes and myths, gods and humans' in B. Bender (ed.) *Landscape, Politics and Perspectives*. Providence, Berg. pp. 281–305.

Cosgrove, D. (2001) *Apollo's Eye*. Baltimore: Johns Hopkins University Press.

Dagognet, F. (1992) 'Toward a biopsychiatry' in J.Crary, and S. Kwinter, (eds) *Incorporations 6*. New York: Zone. pp. 516–41.

Dalby, A. (2000) *Dangerous Tastes. The Story of Spices*. Berkeley: University of California Press.

Davidson, J. (2000) 'Fear and trembling in the mall: women, agoraphobia, and body boundaries' in I. Dyck, N.D. Lewis and S. McLafferty (eds) *Geographies of Women's Health*. New York: Routledge. pp. 213–30.

Deleuze, G. and Guattari, F. (1998) *A Thousand Plateaus*. Minneapolis: University of Minnesota Press.

Donkin, R.A. (1999) *Dragon's Brain Perfume. An Historical Geography of Camphor*. Leiden: Brill.

Ellington, G. (1869) *The Women of New York, or The Underworld of the Great City*. New York: The New York Book Co.

Foucault, M. (2000) *Questions of Method*. London: Tavistock.

Gilroy, P. (1993) *The Black Atlantic: Modernity and Double Consciousness*. London: Verso.

Glendinning, S. (1996) 'Heidegger and the question of animality', *International Journal of Philosophical Studies* 4: 67–86.

Glendinning, S. (2000) 'From animal life to city life', *Angelaki: Journal of the Theoretical Humanities* 5 (3): 19–30.

Halttunen, K. (1982) *Confidence Men and Painted Women: A Study of Middle-Class Culture in America, 1830–1870*. New Haven: Yale University Press.

Harvey, D. (1979) 'Monument and myth', *Annals of the Association of American Geographers* 69: 362–81.

Hoskins, W. (1955) *The Making of the English Landscape*. Harmondsworth: Penguin.

Ingold, T. (1995) '"People like us": the concept of the anatomically modern human', *Cultural Dynamics* 7 (2): 187–214.

Jackson, P, (1989) *Maps of Meaning: An Introduction to Cultural Geography*. London: Unwin Hyman.

Johns, A. (2000) *The History of the Book*. Cambridge: Cambridge University Press.

Johnson, N. (1994) 'Sculpting heroic histories: celebrating the centenary of the 1798 rebellion in Ireland', *Transactions of the Institute of British Geographers* 19: 78–93.

Johnson, N. (1995) 'Cast in stone: monuments, geography and nationalism', *Environment and Planning D: Society and Space* 13: 51–66.

Katz, J. (1999) *How Emotions Work*. Chicago: University of Chicago Press.

Latour, B. (1993) *We Have Never Been Modern*. London: Harvester Wheatsheaf.

Latour, B. (1999) *Pandora's Hope*. Harvard: Harvard University Press.

Law, J. and Mol, A. (2001) 'Situating technoscience: an inquiry into spatialities', *Environment and Planning D: Society and Space* 19: 609–22.

Marks, L. (2000) *The Skin of the Film: Intercultural Cinema, Embodiment and the Senses*. Durham, NC: Durham University Press.

Mitchell, D. (2000) *Cultural Geography: A Critical Introduction*. Oxford: Blackwell.

Newman, C. (1998) *Perfume. The Art and Science of Scent*. Washington: Nature Geographic.

Pearson, K.A. (1997) *Viroid Life: Perspectives on Nietzsche and the Transhuman Condition*. London: Routledge.

Pearson, K.A. (1999) *Germinal Life: The Difference and Repetition of Deleuze*. London and New York: Routledge.

Philo, C. (compiler) (1991) *New Words, New Worlds: Reconceptualising Social and Cultural Geography*. Department of Geography, St David's University College, Lampeter.

Pile, S. (forthcoming) 'Ghosts and the city of hope' in L. Lees (ed.) *The Emancipatory City*. London: Sage.

Pile, S., Brook, C. and Mooney, G. (eds) (1999) *Unruly Cities? Order/Disorder*. London: Routledge.

Rubenstein, J. (1999) *The Cultural Landscape: An Introduction to Cultural Geography*. London: Prentice Hall.

Sauer, C. (1952) *Agricultural Origins and Dispersals*. New York: American Geographical Society.

Sauer, C. (1956) 'The agency of man on earth' in W. Thomas (ed.) *Man's Role in Changing the Face of the Earth*. Chicago: Chicago University Press.

Sharp, J.P., Routledge, P., Philo, C. and Paddison, R. (eds) (2000) *Entanglements of Power: Geographies of Domination/Resistance*, London: Routledge.

Smith, B. (1995) *The Emergence of Agriculture*. New York: Freeman.

Strathern, M. (1999) *Property, Substance and Effect: Anthropological Essays on Persons and Things*. London and New Brunswick: Athlone.

Thévenot, L. (2001) 'Organising complexity', *European Journal of Social Theory* 4: 405–26.

Thomas, N. (1997) 'Anthropological epistemologies', *International Social Science Journal*, 153. Oxford: UNESCO, Blackwell. pp. 333–44

Thrift, N.J. (1996) *Spatial Formations*. London: Sage.

Thrift, N.J. (2000) 'Afterwords', *Environment and Planning D: Society and Space* 18: 213–55.

Thrift, N.J. (2002) 'Summoning Life' in P.J. Cloke, P. Crarg and M. Goodwin (eds) *Envisioning Geography*. London: Arnold.

Till, K. (1999) 'Staging the past: landscape designs, cultural identity and *Erinnerungspolitik* at Berlin's Neue Wache', *Ecumene* 6 (3): 251–83.

Wharton, E. (1986) *The House of Mirth* (1905). New York: Bantam.

Wharton, E. and Codman, O. (1897) *The Decoration of Houses*. New York: Charles Scribner's Sons.

Whatmore, S. (1999) 'Hybrid geographies: rethinking the "human" in human geography' in D. Massey, J. Allen and P. Sarre (eds) *Human Geography Today*. Cambridge: Polity. pp. 22–40.

Whatmore, S. and Thorne, L.B. (1998) 'Wild(er)ness: reconfiguring the geographies of wildlife', *Transactions of the Institute of British Geographers* 23: 435–54.

James VanDerZee, *Couple*, Harlem, 1932 © Donna Mussenden VanDerZee
(Source: Schomburg Center for Research in Black Culture, The New York Public Library.
Photo: courtesy of Donna VanDerZee)

Section I

RETHINKING THE SOCIAL   Edited by Peter Jackson

Introduction: The Social in Question

Peter Jackson

When in 1987 British Prime Minister Margaret Thatcher declared that there was 'no such thing as society', her comments hit a raw nerve with social scientists who had for years suffered a sense of professional anxiety over exactly what constitutes 'the social' and how it might be differentiated from other concepts such as 'the economy' or 'culture'. For Margaret Thatcher, 'society' was a dubious abstraction which was too readily invoked in support of policies that she and her supporters would have defined as socialist. While Mrs Thatcher preferred to speak of individual men and women and in support of nuclear, heterosexual families, her 'no society' rhetoric served as a covert means of promoting economically driven solutions to social problems, with 'market forces' increasingly replacing 'society' at the centre of British political discourse. If nothing else, Thatcher's remarks demonstrate that definitions of the social are politically charged, with more at stake than a question of semantics.

What constitutes 'the social' has always been in question among geographers and other social scientists. For some, it can be defined in terms of hierarchical structures and relatively permanent institutions. For others, it is a much more elusive concept, referring to relatively impermanent interpersonal relations and fleeting affinities, continuously made and remade rather than firmly institutionalized

and historically sedimented. Raymond Williams (1976: 291–3) traces the etymological roots of our current understandings of 'society' and 'the social', demonstrating how a relational view of society as companionship or association (from the sixteenth century) gave way to a more abstract and impersonal understanding of society or social structure, closely related to the state (from the eighteenth century). By the nineteenth century, Williams argues, 'society' was sufficiently objectified to give rise to debates about the relationship between 'the individual' and 'society'. Similar debates continue to resonate today, despite attempts to transcend such dualisms with more sophisticated accounts of the recursive relationship between (social) structure and (human) agency (Giddens, 1985).

Some social scientists, such as John Urry (2000), have begun to doubt the usefulness of the term 'society' and to sketch a 'post-societal' agenda for sociology, acknowledging that sociological categories of class, gender and ethnicity cannot be mapped unproblematically onto the geographical spaces of the nation, region or city. As the flows of people, information, goods and capital are increasingly exceeding the boundaries of individual nations, Urry suggests that metaphors of networks, mobilities and flows are more appropriate than static, bounded notions of 'society'. He also argues that the complex interweaving of society and nature, explored in recent debates about actor network theory and related arguments about 'non-representational theory' (Thrift, 1996), provide further evidence that 'society' is no longer a valid object of sociological discourse. Reversing the conventional logic, theorists like Bruno Latour (2000) have argued that the social should now be seen as that which circulates within the world of things, not what things circulate within.

Social historians such as Patrick Joyce have also put 'the social' in question, challenging our tendency to think of society in reified terms (as in 'social structure' or the 'social system'). Joyce (2001) seeks to replace this solid, ontological sense of 'society' with a more fluid understanding of the social as constituted through particular practices, materialities and embodied beings. As these debates among sociologists and social historians imply, the social sciences have, as disciplinary forms of knowledge, played their own part in the constitution and transformation of the social. And, as Joyce himself points out, geographers have made a significant contribution to this current rethinking of the social.

For geographers, rethinking 'the social' has involved a parallel rethinking of space. Like society, space is no longer adequately theorized in a static or bounded sense but is increasingly understood in relational terms. In Doreen Massey's (1994) work, for example, the emphasis is on the linkages between places at a variety of scales from the local to the global, rather than on a more fixed and bounded sense of place. A progressive or global sense of place is, she argues, about tracing the *routes* which connect different places rather than a nostalgic concern for the *rootedness* of any particular place.

The chapters in this section all approach 'the social' through the lens of the spatial. This does not imply that the social can only be understood through empirically grounded research on particular places at particular times (though an insistence on historical and geographical specificity has much to recommend it). It also involves a commitment to exercising and interrogating our *geographical imaginations* (Gregory, 1994). More broadly, it means that human geographers and other social scientists should expend more intellectual energy in seeking to understand *the spatiality of social life*.

This might seem unremarkable, but for generations social theory has emphasized temporality rather than spatiality. So, in the mid 1980s, it was felt necessary to remind social scientists that society does not take place on the head of a pin, that social relations are underpinned and reproduced through spatial structures, and that there are spatial as well as social divisions of labour (Gregory and Urry, 1985; Massey, 1984). Indeed, the 1980s were characterized by Ed Soja (1989) as involving a 'reassertion of space in critical social theory'.

Since then, across the social sciences, there has been significantly more interest in the complex relationships between society and space. As a result, geographers need no longer be quite so insistent in asserting that 'Geography matters!' (Massey and Allen, 1984), although there is still much to be done to demonstrate 'the difference that space makes' (Sayer, 1985) in particular social situations. From the late 1980s, the intellectual traffic between geography and the other social sciences has been increasingly two-way, with books like David Harvey's *The Condition of Postmodernity* (1989) attracting a remarkably wide audience beyond the discipline of geography. The interdisciplinary field of cultural studies has seen a proliferation of works that explore the significance of spatial metaphors such as borders and margins, spaces and places, centres and margins – though the use of such metaphors is itself contested (Pratt, 1992).

This growing interest in spatiality across the social sciences has been reciprocated by geography's increasing engagement with social theory. Both trends came together in the so-called 'cultural turn' which brought cultural questions of meaning, identity and representation to centre-stage and caused some to question whether political and economic issues had been displaced. Within geography, Nigel Thrift declared the cultural to be virtually hegemonic by 1991 and, a few years later, Andrew Sayer (1994) was warning cultural geographers of the dangers of ignoring 'the economy, stupid'. Charging geographers with neglecting material inequalities in society, Nicky Gregson (1995) felt as though she was writing an obituary for social geography. It was certainly not the case that social inequalities of race, class and gender – the focus of research for generations of social geographers – had suddenly disappeared (far from it). Rather, new concerns with cultural difference and the politics of representation – informed by postmodern, postcolonial and other forms of poststructuralist theory – had taken hold across the social sciences, leading to a relative lack of research on structures of inequality and their associated spatialities. In Gregson's words, questions of meaning, identity, representation and ideology were increasingly replacing studies that were more firmly grounded in an analysis of the material world or concerned with socially significant differences of gender, class, race, sexuality and (dis)ability (1995: 139).

The perceived *evacuation of the social* in social and cultural geography is a serious concern that each of the chapters in this section addresses in a different way. Taken together, however, the chapters provide strong evidence that the 'death of the social' has been prematurely proclaimed. Nicky Gregson's chapter, for example, demonstrates that 'the social' has not simply been replaced by 'the cultural' but is increasingly refracted through the cultural. Gregson demonstrates that multiple social and cultural geographies coexist rather than a single hegemonic social geography being replaced by an equally hegemonic cultural geography. So, too, in Katharyne Mitchell's chapter, the importance of 'grounding' the analysis of hybrid cultures in the materialities of specific times and places is central to her argument. In place of stark oppositions between the social and the cultural, the material and the symbolic, discourse and practice, each of the following chapters seeks to transcend such dualisms. Instead, each chapter looks forward to a geography in which symbolic meanings are embedded in specific material contexts, defined by unequal relations of power, structured by culturally mediated notions of social difference.

The remainder of this introduction focuses on two related issues. The first is intended to further underline the potential for seeing social and cultural questions *through the lens of the spatial*. The second provides further evidence of the value of seeing the social *through the lens of the cultural* (and in so doing demonstrates

the need to transcend all such dualistic ways of thinking).

SEEING THE SOCIAL THROUGH THE LENS OF THE SPATIAL

Our account begins with a period when social geography was in an upbeat and optimistic mood: when geographers were opening up a dialogue with other social scientists about the reciprocal relations between *society and space* (signalled by the journal of that name, launched in 1983). Geographers were no longer in the thrall of the other social sciences, no longer content to be the cartographers of social inequalities that were left to other disciplines to theorize and explain. There was an increasing acceptance that social relations were spatially constituted (signalled by interdisciplinary collections such as Gregory and Urry's *Social Relations and Spatial Structures*, published in 1985). With the cultural turn in human geography, geographers felt a new confidence about contributing to debates in cultural studies, even while we were being encouraged to question our ethnographic cultural authority and examine our own positionality.

The cultural turn, especially through the influence of feminism, also led to a broader understanding of what constitutes 'the political'. What might appear from one perspective as evidence of the end of the social could be read more optimistically from a feminist perspective as a broadening out of the political to include questions of visual and textual representation, the cultural politics of sexuality and the body, and more expansive notions of citizenship. This sense of optimism was, in turn, challenged by voices 'from the margins' of white western feminism which raised uncomfortable questions about the privileges of power including the unexamined 'whiteness' of much current academic practice (hooks, 1992).

While Gregson's chapter explores these issues in the context of recent disability studies in geography, geographies of race and racism are equally revealing of the theoretical ebb and flow of social and cultural geography. The social was traditionally seen in terms of a series of relatively independent 'dimensions' of difference whereby it was implied that inequalities of race, class and gender could be stacked up in some simple additive fashion. These separate 'dimensions' were then mapped, either literally or figuratively, in successive studies of residential segregation. It became increasingly apparent, however, that standard cartographies were unable to deal with new theorizations of the *mutual constitution* of race, class and gender. As Vron Ware (1992) and others argued, gender identities are simultaneously racialized (and vice versa).

Recent work on the geographies of race and racism (as reviewed by Bonnett, 1996; see also Chapter 15, this volume) increasingly acknowledges that we all lead racialized lives rather than assuming that 'race' or 'ethnicity' are terms that only apply to certain minority groups. Rather than focusing on the spatial expression of ethnic or racialized identities, therefore, current research is increasingly exploring *the racialization of space* as an active, constitutive process at a variety of scales from the neighbourhood to the nation and beyond (see Smith, 1989). It is in this sense that Avtar Brah revives the notion of cartography, to argue that what she calls 'diaspora space' is 'inhabited' not only by diasporic subjects but equally by those who are constructed and represented as 'indigenous' (1996: 16). So, too, studies such as Kay Anderson's (1991) analysis of the social and spatial construction of Vancouver's Chinatown illustrate the turn from mapping the geographies of isolated ethnic minorities to studies that explore the role of dominant discourses in the construction of racialized space. Likewise, Jane Jacobs' (1996) account of the postcolonial geographies of urban redevelopment in London demonstrates a relational view of space whereby the view from the financial centre of the City of London (the 'heart of empire') cannot be divorced from an exploration of the margins of racialized inner-city neighbourhoods such as Spitalfields.

A relational view of identity and space is also intrinsic to recent studies of transnationality, explored in Katharyne Mitchell's chapter. Such research focuses on the transnational connections between people, places and things, drawing together a political-economic understanding of capital investment and labour migration with a critical understanding of the hybrid cultures and identities that result from such processes. Indeed, as Roger Rouse's (1991) work on the transnational geographies of Mexico and southern California reveals, transnational labour migration and business development have opened up a series of new spaces that are increasingly encompassing all sections of society, not just those people who are themselves directly connected to transnational migrant communities. In Rouse's words: 'The comfortable modern imagery of nation-states and national languages, of coherent communities and consistent subjectivities … no longer seems adequate … [D]uring the last 20 years, we have *all* moved irrevocably into a new kind of [transnational] social space' (1991: 8, emphasis in the original). Research on transnationality has also repudiated more simple-minded, top-down models of globalization, demonstrating the continued significance of local contexts of consumption and calling forth a more relational view of the global–local nexus.

SEEING THE SOCIAL THROUGH THE LENS OF THE CULTURAL

As well as demonstrating the value of approaching the social through the lens of the spatial, each of the chapters in this section also demonstrates the importance of seeing the *social* through the lens of the *cultural*, rather than seeing social and cultural geographies as mutually antagonistic (one concerned with material inequalities, the other with discursively constructed differences). Recent research on disability (reviewed here by Gregson and by Moss and Dyck) demonstrates the value of tracing the social construction of disability in both its material and its discursive forms. Rather than focusing exclusively on problems of access, where space − defined in terms of barriers and obstacles − is a mere container for human action, recent work has shown how disabled space is actively constituted through countless decisions about planning and design, regulated and reinforced by the power of the state. From this perspective, 'disabling space' is as culturally constructed as 'disability' itself. Some environments are more enabling than others, just as some conditions (such as

HIV/AIDS) are more socially stigmatized. Some disabilities (such as dyslexia) are relatively invisible, while others are more publicly apparent.

While, for many years, geographers and other social scientists adopted a curiously disembodied approach to human subjectivity, recent geographies of self and identity have been increasingly embodied (Pile and Thrift, 1995). Again, however, this is not merely a question of *reinstating corporeality* (as Moss and Dyck insist through their distinction between 'embodied geographies' and 'geographies of embodiment'). Rather, it involves an exploration of the social and spatial construction of our embodied identities. Judith Butler's work has made a key contribution to this debate, arguing that gender difference is not merely a cultural elaboration of pre-discursive, biologically defined sexual difference. According to Butler (1990), sexual difference is itself culturally encoded through the repetitive enactment of gender as a social practice. Gender differences are, in turn, made culturally intelligible through the regulatory grid of compulsory heterosexuality which renders some bodies socially acceptable while making others discursively impossible. Moreover, an emphasis on performativity and practice helps dissolve the boundaries between the discursive and the material, while geographical research on sex–gender differences has underlined the significance of space in providing a hetero-normative context for interpreting the cultural significance of different embodied identities (Bell et al., 1994; Valentine, 1993).

While Moss and Dyck's chapter emphasizes the maturing of qualitative methods in human geography (see Limb and Dwyer, 2001), with an increasing concern for the positionality and reflexivity of the researcher, the chapters in this section should not be read in a triumphalist manner, as a succession of ideas and approaches that demonstrate a steady march of progress towards the heroic present. Rather, the chapters that follow represent a more modest contribution to theoretical enquiry. For, as the previous examples suggest, cherished distinctions between sex and gender, the material and the discursive, the social and the cultural all begin to collapse as soon as our analytical gaze intensifies. Whether the resulting ambiguities and ambivalences are a source of disciplinary anxiety or an opportunity to think in new and unconventional ways raises complex questions about the relationship between power and knowledge. The essays in this section are all motivated by a desire to explore more cultural understandings of the social without abandoning the quest for a more critical politics of difference. Approaching the social through the lens of the cultural and seeing both through the lens of the spatial establishes an exciting and open-ended agenda for future research.

REFERENCES

Anderson, K.J. (1991) *Vancouver's Chinatown.* Montreal and Kingston: McGill–Queen's University Press.

Bell, D., Binnie, J., Cream, J. and Valentine, G. (1994) 'All hyped up and no place to go', *Gender, Place and Culture* 1: 31–48.

Bonnett, A. (1996) 'Constructions of "race", place and discipline: geographies of "racial" identity and racism', *Ethnic and Racial Studies* 19: 864–71.

Brah, A. (1996) *Cartographies of Diaspora.* London: Routledge.

Butler, J. (1990) *Gender Trouble.* London: Routledge.

Giddens, A. (1985) *The Constitution of Society.* Cambridge: Polity.

Gregory, D. (1994) *Geographical Imaginations.* Oxford: Blackwell.

Gregory, D. and Urry, J. (eds) (1985) *Social Relations and Spatial Structures.* London: Macmillan.

Gregson, N. (1995) 'And now it's all consumption?', *Progress in Human Geography* 19: 135–44.

Harvey, D. (1989) *The Condition of Postmodernity.* Oxford: Blackwell.

hooks, b. (1992) *Black Looks: Race and Representation.* London: Turnaround.

Jacobs, J.M. (1996) *Edge of Empire: Post-colonialism and the City.* London: Routledge.

Joyce, P. (ed.) (2001) *The Social in Question.* London: Routledge.

Latour, B. (2000) 'When things strike back: a possible contribution of "science studies" to the social sciences', *British Journal of Sociology* 51: 107–23.

Limb, M. and Dwyer, C. (eds) (2001) *Qualitative Methodologies for Geographers.* London: Arnold.

Massey, D. (1984) *Spatial Divisions of Labour.* London: Macmillan.

Massey, D. (1994) *Space, Place and Gender.* Cambridge: Polity.

Massey, D. and Allen, J. (eds) (1984) *Geography Matters!* Cambridge: Cambridge University Press.

Pile, S. and Thrift, N. (eds) (1995) *Mapping the Subject.* London: Routledge.

Pratt, G. (1992) 'Spatial metaphors and speaking positions', *Environment and Planning D: Society and Space* 10: 241–4.

Rouse, R. (1991) 'Mexican migration and the social space of postmodernism', *Diaspora* 1: 8–23.

Sayer, A. (1985) 'The difference that space makes', in D. Gregory and J. Urry (eds) *Social Relations and Spatial Structures*. London: Macmillan. pp. 49–66.

Sayer, A. (1994) 'Cultural studies and "the economy, stupid"', *Environment and Planning D: Society and Space* 12: 635–7.

Smith, S.J. (1989) *The Politics of 'Race' and Residence*. Cambridge: Polity.

Soja, E. (1989) *Postmodern Geographies*. London: Verso.

Thrift, N.J. (1991) 'Over-wordy worlds?', in C. Philo (ed.) *New Words, New Worlds: Reconceptualising Social and Cultural Geography. Conference Proceedings*. Aberystwyth: Cambrian. pp. 144–8.

Thrift, N.J. (1996) *Spatial Formations*. London: Sage.

Urry, J. (2000) *Sociology beyond Societies*. London: Sage.

Valentine, G. (1993) '(Hetero-)sexing space: lesbian perceptions and experience of everyday spaces', *Environment and Planning D: Society and Space* 11: 395–413.

Ware, V. (1992) *Beyond the Pale: White Women, Racism and History*. London: Verso.

Williams, R. (1976) *Keywords*. London: Fontana.

1

Reclaiming 'the Social' in Social and Cultural Geography

Nicky Gregson

Neil Smith commented recently that 'cultural and social geography have experienced divergent fates in the English-speaking world during the last decades of the twentieth century'. Indeed, he went on to claim that cultural questions now predominate at the research frontiers of the human side of the discipline and that, by comparison, 'social geography has languished' (2000: 25). In this chapter, I want to develop a more nuanced reading of events, to argue that the trajectory of social geography in the English-speaking world, and particularly in Britain, through the 1990s cannot be depicted in the singular, as an increasingly marginal counterpoint to the cultural. Neither is it best understood through the lens of fate and its associated, contrasting academic fashions. Rather, I want to suggest that social geography/ies and cultural geography/ies currently are relational constructions, but in a way which is not just about authorial power within (and ability to define) the discipline. Indeed, I maintain that 'the cultural' is a presence, respectively mobilized to enable specific reconfigurations of 'the social' on the one hand and to permit both a reassertion and an evacuation of the social on the other. Rather than writing out 'the social' then, 'the cultural' has figured as its prism, as a means of its refraction. Correspondingly, I argue in this chapter that, rather than having languished, 'the social' has been simultaneously reconfigured, reasserted and evacuated – a position which immediately exposes the multiplicity of understandings of 'the social' currently copresent within British human geography, and their relative authority/ies.

Developing such a position requires that I cover a deal of ground with considerable rapidity. What it also means though is that neither can

I hope to achieve, nor would I want to claim, an exhaustiveness of coverage. This has risks: in covering so much these arguments will be vulnerable, particularly in their specifics, to charges of oversimplification, glossing and so forth. Nonetheless, this seems a risk worth taking, for two reasons. First, it provides the space to represent British social geography in all its partial, uncertain, provisional and messy complexity, that is as situated within and produced through an incredibly complex interwoven field of power–knowledge. Secondly, this representation helps foreground what we are struggling over, what we mean when we invoke the term 'the social'. While this is rarely debated within human geography in the English-speaking world (at least in recent years), there is a strong case for doing so now; not least because – as I show here – the multiple understandings of 'the social' that exist within the discipline work with very different formulations of how 'the social' connects to society. Indeed, whereas one reading of this relation is grounded in no more than inclusion/ exclusion (and therefore rejection, if not always abjection), for others it goes beyond this and is tied intrinsically to the material conditions of societal reproduction, to the constitution of society through and by economy and polity. Having established these points, in the next two sections I argue that there is a clear case to be made for reclaiming a 'social' that relates directly to the materiality of social life, specifically to the conditions of its organization and reproduction. This involves a return to old(er) questions about the materiality of society and about societal inequalities. The penultimate section offers some suggestions as to how such issues might begin to be reclaimed and reimagined – by taking seriously consumption, material culture and a discursive that connects with the material. This is a tactic not just which is indicative of personal academic investments, but which works by refracting facets of 'the cultural' – specifically material culture – onto a particular reading of 'the social' (see Jackson, 2000). Although few will probably agree wholly, or perhaps even partially, with such a course, my hope and intention here are that such thoughts might at least precipitate a degree of reflection. For – in agreement with Neil Smith – I would argue that 'what constitutes and comprises "the social"' are questions that go beyond the academy to connect again with an older, yet no less pertinent, vision of critical scholarship. Asking these questions might indeed be a necessary radical act for an increasingly self-styled and self-referential critical human geography.

EVACUATING OR RECONFIGURING 'THE SOCIAL'?

A generation ago the future for social geography looked assured. Theoretically, David Harvey's inspirational text *Social Justice and the City* (1973) provided a means of embedding a range of social questions within broader debates about inequalities and redistributional politics. Empirically, much of social geography – at least in the English-speaking world – was characterized by critical and committed analyses of housing, health and education, to take just a few of the most obvious examples. Informed by political economic approaches, and to a more limited degree by feminism, such work exhibited vibrancy and intensity that is evident even now, as witnessed by recent contributions to the gentrification literature or to uncovering geographies of health. Importantly, too, the best of this work was not simply 'radical' in its analysis, but radical in intent too. It was about trying to produce analyses that, potentially, could make a difference to people's everyday lives and life chances. In short, it was work that wore its political colours on its sleeve; and those colours were unashamedly of the left.

So, what happened? Reflecting on the very same question, Neil Smith posits a disciplinary explanation, arguing that social geography was squeezed out – the victim of a pincer movement between political economy and the move to what he labels 'cultural deconstruction'. In the process, he maintains, what had come to be seen as significant lines of social differentiation –

notably the much recited mantra of class, gender and race – came to be recast: as identities, even as subjectivities, to be informed more by a psychoanalytically influenced cultural theory and various versions of anti-foundationalist post-structuralist thinking than by a social theory tainted with materialist associations. Thus, 'the cultural' usurped 'the social', or so the story goes.[1]

Although attractive in its simplicity and persuasive in the way in which the demise of 'the social' is hinged to the progressivist development of the discipline, Smith's argument is problematic in at least two ways: first, because writing (the) history/ies of disciplinary tradition/s is, as Smith himself is well aware (Godlewska and Smith, 1994), a rather more complicated, altogether messier, activity than is given credit here; and secondly, because there is a presumption in this narrative that we all know what is meant by and included within the term 'the social', although it is important to note that this is a very long way from presumed by the end of Smith's commentary. As I go on to argue in this section and in the subsequent one, weaving together these two threads results in a more nuanced reading of the trajectory of 'the social' and in a rather different narrative from a straightforward tale of demise.

In developing my argument here, I want to begin with two of the core tenets of recent writing/s on disciplinary tradition/s: their situatedness and their relation to debates over authority, authorship, inclusion and exclusion (Driver, 1995; Livingstone, 1992; Rose, 1995). In terms of 'the social', these points are suggestive immediately of the coexistence of multiple, rather than singular, understandings of 'the social' – ones which have relative degrees of authority within the discipline, depending on their relation to prevailing and/or fashionable theoretical canons. This we can see, not least when we look back to the 1970s, when political-economy-inspired readings of 'the social' held sway over, yet were copresent with, more empiricist interpretations, characterized notably by the then plethora of segregation studies (Peach et al., 1981; see Jackson and Smith, 1984). Correspondingly, what this should alert us to in the contemporary period is the coexistence of yet further representations of 'the social'; and that, rather than having languished, social geography might have been reconfigured.

When we look at a range of work produced through the 1990s by geographers, we can see that such suggestions are indeed borne out. Studies of sexuality, (dis)ability, children's geographies, parenting and youth, for example, are all indicative of the diversity of research currently being conducted under what could be termed a broadly 'social' umbrella.[2] Without doubt, this work has been critical in countering many of the unconsidered yet highly troubling assumptions lurking within much previous social geography, notably its ableism, its heterosexism and its adult-centredness. Yet – and this is the big point of contention – such work has very different, frequently implicit, understandings of 'the social' from those which shaped and continue to shape readings informed by political economic accounts. These differences are worth spelling out here, for they are at the heart of understanding what is at stake when claims as to the demise and/or evacuation of 'the social' are made. Consequently, and by way of illustration, I use the growing body of work on (dis)ability to begin to tease out the vision/s of 'the social' which are being presumed here.[3]

Research on (dis)ability in geography comprises a vibrant and burgeoning field, characterized by a proliferation of perspectives (behaviouralist, historical materialist, phenomenological, Foucauldian), strong disagreements over appropriate terminologies, and connections which span the full gamut of knowledge divisions: from medicine to architecture, planning and design, to cultural studies and psychology.[4] Empirically, it encompasses studies as diverse as: mental health and illness (in itself a disputed presence); disability in the workplace and employment; the lived experience of disability; technology; activism; and city and housing design. Methodologically, it features profound distinctions: between writing on, for, with or about, to writing by and autobiographical writing. As such, even to attempt a (necessarily) condensed overview such as this seems invidious. Yet, as most reviews by those working directly in the field point out, there is a clear trajectory shaping these studies – a narrative of progress that is used to relate the growing sophistication of the field. It is on these that I draw here, because they are most suggestive of how 'the social' has been reconfigured.

Almost all of these reviews begin by reiterating the basic distinction between approaches which centre the disability of the individual (a view which foregrounds pathology, attention and care), and those which understand and represent disability as produced by society: that is, as the social processes which work to exclude and/or disadvantage those with disabilities. For some, and I'm thinking here particularly of the work by Brendan Gleeson (1996; 1997), the concern has been specifically to connect disability with political economy; to expose how the development of

advanced capitalism has progressively devalued the labour power of those with disabilities (see Hall, 1999). Yet, as is clear from a recent edited collection (Butler and Parr, 1999) and from recent papers, there is a growing dissatisfaction with this position; specifically over its denial of the significance of impairment, its exclusion of pain, fatigue and depression; in short, a dissatisfaction with its exclusion from analysis of the very materiality of the disabled body (Hall, 2000). Correspondingly, and bolstered by the more general turn to the bodily within geography,[5] it is the messy, 'undisciplined' matter of disabled bodies which we can anticipate to provide a focal point for much future research in this area.

What this might mean in terms of understandings of 'the social' is less transparent, but there are nonetheless several pointers within the literature. In a recent paper, for example, Edward Hall argues for an 'embodied' and 'biological' approach to impairment, 'which sees the body as social and society as bodily' (2000: 24), whilst in the introduction to their recent collection, Ruth Butler and Hester Parr (1999) centre the notion of 'mind and body spaces'. Although different, both arguments emphasize a corporeality which is the site of social inscriptions and which defines and produces particular social relations and day-to-day lived geographies (clinics, surgeries, drop-in centres, parks) as well as identities and subjectivities. There is a sense here then – although this is nowhere made explicit – of a social which is understood in terms of the everyday; as routine, regulated, resisted even, but individual-centred. Yet, simultaneously, this is a social that is located in and defined by the materiality of specific (disabled) bodies. This, in turn, is suggestive of a social located more in the politics of difference than in a politics of inequality; a view confirmed by Butler and Parr's portrayal of Gleeson's work as 'materialistic', and from which they appear to wish to distance themselves.

More generally, looking across this literature and elsewhere within recent social geographical work, there are four points that I would pull out as comprising key components in how 'the social' is increasingly being understood. Two are presences, in the sense that they are positively talked up in the literature; and two are significant absences.

The first of these components is its body-centredness. This we can see from the way in which social geography's subjects are increasingly defined through their relation to the bodily and from the frequently asserted claim that the body is social and that society has to be related to the bodily.[6] Illustrative of the more widespread move to centre the body within social theory and the social sciences more generally (Butler, 1990; 1993; Crossley, 1995; 1997; Featherstone et al., 1991; Gatens, 1996; Grosz, 1994; Shilling, 1993; 1997; Turner, 1996; though see Witz, 2000), this work also signals a significant reorientation of the scale/s through which 'the social' is being understood and interrogated within social and cultural geography. Indeed, this bodily 'social' is predominantly represented as located within the 'scales' inhabited by individual bodies, notably homes and neighbourhoods. This is very different from definitions of 'the social' that privilege other geographical scales, for instance nation-states or (western) cities, or indeed from readings of the body that connect to governance, regulation and citizenship. Just how different can be seen when we think, for example, about the various ways in which these scales connect to materiality. Here, we can posit a key distinction between the materialities of bodies – which concern the flesh, bodily surfaces and boundaries, fluids, dirt, matter-out-of-place, pollution and so on – and the materialities of neighbourhoods or cities, for instance. For with the latter, matter – the material – is not just 'fixed' in space (which is an important enough distinction from materialities of the body) but is woven together and understood through various social relations – but particularly of exchange and use – as resources: that is, as things, goods, services or commodities which can be used and/or bought/sold.[7] Consequently, materiality at these scales is more likely to be thought about and articulated in relation to bricks and mortar, concrete and steel (housing, schools, health centres, shops, places of work, etc.) and is more likely too to be connected to questions of disadvantage and deprivation, to be discussed in terms of inequality/ies in access/ provision, life chances and so on.[8] Reconfiguring the scale of 'the social' has important effects in terms of how materiality is understood, and to what it relates. In short, it has significant ontological implications, which need to be recognized.

A second core component in this social geography follows on from this. As I have shown above, this bodily social is one which appears to be being constructed primarily as pertaining to and defining of the person and of particular embodiments, be these transitional, temporary or permanent states of being. Furthermore, the situation of this bodily social – its key geographies – is principally within 'everyday' landscapes, places and institutions. Now this is frequently but a short step, albeit not a necessary one, from a social that is understood in terms of embodiments as experienced and lived, that is personally by individuals. Some of the difficulties with such a position have been well documented, notably in

relation to the unproblematized relation which this posits between knowledge and experience, and the ease with which such a position comes to represent individuals as (the) victim/s. But what I find particularly intriguing, and to an extent depressing, here is the way/s in which individuals come to represent and be represented as the effect of 'the social'. Rather than addressing the varied sociality/ies of particular embodiments, particularly the way/s in which collective mobilization, campaigning and activism are seeking to challenge social exclusion (Chouinard, 1999), much of this geographical work still tends to begin and end with the individual's experience of exclusion – a comment which is particularly apposite to some of the research on disability (Dyck, 1999; Moss, 1999).

This leads me to a third component of this social geography: an absence. What is missing from so much of this work is a sense of the social which goes beyond the individual (or groups of individuals who are socially defined); which hinges its imagining of the social to an understanding of society and the conditions of its organization and reproduction. To expand: when we look at a great deal of current 'social' geographical literature, the understanding of society which it appears to operate with is one which is grounded in relatively imprecisely specified notions of inclusion/exclusion and marginalization, which frequently retreat – often imperceptibly – to little more than analyses of 'the excluded' and/or 'marginalized'. Much of the disability literature provides a case in point: in concentrating on specific individuals with particular disabilities, this work defines itself simultaneously against the able-bodied and in terms of the excluded and/or marginalized group. Similarly, much the same comment was made in respect of earlier work on sexuality, with its focus on gay, lesbian and bisexual identities, rather than on the processes, structures and institutions of heterosexism and homophobia, and it is also a charge which could be levelled at some work on children and youth. Conceptually, what this means is that such research ends up reinstating the very oppositions which it seeks to challenge, and that 'the excluded' are defined by, and remain trapped within, their representation as specific instances of exclusion – a situation which has been explored theoretically elsewhere in connection with poststructuralist feminist thinking by Gillian Rose (1993). What concerns me more, however, in this context is the lack of comment that the inclusion/exclusion opposition all too frequently appears to merit within much of this social geography, and particularly the lack of reflection on what this might presume about how societal reproduction is being understood.

A notable exception to this tendency is, of course, David Sibley's *Geographies of Exclusion* (1995), a text that is justifiably widely cited in this literature. But what intrigues me about so much of this positive citation is that it ignores the crux to Sibley's arguments, which is to locate his understanding of exclusion not at the level of particular excluded social groups but in object relations theory, and in Kristeva's notion of abjection. His reading of 'the social' then (and societal reproduction) is grounded very strongly – and transparently – in a psychoanalytic reading, which means that 'the social' becomes little more than the effect of fear, rejection, expulsion, loss and so forth. This, of course, is very different from a historical materialist reading of exclusion, which is located primarily in relation to (non-participation within) the labour force. All of which is to argue that it is not enough merely to cite inclusion, exclusion and/or marginalization. Rather, these in themselves are no more than descriptive terms which require theoretical (and not just empirical) work to expose the nature of their connection/s to 'the social' and to society.

Fourthly and relatedly, I would highlight what for me is another striking absence, the seeming reluctance to locate this 'social' within any arena that might be labelled 'economic'. Indeed, it would be very easy to get the impression from reading this social geography literature that it operates with a sense of the social that is counterposed to an unspoken, unarticulated 'economy'. For example, the 'spaces' of this work are bodies, homes, neighbourhoods, 'public' spaces (streets, squares, parks) and rarely those of paid employment and unpaid 'work'. And it is unusual for this research to make reference to the ways in which these same spaces, and particularly the home and the body, are permeated with and/or shaped by the worlds of work – in the ways, say, that much feminist economic work has attempted to do (Hanson and Pratt, 1995; Massey, 1995; McDowell and Court, 1994). In addition, it is worth noting that many of the subjects of this social geography are connected by their marginalization and/or exclusion (direct or indirect) from legal employment (children; youth; the elderly; people with disabilities) and that at least several areas of empirical interest are represented as in some way 'cultural' – for instance, 'cultures' of parenting and/or mothering, youth 'cultures'. For me, this absence is the area more than any other within this social geography which signals the influence of the cultural within/to its reimagining. Not only is it possible to infer in all this a positive identification with 'the cultural' – which we might note remains tacit in its reading of 'the cultural' – but also the denial of certain materialities and social relations seems to imply an acceptance of the (false) opposition drawn in much recent geographical literature between economy and culture (Gregson et al., 2001; Ray and Sayer, 1999; Sayer, 1997). Hence, at the same time as it is taking its cues from arguments about the presence/absence of the body in social theory, it is the prism of the cultural which appears to have exerted a critical force in the contemporary reimagining of the social within social and cultural geography.

So, and at the risk of oversimplifying and eliding considerably, my contention is that rather than having languished through the 1990s, social geography has been reconfigured and that this is a reconfiguration which has used an alliance with 'the cultural' to define in/out its field. This reconfiguration has been primarily body-centred and is (mostly) about individual experiences. It is frequently, though not exclusively, divorced from a notion of society which continues to take seriously needs, values and the pattern and provision of goods and services. This, of course, is at considerable odds with political economic interpretations of 'the social', in which the (re)production of social life, its conditions, possibilities and limits within capitalist, market-led and increasingly neoliberal regimes, figures centrally. Small wonder then that, from the latter perspective, all this reads like an evacuation of the social. But what I would also want to argue is that evacuation is not simply a product of reconfiguration. Rather, and concurrent with these developments, significant changes have occurred within those strands of social geography informed by political economy. These point to a different understanding, an evacuation of 'the social' from within. It is to these that I turn in the following section.

OLD WOR(L)DS, NEW WOR(L)DS: ON EVACUATION FROM WITHIN

Talking and thinking explicitly in terms of material inequalities used to be familiar terrain in social geography. Grounded in the historical materialist tradition, materiality was widely and unproblematically regarded as comprising the basic conditions necessary for the reproduction of social life on both a daily and a generational basis (namely shelter, food, income, health, education, employment, welfare). Inequalities in resource distribution, specifically their connections with class, were regarded as of primary empirical interest, and, more theoretically, these inequalities were understood to be produced by the workings of specific structures – notably capitalism – whose key, defining relations were conceptualized as fundamentally unequal and exploitative (capital–labour being the prime instance), and to be regulated by the state, specifically by particular forms of welfare regime.[9]

In the context of the UK it was the regulatory matrix that provided the critical theoretical and empirical content for much of social geography. Hence the focus on the state: on housing, health and education (collective consumption); on (un)employment; and on the social divisions and inequities which state policies both attempted to counter and in turn generated and/or reinforced. The rise of the new right through the 1980s, specifically via privatization, the development of markets in housing, health, transport and increasingly in education, combined with the 'hollowing out' of the state – neoliberalism in other words – changed all this, in profound ways. One of the most far-reaching of these in terms of its effects was in the language of 'the social'. In its most extreme form this attempted to dispense with the connections between individual and society, with – to paraphrase an infamous sound bite – society being recast as no more than individuals (and families, of the nuclear, heterosexual sort). More generally though, these transformations saw the reconfiguration of the individual–society relation through the market, and consequent changes in patterns of talk, with increasing reference to customer/client, choice, opportunity and so on. This meant, in turn, that thinking and talking in terms of inequalities became tantamount to unthinkable and unspeakable, certainly within the project of an emerging New Labour in the mid 1990s. Lack of choice and/or opportunity maybe, but to highlight the inequities produced through the market would have been to cut to the heart of the neoliberalism project – to risk reasserting an Old Labourism in the brave new world of New Labour.

For social geographers working in the political economy tradition, these changes have proved hard to negotiate, involving as they do not just radical transformations in the provisioning of traditional areas of inquiry (housing, health, education and so on), but fundamental political change too as New Labour's commitment to neoliberalism has become increasingly apparent. To this we need to add the challenge posed by the rise to prominence of poststructuralist thinking within the academy generally and social and cultural geography more particularly. The key question for many has become how (or indeed whether) to intervene in all this, both academically and politically.

For some social geographers working in these areas the response has become increasingly transparent: to reassert the primacy of (an unreconstructed) class analysis and political economy. In general terms this course depends on drawing clear lines of opposition between the latter and relatively crude understandings of poststructuralist thinking, which – equally problematically – are often elided with 'the cultural turn'. More specifically though, reassertion has prompted moves to expose the effects of market provision, pointing to how these connect with the development of two-tier markets in health and education (based on ability to pay) and – rather more implicitly – to their different effects on their respective consumers (see, for example, Mohan, 1995; 2000). Aside from its revisionism, the importance of this tactic for my purposes here lies in what it has to say, or rather doesn't, about inequality. Inequality, if it is mentioned at all, is typically couched in empirical terms, as an effect of distribution, as the outcome of differences in service delivery and/or provision which maps into particular social categories (typically subsets of the elderly, women, ethnic minorities). That these differences have largely predictable geographies (inner city versus suburban; rural versus urban; north versus south (-east)) almost goes without saying. What has gone here though is the connection to older theoretical understandings of inequality and its regulation. In part, I think, this is inevitable – at least if one remains theoretically within the parameters of conventional radical political economy. Indeed, the transformations that have been wrought in service delivery/provision have meant that, increasingly, the core relation for so much of this work is defined by the market and enacted through exchange. Private health care and education provide perhaps the most fully developed instances of this, but so pervasive is the market relation that the social relations of exchange (in particular the emphasis – both discursively and in practice – on the customer/client) permeate even those areas currently outwith fully developed commodity relations (state education and parts of the NHS). Notwithstanding the differences experienced by consumers, to see such relations as exploitative – and hence couch them in terms of inequalities as classically understood – would be misplaced. Yet what this has meant is that inequality has receded from view even in the analyses of those whom one might most expect to make reference to it.

For other social geographers though, the response to changing times has been rather more equivocal. Indeed, for many working in social geography the emphasis has switched more to the analysis of issues with a less overt politics. A case in point, and possibly one of the best instances of this tendency, is the clutch of work on social polarization that appeared during the mid 1990s (Dorling and Woodward, 1995; Hamnett, 1994; 1996; Hamnett and Cross, 1998; Williams and Windebank, 1995; Woodward, 1995) about which I want to say a little more.

Taking much of its inspiration from the arguments of Saskia Sassen (1991), from the outset this debate has had strong connections with two of the core concerns of social geography: the changing nature of the class structure and its mediation by particular welfare regimes. Nonetheless, the primary issues for debate in this field revolve around polarization as pattern, specifically key measurement questions of gaps in income and/or income differentials and change/s in the occupational class structure; polarization as a characteristic of globalization (the New York–London etc. comparison); and polarization at the local scale. Occasionally within all this there is some discussion of causal processes, particularly in local studies or, for instance, in Breugel's (1996) insistence on the importance of transformations in gender relations to polarization. But the emphasis remains on the particular form of polarization – witness the prevalence of eggs, dumb-bells, hourglasses and onions as the prevailing metaphors.

This aside, one of the intriguing facets of the polarization debate is the way in which it engages with inequality. Running through much of this literature is a thread that connects polarization to inequality. At times this is explicit: polarization is itself seen to expose social inequality (Woodward); alternatively, income differentials are represented as inequalities, which in turn are elided with polarization (Hamnett). This suggests that polarization is seen to be revealing of inequality; that it maps it, charts it, makes it visible. But inequality is primarily understood in terms of observable, measurable income differences and/or differentials. This raises the question of whether polarization is poverty research in another (more acceptable?) guise – discursively recast, and in ways which, by focusing attention on the façade of polarization, work simultaneously to mask/obscure and to deflect attention away from the existence of poverty.

Yet for much of the time within the polarization debate the thread which elsewhere connects or elides polarization and inequality goes unmentioned. Specifically this happens where attention turns to class, and is connected intrinsically with the emphasis afforded to the occupational class structure. With its focus on a categorical, rather than relational, understanding of class this is – I think – inevitable. For discussions of numbers of professionals managers and so forth and their relative growth, even comparative ones, leave little room for inequality. Witness how bizarre, not to mention inappropriate, the prospect of talking even in terms of income inequalities (rather than the weaker differences) amongst and between professionals and managers appears. Yet the question this also raises is to what extent a categorical representation of class might also signal a tacit erasure – a means of enabling a shift away from relational understandings with their intrinsic connections with inequality. I say this because such a manoeuvre has its attractions. Specifically, it allows work on social polarization to connect with the core concerns within class analysis for much of the 1990s, namely the middle classes, and because relational inequality is so transparent a feature of polarization (in the sense that one class is frequently directly employed by the other to service facets of social reproduction: Gregson and Lowe, 1994) that to focus exclusively on middle-class occupational change seems at best a deflection, at worst a recasting to fit the times. It is a fashionable way of doing what might otherwise risk being dubbed as deeply unfashionable.

More broadly, these presences and silences around inequality within the polarization literature are indicative of a profound ambiguity: about what inequality itself might mean; about how we might talk about this; and about how this might or might not figure within a reconfigured class analysis. Polarization seems to be a way of occupying this gap. Indeed, for its protagonists it would seem to be an acceptable way in which to talk about inequality – as an effect, observable, measurable, controllable, tolerable even (within limits), rather than as a relation, with all those messy connections with power (and with problematic, not to mention unfashionable, words like domination, exploitation and oppression). It is a way of talking about poverty (and class) without mentioning the(ir) name(s), in short an avoidance that speaks. This suggests that 'the social', as once understood and talked about, has been evacuated from within, so much so that it is little more than a trace, even in areas of research where one might most expect to find its continued presence.

As the previous two sections have shown, reconfiguration and evacuation have had profound effects on social and cultural geography. Indeed, currently it is possible to discern not just the copresence of political-economic and psychoanalytically inspired readings of 'the social' alongside empirical readings defined by the identification of particular 'excluded' social groups, but a social which is located in the body, as well as another which continues to concern itself with questions of class and poverty, even if it doesn't quite speak their names. There is much more that could, and probably should, be included here: on race, on gender, on ethnicity, on sexuality. But my point in formulating this discussion in this particular way has been to position and situate, to site 'the social' within a field of power–knowledge which, at least as this appears to be being played out, is increasingly foregrounding different readings of materiality – one embodied, the other more conventionally materialist in that it centres the conditions of societal reproduction. Moreover, and as befits this locatedness, all too frequently these materialities are being represented and talked about as if they are oppositional. Intriguingly, it is various representations of 'the cultural' that are mobilized to enable this. Again, what this points to is the importance of siting discussions of 'the social' within a power–knowledge framework. For it is an association with 'the cultural' on the one hand (often explicitly against political economy) which confers discursive authority on a particular (bodily) social, and an explicit disassociation on the other (on the part of those wishing to reassert political economy) which is used by the former to define the latter out, to reposition it (I would argue falsely) within 'the economic'. This establishes, I think, that 'the social' has neither languished nor been marginalized by 'the cultural', but rather is itself being contested through it.

Yet, as the discussion thus far also shows, in the course of this positioning, repositioning and reconfiguration, the retreat from inequality as a theoretical concern has become increasingly apparent. This, as I have shown, is as much an effect of evacuation from within as it is of being defined 'out'. The question that perhaps ought to be posed within social and cultural geography is the normative one: should this continue to be so? Do we want – for whatever motivations – to lose sight of a social and cultural geography that connects with inequality and the politics of distribution? My answer to this would be a definite 'no', and for the following reasons.

Theoretically, thinking about 'the social' cannot be divorced from ontological questions, principally those that concern the constitution of individual–society in time–space. This means that 'the social' is about how (and why) individuals relate to and with others, in particular routinized, regulated and frequently institutionalized (if negotiated and/or contested) ways, and with what critical effects – a position which admits to the importance of the politics of identity and difference. But, this also means that 'the social' cannot be separated from society; and because it is about this it is also, inevitably and simultaneously, about needs, resources, power, justice, rights, values and normativity, and societal reproduction. Moreover, I would want to insist, since our situatedness remains primarily within capitalist society/ies which continue to be organized on territorial lines as nation-states, that this 'social' is still about inequalities (and their regulation). This is not to say, as so much of contemporary social and cultural geography seems to presume, that inequality has to be interrogated from a revisionist political economy position.

In what remains of this chapter then, and with the intention of opening up rather than foreclosing debate, I want to make a number of suggestions regarding how such a social might be reclaimed within social and cultural geography. Specifically, I want to focus on the need to consider forms of materiality that connect with the conditions which shape the reproduction of social life, and to think through how inequality itself might begin to be reimagined. This, I argue, requires that we address issues of consumption and of material culture, and that we take seriously a discursive that connects with materiality.[10]

RECLAIMING 'THE SOCIAL': SOCIETAL REPRODUCTION, CONSUMPTION CULTURE/S AND RETHINKING INEQUALITY

In addressing these questions it would seem imperative to begin by returning to a fundamental question: how we think about the materiality of societal reproduction. Now, whilst various authors continue to argue the importance of previous understandings' grounding in the basic conditions necessary for the reproduction of human life (food, shelter and so on: see, for example, Doyal and Gough, 1991; Sen, 1992), others, and particularly those concerned with consumer culture/s, emphasize the impossibility of distinguishing basic needs from the cultural conditions of their production, from intersubjective meanings and institutionalized norms (Slater, 1997). Discussions of the 'consumerist west' have seen 'needs' reconfigured through a particular form of consumption culture, specifically through our ability/ies to satisfy needs (and understand them) through our relationship/s to particular goods and services. So, for instance, we tend to think of housing no longer primarily as shelter but as an expression of identity and self; likewise with food and clothing. Moreover, and in agreement with many of those who have worked recently on consumption issues in the First World, I would argue that people – particularly those located in the First World – increasingly relate to the world as consumers of goods and services, rather than as their producers, and that for many it is consumption which provides the primary set of motivations (Bauman, 1998). This has profound implications for how we think about societal reproduction. In short, it suggests that rather than thinking about this in terms of basic material need/s, we have to rework our understanding of materiality through consumer culture/s. Moreover, we need to take objectification seriously, and to connect the general desire for goods which is consumption to a desire for specific sorts of goods – to the notion that our relationship to goods is itself refracted through the ways in which some things/goods matter more than others (Miller, 1987; 1998b).

Such arguments, of course, also have implications for how we might think about issues of inequality. Globally, the importance of the relationship between First World consumption patterns and Third World producers' poverty has long been recognized within analyses of uneven development. Particularly apparent in the production/consumption of food and clothing, this unequal relation lies behind campaigns against particular retailer/manufacturers' practices (Nike and Gap, for instance), consumer boycotts and ethical consumption and/or fair trade agreements (Klein, 2001). But what this relationship also means is that inequality in its classic materialist sense – that is, linked intrinsically to the social relations of production – has been displaced geographically; that the critical effects of First World consumption patterns are located elsewhere, distant and frequently masked; and that inequalities themselves in these First World societies are increasingly being understood exclusively in relation to goods and their consumption, that is, through consumer culture/s. This we can see from various studies of the meaning of poverty in the UK, most of which document the importance of specific consumer items to household reproduction. Hence, whilst videos, TVs, satellite dishes, mobile phones and children's toys might not constitute life-supporting goods in the strict materialist sense, that they are talked about as needs by the majority of people does actually matter. Being without these things (and frequently in the case of children, without specific toys and/or clothes), through force of circumstance/s rather than choice, is – in this culture – exclusionary. It prevents people from participating in society in ways which are considered to be the norm and which are institutionalized as normative, which in a consumer society is read and talked about as inequality.

I make these points not just because my personal conviction is that both societal reproduction and inequality have to be rethought through consumption culture/s, but for a further set of reasons. Because, notwithstanding the laudability of the objectives of many consumer campaigns, such analyses seem to me to be remarkably thin and/or naive in terms of their understandings of First World consumption cultures. Indeed, they seem to proceed from the assumption that improved consumer knowledge connected to a (re)moralized pattern of exchange/trade will suffice to effect radical change/s in patterns of First World purchasing. What this manages to overlook, however, is the immense complexity of First World consumption. For example, the imperatives behind much household provisioning on an everyday (as opposed to 'gift') basis in the UK at least are about thrift, value (for money) and saving. Consequently, to spend more on basic food goods for ethical and/or moral reasons is not just a choice which is only possible on income grounds for the middle classes but one which challenges some of the primary meanings of consumption as practised within this consumer culture. To take another example, consider branded clothing. The brand is critical for the ways in which it brings together both identification and distinction; its meaning is not simply about its purchase but about the wearing and the specific socialities which this defines in/out (Lury, 1996). Not to buy a particular branded item then (kids' trainers would be a good instance) can (and does) have huge critical effects – on friendship patterns, bullying, individual social and psychological development and so on. Whilst for adults, rather than condemning out of hand the stereotype of the Levi's/Nike-clad, McDonald's-eating, Coca-Cola-swigging male or 'his' female counterpart chasing the cheap brands in Matalan or at Cribb's Causeway, we need to think hard about the social securities and certainties which these brands confer. We need to appreciate that to enact social difference through the consumption of goods is dependent on particular identities and subjectivities, which are middle class.

Having said something about how we might rethink the material conditions of societal reproduction through consumption culture/s, I want to say rather more first about how we might approach analyses of inequality within consumer culture/s and secondly about their connection to key social categories.

If, as I argued above, we are now in a situation where inequality in the strict, exploitative, materialist sense has been displaced from First World, societies to the Third World, and if 'inequality' in the First World (and almost certainly elsewhere) is increasingly being understood through consumption culture/s, then it seems to me that we have to take both the discursive – talk – and material culture seriously in any reclamation of inequality. Inequality has different, situated meanings, ones that are articulated in and through different consumption cultures. Consequently, how people talk about things – particularly in consumer society/ies – is critical not just to their understanding of things, goods and their pattern/s of provision and use but to their constitution of particular social relations, to how they construe their relation to society and to their understanding of inequality. Already from some of the preceding discussion we can begin to see how this might happen: specifically, how 'inequality' within a capitalist consumer society comes to be regarded in terms of the presumed right of individuals to buy and consume particular things. But a further instance helps make the point even more strongly, showing how this right to buy and consume is about consuming in particular ways, which themselves are about enacting key socialities through talk.[11] The example I want to use here is the mobile phone.

The volume of mobile phone sales in the UK over the Christmas 2000 period was estimated at some 5 million handsets. To question mobile phones as a basic material need within this society however, or indeed even to get embroiled in a debate on such terms, is to miss the point that this gadget has come to be talked about in the UK as necessary, for children and adults alike. Not having one, through force of circumstance rather than through choice, is talked about not simply as exclusionary within this consumer culture but as an inequality, as an inability to satisfy what is seen as a basic need through possessing a particular good. This has more than a little to do with its connection to practices of talk. Talk, particularly the new and different forms of talk enabled by mobile phone technologies, has been critical to the identification of mobile phones as things that matter. Routine safety and/or emergency talk ('I've arrived', 'I'm back', 'There's a holdup', 'The train's delayed'), quick/query talk ('the train's left Darlington, pick me up in 10 minutes'; 'they haven't got any tuna steaks but they have got salmon') and text messaging are all novel forms of talk/communication, in the sense that they have no prior equivalents, particularly in terms of their ability to transcend spatial fixity – in this case by enabling talk between spatially dislocated yet mobile individuals. But more than this, their importance lies in the way/s they allow for the enactment of key socialities. Providing a child with a mobile phone is simultaneously about being the 'good', safety-conscious, concerned and aware parent, and about facilitating what children construe as necessary ways of peer group communication (witness current playground use). Equally, quick/query talk between adults is as much about the reproduction in practice of significant social relations – particularly with the significant other – as it is about the content of the talk. But what this means in turn is that these forms of talk have become ways in which key social relations are practised: increasingly, some of the primary ways in which significant social relations are constituted. So, 'being without' is not just a matter of being without a particular thing that matters; in this case it is about not being able to practise ways of social talk that are becoming increasingly routinized as normative modes of communication. This then is where 'inequality' goes beyond a particular good to being about how the practices that surround a particular good constitute inequalities in the constitution and reproduction of social relations. Where this takes us, perhaps, is to the situation amongst young people in Japan, where it is argued that the act of communicating – indeed, being continually available for contact – rather than talk itself, assumes primary significance and where social relations are argued to require mobile phones to be both formed and maintained (*The Guardian*, 10 May 2001, quoting sociologist Hisao Ishii).

A second integral component in all this is to rethink how key social categories might be brought into analyses of inequality within consumer societies. Made inherently problematic by increasingly sophisticated accounts of significant social differences, the relationship between key social categories and inequality is messy, theoretically and empirically. For, if inequality/ies are primarily understood in relation to consumption cultures then it follows that these are always in a process of production, and that their relation to social categories is much the same, that this can be neither read off nor presumed. And yet, empirically, it is clear that particular social groups – notably the elderly, single-parent households, certain ethnic 'minority' groups, benefit-dependent households – are, simply as an effect of income, more likely to be unable to relate to goods and services in way/s which conform with culturally specific definitions of 'need'. We know then that certain groups are more likely to be 'disadvantaged consumers' than others. But quite what the effect/s of this – socially, economically and individually – might be is unclear. What, for example, will be the effect/s of transformations in online shopping? Are inequalities in relation to certain goods as significant for some disadvantaged consumers as for others?

These are just some of the critical questions raised by rethinking inequality through consumption cultures, but what these arguments also begin to suggest is the need to rethink, too, old assumptions about the identification of inequalities with specific class, race and gender differences. The category 'disadvantaged consumer' is heterogeneous: it cross-cuts these divisions and differences and disrupts their pre-eminence. Perhaps this is no bad thing. But, if we switch for a moment to thinking about consumption as the work of shopping (admittedly a much narrower definition than consumption cultures), then – as many studies continue to show – it is women upon whom the onus of this work rests, particularly when it comes to weekly food shopping, gift buying and so on; and it is women who continue to sacrifice themselves as consumers within the household, buying for themselves last, if at all. This is something that needs careful thought. So, whilst at one level, as Miller has argued, this is about shopping as love, at others – particularly in relation to routine food shopping and for that matter its end point in meal preparation – it is about inequalities in domestic labour. These differences matter, and increasingly so in a world in which women's relation to the labour market is undergoing radical change. The question that perhaps needs to be asked here though is whether it is helpful to continue to think about this in terms of inequality. My answer to this would be equivocal. At one level, yes, I think it is. That men in general do less of this consumption work is a problem for (many) women; it is part of what used to be called the 'double day'. But to progress requires that we break out of thinking in terms of categorical identifications and inequality, for identifying this situation as a gender inequality offers few insights into how things might be changed. Instead, I would urge that it is precisely here, amidst the material, that anti-foundationalist thinking can exert its most radical effects. For (re)reading these inequalities as regulated performance/s of gender enables them to be seen as just this – as performances which can be disrupted and enacted differently.

Having said this much, I want to return finally in this section to make a few very general observations concerning those previous concerns of social geographers: housing, education, health care and so on. No longer central concerns, particularly for a reconfigured social geography,

these remain nonetheless critical components within societal reproduction, and the manner of their provision and use is not just a policy consideration but a key indication of the organizing principles of any society. Specifically then, I want to suggest – indeed urge – that their analyses be retrieved; that they not be abandoned to a social geography that is motivated by a revisionist political economy. I think this can be achieved by connecting them with contemporary research on consumption culture/s, material culture and the discursive. I say this because, whilst there have been many fine studies of provision by social geographers, these studies remain that: analyses which look at the nature and geographies of these services up to the point of delivery. Looking at these services as they are consumed, as commodities and/or as goods, would seem to have additional advantages. One, possibly the most significant in terms of its critical purchase, would be that it would allow us to see if certain services and/or goods are seen by consumers and understood to be significantly different in their meaning/s to others. If they are, in whatever ways, then this is important. Is health care (and education for that matter) still regarded as a basic right, notwithstanding its increasing permeation with commodity relations? What about housing? Might this be understood differently? Might this have been more readily incorporated into the commodity form, and if so, why? Moreover, thinking in terms of how discursive power works through organizations – to connect with practice, or not – clearly has potential in terms of enhancing understanding/s of service provision and delivery. We might envisage, for example, research which brought together critical discourse analyses of various forms of service provision and accounts of how such services are consumed by particular social groups – of the production and consumption of health care, education and so on. This means that inequalities would be understood not as the product of structures and subject position/s defined by structures, but as the critical effect of a materialized discursive.

SPECULATIONS ABOUT 'FUTURE TALK'

Given my primary intentions in this chapter, to end on a concluding note would be a futile gesture. Yet, the editors' response to an earlier draft 'encourages' me 'to be more speculative (about future orientations)' here, where convention decrees I conclude. This poses me significant difficulties with this chapter, notably that it has the tendency to ring fence rather than enable the framing of future 'debate'. Moreover, speculations about 'the future' have the added drawback of both positioning their authors in the unenviable and untenable position of all-seeing clairvoyant, and writing out the way in which academic commentary is always situated in the conditions of its production. So, in moving to 'speculate about futures' I'm only going to go part way down the cooperation road. At this point, therefore, I suspend the conventions of academic writing practice – by admitting into the text both comments on a previous draft of this chapter and my responses to them. In so doing then, I'm choosing to work with a tactic which highlights how dialogic talk shapes practice in the constitution of academic texts, and which resonates with some of my earlier comments about the importance of talk.

First then, and in speculative mode: I hope that what I have labelled here as 'reclaiming the social' is seen to matter, at least in some quarters, and that some of my suggestions regarding potential empirical foci receive a degree of consideration. Moreover, I would hope that 'reclamation' is accompanied by transparent debate within social and cultural geography about three issues: (1) what we mean by 'the social', and – perhaps even more importantly – how this connects with and to society and societal reproduction, and therefore with economy and polity as well as culture; (2) what vision/s of society we have – for example, whether this is (still) about commitments to ameliorate and/or eradicate inequalities through redistribution, or is based on equality of opportunity – and whether these (still) construe themselves as broadly left; and (3) the role we attach to 'the academic' – simply commentator or critic. All of these are questions which have been largely sidelined or placed backstage in recent formulations of the social. Particularly within the ascendant 'reconfiguration' literature, these would signal a re-evaluation of the use of the much-used, and therefore potentially abused, label 'critical' in human geography.

Secondly, and in response to another fragment of 'talk' – yet less conventionally authoritative – 'several people … felt that the chapter was a bit "too British"', the implication being to do something about this by referring to other places too. My response to this is both critical and resistant. Indeed, I want to assert that this chapter is intentionally and explicitly 'British', for two reasons. First, as I have written elsewhere with others (Gregson et al., 2001), it is about time that 'we' British started to acknowledge the partialities of our knowledge, and to think about resisting the

power geometries which permit us to frame, shape and know others through ourselves. Secondly, any engagement with 'the social' requires an engagement with the spaces of the social. These, at least at the moment, remain in part the concern of nation-states. So, when we talk – as I have done here in places – about facets of societal reproduction such as health, education and so forth, we have to look inevitably at situations within (and between) nation-states. Reverting to speculative mode though, societies are not necessarily closed, bounded spaces sealed by borders; they are relationally constituted too, in ways that connect and blur spaces to produce multicultural, often diasporic, identities. This is indeed pretty much the orthodox narrative within social and cultural geography, and one which is located within (although this is rarely acknowledged) the context of its production – largely metropolitan, largely British and North American. But there are other places in the world where forging connections between societies and space is about more homogenizing, bounding practices, violence even – the Balkans (Bosnia, Kosovo, Macedonia), Palestine, Rwanda to mention just three examples. These places however, or at least writing about them, remain seemingly the preserve of political and economic geographers; they are places which social (and for that matter cultural) geographers have been apparently content to leave alone. My hope is that 'the future' – triggered by a reclamation of 'the social' – might see a change to this insularity as well as an explicit acknowledgement (as here) of the partialities and particularities of the spaces and practices we write about.

NOTES

1 An addition to the standard disciplinary history is provided by the renaming of the Social Geography Study Group as the Social and Cultural Geography Study Group of the RGS-IBG. All naming has a politics, and this relabelling as social *and* cultural seems, in retrospect, to have been a key defining moment. It is immediately suggestive of ambiguity: is this 'and' about separation, combination or relation? At the same time, the 'and' conceals the power relations between the two terms, which, there can be little disputing, have seen 'the cultural' (with the authority conferred by the cultural turn) exercise discursive power over 'the social'.

2 See, for example, Aitken (1994), Aitken and Herman (1997), Bell and Valentine (1995), Bingham et al. (1999), Holloway (1998), Holloway and Valentine (2000), Malbon (1999), Skelton and Valentine (1997), Valentine (1996; 1997).

3 My selection of work on (dis)ability is in no sense meant to be either privileging this work or singling it

out for special criticism. On the contrary, as I make clear later, whilst the importance of this work lies in the way in which it signals some of the core concerns of the contemporary moment and in the way in which it speaks to a new generation of social geographers, it is in no sense unique in its more general characteristics. In addition, using (dis)ability as an exemplar has the added advantage of connecting to the concerns of the next chapter in this section.

4   See, for example, Butler (1994), Butler and Bowlby (1997), Butler and Parr (1999), Chouinard (1997), Chouinard and Grant (1995), Dear et al. (1997), Dyck (1995), Gleeson (1996; 1997), Golledge et al. (1979; 1991), Hall (2000), Imrie (1996a and b), Lovett and Gattrell (1988), Parr (1999).

5   See, for example, Dorn and Laws (1994), Duncan (1996), Johnston (1996), Longhurst (1995; 2000a; 2000b), McCormack (1999), McDowell and Court (1994), Mowl et al. (2000), Nast and Pile (1998), Pile (1996), Simonsen (2000).

6   Besides disability research, recent work on children, youth and the elderly is also indicative of this tendency, as is that on pregnant bodies, breast-feeding and body work. It might be argued that work on subjects defined by age is a less clear-cut case of the bodily; that age is no more than a social construction, and that the categories of children, the elderly and so on are no more than ones which society inscribes on particular types of human bodies. But the point is that these inscriptions depend on specific physiological and mental changes and/or transformations (grey hair, memory loss, small bodies, prepubescent bodies, physical frailty, etc.) to make sense.

7   This is not to say that bodies cannot be bought and sold: prostitution, surrogacy and organ exchange provide various instances of precisely this. Nonetheless, certain forms of such exchange remain the exception rather than the norm, and the commodification of body parts – if not certain female bodies – is the subject of intense ethical debate.

8   This is not to deny that cities and neighbourhoods cannot be conceptualized differently and in ways more accordant with body theory/ies – as networks and flows, for example. But it is to question the merits of such thinking in relation to such objects.

9   See, too, the feminist critique of this position: for example, Jones and Jonasdottir (1988), McIntosh (1978), Watson (1990), Wilson (1977).

10 In arguing this line I am drawing on some of the core tenets within my own recent work, as well as on that of others working in the consumption field (Gregson and Crewe, 1997a; 1997b; Miller, 1998a; Miller et al., 1998; Slater, 1997). It is also perhaps worth noting that this type of work is seldom positioned within social geography, being more commonly represented as either cultural or economic in orientation. As well as providing further grist to previous arguments, my intention in developing this position here is to point to the way/s in which a materialist reading of society should not be equated automatically with a revisionist political economy position. Such assumptions

seem to me currently to bedevil much of the reconfigured social geography literature, and work to inhibit hinging 'the social' to a wider understanding of societal reproduction.

11 For a rather different take on talk, which concentrates on the differences between talk and text, see Laurier (1998). For work discussing the importance of talk within organizations, see Boden (1994), Cameron (2000), du Gay (1996) and Leidner (1993).

REFERENCES

Aitken, S. (1994) *Children's Geographies.* Washington, DC: Association of American Geographers.

Aitken, S. and Herman, T. (1997) 'Gender, power and crib geography: transitional spaces and potential spaces', *Gender Place and Culture* 4: 63–88.

Bauman, Z. (1998) *Work, Consumerism and the New Poor.* Buckingham: Open University Press.

Bell, D. and Valentine, G. (eds) (1995) *Mapping Desire.* London: Routledge.

Bingham, N., Valentine, G. and Holloway, S. (1999) 'Where do you want to go tomorrow? Connecting children and the Internet', *Environment and Planning D: Society and Space* 17: 655–72.

Boden, D. (1994) *The Business of Talk: Organizations and Action.* Cambridge: Polity.

Breugel, I. (1996) 'Gendering the polarization debate: a comment on Hamnett's social polarization, economic restructuring and welfare regimes', *Urban Studies* 33: 1431–9.

Butler, J. (1990) *Gender Trouble.* London: Routledge.

Butler, J. (1993) *Bodies that Matter.* London: Routledge.

Butler, R. (1994) 'Geography and vision-impaired and blind populations', *Transactions of the Institute of British Geographers* 19: 366–8.

Butler, R. and Bowlby, S. (1997) 'Bodies and spaces: an exploration of disabled people's experiences of public space', *Environment and Planning D: Society and Space* 15: 411–33.

Butler, R. and Parr, H. (eds) (1999) *Mind and Body Spaces: Geographies of Illness, Impairment and Disability.* London: Routledge.

Cameron, D. (2000) *Good to Talk? Organizations in Action.* London: Sage.

Chouinard, V. (1997) 'Making space for disabling differences: challenging ableist geographies', *Environment and Planning D: Society and Space* 15: 379–87.

Chouinard, V. (1999) 'Body politics: disabled women's activism in Canada and beyond', in R. Butler and H. Parr (eds) *Mind and Body Spaces.* London: Routledge. pp. 269–94.

Chouinard, V. and Grant, A. (1995) 'On being not anywhere near "the project"', *Antipode* 27: 137–66.

Crossley, N. (1995) 'Merleau Ponty, the elusive body and carnal sociology', *Body and Society* 1: 46–63.

Crossley, N. (1997) 'Corporeality and communicative action: embodying the renewal of critical theory', *Body and Society* 3: 17–46.

Dear, M., Gaber, L.Takahashi, L. and Wilton, R. (1997) 'Seeing people differently: the socio-spatial construction of disability', *Environment and Planning D: Society and Space* 15: 455–80.

Dorling, D. and Woodward, R. (1995) 'Social polarization 1971–91: a micro geographical analysis of Britain', *Progress in Planning* 45: 1–66.

Dorn, M. and Laws, G. (1994) 'Social theory, body politics and medical geography: extending Kearns's invitation', *Professional Geographer* 46: 106–10.

Doyal, L. and Gough, I. (1991) *A Theory of Human Need*. Basingstoke: Macmillan.

Driver, F. (1995) 'Submerged identities: familiar and unfamiliar histories', *Transactions of the Institute of British Geographers* 20: 410–13.

du Gay, P. (1996) *Consumption and Identity at Work*. London: Sage.

Duncan, N. (ed.) (1996) *Bodyspace: Destabilising Geographies of Gender and Sexuality*. London: Routledge.

Dyck, I. (1995) 'Hidden geographies: the changing lifeworlds of women with multiple sclerosis', *Social Science and Medicine* 40: 307–20.

Dyck, I. (1999) 'Body troubles: women, the workplace and negotiations of a disabled identity', in R. Butler and H. Parr (eds) *Mind and Body Spaces*. London: Routledge. pp. 119–37.

Featherstone, M., Hepworth, M. and Turner, B. (eds) (1991) *The Body: Social Process and Cultural Theory*. London: Sage.

Gatens, M. (1996) *Imaginary Bodies: Ethics, Power and Corporeality*. London: Routledge.

Gleeson, B. (1996) 'A geography for disabled people?', *Transactions of the Institute of British Geographers* 21: 387–96.

Gleeson, B. (1997) *Geographies of Disability*. London: Routledge.

Godlewska, A. and Smith, N. (eds) (1994) *Geography and Empire*. Oxford: Blackwell.

Golledge, R., Parnicky, J. and Rayner, J. (1979) 'An experimental design for assessing the spatial competence of mildly retarded populations', *Social Science and Medicine* 13D: 292–5.

Golledge, R., Loomis, J., Flury, A. and Yang, X. (1991) 'Designing a personal guidance system to aid navigation without sight: progress on the GIS component', *International Journal of GIS* 5: 373–95.

Gregson, N. and Crewe, L. (1997a) 'The bargain, the knowledge and the spectacle: making sense of consumption in the space of the car boot sale', *Environment and Planning D: Society and Space* 15: 87–112.

Gregson, N. and Crewe, L. (1997b) 'Performance and possession: rethinking the act of purchase in the light of the car boot sale', *Journal of Material Culture* 2: 241–63.

Gregson, N. and Lowe, M. (1994) *Servicing the Middle Classes*. London: Routledge.

Gregson, N., Simonsen, K. and Vaiou, D. (2001) 'Whose economy for whose culture? Moving beyond oppositional talk in European debate about economy and culture', *Antipode* 33: 617–46.

Grosz, E. (1994) *Volatile Bodies: Towards a Corporeal Feminism*. London: Allen and Unwin.

Hall, E. (1999) 'Workspaces: refiguring the disability–employment debate', in R. Butler and H. Parr (eds) *Mind and Body Spaces*. London: Routledge. pp. 138–54.

Hall, E. (2000) 'Blood, brain and bones: taking the body seriously in the geography of health and impairment', *Area* 32: 21–9.

Hamnett, C. (1994) 'Social polarization in global cities, theory and evidence', *Urban Studies* 31: 401–24.

Hamnett, C. (1996) 'Social polarization, economic restructuring and welfare state regimes', *Urban Studies* 33: 1407–30.

Hamnett, C. and Cross, D. (1998) 'Social polarization and inequality in London', *Environment and Planning C: Government and Policy* 16: 659–80.

Hanson, S. and Pratt, G. (1995) *Gender Work and Space*. London: Routledge.

Harvey, D. (1973) *Social Justice and the City*. London: Arnold.

Holloway, S. (1998) 'Local childcare cultures: moral geographies of mothering and the social organization of pre-school education', *Gender Place and Culture* 5: 29–54.

Holloway, S. and Valentine, G. (eds) (2000) *Children's Geographies*. London: Routledge.

Imrie, R. (1996a) *Disability and the City*. London: Chapman.

Imrie, R. (1996b) 'Ableist geographies, disableist spaces: towards a reconstruction of Golledge's "Geography and the disabled"', *Transactions of the Institute of British Geographers* 21: 397–403.

Jackson, P. (2000) 'Rematerialising social and cultural geography', *Social and Cultural Geography* 1: 9–14.

Jackson, P. and Smith, S. (eds) (1984) *Exploring Social Geography*. London: George Allen and Unwin.

Johnston, L. (1996) 'Flexing femininity: female body builders refiguring "the body"', *Gender Place and Culture* 3: 327–40.

Jones, K. and Jonasdottir, A. (eds) (1998) *The Political Interests of Gender*. London: Sage.

Klein, N. (2001) *No Logo*. London: Flamingo.

Laurier, E. (1998) 'Geographies of talk: "Max left a message for you"', *Area* 30: 36–45.

Leidner, R. (1993) *Fast Food, Fast Talk: Service Work and the Routinization of Everyday Life*. Berkeley: University of California Press.

Livingstone, D. (1992) *The Geographical Tradition*. Oxford: Blackwell.

Longhurst, R. (1995) 'The body and geography', *Gender Place and Culture* 21: 97–106.

Longhurst, R. (2000a) 'Corporeographies of pregnancy: bikini babes', *Environment and Planning D: Society and Space* 18: 453–72.

Longhurst, R. (2000b) *Bodies: Exploring Fluid Boundaries*. London: Routledge.

Lovett, A. and Gatrell, A. (1988) 'The geography of spina bifada in England and Wales', *Transactions of the Institute of British Geographers* 13: 288–302.

Lury, C. (1996) *Consumer Cultures*. Cambridge: Polity.

McCormack, D. (1999) 'Body shopping: reconfiguring geographies of fitness', *Gender Place and Culture* 6: 155–77.

McDowell, L. and Court, G. (1994) 'Performing work: bodily representations in merchant banks', *Environment and Planning D: Society and Space* 12: 727–50.

McIntosh, M. (1978) 'The state and the oppression of women', in A. Kuhn and A.Wolpe (eds) *Feminism and Materialism*. London: Routledge.

Malbon, B. (1999) *Clubbing: Dancing, Ecstasy and Vitality*. London: Routledge.

Massey, D. (1995) 'Masculinity, dualisms and high technology', *Transactions of the Institute of British Geographers* 20: 487–99.

Miller, D. (1987) *Material Culture and Mass Consumption*. Oxford: Blackwell.

Miller, D. (1998a) *A Theory of Shopping*. Cambridge: Polity.

Miller, D. (1998b) 'Why some things matter', in D. Miller (ed.) *Material Cultures*. London: UCL Press.

Miller, D., Jackson, P., Thrift, N., Holbrook, B. and Rowlands, M. (eds) (1998) *Shopping, Place and Identity*. London: Routledge.

Mohan, J. (1995) *A National Health Service? The Restructuring of Health Care in Britain since 1979*. New York: St Martin's.

Mohan, J. (2000) 'Geographies of welfare and social exclusion', *Progress in Human Geography* 24: 291–300.

Moss, P. (1999) 'Autobiographical notes on chronic illness', in R. Butler and H. Parr (eds) *Mind and Body Spaces*. London: Routledge. pp. 155–66.

Mowl, G., Pain, R. and Talbot, C. (2000) 'The ageing body and the homespace', *Area* 32: 189–97.

Nast, H. and Pile, S. (eds) (1998) *Places through the Body*. London: Routledge.

Parr, H. (1999) 'Delusional geographies: the experiential worlds of people during madness/illness', *Environment and Planning D: Society and Space* 17: 673–90.

Peach, C., Robinson, V. and Smith, S. (eds) (1981) *Ethnic Segregation in Cities*. London: Croom Helm.

Pile, S. (1996) *The Body and the City*. London: Routledge.

Ray, L. and Sayer, A. (eds) (1999) *Culture and Economy after the Cultural Turn*. London: Sage.

Rose, G. (1993) *Feminism and Geography*. Cambridge: Polity.

Rose, G. (1995) 'Tradition and paternity: same difference?', *Transactions of the Institute of British Geographers* 20: 414–16.

Sassen, S. (1991) *The Global City: New York, London, Tokyo*. Princeton: Princeton University Press.

Sayer, A. (1997) 'The dialectic of culture and economy', in R. Lee and J. Wills (eds) *Geographies of Economies*. London: Arnold. pp. 16–26.

Sen, A. (1992) *Inequality Re-examined*. Oxford: Clarendon.

Shilling, C. (1993) *The Body and Social Theory*. London: Sage.

Shilling, C. (1997) 'The undersocialised condition of the (embodied) agent in modern sociology', *Sociology* 31: 737–54.

Sibley, D. (1995) *Geographies of Exclusion*. London: Routledge.

Simonsen, K. (2000) 'The body as battlefield', *Transactions of the Institute of British Geographers* 25: 7–10.

Skelton, T. and Valentine, G. (eds) (1997) *Cool Places: Geographies of Youth Culture*. London: Routledge.

Slater, D. (1997) *Consumer Culture and Modernity*. Cambridge: Polity.

Smith, D. (2000) 'Moral progress in human geography: transcending the place of good fortune', *Progress in Human Geography* 24: 1–18.

Smith, N. (2000) 'Socialising culture, radicalising the social', *Social and Cultural Geography* 1: 25–8.

The Guardian (2001) 'Hooked-up babes', 10 May.

Turner, B. (1996) *The Body and Society*, 2nd edn. London: Sage.

Valentine, G. (1996) 'Angels and devils: moral landscapes of childhood', *Environment and Planning D: Society and Space* 14: 581–99.

Valentine, G. (1997) '"My son's a bit dizzy", "My wife's a bit soft": gender, children and cultures of parenting', *Gender Place and Culture* 4: 37–62.

Watson, S. (ed.) (1990) *Playing the State*. London: Verso.

Williams, C. and Windebank, J. (1995) 'Social polarisation of households in contemporary Britain', *Regional Studies* 29: 723–8.

Wilson, E. (1977) *Women and the Welfare State*. London: Tavistock.

Witz, A. (2000) 'Whose body matters? Feminist sociology and the corporeal turn in sociology and feminism', *Body and Society* 6: 1–24.

Woodward, R. (1995) 'Approaches towards the study of social polarisation in the UK', *Progress in Human Geography* 19: 75–89.

2

Embodying Social Geography

Pamela Moss and Isabel Dyck

INTRODUCING THE BODY IN SOCIAL GEOGRAPHY

Fascination with the body, or what Kirsten Simonsen (2000: 7) calls 'body fixation', is burgeoning in the English-speaking academy. Indeed, 'the body' is no longer a construct with a singular corresponding material entity. Rather, 'the body' refers to an abstract 'thing' associated with multiple and varied discursive formations that inscribe corporeal vessels signifying human being(s). In this sense, there is no single, universal *body*; there are only multiply differentiated *bodies*. And it is the processes that differentiate bodies that are holding the attention of scholars throughout the humanities and social sciences. One of the most popular notions of differentiation as a process is *embodiment*, most often cast as lived experience. Scrutiny of 'the body' as myriad discursive formations, 'bodies' as concrete entities, and embodiment as lived experience involves exploring links between conceptualizations of the body (not 'the body') and states of bodily being, bodily experiences and bodily activities. Together, body and embodiment are key concerns in theorizing human experience, subjectivity and the relations of power through which difference is constructed and regulated. In geography, space and place necessarily are central in analyses that take issues about the body seriously, thus contributing to spatializing and *embodying* social geography.

Several coalescing interests are feeding into the attractiveness of body and embodiment as sites for theory building, not only about bodies but also about knowledge arising from bodily experiences, and are often cited as justifications for studying the body. It makes sense to identify some of these trends so that we can better understand the context within which theories of the body emerge. Social change at a variety of scales – local, regional, global – is shaping the way bodies and societies articulate. Rising costs of healthcare, linked both implicitly and explicitly to enormous financial investments in biotechnology, set parameters for determining what is, can be and should be a 'healthy' body. Enhanced technological and pharmaceutical control of bodily functions, with regard to organ replacement, performance-enhancing drugs and fertility procedures, are redefining limits to the capacities of bodies. Improvements in global goods distribution networks continue to deplete natural and human resources and threaten the social and economic welfare of people in the south. In these contexts, social change is transforming how bodies exist within, relate to and constitute societies. Interest in such social change enhances curiosity about the multiple facets of physical and cultural bodies – both the materiality of actual bodies and their discursive inscriptions.

Rapid economic change throughout the last quarter of the twentieth century transformed the nature of paid work, unpaid work and leisure. Demands of manual labour have moved from muscle, mass and strength to dexterity, flexibility and endurance. Factory floors are now corded-off machine areas where workers, positioned spatially at the margins, oversee production from glassed-in cubicles, wherein the machines' controls are safely encased. With the explosion of the tertiary and quaternary sectors, cubicles are the most common spatial configuration leading to a more atomistic labour process, intensifying the alienation of workers. Concurrent with these drastic changes in paid work, unpaid work associated with social reproduction has too transformed. Commodified forms of caring work in prenatal care and birthing, daycare and eldercare have replaced traditional kinship, familial and community ties. Commodification of conventional household tasks, as for example cleaning, shopping, dog walking, gardening and plant watering, through small and franchised firms and accessible internet access have changed the running of households. Together these shifts have opened up leisure time as never before, permitting many people to pursue idealized forms of the body, or 'the body beautiful,' through exercise regimes, fitness machines, diets and surgery.

Links between the materiality of bodies and their cultural representations through varied media are nowhere clearer than in reports about the emergence of new diseases and 'scientific breakthroughs' in biomedicine. Extensive coverage in popular media about deadly viruses over which 'science' has little (as in HIV) or no (as in the Ebola virus) control heightens apprehension over ways in which bodies can, do and should relate. Intense publicity about diseases and syndromes affecting immune, neurological and endocrine systems, as for example Creuzfeldt-Jacob disease, incite disquiet among those who try to control outbreaks, such as public health officials or international customs agencies, disdain of those people developing the disease or syndrome, and dread among the rest who can only hope they do not have to deal directly with such horrific bodily conditions. The social implications of the complete mapping of the human genome, narrowing the gap between science fiction and reality, have yet to be fully drawn out, but certainly such a mapping forces a reconsideration of the relationship between capital and bodies. Take for instance the full-scale purchase of Iceland's gene pool by DeCODE Genetics, or the recent relaxation of human cloning laws in Britain for medical research: both stretch the notion of what denotes individual and collective ownership of bodies and body parts and what aspects of the body are consumable.

In conjunction with social and economic change and the increased presence of biomedicine for public consumption, shifts in intellectual movements have been a factor in boosting the appeal of investigating body as a site for creating, building and expressing theory. The postmodern challenge to take a sceptical view of the grand narratives of modernist thinking spawned numerous critiques of objectivity, identity, truth and reason. The notion that the body is separate from the mind has permeated western thinking about the body since René Descartes published his famous utterance, *cogito, ergo sum*. Challenges to this particular dualism of valuing mind over body, as well as other manifestations of binary thinking such as binaries set up to value males over females, masculinity over femininity, and culture over nature, are now well ensconced in academic literature on the body. Theories of the body that emphasize the multiplicity of bodily shapes, functions, inscriptions and meanings fit quite nicely with postmodernist claims of the world as partial, fragmented, contingent, unstable and diverse, while critically challenging submerged voices located within bodily being, experience and activity.

Within this milieu of ever more complex relationships among bodies and social change, biomedicine and intellectual movements, we can place the surfacing and substantiation of body and embodiment in social geography. In this chapter, we chart the course of body and embodiment in geography, not chronologically, but through a combination of intellectual passages and spatialized embodiments, or particular 'types' of bodies depicted in the literature. In what follows, we first differentiate between theorizing body and pursuing embodiment as two key problematics framing the positioning of the body within social and cultural geography. We then delineate three intellectual passages, or enactments, of social and cultural geographical concerns involving power, identity and difference by considering bodies and spaces. By sorting through geographical works about body and embodiment, we show what geographers have been interested in pursuing, mostly in the 1990s. Next we provide a critical take on embodiment and demonstrate how we have taken up some of these concerns in our own work on women with chronic illness. We close with comments on the contributions and contestations geographers have been able to make about social geography because of their engagement with the body as well as what they could possibly say in the future.

SOCIAL GEOGRAPHIES OF BODIES AND EMBODYING SOCIAL GEOGRAPHY

Not all geographical studies of the body are studies about embodiment, nor do they necessarily invoke embodiment as an analytical category. In order to understand how the body expressly manifests in geography, we need to distinguish between a social geography of the body and an embodied social geography, paralleling concerns in sociology about the disembodied nature of knowledge that is not grounded in lived experience (Williams and Bendelow, 1998). Social geographies of bodies can, among other things, describe the nexus of personal and collective experiences of social, built and natural environments; tease out the constituent processes of individual and collective identities in relation to power; and explore the possibilities of bodily activities in specific spaces. The body, too, can be 'recovered' from classical works, as for example, Marx's 'sensual body' in 'Private Property and Communism', the 'alienated body' in 'Estranged Labour' and the 'labouring body' in *Capital* (Marx, 1964; 1967). This recovery project, possibly (and perhaps wrongly) criticized for salvaging modernist thinkers in postmodern times, is fully participating in constituting a truly embodied social geography wherein the body itself is both the subject of theory and a site for theorizing the social.

The same can be said for works in geography. Gill Valentine (1999: 329) makes this point when she identifies David Seamon's (1979) early work that looks at the ways through which bodies move through everyday spaces. Seamon's work is representative of one strand of 'body work' of the humanist tradition that emerged in contradistinction to Sauerian cultural geography. These humanist interpretations of lifeworlds and even some of the early auto/biographical writings in geography could be considered 'body work' in that they try to account for the way in which the body 'fits' into society (for example, Billinge et al., 1984; Buttimer, 1976; Buttimer and Hägerstrand, 1988; Eyles, 1985; Pred, 1981; Sauer, 1931). As Valentine (1999: 329) further argues, echoing Chris Shilling (1993: 9), the body clearly has been *in* geography, but such a claim has not been made apparent. These spatialized 'bodies' have been 'hidden' through subsequent iterations of other, perhaps more immediately relevant at the time, theoretical points/spaces of any one piece, resulting in readings that did not emphasize the body, its context or the implications of knowledge arising out of bodily activities. While not merely objects of study, the bodies in these types of geographies are still discrete entities that engage external environments, are end products of identity formation processes and perform through pre-scripted norms. This social geography of the body, or social geographies of bodies, are important in detailing connections among bodies, spaces and places.

In contrast to a social geography of the body, an embodied social geography is not simply a geography interested in exploring the body as a discrete entity. Separating bodies as outcomes from their constitutive material and discursive processes severs the living connection between bodies and those things that create, make up and sustain bodies themselves. As well, an embodied social geography is concerned with constructing knowledge that theorizes *from* bodies, privileging the *material* ways in which bodies are constituted, experienced and represented. Embodied knowledge, as a situated knowledge (after Haraway, 1988), challenges abstractions that are divorced from materiality and the spatially specific manner in which power is exercised and contested in society. Embodying social geography is therefore both a methodological and an epistemological matter as well as being closely concerned with theorizing body through spatial lenses of geography as a discipline. Theorizing body, indeed, is crucial to how we understand 'lived experience' and 'situated bodies', so that in pursuing the notion of an embodied social geography we need to retain at the centre of discussion the interaction between the body's analytic categories, the empirical constructs used to describe bodily experience and activity, and the theoretical understanding of the body and how it articulates with society.

Body and embodiment are often unproblematically connected in analysis. In order to problematize the link between the two, it is important to discern how the body is conceptualized from various theoretical, non-unified perspectives and how embodiment is used to denote constituent aspects of the body, including identity, power and the materiality of the body itself. We conceive the body as a material entity that is complexly constitutive of bodily notions, ideas and inscriptions. We think of embodiment as lived spaces where bodies are located corporeally and conceptually, concretely and metaphorically, materially and discursively. This means *being* simultaneously part of bodily forms, their social constructions and the materialization of their constitutive interaction. For example, becoming chronically ill while employed ignites a series of activities and events, some of which appear boundless as in the body that is ill, and some that have in a sense been pre-scripted as in the diagnosis and official responses to ill employees. Chronic illness as it manifests through the body as, for instance, fatigue and pain mediates the way in which the manner of specific work tasks get completed and the time in which it takes to complete the tasks. But abiding the bodily sensations of chronic illness is not 'enough' justification for a worker to reorganize her paid work beyond that which is acceptable within the range of the expected. A biomedical diagnosis, as a conventionally accepted inscription of disease, is required as 'proof' of an employee's bodily experience. A diagnosis assists in spelling out what is at stake prognosis-wise with a particular constellation of bodily sensations, known as symptoms. From this biomedical inscription, the worker with the 'ill body' regains some sense of limits on her 'out of control' body, interpolates the cultural representations of illness and disease with her bodily experiences, resulting in a shift in bodily activities, including treatment, care and identity, and begins thinking of herself as 'chronically ill'. At the same time, the diagnosis legitimates the worker's bodily sensations to the employer so that the employer can plan for and take action either with the worker, the workplace or both. In negotiating these inscriptions, both the worker and the employer (re)constitute working relationships in the workplace and the workplace itself.

In formulating our ideas about embodying social geography, we focus on works that assist in supporting our own notion of embodiment that contribute to a nuanced understanding of how bodies are complicit in creating an embodied knowledge upon which an embodied social geography can be based. The body *is* clearly central in thinking about embodiment, and for us to think of either without the other dissolves those textured connections replete with histories of articulations of identities and power within their spatialized contexts. However, as we will discuss throughout the chapter, the distinction between theorizing body and working with embodied knowledge needs to be held in tension because, with this tension, we can identify ways to understand the body innovatively that challenge conventional, static bodily conceptions and poke at the confining borders of binary thinking. Furthermore, thinking through an embodied social geography such as the one we propose includes a concern with political engagement as integral to, not separate from, methodological, epistemological and theoretical work.

BODIES IN THE SOCIAL SCIENCES AND GEOGRAPHY

As with other social sciences, the upsurge of interest in the body and issues of embodiment in geography have focused on various lines of inquiry. A move from modernist to postmodernist concerns has presented the body as a central theoretical problematic. Leading postmodern thinkers used the body distinctively as a site for extrapolating specific points about human experience, subjectivity and the relations of power. Michel Foucault's work on the body concentrates on the discipline of the body within the context of the way power operates in society,

as for example, through prisons, medical clinics and normative expressions of sexuality (Foucault, 1973; 1977; 1978). Gilles Deleuze and Félix Guattari (1983; 1987) transfixed theorists across disciplines with their introduction of 'the body without organs' and reacquainted psychoanalysts with new, revolutionary ideas about the link between the body and society. Although they are wide-ranging in their scope of influence on body and space, engagement with these specific works by geographers seems to be somewhat limited (though not exhaustive, some examples include Doel, 1995; Driver, 1985; 1993; *Environment and Planning D: Society and Space*, 1996; Matless, 1991; Philo, 1992; 2000). These works, too, tend to focus on how the representations of mechanisms and technologies define bodies and on whether there are possibilities of providing a framework for more critical work about the body in human geography.

However, it is within feminism and queer studies that the methodological, epistemological and theoretical project of embodiment has been the most influential, and probably the most effective, in challenging binary thinking through the body, contributing to theory about the body, and demonstrating the materiality of discourse on the body – both within and outside geography. Philosophers such as Jana Sawicki, Elspeth Probyn and Elizabeth Grosz are prominent examples of how feminists and queer theorists have pushed the bounds of binary thinking beyond static dichotomies. Sawicki (1991) decentres power and shows how ideas attached to the binary of masculinity and femininity are not useful for feminist politics. She argues that exaltation of the lesser valued dyad prevents the possibility of new configurations of power and knowledge, particularly as they relate to reproductive technologies. Probyn (1993) offers an innovative destabilization of the fragile nature of the social constructions of sex and gender. Her analysis promotes a more complex understanding of how gender itself is precariously positioned within multiple sets of power relations based on sex, race and sexuality and how then these intersect with each other to produce identities and subjectivities, especially with reference to the cultural forms these identities and subjectivities take. Grosz (1994; 1995) entwines instability of identity with the unpredictability of bodily functions and activities in her call for a more material and less deterministic framework for a feminist politics. Her work undertakes to delink stable binary pairs, such as discourse and materiality, culture and nature, and presence and absence, and resituate them so that norms, rules and expectations relating to each are redefined as unstable and thus open to the possibility of change, whether it be with styles of architecture, libidos or cities.

These types of feminist and queer works, in addition to feminist contributions in the field of ethics (see Diprose, 1994; Gatens, 1993; Shildrick, 1997; Weiss, 1999), suggest that the body as a site for theory development is indeed abundant. Such is the case for other social science disciplines. In sociology, bringing the body 'back in' (signalled by key works such as Frank, 1989, and Zola, 1991) spawned even more feminist interest in various bodies: the surgically altered body (Davis, 1995); the body under surveillance (Howson, 1998); and the body of gender difference (Lindeman, 1997). Feminist anthropologists, too, have explored understandings of gender and the body in terms of scientific knowledge, binary thinking, health and cultural activities (see, for example, Aalton, 1997; Martin, 1992a; 1992b; 1994). As well, feminist anthropologists are providing embodied alternatives to traditional ethnographic fieldwork, though not necessarily explicitly locating the work within the body literature (see, for example, Barillas, 1999; Dossa, 1999; Williams, 1996). In psychology, outside the predominance of a realist, empirically driven perspective, feminists are pursuing more contextual explanations of the body. Negotiating the discursive and material binary, feminist psychologists and psychoanalysts interested in queer theory are posing possible alternatives to conventional counselling and political action (see, for example, chapters in Ussher, 1997a, especially Squire, 1997; Stoppard, 1997; and Ussher, 1997b; 1997c).

Geography is much like these other social science disciplines. Geographers have demonstrated that bodies have a history and a geography (Nast and Pile, 1998; Teather, 1999). Questions of body and embodiment are being pursued by those working in the critical vein of the discipline – feminists, socialists, Marxists, queer theorists, psychoanalytic theorists and poststructuralists. Social and cultural geographers have engaged with the body and embodiment historically, economically, culturally and in health. Feminist, poststructural and psychoanalytic work with the influence of cultural studies have been central in the deconstruction of a unitary body (as well as subject, subjectivity and identity; see Section 5 in this volume), bringing to the forefront power, gender and sexuality in analyses of body and embodiment. The materiality of the body has been counterposed against the body as a surface of inscription, as a readable text. Although there has been less engagement with classed and racialized bodies, the body in

economic production in Marxist analyses and the role of 'whiteness' in the construction of the other in postcolonial studies are increasing.

These are the types of works we are interested in reviewing to show how geographers are embodying social geography. How have geographers used geographical concepts, such as space, place and scale, in constructing understandings of body and embodiment? The body as a *scale* of investigation, whether in geopolitics, urban studies or studies of the natural environment, has implications for methodology in that studies of articulations of the body with society may presuppose particular types of research, as for example, ethnographic methods for understanding everyday life or depth interviews for bodily inscriptions. How do geographers break the confines of scale? Or do they need to? A focus on lived experience through spatiality grounds investigations of the body in the everyday, bringing abstract subjects/bodies into the materiality of life in specific spaces and locations. Holding theorizing body in tension with an embodied knowledge poses additional questions not only in the sense of how these works have contributed to an *embodied* geographical knowledge, but also in how working with embodied knowledge contributes to understanding and explaining how bodies exist in various contexts.

INTELLECTUAL PASSAGES

Emerging through this tension between creating theory about the body and working from embodied knowledge are three key intellectual passages, around which we group our review of work about the body in social and cultural geography: examining the body as a site of regulation, oppression and control; delving into embodied subjectivity and spatiality; and challenging binary categories through problematizing the body. Much like Eleonore Kofman's (1994) assertion that research agendas brought forward in the 1980s remain unfinished, we propose a similar heuristic device referring to an unfinished grouping of body works as intellectual passages.

The body as a site of regulation, oppression and control

Many works in social and cultural geography draw on the notion that categories like gender, race, sexuality and disability are not to be taken for granted, but instead are to be thought of as constructed through social relations of power, via interaction, negotiation and constitution of specific bodies. Although not explicitly theorizing the body, early work in the 'new' cultural geography (see, for example, Jackson, 1989; Jackson and Penrose, 1992) collectively spelled out what is infeasible about understanding the body through unproblematized categories of power, both analytically and practically. Of central concern then is the intersection of oppressions arising out of regulatory mechanisms that control the body to varying degrees and the spatial manifestations of the particular ways these relations of power play out. Through conceiving analytical categories and empirical constructs as socially constructed, theorists have been able to demonstrate how multiple sets of power relations mutually constitute identities in such a way as to ground difference not in the manifestations of bodily being, as for example through sight, DNA or biology, but in the processes through which difference comes to be recognized as difference, as for example through the way reproductive labour is shaped by both gender and race (see, for example, Peake, 1995). By drawing out material processes related to production, theorists have been able to demonstrate how spaces, saturated with power organized around both oppressions and imaginations, are produced and reproduced in different contexts and time periods (see Gleeson, 1999 on disability and the production of urban space). Another aspect of the notion of socially constructed categories is the regulation of bodies through the same sets of relations that signal, denote and deploy oppression. Bodies act as the link between everyday activities and the larger organization of social power, and are deeply embedded within the ways people negotiate power through social relations. Regulatory mechanisms control the range of bodily activities as well as the bodies themselves, thus producing bodies that are constantly under surveillance either by the self or by society, either concretely or discursively, as for example in cases of anorexia nervosa (Bray and Colebrook, 1998) and in 'white privilege' in the university classroom (Sanders, 1998).

Embodying subjectivities and spatialities

With the upsurge of interest in the body came an associated rise in studies of identity and subjectivity. Understanding the body became closely related to understanding the subject, and vice versa. One of the central themes of inquiry into subjectivity influenced heavily by feminist and poststructuralist thinking is the notion of 'becoming', in lieu of a 'pre-existing person which is then channelled into diverse forms' (Price and Shildrick, 1999: 79). In diversity, then, while always in a state of becoming, the body never reaches a point of completeness or stasis. Ongoing external coercion, self-surveillance and agency are three primary modes of sustaining a state of becoming (Bordo, 1993). Problems crop up however when differentiating subjectivities for a specific politics, as for example feminist, queer or disability politics. Recognition of a marginalized individual or collective identity as a basis for a politics often involves glossing over diversity for the sake of a seemingly unified political stance, as for example in diminishing the importance of race politics in feminist movements or of sexual politics in workers' movements. This unintentional exclusionary politics happens discursively as well. For example, some young women may find it easier to claim their sexuality and be part of the lesbian, gay, bisexual and transgendered movement than to claim to be feminist in the more politically general women's movement. Empirical categories used to distinguish ability, age, class, ethnicity, gender, nationality, race, sex and sexuality already have within them a negation of dominant categories and an exclusion of other categories, unless hyphenated (see Visweswaran, 1994: 119, for a discussion of hyphenated identities). Thus when claiming, for example, femininity as resistance or transgression, the binary upon which femininity is inevitably linked ends up being supported, reinforcing then the hegemony of masculinity.

Geographical explorations into the body, subjectivity and, subsequently, embodied subjectivity have tried to embrace differentiation while keeping at bay claims that recognition of difference enhances pre-existing dominance (see, for example, chapters in Pile and Thrift, 1995). One strategy has been to elaborate the historical and geographical specificity of particular bodies. For example, Julia Cream (1995) meticulously distinguishes bodies as they relate to the introduction of the 'pill'. While engaged in the same act – that of taking the 'pill' – different women are at the same time contracepted for birth control, liberated for sexual freedom and exploited for the advancement of science. Spatializing embodied subjectivities like those that Cream identified can take many forms. Annabel Cooper et al., (2000) investigated the links between toilet provision for women in New Zealand between 1860 and 1940. They concluded that the more women became part of public life, the more private toilets became. Although not specific to their argument, the shift in the spatial regulation of women's bodies would necessarily have an impact on the constitution of these women's identities, or so the theory goes. In a geographical reading of Foucault's *The Birth of the Clinic* (1973), Chris Philo (2000) reintroduces Foucault's interlocking 'spatializations', rooted in the intensification of medicine after the French Revolution as well as the shifts in attitudes toward medicines in France in the eighteenth and nineteenth centuries. Even though Philo argues that this work is inspirational *in spite of* its historical and geographical specificity, we would argue that Foucault's work is compelling precisely *because of* its specificity. Considering past configurations of spatialities of subjectivities is but one strategy geographers use to embody subjectivities and spatialities. Other strategies include transforming space through discourse on the body (Robinson, 2000), unsettling space through uncertain bodily boundaries (Wakeford, 1998), transgressing space through bodily movement (Longhurst, 2001) and imagining space through socially imposed exile (Munt, 1998).

Challenging binary categories through problematizing the body

The body has become a lucrative site of theoretical attempts to overcome various binary categories, such as male/female, masculine/feminine, culture/ nature and discourse/materiality. By problematizing the body through conceptualizing it as multifaceted with multiple sets of complex links into itself and to the external world, geographers are challenging dichotomous thinking in ways that cast difference, self and subjectivity as fluid categories resistant to either overdetermination of structural imperatives or unrestricted agency with full free will (another binary). One particularly popular approach to understanding the body in various contexts is understanding categories of difference as performative, 'constituting the identity it is purported to be' (Butler, 1990: 25). Elaborating performativity even outside theatrics has proven to be useful in courting the notion that identities, for example, are imitative, caught up in a cycle of mimicry outside a pre-given state of being, with space within their reiteration for normative and resistant acts. Thus, differentiation and related processes such as subjectivation, sexualization and racialization of the self are set up within a space with no fixed representation. Representations themselves are fluid, imitative, mimicked and reiterative. Earlier spatial engagements of performativity focused on understanding bodies in space and their meanings. For example, Linda McDowell and Gill Court (1994) looked at how gendered representations abound in the finance industry. They showed how gender

performances of 'working bodies' in banks have shifted within the previous, financially volatile decade. In another early work, David Bell et al. (1994) contrasted hyper-sexualized spaces of gay skinheads and lipstick lesbians with the predominant, normalizing heterosexual spaces of everyday life. In performing hyper-gendered roles in hegemonic spaces, bodies can continually reconstitute meanings of their identities and subvert meanings of everyday spaces. More recent studies in performativity tend to draw more widely on performance theorists and promote a more nuanced appreciation of the insight such a theoretical framework can offer. Rather than engaging in the closely associated gender, sex and sexuality studies of performativity, Nicky Gregson and Gillian Rose (2000) argue that spaces, too, need to be thought of as performative. To illustrate their point, they expand topically into understanding the narratives of a group of community arts workers and the phenomenon of observing and participating in car-boot sales while simultaneously highlighting academic performances related to research and writing. In contrast to the relatively open spaces of community arts workers and car-boot sales, Teresa Dirsuweit (1999) maps a space heavily controlled through surveillance. She examines the coded sexual performances of prisoners in a South African women's prison expressing same-sex sexuality and engaging butch–femme roles. She found that performances of gender and sexuality were space-specific. Performativity matters in studies of subjectivity in that it has the potential to break open binary categories by mediating interpretations of bodily activities and demonstrating how space is intermingled with categories, constructs and expressions of the complexities of the manifestations of identity.

GEOGRAPHIES OF BODIES AND EMBODIED GEOGRAPHIES

These intellectual passages within geography – examining the body as a site of regulation, oppression and control; delving into embodied subjectivity and spatiality; and challenging binary categories through problematizing the body – are manifested in a wide range of work on the body, in the sense of both geographies of bodies and embodied geographies. For the most part, geographers have been preoccupied with trying to tease out the complexities of linkages among bodies and spaces, in the sense both of theorizing body and of recognizing and utilizing knowledge emerging from bodies. These complexities arise out of empirical issues, such as understanding the body in the context of labour, and theoretical ones, such as explaining how difference comes to be constituted through specific social practices. As a way to introduce some of the complexities geographers deal with, we now look at two specific 'bodies' that assist in showing how the intellectual passages show up in 'body work'. Like our previous categorization, we offer these two types of 'bodies' heuristically, as a way to sort the works about the body in geography, and refuse to claim that they or the works within them are exhaustive of the 'body work' in geography.

Economic bodies

Given long-standing interests in social relations of labour, it comes as no surprise that geographers are exploring economic bodies. It is, however, unexpected that this bodily interest was relatively late in coming into print, waiting until the late 1990s to appear. In what can be seen as a recovery of the body in classical analysis, David Harvey (1998) argues that the body is at once a site of political-economic contestation and of the very forces that construct it. Bodily practices that arise out of engaged labour (which sometimes create unhealthy bodies contradicting capital's needs for healthy bodies) and the circulation of variable capital are a means to transform relations of production and to create an emancipatory politics. Although cast as sensual, the body remains an entity unto itself, without specifying how the body is constituted through labour.

In a sympathetic critique, Felicity Callard (1998) situates her arguments at the juncture of Marxist and queer theory and suggests rethinking the connections between the labouring body and the corporeal configurations that give rise to particular *meanings* of the body. This refocus on the meaning *and* materiality of bodies shows how analysts can move through dichotomies organizing our thinking about labouring bodies. Deborah Leslie and David Butz (1998) problematize injured bodies by linking the spatial organization of a specific cybernetic labour process with a critique of neoliberal economic discourse. They argue that restructuring occurs at multiple geographical scales, including the body, setting up bodies, made vulnerable through an imposition of an unfamiliar set of material and discursive practices on existing bodies, at high risk for repetitive injuries. These injuries then take on specific meanings depending on the context within which the body with the injury appears – on the assembly line, in union meetings or in non-working

places like ski resorts – thus inscribing the body with the contradictions of the labour process. These inscriptions, read off these injured bodies in ways that seek to discipline, control and manage their bodies, can be points of departure for resistance and struggle in the workplace.

Bodies in consumption, both as consumers and as consumable, bring forward issues involving measurements against some idealized form of the body, resistance to hegemonic constructions of identities, and senses of subversion and conformity to hegemonic ideals of the body. Examining the body as a consumable good discursively bleeds into the materiality of the body and bodily activities. For example, with regard to body building and consumption, Derek McCormack (1999) intertwines the notions of consumer and consumed and explores the 'fitness' geographies produced by the fitness manufacturer, NordicTrack. He perceptively sorts through the complex co-configurations of body and machine in discourses of fitness. Through an exploration of the meaning and materiality of body fitness, he muddies the binaries of human/ non-human, male/female and nature/culture. Exploring the boundaries of bodies and spaces, Joyce Davidson (2001) also addresses the notions of consumer and consumed in her study of women with agoraphobia. She makes the case that 'going shopping', based on sensory stimulation with colourful displays, background music and scents, is often too intense for women fearing public consumption places like malls and supermarkets. But this is not just a matter of sensory overload; rather, she argues that consuming is based on a recognition of differentiating one's own self from the selves presented as part of the experience of consumption. Yet experiences of women with agoraphobia are disconcerting because their senses of themselves are disparate, fragmented and unwieldy in specific spaces of consumption, such as shopping malls, which locate women in a contradictory subject position as both consumer and one to be consumed by and for others.

These examples of economic bodies demonstrate how rooting an analysis in a binary – in these cases that of discourse and materiality – assists in unsettling other dichotomous categories. These types of work build an embodied geography by interrogating how specific social and cultural practices constitute subjectivities through bodily acts and their meanings.

Bodies as the intersection of oppressions

As part of the wider trend in body studies, social and cultural geographers have engaged the body as an intersection of oppressions. Because of geographers' extensive work on identity, difference and power, the body in relationship to power and subjectivity is probably the most popular type of body interrogated in geography, especially in regard to the intersection of gender, 'race', sexuality and ability, and to a lesser extent age, citizenship and nationality. Together these studies of power and the body tend to focus on spatial responses to everyday phenomena, the performance of identity in an array of contexts, and the destabilization of categories in terms of dominance and privilege. In what could be a far-reaching analysis of an emancipatory politics, Joanne Sharp (1996) explores how the relationship between gender and nationhood, its constitution and the contingency of its material manifestation, might contribute to a radical democratic feminist politics (following Mouffe, 1992). She argues that denaturalizing the constitutive process of dominance and subjugation is useful in supporting a radical democracy. The move to denaturalize any oppressive arrangement of the deployment of power speaks to how specific bodies find spaces to resist. Theorizing the body as a site of both oppression and resistance can access the intersection of multiple oppressions as well as the intersection of meaning and materiality (Moss and Dyck, 1996: 749).

What is missing analytically in many studies of the intersection of oppressions, as Kristen Day (1999) has pointed out, is scrutiny of the 'whiteness' of various constructions of oppression and the complexities of 'race'. 'Race', so denoted to emphasize the socially constructed nature of the category, should be very much a part of body work in social and cultural geography (see Bonnett and Nayak, Chapter 15 in this volume). Understanding bodily experience in terms of 'race' enhances embodied knowledge especially because '"racial" knowledge about the "other" is what provides the contradictory experience of "race"' in everyday life (Nkweto Simmonds, 1999: 57). Embodied knowledge is perhaps the most effective way to challenge the hegemony of 'whiteness' so prevalent in the west. At stake in the production of academic knowledge and, more widely, in society is the denaturalization of difference (after Kobayashi and Peake, 1994) and the deinternalization of dominance (after Chouinard and Grant, 1995). Although not explicitly addressing either bodies or knowledge grounded in bodily experience, social and cultural geographers nevertheless include racialized bodies in their interpretations of 'race' and spatial politics. In a nuanced study of collective politics, Laura Pulido (1997) complicates the category 'woman' in feminist theory and politics by questioning whether or not the subaltern can speak. In a study of environmental justice activist women's groups, she showed how low-income women of colour chose to show a united front within their respective racialized communities in lieu of drawing on 'woman' and feminism. She goes on to argue that academics and other activists tend to identify these activists in terms of gender rather than 'race', further demonstrating the lack of political astuteness of people invoking feminist theory and practising feminism. This theme of misreading racialized identities appears in Alastair Bonnett's (1996) work scrutinizing 'white' identities in anti-racist discourse. He argues that through the reification of 'white' identities, white anti-racists have failed to consider 'whiteness' fully as a contingent, socially constructed category, and this has resulted in a mixture of white confessions, white guilt and whites trying to understand the 'non-white' as different. With this focus on privilege, analysts have provided insight into the ways power operates diffusely, subtly and hegemonically.

As in feminist and queer theory, in cultural geography much of the theorizing of both the body itself and knowledges emerging from the body appears in studies of space and sexuality. Although not all social or cultural geographic works in the area of sexuality deal with bodies (see recent reviews by Binnie and Valentine, 1999 and Hubbard, 2000), works specifically on the sexualized body invoke and develop arguments that destabilize dichotomous categories while at the same time emphasizing that the body as an intersection of oppressions is also a site of resistance. For example, Ki Namaste (1996), in her study of 'genderbashing' in Montréal, argues that the fusion of gender and sexuality has distinct implications, often with severe consequences in public spaces. Women and men who transgress the socially accepted norms of gendered and sexualized limits are most at risk of assault. Scrutiny of the intersection of sexualized subject positions of identity formation also includes examination of the ways in which power differentiates, oppresses and marginalizes particular people within sexualized groups – whether they be part of the compulsory heterosexuality that regulates public and private spaces (see Nast, 1998) or already identifiable marginalized groups, as for example lesbians, gay men and bisexual individuals.

Much of the explicit 'body work' in social and cultural geography has been with disabled, impaired and ill bodies. Discontent over research methods involving disease mapping in epidemiological studies produced calls for an *embodied* medical geography (Dorn and Laws, 1994).

Many studies are located at the nexus of body politics, knowledge/power in biomedicine and the instability of categories used in analysis and description. Disability studies in social and cultural geography focus on conceptualizing disability as a social construct, one that moves away from individualized definitions of ability that lay blame on the deficiencies of the individual toward explaining disability in terms of the social organization of society, place and access to resources and their allocation (see Gleeson, 1999). Explicit theorizations of the body accompanied this increased interest in disability studies as endeavours to understand difference, identity and the deployment of power (see, for example, Butler and Bowlby, 1997; Butler and Parr, 1999; Chouinard, 1999; Dorn, 1998; Moss and Dyck, 1996). Ruth Butler (1999) illustrates this point in her examination of the political connections between sexuality, space and disability. She points out that lesbian, gay men and bisexual individuals are marginalized within the disabilities movement just as disabled people are marginalized within the gay movement. For Butler, recognition of difference within communities of oppressed and marginalized persons is important in effecting political change for any one group.

Spatialized configurations of gendered, 'raced,' sexualized and disabled bodies may further the understanding of the intersection of regulation, oppression and control of bodies because, in all these examples, theorizing the body stems from experiences of *embodied* subjectivities.

CRITICALLY RETEXTURING BODIES

Having worked through various geographical approaches to theorizing body and embodiment in different topical areas, it is clear that subdisciplinary borders are crossed and recrossed. The influence of feminism and poststructural thought in particular bring common theoretical problematics to queries and concerns in subfields of social and cultural geography, whether they be about economy, nationhood, sexuality or disability. As these works show, these crossings can usefully challenge binary thinking by opening up categories for analysing embodied subjectivity, a concern that must be central in understanding the constitution of social divisions, their intersections in carving out social experience, and the ways in which powerful discourses frame both subject positions and forms of resistance. As we would expect, the strength of geography's contribution to theorizing body as well as to embodied knowledge is through attention to the

spatiality of embodiment and its materiality. Pile and Thrift's (1995) metaphor of the body as a 'spatial home' of the subject emphasizes the body's movement in time and across space in particular configurations of relations and distributions of power as they coalesce in (and through the continuous process of creating) 'place' (see Massey, 1993).

Places also are encoded with meanings that signal 'out of placeness' (after Cresswell, 1996) or inclusiveness, which are embodied through, for example, classed, 'raced' and gendered performances which reconstitute, or alternatively transgress, such place meanings. Transgression may be contested, as in the use of threat and violence, or alternatively through self-surveillance in fear or acceptance of dominant norms. If 'successful', transgression acts as a transformative politics adding to the multiplicity of meanings of a particular place. As McDowell emphasizes, 'Social, economic and political structures are crucial in defining and maintaining not only a particular urban form but also particular versions of acceptable bodies' (1999: 66). Engaging feminism and poststructural thought, especially queer theory, opens up space for alternative voices to the heterosexual male, white, able-bodied, middle-class norm predominant in 'writing' the landscape and social organization of the greater part of twentieth-century society and, in association, its scholarship. In geography these approaches also firmly ground the subject/body in the concrete spatiality of everyday life. Work pointing to the fluidity of bodies and identities, the multiple performances of identity in the specificities of place, and the conceptualization of a 'third space' through which identities may be reformulated, provide an intellectual space from which to challenge binary and 'fixed' social categories.

Thinking of the body as simultaneously and mutually constituted as corporeal and discursive is crucial to effecting this challenge of unsettling binaries while expanding the materiality of spatialities. Despite health geography's relatively late takeup of theoretical concerns about body and embodiment, it is perhaps these nascent works that can be particularly useful in demonstrating how the reassertion of the body as non-essentialist generates space for a body politics that recognizes the mutability of bodies and the intersectionality of multiple differences. In an approach we broadly term a feminist materialist perspective, we, like other social and cultural geographers, have drawn upon ideas of a discursively produced body which is constantly lived through both its materiality and its representations in particular environments (after

Grosz, 1994; 1995). Such environments have a concrete spatiality with encoded meanings that may be reproduced or resisted through the ongoing negotiation of everyday life in homes, workplaces, communities and nation-states in the context of audiences of the repeated performativity of identity (see Butler, 1990: 25 for the *type* of performativity we refer to here).

However, like Lise Nelson (1999), we are leery of grasping onto an uncritical acceptance of Judith Butler's (1990) notion of performativity. On the one hand, 'by interrogating implicit norms within enunciations of "identity" and recognizing it as a process of identification, something that is *done over and over* instead of something that is an *inherent* characteristic', Butler's notion of 'performativity opens up new terrains of analysis' (Nelson, 1999: 339, emphasis in original). On the other hand, according to Nelson, Butler inevitably has to invoke the notion of conscious agency in order to overcome the contradiction of intent and action in her theorization of performativity. Nelson proposes that conceiving identity formation as an 'iterative, non-foundational process', one that articulates with intentional human practice, is preferable to the dichotomous conceptualization in Butler's work: 'between the masterful humanist subject and the subject as a node in the power discourse matrix, a site of compelled repetition of hegemonic identities' (1999: 333, 347). An uncritical acceptance will no doubt reproduce the same weaknesses and constrictions of performativity, much like Gillian Rose's (1997) critique of 'transparent reflexivity'.

Unlike Nelson, though, we do not think it necessary to enter (back) into discussion of the intentionality of human action and agency as the basis for a materially embodied critique of Butler's theory of performativity. Butler's refusal of any pre-discursive sex assignment sets up a non-biologically determined framework that could demonstrate the power of discourse not only to *shape* but also to *read* bodies. Identifying and elaborating a process – performativity – through which discourses inscribe onto the body particular renditions of sex as well as through which we can make bodies culturally intelligible is the strength of Butler's work in the context of theorizing body and embodiment (see also Butler, 1993). This (purposeful?) overemphasis of discourse, although useful in understanding how the body relates to society and even strategic in practising a feminist body politics, eclipses the power of materiality not only to *shape* but also to *read* bodies. Actual bodies as concrete entities – or our corporeal embodiments – are locales of human being(s) and media through which people exist, act and experience environments. Reading

other bodies inevitably relies on the materiality of the particular bodies doing the reading. When dealing with body and embodiment, avoiding traps of binary thinking – whether it be not resurrecting the structure/agency debate, the nature/nurture argument, the subject/object controversy or the discourse/materiality divide – is vital because the body, viewed as a nexus, is where binaries exist in tandem. Hence, drawing on Butler's notion of performativity as reiterative, routinized behaviour is as useful as Sawicki's, Probyn's and Grosz' attempts at destabilizing binary notions of the body.

In our own work, we find the variably ill body a useful example to demonstrate the intertwining of the corporeality and constancy of performance – of multiple positions of gender, 'race', class, sexuality, ability and age, for example – in establishing or attempting to maintain, or alternatively resisting a destabilization of, a cohesive identity. In our work with women diagnosed with multiple sclerosis (MS), myalgic encephalomyelitis (ME) or rheumatoid arthritis (RA), we found that many women experience changes in their corporeality that threaten their previous abled and gendered identities. As a way to deal with these 'new', sometimes well-scripted notions of their body as ill and/or diseased, the women were forced to renegotiate both their social and physical environments as well as the representations of their body as ill and/or diseased. These negotiations, with varying degrees of 'success', took place within a web of power relations predicated on biomedical discourses and practices which continually inscribe the women as 'deviant'. In Foucauldian terms such inscription acts to mediate access to resources and use of space through regulation and self-surveillance, for the diagnosis and prognosis set expectations of what a woman might be able to accomplish in day-to-day life, whether corporeally 'accurate' or not. Yet the unpredictability of much chronic illness makes the re-establishment of identity through routine material practices and performance difficult as the body slips between hegemonic notions of what being well and being ill are.

We explored women's experiences of chronic illness from their accounts of living with a body constructed as 'deviant' through biomedical representational and material practices (see Moss and Dyck, 1996; 1999a). Rather than 'fixed' – in terms of capacity to perform 'as usual' – the women's bodies were sometimes ill, sometimes well, and sometimes both ill and well at the same time (Moss, 2000). Uncertainty prevailing in what women might expect to be able to do and in how others in positions of power might perceive them, such as employers and disability insurers, placed women in precarious positions with regard to maintaining a job or securing financial support (Moss and Dyck, 2001). The women actively negotiated the dissonance between previous able identities and/or current instability of corporeal capacities in *particular* settings with gendered performance expectations, whether home or workplace (Dyck, 1999). Strategies the women used varied widely but most included ways of navigating physical environments that had become 'hostile', in the sense of being difficult to traverse, use or exist within, and ways of concealing corporeal limitations in workplaces where there was fear of reprisal – such as dismissal or stigmatization – if a woman's illness happened to be revealed. In each case, class was an important dimension of experience and one would expect that 'race' and other devalued differences would also intersect in women's embodied experience and knowledge of being chronically ill, although most of the women in our studies were heterosexual and white.

Our studies show that it is through the body – which had become a problem for these women through unsettling constancies, capacities and materialities – that the women think through and talk about their lives in perhaps a more explicit way than those with an able and non-chronically ill body which has not been reconstructed as 'deviant'. The *transition* between states of bodily being heightens the perception of change and intensifies the bodily experiences of that change so that when recounting, women were readily forthcoming with details about how they deal with their illness in a number of venues (Moss and Dyck, 1999b). In a sense, they had yet to naturalize or internalize their 'deviance', resulting in richly, and sometimes garishly, textured descriptions of their bodies. This 'being in transition' highlights the *fluidity* of bodies and associated inscriptions, through specifying particular spaces encoded with performance expectations and particular strategies developed to manage discrepancies between performance and appearance. 'Being in transition' also brings into focus bodily boundaries in identity formation or transformation – ones that differentiate specific bodily aspects of identity as well as differentiate bodies from each other. Like most boundaries, these bodily boundaries, too, are created, reproduced, resisted and transgressed (see Moss, 1999).

CONTRIBUTIONS AND CONTESTATIONS

The sum of work in social and cultural geography concerned with theorizing body and embodiment in various ways enables us to begin thinking about knowledge, 'being' and identities beyond the confines of binaries in two main ways. First, methodological approaches that focus on narrative accounts and other qualitative methods provide ways of producing knowledge that creates space for *embodied knowledge* to emerge from bodily being, experiences and activities from a plurality of voices *and* places. This in itself challenges an understanding of the world based on power-laden categories that cast frameworks of knowledge in relation to normative depictions of who and which groups constitute the centre and margin, the self and other. Taking up difference entails a reworking of boundaries and borders and a destabilizing of unitary taken-for-granted categories through the performance of multiple ways of *being* gendered, classed, 'raced', sexed, sexualized and abled. Because identities are embodied, audiences respond to transgressive and conforming performances in the specificities of the spaces where these performances are enacted and observed.

Second, theorizing body as material and discursive with each constituting the other, and grounded in the particular space/time organization of quotidian life, is useful in thinking through challenges to binary categorization. Examining this grounding empirically, geographers are building a mass of studies where contestations, resistances and/or transgressions of dominant and normative depictions of gender, class, 'race', sex, sexuality and ability are unsettled. In grounding the abstract subject/body in the materiality of everyday life such unsettling, and the success or otherwise of its myriad implications, provides room for multiple and wide-ranging audiences, places and cultural contexts. Feminism and queer theory have been insightful in suggesting ways of thinking beyond binaries and incorporating corporeality without falling into the perils of biological reductionism or essentialism, while theorizing body and embodiment in ways that demonstrate the non-viability of understanding social relations based on simplistic associations of biology, behaviour and capacities for action and thinking.

Throughout this chapter we have pointed out the difference between a geography of the body and an embodied geography while holding in tension theorizing body and working with embodied knowledge. Although creating geographies of bodies is useful in counting, accounting for and recounting people's lives, they are somewhat limited in explaining and understanding how multiple sets of power relations mutually constitute individual persons and groups of people or how the mutability of bodies and fluidity of subject/body is integral to (re)constituting embodied subjectivities (see the exchange on accessibility issues in disability studies in geography between Golledge, 1993; 1996; Gleeson, 1996; and Imrie, 1996). For example, there is no doubt that accessibility maps contribute to knowing both what makes a public toilet inaccessible and where accessible toilets are located. However, it might make more sense to understand these maps in the context not only of how the experience of mobility restrictions or incontinence shape and are shaped by the layout of toilets and their locations, but also of how the subjectivities of women and men are constituted by the experience of the accessibility of toilets through multiple social relations of power. By integrating real, deviant bodies into our understandings of geography, geographies of the body can be transformed into embodied geographies, ones that problematize the body complexly while drawing on varied types of embodied knowledge.

NOTE

We would like to thank Karen Boyes for her assistance in gathering literature for this chapter and Peter Jackson for his constructive comments.

REFERENCES

Aalton, A. (1997) 'Performing the body, creating culture', in K. Davis (ed.) *Embodied Practices: Feminist Perspectives on the Body.* London: Sage. pp. 41–58.

Barillas, M.O. (1999) 'Colonial and post-revolutionary discourses and Nicaraguan feminist constructions of *Mestiza*: reflections of a cultural traveller', in R. Bridgman, S. Cole and H. Howard-Bobiwash (eds) *Feminist Fields: Ethnographic Insights.* Peterborough, ON: Broadview. pp. 196–211.

Bell, D., Binnie, J., Cream, J. and Valentine, G. (1994) 'All hyped up and no place to go', *Gender, Place and Culture* 1: 31–48.

Billinge, M., Gregory, D. and Martin, R.L. (eds) (1984) *Recollections of a Revolution.* London: Macmillan.

Binnie, J. and Valentine, G. (1999) 'Geographies of sexuality – a review', *Progress in Human Geography* 23 (2): 175–87.

Bonnett, A. (1996) 'Anti-racism and the critique of "white" identities', *New Community* 22 (1): 97–110.

Bordo, S. (1993) 'Feminism, Foucault and the politics of the body', in C. Ramazaglù (ed.) *Up Against Foucault: Explorations of Some Tensions between Foucault and Feminism.* London: Routledge. pp. 179–202.

Bray, A. and Colebrook, C. (1998) 'The haunted flesh: corporeal feminism and the politics of (dis)embodiment', *Signs: Journal of Women in Culture and Society* 24 (1): 35–67.

Butler, J. (1990) *Gender Trouble: Feminism and the Subversion of Identity.* London: Routledge.

Butler, J. (1993) *Bodies that Matter.* London: Routledge.

Butler, R. (1999) 'Double the trouble or twice the fun? Disabled bodies in the gay community', in R. Butler and H. Parr (eds) *Mind and Body Spaces: Geographies of Illness, Impairment and Disability.* London: Routledge. pp. 203–20.

Butler, R. and Bowlby, S. (1997) 'Bodies and spaces: an exploration of disabled people's experiences of public space', *Environment and Planning D: Society and Space* 15: 411–33.

Butler, R. and Parr, H. (eds) (1999) *Mind and Body Spaces: Geographies of Illness, Impairment and Disability.* London: Routledge.

Buttimer, A. (1976) 'Grasping the dynamism of the lifeworld', *Annals of the Association of American Geographers* 66: 277–92.

Buttimer, A. and Hägerstrand, T. (eds) (1988) *Geographers of Norden: Reflections on Career Experiences.* Lund: Lund University Press.

Callard, F. (1998) 'The body in theory', *Environment and Planning D: Society and Space* 16: 387–400.

Chouinard, V. (1999) 'Body politics: disabled women's activism in Canada and beyond', in R. Butler and H. Parr (eds) *Mind and Body Spaces: Geographies of Illness, Impairment and Disability.* London: Routledge. pp. 269–94.

Chouinard, V. and Grant, A. (1995) '"On not being anywhere near the 'project': revolutionary ways of putting ourselves in the picture', *Antipode* 27: 137–66.

Cooper, A., Law, R., Malthus, J. and Wood, P. (2000) 'Rooms of their own: public toilets and gendered citizens in a New Zealand city, 1860–1940', *Gender, Place and Culture* 7: 417–33.

Cream, J. (1995) 'Women on trial: a private pillory?', in S. Pile and N. Thrift (eds) *Mapping the Subject: Geographies of Cultural Transformation.* London: Routledge. pp. 158–69.

Cresswell, T. (1996) *In Place, Out of Place: Geography, Ideology, and Transgression.* Minneapolis: University of Minnesota Press.

Davidson, J. (2001) 'Fear and trembling in the mall: women, agoraphobia, and body boundaries', in I. Dyck, N. Davis Lewis and S. McLafferty (eds) *Women's Health Geographies.* London: Routledge. pp. 213–30.

Davis, K. (1995) *Reshaping the Female Body: The Dilemmas of Cosmetic Surgery.* London: Routledge.

Day, K. (1999) 'Embassies and sanctuaries: women's experiences of race and fear in public space', *Environment and Planning D: Society and Space* 17: 307–28.

Deleuze, G. and Guattari, F. (1983) *Anti-Oedipus: Capitalism and Schizophrenia.* Minneapolis : University of Minnesota Press.

Deleuze, G. and Guattari, F. (1987) *A Thousand Plateaus: Capitalism and Schizophrenia.* Minneapolis: University of Minnesota Press.

Diprose, R. (1994) *The Bodies of Women: Ethics, Embodiment and Sexual Difference.* London: Routledge.

Dirsuweit, T. (1999) 'Carceral spaces in South Africa: a case study of institutional power, sexuality and transgression in a women's prison', *Geoforum* 30: 71–83.

Doel, M. (1995) 'Bodies without organs: schizoanalysis and deconstruction', in S. Pile and N. Thrift (eds) *Mapping the Subject: Geographies of Cultural Transformation.* London: Routledge. pp. 226–40.

Dorn, M. (1998) 'Beyond nomadism: the travel narratives of a "cripple"', in H.J. Nast and S. Pile (eds) *Places through the Body.* London: Routledge. pp. 183–206.

Dorn, M. and Laws, G. (1994) 'Social theory, body politics, and medical geography: extending Kearns's invitation', *The Professional Geographer* 46: 106–10.

Dossa, P. (1999) 'Narrating embodied lives: Muslim women in the coast of Kenya', in R. Bridgman, S. Cole and H. Howard-Bobiwash (eds) *Feminist Fields: Ethnographic Insights.* Peterborough, ON: Broadview. pp. 157–72.

Driver, F. (1985) 'Power, space and the body: a critical assessment of Foucault's *Discipline and Punish*', *Environment and Planning D: Society and Space* 3: 425–46.

Driver, F. (1993) 'Bodies in space: Foucault's account of disciplinary power', in C. Jones and R. Porter (eds) *Reassessing Foucault: Power, Medicine and the Body.* London: Routledge.

Dyck, I. (1999) 'Body troubles: women, the workplace and negotiations of a disabled identity', in R. Butler and H. Parr (eds) *Mind and Body Spaces: Geographies of Illness, Impairment and Disability.* London: Routledge. pp. 119–37.

Environment and Planning D: Society and Space (1996) Special Issue devoted to Deleuze and Guattari, 14: 379–499.

Eyles, J. (1985) *Senses of Place.* Warrington: Silverbrook.

Foucault, M. (1973) *The Birth of the Clinic: An Archaeology of Medical Perception.* New York: Vintage, 1994.

Foucault, M. (1977) *Discipline and Punish: The Birth of the Prison.* New York: Vintage, 1979.

Foucault, M. (1978) *History of Sexuality: An Introduction,* Vol I. New York: Vintage, 1990.

Frank, A.W. (1989) 'Bringing bodies back in: a decade review', *Theory, Culture and Society* 7: 131–62.

Gatens, M. (1993) *Imaginary Bodies: Ethics, Power and Corporeality.* London: Routledge.

Gleeson, B. (1996) 'A geography for disabled people?', *Transactions of the Institute of British Geographers* 21: 387–96.

Gleeson, B. (1999) *Geographies of Disability.* London: Routledge.

Golledge, R.G. (1993) 'Geography and the disabled: a survey with special reference to vision impaired and blind populations', *Transactions of the Institute of British Geographers* 18: 63–85.

Golledge, R.G. (1996) 'A response to Gleeson and Imrie', *Transactions of the Institute of British Geographers* 21: 404–11.

Gregson, N. and Rose, G. (2000) 'Taking Butler elsewhere: performativities, spatialities and subjectivities', *Environment and Planning D: Society and Space* 18: 433–552.

Grosz, E. (1994) *Volatile Bodies: Toward a Corporeal Feminism.* Bloomington: Indiana University Press.

Grosz, E. (1995) *Space, Time and Perversion: Essays on the Politics of the Body.* London: Routledge.

Haraway, D. (1988) 'Situated knowledges: the science question in feminism and the privilege of partial perspective', *Feminist Studies* 14: 575–99.

Harvey, D. (1998) 'The body as an accumulation strategy', *Environment and Planning D: Society and Space* 16: 401–21.

Howson, A. (1998) 'Embodied obligation: the female body and health surveillance', in S. Nettleton and J. Watson (eds) *The Body in Everyday Life.* London: Routledge. pp. 218–40.

Hubbard, P. (2000) 'Desire/disgust: mapping the moral contours of heterosexuality', *Progress in Human Geography* 24 (2): 191–217.

Imrie, R. (1996) 'Ableist geographers, disablist spaces: towards a reconstruction of Golledge's "Geography and the disabled"', *Transactions of the Institute of British Geographers* 21: 397–403.

Jackson, P. (1989) *Maps of Meaning.* London: Routledge.

Jackson, P. and Penrose, J. (eds) (1992) *Constructions of Race, Place and Nation.* Minneapolis: University of Minnesota Press.

Kobayashi, A. and Peake, L. (1994) 'Unnatural discourse: "race" and gender in geography', *Gender, Place and Culture* 1: 225–43.

Kofman, E. (1994) 'Unfinished agendas: acting upon minority voices of the past decade', *Geoforum* 25: 429–43.

Leslie, D. and Butz, D. (1998) '"GM suicide": flexibility, space, and the injured body', *Economic Geography* 74 (4): 360–78.

Lindeman, G. (1997) 'The body of gender difference', in K. Davis (ed.) *Embodied Practices: Feminist Perspectives on the Body.* London: Sage. pp. 73–92.

Longhurst, R. (2001) *Bodies: Exploring Fluid Boundaries.* New York: Routledge.

Martin, E. (1992a) 'The end of the body?', *American Ethnologist* 19 (1): 120–38.

Martin, E. (1992b) *The Woman in the Body: A Cultural Analysis of Reproduction,* 2nd edn. Boston: Beacon.

Martin, E. (1994) *Flexible Bodies: The Role of Immunity in American Culture from the Days of Polio to the Age of AIDS.* Boston: Beacon.

Marx, K. (1964) *Economic and Philosophic Manuscripts of 1844.* New York: International.

Marx, K. (1967) Capital, Vol. 1. New York: International.

Massey, D. (1993) 'Power-geometry and a progressive sense of place', in J. Bird, B. Curtis, T. Putman, G. Robertson and L. Tickner (eds) *Mapping the Futures: Local Cultures, Global Change.* London: Routledge. pp. 59–69.

Matless, D. (1991) 'An occasion for geography: landscape, representation and Foucault's corpus', *Environment and Planning D: Society and Space* 10: 41–56.

McCormack, D. (1999) 'Body shopping: reconfiguring geographies of fitness', *Gender, Place and Culture* 6: 155–77.

McDowell, L. (1999) *Gender, Identity and Place: Understanding Feminist Geographies.* London: Polity.

McDowell, L. and Court, G. (1994) 'Performing work: bodily representations in merchant banks', *Environment and Planning D: Society and Space* 12: 727–50.

Moss, P. (1999) 'Autobiographical notes on chronic illness', in R. Butler and H. Parr (eds) *Mind and Body Spaces: Geographies of Illness, Impairment and Disability*. London: Routledge. pp. 155–66.

Moss, P. (2000) '"Not quite abled" and "not quite disabled": experiences of being "In Between" ME and the academy', *Disability Studies Quarterly* 20 (3): 287–93.

Moss, P. and Dyck, I. (1996) 'Inquiry into environment and body: women, work and chronic illness', *Environment and Planning D: Society and Space* 14: 737–53.

Moss, P. and Dyck, I. (1999a) 'Body, corporeal space and legitimating chronic illness: women diagnosed with ME', *Antipode* 31: 372–97.

Moss, P. and Dyck, I. (1999b) 'Journeying through M.E.: identity, the body and women with chronic illness', in E.K. Teather (ed.) *Embodied Geographies: Spaces, Bodies and Rites of Passage*. London: Routledge. pp. 157–74.

Moss, P. and Dyck, I. (2001) 'Material bodies precariously positioned: working women diagnosed with chronic illness', in I. Dyck, N.D. Lewis and S. McLafferty (eds) *Geographies of Women's Health: Place, Diversity and Difference*. London: Routledge. pp. 231–47.

Mouffe, C. (1992) 'Feminism, citizenship and radical democratic politics', in J. Butler and J.W. Scott (eds) *Feminists Theorize the Political*. London: Routledge. pp. 369–84.

Munt, S.R. (1998) 'Sisters in exile: the lesbian nation', in R. Ainley (ed.) *New Frontiers of Space, Bodies and Gender*. London: Routledge. pp. 3–19.

Namaste, K. (1996) 'Genderbashing: sexuality, gender, and the regulation of public space', *Environment and Planning D: Society and Space* 14: 221–40.

Nast, H.J. (1998) 'Unsexy geographies', *Gender, Place and Culture* 5: 191–206.

Nast, H. and Pile, S. (eds) (1998) *Places through the Body*. London: Routledge.

Nelson, L. (1999) 'Bodies (and spaces) do matter: the limits of performativity', *Gender, Place and Culture* 6: 331–53.

Nkweto Simmonds, F. (1999) 'My body, myself: how does a black woman do sociology?', in J. Price and M. Shildrick (eds) *Feminist Theory and the Body: A Reader*. New York: Routledge. pp. 50–63.

Peake, L. (1995) 'Toward an understanding of the interconnectedness of women's lives: the "racial" reproduction of labour in low-income urban areas', *Urban Geography* 16: 414–39.

Philo, C. (1992) 'Foucault's geography', *Environment and Planning D: Society and Space* 10: 137–61.

Philo, C. (2000) '*The Birth of the Clinic*: an unknown work of medical geography', *Area* 32: 11–20.

Pile, S. and Thrift, N. (eds) (1995) *Mapping the Subject: Geographies of Cultural Transformation*. London: Routledge.

Pred, A. (1981) 'Social reproduction and the time-geography of everyday life', *Geografiska Annaler* 63B: 5–22.

Price, J. and Shildrick, M. (eds) (1999) *Vital Signs: Feminist Reconfigurations of the Bio/logical Body*. Edinburgh: Edinburgh University Press.

Probyn, E. (1993) *Sexing the Self: Gendered Positions in Cultural Studies*. London: Routledge.

Pulido, L. (1997) 'Community, place, and identity', in J.P. Jones III, H.J. Nast and S.R. Roberts (eds) *Thresholds in Feminist Geography: Difference, Methodology, Representation*. Latham: Rowman and Littlefield. pp. 11–28.

Robinson, J. (2000) 'Feminism and the spaces of transformation', *Transactions of the Institute of British Geographers* 25: 285–302.

Rose, G. (1997) 'Situating knowledges: positionality, reflexivities and other tactics', *Progress in Human Geography* 21: 305–20.

Sanders, R. (1998) 'Introducing "white privilege" into the classroom: lessons from finding a way', *Journal of Geography* 98: 169–75.

Sauer, C. (1931) 'Cultural geography', in P. Wagner and M. Mikesell (eds) *Readings in Cultural Geography*. Chicago: University of Chicago Press. pp. 30–4.

Sawicki, J. (1991) *Disciplining Foucault: Feminism, Power, and the Body*. London: Routledge.

Seamon, D. (1979) *The Geography of the Lifeworld*. London: Sage.

Sharp, J.P. (1996) 'Gendering nationhood: a feminist engagement with national identity', in N. Duncan (ed.) *BodySpace: Destabilizing Geographies of Gender and Sexuality*. London: Routledge. pp. 97–108.

Shildrick, M. (1997) *Leaky Bodies and Boundaries: Feminism, Postmodernism and (Bio)Ethics*. London: Routledge.

Shilling, C. (1993) *The Body and Social Theory*. London: Sage.

Simonsen, K. (2000) 'Editorial: the body as battlefield', *Transactions of the Institute of British Geographers* 25 (1): 7–9.

Squire, C. (1997) 'AIDS panic', in J.M. Ussher (ed.) *Body Talk: The Material and Discursive Regulation of Sexuality, Madness and Reproduction*. London: Routledge. pp. 50–69.

Stoppard, J.M. (1997) 'Women's bodies, women's lives and depression: towards a reconciliation of material and discursive accounts', in J.M. Ussher (ed.) *Body Talk: The Material and Discursive Regulation of Sexuality, Madness and Reproduction*. London: Routledge. pp. 10–32.

Teather, E.K. (eds) (1999) *Embodied Geographies: Spaces, Bodies and Rites of Passage*. London: Routledge.

Ussher, J.M. (1997a) *Body Talk: The Material and Discursive Regulation of Sexuality, Madness and Reproduction*. London: Routledge.

Ussher, J.M. (1997b) 'Framing the sexual "Other": the regulation of lesbian and gay sexuality', in J.M. Ussher (ed.) *Body Talk: The Material and Discursive Regulation of Sexuality, Madness and Reproduction*. London: Routledge. pp. 131–59.

Ussher, J.M. (1997c) *Fantasies of Femininity: Reframing the Boundaries of Sex*. London: Routledge.

Valentine, G. (1999) 'A corporeal geography of consumption', *Environment and Planning D: Society and Space* 17: 329–51.

Visweswaran, K. (1994) *Fictions of Feminist Ethnography*. Minneapolis: University of Minnesota Press.

Wakeford, N. (1998) 'Urban culture for virtual bodies: comments on lesbian "identity" and "community" in San Francisco Bay Area cyberspace', in R. Ainley (ed.) *New Frontiers of Space Bodies and Gender*. London: Routledge. pp. 176–90.

Weiss, G. (1999) *Body Images: Embodiment as Intercorporeality*. London: Routledge.

Williams, B. (1996) 'Skinfolk, not kinfolk: comparative reflections on the identity of participant-observation in two field situations,' in D.L. Wolf (ed.) *Feminist Dilemmas in Fieldwork*. Boulder: Westview. pp. 72–95.

Williams, S.J. and Bendelow, G. (1998) *The Lived Body: Sociological Themes, Embodied Issues*. London: Routledge.

Zola, I. (1991) 'Bringing our bodies and ourselves back in: reflections on a past, present, and future "medical sociology"', *Journal of Health and Social Behavior* 32 (March): 1–16.

3

Cultural Geographies of Transnationality

Katharyne Mitchell

WHAT IS TRANSNATIONALITY?

The idea of transnationality is an important one in contemporary theory for a variety of reasons. Perhaps the foremost concern of scholars today is to make connections across and between entities that were formerly theorized as discrete and autonomous. The term 'transnational' with the emphasis on the *trans* allows this type of relational theorizing. Furthermore, it encourages it in a number of different arenas – from examinations of the interactions and literal back-and-forth movement of goods, people and ideas across national borders, to the theoretical suppleness of poststructuralist thought across containing and linear narratives and disciplinary confines.

Theorizing transnationality, considered within this broad rubric, is inherently transgressive. The emphasis on relations *between* things and on movements *across* things forces a reconceptualization of core beliefs in migration and geopolitical literatures, which formerly emphasized state-centric narratives and territorially defined national borders. It forces a rethinking of economic categories, which exclusively privileged highly abstracted global forces such as capitalism. It also forces a rethinking in broader areas of epistemological inquiry, including questions of identity, subjectivity formation and foundational beliefs about space and time. In the best work on transnationality, the border crossings evident in rethinking these literatures of the past decade have come together in a fruitful dialogue. Thus, issues related to the changing relationship of the nation *vis-à-vis* global economic and political forces, and the shifting understandings of

categories and foundational cultural narratives, are interrelated in complex and interesting ways. In the following sections I examine some of the ways that thinking 'transnationally' problematizes older categories, then go on to discuss contemporary transnational research in cultural geography and some of the questions and dilemmas that it raises.

THE CONDITION OF TRANSNATIONALITY

In a prominent discourse on the nature of modernity, Marshall Berman (1982) defined *modernization* as a process of change characterized by a number of key forces, especially the urbanization, industrialization and bureaucratization that were occurring with accelerated intensity in Europe in the latter half of the nineteenth century. He defined *modernism* as the cultural fallout of these processes – the expression of change evident in the art, music, architecture and literature of that time. *Modernity*, for Berman, was the overall experience of change that was felt by both urban and rural residents as the landscape literally shifted beneath their feet.

These divisions are useful for considering transnationality as well, as the term has become ubiquitous across disciplines and often loses its coherence as a result. *Transnational* and *transnationalism* are terms favoured by cultural studies and literary theorists and generally refer to the poststructural concepts of in-betweenness or ambivalence, especially with reference to the nation. They are also frequently used to denote cultural blending of various kinds, invoking a form of syncreticism in art or creolization in relation to identity (Hannerz, 1996; Vertovec, 1999). Often these terms are used as textual markers, abstracted away from the material relations of a particular time and place. In response to this ahistorical and ageographical type of theorizing, most Marxist geographers, and others interested in both transnational concepts and political economy, have called for a more grounded theory, especially with reference to economic processes (Jackson and Crang, 2000; Mitchell, 1997a; Nonini and Ong, 1997).

In earlier work, following the rapid rise in popularity of the term 'postmodernism', for example, the pointed reference to the necessity of theorizing epistemological concepts alongside economic shifts was clearly the intent behind the title of Harvey's influential book *The Condition of Postmodernity* (1989). In direct response to the epistemological celebrations of the end of metanarratives and the death of the individual subject found in Lyotard's *The Postmodern Condition* (1984), Harvey set out to show the reader that many of these reputedly liberatory positions were connected to an increasingly polarizing and debilitating form of capitalism. In this regard, terms such as the 'condition of transnationality' or 'transnationalization' are perhaps more apropos of describing the changing interactions, subjectivities and narratives that are directly related to specific material processes, particularly the global restructuring of the economy under late capitalism.

In the last three decades a number of forces have constellated to create a highly integrated and interdependent international system. The on-going process in which this interdependency is produced and maintained is broadly referred to as 'globalization', a term with such a large tent-like structure that many things have crept to shelter under it.[1] Transnational research in geography is linked with globalization studies through a shared focus on changes in the systems of world governance, including the proliferation of regional and cross-border trade agreements and pacts, the end of the Cold War and the beginning of a 'new world order', and the ongoing impacts of neocolonialism and neoliberalism. They are also linked through the emphasis on shifts in the nature of capitalism as a global socio-economic system, especially the changing geographies of production, the changing flows of capital and labour, the rise of new kinds of networks and commodity flows, and the increasing polarization of wealth on both

macro (geographic) and micro (household) scales.

A key feature of contemporary capitalism is a far greater globality and flexibility in regimes of accumulation (Harvey, 1989). These include the deterritorialization of finance (Corbridge et al., 1994; Roberts, 1994), the geographical fragmentation of production systems (Dicken, 1992) and the strategic elasticity of institutions (Herod et al., 1998). This flexibility is also manifested in the increasing movements of people in extended social fields across borders, and in the consciousness, socio-cultural expressions and identity formation of both these transnational migrants and those who have remained locally based (Rouse, 1991; Silvey and Lawson, 1999). As Sassen (1988) showed over a decade ago, the growing flexibility of capital and the flows of people are intricately interlinked. Thus geographers interested in the social formations of transnationalism, including cultural expressions of in-betweenness, or disruptions of national narratives, must necessarily examine the broad global economic context in which these transnational processes are occurring.

For example, in the mid 1980s a new category of immigration, called the Business Immigration Programme, was initiated by the Canadian government in order to attract wealthy investors and entrepreneurs to settle and invest in Canada. The programme was greatly expanded following a recession in 1981–2, and was explicitly designed as a method of kick-starting the economy through the influx of capital and people with business expertise, and also through the expansion of ties with the booming Asia–Pacific region. The programme was one of numerous measures introduced by the federal government in the 1980s that were part of a broad neoliberal strategy of easing the national and international circulation of capital and reducing the role of government in many sectors of the economy. The main targets of the programme were Hong Kong Chinese business people who were leaving the colony as a result of its impending transfer to control by the People's Republic of China in 1997. Hong Kong led as the primary source of immigrants within this programme for a number of years.

Being cognizant of this global and national economic context is crucial for understanding the types of cultural disputes and political struggles that occurred in Canada following the rapid influx of so many wealthy immigrants from Hong Kong. In my own research, which centred on the city of Vancouver, British Columbia, I investigated the many socio-cultural clashes over landscape design, house demolitions, house size, architectural style, tree removal, downzoning and other points of conflict involving the recent immigrants from Hong Kong and the long-term residents of the neighbourhoods. Although the clashes were often represented as primarily about race and racism, there was a complex, multi-dimensional layering to the cultural conflicts that was deeply bound up in the economic context in which the immigrants arrived and in which the struggles occurred.

As a result of the popularity of the Business Immigration Programme, combined with the easing of restrictions over banking and investment in Canada (owing to a number of neoliberal federal policies), numerous Hong Kong property investors became interested in Vancouver as a site of investment. This interest snowballed rapidly, and during the late 1980s the Vancouver property market was so hot that numerous developments that were controlled by Hong Kong developers sold out in Hong Kong in a matter of hours, before Vancouver buyers had a chance to make a purchase. At the same time, many houses in older, staid neighbourhoods were being demolished to make way for houses two or three times their size. These new, so-called 'monster' houses were targeted mainly for purchase by the new Hong Kong immigrants.

As a result of government policies designed to rejuvenate the business sector with an influx of capital and to integrate Vancouver and Canada into the global economy, many cities and regions in the country went through major transformations in form and atmosphere. The changes in some neighbourhoods were so rapid and all-encompassing that many older residents felt a complete loss of control and alienation in neighbourhoods they had lived in most of their lives. At the same time, owing to the stipulations of the Business Programme, which required that those migrants arriving as 'entrepreneurs' must establish a business in Canada, many recent immigrants felt it necessary to live part-time in Canada and part-time in Hong Kong in order to sustain successful business operations in both places. (At this time the Hong Kong economy was booming, while the Canadian economy was comparatively stagnant.) The evident 'trans-national' mobility of the immigrants made them appear as 'sojourners' to many of the older Vancouver residents, who then questioned their allegiance to the neighbourhoods and to the nation. Many battles over the neighbourhoods' redevelopment then ensued.

Thus although racism was clearly a factor in the socio-cultural struggles between the trans-national immigrants and the long-term Vancouver residents, it was just one factor among many. Understanding the economic and geopolitical contexts of Canadian federal immigration policy, of the perceived imperative of global economic integration, of the attraction of the booming economic region of Asia in the mid 1980s, of the rise of global property markets, of excess capital accumulation in Hong Kong, and of Hong Kong's transition to Chinese control, is a necessary starting point for any discussion of Vancouver's urban change, social formations, political alliances or cultural struggles. The concept of transnationalism should recast our understanding of the social and the cultural by showing how they are always bound up with the economic and the political at a number of different scales.

A second important feature of contemporary capitalism that affects the cultural geography of transnationality is the rise of networks. Networks are webbed structures through which goods, information, capital and people flow in a multi-directional manner. They are also forms of corporate governance that are based less on hierarchical models and more on flexible nodes and vertices linked together through both formal and informal relationships such as subcontracting (Thrift and Olds, 1996). Social networks are based on affinities between people (such as ethnicity or college ties) and much of the work on the social adaptation of immigrants focuses on the ways in which the intimate connections between immigrants greatly affect their access to certain kinds of residence and employment opportunities, business information and/or credit (Light and Karageorgis, 1994; Portes and Manning, 1986). Transnational migration within the contemporary global economy may accelerate the extension of these social networks, including ethnic ties, across space in a manner that has implications for capitalist articulations globally as well as locally (Mitchell, 1995; Mitchell and Olds, 2000). Transnationalism also has implications for the ways in which local networks operate *vis-à-vis* immigrant adaptation and in terms of commodity flows and global consumerism (Jackson, 1999).

It has been evident that advances in the technological arena have facilitated and accelerated globalization processes. Many scholars have argued further that new technologies such as telecommunications and computing have actually enabled global restructuring to take place. Castells (1989; 1996), for example, proposes a two-way process: the global restructuring of capitalism conditions the production and application of new information technologies, and the new technologies enable this kind of restructuring to occur. In these types of arguments, the new technologies are given causal weight and are seen to influence not just an acceleration of international linkages, but the ways in which society itself is structured. The local impact of the new information processes is, for Castells, immense, as an abstracted 'space of flows' begins to dominate more local, place-based processes and understandings.

Research that connects transnational cultural geographies with economic processes, particularly those focusing on the rise of the network society and the shift to regimes of flexible accumulation, is imperative in order to understand the structural factors in how globalization works. For example, in the past, the experience of international migration for a majority of migrants involved a permanent move from one nation-state to another. Although communication and even periodic visits might have continued between the state of origin and that of the migrant's destination, the time, expense and difficulties associated with these international connections made them relatively infrequent. But in the contemporary period of rapid and inexpensive jet flights, satellite television, internet connections, international banking and cheap and direct telephone lines, the ability to remain linked on a semi-permanent and nearly instantaneous basis with two or more places became not just possible but relatively easy.

This shift has had remarkable implications for both economic networking and cultural meaning. Migrants are likely to retain important relationships binationally and to orchestrate economic, political and social decisions via complex networks across space. Cultural meanings as well as financial remittances flow along these newly forming networks, facilitated by the new telecommunications technologies. As people move and communicate in new kinds of ways this, in turn, affects the flow of money and credit and also the movement and manipulation of commodities. (Demand for certain kinds of foodstuffs from countries of origin, for example, has created new markets for products and services.) Transnational migrants are important agents of economic transformation in themselves, and their activities and practices in numerous spheres have had a major economic impact on local, regional and global arenas. Thus, how globalization operates in terms of the expansion and diffusion of capitalism is intricately connected with the social formations and cultural identities of these transnational labourers and business people on the move.

Until recently, however, much of contemporary research on globalization has emphasized a kind of transnationalization 'from above', and neglected the contextual nature of these economic and cultural interlinkages. For example, research that focused on increasing globalization related just to the general expansion of capitalism worldwide often relied on an extremely narrow and homogeneous vision of the workings of capitalism. Capitalism, money and information were assumed as standards that were self-referencing and all-encompassing. The origins of these processes receded from view, and their power and ability to expand and diffuse took on the characteristic of the self-evident.

Transnational theory which foregrounds postcolonial and poststructuralist frameworks seeks to question the limits of ahistorical and homogenizing narratives such as these, and enables new kinds of questions concerning so-called global processes. In studies of capitalist formation, for example, contemporary research focuses on the specific configurations of differing economic systems within their own geographical and historical contexts. Thus, instead of negatively framed questions such as the classic query in sociology, 'why did capitalism not develop in China?', the organization of Chinese commerce must be explained in its own terms (Hamilton, 1985: 65). In conceptualizations of this type, wherein the historical dynamics of culture and systems of organizational authority are given substantial theoretical weight, singular visions of the nature of capitalism and/or of the 'natural' path of capitalist development are effectively avoided. Economic and socio-cultural activity can then be viewed as mutually interactive, as intertwined rather than causal (Mitchell, 1995: 365).

Knowledge of local economic variation moves beyond a cultural 'embeddedness' approach to investigations of alternative originations and evolutions. What constitutes capitalism as a socio-economic system varies greatly by time and place and affects how capitalist practices are both implemented and understood. Furthermore, through hegemonic struggles over capitalist meanings and practices, local economic regimes provide the seeds of future variation for *global* economic change. Blim (1996: 88), for example, has discussed the highly heterogeneous forms of property ownership and the varied understandings of private and collective property rights in contemporary China. Chinese private property rights exist in a hybridized state of formal and informal, stable and unstable, that is unique to China at this particular historical juncture. The current struggles between China and the United States over the normalization (American-style) of these property rights (particularly in the area of patents and intellectual copyrights) can be seen as a hegemonic struggle over the meaning of capitalism between two varied economic regimes. How this ideological struggle over the 'nature and norms of capitalism' plays out will inevitably and powerfully affect world economic standards and the capitalist cycle of accumulation in the long term.

Just as a singular understanding of capitalism remains limited, so too does a homogeneous vision of information. The idea that new information technologies will lead to a domination of local cultural identities relies largely on a fetishized understanding of information technology – one in which the actual social relationships necessary for the effective exchange of information are elided. Information itself is culturally specific, as it relies on a shared system of meanings between the sender and the receiver. As with linear, top-down understandings of capitalism, the top-down view of information as a process controlled primarily through access to technology and by bureaucratic power structures reduces the social and cultural complexity of its actual originations and transnational evolutions. Furthermore, as Smith (1996: 69) has pointed out, the supposedly placeless 'spaces of flows' are always paralleled, albeit sporadically and unevenly, by a 'deepening *spatial fixity*' – one that is linked with strategic sites around the globe. These sites are not irrelevant to the exchange of information, nor can they be conceptualized solely as the privileged, 'switched-on' areas of the world. Their location in cities such as New York, Los Angeles, Tokyo and London is a location in history as well as in space; it is that contextual location that impacts on the message and infects the system with particular cultural meanings.

THE GEOPOLITICS OF THE TRANSNATIONAL

Transnational theory, which emphasizes the limits of homogenizing narratives such as those that privilege a singular, western-centric vision of capitalism or information, also forces a reconceptualization of formerly static categories and conceptual 'power' containers such as 'the nation'. State-centric narratives of containment, the bread and butter of political science discussions in the past, have given way to more nuanced studies of the 'boundary-drawing practices and performances that characterize the everyday life of states' (Ó Tuathail and Dalby, 1998: 3). Through its emphasis on relationships and interactions rather than static formations and containment, transnationality encourages new ways of envisioning the nation, the state and the hyphenated properties and relations of the nation-state.

The bulk of geopolitical theory in the past focused on state autonomy, the borders and 'outsides' of the state, and interstate relations (Walker, 1993). Cold War rhetoric employed the language of power strategies and containment and the spaces of the nation were described in terms that suggested both their inevitability and their naturalness. National landscapes were depicted as receding nostalgically into the past and extending indefinitely into the future in a timeless yet necessary manner. The *nation* was seen as the best scale of analysis for theorizing both world politics and local politics. Furthermore, the practice of geopolitics itself was perceived to be neutral and objective (Ó Tuathail, 1996; Ó Tuathail and Dalby, 1998).

Transnational theory that borrows from the conceptual framework of poststructuralism forces a continual rethinking of these older categories and general geopolitical assumptions. Utilizing scales of analysis other than the nation, for example, allows the posing of numerous global, local and regional questions that cannot be solved or discussed within a national politics (Beck, 1998: 29). Furthermore, the geopolitics of transnationality opens up questions of scale through the emphasis on scale as produced (rather than given) and on the interrelationship between scales as of crucial import. As numerous scholars such as Cox (1993), Smith (1995), Brenner (1998) and Katz, Newstead and Sparke (Chapter 26 in this volume) have shown, scale is a central component of capitalist restructuring. Scales are constantly reworked in conjunction with territorial organization and the crises associated with global capitalist restructuring. Brenner writes, 'spatial scales constitute a hierarchical scaffolding of territorial organization upon, within, and through which the capital circulation process is successively territorialized, deterritorialized and reterritorialized' (1998: 464).

Transnational theory also emphasizes the production of boundaries and the ongoing performance of the nation as integral to the everyday practices of states. 'In contrast to conventional geography and geopolitics, both the material borders at the edge of the state and the conceptual borders designating this as a boundary between a secure inside and an anarchic outside are objects of investigation' (Ó Tuathail and Dalby, 1998: 3). Drawing from the works of postcolonial critics such as Homi Bhabha, theories of the *trans* in transnational constantly bring into play the ambivalences within the nation; they highlight the *process* of nation-making, especially the ways in which an 'irredeemably plural modern space' (Bhabha, 1994: 140) is inexorably reworked into a politically unified space through national narratives and state practices.

Borders are, of course, differentially porous, varying not just by nation or by political regime, but by types of flows and by particular narratives of the nation transpiring at differing moments. In geopolitical research states are often assumed as the *de facto* containers of the nation in ways that elide the ongoing processes of nation-building. Rather than an emphasis on state-making as a process and the state as a set of routinized institutional norms and practices as Timothy Mitchell (1991) has outlined, the state is conceived as a complete and circumscribed entity. In the contemporary period of globalization, however, the role and boundaries of the state are in constant flux. The state is not pre-eminent, nor has it lost its general viability; rather, there is an insistent tension between the project of the modern nation-state and its ideological control over the circulation of both its citizens and their capital in diaspora.

Another of the assumptions within conventional geopolitical theory is the natural correspondence between the state and its territory. In path-breaking books such as Basch et al.'s *Nations Unbound* (1994), and in other empirical research on transnational migration, these 'necessary' correspondences are problematized. As Luis Guarnizo (1994; 1997; 1998) has shown in his research on Mexico and the Dominican Republic for example, and Sarah Mahler (1998) for El Salvador, foreign revenue from migrant remittances has spurred state interest in reaching beyond the territorial borders of the nation to maintain connections with migrants working abroad. The state's attempt to capture these migrants and their capital remittances has included positive incentives such as various health and welfare benefits, property rights, voting rights and even rights to dual citizenship. In this scenario nationalism is not based on an understanding of the modern state as a territorial container, nor are citizenship rights granted through one's location in the spaces of the nation-state. Rather, state practices and the rights and responsibilities of state citizenship expand and contract with prevailing transmigration currents and capital flows. As Wakeman rightly discerns, the 'loosening of the bonds between people, wealth and territories has altered the basis of many significant global interactions, while simultaneously calling into question the traditional definition of the state' (1988: 86).

Tensions *within* the state and between state practices and the performance of national narratives can also be exposed in transnational research. Moving away from the conceptualization of the state as a monolithic black box, it is immediately possible to see the potential for rivalries, conflicting priorities and differing agendas between different bureaucratic structures lodged within the tiers of state governance. These tensions often become manifest around issues of migration, where internal conflicts, even between individual politicians and parties, are rife. With regard to transnational migration research, the multitude of ways in which state immigration laws, citizenship statutes and informal policies come into conflict with national narratives of territorialization or multiculturalism or timelessness are also brought to the surface. Thus the insistent tension between the nation and global forces is mirrored in many respects by the tensions evident in the hyphen between the nation and the state.

Finally, as Hyndman (1997: 150) shows in her work on cross-border humanitarian organizations and the 'refugee industry', the heretofore common theoretical analyses which foregrounded core–periphery and centre–margin binaries are no longer adequate for addressing the dialectical and ever-shifting relationships between refugee flows and the flows of humanitarian assistance. Emphasizing a 'transnational politics of mobility', Hyndman argues that static geopolitical divisions of north and south or First and Third Worlds are unsuccessful in capturing the contemporary dynamics of refugee movement or of the ways in which supranational institutions such as the Office of the United Nations High Commissioner for Refugees (UNHCR) are implicated in this movement.[2]

THE TRANSNATIONAL SOCIAL FIELDS OF MIGRATION

As a postmodern architectural style of fragmentation, historical reference and pastiche jump-started the theoretical conceptualizations of postmodernism, so transnational migration theories kicked off the recent wave of thinking about transnationalism. A number of migration scholars working in the 1980s conducted empirical work on immigrants in the US and then traced them back to their sending countries. After researching the movements of the migrants over time, it became apparent that older migration frameworks emphasizing either a unilinear migration pattern (movement from one place to another and settlement there) or circular migration (brief circular movements with an eventual return 'home') were simply inadequate in describing all of the migration flows that were occurring.

In the former framework of unilinear migration, migrants leave one contained and defined spatial territory, cross one or more borders, and arrive in another identifiable national space, to which they must then assimilate. This way of thinking began to be challenged with the new social morphology of movement within transnational theory. Borrowing from some of the ideas of poststructuralist theory, especially the critique of boundedness and fixity, and the celebration of mobility and hybridity, several theorists postulated new ways of thinking about migration that could supplement pre-existing theories.[3] According to Roger Rouse,[4] the assumptions implicit in many of the older theories needed to be addressed, including a moral undertone in US migration research which seemed to infer that migrants could and *should* only be truly involved in one place. Rouse links this assumption to the 'dreamscape of the bourgeois state' where the territorial state and citizenship are forever locked together, and the image of the national landscape is as a separate, territorially bounded entity that remains constant through time.

Transnational theory in migration research developed a less rigid view of sending and receiving nations and of migrants' relationships to them. The idea of 'social fields', advanced initially by Glick Schiller et al. (1992), captured a dynamic of migration and of migrants' lives that was multilocal – the idea that social 'places' could be tightly woven together across borders and across space (see also Guarnizo, 1994; Rouse, 1991). They, along with a growing number

of others, argued that with the technological innovations of electronic banking, telephones, computers and jet planes, physical distances could be collapsed, and it was possible for migrants to participate simultaneously in the events and activities of different national sites. In numerous empirical case studies, migrants were shown to live fully and actively in two nations, and often to conduct business, political and family affairs in both.

In one example of political activities crossing the borders of nations, Basch, et al. (1994: 2) relate the story of Grenada's ambassador to the United Nations, who was also a leader of New York's West Indian community for a number of years, and took a major role in electing mayors of New York City. In a different example, they discuss the formation of a 'tenth' department within Aristide's Haitian regime, composed entirely of overseas Haitians. A third, frequently cited example from Sarah Mahler's (1998) research is of the odd case in which the El Salvadoran government provided free legal assistance to political refugees in the US. Owing to the great economic importance of their remittances home, the government provided aid to those citizens who were fleeing its own oppressive regime (see also Vertovec, 1999: 455). In all of these cases, both the state and its citizens stretched beyond conventionally defined territorial borders and engaged with each other and with national affairs from a multitude of locations.

This spatial stretching is often theorized as a form of national 'deterritorialization'. However 'respatialization' may be a more accurate term, as it indicates a reworking of spatial arrangements that had previously been more localized. The respatialization of the state and of individuals' lives does not suggest either a loss of power or a liberatory process for either entity. The particular state practices of a particular nation may become more or less extensive and far-reaching, more or less controlling and disciplining, and more or less powerful depending on the conditions in which that nation was and is inserted into the global regime of flexible capitalism. Similarly, a transnational mode of life for an individual migrant cannot be necessarily equated with freedom (from family or state), but must always be examined in context. As Guarnizo (1997) has shown *vis-à-vis* the experience of Dominican migrants, for example, the implications of the transnational movements back and forth between the US and the Dominican Republic are vastly different for men and for women; the experience of living and working in a social field across borders serves to reinforce male authority in many instances, and thus is not experienced as 'liberatory' for numerous women (see also Glick Schiller, 1997; Jones-Correa, 1998).

POSTSTRUCTURAL SPACES OF THE TRANSNATIONAL

Epistemological transgressions employing the rhetoric of transnationalism have focused on the contestation of linear and containing understandings of time and space and on singular and homogenizing narratives of processes such as capitalism, culture and modernity. Numerous scholars have also celebrated new anti-essentializing concepts of subjectivity that emphasize plurality, mobility, hybridity and the margins or spaces 'in between'. As discussed above, many of the questions raised by these movements across the borders of prior theoretical assumptions can be related to understandings of the literal movements of people and goods across the borders of the nation-state.

In anthropology, scholars such as Clifford (1992) have introduced the concept of the informant as traveller, where the relations of movement and displacement are foregrounded over those of dwelling and local, confined knowledge. In this view, culture is best understood when its locus is a place of movement or a 'site of travel' rather than a controlled space such as a 'site of initiation and inhabitation' (1992: 101). Similarly, Arjun Appadurai's (1988) concerns about the previous privileging of the local and the representational in western analyses of 'native' peoples has drawn him toward a celebration of deterritorialization in his discussion of disjuncture and difference in the new cultural mediascapes of late capitalism (Appadurai, 1990). Here he seeks to escape the 'metonymic freezing' of people's lives in western anthropological discourse through an emphasis on historical mobility and ongoing displacement.

Other scholars interested in questions of identity and the constitution of subjectivity herald the ways in which new cross-border movements have facilitated the production and reworking of multiple identities, dialogic communications and syncretic cultural forms. Perhaps the most famous celebrant of the spaces of in-betweenness and hybridity is Homi Bhabha (1994), who depicts the spaces of the margins as the privileged location from which to make consequential interventions in hegemonic narratives of race and nation. Theories privileging the liminal and the hybrid have effectively destabilized many prior assumptions of purity, authenticity and local and fixed subjectivities. They have also raised important questions relating to the homogenizing and western-based provenance of both historical structural and neoclassical accounts of globalization processes. This transgressive work, however, manifests certain kinds of limits as well.

The destabilization of linear and/or essentializing narratives has been an important first step toward opening up alternative ways of theorizing subjectivity and the social (for example, Laclau and Mouffe, 1985). Recently, this work has also been applied to theories of the economic (Gibson-Graham, 1995). Although destabilizing concepts of capitalism allows for new ways of examining different hegemonic constructions and negotiations over the meaning of capitalism (such as discussed in the contemporary case of China and the United States), it can also present certain theoretical black holes. The most obvious concern is the proverbial problem of throwing out the baby with the bathwater. If the term 'capitalism' is deprived of associated meanings such as capital accumulation, class relations, surplus value, dynamism and crisis, then it is voided of explanatory potential, or indeed of any meaning at all. 'To theorize capitalism itself as different from itself (as having, in other words, no essential or coherent identity) is to multiply (infinitely) the possibilities of economic alterity' (1995: 279). Unfortunately this kind of subtle theoretical finesse of infinite multiplicities also multiplies the possibilities for the subject itself to disappear. While it is clearly imperative to avoid research which relies on a singular conception of capitalism, it is also theoretically dubious and politically dangerous to completely empty the term of all essential meanings. If, as Gibson-Graham argues, 'there is no underlying commonality among capitalist instances, no essence of capitalism like expansionism or power or profitability or capital accumulation, then capitalism must adapt to (be constituted by) other forms of economy as much as they must adapt to (be constituted by) it' (1995: 279). Yet transnational research shows this kind of mutual constitution to be unlikely. While it is absolutely crucial to examine the different forms that capitalism takes in different contexts, and to theorize the ways in which contemporary struggles over the hegemonic meanings of capitalism lead to changes in its global operations, it is also imperative to maintain a knowledge of the structural principles undergirding a system that infects and is infected by every other system in an *unequal* exchange. Without this, the power relations evident in every facet of transnational contact – between states, institutions and people – become lost. And without an understanding of power relations, one that transcends purely local knowledges, the possibility of political and economic resistance is greatly diminished.

A second area in which the limits of the liminal emerge is in the discussion of cultural mobility. In numerous celebratory representations of 'new' transnational cultures and hybrid subject positions, the powerfully oppressive socio-economic forces underlying the changes are neglected, as are many of the people caught within them. As bell hooks (1992) has noted of Clifford's somewhat playful evocation of travel and 'hotel lobby' culture, the actual, terrorizing experience of border crossings for many people of colour is effectively ignored. On-the-ground experiences are relegated to a secondary position – if included at all – in the general rush to proclaim the beneficial potential of hybrid forms, 'third' spaces, and state-sponsored drives toward increasing cultural diversity and mutual tolerance. In an era of global capitalism, the heralding of subject positions 'at the margins' too often neglects the actual marginalization of subjects. And positive readings of the forces of deterritorialization inadequately address 'the powerful forces of oppression unleashed by them' (Visweswaran, 1994: 109).

A different set of problems arises when theorists herald hybridity precisely because they believe it to be the only space of resistance left to the marginalized – particularly those marginalized by the exclusionary forces of nationalism. According to scholars like Bhabha (1994), subjects located 'in between' nations or subject positions are best positioned to resist hegemonic narratives of race or nation; it is from a place at the margins that substantial interventions in the ongoing production of the nation can be launched. Although the potential for resistance is clear, the celebratory theoretical assumption of a progressive politics of intervention is not always borne out in empirical transnational studies. For example, research on contemporary Chinese businessmen has shown that various kinds of diasporic, deterritorialized and hybrid subject positions have been used strategically for economic gain. In other words, strategic self-fashioning in liminal and partial sites can be used for the purposes of capital accumulation quite as effectively as for the purposes of intervention in hegemonic

narratives of race and nation (Mitchell, 1997b; see also Ong, 1993; 1999).

It is this problem of a frequent disregard for grounded empirical work that limits many epistemological inquiries into transnational processes. Theorizing global processes with new conceptual tools enables alternatives to the 'globalization-from-above' model. But without 'literal' empirical data related to the actual movements of things and people across space, theories of anti-essentialism, mobility, plurality and hybridity can quickly devolve into terms emptied of any potential political efficacy. Through geographically informed research and theoretically nuanced understandings of difference and alterity, the difficult questions related to borders and identities will be forced to the surface, even if they remain partially unanswered and unanswerable.

TRANSNATIONAL RESEARCH IN CULTURAL GEOGRAPHY

The practices and meaning of location, especially the reconstruction of place in the contemporary period of 'fast' capital, are common themes in cultural geography. In what way is the 'condition of transnationality' affecting the culture of location and the consciousness of people on the move, as well as those unable or unwilling to move? What are the transnational practices and imaginings that have been galvanized by global shifts (Ong, 1999)? How does the extension of 'multiple ties and interactions' (Vertovec, 1999) across borders impact on the cultural geographies of individuals and of nations?

In recent years, numerous scholars have begun to examine the socio-spatial dynamics of cross-border economies and lives. Peter Jackson and Phil Crang (2000), for example, trace the paths of commodity flows and their cultural implications. They are interested in the ways that notions of commerce and the traffic in things can be conceptualized in relation to transnational theory. For them, a focus on commodity culture can provide a significant lens with which to examine a multitude of locations and movements between specific sites rather than the more familiar movements of migrants back and forth across the globe. Aiming to help 'ground' transnational discourse yet avoid fixing it in a simplistic way, they examine the 'refiguring of the spaces of culture' through the links between the multiple sites of commodity consumption and production.

Research on the impacts of transnational movement for families and particularly for women is also a growing area of research. Yeoh and Willis (1999), for example, examine the ways in which transnational business networks and the regionalization process in Singapore draw on gendered ideologies and have major implications for the ongoing constructions of women's and men's identities in Asia. As it is men who are usually involved in businesses that span national borders, it is generally the women in the family who are left 'at home' to protect the family and the meanings of hearth and home. Drawing from extensive interviews of economic migrants, Yeoh and Willis found that 'male and female family members often attempt to resolve the tensions of being "home" and "away" through a transnational gender division of labour: while male labour is deployed in spearheading and driving the external economy and lubricating its wings abroad, privatised female labour shores up the home front and nourishes the nation's "heartware"'.

This pattern is reversed in interesting ways in the case of Hong Kong business migrants to Vancouver, where it is generally the woman who is left in the 'new' home of Canada, while the man continues to conduct business transnationally, often retaining both a home and a mistress in Hong Kong. This pattern became so common in the 1980s that the term *tai kong ren* was coined to describe the phenomenon. Tellingly, the term carries a *double entendre*, meaning both 'astronaut' and 'empty wife' in Mandarin (Mitchell, 1993; Ong, 1993). In both of these cases, the woman serves as the bridge between a home/family life and the economically productive world of the high-flying transnational male. The highly gendered split between public and private and between productive and reproductive is made even stronger and deeper through the literal spatial separation of international borders (Yeoh and Willis, 1999).

One major focus of inquiry over the past decade has been the impact of transnational migration and capital flows on the cultural politics of the nation. Anthropologists such as Ong (1999), Basch et al. (1995) and Appadurai (1996) examine the ways in which cross-border flows deeply transform the meaning of belonging, creating imagined communities at large. Mitchell (1993; 1997c) also traces the shifting discourses around national narratives such as multiculturalism, the public sphere and democracy in the context of global restructuring and increasing transnational flows. In all of these works there is an interest in connecting economic and cultural processes, especially the shift from mass industrial production to the flexible accumulation regime of late capitalism, with shifts in the cultural construction of both the nation and its citizens in motion.

In work on the wealthy Hong Kong Chinese transnational migrants to Vancouver, for example, Mitchell (2002) shows the ways in which this migrant class has disrupted long-standing assumptions of community and made manifest the tensions in the hyphen between *nation* and *state*. As the nation is primarily concerned with the formation of an affective community, the narratives associated with this formation are those of place, rootedness, values, communitarianism, territory, etc. These narratives are shown to be contradictory with the actual practices of the Canadian state, which seeks to entrench a neoliberal agenda of laissez-faire capitalism and the attrition of governance. The cultural narratives of the nation are those of settled places and consciousness of place (*genius loci*), while state practices in the 1980s encouraged circulation, speed and penetration of place (through the entry of fast capital). Wealthy transnational migrants entering the country within the 'business' category of immigration make manifest this contradiction via their literal embodiment of capital. As capitalist 'trans'-migrants, perceived to be riding into the city of Vancouver and the country at large on a 'tidal wave of capital',[5] and frequently represented as sojourners moving back and forth across the Asia–Pacific with little or no allegiance to 'place', these migrants make manifest the contradiction between the discourse of nation and the actions of the state in an extremely striking manner.

Global communication, especially media dissemination, has become so pervasive and so persuasive that many now postulate the formation of either a global monoculture or a 'global ecumene' (Hannerz, 1996). Most cultural geographers, however, eschew either apocalyptic thinking or epistemological celebrations abstracted from economic history and the everyday practices of social and cultural reproduction. Although the concept of transnational cultural spaces was initially the provenance of cultural studies critics and anthropologists, cultural geographers have begun to investigate the implications of transnational networks on the form and meaning of house and home (Mitchell, 1998; Vasile, 1997), on the spaces of neighbourhood and culture (Featherstone and Lash, 1999; Walton-Roberts, 1999) and on the ways in which capital connections and transnational business ties affect cultural consciousness and vice versa (Dwyer and Prinjha, 2000; Jackson and Crang, 2000).

Cultural, transnational research is primarily ethnographic, and must be multisited and spatial in order for the interconnections between places, people and events to become evident. Unlike most of the work in literary criticism, geographical research draws on current epistemological understandings as well as grounded empirical research. Cultural geographies of transnationality examine the embodied movements and practices of migrants and/or the flows of commodities and capital, and analyse these flows with respect to national borders and the cultural constructions of nation, citizen and social life. It is the interweaving of economic and cultural processes and theory in cross-border studies which allows for this rich interplay and in which the best scholarship in transnational studies continues to be conducted.

In future work on transnationalism, there are several areas in which research could fruitfully expand. The conceptualization of the economic effects of transnational migration, for example, has remained somewhat limited because of the overweening focus on the impact of remittances. Although global remittances are monumental in scale (somewhere around $71 billion worldwide) and have a crucial importance to certain countries such as India and El Salvador, the overall economic effects of transnational mobility are probably far greater even than this. Clearly there are numerous multiplier effects associated with the accelerated movements of people and goods. Research investigating the shifting policies and practices of corporations, especially those involved in banking, insurance, money transfer (such as Moneygram or Western Union) and other international services, would begin to tease out some of the tremendous economic ramifications of an increasing transnationalism.

Similarly, contemporary work on the political effects of transnationalism could be usefully broadened. Currently research focuses primarily on the scale of the nation-state. Of equal interest, however, are questions related to the politics of other scales, and of the interrelations between scales. What, for example, is the impact of transnational migration on conceptions of education, particularly how children should be educated to become 'citizens' of a particular nation? In England, Canada and the United States,

debates are now raging over the purpose of education, as well as how education should be best delivered. Should students be educated to become effective global workers or to participate actively in the national arena? Is the national government the best provider of education, or should it be opened to market, perhaps international, forces? These debates are clearly tied up with the condition of transnationality, and research in this vein would shed greater light on institutional politics that are tightly intertwined with those of the nation, but yet which operate on different scales (see, for example, Mitchell, 2001).

Other issues associated with social reproduction include those of health and urban consumption more generally. Do migrants demand the same services or products? Do markets and government programmes shift as a result of differing needs and demands? Do patterns of consumption alter, and do these impact on commodity chains and production processes more generally? Does capital follow transnational actors? Do they open up new market niches for specific kinds of 'ethnic' products or services? These are the kinds of questions that will begin to uncover the embedded activities and ramifications of transnational movements in specific, contextually grounded sites. In the future, transnational research should begin to cast a wider net in order to capture the nuanced cultural changes associated with the new networks, transactions and socio-cultural interactions that are occurring in the contemporary period of accelerated movements and exchange.

NOTES

Parts of this chapter draw on an earlier essay on transnationality published in *Antipode* (Mitchell, 1997a).

1 There are literally hundreds of contemporary books on globalization, but few of these provide a particularly spatial emphasis. The best studies of globalization written from a specifically geographical perspective include Agnew and Corbridge (1995), Amin and Thrift (1994), Cox (1997), Dicken (1992), Herod et al. (1998), Olds et al. (1999).

2 For other work on transnational organizations and their implications for geopolitical theory see Drainville (1998) and Kriesberg (1997).

3 For some of the early work in this vein, see especially Basch et al. (1994), Glick Schiller et al. (1992), Guarnizo (1994), Kearney (1991), Rouse (1991).

4 From personal conversation and guest lecture (2 October 2000).

5 The allusions to water and waves in reference to the Hong Kong immigration into Vancouver in the 1980s were frequent in media and popular books of the time. See, for example, *China Tide* by Margaret Cannon in 1989, which chronicled Hong Kong investment in Vancouver. See also 'Asian capital: the next wave' in *B.C. Business*, July 1990; 'Tidal wave from Hong Kong' in *B.C. Business*, February 1989; and 'Hong Kong capital flows here ever faster' in the *Vancouver Sun*, 8 February 1989.

REFERENCES

Agnew, J. and Corbridge, S. (1995) *Mastering Space: Hegemony, Territory and International Political Economy*. New York: Routledge.

Amin, A. and Thrift, N. (1994) *Globalization, Institutions, and Regional Development in Europe*. Oxford: Oxford University Press.

Appadurai, A. (1988) 'Putting hierarchy in its place', *Cultural Anthropology* 3 (1): 36–49.

Appadurai, A. (1990) 'Disjuncture and difference in the global cultural economy', *Public Culture* 2 (2): 1–24.

Appadurai, A. (1996) *Modernity at Large: Cultural Dimensions of Globalization*. Minneapolis: University of Minnesota Press.

Basch, L., Glick Schiller, N. and Szanton Blanc, C. (1994) *Nations Unbound*. New York: Gordon and Breach.

Basch, L., Szanton Blanc, C. and Glick Schiller, N. (1995) 'Transnationalism, nation-states, and culture', *Current Anthropology* 36 (4): 683–6 .

Beck, U. (1998) 'The cosmopolitan manifesto', *New Statesman* 20: 28–30.

Berman, M. (1982) *All That Is Solid Melts into Air: The Experience of Modernity*. New York: Penguin.

Bhabha, H. (1994) 'DissemiNation: time, narrative and the margins of the modern nation', in *Nation and Narration: The Location of Culture*. New York: Routledge.

Blim, M. (1996) 'Cultures and the problems of capitalisms', *Critique of Anthropology* 16 (1): 79–93.

Brenner, N. (1998) 'Between fixity and motion: accumulation, territorial organization and the historical geography of spatial scales', *Environment and Planning D: Society and Space* 16: 459–81.

Cannon, M. (1989) *China Tide: The Revealing Story of the Hong Kong Exodus to Canada*. Toronto: Harper and Collins.

Castells, M. (1989) *The Informational City*. Oxford: Blackwell.

Castells, M. (1996) 'The net and the self', *Critique of Anthropology* 16 (1): 9–38.

Clifford, J. (1992) 'Traveling cultures', in L. Grossberg, C. Nelson and P. Treichler (eds) *Cultural Studies*. New York: Routledge.

Corbridge, S., Martin, R. and Thrift, N. (1994) *Money, Power and Space*. Oxford: Blackwell.

Cox, K. (1993) 'The local and the global in the new urban politics: critical view', *Environment and Planning D: Society and Space* 11: 433–48.

Cox, K. (1997) *Spaces of Globalization: Reasserting the Power of the Local*. New York: Guilford.

Dicken, P. (1992) *Global Shift: The Internationalization of Economic Activity*. New York: Guilford.

Drainville, A. (1998) 'The fetishism of global civil society: global governance, transnational urbanism and sustainable capitalism in the world economy', in M. Smith and L. Guarnizo (eds) *Transnationalism from Below*. New Brunswick: Transaction.

Dwyer, C. and Prinjha, S. (2000) ' Transnational spaces of commodity culture'. Paper presented at the Association of American Geographers, Pittsburgh.

Featherstone, M. and Lash, S. (1999) *Spaces of Culture: City, Nation, World*. London: Sage.

Gibson-Graham, J.K. (1995) 'Identity and economic plurality: rethinking capitalism and capitalist hegemony', *Environment and Planning D: Society and Space* 13: 275–382.

Glick Schiller, N. (1997) 'The situation of transnational studies', *Identities* 4 (2): 155–66.

Glick Schiller, N., Basch, L. and Szanton Blanc, C. (1992) 'Transnationalism: a new analytical framework for understanding migration', in N. Glick Schiller, L. Basch and C. Szanton Blanc (eds) *Towards a Transnational Perspective on Migration: Race, Class, Ethnicity and Nationalism Reconsidered*. New York: New York Academy of Sciences. pp. 1–24.

Guarnizo, L. (1994) '*Los Dominicanyorks*: the making of a binational society', *Annals of the American Academy of Political and Social Science* 533: 70–86.

Guarnizo, L. (1997) 'The emergence of a transnational social formation and the mirage of return migration among Dominican transmigrants', *Identities* 4 (2): 281–322.

Guarnizo, L. (1998) 'The rise of transnational social formations: Mexican and Dominican state responses to transnational migration', *Political Power and Social Theory* 12: 45–94.

Hamilton, G. (1985) 'Why no capitalism in China: negative questions in historical, comparative research', in A. Buss (ed.) *Max Weber in Asian Studies*. Netherlands: Brill. pp. 65–89.

Hannerz, U. (1996) *Transnational Connections: Culture, People, Places*. London: Routledge.

Harvey, D. (1989) *The Condition of Postmodernity*. London: Blackwell.

Herod, A., Ó Tuathail, G. and Roberts, S. (1998) *Unruly World? Globalization, Governance and Geography*. London: Routledge.

hooks, b. (1992) 'Representing whiteness in the black imagination', in L. Grossberg, C. Nelson and P. Treichler (eds) *Cultural Studies*. New York: Routledge.

Hyndman, J. (1997) 'Crossing borders', *Antipode* 29 (2):

Jackson, P. (1999) 'Commodity cultures: the traffic in things', *Transactions of the Institute of British Geographers* 24: 95–108.

Jackson, P. and Crang, P. (2000) 'Transnationalism and the spaces of commodity culture'. Paper presented at the Association of American Geographers, Pittsburgh.

Jones-Correa, M. (1998) 'Different paths: gender, immigration and political participation', *International Migration Review* 32: 326–49.

Kearney, M. (1991) 'Borders and boundaries of state and self at the end of empire', *Journal of Historical Sociology* 4 (1): 52–73.

Kriesberg, L. (1997) 'Social movements and global transformation', in J. Smith, C. Chatfield and R. Pagnucco (eds) *Transnational Social Movements and Global Politics*. Syracuse: Syracuse University Press. pp. 3–18.

Laclau, E. and Mouffe, C. (1985) *Hegemony and Socialist Strategy: Towards a Radical Democratic Politics*. London: Verso.

Light, I. and Karageorgis, S. (1994) 'The ethnic economy', in N. Smelser and R. Swedburg (eds) *The Handbook of Economic Sociology*. Princeton: Princeton University Press. pp. 467–71.

Lyotard, J. (1984) *The Postmodern Condition: A Report on Knowledge*. Minneapolis: University of Minnesota Press.

Mahler, S. (1998) 'Theoretical and empirical contributions toward a research agenda for transnationalism', in M. Smith and L. Guarnizo (eds) *Transnationalism from Below*. New Brunswick: Transaction. pp. 64–100.

Mitchell, K. (1993) 'Multiculturalism, or the united colors of capitalism?', *Antipode* 25: 263–94.

Mitchell, K. (1995) 'Flexible circulation in the Pacific Rim: capitalisms in cultural context', *Economic Geography* 71 (4): 364–82.

Mitchell, K. (1997a) 'Transnational discourse: bringing geography back in', *Antipode* 29 (2): 162–79.

Mitchell, K. (1997b) 'Different diasporas and the hype of hybridity', *Society and Space* 15: 533–53.

Mitchell, K. (1997c) 'Conflicting geographies of democracy and the public sphere in Vancouver, B.C.', *Transactions of the Institute of British Geographers* 22 (2): 162–79.

Mitchell, K. (1998) 'Fast capital, modernity, race and the monster house', in R. George (ed.) *Burning Down the House: Recycling Domesticity*. New York: Westview. pp. 187–212.

Mitchell, K. (2001) 'Education for democratic citizenship: transnationalism, multiculturalism, and the limits of liberalism', *Harvard Educational Review* 71 (1): 51–78.

Mitchell, K. (2002) 'Transnationalism and the Politics of Space'. Unpublished manuscript.

Mitchell, K. and Olds, K. (2000) 'Business networks and the globalization of property markets in the Pacific Rim', in K. Olds and H. Yeung (eds) *The Globalization of Chinese Business Firms*. Oxford: Oxford University Press. pp. 147–73.

Mitchell, T. (1991) 'The limits of the state: beyond statist approaches and their critics', *American Political Science Review* 86 (4): 77–96.

Nonini, D. and Ong, A. (1997) 'Chinese transnationalism as an alternative modernity', in A. Ong and D. Nonini (eds) *Ungrounded Empires: The Cultural Politics of Modern Chinese Transnationalism*. London: Routledge. pp. 3–33.

Olds, K., Dicken, P., Kelly, P., Kong, L. and Yeung, H. (1999) *Globalization and the Asia Pacific: Contested Territories*. London: Routledge.

Ong, A. (1993) 'On the edge of empires: flexible citizenship among Chinese in diaspora', *Positions* 1 (3): 745–78.

Ong, A. (1999) *Flexible Citizenship: The Cultural Logics of Transnationality*. Durham: Duke University Press.

Ó Tuathail, G. (1996) *Critical Geopolitics: The Politics of Writing Global Space*. Minneapolis: University of Minnesota Press.

Ó Tuathail, G. and Dalby, S. (1998) *Rethinking Geopolitics*. New York: Routledge.

Portes, A. and Manning, A. (1986) 'The immigrant enclave: theory and empirical examples', in S. Olzak and J. Nagel (eds) *Comparative Ethnic Relations*. New York: Academic. pp. 47–68.

Roberts, S. (1994) 'Fictitious capital, fictitious spaces? The geography of offshore financial flows', in S. Corbridge, R. Martin and N. Thrift (eds) *Money, Power and Space*. Oxford: Blackwell. pp. 91–115.

Rouse, R. (1991) 'Mexican migration and the social space of postmodernism', *Diaspora* Spring: 8–23.

Sassen, S. (1988) *The Mobility of Labor and Capital: A Study in International Investment and Labor Flow*. Cambridge: Cambridge University Press.

Silvey, R. and Lawson, V. (1999) 'Placing the migrant', *Annals of the Association of American Geographers* 89: 121–32.

Smith, N. (1995) 'Remaking scale: competition and cooperation in prenational and postnational Europe', in H. Eskelinen and F. Snickars (eds) *Competitive European Peripheries*. Berlin: Springer. pp. 59–74.

Smith, N. (1996) 'Spaces of vulnerability', *Critique of Anthropology* 16 (1): 63-77.

Thrift, N. and Olds, K. (1996) 'Refiguring the economic in economic geography', *Progress in Human Geography* 20 (3): 311–37.

Vasile, E. (1997) 'Re-turning home: transnational movements and the transformation of landscape and culture in the marginal communities of Tunis', *Antipode* 29 (2): 177–96.

Vertovec, S. (1999) 'Conceiving and researching transnationalism', *Ethnic and Racial Studies* 22 (2): 447–462.

Visweswaran, K. (1994) *Fictions of Feminist Ethnography*. Minneapolis: University of Minnesota Press.

Wakeman, F. (1988) 'Transnational and comparative research', *Items* 42 (4): 85–7.

Walker, R. (1993) *Inside/Outside: International Relations as Political Theory*. Cambridge: Cambridge University Press.

Walton-Roberts, M. (1999) 'Returning, remitting, reshaping: non-resident Indians and the transformation of society and space in Punjab, India'. Paper presented at the Association of American Geographers Conference, Pittsburgh.

Yeoh, B. and Willis, K. (1999) '"Heart" and "Wing": nation and diaspora: gendered discourses in Singapore's regionalisation process', *Gender, Place and Culture*.

(Photo: Steve Pile)

Section 2

THE CULTURE OF ECONOMY   Edited by Trevor J. Barnes

Introduction: 'Never Mind the Economy. Here's Culture'

Trevor J. Barnes

Culture is not a decorative addendum to the 'hard world' of production and things, the icing on the cake of the material world. (Hall, 1988)

It is possible to argue that economic geographers have become some of the leading exponents of cultural geography. (Thrift, 2000a: 692)

[P]unk became real culture ... [making] ordinary social life seem like a trick, the result of sado-masochistic economics. (Marcus, 1989: 69)

I got the Sex Pistols' album 'Never Mind the Bollocks. Here's the Sex Pistols' the first day it was for sale in October 1977. I bought it at the HMV store on Oxford Street in London. It came in plain brown wrapping to prevent upright Londoners from swooning at the sight of vulgar language. Two tracks from it were in effect banned by the BBC and other radio stations (although John Peel played them): the Pistols' contribution to the Queen's Silver Jubilee celebrations, 'God Save the Queen', and their demonic version of national political analysis, 'Anarchy in the UK'. I never saw them perform live, but watched them on TV throw around chairs as well as four-letter words on Bill

Grundy's LTV 'Today' programme, listened to my conservative (and Conservative) aunts with whom I lived tell me that they were an 'abomination' and a 'disgrace' (and which consequently immensely magnified them in my estimation), and walked with my friends on a Saturday afternoon down King's Road, Chelsea, outwardly sneering at the punk fashion scene around me and which the Pistols exemplified – ripped jeans and T-shirts, green- and red-dyed hair, Doc Marten boots, and the ubiquitous use of safety pins for tethering things that should never be tethered – but inwardly admiring, and secretly wishing to join in.

The Pistols were a cultural revelation. Their ferocious energy and sound of 'broken glass and rusty razor blades' (Savage, 1993: 206) were the perfect antidote to the bloated, self-indulgent, and anodyne music of such groups as the Eagles or Genesis that characterized the first part of the decade, and to whom I was subjected as a teenager. As Savage writes, 'At a time when songs generally dealt with the pop archetypes of escape or love, the Sex Pistols threw up a series of insults and rejections, couched in a new pop language that was tersely allusive and yet recognisable as everyday speech' (1993: 206). That language, along with the Pistols' clothing, hair style, body piercing, snarling, and swearing all seemed for me at the time a seductive oppositional youth culture and way of life that uncannily matched the mid to late 1970s England of strikes, discontent, and resentment in which I lived. I even thought they might have something to say to geographers, and cheekily titled one of my third-year undergraduate papers 'Never Mind the Truth. Here's the Ideologues', a first foray into the world of intellectual anarchy.

For this section of the book, and for my own editorial introduction, I've again cheekily misappropriated the Pistols' title, and on their own silver jubilee. (Will anyone sing 'God Save the Pistols?') My contention is that the history of the band and their record raise the same issue that is at the centre of the four contributions in this section of the *Handbook*: the relationship between culture and economy. This might appear a stretch even for a book in cultural geography. But in both popular and academic treatments of the Sex Pistols, what emerges is a tension between them as a voice of culture and as an economic commodity. On the one hand, they represented a distinct 'break in the pop milieu ... nothing like it had been heard in rock 'n' roll before and nothing like it has been heard since' (Marcus, 1989: 2–3). As the Radio One DJ John Peel put it, 'You went to the gigs and there was a feeling that you were participating in something that had come from another planet, it seemed so remarkable it was happening at all' (quoted in Marcus, 1989: 41). If the hallmarks of culture are innovation, new forms of language, and changed values and ways of life, the Sex Pistols were real culture. On the other hand, the Pistols from their very creation were of making money, of selling product, of generating 'filthy lucre', of being part of 'the great rock'n'roll swindle' (Mitchell, 2000: 68). It was not for nothing that they recorded with on-the-run great train robber Ronnie Biggs, that Johnny Rotten engaged in an eight-year legal suit with Malcolm McLaren, the Pistols' manager, to recover unpaid royalties (Lydon, 1994: Chapters 19–20), or that McLaren himself coined the slogan 'cash from chaos'.

The four contributors to this section of the book – Linda McDowell, Adam Tickell, Meric Gertler, and Don Slater – see the same kind of tension between culture and economy that I am claiming for the Sex Pistols playing out

in different parts of economic geography, respectively, in production, labour markets, finance, and consumption. This is new. Until the recent past, culture was a dirty term within the discipline; its use met with the same word that the London Metropolitan Police were so anxious to conceal from the Sex Pistols' album cover. One result, as Thrift puts it, was that 'by the 1980s economic geography was in a pretty moribund state, at risk of boring its audience to death' (2000a: 692). It was as if the Eagles and Genesis had left the world of pop music, and taken up home in economic geography. But things are changing. During the 1990s, economic geographers began opening up a Pandora's box of culture, to use Thrift's (2000a) metaphor. And once opened there is no closing it again. Furthermore, as also in that original myth, what remains after the lid is off is hope. The same holds true here. In this case, the hope as economic geography engages culture is for a vibrant, energetic, and edgy discipline, a punk economic geography.

In this editorial introduction, I begin by briefly reviewing some of the different positions on the culture versus economy issue, and then examine how they have been worked out in the discipline. There are no easy solutions. Almost everyone except for a 'paid-up member of the Khmer Rouge' (Eagleton, 1995: 35) thinks that it is not either/or but both/and when it comes to culture and economy. But the difficult issue is their precise relation and theorization.

CULTURE VERSUS ECONOMY

Terry Eagleton's (2000) 'manifesto' on culture begins with the term's tangled and ambivalent etymological meaning. At first culture 'denoted a thoroughly material process', that of cultivating the land, of using brawn, skill, and material resources to put food on the table (2000: 1). Later, though, the word is 'metaphorically transposed to affairs of the spirit' (2000: 1), becoming a Bach fugue, a Botticelli portrait, a Balzac novel.

> The word [culture] thus charts within its semantic unfolding humanity's own historic shift from rural to urban existence, pig-farming to Picasso, tilling the soil to splitting the atom. In Marxist parlance, it brings

together both base and superstructure in a single notion. (2000: 1)

Culture and economy are yoked from the beginning, and the supposed opposition between a basic, brute materialist logic, and an ethereally refined non-materialist one, is false. Eagleton's reference to Marxism is also useful. Marxism has been the main forum in which debates about that yoking have been staged, at least in the post-war period, and certainly in economic geography.

That Marxism has played that role is unsurprising. Neoclassical economics, which offers the principal (and orthodox) alternative interpretation of the economy, has no truck with culture, reducing it to the 'colorless blanket' of utility-maximizing rational agents (Georgescu-Roegen, 1968: 264). As Margaret Thatcher might have said, 'There is no such thing as culture.' In contrast, culture is there from the beginning in Marx's analysis. His most succinct and perhaps best-known theoretical statement is found in the Preface to A Contribution to the Critique of Political Economy (1859). There he writes, 'the mode of production of material life conditions the social, political and intellectual life process in general. It is not the consciousness of men that determines their being, but, on the contrary, their social being that determines their consciousness' (Marx, 1904: Preface).

It is around those two sentences that an academic interpretive industry of Fordist proportions has agglomerated. The 'classical' interpretation is of economic determinism: culture as a set of 'social, political and intellectual processes in general' is irrevocably determined by the economy, 'the mode of production'. Culture is thereby reduced to an epiphenomenon, performing the functional role of ideological smokescreen for an oppressive capitalist class bent on immiserating the proletariat. While in the past this position may have had some currency, there is not much evidence of it now, at least in geography. Former dyed-in-the-wool, classical Marxists such as Dick Peet are now searching for 'the cultural source of economies', urging the use of 'cultural terms such as symbol, imaginary, and rationality ... to understand crucial economic processes' (2000: 1215, 1213). As Peet writes, 'In a phrase I never thought I would

say, political economy should become cultural economy' (2000: 1231). Or Neil Smith, who in his earlier days trumpeted 'the universalization of value in the form of abstract labour' (1984: 82), now says, '"Back-to-class" in any narrow sense is its own self-defeating cul-de-sac' (2000: 1028), and it is necessary 'to find a way of integrating class into the issues of identity and cultural politics' (2000: 1011).

David Harvey is maybe one of the few holdouts, although his own position has never been straightforward. At the very least there is a disjuncture between the prefaces and introductions to his books, which are lithe and limber, with references to novelists, popular culture, and the cultural situation of Harvey himself, and the body of the text that follows, which often goes in for categorical statements about the paramount importance of the economy. His latest book, Spaces of Hope (2000: Chapter 1), gives both perspectives in the same introductory chapter, 'The Difference a Generation Makes'. Harvey provides a wonderfully evocative account of his own shifting cultural position as a university teacher running an annual seminar since 1971 on Marx's Capital (Volume 1) in American and British universities. But the thrust of his argument is against just such a cultural positioning, and more broadly, against 'cultural analysis [which has] supplanted political economy (the former, in any case, being much more fun than being observed in the dour world and crushing realities of capitalist exploitation)' (2000: 5). For Harvey (2000: 7) it is those 'crushing realities' that demand our attention, and his list of the most important – 'fetishism of the market', 'the savage history of downsizing', 'technological change', 'weakened organized labour', and an 'industrial reserve army' – make it clear that it isn't going to be fun. We need to roll up our sleeves, and be prepared for some serious work. No more lolly-gagging and fripperies, no more 'Holidays in the Sun' (Sex Pistols, 1977).

That said, there are places where Harvey offers a softer position, recognizing the autonomy and importance of the cultural sphere. In fact, within the Marxist canon there is a continuum of softer positions, which vary from opening the economy to culture just a crack to engaging in a full-blown cultural analysis where the economy is barely present, if it is present at all. At one end, and closest to classical Marxism, are analyses by writers such as Antonio Gramsci ('cultural hegemony') or Louis Althusser ('determination in the last instance') who, while recognizing that culture is not utterly determined by the economy, keep it on a short leash, 'characteristically analys[ing it] *in relation* to class structures and the social hegemony of economically dominant groups' (Bradley and Fenton, 1999: 114).

Softer still, and occupying a middle ground, are people associated with the founding of cultural studies such as Raymond Williams ('structure of feeling'), Richard Hoggart ('the felt quality of life'), and later Stuart Hall ('Marxism without guarantees'), and discussed by a number of contributors to this volume. The importance of this group is in attempting to hang on to class analysis and the economy, and also recognizing values, ways of life, and emotional and political commitments that lay outside: hence, for example, Williams' phrase the 'structure of feeling' that connotes the 'doubleness of culture … [as both] material reality and lived experience' (Eagleton, 2000: 36). Also somewhere in this middle range, but from a different intellectual lineage, is Andrew Sayer's (1997; and Ray and Sayer, 1999) position informed by a critical realist philosophy. Using realism as a scalpel, Sayer pares away the superfluous and contingent, revealing the precise meanings and limits of the concepts 'economy' and 'culture', showing that while they are not synonymous, they are not antonyms either (Ray and Sayer, 1999: 4). Rather, economy and culture interact according to their respective logics – culture as 'dialogical', economy as 'instrumental' (Sayer, 1997: 25) – producing complex effects that must be continually scrutinized conceptually and empirically. One final example of work carried out in this intermediate terrain is Nancy Fraser's (1995; 1999), influencing the new Neil Smith (2000; also see McDowell, 2000a). Economy and culture for Fraser are transposed into two seemingly quite different claims for social justice, respectively the politics of redistribution and the politics of recognition. The trick though, as in all of this work, is to have both within 'a single comprehensive framework' (Fraser, 1999: 26), which is what she attempts.

Finally, at the other end of the continuum are postmodern or poststructural approaches that not only begin to disassociate the economy from culture, but in their 'more extreme versions' reduce and desocialize culture to 'no more than a free play of texts, representations and discourses' (Bradley and Fenton, 1999: 114). In economic geography, the best example is J.K. Gibson-Graham's (1996) work that deploys a radical version of Althusserian overdetermination in which everything causes everything else. In brief, Gibson-Graham's argument is that one of the mistakes of political economy is to assume a single, unified capitalist totality: that is, to treat it as an inviolable constant, rather than as a particular kind of discourse. Once the discursive nature of capitalism is recognized, political possibilities and strategies for change suggest themselves. To enact them, however, requires understanding how the idea of capitalism became hegemonic in the first place. For Gibson-Graham, drawing on a number of poststructuralist writers, it is because knowledge of the economy is approached from a particular cultural slant: heroic, essentialist, and masculinist. Change that slant and we will have *The End of Capitalism (As We Knew It)*.

In sum, this review provides only the thinnest of glosses. There are other approaches within economics for interpreting culture that I have not discussed, such as post-Keynesianism, evolutionary economics, or institutionalism (in this light, see Ron Martin and Peter Sunley's 2001 reply to Ash Amin and Nigel Thrift, 2000, castigating them for ignoring the panoply of heterodox theoretical economic traditions available for reconstructing economic geography). Or again, there is the Weberian take on culture and economics bound up with 'social status' and 'life chances'. Or the functionalist one of Talcott Parsons, and worked out in terms of imperatives for social integration involving among other things cultural 'latency' and economic 'adaptation'. And recently, even the pragmatist American philosopher Richard Rorty, hitherto an aggressive champion of the cultural, has got into the act, writing that 'the soul of history *is* economic' (1999: 227). This is from a man who also argues that 'the sheer clumsiness' of attempts to use 'a problematic coming from the Marxist tradition' when dealing with contemporary problems is the most persuasive reason for doubting … that we must read and reread

Marx' (1999: 221). Clumsiness, though, hardly describes the various attempts I reviewed by Marxists to deal with culture and economy. They are creative and innovative, just like culture itself. But the task of delineating clear links between culture and economy is daunting. Raymond Williams (1976) said that culture alone is 'one of the two or three most complex words in the English language'. And once joined with economy the combination becomes dense and tangled. Perhaps the way forward is not the single road of grand theoretical statement, but paths that are more piecemeal, less defined and limited, and which join bits of empirical study and modest cultural theory. As Harriet Bradley and Steve Fenton write: 'The relationship between culture and economy ... cannot be deduced from abstract principles, but can only be elucidated in specific contexts' (1999: 122). This is not to forget the larger question, but to bracket it to enable a 'close dialogue' between economy and culture, as Linda McDowell (2000b: 16) puts it. It is precisely this kind of 'close dialogue' that characterizes the work in which economic geographers have been engaged, and to which I will now turn.

CULTURE AND ECONOMIC GEOGRAPHY

Issues of culture implicitly run through Anglo-American economic geography from the beginning (Barnes, 2001a). Early texts like George Chisholm's *Handbook of Commercial Geography* (1889) or J. Russell Smith's *Industrial and Commercial Geography* (1913), while replete with trade figures, production statistics, and maps of economic specialization, were also about ways of life, values and beliefs, and material artefacts (Chisholm's 'commodities' and Smith's 'industrial products'), that is, they were about culture. Even during the 1960s and early 1970s when economic geography was 'spatial science', and mimicking neoclassical economics, culture still made a difference. For example, one of the founding texts of spatial science was the German location theorist August Lösch's *The Economics of Location* (1954) (Barnes, 2001b), framed in terms of the spartan landscape of geometrical axioms and differential equations. But behind that formalism was culture, in this case Lösch's own lifeworld of pre-Second World War Swabia. Lösch says as much in his Preface:'my youthful experience in a little Swabian town constitutes the real background of this book ... [and] my original experience there confirms my final theories' (1954: xv). This is picked up by Peter Gould who writes, '[Lösch's] landscape is not just geometry, but is inhabited by people joined by a complexity of social relations, not the least of which may be a deep sense of rootedness, of *Bodenständigkeit*, in the region itself' (1986: 15). You can take Lösch out of Swabian culture, but you can't take Swabian culture out of Lösch. It goes all the way down even into mathematical symbols and precisely drawn figures (for other examples, see Barnes, 2001b).

Of course, this is revisionist history, and I am not suggesting that economic geographers at the time articulated their concerns in the vocabulary of culture. Quite the opposite. But it indicates that culture was always present at least implicitly within the discipline. The first explicit introduction is in Doreen Massey's book *Spatial Divisions of Labour* (1984). There she builds into the very spatial process of capitalist accumulation a role for local culture, conceived historically and geographically as the sum of the sedimented layers of past interactions between rounds of investment and cultural characteristics of place (Warde, 1985). She uses the example of South Wales, and represents its history over the twentieth century as a reciprocal relationship between public and private investment and the culture of that region. Constituting that culture are masculinism, a family structure of male patriarchy, a set of religious beliefs and practices especially around the Methodist Church, a left-wing politics associated with both trade unionism and the British Labour Party, and strong, tight-knit local communities of relatively isolated single-industry towns (Massey, 1984: Chapter 5).

Although the significance of Massey's book was quickly appreciated, it was a long while before its cultural sensibility was widely taken up. Admittedly, the locality project that followed closely on its heels gave the promise of culture, but many of the studies it generated turned out as traditional and often narrowly conceived empirical analyses of local labour markets, and saying very little about culture (Cooke, 1989). That the cultural part of Massey's book remained underemphasized was in part because of the continuing dominance of political economic approaches in economic geography such as regulationist theory (Tickell and Peck, 1992), or Scott and Storper's (1992) framework of flexible production, or Harvey's (1989) ideas of space–time annihilation, all of which gave primary prominence to the economy and relegated culture to at best a secondary role. This began to change in the early 1990s as ideas from cultural studies, the British version as well as the American, which tended to be more postmodern and poststructural, entered the discipline. Furthermore, this coincided with a realization by economic geographers and others of a change in the very nature of the economy as it defined itself through culture, and which affected the nature of goods produced and sold, the behaviour and choices of consumers, and the very internal workaday operations of private firms. Here Lash and Urry's (1994) book on 'economies of signs', Beck's (1992) work on 'reflexive modernization' and, in geography, Thrift's (1997) writings on 'soft capitalism' were signal contributions. There was a recognition that the economy operated as a discursive construction for everyone within it, and therefore it was susceptible to the tools of cultural analysis. As Thrift says, 'capitalism seems to be undergoing its own cultural turn as increasingly … business is about the creation, fostering, and distribution of knowledge' (1997: 30).

Good reviews of economic geography's subsequent 'cultural turn' already exist (Crang, 1997; Thrift, 2000a), and the four chapters that follow also provide assessments of their respective subareas. As a result, let me only briefly highlight five substantive areas of writing to provide a flavour of the burgeoning literature in this field. The first is on labour markets and work. Often drawing upon poststructural and postcolonial feminist theory, there is an emphasis on the close relation among cultural performance at work centred on the body, the places and spaces in which that work is carried out, and the material consequences (Crang, 1994; Hanson and Pratt, 1995; Leslie and Butz, 1998; McDowell, 1997;

Pratt, 1999). The second is on the development and use of hi-tech. The stress here is on the significance of institutional embeddedness, close personal contacts that demand proximity, tacit knowledge, and shared cultural assumptions. Without this particular kind of culture of production, goes the argument, geographical phenomena like learning regions, hi-tech centres, and networks of association would never emerge: it is part of their very constitution (Cooke and Morgan, 1998; Gertler, 1997; 2001; Saxenian, 1994; Storper, 1997). The third is on the financial sector, and other high-level, information-based service sectors such as business consultancy. Given the need for high degrees of both personal interaction and interpretive skills in these activities, research focuses on the culture of those who interact – their ethnicity, gender, values, and beliefs – the places of interaction (often the cores of world cities), and the semiotic and discursive strategies used for interpretation (Clark and O'Connor, 1997; Leyshon and Thrift, 1997; McDowell, 1997; Thrift, 2000b; Tickell, 1996). The fourth is about the corporation. Accentuated here are the various discourses, different and even contradictory, that shape high-level management culture, and, in turn, influence the course of the firm. As O'Neill and Gibson-Graham put it, 'business is the process of talk' (1999: 15), and for talk, even corporate talk, to make sense and have effects it must be understood within specific cultural practices (Marcus, 1998; O'Neill and Gibson-Graham, 1999; Schoenberger, 1997; 1999). The last area is around consumption. Much of this work is concerned with reconceiving the consumer as an actor by moving away from the models of rational utility maximization and the consumer as dupe of the market to approaches that allow a cultural sensitivity in understanding the nature of the goods consumed (commodities as signs), the motivations for consumption, and the places in which they are bought (leading to a new geography of retailing: Glennie and Thrift, 1996; Goss, 1999; Gregson and Crewe, 1997; Jackson et al., 1998)

Whether Thrift is right that economic geographers are now 'the leading exponents of cultural geography' (2000a: 692) is in some sense moot. What is astounding is that economic geographers are doing cultural geography at all. While culture was always implicit within the discipline, it required a radical transformation to make it explicit. Of course, there are critics like Harvey (2000) and Storper (2001) who argue that the focus on culture distracts too much from 'the "hard world" of production and things' (Hall, 1988), and economic geographers would be better off if they devoted their energies to them. But what emerges from the literature I have reviewed is the inseparability of culture from that 'hard world'. It is not something to be detached and put to one side while serious work is first devoted to 'production and things'. It is more complicated, and no less serious. It is not let's do the serious thing first by examining the economy and then if there is time have 'fun' with culture, but it is doing both together.

CONCLUSION

Throughout this chapter, I've often used the two terms 'economy' and 'culture' as if they are self-sufficient, separate, and centred. They are not. One of the intellectual impulses behind the cultural turn in economic geography is to undermine dualities, and the dualism of culture and economy is one that should go. The hope is for a world in which the very distinction between economy and culture is no longer important. Such a reorientation, though, is difficult, and disorientating, because familiar conceptual handholds for understanding are taken away.

This leads back to the Sex Pistols. Their music was about disrupting traditional categories, about not fitting into one conceptual box or another. Graham Lewis of *Wire* says of punk rock, 'it was a deconstruction, it was a piss-take of Rock music. The structures were Rock'n'Roll, taken apart and put together in different ways. This is how they go, but not quite. They *swerve*' (quoted in Savage, 1993: 329). As a result, the Pistols didn't look like, sound like, or write songs like any musicians before them; indeed, some would say they weren't musicians at all. Bernard Sumner, lead singer for New Order, and formerly a member of Joy Division, said after seeing the Pistols, 'They were terrible. I thought they were great. I wanted to get up and be terrible too' (quoted in Marcus, 1989: 7). It's the same with the cultural turn in economic geography. It attempts to undo formerly fixed conceptual categories of economic geography, and put them together again in different ways, and add new ones as well; it swerves. Furthermore, just as there was experimentation, of trying things out – Sid Vicious said, 'you just pick a chord, go twang, and you have music' – and a do-it-yourself approach to punk, the same applies to the new culturally informed approach to economic geography. It won't always work, it will be 'terrible,' but it will be 'great' as well. For some punk economic geography, read on.

NOTE

I am grateful to a number of people for the comments they made on this essay, and which greatly improved it: Steve Pile, Hugh McDowell, who after reading it thought I was 'a sad old git', which I take to be a punk compliment, and the polymath Adam Tickell. who set me straight on both new wave music facts and regulationist theory.

REFERENCES

Amin, A. and Thrift, N. (2000) 'What kind of economic theory for what kind of economic geography?', *Antipode* 32: 4–9.

Barnes, T.J. (2001a) 'In the beginning was economic geography: a science studies approach to disciplinary history', *Progress in Human Geography*, 25: 455–78.

Barnes, T.J. (2001b) 'Location, location, location: from the old location school to Paul Krugman's "new economic geography"'. Humboldt Lecture, Catholic University of Nijmegen, the Netherlands (http\\:www.kun.nl/socgeo/colloquium/index.hmtl).

Beck, U. (1992) *Risk Society: Towards a New Modernity*. London: Sage.

Bradley, H. and Fenton, S. (1999) 'Reconciling culture and economy: ways forward in the analyses of ethnicity and gender', in L. Ray and A. Sayer (eds) *Culture and Economy after the Cultural Turn*. London: Sage. pp. 112–34.

Chisholm, G.G. (1889) *Handbook of Commercial Geography*. London: Longman, Green.

Clark, G. and O'Connor, K. (1997) 'The informational content of financial products and the spatial structure of the global finance industry', in K. Cox (ed.) *Spaces of Globalization*. New York: Guilford.

Cooke, P. (1989) *Localities: The Changing Face of Urban Britain*. London: Unwin Hyman.

Cooke, P. and Morgan, K. (1998) *The Associational Economy: Firms, Regions and Innovations*. Oxford: Oxford University Press.

Crang, P. (1994) 'It's showtime: on the workplace geographies of display in a restaurant in South East England', *Environment and Planning D: Society and Space* 12: 675–704.

Crang, P. (1997) 'Introduction: cultural turns and the (re)constitution of economic geography', in R. Lee and J. Wills (eds) *Geographies of Economies*. London: Arnold. pp. 3–15.

Eagleton, T. (1995) 'Marxism without Marxism', *Radical Philosophy* 73: 35–7.

Eagleton, T. (2000) *The Idea of Culture*. Oxford: Blackwell.

Fraser, N. (1995) 'From redistribution to recognition? Dilemmas of justice in a "postsocialist" age', *New Left Review* 212: 68–93.

Fraser, N. (1999) 'Social justice in the age of identity politics: redistribution, recognition and participation', in L. Ray and A. Sayer (eds) *Culture and Economy after the Cultural Turn*. London: Sage. pp. 25–32.

Georgescu-Roegen, G. (1968) 'Utility', in the *International Encyclopaedia of Social Sciences*, vol. 16. New York: Macmillan, Free. pp. 236–67.

Gertler, M.S. (1997) 'The invention of regional culture,' in R. Lee and J. Wills (eds), *Geographies of Economies*. London: Arnold. pp. 47–58.

Gertler, M.S. (2001) 'Best practice? Geography, learning and the institutional limits to strong convergence', *Journal of Economic Geography* 1: 5–26.

Gibson-Graham, J.K. (1996) *The End of Capitalism (As We Knew It): A Feminist Critique of Political Economy*. Oxford: Blackwell.

Glennie, P. and Thrift, N.J. (1996) 'Consumers, identities and consumption spaces in early-modern England', *Environment and Planning A* 28: 25–45.

Goss, J. (1999) 'Once upon a time in the commodity world: an unofficial guide to the mall of America', *Annals of the Association of American Geographers* 89: 45–75.

Gould, P. (1986) 'August Lösch as child of his time', in R.H. Funck and A. Kuklinkski (eds) *Space–Structure–Economy: A Tribute to August Lösch*. Karlsruhe: von Loeper. pp. 7–19.

Gregson, N. and Crewe, L. (1997) 'The bargain, the knowledge and the spectacle: making sense of consumption in the space of the car boot sale', *Environment and Planning D: Society and Space* 15: 87–112.

Hall, S. (1988) 'Brave new world', *Marxism Today* 24–29 October.

Hanson, S. and Pratt, G. (1995) *Gender, Work and Space*. London: Routledge.

Harvey, D. (1989) *The Condition of Postmodernity: An Enquiry into the Origins of Cultural Change*. Oxford: Blackwell.

Harvey, D. (2000) *Spaces of Hope*. Berkeley: University of California Press.

Jackson, P., Miller, D., Holbrook, B., Thrift, N.J. and Rowlands, M. (1998) *Consumption, Place and Identity*. London: Routledge.

Lash, S. and Urry, J. (1994) *Economies of Signs and Space*. London: Sage.

Leslie, D. and Butz, D. (1998) '"GM Suicide": flexibility, space and the injured body', *Economic Geography* 74: 360–78.

Leyshon, A. and Thrift, N.J. (1997) *Money/Space: Geographies of Monetary Transformation*. London: Routledge.

Lösch, A. (1954) *The Economics of Location* (1940), 2nd edn. New Haven: Yale University Press.

Lydon, J. (1994) *Rotten: No Irish, No Blacks, No Dogs. The Authorised Autobiography of Johnny Rotten of the Sex Pistols, with Keith and Kent Zimmerman*. New York: Picador.

Marcus, G. (1989) *Lipstick Traces: A Secret History of the Twentieth Century*. Cambridge, MA: Harvard University Press.

Marcus, G.E. (ed.) (1998) *Corporate Futures: The Diffusion of the Culturally Sensitive Corporate Firm*. Chicago: University of Chicago Press.

Martin, R. and Sunley, P. (2001) 'Rethinking the "economic" in economic geography: broadening our vision or losing our focus?', *Antipode* 33: 148–61.

Marx, K. (1904) *A Contribution to the Critique of Political Economy*, trans. from the second German edition by N.I. Stone. Chicago: Kerr.

Massey, D. (1984) *Spatial Divisions of Labour: Social Structures and the Geography of Production*. London: Macmillan.

McDowell, L. (1997) *Capital Culture: Gender at Work in the City*. Oxford: Blackwell.

McDowell, L. (2000a) 'Economy, culture, difference and justice', in I. Cook, D. Crouch, S. Naylor and J.R. Ryan (eds) *Cultural Turns/Geographical Turns: Perspectives on Cultural Geography*. Harlow: Prentice Hall. pp. 182–95.

McDowell, L. (2000b) 'Acts of memory and millennial hopes and anxieties: the awkward relationship between the economic and the cultural', *Social and Cultural Geography* 1: 15–30.

Mitchell, D. (2000) *Cultural Geography: A Critical Introduction*. Oxford: Blackwell.

O'Neill, P. and Gibson-Graham, J.K. (1999) 'Enterprise discourse and executive talk: stories that destabilise the company', *Transactions, of the Institute of British Geographers* 24: 11–22.

Peet, R. (2000) 'Culture, imagery and rationality in regional economic development', *Environment and Planning A* 32: 1215–34.

Pratt, G. (1999) 'From Registered Nurse to registered nanny: diverse geographies of Filipina domestic workers in Vancouver, BC', *Economic Geography* 75: 215–36.

Ray, L. and Sayer, A. (1999) 'Introduction', in L. Ray and A. Sayer (eds) *Culture and Economy after the Cultural Turn*. London: Sage. pp. 3–24.

Rorty, R. (1999) *Philosophy and Social Hope*. Harmondsworth: Penguin.

Savage, J. (1993) *England's Dreaming: Anarchy, Sex Pistols, Punk Rock and Beyond*. New York: St Martin's.

Saxenian, A.L. (1994) *Regional Advantage. Culture and Competition in Silicon Valley and Route 128*. Cambridge, MA: Harvard University Press.

Sayer, A. (1997) 'The dialectics of culture and economy', in R. Lee and J. Wills (eds) *Geographies of Economies*. London: Arnold. pp. 16–26.

Schoenberger, E. (1997) *The Cultural Crisis of the Firm*. Oxford: Blackwell.

Schoenberger, E. (1999) 'The firm in the region and the region in the firm', in T.J. Barnes and M.S. Gertler (eds) *The New Industrial Geography: Regions, Institutions, and Regulation*. London: Routledge.

Scott, A.J. and Storper, M. (1992) 'Regional development reconsidered', in H. Ernste and V. Meier (eds) *Regional*

Development and Contemporary Industrial Responses: Extending Flexible Production. London: Bellhaven. pp. 3–24.

Smith, J.R. (1913) *Industrial and Commercial Geography.* New York: Holt.

Smith, N. (1984) *Uneven Development.* Oxford: Blackwell.

Smith, N. (2000) 'What happened to class?', *Environment and Planning A* 32: 1011–32.

Storper, M. (1997) *The Regional World: Territorial Development in a Global Economy.* New York: Guilford.

Storper, M. (2001) 'The poverty of radical theory today: from the false promises of Marxism to the mirage of the cultural turn', *International Journal of Urban and Regional Research* 25: 155–79.

Thrift, N.J. (1997) 'The rise of soft capitalism', *Cultural Values* 1: 29–57.

Thrift, N.J. (2000a) 'Pandora's box? Cultural geographies of economies', in G.L. Clark, M.P. Feldman, and M.S. Gertler (eds) *The Oxford Handbook of Economic Geography.* Oxford: Oxford University Press. pp. 689–704.

Thrift, N.J. (2000b) 'Performing cultures in the new economy', *Annals, Association of American Geographers* 91: 674–92.

Tickell, A. (1996) 'Making a melodrama out of a crisis: reinterpreting the collapse of Barings Bank', *Environment and Planning D: Society and Space* 14: 5–33.

Tickell, A. and Peck, J. (1992) 'Accumulation, regulation and the geographies of post-Fordism: missing links in regulationist research', *Progress in Human Geography* 16: 190–218.

Warde, A. (1985) 'Spatial change, politics and the division of labour', in D. Gregory and J. Urry (eds) *Social Relations and Spatial Structures.* London: Macmillan. pp. 190–212.

Williams, R. (1976) *Keywords: A Vocabulary of Society and Nature.* London: Fontana.

4

Cultures of Labour – Work, Employment, Identity and Economic Transformations

Linda McDowell

Twenty-five years ago the sociologist Ray Pahl (1984) suggested that the changing nature of work was one of the key issues facing industrial societies. At the start of the new millennium his claim retains its relevance as the future of work – both waged and unwaged – is a key issue of debate across the social sciences as well as in more popular arenas. There seems to be a growing anxiety about the nature of work, a sense that it is becoming less significant and less certain in providing meaning, as well as income, in advanced industrial societies – a less central element in the social construction of identity.

There is no doubt that in the last three decades or so, in the advanced industrial west, there have been huge changes in the nature, form, distribution and location of waged work – changes so significant that a common claim has developed across the disciplines that there has been a transformation in these industrial societies 'since around 1973' (Harvey, 1989). The terms used to distinguish the complex set of changes that comprise the transformation vary: they include a suggested shift from a modern to a postmodern society, from a first to a second or reflexive modernity, and from a Fordist to a post-Fordist or a risk society. In each description of the transformation, however, changes in the nature of work and employment are a key defining characteristic. In the modern or Fordist regime that dominated economies such as the USA and the UK after the Second World War, mass production, mass labour and mass consumption resulted in a uniform or standard and masculinized working class, whose entitlement to work and to a living wage was supported by collective bargaining, relatively strong trade unions, Keynesian macro-economic policies and state welfare

provision. In this period paid work was the prime source of identity for men who, in their role as breadwinners, were expected to support their dependants, usually women and children. This world was one of relative stability, at least for the more skilled members of the proletariat, characterized by a high degree of temporal and spatial standardization in, for example, work contracts and pay rates. Spatial stability was enhanced through local links between the cultures of work and the workplace, and those of family and community. Consequently geographically specific local cultures and ways of living were discernible, where industrial traditions based on specific industries were deeply implicated in the development of spatially identifiable family obligations, class and gender relations and political beliefs (McDowell and Massey, 1984; Thrift and Williams, 1987).

These older ways of living have disintegrated in the face of economic restructuring and global shifts in the nature and distribution of work since the mid 1970s. Work – in the sense of waged labour – has, according to key theorists, become flexible, destandardized, detraditionalized and individualized (Bauman, 1998; Beck, 1992; Beck et al., 1994; Lash and Urry, 1994; Sennett, 1998). The old certainties of Fordism have been replaced by a fragmented and plural employment system characterized by 'highly flexible, time-intensive, and spatially decentralized forms of deregulated paid labour' (Beck, 2000: 77). This new system is based on networks rather than bureaucratic hierarchies in knowledge-based or informational economies in which highly skilled and individualized workers, able to take risks, construct mobile portfolio careers, and less skilled workers become increasingly redundant or replaceable. In western societies, the old institutions of the welfare state that had been such significant contributors to the reproduction of a stable working class *in situ* have been replaced by workfare policies to 'encourage' employment participation by ever-larger numbers of workers, exhorted to 'get on their bikes' to seek employment. The new knowledge societies are mobile and fast-moving – societies in which risk takers may be rewarded but where conventional patterns of institutional loyalty have lost their significance (Sennett, 1998). While, for the affluent, new forms of work may bring high rewards and opportunities to construct 'lifestyle identities' through work but also through the purchase of a range of key consumer goods, for the less skilled and less able the risk society entails not mobility and excitement but instead growing insecurity and uncertainty. But for both these groups in an increasingly polarized workforce, it is argued, employment *per se* seems to have lost its centrality in the construction of a sense of self and identity.

In this chapter then I want to interrogate these claims about the nature and consequences of economic transformation, assessing their relevance, their generality and their association with changing spatial divisions of labour. I want to set the debates about the changing meaning of work and identity in the context of wider changes in the global proletariat. In the pages that follow, I contrast the post-war era, when employment was essentially a local matter, with the present day, when work has become for many a less certain part of life. My aim, as a geographer, is to link social to spatial changes in this examination of the ways in which cultures of labour – the meanings of work for individuals and groups – have changed as global economic restructuring has had a differential and uneven impact on the space economies of different nations. While the main focus is at the scale of the locality and the nation-state, I look at the ways in which culture *per se* has become a central issue for global corporations, employing increasingly diverse workforces as they extend their global reach. At all spatial scales, however, from the organizational to global restructuring, in economies in which the exchange of knowledge and ideas has become more significant than the production and exchange of material objects, 'culture', in the sense of meanings and symbols, has come to play a growing part in understandings of the new space economy (Lash and Urry, 1994). I want to start therefore by looking at the consequences of globalization for the composition of the working class to provide a wider context for evaluating specific claims about the declining significance of waged work for individual and group identities.

WORK, IDENTITY AND THE IMPACT OF GLOBALIZATION

Work, in its widest sense, is all those activities that are central to our material existence and our place in the world. Work provides sustenance, goods for exchange and, in most societies, income. It makes life possible not only for workers themselves but also for all those people who are considered too young, too old or too weak or incapable of working. The activity of working – labouring – might therefore be defined as the application of human effort in order to transform material resources into goods for the use of individual workers and their households or for exchange with others. It takes many forms – waged and unwaged, illegal, informal and voluntary – and occurs in a wide range of locations – in the home, in the community and in specialized locations such as factories and offices. Over the twentieth century, a larger and larger proportion of all work in the world was undertaken as part of capitalist wage relations. Since urban industrialization in the west and its spatial extension, selling labour power in the market has become the main way of making a living for most of the world's population. Thus at the start of the new millennium the size of the global proletariat is larger than it has ever been before in world history. The penetration of capital throughout the world and the dominance of neoliberal economic and social policies are leading to the formation of a global proletariat on a previously unknown scale. And, as Panitch and Leys (2000) have argued, this global working class is an increasingly complex and differentiated one. While the old proletariats of the industrial 'west' are experiencing deindustrialization, work intensification, casualization and job insecurity, in the Third World, where the 'golden age' of Fordist social and economic regulation never existed, industrialization is associated with a growing urban proletariat working long hours for low pay, the exploitation of child labour and the growing participation of women, the denial of union rights and often state repression. As Panitch and Leys point out, 'the working conditions, pay and social rights of the emerging labour forces [in the Third World] share much in common with those of the core capitalisms earlier in the twentieth century' (2000: ix). Similarly, David Harvey (2000) has argued that Marx's analysis of the social conditions of nineteenth-century industrial society has a new relevance at the beginning of the twenty-first century as gross exploitation of workers characterizes more and more societies. At the same time, however, as economic inequalities are increasing, questions about the cultural meanings of work and its diversity have also become more important.

But work, as well as being a necessary part of earning a living, also bestows status, companionship, forms of solidarity with co-workers and a sense of meaning and identity on those who labour in particular ways in different organizations, institutions and localities. As more and more people are drawn into waged labour their connections to each other, to the organizations in which they labour and to the locality are also changing. Work, which was once a local affair in which people tended to be employed in the locality in which they lived, often in locally owned firms, now links people across increasingly extended spaces, regions and nations, sometimes involving physical movement across space, of both labour and capital, but also linking workers in particular locations into new networks of ownership. The new global proletariat is therefore increasingly complex and diverse. It not only combines what Panitch and Leys (2000) term the old and new working classes but also mixes them up spatially, bringing them into physical contact with each other as well as connecting them through ownership patterns. Labour, as well as capital, has become more mobile. Thus in the core economies of the old industrial west, for example, there is a growing reliance on migrant labour from the Third World to run key urban services as well as work in sweated conditions in basic manufacturing industries such as textiles, clothing and electronics. Parts of these same industries, however, have relocated to the border regions in Third World economies, to export processing zones in South East Asia or the maquiladoras of the US/Mexico border, for example, where labour costs are lower. Thus the old and the new working classes are spatially contiguous in western metropolises but also spatially differentiated by the dispersal of workers in a particular sector, or employees of a single multinational company, across the spaces of national economies, raising new questions for managers and for labour organizers. In the 'new economy' too – in the financial services and information processing industries, for example – the geographic reach of contemporary capitalist organizations has expanded and so managers, workers and organizers, whether in 'old' or 'new' sectors or economies, have to cope with cultural and linguistic diversity, whether in negotiating agreement and compliance or in organizing or defusing resistance. Significant social, local and national differences in customs – beliefs and cultures among a workforce that is increasingly diverse, as women, children, rural to urban migrants, ethnic minorities, refugees, asylum seekers and economic migrants enter labour markets previously dominated, in the west at least, by men – mean that 'cultural' understanding and connections have a growing salience in 'economic' organization. Divisions of labour now cross, or are negotiated over, diverse and multiple cultural and linguistic spaces and so new ways of drawing in and constructing co-workers and of managing cultural differences among them are important in multinational spaces. Globalization is therefore not an abstract process, based on undifferentiated labour power, but is affected by as well as affects social and cultural processes, the meanings of waged work, and the subjectivities of workers themselves. I want, then, to look at how culture has been defined and its relationship to current analyses of this transition to a more complex and differentiated global workforce.

DEFINING CULTURE: FROM A FIXED TO A MUTABLE DEFINITION

Sociologists, economic anthropologists and historians have long recognized that work in all its various forms and settings plays a central role in shaping people's view of the world and their sense of themselves as individuals and members of a group. As the influential historian E.P. Thompson (1967; 1980; 1991) has documented, the customs and cultures of the workers in early modern England were transformed by industrialization as attempts to synchronize labour and impose the discipline of clock time were resisted with greater or lesser degrees of success in different sectors and different locations. The definition of culture in many of the earlier studies of work is now being challenged, even though earlier analysts recognized that its definition was a complex and contested question (Williams, 1976; 1981). In general, it was associated with the development of particular ways of life, customs and social meanings in a clearly defined place or locality, be this an organization or a community. In recent writings, however, whether by social theorists, anthropologists, cultural and literary theorists, historians, sociologists or geographers, the concept of culture has been the subject of innovative redefinition, in part influenced by the very flows of people, money and ideas across space outlined above, which have disrupted previous connections between territory, cultural beliefs and customs. Thus, as social anthropologist George Marcus has argued, the idea of culture has moved from a sense of a whole, integrated, self-contained social group and way of life to a sense of an entity, that while still defining a coherent group or community, is highly mutable, flexible, open to shaping from many directions at once in its changing environment, and, most importantly, a result of constructions continuously debated and contested among its highly independent, even unruly, membership. (1998: 6)

Influenced by this redefinition of culture as mutable and riven by the social relations of power and conflict, an exciting and vibrant new field of labour geographies (Herod, 1997; Martin, 2000; Wills, 1998) has begun to be developed, as well as interdisciplinary work by economic sociologists, anthropologists and organization theorists focusing on the analysis of labour and organizations as socially constructed entities. The ways in which class, gender, ethnic and place-based differences, as well as local habits, customs and cultures, are used to assemble, differentiate, control and reward workers, who are drawn into and expelled from the labour market in particular ways in different places and industrial sectors, as well as the patterns of worker resistance and struggles against exploitation, are being analysed and explained in the growing number of these studies that insist on the social and cultural constitution of labour power and organizations (Grahber, 1993; Grint, 1998; Martin, 2000; Massey, 1984; Peck, 1996; Sayer and Walker, 1992; Smelser and Swedberg, 1994; Zukin and DiMaggio, 1990). In this work, more complex and contested notions of labour are common. There are, for example, a growing number of interesting case studies by geographers and others focusing on the social meaning of waged work and the identity of workers, some of them, but not all, feminist in inspiration and many drawing on poststructuralist analyses of the social and discursive construction of subjectivity (Casey, 1995; du Gay, 1996; Hanson and Pratt, 1995; Leslie and Butz, 1998; Massey, 1995; Pringle, 1998; Wajcman, 1998; Wright, 1994). Thus, work on gendered identities, and on pleasure, desire and discipline in the workplace, has become part of explorations of what constitutes class identity. Here ideas about performance, embodiment and aesthetic labour are crucial and have been significant in finding a way to include the cultural construction of difference into what seemed initially hostile economic debates. Through such analyses it might be possible to find a way through the sterile juxtaposition of economic versus cultural questions/approaches that has tended to dominate geographical debates as well as the equality/difference distinction that has dominated feminist theorizing and politics

(McDowell, 1999; 2000; Phillips, 1999; Sayer, 2000; Segal, 1999; Thrift, 2000a).

Rather than pursue the theoretical implications of overcoming these dichotomous distinctions, I want now to return to an assessment of the late-twentieth-century transformation in work. My main focus will be on the old industrial economies of the west, not only because this is where I undertake my own empirical research but also because the current debates about the nature of work in 'new' knowledge-based economies and their implications for cultural identity are currently western-centric. First, then, let us turn to a set of assumptions about the meaning of work in the Fordist era to provide the basis for a comparison and evaluation of the debates about transformation.

POST-WAR ASSUMPTIONS ABOUT WORK AND THEIR DISRUPTION

Disruptions 1 and 2: work and masculinity; leisure and work

Before the late twentieth century, in the long decades following industrial capitalism, waged work was, in general, imbued with the attributes of masculinity. This association took a distinctive binary form, linked to class position. For working-class men who laboured in manual occupations, work was represented as an heroic embodied struggle against both the material world and the owners of land and capital. For middle-class men, by contrast, work consisted of rational disembodied cerebral thought. The general association of work with men, however, was widespread and taken for granted. Women, in contrast, were on sufferance in the labour market, constructed by their femininity as out of place in the public arena of the workplace, except in the segregated ghettos of female employment. Here the stereotypical attributes of femininity – caring, empathy and dexterity – apparently fitted them for less prestigious and less well-paid work. In the main throughout the century from 1850, it was assumed that women's proper place was in the home, dependent on the waged labour of their menfolk, despite well-documented exceptions to this belief (Bradley, 1989; McDowell and Massey, 1984). This assumption, as numerous feminist scholars have documented in detail, not only affected the patterns of labour market participation and remuneration but also structured appointment and promotion policies, the cultural assumptions of firms and organizations, the nature of the labour movement and everyday forms of interaction in the workplace, behaviour on the shopfloor and in the office as well as entry to the clubs and meetings of professional associations and the trade unions, albeit with important local and regional variations depending in part on the nature of dominant forms of employment and the history and traditions of the locality (see, for example, Walby, 1989). But work was perceived and theorized as an issue about men, about class rather than gender (Grint, 1998), as the conventional division of labour that developed as an ideal in Victorian industrial capitalism continued to dominate the structures and institutions of twentieth-century western economies. And work was taken for granted as both a right and a duty of men, indeed as an essential attribute of their very masculinity. It was their rightful place in the world of work that distinguished men from women.

In the second half of the twentieth century, however, these assumptions and relationships began to be disrupted both in theoretical debate and in empirical shifts. First, a much heralded and welcomed reduction in the significance of employment was identified by sociologists as a new, more leisured and increasingly middle-class workforce began to replace the traditional working class (Bell, 1960; 1973). However, only a decade or so later analysts began to regret the apparent demise of the working class (Gorz, 1982), ruing the consequences for left-wing politics (Hobsbawm, 1981). By the 1990s, a different set of regretful consequences began to be documented. The traditional solidarity and stability of local working-class culture, which had been replaced by a more mobile and ambitious middle class without roots, apparently with adverse consequences for both individuals and society (Bauman, 1998; Beck, 2000; Sennett, 1998; 1999) as an aestheticized bourgeois neglected its familial and communal responsibilities (Putnam, 2001). In all instances, however, these grand statements turned out to be less a general trend than specific changes in the place of men in the labour market and the place of work in some men's lives. For women, for example, the last decades of the twentieth century marked growing, rather than declining, significance of waged labour in their lives (Crompton, 1999) and internationally, of course, as I outlined earlier, the size of the global working class is at its maximum. Fears of the end of the working class are a specifically western preoccupation.

But for a time, between 30 and 40 years ago, in the industrial economies of the west, it did seem that employment was set to become less significant, both as a filler of time and as a measure of social meaning. Leisure, and activities associated with consumption, it was argued, would replace work or production as the centre of life and meaning. In 1961, for example, Ferdynand Zweig discerned the emerging outlines of a new leisure society, in which individuals would work less and have more time for leisure and pleasure, producing a more civilized view of the future in which hardship and want, at least for the majority, would diminish in an increasingly affluent society. Some years later, developing the theme of growing affluence, Goldthorpe and his coworkers (1968), in a study of car workers in Luton, also intimated that family life and domestic concerns had as much significance for the newly affluent male workers as their working lives. The commitment to workplace activities and to solidaristic forms of organization, they suggested, would decline, despite their predictions being almost immediately challenged by industrial unrest in the car plant where they undertook their empirical research.

Indeed, as with so many social predictions, reality turned out to be different. At the beginning of a new century, waged work has come to play an even more significant part in the space economy, the social structure, and the pattern of life chances and income inequalities than it did in previous decades in advanced industrial societies such as the USA and the UK. For growing numbers of people, for men as well as for women, in the old industrial economies, as well as in the Third World, waged work is now more central to their material existence and to their sense of selves as an individual and a member of a local community than it has been for many years. In Great Britain, as well as the USA, for example, not only does waged work endow income and social status but, increasingly, participation in the labour market is regarded as an essential aspect of full membership of the nation-state and as the legitimate way of gaining eligibility to social and income benefits from the state. In these societies, individuals are expected to be in work and indeed, larger than ever numbers of the population are employed. Participation rates in the UK, for example, are currently at 75 per cent of all those of working age and the Labour government, now in its second term, has plans to increase this proportion. Employment participation in the old industrial economies of the west is also increasingly varied and undertaken in a wide range of circumstances (Cully et al., 1999; Greg and Wadsworth, 1999). For example, working hours have become less predictable, as has participation over the life cycle.

In Great Britain, the old assumption that employment entails a job for life, often for a single employer, working in the main steady and regular hours – an assumption that was never as strong in the USA and which, of course, in Great Britain only ever applied to a minority of 'workers', mainly middle-class men, 'career' women and the labour aristocracy – has been displaced. Employment for many is now uneven, contingent, casualized and insecure (Gallie et al., 1998). Similarly, the idea that work should be more significant in the lives of men than women, as male breadwinners brought home the bacon for their dependants, has been displaced. More and more women are entering the labour market, 'encouraged' by the social and economic policies of governments that emphasize the responsibilities of individuals rather than the mutual needs of a family or household. Unlike earlier periods in the twentieth century, motherhood is no longer regarded as an acceptable alternative to waged work. Further, even in the industrial west, where participation in education has lengthened, growing numbers of children and young people are now employed while they are still undertaking education and training (Mizen et al., 1999). This redistribution of employment has had a significant impact on the nature and meaning of work, the conditions under which people labour, the social relationships in the workplace and people's sense of themselves as members of a community.

Disruption 3: work and local cultures

As I argued earlier in this chapter, the nature of work, its organization, social structure and status, are also related to changes in the organization of production, control, ownership and location, as well as to the participation and social characteristics of workers. The growing concentration of ownership into the hands of a small number of multinational corporations operating across the boundaries of the nation-state, and the consequent loss of power both by the state and by workers and their organizations (Dicken, 1998; Held et al., 1999), have recast the relationships between firms and corporations and their workforce, leading to a disruption in the spatial associations between work and localities. Nation-states, eager to attract and retain inward investment by increasingly footloose capital, have adopted policies of financial and labour market deregulation, and, in the US and the UK *par excellence*, the benefits of labour flexibility are trumpeted. Labour power, which is seldom as footloose or flexible as capital, has been left unprotected against the demands for ever-increasing efforts to cut costs and increase

profits, and so the old industrial proletariats have seen their livelihoods disappear and their living standards plummet as capital draws in and exploits the attributes of the new industrial working class. The fate of these old working-class communities, discarded by mobile capital, has been well documented in the literatures of deindustrialization (Hudson, 2000; Lawless et al., 1998; Martin and Rowthorn, 1986). What is clear is that the links between workplace cultures – the attitudes, beliefs and customs that develop in a workplace – and a specific locality have unravelled, or rather been reconstituted at different spatial scales, as organizations rethink their structures of regulation and control (Castells, 2000a and b; Marcus, 1998; Standing, 1999; Yeung, 1998; 2000). The strong association between place and a dominant industry that was a distinctive feature of the space economy of the first industrial nations is disappearing.

Until perhaps some time in the 1970s, in a country such as Britain, there were evident associations between place and industrial structure and customs, as well as voting patterns, that meant that class mapped onto space in specific ways (Thrift and Williams, 1987) and produced the sense of a local culture that Raymond Williams defined as 'a structure of feeling'. But economic restructuring and its correlate, the growing dominance of service employment, have altered these regional associations as the economic geography of the nation perhaps becomes less distinctive. Service sector employment tends to be less certain, often undertaken on a casual or part-time basis, and also less spatially distinctive, at least in the low-income parts of the sector. Labour turnover is higher, which also alters older patterns of loyalty to an employer. New forms of work and new patterns of ownership have also recast the nature of workplace cultures. The rites and symbols that construct forms of high-tech work or professional services employment, for example, as distinctive are different from those that defined the solidaristic community of miners or steel makers, now only visible in romanticized representations in movies such as *The Full Monty, Brassed Off* or *Billy Elliott*. In new forms of work, professional loyalties rather than place-based connections are often more important and job mobility is greater. In the localities formerly dominated by older forms of work, workplace and local customs were elided as the ties formed between predominantly male workers also influenced their patterns of leisure activities and political organization outside work (Beynon, 1975; Beynon and Blackburn, 1984; Cooke, 1988; Dennis et al., 1956; McDowell and Massey, 1984; Thompson, 1980).

At the beginning of the new century, this elision of work and leisure is perhaps beginning to emerge in new forms, less politicized than the old. In, for example, the currently dominant new types of work in the service sector, a culture of presentism is evident that requires that long hours are spent in the workplace engaged in a range of work-related activities (Budd and Whimster, 1993; Hochschild, 1997; Lewis, 1989). Similarly these new forms of aestheticized work (Bauman, 1998) demand particular embodied performances from employees that often result in workers spending their leisure hours working to produce a particular bodily form and appearance. The merchant bankers whom I studied in the mid 1990s, for example, were more likely to spend their leisure in the gym at work or in escorting clients to sporting or cultural events than in locally based community activities, still eliding work and leisure but in a different way than in older industrial occupations (McDowell, 1997).

In the next sections of this chapter, I shall illustrate these claims about the transformation of workplace cultures and identity through a review of case studies of particular sectors or occupations at different periods of time from the social sciences literature. In particular, I want to explore the implications for workers of the growing significance of waged work in all aspects of their lives and yet its increasing uncertainty. I shall assess the contentions of some of the key claims of theorists of labour market change and transition, including more pessimistic claims that work has become less certain, corroding the sense of pride and identity of workers (Sennett, 1998), and more optimistic claims of the advantages of the new 'detraditional' working patterns and organizations (Beck et al., 1994; Lash and Urry, 1994). In both cases, however, I want to caution against too radical an assertion of a total transformation in the nature of work. The old Fordist pattern of full-time attachment for industrial and white-collar workers alike, as the sociologist of work Ray Pahl (1984; 1989) has consistently argued, was always a temporally and spatially specific phenomenon, dominant for perhaps only a 30-year period from the 1940s. Furthermore, exaggerated claims about the end of the working class (Gorz, 1982), the end of work (Rifkin, 1996), even the brave new world of work (Beck, 2000) need careful and empirical interrogation. For some workers, perhaps, there is a brave new world in sight, but for many others, trapped in the least prestigious parts of the new economy, the new patterns, as I suggested above, seem to reflect older forms of labour market attachment and exploitation.

FORDIST CULTURES OF WORK: LABOURING HEROES IN THE POST-WAR DECADES

Many of the post-war studies of the worlds of work were undertaken by sociologists. Anthropologists had not yet returned from 'the field' in foreign lands to cast a critical eye on their compatriots, and geographers were interested in a wider spatial focus – the region or the nation-state rather than singular workplaces. Fifty years later, scholars in all these disciplines seem united in their interest in a wide range of issues about the cultures of labour in and across a wide range of workplaces and sites of production, attempting to document the links between global and local processes and patterns (see, for example, Burawoy, 2000, for a review of this changing focus). In the first decades after the Second World War, however, the most common method was a single-site ethnography, often presented as a typical example of the nature of work, rather than as a single case study set in its own particular historical and geographical context. Numerous studies were undertaken of the customs, rites and rituals found among men working in heavy and dangerous occupations in primary and manufacturing industries, based mainly on studies within a plant or a firm. The ways in which shared risks led to a specific sense of camaraderie among, for example, miners, deep sea fishermen or steel makers were examined, as well as the tedium and boredom of working on the assembly line in the expanding car plants, components industries, meat-packing plants, white goods assembly and so forth of the early post-war decades (Beynon, 1975; Dennis et al., 1956; Goldthorpe et al., 1968; Halpern, 1996; Stull, 1997). These workplace ethnographies also uncovered the multitude of ways in which workers manipulated and resisted workplace discipline, through horseplay, having a laugh, covering for each other's absences, making preposterous demands through to prolonged periods of serious industrial action, as well as some of the divisions between workers, on the basis of ethnicity for example. In their respective studies of life on the line in a car plant in the UK and USA, Huw Beynon (1975) and Ben Harmer (1992) showed how men tried to survive the noise, monotony and repetition of the assembly line by everyday acts of resistance and mutual support.

As Thompson emphasized in his historical analyses, the notion of culture, or customs, a concept synonymous with culture, includes a bundle of attributes – 'rites, symbolic modes, the cultural attributes of hegemony, the inter-generational transmission of custom and custom's evolution within historically specific forms of working and social relationships' (1999:13). Although these customs are spatially variable, the lineaments of certain general changes might be discerned. Even in the immediate post-war period in Great Britain, the relationships between workplace cultures and local communities were beginning to be transformed. Deferential attitudes to paternalistic employers – the textile mill owners, the coal and steel barons of northwest and north-east England – had weakened in the Depression and especially during the Second World War, and the election of the post-war Labour government accelerated this change. Similarly the links between a local neighbourhood, forms of mutual organization and benefit societies and the church had weakened, to be replaced by growing allegiances to trade unions, political party and the welfare state (Clarke, 1996; Obelkevich and Catterall, 1994). Thus the 'localness' or geographical specificity of workplace institutions began to diminish; national-level institutions, systems of regulation, control and bargaining become more important; and regional variations in, for example, conditions of work and remuneration levels decreased.

The period between the end of the Second World War and the end of the 1960s was also, for most working-class men in Britain, one of relatively secure work and, as I noted above, the development of new forms of association and leisure interests. It was widely assumed that the so-called embourgeoisement of the working class would become more widespread, as workers developed typically middle-class consumption habits and behaviours, with, for example, the rise in home ownership and the widespread possession of a range of consumer durables. This shift in consumption patterns, it was argued, would be reflected in shifts in voting patterns, especially in the realignment of older class associations, in a reduction in industrial conflict and in new forms of companionate marriage and shared leisure activities within families (Franklin, 1985; Summerfield, 1994; Young and Willmott, 1973).

Growing numbers of women entered the labour market in this period, in part to support the increased costs of the rising living standards but also to meet labour shortages in both the public sector as the institutions of the welfare state expanded and in new manufacturing industries producing consumer durables. Women moved into both these sectors in growing numbers, often employed on a part-time basis. Their presence in the workforce, especially on the factory floor, led to a theoretical and empirical

challenge to the assumptions about workplace cultures and to new work uncovering the ways in which labour segmentation by gender and race is reinforced by shop and factory rituals and everyday behaviours. In a study of social relationships in a Leicestershire hosiery factory in the early 1980s, for example, Sallie Westwood (1984) documented in fascinating detail the ways in which women breached the traditional distinctions between home and work as separate spheres. Based on different traditions and practices to male workers, these women introduced reminders of home into the workplace, personalizing their benches for example, swapping food and recipes, undertaking sewing for their own families in slack periods and insisting on the celebration of birthdays, engagements and weddings by workmates.

There are now a rapidly expanding number of studies of women's work in different circumstances on assembly lines and in factories, exploring women's customs and workplace cultures both in the older industrial economies (Cavendish, 1983; Cockburn, 1986; Glucksmann, 1990; Redclift and Sinclair, 1991; West, 1982) and increasingly across the newly industrializing societies (Chant and McIllwaine, 1995; Elson and Pearson, 1981; Faulkner and Lawson, 1991; Fuentes and Ehrenreich, 1983; Hsiung, 1996; Jackson and Pearson, 1998; Lee, 1997; Mies, 1986; Momsen and Kinnaird, 1993; Ong, 1987; Pearson and Mitter, 1993). In this work the interconnections between women's position in the home and locality and the ways in which they influence both the assembly of female labour forces and the social relationships between women and between men and women in the workplace are explored. In these studies the ways in which the cultural assumptions and symbolic meanings about work and gendered identities are explored, demonstrating in fascinating empirical detail how cultural attitudes and workers' gendered subjectivities are an essential part of developing an understanding not only of local labour markets and labour processes therein but also of the larger-scale nature of globalization. In a study of women working in the information processing industry in Barbados, for example, Freeman (2000) has vividly demonstrated the importance of a cultural analysis of the production process.

There are also growing numbers of feminist studies of the wide variety of women's work in the service sector, in both highly paid, professional jobs and more typical 'women's jobs' in caring and servicing roles (England and Stiell, 1997; Greed, 1991; Gregson and Lowe, 1994; Halford et al., 1997; Moss, 1997; Pratt, 1997;

Pringle, 1989; 1998), which also document the ways in which the social meaning of work and the cultures of production are connected to gender divisions of labour. It has been argued, for example, that women managers tend to be less bureaucratic than male managers, challenging traditional structures, although not always successfully changing them (Marshall, 1984; 1995). In studies of 'caring' occupations, analysts have shown how the management of emotions, or the emotional relations involved in caring for children or for the elderly, affects relationships not only between employees and service consumers but also between employees and employers. Interesting new work about the conflicts between, for example, women who employ domestic workers, often to facilitate their own participation in waged work, is exploring the limits to classic accounts of the exploitation of workers by employees (Anderson, 2000).

NEW FORMS OF WORK, NEW CULTURES OF LABOUR?

In this section, I want to turn to a set of theoretical arguments about the links between the transformation of work, social identity and cultures of labour at the end of the twentieth century that unfortunately have, in the main, ignored the excellent empirical studies of the culture of different types of work reviewed above.

One of the most significant features of contemporary advanced industrial economies is the predominance of employment in the service sector, and there is now a huge literature delineating the shift from manufacturing dominance, the spatial distribution of services, conditions of employment and the nature of work in different workplaces, as well as a literature about new forms of 'flexible' industrial production (see, for example, Aglietta, 1979; Allen, 1992; Amin, 1994; Christopherson, 1989; Daniels, 1995; Lash and Urry, 1994; Lipietz, 1987; Piore and Sabel, 1984; Pollert, 1988; Standing, 1999). One of the common lines of agreement in this vast array of studies is that service-dominated economies are marked by significant and growing labour market inequalities (Bauman, 1998; Pinch, 1993; Sassen, 1991). Despite debates about the extent and causes of this inequality, it seems that a service-dominated economy is one that takes a bifurcated form, with the most rapid expansion of employment occurring at the top end in well-paid occupations that demand educational and professional credentials and, at the bottom end, often entry-level jobs and occupations which are poorly paid,

unskilled and offering little job security and few work-related benefits (Fine and Weiss, 1998; Nelson and Smith, 1999; Newman, 1999).

Even in the most well-paid occupations of the new service economy, job security and permanent employment are becoming less common. Older hierarchical and bureaucratic institutional structures with almost guaranteed progression and promotion are increasingly being displaced by new forms of internal organization, often based on team work. Middle management is being replaced by 'horizontal' groups and 'empowered' employees, for example, and promotion is often related to individualized performance-based measures. In both older and newer high-status occupations – banking and finance, dot.com companies, the legal profession, business services – the products being exchanged increasingly consist of information and advice. Indeed the social theorist Castells (2000a) has identified the onset of a new form of capitalism that he terms informationalism or informational capitalism in which workers are either 'self-programmable labour' – highly educated and able to retrain and adapt to new tasks and processes – or 'generic labour' – exchangeable, disposable and usually unskilled (Castells, 2000b).

Work as a performance?

In the new informational economy, designated rather fancifully by others as 'weightless' (Coyle, 1997) or as 'living on thin air' (Leadbeater, 1999), work in the elite occupations has become a matter of producing a convincing performance, rather than being based on clearly defined rules and practices. Work, in other words, has become an elaborate game of pretence and a spectacle in which, as the sociologist Zygmunt Bauman has suggested, 'bosses do not really expect employees to believe that they mean what they say – they wish only that both sides pretend to believe that the game is for real, and behave accordingly' (1998: 35). Thus work itself is as much about the cultural production of employees as about the material production of goods and services. In his aggregate assessment of western economic change, Bauman has also argued that as the nature of work itself has changed for many to become discontinuous and flexible, employment no longer acts as the basis for the building of a lifelong identity in the same way as it did in an earlier era – whether this era is termed modernity or Fordism. Previously, Bauman claims, 'the fixed itinerary of work-career and the prerequisites of lifelong identity construction fit each other well' (1998: 27) (although it is important to remember that this coincidence or fit was for men in the main, and among the working class only for the labour aristocracy). In the new world, Bauman argues, identity is constructed in the sphere of consumption, through lifestyle purchases, in a society in which aesthetic ideals rather than ethical norms dominate. In the following passage Bauman spells out the implications of this shift for employment:

> The status occupied by work, or more precisely by the job performed, could not but be profoundly affected by the present ascendancy of aesthetic criteria. Work has lost its privileged position – that of an axis around which all other effort at self-constitution and identity-building rotate. But work has also ceased to be a focus of particularly intense ethical attention in terms of being a chosen road to moral improvement, repentance and redemption. Like other life activities, work now comes first and foremost under aesthetic scrutiny. Its value is judged by its capacity to generate pleasurable experience. Work devoid of such capacity – that does not offer 'intrinsic satisfaction' – is also work devoid of value. (1998: 32)

For the elite, in high-status occupations, 'the line dividing vocation from avocation, job from hobby, work from recreation' has been effaced, lifting 'work itself to the rank of supreme and most satisfying entertainment. An entertaining job is a highly coveted privilege' (1998: 34). And so, Bauman notes, 'workaholics with no fixed hours of work, preoccupied with the challenges of their jobs twenty-four hours a day and seven days a week, may be found today not among the slaves, but among the elite of the lucky and successful' (1998: 34).

Bauman's analysis has parallels with the arguments about the elision of work and leisure mentioned earlier in this chapter and with the expanding number of recent studies that insist on the significance of an embodied performance in the workplace (Acker, 1990; Adkins, 1995; du Gay, 1996; Hochschild, 1983; Kerfoot and Knights, 1996; Leidner, 1993; Leslie and Butz, 1998; McDowell, 1997; Pringle, 1989; 1998). While this erasure of the division between work and 'life', employment and consumption, may be the case for an elite, for the masses in post-Fordist economies, Bauman argues, work has become increasingly meaningless, boring, without worth, with no security or corresponding commitment. These differences between types of work are now more obvious without a work ethic that emphasizes the dignity of labour for all workers, and which once conveyed a message of equality of respect between men, despite evident differences in their rewards and conditions. Thus, in Bauman's view, work itself has lost the inherent value that it once possessed. These are grand claims and need substantiation or modification through careful empirical analysis of the meaning of work in different occupations at all levels in the labour market. In a fascinating study of the skills developed through low-wage work in the fast-food sector in New York City, anthropologist Katherine Newman (1999) persuasively argues that these jobs not only inculcate work discipline but also provide workers with a sense of self-worth and respect. While not denying the often exploitative conditions of employment, it is demeaning to dismiss 'McJobs' as intrinsically worthless and Bauman's assertions need to be tested against employees' own opinions of their working lives.

The corrosion of character?

A rather similar argument about the loss of the inherent value of employment in 'flexible' capitalism has been made by the sociologist Richard Sennett (1998). Like Bauman, Sennett suggests that, in the brave new world of a new capitalism characterized by risk, flexibility, networking and short-term team work, the ability to reinvent oneself and to construct a convincing performance is a crucial attribute of success. However, in Sennett's view, this essential characteristic leads to a destructive corrosion of a sense of self-worth, and the loss of trust and integrity which were valued by an earlier generation of both employees and employers. Long-term commitment on both sides has been destroyed by new institutional and labour market practices in which it is the short term that matters – for both profits and employment. Consequently the senses of linear time and cumulative achievement that marked the lives of the 'decent' working class in previous (post-war) decades have been replaced by uncertainty and, Sennett suggests, a loss of connection to locality and community.

Both Bauman and Sennett, in my view, fail to recall Ray Pahl's (1984) warning about the transitory nature of the Fordist era, reading off its characteristics as a singular ideal that no longer exists. They fail to recognize the extent of the variations in earlier eras in the nature of attachment to the labour market: most women and a large proportion of the male working class faced uncertain and transitory attachment to the labour market throughout most of the twentieth century. There is also a noticeable lack of consideration of the variability and multiplicity of the ways in which new identities are being constructed in the labour market. Neither theorist, for example, looks at gender differentiation, despite their assertion that many men's lives are changing for the worse, or at spatial variations. As I have already suggested, feminist theorists, while not wanting to celebrate the features of the new capitalism, in which there is no doubt about the extent of inequalities, have begun to produce an exciting body of new theory and empirical documentation of the ways in which organizational practices and everyday behaviours are part of the fluid and multiple construction of workers' identities. The theoretical insistence of postmodern and poststructural theorists that identity is provisional, fluid and discursively constructed coincided with a new set of issues in the organization and restructuring of work. The rising numbers of women entering the labour force of almost every industrial and industrializing nation (with the exception of the former socialist societies in the early 1990s) challenged, for example, the association of waged labour with men and masculinity and the workplace as either a rational, unemotional and disembodied sphere or an arena for the display of heroic masculinity and bodily strength. The ways in which assumptions of masculine superiority and feminine inferiority dominated workplace practices of recruitment and promotion, reward structures and daily social interactions have been investigated and challenged in a range of organizations (Acker, 1990; 1992; 1998; Adkins, 1995; Kerfoot and Knights, 1996; 1998). In recent work on industrial economies, interesting new analyses that link class, ethnicity and gender together and show the multiplicity of the ways in which gendered performances at work might challenge binary distinctions between masculinity and femininity and the sexed body are beginning to provide new ways of thinking about the social construction of identity (Lamphere et al., 1997).

Detraditional work?

A rather different analysis of the consequences of new forms of work, especially in the professions and in elite occupations, which also draws on postmodern notions about difference, diversity and performance, is to be found in the work of theorists such as Giddens (1991), Lash (1994) and Beck (1992; 2000), all of whom have made significant contributions to debates about the contemporary features of 'reflexive modernity' and the 'risk society'. Unlike Sennett, these theorists emphasize the new opportunities that exist in the breakdown of traditional structures in the workplace, such as rigid bureaucracies. Instead of workplaces being dominated by hierarchical

relationships where success is dependent on status and experience, these theorists identify an intensification in processes of individualization. By this term, they mean the dominance of processes in which individuals are required to create their own self-identities as individuals and new forms of authority at work. Here there are clear parallels with Bauman's notions about the significance of performance in the workplace.

These theorists of reflexive modernity go further, however, in suggesting that the intensification of individualization actually challenges, and even breaks down, existing social forms such as class, status and gender (Beck and Beck-Gersheim, 1996), untying individuals from the rules and norms that dominate modern institutional forms such as industrial organizations. Consequently, according to Beck (1992), the significance of social class and gender in influencing individuals' positions in the labour market have or will become less important than individual performance and the social construction of a particular identity. Thus workers with standard contracts guaranteeing (perhaps lifetime) employment are being replaced by workers on contracts where income and temporary security are linked to the ability to perform (Beck and Beck-Gersheim, 1996). According to Beck (1992), 'do-it-yourself' workers have to stress their individuality and uniqueness to sell themselves to employers. While these claims may have some purchase in certain sectors of the new knowledge economy – perhaps in the now rapidly deflating dot.com companies where image is all – here too detailed empirical work in a wide range of sectors will be essential to assess the extent to which the class and gender inequalities of 'modern' organizations are being dislocated. Adkins (1999; 2000), for example, in an initial evaluation, is sceptical of the claims about the growing insignificance of gender as a key social division in contemporary labour markets and organizations in most industrial and industrializing nations. Certainly, current evidence both from aggregate analyses of the continuing significance of gender in job segregation and income inequality (Cully et al., 1999; Forth, 2001; Gallie, 2001; Gallie et al., 1998) and from case studies undertaken within organizations (Crompton, 1999; Halford et al., 1997; McDowell, 1997; Pringle, 1998; Wajcman, 1998), which reveal the continuing significance of discrimination in everyday social practices at work and in the cultural traditions that support such discrimination, cast considerable doubts on the generality of Beck's claims.

ORGANIZATIONAL CULTURES

In this final substantive section, I want to change the emphasis from labourers *per se* to the new and interesting work about organizational cultures. Here one of the key foci of research is about the ways in which corporations operating in the new, globalizing economy endeavour to construct a sense of corporate belonging across increasingly diverse geographical spaces and among increasingly differentiated labour forces (Thrift, 2000b). While labour historians, radical sociologists, feminist analysts and some geographers have had a long-standing interest in the working class, more recently there has been a new and growing emphasis in the social science literature about work and organizations on middle-class workers, on managers and, especially, on the notion of corporate culture. As the anthropologist George Marcus (1998) noted, there has been a parallel 'corporate interest in culture and a cultural interest in the corporate' (1998: 3), mirroring in part the growing dominance of multinational corporations (Dicken, 1998). Analyses of the culture of corporations have, of course, a long history in the social sciences, albeit a shorter one in geography. However in the tradition of work on the modern corporation that reaches back over the twentieth century, the dominant concerns have tended to be with the rational bases of organizational behaviour, scientific management and economic modelling. It has only been in the last 20 years or so that new questions about social interaction, power and social control in the workplace have become important. There is a new interest, for example, in norms, values, collective ethos, organizational culture, authority and power in interpersonal relations, and issues about ethics and social responsibility in corporations among both the owners and managers of these corporations and those for whom they are a research subject. The idea of an economy and an economic organization as a social institution, flexibly constructing itself through symbols, conventions and rules, has come to dominate recent work. And, as economic sociologists Neil Smelser and Richard Swedberg have noted, 'the particular perspectives of social networks, gender and cultural context have also become central' (1994: 3). A rapidly expanding literature, both academic texts and practical manuals, currently addresses questions about corporate culture, in association with new forms of management practices and the growth in importance of human relations departments within corporations. From new forms of organization

including networks to replace hierarchies, through issues about the feminization of management and new concerns with work/life balance, to new dress codes and 'dressing down' days, the social and cultural formation of corporations and firms is a key area of investigation and innovation.

There seems to be a number of reasons for this new emphasis. As I argued earlier, the concept of culture itself has been subject to innovative redefinition leading to new emphasis on diversity and change. But the shift also reflects the reorientations of corporations attempting to manage change both in a period of greater risk and uncertainty and across diverse geographical spaces. By inculcating a strong sense of corporate culture, organizations hope to increase control over their diverse labour forces and to be able to manage the uncertainties of social change within their own organizational boundaries. The increasing dominance of transnational corporations that operate across national boundaries has paradoxically necessitated closer attention both to the significance of cultural differences between sites/nations and to the establishment of greater uniformity in cultural practices within the organization. Interesting new areas for analysis that straddle the older boundaries between economic and cultural geography thus arise, about transnational business cultures, global discourse and cosmopolitan workforces (Hannerz, 1990) on the one hand, and about managing diversity in local, regional and national needs, cultural attitudes and different business practices on the other. A new emphasis, shifting from the culture within an organization at 'home' to one that is developed across boundaries, is an essential element of developing work on economic globalization. Thus, as Hannerz noted, the rise of transnational organizations has led to 'cultural work' – what he terms the rise of 'a culture shock prevention industry' (1990: 108) that may include, for example, cross-cultural training programmes. Interestingly Hannerz suggests that a 'culture of critical discourse' may develop across a global corporation, that is 'reflexive, problematising, concerned with metacommunication' (1990: 109). It is also

> generally expansionist in its management of meaning. It pushes on and on in its analysis of the order of ideas, striving towards explicitness where common sense, as a contrasting mode of meaning management, might come to rest comfortably with the tacit, the ambiguous and the contradictory. In the end, it strives towards mastery. (1990: 109)

It may be, however, that in the end such a meta-narrative will fail to accommodate the diversity and difference that exist in business practices that straddle several different cultural contexts. Interesting new research questions about the practices and relative success of this 'cultural work' need to be investigated in a range of industries and locations.

It is becoming clear, however, that global corporations increasingly and deliberately attempt to create a set of myths and stories to constitute an identifiable culture, as well as their practice of developing and marketing 'brand' identities and loyalties associated with particular lifestyles and behaviour (Klein, 2000). In a fascinating glimpse into the inner workings of Shell, Davis-Floyd (1998) has told the story of how a Professor of English, Dr Betty S. Flowers, was employed in a consultant capacity to serve as the editor for the myths that Shell was consciously creating, the stories they wanted to write about the future as well as the past. Backed up by international and comparative data collected by a team of 20 economists, Flowers produced a series of scenarios to be used to teach managers to

> think mythologically and causally, to see every major local and world event as potentially located in a story, and to make on the spot business and policy decisions based on what they know that story would lead to if allowed to play itself out. (1998: 142)

Although Shell had used alternative scenarios for future planning for decades, it was felt that they had not been complex enough to encompass the diverse set of circumstances that managers increasingly had to face.

In her report, Betty S. Flowers defined corporate culture as myth and narrative, suggesting that 'a myth is a story that organizes experience through telling something explicitly about meaning – where we're going, where we came from, or who we are' (Flowers in Davis-Floyd, 1998: 146). But stories that corporations and institutions tell about themselves may, of course, fail as well as succeed in their aim. Two recent studies have illustrated the problems of continuing to cling to orthodox liberal narratives with their emphases on rational decision-making and competitive behaviour in conditions of uncertainty and rapid economic change. In a path-breaking study, economic geographer Erica Schoenberger (1996) analysed the decisions made by key management figures in two classic US manufacturing firms (Rank Xerox and Lockheed) that were losing their competitive edge throughout the 1980s and 1990s. As she showed, the culture of these firms and, especially, the socialization of their key personnel prevented them fully understanding the new economic circumstances that were causing problems for their organization.

The second example is economist Robert Shiller's (2000) analysis of the huge surge in the value of new internet companies and the hype that surrounded them at the end of the 1990s. Shiller's analysis is condemnatory, demolishing claims of a rational evaluation by investors of the prospects for the new e-economy. Instead, he insisted,

> the market is high because of the combined effect of indifferent thinking by millions of people, very few of whom felt the need to perform careful research on the long-term investment value of the aggregate stock market, and who are motivated substantially by their own emotions, random attention and perceptions of conventional wisdom. (2000: 2)

This claim may be intended as criticism of the general public rather than of Shiller's professional economist peers or institutional investors, but the infamous collapse a year or so earlier of the US firm Long Term Capital Markets, headed by Nobel laureate economists, had already severely dented the reputation of neoliberal approaches to risk and investment. The current commentary accompanying the fall in the value of high-technology shares tends to support Shiller's arguments about the emotional basis of investment decisions.

It is interesting that, in the company of David Harvey, that former bastion of neoliberal orthodoxy *The Economist* magazine suggested recently that *Das Kapital* was a more accurate guide to understanding the operation of capitalism in turn-of-the-millennium economies than conventional economic texts. Marcus (1998) has also linked earlier aggregate analyses of capitalism to the new interest in culture within corporations, arguing that this interest is a reflection of the recent reminders through restructuring that capitalism is, as Schumpeter argued, a process of creative destruction. Marcus suggests that

> the double-edged quality of the term *creative destruction* itself captures well the ideological and cognitive work that cultural discourse currently does for corporations: *creative* counterbalances and gives positive value to a process that is undeniably *destructive* with considerable human costs and displacements implied. (1998: 10, original emphases)

Thus a rhetoric of cultural change is often adopted by the very managers whose jobs will be forfeited because of it. As I have argued throughout this chapter, in capitalist societies the nature of work and the characteristics of those who labour for a living are neither fixed nor permanent; the type of work undertaken, by whom and under what conditions have undergone radical shifts in the last decades of the twentieth century.

CONCLUSIONS

I have ranged so widely in this chapter that it is difficult to end with a set of concise conclusions about the significance of the changing nature and distribution of work and the diverse set of theoretical approaches to its analysis. I want therefore in this conclusion to focus in the main on a set of issues about policy and political responses to contemporary shifts rather than to attempt a summary. I shall address three sets of issues at different spatial scales from inside the organization to international cooperation.

At the scale of the organization, perhaps one of the most significant implications of the expanding body of work about social identity and workplace experiences has been for theoretical and practical conceptions of social justice and for policies to implement more equal treatment between diverse workers. There is a growing recognition that issues about embodiment, weight, age, sexuality, physical ability and skin colour as they affect workplace performances and evaluations are economic, and not merely socio-cultural, issues. Cultural attributes are thus part of the basis of economic discrimination (Fraser, 1997; Turner, 1996; Young, 1990). This argument has opened up a series of questions about how to effect greater equality in the workplace through recognizing difference and so introducing specific policies for different groups. Examples being introduced include: ideas about workplace mentoring; policies to ensure that appraisal and promotion schemes discriminate evenly rather than assuming workers have no dependants; and wider policies based on ethnic and gender audits to identify possible areas of unequal treatment. In the economy as a whole, some nation-states have accepted that to facilitate growing workplace participation for all individuals, regardless of age and status, new forms of what have become called work/life balance policies are necessary in order to ensure that responsibilities that were once accepted by the family or supported by welfare provisions for dependants, which are now being cut in neoliberal states, are still able to be carried out as more and more households include dual or multiple workers. As well as policies that explicitly recognize women's maternal responsibilities, some states have accepted that to shift stubborn gender inequalities, policies that facilitate men's participation in the domestic sphere are also needed. Paternity and parental leave provisions have been introduced in, for example, many of the European Union states, although progress is slow and uneven.

While local and national policies may have a limited impact as global capital constructs new geographies of investment and disinvestment and the global proletariat is increasingly at the mercy of decisions taken in the headquarters of global corporations, it is clear that policies to protect workers from their immediate consequence are also crucial. While Britain has chosen to emphasize the 'flexibility' of its workforce with extremely limited protection against redundancy, France, for example, has adopted a more protectionist course, and yet there seems to be little difference in their relative success at attracting inward investment. In the face of global capitalism, however, new forms of workers', consumers' and ecological movements that traverse spatial and cultural differences in the same way as capitalism are becoming more important, aided by the technological innovations that lie behind the emerging knowledge and network societies. In these movements, social and cultural beliefs about equity and justice that also accept diversity and build on it are an essential aspect of the challenge to the profit motive and economic 'rationality'. Indeed, new areas of theoretical and empirical investigation developing theoretical approaches and methodologies that are more usually the province of social and cultural geographers are providing part of the impetus to look in different ways at the uneven effects of economic restructuring, economic inequality, spatially uneven development and the cultures of particular ways of labouring. Perhaps a challenge that remains is to develop analyses that connect these new approaches to political struggles against the unequal impacts of labour market restructuring and economic change between and across geographic scales.

Analyses of the future of work that celebrate the relative freedoms of highly skilled 'detraditional' workers in new knowledge economies, but neglect the consequences for the growing global working class who labour under conditions of increasing exploitation, are an inadequate response to the enormous implications of the new ways of working that are emerging in the twenty-first century. It seems clear to me that a combination of materialist and cultural perspectives is necessary to understand the complex ways in which diverse labour forces are constructed and cultures of production are produced and maintained within organizations and localities. The central role played by local women workers in the extension of new modes of economic flexibility in the development of a global proletariat, for example, cannot be explained without an understanding of the spatially variable gendered practices and ideologies that influence labour market behaviour. History and local specificity are important in understanding how globalization is affecting workers 'on the ground', just as the analysis of the abstract structures of capitalist exploitation enables continuities and similarities between workers and places to be understood. Perhaps a further challenge lies in exploring how to hold together theories of diversity and difference and local particularities with larger-scale theories of economic inequality.

REFERENCES

Acker, J. (1990) 'Hierarchies, jobs, bodies: a theory of gendered organisations', *Gender and Society* 4: 139–54.

Acker, J. (1992) 'Gendering organisational theory', in A. Mills and P. Tancred (eds) *Gendering Organisational Analysis*. London: Sage. pp. 248–60.

Acker, J. (1998) 'The future of gender and organisations: connections and boundaries', *Gender, Work and Organisation* 5: 195–206.

Adkins, L. (1995) *Gendered Work: Sexuality, Family and the Labour Market*. Buckingham: Open University Press.

Adkins, L. (1999) 'Community and economy: a re-traditionalisation of gender?', *Theory, Culture and Society* 16: 119–39.

Adkins, L. (2000) 'Objects of innovation: post occupational reflexivity and re-traditionalisations of gender', in S. Ahmed, J. Kilby, C. Lury, M. McNeil and B. Skeggs (eds) *Transformations: Thinking through Feminism*. London: Routledge. pp. 259–72.

Aglietta, M. (1979) *A Theory of Capitalist Regulation*. London: New Left.

Allen, J. (1992) 'Services and the UK space economy', *Transactions of the Institute of British Geographers* 17: 292–305.

Amin, A. (1994) *Post-Fordism: A Reader*. Oxford: Blackwell.

Anderson, B. (2000) *Doing the Dirty Work*. London: Zed Press.

Bauman, Z. (1998) *Work, Consumerism and the New Poor*. Buckingham: Open University Press.

Beck, U. (1992) *Risk Society: Towards a New Modernity*. London: Sage.

Beck, U. (2000) *The Brave New World of Work*. Cambridge: Polity.

Beck, U. and Beck-Gersheim, E. (1996) 'Individualisation and precarious freedoms: perspectives and controversies of a subject-oriented sociology', in P. Heelas, S. Lash and P. Morris (eds) *Detraditionalization: Critical Reflections on Authority and Identity*. Oxford: Blackwell.

Beck, U., Giddens, A. and Lash, S. (1994) *Reflexive Modernisation: Politics, Tradition and Aesthetics in the Modern Social Order*. Cambridge: Polity.

Bell, D. (1960) *The End of Ideology*. Glencoe: Free Press.

Bell, D. (1973) *The Coming of Post-industrial Society: A Venture in Social Forecasting*. New York: Basic.

Beynon, H. (1975) *Working for Ford*. Wakefield: EP.

Beynon, H. and Blackburn, R. (1984) 'Unions: the men's affair?', in J. Siltanen and M. Stanworth (eds) *Women and the Public Sphere*. London: Hutchinson.

Bradley, H. (1989) *Men's Work, Women's Work*. Cambridge: Polity.

Budd, L. and Whimster, S. (eds) (1993) *Global Finance and Urban Living*. London: Routledge.

Burawoy, M. (2000) *Global Ethnography: Forces, Connections and Imaginations in a Post-modern World*. Berkeley: University of California Press.

Casey, C. (1995) *Work, Self and Society: After Industrialism*. London: Routledge.

Castells, M. (2000a) *The Information Age: Economy, Society and Culture*, rev. edn, 3 vols. Oxford: Blackwell.

Castells, M. (2000b) 'Materials for an exploratory theory for a network society', *British Journal of Sociology* 51: 5–24.

Cavendish, R. (1983) *Women on the Line*. Basingstoke: Macmillan.

Chant, S. and McIllwaine, C. (1995) *Women of a Lesser Cost: Female Labour, Foreign Exchange and Philippine Development*. London: Pluto.

Christopherson, S. (1989) 'Flexibility in the US service economy and the emerging spatial division of labour', *Transactions of the Institute of British Geographers* 14: 131–43.

Clarke, P. (1996) *Hope and Glory: Britain 1900–1990*. Harmondsworth: Penguin.

Cockburn, C. (1986) *Machinery of Dominance*. London: Pluto.

Cooke, P. (ed.) (1988) *Localities: The Changing Face of Urban Britain*. London: Unwin Hyman.

Coyle, D. (1997) *The Weightless World*. Oxford: Capstone.

Crompton, R. (ed.) (1999) *Restructuring Gender Relations and Employment*. Oxford: Oxford University Press.

Cully, M., Woodland, S., O'Reilly, A. and Dix, G. (1999) *Britain at Work as Depicted by the 1998 Workplace Employee Relations Study*. London: Routledge.

Daniels, P. (1995) 'Services in a shrinking world', *Geography* 80: 97–110.

Davis-Floyd, R.B. (1998) 'Storying corporate futures: the Shell scenario', in G.E. Marcus (ed.) *Corporate Futures*. Chicago: University of Chicago Press. pp. 141–79.

Dennis, N., Henriques, F. and Slaughter, C. (1956) *Coal is Our Life*. London: Eyre and Spottiswood.

Dicken, P. (1998) *Global Shift*, 3rd edn. London: Guilford.

Du Gay, P. (1996) *Consumption and Identity at Work*. London: Sage.

Elson, D. and Pearson, R. (1981) 'Nimble fingers make cheap workers: an analysis of women's employment in Third World export manufacturing', *Feminist Review* 7: 87–107.

England, K. and Stiell, B. (1997) '"They think you are as stupid as your English is": constructing foreign domestic workers in Toronto', *Environment and Planning A* 29: 195–215.

Faulkner, A. and Lawson, V. (1991) 'Employment versus empowerment: a case study of the nature of women's work in Ecuador', *Journal of Development Studies* 27: 16–47.

Fine, M. and Weiss, L. (1998) *The Unknown City: The Lives of Poor and Working Class Young Adults*. Boston: Beacon.

Forth, J. (2001) 'The gender pay gap: new evidence from the Workplace Employee Relations Survey'. Unpublished paper available from the author at the National Institute for Economic and Social Research, Smith Square, London SW1P 3HE.

Franklin, M. (1985) *The Decline of Class Voting in Britain*. Oxford: Clarendon.

Fraser, N. (1997) *Justice Interruptus: Critical Reflections of the 'Postsocialist' Condition*. London: Routledge.

Freeman, C. (2000) *High Heels and High Tech in the Global Economy*. London: Duke University Press.

Fuentes, A. and Ehrenreich, B. (1983) *Women in the Global Factory*. Boston: South End.

Gallie, D. (2001) 'Skills change and the structure of the labour market: class, gender and unemployment'. Paper presented at a Conference on Disadvantage in the Labour Market, National Institute of Economic and Social Research, 15 June (paper available from the author at Nuffield College, Oxford).

Gallie, D., White, M., Cheng, Y. and Tomlinson, M. (1998) *Restructuring the Employment Relationship*. Oxford: Oxford University Press.

Giddens, A. (1991) *Modernity and Self Identity: Self and Society in the Late Modern Age*. Cambridge: Polity.

Glucksmann, M. (1990) *Women Assemble: Women Workers and the New Industries in inter-War Britain*. London: Routledge.

Goldthorpe, J., Lockwood, D., Bechhofer, F. and Platt, J. (1968) *The Affluent Worker: Industrial Attitudes and Behaviour*. Cambridge: Cambridge University Press.

Gorz, A. (1982) *Farewell to the Working Class*. London: Pluto.

Grahber, G. (ed.) (1993) *The Embedded Firm: The Socio-Economics of Industrial Networks*. London: Routledge.

Greed, C. (1991) *Surveying Sisters: Women in a Traditional Male Profession*. London: Routledge.

Greg, P. and Wadsworth, J. (ed.) (1999) *The State of Working Britain*. Manchester: Manchester University Press.

Gregson, N. and Lowe, M. (1994) *Servicing the Middle Classes*. London: Routledge.

Grint, K. (1998) *The Sociology of Work*. Cambridge: Polity.

Halford, S., Savage, M. and Witz, A. (1997) *Gender, Careers and Organisations*. Basingstoke: Macmillan.

Halpern, R. (1996) *Meatpackers: An Oral History of Black Packing-House Workers and their Struggle for Racial and Economic Equality*. London: Prentice Hall.

Hannerz, U. (1990) 'Cosmopolitans and locals in a world culture', in M. Featherstone (ed.) *Global Culture: Nationalism, Globalization and Modernity*. London: Sage.

Hanson, S. and Pratt, G. (1995) *Gender, Work and Space*. London: Routledge.

Harmer, B. (1992) *Rivethead*. New York: Basic.

Harvey, D. (1989) *The Condition of Postmodernity*. Oxford: Blackwell.

Harvey, D. (2000) *Spaces of Hope*. Edinburgh: Edinburgh University Press.

Held, D., McGrew, A., Goldblatt, D. and Perraton, J. (1999) *Global Transformations: Politics, Economy and Culture*. Cambridge: Polity.

Herod, A. (1997) 'From a geography of labour to a labour of geography: labour's spatial fix and the geography of capitalism', *Antipode* 29: 1–31.

Hobsbawm, E. (1981) 'The forward march of labour halted?', in M. Jacques and F. Mulhearn (eds) *The Forward March of Labour halted?* London: New Left.

Hochschild, A. (1983) *The Managed Heart: Commercialization of Human Feeling*. Berkeley: University of California Press.

Hochschild, A. (1997) The Time Bind: *When Work Becomes Home and Home Becomes Work*. New York: Holt.

Hsiung, P.-C. (1996) *Living Rooms as Factories: Class, Gender and the Satellite Factory System in Taiwan*. Philadelphia: Temple University Press.

Hudson, R. (2000) *Production, Places and Environment*. London: Prentice Hall.

Jackson, C. and Pearson, R. (eds) (1998) *Feminist Visions of Development*. London: Routledge.

Kerfoot, D. and Knights, D. (1996) '"The best is yet to come?" The quest for embodiment in managerial work', in D. Collinson and J. Hearn (eds) *Men as Managers: Managers as Men*. London: Sage.

Kerfoot, D. and Knights, D. (1998) 'Managing masculinity in contemporary organizational life: a "man" agerial project', *Organization* 5: 7–26.

Klein, N. (2000) *No Logo*. London: Harper Collins.

Lamphere, L., Ragone, H. and Zavella, P. (eds) (1997) *Situated Lives: Gender and Culture in Everyday Life*. London: Routledge.

Lash, S. (1994) 'Reflexivity and its doubles: structures, aesthetics, community', in U. Beck, A. Giddens, and S. Lash (eds) *Reflexive Modernization: Politics, Tradition and Aesthetics in the Modern Social Order*. Cambridge: Polity. pp. 110–73.

Lash, S. and Urry, J. (1994) *Economies of Signs and Space*. London: Sage.

Lawless, P., Martin, R. and Hardy, S. (eds) (1998) *Unemployment and Social Exclusion: Landscapes of Labour Inequality*. London: Kingsley.

Leadbeater, C. (1999) *Living on Thin Air: The New Economy*. Harmondsworth: Penguin.

Lee, C.K. (1997) 'Factory regimes of Chinese capitalism: different cultural logics in labor control', in A. Ong and D. Nononi (eds) *Ungrounded Empires: The Cultural Politics of Modern Chinese Transnationalism*. London: Routledge. pp. 115–42.

Leidner, R. (1993) *Fast Food, Fast Talk: Service Work and the Routinization of Everyday life*. Berkeley: University of California Press.

Leslie, D. and Butz, D. (1998) '"GM Suicide": flexibility, space and the injured body', *Economic Geography* 74: 360–78.

Lewis, M. (1989) *Liar's Poker: Two Cities, True Greed*. London: Hodder and Stoughton.

Lipietz, A. (1987) *Mirages and Miracle: The Crises of Global Fordism*. London: Verso.

Marcus, G. (ed.) (1998) *Global Futures*. Chicago: University of Chicago Press.

Marshall, J. (1984) *Women Managers: Travellers in a Male World*. Chichester: Wiley.

Marshall, J. (1995) *Women Managers Moving On: Exploring Career and Life Choices*. London: Routledge.

Martin, R. (2000) 'Local labour markets: their nature, performance and regulation', in G. Clark, M. Feldman and M. Gertler (eds) *The Oxford Handbook of Economic Geography*. Oxford: Oxford University Press. pp. 455–76.

Martin, R. and Rowthorn, B. (eds) (1986) *The Geography of Deindustrialisation*. London: Macmillan.

Massey, D. (1984) *Spatial Divisions of Labour*. London: Macmillan.

Massey, D. (1995) 'Masculinity, dualisms and high technology', *Transactions of the Institute of British Geographers* 20: 487–99.

McDowell, L. (1997) *Capital Culture: Gender at Work in the City*. Oxford: Blackwell.

McDowell, L. (1999) 'Economy, culture, difference and justice', in I. Cook, D. Crouch, S. Naylor and J. Ryan (eds) *Cultural Turns/Geographical Turns: Perspectives on Cultural Geography*. London: Prentice Hall. pp. 182–95.

McDowell, L. (2000) 'Acts of memory and millennial hopes and anxieties: the awkward relationship between the economic and the cultural', *Social and Cultural Geography* 1: 15–30.

McDowell, L. and Massey, D. (1984) 'A woman's place', in D. Massey and J. Allen (eds) *Geography Matters*. Cambridge: Cambridge University Press.

Mies, M. (1986) *Patriarchy and Accumulation on a World Scale: Women in the International Division of Labour.* London: Zed.

Mizen, P., Bolton, A. and Pole, C. (1999) 'School-age workers: the paid employment of children in the UK', *Work, Employment and Society* 13: 423–38.

Momsen, J. and Kinnaird, V. (1993) *Different Voices, Different Places*. London: Routledge.

Moss, P. (1997) 'Spaces of resistance, spaces of respite: franchise housekeepers keeping house in the workplace and the home', *Gender, Place and Culture* 4: 179–96.

Nelson, M. and Smith, J. (1999) *Working Hard and Making Do: Surviving in Small Town America*. Berkeley: University of California Press.

Newman, K. (1999) *There's No Shame in My Game: The Working Poor in the Inner City*. New York: Knopf and Russell Sage Foundation.

Obelkevich, J. and Catterall, P. (eds) (1994) *Understanding Post-war Society*. London: Routledge.

Ong, A. (1987) *Spirits of Resistance and Capitalist Discipline: Factory Women in Malaysia*. New York: State University of New York.

Pahl, R. (1984) *Divisions of Labour*. Oxford: Blackwell.

Pahl, R. (1989) *On Work*. Oxford: Blackwell.

Panitch, L. and Leys, C. (eds) (2000) *Working Classes, Global Realities*. London: Merlin.

Pearson, R. and Mitter, S. (1993) 'Employment and working conditions of low skilled information processing workers in less developed countries', *International Labour Review* 132: 53–69.

Peck, J. (1996) *Work Place*. London: Guilford.

Phillips, A. (1999) *Which Equalities Matter*. Cambridge: Polity.

Pinch, S. (1993) 'Social polarisation in Britain and in the United States', *Environment and Planning A* 25: 779–96.

Piore, M. and Sabel, C. (1984) *Second Industrial Divide: Possibilities for Prosperity*. New York: Basic.

Pollert, A. (1988) 'Dismantling flexibility', *Capital and Class* 34: 42–75.

Pratt, G. (1997) 'Stereotypes and ambivalence: the construction of domestic workers in Vancouver, British Columbia', *Gender, Place and Culture* 4: 159–78.

Pringle, R. (1989) *Secretaries Talk*. London: Verso.

Pringle, R. (1998) *Sex and Medicine*. Cambridge: Cambridge University Press.

Putnam, R. (2001) *Bowling Alone*. New York: Simon and Schuster.

Redclift, N. and Sinclair, T. (1991) *Working Women: International Perspectives on Labour and Gender Ideology*. London: Routledge.

Rifkin, J. (1996) *The End of Work*. New York: Putnam.

Sassen, S. (1991) *The Global City*. Princeton: Princeton University Press.

Sayer, A. (2000) 'Critical and uncritical turns', in I. Cook, D. Crouch, S. Naylor and J. Ryan (eds) *Cultural turns/Geographical Turns: Perspectives on Cultural Geography*. Harrow: Prentice Hall. pp. 166–81.

Sayer, A. and Walker, R. (1992) *The New Social Economy: Reworking the Division of Labour*. Oxford: Blackwell.

Schoenberger, E. (1996) *The Culture of the Firm*. Oxford: Blackwell.

Segal, L. (1999) *Why Feminism?* Cambridge: Polity.

Sennett, R. (1998) *The Corrosion of Character: The Personal Consequences of Work in the New Capitalism*. New York: Norton.

Sennett, R. (1999) 'Growth and failure: the new political economy and its culture', in M. Featherstone and S. Lash (eds) *Spaces of Culture*. London: Sage. pp. 14–26.

Shiller, R. (2000) *Irrational Exuberance*. Cambridge, MA: MIT Press.

Smelser, N. and Swedberg, R. (eds) (1994) *The Handbook of Economic Sociology*. Princeton: Princeton University Press.

Standing, G. (1999) *Global Labour Flexibility*. Basingstoke: Macmillan.

Stull, D. (1997) 'Knock 'em dead: work on the killfloor of a beefpacking plant', in L. Lampere, H. Ragone and P. Zavella (eds) *Situated Lives: Gender and Culture in Everyday Life*. London: Routledge. pp. 311–36.

Summerfield, P. (1994) 'Women in Britain since 1945: companionate marriage and the double burden', in J. Obelkevich and P. Catterall (eds) *Understanding Post-war British Society*. London: Routledge. pp. 58–72.

Thompson, E.P. (1967) 'Time, work-discipline and industrial capitalism', *Past and Present* 38:

Thompson, E.P. (1980) *The Making of the English Working Class*. Harmondsworth: Penguin.

Thompson, E.P. (1991) 'Customs in Common'. London: Merlin. First published in 1967 in *Past and Present*, 38: 352–403.

Thrift, N. (2000a) 'Pandora's box? Cultural geographies of economies', in G. Clark, M. Feldman and M. Gertler (eds) *Handbook of Economic Geography*. Oxford: Oxford University Press.

Thrift, N. (2000b) 'Performing cultures in the new economy', *Annals of the Association of American Geographers* 90: 674–92.

Thrift, N. and Williams, P. (1987) *Class and Space*. London: Routledge.

Turner, B. (1996) *The Body and Society: Explorations in Social Theory*, 2nd edn. London: Sage.

Wajcman, J. (1998) *Managing Like a Man: Women and Men in Corporate Management*. Cambridge: Polity.

Walby, S. (1989) *Theorising Patriarchy*. Oxford: Blackwell.

West, J. (1982) *Women, Work and the Labour Market*. London: Routledge and Kegan Paul.

Westwood, S. (1984) *All Day, Every Day*. London: Pluto.

Williams, R. (1976) *Keywords*. London: Fontana.

Williams, R. (1981) *Culture*. London: Fontana.

Wills, J. (1998) 'Taking on the CosmoCorps? Experiments in transnational labour organisation', *Economic Geography* 74: 111–30.

Wright, S. (1994) *Anthropology of Organisation*. London: Routledge.

Yeung, H. (1998) 'The socio-spatial constitution of business organisations: a geographical perspective', *Organisation* 5: 101–28.

Yeung, H. (2000) 'Organising "the firm" in industrial geography. 1: Networks, institutions and regional development', *Progress in Human Geography* 24: 321–35.

Young, I. M. (1990) *Justice and the Politics of Difference*. Princeton: Princeton University Press.

Young, M. and Willmott, P. (1973) *The Symmetrical Family*. London: Routledge and Kegan Paul.

Zukin, S. and DiMaggio, P. (eds) (1990) *Structures of Capital: The Social Organisation of the Economy*. Cambridge: Cambridge University Press.

Zweig, F. (1961) *The Worker in an Affluent Society*. London: Heinemann.

5

Cultures of Money

Adam Tickell

For a period during the 1980s, the *New Republic*, an American liberal magazine, ran an occasional series of articles under the collective heading of 'The Money Culture'. In an idiosyncratic manner, this column charted the transformation of the imagined financier from the conservative banker in a sober suit to the brash trader wearing red suspenders. This was the era of Ivan Boesky, the market wizard who was subsequently exposed for dealing using privileged inside information, and Gordon Gecko, Boesky's fictional counterpart in the film *Wall Street*. It was also an era where governments were sweeping away the regulatory creations forged in the rubble of the 1930s financial catastrophe in the United States and the cosy social environment in the City of London. If finance was reaching a pivotal position in the popular imagination, this reflected both its (re)emergence at the core of the American and British economies and its regulatory transformation. If anything, the money culture has become more pervasive in the interim. As US stock markets powered ahead during the 1990s, television channels proliferated dedicated to charting the minutiae of price changes and giving anodyne stock tips; individual shareholding (both directly and through pension and mutual funds) expansively grew across western countries (Clark, 2000); brokerage firms set up telephone and then internet arms for individuals trading from home; and governments increasingly appeared to take heed more of the reaction of 'the markets' to their economic policies than of their electorates.

And yet, for all its pervasive influence on social life in capitalist countries, cultures of money have been of marginal interest to social scientists other than economists. Money and finance are somehow too hard, too boring, too technical for non-economists. Another, contextually richer, vein of writing on money is to be found in the mass of popular accounts of individual financial institutions or events, usually written by journalists and former and current financiers. Not only do such writers enjoy unrivalled access to their research objects (people involved in finance from floor trader to chief executive are far happier talking to the *Financial Times* or *Wall Street Journal* than some unknown academic), but they often have developed an intuitive feel for the dynamics of part of the industry. For example, Michael Lewis' beautifully written recollections of life in Salomon Brothers on Wall Street in the 1980s evoke the culture of the time and do as much to expose the gendered politics of financial traders as theoretically nuanced accounts by social scientists who (inevitably) rely upon ethnographies and interviews (Abolafia, 1996; McDowell, 1997). Similarly, whilst the collapse of Barings Bank was the subject of numerous analyses (Fay, 1996; Leeson, 1996; Tickell, 1996; 2001), the definitive account of the events was written by two *Financial Times* journalists (Gapper and Denton, 1996) who drew upon the official reports into the affair and extensive discussions with almost all of the principal participants.

All this said, anthropologists, sociologists and geographers have produced a theoretically robust and conceptually rich set of approaches that develop a distinctive cultural economic geography of money and finance. Such an approach rejects the implicit dualism which sees the cultural and the economic as analytically distinct arenas. Instead of focusing on finance as a solely economic entity, a culturally inflected analysis explores the ways that institutions, discourses, representations and symbols interact with more material processes and forms in producing and reproducing money and finance. As Mitchell, whose cultural geography remains heavily materialist, puts it, 'Cultural geography is precisely the study of how particular social relations intersect with more general processes, a study grounded in the production and reproduction of actual places, spaces and scales *and* the social structures that give those places, spaces and scales meaning' (2000: 294). A cultural geography of money, therefore, examines the processes and practices that constitute money and finance and explores the interweaving between narrative and material practice. This chapter explores the culture of finance in three ways: first, how money is culturally and economically constituted; second, how the financial industry is spatially and culturally formed; and third, how finance represents a political, cultural and economic project. In the light of this, the chapter concludes with a reassessment of the potential for recent work from within the regulation school to live up to its earlier promise of integrating economic analysis with a more robust understanding of cultural change.

THE CULTURAL GEOGRAPHY OF MONEY

Although money occupies an unhealthy and pervasive position in western societies, its very nature remains both peculiar and multifaceted. It is at once a container of value; a universal form of measure; a medium of exchange (i.e. it is the commodity by which actors exchange other commodities, obviating the need for barter); and a store of value (that is, it allows actors to retain value created in one time and space indefinitely). It is also variable: there is not one international money but a plethora of national currencies, which means that for all of its absolute qualities, money is relative and the value stored and exchanged can be 'devalued' and lost (Harvey, 1982; Swyngedouw, 1996). Further, Altvater (1993) shows that it is only possible to make sense of money when it is in motion, that is, in constant processes of circulation and transformation.

Although Ron Martin (1999) argues that Alfred Weber intended to write a companion to his volume on industrial location, which would have put money on an equal footing to industry, the first serious treatment of money and finance within geography came firmly from the political economic tradition. Marx, after all, had observed that although money was a basic unit in pre-capitalist societies, the accumulation of value through the money form lay at the heart of capitalism. As Fine and Lapavistas put it in their faithful rendition of Marx's analysis: 'Money could

penetrate pre-capitalist societies, maintaining a marginal position within them, but could also potentially exercise disruptive and antagonistic influences on the essential relations of capitalist production. In contrast, capitalism is a society premised on the creation of value (and surplus value), rendering the role of markets and money fundamental to capitalist reproduction' (2000: 367). Therefore, in his substantial re-examination of Marx's *Capital,* Harvey (1982) demonstrates the ways that money creates and transforms geographic space. As a commodity, money obscures the social relations that underlie its existence, and allows activities separated by both time and space to be linked together, contributing to the homogenization of economic spaces (Leyshon, 1996).[2]

If money in general allows for the linkage of space and time, within the past 30 years new forms of money have emerged that raise important theoretical and empirical questions about the nature of money itself and have unsettled Marxian analyses. Here, I focus on two, very different, forms of money to illustrate this: internationalized derivatives and localized currency schemes. In each case, it is possible to construct an economic rationality for their existence: financial derivatives give firms a degree of certainty in volatile economic environments, whilst local currency schemes allow individuals to engage in economic transactions while bypassing the formal economy. However, these economic rationalities have a limited purchase on the processes involved. Instead, it is important also to explore how economic relations are formed and transformed by reference to geographical, cultural and political practices, and vice versa. For both internationalized and localized money, then, a culturally sensitive analysis undermines economic determination (however final the instance may be) and illustrates that culture inflects money at all spatial scales, from the most international to the very local.

The early 1970s saw the end of an international regime – known as the Bretton Woods system – where exchange rates were largely fixed against each other and interest rates were relatively stable (Altvater, 1993; Bordo and Eichengreen, 1993; Leyshon and Tickell, 1994; Walter, 1992). Occurring contemporaneously with the collapse of Fordism (Aglietta, 1979), western economies faced significant inflationary pressures and macro-economic instability that created new uncertainties in the business environment for banks and corporations. One response to these monetary pressures was the 'invention' of financial derivatives in Chicago in 1972, a moment which heralded a significant shift in the history of financial capitalism (Tickell, 2000b).[3] Derivatives are products that allow firms to manage risk, for example, by fixing interest rates for a given period, or by agreeing a price at which foreign exchange can be bought at a particular point in the future.

Yet, while derivatives may appear simply to be an economic tool that institutions and, indeed, individuals use to manage their financial transactions, the instruments are much more than that. First, although developed as a tool for risk management and diversification on the part of individual actors, their spread and ubiquity mean that they have paradoxically increased risk in the global financial system, as they have coincided with and stimulated a risk-taking culture in financial markets. Although this culture cannot be divorced from broader changes in social attitudes towards risk, derivatives products have changed the attitudes of both financial institutions and regulators towards the nature of money. At the extreme, these changes have led to high-profile disasters such as in the collapse of Barings Bank in 1995 after a trader who bet badly in the markets began fraudulently to misrepresent and hide his losses (Gapper and Denton, 1996; Tickell, 1996). While the loss of Barings was little mourned – because the bank had been given away by the owners to a charitable foundation in the 1960s, it was hard to identify any losers from its collapse – a better indication of the pervasive ways in which the growth of derivatives signalled a broader change in the money culture was to be found in the events in municipalities in London and California. In both Orange County and Hammersmith and Fulham, local authority treasurers disastrously traded derivatives in attempts to overcome revenue shortfalls, leading to bankruptcy in California and a declaration from the courts in the UK that local authority trading for profit was illegal (and therefore that the debts that they incurred were unrecoverable). Yet, while responsibility for the losses remains with the traders concerned, I have argued elsewhere that the risk-taking behaviour reflects broader changes in the discourses and cultures of the financial community (Tickell, 1998).

Second, and more broadly, derivatives change and transform both geography and the nature of money. Leyshon (1996) shows how interest rate swaps, a relatively simple form of derivative, only exist because geographical differences in perceptions of risk exist in different financial markets. More importantly, they allow financial markets to overcome the 'problem' of the embedded nature of financial markets, that is, that local knowledges mean that financiers are better able to judge local risk and price it accordingly. This means that derivatives tend to flatten differences in prices in different financial markets. Financial markets are converging in other ways too: mergers and competitive changes mean that a smaller and smaller number of transnational financial institutions are dominating the industry; cooperation agreements between financial markets are reducing the informational and financial costs of dealing overseas; international agreements on the supervision and regulation of financial institutions are leading to an unprecedented harmonization; credit cards are becoming international money equivalents; and accounting standards are regularizing around an (Americanized) international norm (Clark et al., 2001; Previts and Merino, 1999; Tickell, 2000a). Leyshon argues that this means that 'the distinctiveness of national financial space is being eroded, reflecting the empowering of financial capital over space and the disempowering of other economic actors' (1996: 77).

Third, Pryke and Allen (2000) have recently argued that derivatives represent a new, self-referential form of money which, following Rotman (1987), has reversed the relationship between trade and finance. In constructing this claim, Pryke and Allen draw upon Simmel's (1990) account that money is more than a store of value and a means of exchange. It is simultaneously a sociological phenomenon that embodies and entails a belief in the prevailing social order and requires relations of trust rather than simply a set of atomistic market transactions. For Simmel, and perhaps most extensively developed and re-articulated by Dodd, money possesses no *intrinsic* qualities to determine how and why it is to be used:

> Money's indeterminacy is its sole distinguishing feature … wherever and whenever it is used, [money] is not defined by its properties as a material object but by symbolic qualities generically linked to the ideal of unfettered empowerment. This is an ideational feature of money and monetary transformation which, as Simmel has shown, has far-reaching cultural and economic consequences … The abstract properties of money are defined by its symbolic features. Those properties are not, however, reducible to such features. Implicit in the use of money is a set of assumptions about its re-use elsewhere and in the future. In other words, actions involving the use of money have an implicit spatial and temporal orientation. (Dodd, 1995: 152)

It follows from this that changes in money have the potential to bring about fundamental social transformations: 'the significance of money in a given period is first of all illustrated by the fact that a *change* in monetary circumstance brings about a change in the pace of life' (Simmel, 1990: 498, quoted in Pryke and Allen, 2000: 270). Pryke and Allen argue that, with their capacity to flatten space and compact time, derivatives are symbolic of a monetized world: 'At the operational heart of this money form, moreover, lies an idea of money that involves the recoding of time – space, the fostering of a new imaginary ... Derivatives emerged as a symbol of a new money culture designed to deal with this new risk awareness. Linked to the computer in such a financial world, this money form reenergized the idea of what money could do, rapidly transporting this new money sign into a growing number of everydays' (2000: 282).

Yet, this leaves us with something of a paradox. In their trenchant critiques of Richard O'Brien's (1991) thesis that finance has, at least theoretically, escaped the locational constraints of space, both Clark and O'Connor (1997) and Martin (1994) show that the overwhelming majority of financial trading remains local. Similarly, drawing upon extensive analysis of the ways in which money is actually used, Viviana Zelizer (1994; 2000) has argued that social theory has consistently been wrong in its claims that money flattens and homogenizes space. While money *is* theoretically impersonal and, once exchanged, the previous owner does not influence its use, Zelizer argues that in fact people inscribe money with values and meanings that vary over time and space. There are qualitative distinctions between different types of 'special monies': domestic money, gift money, institutional money and sacred money. For each of these, the different cultural and social settings in which they exist exert controls, restrictions and distinctions in the uses, users, allocation, regulation, sources and meanings of money (Zelizer, 1998). This means that, even as finance internationalizes,

> Money has not become the free, neutral, and dangerous destroyer of social relations. As the world becomes more complex, some things do of course standardize and globalize, but as long-distance connections proliferate, for individuals everywhere life and its choices become more, rather than less intricate. As the case of domestic money illustrates, earmarking currencies is one of the ways in which people make sense of their complicated social ties, bringing different meanings to their varied exchanges. (1998: 66)

There are two explanations for this paradox. First, as I show below, social interaction within financial markets gives local traders a greater knowledge base than external ones. Second, while the majority of trading is local, the direction of change is for greater harmonization and the flattening of economic and regulatory space.

If international finance has fostered an understanding of the money form which allows for more complex relationships with the material economy than political economic theories allow for and that change the nature of geographic space, so too does *local* finance. Local currency systems, initially developed in British Columbia in the early 1980s but which have since become popular across the US, Canada, the UK and Australia, have as their essential aim the development of a unit of exchange that is both generated and spent within a local community in an attempt to ground circuits of economic and social reproduction within a locality (Lee, 2000; Pacione, 1999). In the relatively small number of successful examples, such as in Tomkins County, Ithaca, the schemes have supported the generation of local employment and appear to be a mechanism with which to ground money, to make finance local once again. In some accounts, local currency schemes are seen as a potential force to undermine the effects of global financial flows and economic integration. Pacione, for example (see also Thorne, 1996; Williams, 1996; Williams and Windebank, 1998), has argued that although 'a local currency cannot insulate the local economy from the negative effects of globalisation ... it can afford a degree of protection against the spatially insensitive currents of the international financial system' (1999: 70). It is important, however, to take such claims with a healthy pinch of salt. Not only do local currencies remain isolated but the most successful examples have been in small, relatively affluent communities.

For Roger Lee (2000), local currencies are less economic interventions than politico-cultural ones; they are profoundly radical acts which challenge naturalized truths about the immutability of global finance. Local currencies challenge alienated, aspatial conceptions of money but they do far more than that: 'They say, simply, that resistance is possible, that truths may be reconstituted, and that alternatives might not only be envisioned but that they are accessible and may be practised in day-to-day geographies even when they are structurally "impossible" ... they are micropolitical practices which cannot be reduced either to [local] geographies or to responses to exclusion from social reproduction. They are, rather, acts of resistance to dominant discourses and relations of force' (2000: 1006; also Maurer, 2000).

We should, however, resist the temptation to see the effect of new forms of money as being entirely novel. Indeed, the acceptance of paper money had equally transformative impacts upon economic and social life in the emergent states of

North America. Emily Gilbert (1998), for example, explores the iconography of nineteenth-century Canadian banknotes in an attempt to reconcile what she identifies as money's simultaneous existence as both a symbol and a thing (Ganssman, 1988). Gilbert shows how paper money moved from being a distrusted commodity to being a taken-for-granted equivalence, and how the images and symbols chosen underwrote this process:

> The iconography of banknotes performs a kind of alchemy, transforming commodities or values into their equivalents, and investing in pieces of paper values that it did not possess of its own accord. Exploring the images of paper money with reference to the social and cultural practices in which these notes are exchanged illustrates the ways in which money involves a displacement of values, not only … of economic values, but of social and cultural values. (1998: 76; see also Helleiner, 1999)

Just as new forms of money in the 1980s and 1990s led to a reconfiguration of space, in Gilbert's account, so the emergence of paper money in Canada both situated colonists within personal, national and imperial geographies and reflected the economic and cultural practices that specified the colonial space.

It is clear, then, that an understanding of the geography of money is transformed with a sensitivity to its cultural constitution. At all spatial scales, the interplay between cultural and economic processes problematizes accounts that simply see money as a unit of exchange. However, culturally sensitive accounts have had a greater impact on recent analyses of the financial sector, and it is to this that I now turn.

THE CULTURAL ECONOMY OF FINANCE

At one level, as Clark (1998) argues, financial markets are rational, functional networks of relationships and transactions that are integrated by information channels. In other words, financial markets correspond with the entities beloved of neoclassical economists that reflect and show rationally derived prices for commodities. While it has been the dominant academic discourse on finance, mainstream economics has been reluctant to discard models based on rational economic actors in order to understand the dynamics of financial markets. Yet, as financial markets are networks of real people in real places, they are as subject to the vicissitudes of human behaviour as any other sphere. Consider the example of small

speculators who routinely and annually lose approximately 20 per cent of their stake in futures market trading but who continue to trade (Zeckhauser et al., 1991). In attempting to explain the dynamics of financial markets more effectively than utility maximizing models, behavioural economists (Thaler, 1993; Tversky and Kahneman, 1981) argue that a marriage between economics and psychology is necessary: 'research on individual decision-making is highly relevant to economics whenever predicting (rather than proscribing) behavior is the goal. The notion that individual irrationalities will disappear in the aggregate must be rejected' (Russell and Thaler, 1985: 1080–1). Empirically, behavioural economists have shown how even the most sophisticated and professional financial traders make illusory correlations, believe that unusual and unsustainable trends are likely to last indefinitely, and place too much emphasis on recent events. Similarly, O'Barr and Conley's (1992) analysis of pension fund managers concluded that each pension fund has a unique culture and argued that the stories told by the managers are akin to creation myths identified by classical anthropology. This culture creates common understandings of how financial markets work, and of how economic data should be interpreted which, in turn, drives investment decisions. Furthermore, even if financial professionals believe that, for example, the stock prices are unsustainably high, financial market dynamics may exert powerful disciplinary pressure.

One of the few mainstream economists to take ideas of behaviour and culture seriously is Robert Shiller, whose *Irrational Exuberance* (2000) powerfully demonstrated that stock values during the late 1990s relied as much on investor sentiment and herding behaviour as they did on any 'objective' analysis of underlying market value:

> The market is high because of the combined effects of indifferent thinking by millions of people, very few of whom felt the need to perform careful research on the long term investment value of the aggregate stock market, and who are motivated substantially by their own emotions, random attentions and precepts of conventional wisdom. (2000: 203; see also Shiller, 1997)

While the subsequent collapse of the dot.com boom may retrospectively undermine the prescience of Shiller's analysis, mainstream economists had completely forgotten the lessons learnt during the decade following the 1929 Wall Street Crash (Galbraith, 1975; Kindleberger, 1978) and were confidently predicting that US stock prices could yet triple beyond their historic highs (for example, Glassman and Hassett, 1999).[4] Even though behavioural finance remains on the relative margins of financial economics – hostility to theories of finance that are not easily quantifiable or do not fit into conventional utility models of human behaviour continues (for example, Rubinstein, 2001) – it nevertheless remains significant that economists are taking seriously the idea that humans may behave in economically irrational ways.[5] However, as Froud et al. argue, these gains are still modest in that they represent 'only the substitution of one scientism for another within a sub-discipline which continues to be narrowly preoccupied with explaining stock prices and remains largely indifferent to social and economic context' (2000: 69).

An alternative response to neoclassical approaches is to understand financial markets as embedded, as spatially bounded entities (Fligstein, 1990; 1996; 2001). As Mitchel Abolafia argues, finance, 'is not a world that can be explained in terms of individual *homines economic*, independently maximizing their utility. It is also not a world of unbridled competitive abandon, but rather a world in which powerful actors create systems to restrain themselves and others' (1996: 12). In a careful ethnography of some of the different markets collectively known as Wall Street, including bond markets, futures markets and the New York Stock Exchange, Abolafia argues that the nature of financial markets creates particular cultures which inscribe and delimit the strategies of market actors. In the bond markets, for example, Abolafia identifies deceptive and opportunistic traits which are a broadly accepted feature, exacerbated by information overload, unrestrained as a consequence of regulatory liberalism and underwritten by employing institutions which expect traders to be self-reliant, calculated risk-takers who value, above all else, money:

> Money is more than just the medium of exchange; it is a measure of one's 'winnings'. It provides an identity that prevails over charisma, physical attractiveness, or sociability as the arbiter of success and power on the bond-trading floor. The top earning trader is king of the mountain. (1996: 30)

Yet, financial markets are not anarchic and uncontrolled: they are frequently high-trust environments with mechanisms of self-regulation that can, but do not *necessarily*, underwrite probity. Simply speaking, the long-term viability of financial markets requires that behaviour which undermines market integrity (whether this is legal or not) must be restrained. Some of this restraint will be provided by formal state regulation but, as Abolafia (1996) stresses, trader behaviour is also controlled by internal self-interest (while opportunistic behaviour may be acceptable, in markets where such opportunism is not culturally approved it is likely to be a self-limiting activity). Such internal controls exist within a set of formal and informal ritualized social arrangements, such as the setting of standards or the monitoring of aberrant behaviours.

Yet there have, of course, been many cases where systematic and deliberate fraud has undermined the integrity of financial firms and markets and, *in extremis*, the integrity of the financial system (for example, Tickell, 1996; 1998). When an individual crisis hits prominence, media accounts and official enquiries often foreground the failings of individuals, underestimating the wider causes of failure. As Passos argues,

> accounts labelled as 'conspiracy theories' – even if true – are easily discredited in public discourse, become fictionalised in commercial books and thus have no real impact … They [also] imply a bad apple theory, consistent with the … culture of individualism and attribution of both success and failure to particular people – which clouds the systemic risk of similar disasters in future. So if we catch the bad guys the problem is considered solved and structural conflicts are overlooked. (1996: 810)

For Stanley, corrupt financial cultures are not some aberration to be explained by reference to the failings of individuals or firms, they are written into the code of the new financial model. In the case of fraudulent activity on the part of a British investment bank, it was not only that investment bankers broke the law, or that their managers did not understand the new environment in which they were working. More significantly, the regulatory authorities' behaviour gave 'rise to a suspicion of collusion: not so much a design weakness but a weakness by design … The erosion of the boundaries between legality and illegality in terms of financial transactions [is related] to the economic aspirations of neo-liberalism and the imperative of deregulation within a strong state' (1996: 93; see also Tickell, 1996; 2001).

Making the transition from showing that financial markets are culturally constituted to demonstrating that this makes a difference to, for example, the price and efficiency of these financial markets is a complex and underdeveloped task. As Muniesa puts it: 'contemporary financial markets appear to be … the best place to try to discuss the relevance of a sociological analysis. How [then] to deal with the hard content of those markets without doing economics?' Furthermore, theories of financial market cultures overwhelmingly explore the ways in which cultures develop in face-to-face environments. The mediating role of traders is, however, under pressure from the technological transformations sweeping the sector, and floor traders are a historical institution in most of the main financial markets in Europe and increasingly too in North America. For economists, this means that markets are becoming more efficient and more rational without intermediaries to provide distortions (for example, McAndrews and Stefanadis, 2000).

On closer inspection, however, although auto-mated exchanges do change the forms of inter-action in financial markets, Fabian Muniesa's (2001) careful examination of trading algorithms shows that they are negotiated between market administrators, traders and firms in the light of their pre-existing views of the world and compu-tational constraints, economists' theoretical models and statistical evidence. As such, Muniesa argues that even this most transparent and efficient of all economic markets is per-formed and, following Callon (1998), framed. Muniesa's analysis of finance here is an impor-tant contribution because he shows that the market is transformed by these frames. Appar-ently neutral algorithms – which are the outcomes of negotiations between key actors – are used by traders seeking the 'best price' in order to inform their decisions. These decisions in turn play a part in reconstituting the frame. The prices of financial products are the outcome not only of the processes of supply and demand but also of the parameters built into the frame. Economic rationality and even the transparency of financial markets, then, are attributes of rules, protocols and frames rather than reflecting relatively simple economic 'laws'. Therefore,

> The complexity and heterogeneity of trading architec-tures show how 'market behaviour' cannot be reduced to a schematised version of what traders have got in their heads. It has to deal also with engineering, knowl-edge and architectural frameworks ... Prices are then 'performed' within this frame: they are the result of translations, negotiations and efforts of all kinds that give them their specific form ... Trading architectures perform economic categories. (Muniesa, 2001: 290)

Paradoxically, however, given the emphasis in sociological and anthropological research on over-coming the shortcomings of neoclassicist appro-aches to finance, Preda claims that the collective impact of emphasizing networks, trading commu-nities and framing means that the 'human' aspect of financial markets has been neglected:

> financial sociologists, lured by the suave, sophisticated smell of the Eau d'ANT [actant network theory]

manufactured in Paris ... have given plenty of attention to the trading floors' cognitive processes and epistemic arrangements. But, somehow, miraculously, human actors have been lost on the way. Setting the focus on the implementation and working of exchange algo-rithms has somehow led to brushing aside the role of gossip, clique-building, asymmetric information and personal connections which, alas are only too human. (2001: 17)

While Preda's claim is rhetorically strong, it is self-consciously overemphatic (Abolafia, 1996; Boden, 1993; Boden and Molotch, 1993; French, 2000). The geographically constituted nature of financial centres illustrates this. In his various analyses of international financial centres, and in particular the City of London, Nigel Thrift (1987; 1994; 2001; Amin and Thrift, 1992; Leyshon and Thrift, 1997; Thrift and Leyshon, 1992) shows how financial markets have cultures based upon information, expertise and contacts (see Davis and Greve, 1997; Granovetter, 1985). First, international financial centres are centres of representation for the global financial services industry where research, analysis and informa-tion processing occur; second, as they occupy a privileged position in the global financial knowl-edge structure, international financial centres are where new products are created, tested and trans-mitted; and third, such centres are loci of social interaction, even in a technologically advanced industry, which underscores the dynamism and trust environments necessary for fluid financial markets.

However, while financial centres may rely upon trust cultures and embedded knowledges, the overwhelming hallmark of the financial market culture is its dynamism and adaptability. Furthermore, while financial markets may be capable of some limited self-regulation, more than in most other socio-economic formations real government intervention, or the threat of it, is a persistent feature. The recent history of the City of London is instructive. Throughout the nineteenth and most of the twentieth century the City operated as something of a 'club' where many of the 'rules' of the international financial system were set and policed. This was a com-munity sustained by an unrivalled knowledge structure (Thrift, 1994), and close-knit ties between individuals maintained the City's coher-ence (Courtney and Thompson, 1995). Until the internationalization and transformation of finance in the 1970s and, particularly, the 1980s, the City's community was drawn from a narrow, upper-class social stratum which both 'strove towards endless expansion ... so as to gain com-petitive advantage over rivals, and ... tried to enlist non-economic power to regulate the system and to give monopolistic advantages to members' (Amin and Thrift, 1992: 581). The regulatory analogue to this community was the Bank of England, a central bank that retained considerable operational autonomy even after it was half-heartedly nationalized in 1946. For the majority of the twentieth century, the Bank of England's approach reflected its social and cultural affinities with the community it supervised, and, at key moments, it was difficult to determine whether the Bank operated as the government's enforcer within the City or the City's representative in government. Certainly, the Bank encouraged informal, uncodified regulation and self-regulation on the part of a coherent group of bankers (Moran, 1991; Tickell, 2001).[6] During this period, the regulation of Britain's banking community rested upon strong social ties, cultural understandings of acceptable behaviour and the City's monopolistic and competitive advantages. Critically, this regulatory form was strongly spatially constituted: banks were all headquartered within a tightly defined geographical space in the City proscribed by the Bank of England (Amin and Thrift, 1992; Davis and Greve, 1997).

The City of London today is a very different environment to its nineteenth-century predecessor. The informal, culturally specific regulatory approach of the Bank of England has been replaced by a more codified, rules-based system (Tickell, 2001), traditional social structures have been replaced by more reflexive and culturally diverse groups of professionals, and trust structures have been reconstituted through relationships. As Thrift puts it: 'The formal gavottes of the Old City have therefore become much more complicated dances: "the first thing is self"' (1994: 348). Yet, the City remains one of the world's three most important financial centres (alongside New York and Tokyo) because, as Amin and Thrift (1992; Leyshon and Thrift, 1997; Thrift, 1994) point out, the City has been able to adapt to, and thrive on, change.

For all the theoretical sophistication of recent sociological and anthropological literature on finance, analysis of the cultural constitution of finance remains largely gender blind (exceptions being, for example, Halford and Savage, 1995; Halford et al., 1997). Yet, this is an industry where homosocial trust environments have traditionally been *male* environments. Indeed, the very language of finance is masculinist: successful traders at Salomon Brothers during the 1980s were given the honorific 'Big Swinging Dick' (Lewis, 1989), while the Bank of England's nickname, 'The Old Lady of Threadneedle

Street', invokes not only the antiquity of the institution but also the spinster aunt so beloved and derided by the English upper classes. Nevertheless, as Thrift (1994) points out, it is important to recognize that just as the old homosocial City is being replaced by a newly reflexive managerial and professional cadre, so too are the old proscriptions on women working in senior positions. At one level, the uniquely male financial environments in London, New York, Chicago and so on are being undermined by the twin pressures of meritocracy and litigation. For all of its institutionalized sexism, finance is an intensely competitive business where many of the old ties which bound it are breaking down.[7]

The most substantial geographic treatment of the gendered nature of finance is McDowell's (1997; 2001; McDowell and Court, 1994a; 1994b; 1994c; see also Jones, 1998) analysis of the City of London. In the City, McDowell argues, women are underrepresented at senior levels, less well paid than men for substantively equivalent jobs, and frequently socially and occupationally excluded. However, this is not the outcome of simple, deliberate sexism where men actively seek to exclude women from the job (although there are vestigial remains of an era where such behaviour was the norm), and our explanations must reflect the complexity of the process. Three sets of literature illuminate the account. First, Butler (1987) elaborates the ways in which constantly repeated acts congeal over time to naturalize and construct a learnt and performed gender (which can, nevertheless, be subverted). Second, accounts of the relationship between masculinity and forms of power (Connell, 1987; Roper, 1994) argue that there are different masculinities which embody relations of domination, alliance and subordination. Finally, there is work on the economic sociology of embeddedness (Granovetter, 1985; Zukin and DiMaggio, 1990). These literatures allow McDowell to theorize gender in the *workplace* as being a performed activity, within a broader context where *investment banking* has forms of hegemonic masculinity (including long-standing paternalistic masculinities and aggressive trading cultures which subordinate both femininities and alternative masculinities). This contributes to a situation where women's bodily appearances clash with the dominant work culture, transgressing accepted organizational norms. Consequently, women's subordinate position within the City reflects embedded social actions and rationalities, dominant masculinities within specific institutional contexts (different parts of a bank and different banks) and performative actions.

THE DISCURSIVITY OF FINANCE

Critical to understanding the cultural geographies of finance is a recognition of interpretive power struggles, where different sets of scripts and discourses conjure up different economic worlds. This is clearly evident in social scientific academic accounts of financial phenomena (see Leyshon and Tickell's 1994 unpacking of alternative historical geographies and political economies of the post-Second World War financial order). It is also apparent in the ways in which economists' analyses spill over into the object of their study: economists are not some neutral arbiters or dispassionate modellers of the system. Just as economists in general have framed and transformed both the *analysis* of the economy and its real *mechanisms* (for example, Callon, 1998; McCloskey, 1985),[8] so too have they altered the terrain of contemporary financial markets. Economic theories and practices have transformed the derivatives markets (Bernstein, 1993), and the events at Long Term Capital Management (LTCM) in 1998 provide a perfect illustration of the interrelationship between the framers of financial architectures and the architecture itself. LTCM was set up as a hedge fund that exploited marginal differences in prices in different financial markets with an aggressive use of leverage (or borrowed money). During August 1998, companies trading with LTCM took fright after Russia announced a partial default of some of its debts, precipitating a run on LTCM and fears that the integrity of the global financial system would be undermined (see, for example, de Goede, 2001; President's Working Group on Financial Markets, 1999). Much of the mainstream commentary on the collapse of LTCM emphasized that the company had been founded by two of the three Nobel economists who had devised the original pricing model for derivatives, suggesting that economists should refrain from engaging with the practices they describe. However, MacKenzie's analysis shows that the relationship between finance and finance theory is an evolving and codetermining one:

> The dominant tendency, over the last thirty years, of ... the 'financial innovation spiral' has been to increase the truth of finance theory's typical assumptions ... LTCM's fate has provoked some anti-intellectual nonsense. Mathematical finance is part of the modern world. The techniques developed out of the research of Black, Scholes and Merton continue to work perfectly well in millions of transactions daily, and their abandonment would be unthinkable folly. Yet we must also remember that finance theory describes not a state of nature but a world of human activity, of beliefs and of institutions. Markets, despite their thing-like character, their global reach, and their huge volumes, remain social constructs, and the feedback loops that constitute them are intricate, knotted and still far from completely understood. (2000: 1, 5)

Finance, then, has the capacity to act upon discursive constructs with a speed and efficacy that academics find it difficult to comprehend.

An intriguing and sophisticated account of the interplay between the cultures, economics and spaces of finance is found in Anna Tsing's (1999) exploration of the events surrounding Bre-X. This was a Canadian gold mining company that claimed to have found significant deposits of gold in a 'new' area of Indonesia. The 'discovery' had a series of major impacts: the destruction of protected forests; claims and counterclaims as to whom the land on which the discovery was found actually belonged; corrupt relationships with government officials; and soaring share values for the company (which rose from 51 Canadian cents in 1993 to C\$286.5 by May 1996). By early 1997 the company's geologist had mysteriously fallen from a helicopter (his body was never found, and it was never clear whether the fall was an accident, suicide or murder) and the company's gold discovery was shown to be illusory. Tsing shows how this story allows us to expose a series of projects embedded within the discourses of global finance (see also de Goede, 2001; Tickell, 2000a). While many understandings of finance are that it has become detached from the material world (money moves instantaneously as bits of information) and from geographic space (finance is globalized, trading occurs 'around the clock'), Tsing argues that these undervalue the complexity of what is going on. On the one hand, the apparent disjuncture between money as sign and the 'real' world of money is a reflection of a 'conjuring aspect' of finance, which makes things appear to be real. On the other, finance is both performative and discursive in nature:

> The conjuring aspect of finance interrupts our expectations that finance can and has spread everywhere, for it can only spread as far as its own magic. In its dramatic performances, circulating finance reveals itself as both empowered and limited by its cultural specificity. Contemporary masters of finance claim not only universal appeal but also a global scale of deployment. What are we to make of these globalist claims, with their millennial whispers of a more total and hegemonic worldmaking than we have ever known? Neither false ideology nor obvious truth, it seems to me that the globalist claims of finance are also a kind of conjuring of a dramatic performance. In these times of heightened

attention to the space and scale of human undertakings, economic projects cannot limit themselves to conjuring at different scales – they must conjure the scales themselves. In this sense, a project that makes us imagine globality in order to see how it might succeed is one kind of 'scale-making project' … By letting the global appear homogeneous, we open the door to its predictability and evolutionary status as the latest stage in macronarratives. (Tsing, 1999: 119)

The appearance of homogeneity is underwritten by the ways in which international investment works, obscuring the ability of 'investors' (often simply a synonym for short-term speculators) to distinguish between companies with long-term potential and 'those that are merely good at being on stage' (1999: 127). While finance may appear to be global, particular investments embody strategies at the scales of the global (finance capital); the national state (in the case of Bre-X, corruption on the part of both Canadian executives and local officials); and the region (the redefinition of forest lands with an existing population as the frontier, in much the same way as the American west became an unpeopled frontier, in the nineteenth century). These are scalar projects which become tangentially linked at particular moments and whose copresence is strengthening. Yet, each of these projects should be recognized as unpredictable and specific rather than inevitable and ubiquitous:

the national specificity of attraction to investments disappears in the excitement of commitments to globalism in the financial world. When one thinks about finance in the Bre-X case, there was nothing worldwide about it at all; it was Canadian and US investment in Indonesia. Yet it is easy to assimilate this specific trajectory of investment to an imagined globalism to the extent that the global is defined as the opening-up process in which remote places submit to foreign finance. Every time finance finds a new site of engagement, we think the world is getting more global. In this act of conjuring, *global* becomes the process of finding new sites. (1999: 142)[9]

CONCLUSIONS

Research on the cultures of money has flowered at precisely the moment when finance has reached supremacy in the material and discursive constitution of Anglo-American capitalisms, and the developing literatures from France and Germany are indicative of the emerging place of finance in those countries too. It has also occurred during a period when the economic as an object of analysis has become subsumed within the majority of the social sciences.

Explanations for this are not hard to find: much orthodox economics and Marxism treated culture as an irrelevance; economic rationalities do not approach an adequate understanding of lives and economic practices which are socially and culturally constituted and embedded; left social scientific enquiry falls on stony ground in a world where economic theories remain dominated by liberalism; 'the study of economics has become devalued in the sense that moral values have been expelled from consideration' (Sayer, 1999: 54); and so on. The result, however, is that something of an imbalance has arisen, somewhat acerbically summed up by Thrift's claim that, '"Cultural" analysis has become more and more sophisticated but it is mixed in with a level of "economic" analysis which rarely rises above that of anyone who can read a newspaper' (1999: 35). And yet money is not just another commodity to be analysed, it is the essence of contemporary capitalism. It is, as Swyngedouw polemicizes,

one of the most powerful signs in a world of almost complete commodification … But surely, it is not just a sign and a metaphor ready for deconstructive enquiry. Money incorporates also, and arguably foremost, direct bodily power. Starvation in Sudan or in the homeless shelter of London's South Bank, the plight of the unemployed or the summits of economic and political power show the powers of money in their most repressive, violent, subordinating or, as the case may be, empowering and emancipatory capabilities. (1996: 138; see also McMurtry, 1999)

This means that the most effective cultural geographies of money need to jettison unhelpful distinctions between the separate spheres of economy and culture, and develop a more sophisticated approach to the integration of cultural and economic explanations. One possible avenue is to develop the research project of the regulation school of economists. Regulation theory now occupies a footnote in the recent history of geography but was originally developed by formerly Althusserian economists (notably, Aglietta, 1979; Lipietz, 1983) as a means of integrating macro-economic analysis and a sensitivity to macro-social and macro-cultural forms in order to explain the medium-run stability of capitalism in the light of the mode's severe tendencies towards crisis and instability. The paradigmatic example of regulation school analysis was of the period after 1945 when mass production, mass consumption, largely nationally oriented economic systems, the emergence of mass consumer finance, household structures, Taylorism, the gender division of labour and a (limited) welfare safety net coalesced in the form of the long Fordist – Keynesian boom

(Aglietta, 1979; Jessop, 1995; Tickell and Peck, 1992). In its most sophisticated variants, then, regulation theory was always sensitive to society and culture.

As regulation theory was adopted and developed by social scientists beyond the original group, it tended to focus on the state, the spatialization of the economic and – in cruder variants which did much to discredit the approach – the supposed 'post-Fordist' mode of growth. A re-engagement of culturally sensitive analysis with regulation theory would build upon a dialogue that Michel Aglietta has been engaged in with anthropologists since the early 1980s (Aglietta and Orléan, 1982; 1998; see Grahl, 2000, for an English language review). Aglietta's account is distinctive because not only is it economically literate and resistant to liberal fallacies, but he embeds his theories within the political, social and cultural and vice versa: 'Far from being an appendix of the real economy, finance is the nervous system of the economy as a whole' (1988: 113; quoted in Grahl, 2000). In recognition of this, some of the original architects of regulation theory have begun to explore the capacity for a new economic-social model to coalesce and cohere: that of a financialized mode of capitalism (for example, Aglietta, 1998; Boyer, 2000a; 2000b; see also Froud et al., 2000). This entails subjecting to critical enquiry the micro-, meso- and macro-economic and cultural changes where finance has appeared to impose its logic. Can we, for example, talk of a financialized mode of growth? Is such a mode economically and/or politically sustainable? How do imperious cultural norms stabilize and transform finance?

Such a venture is not unproblematic. While regulation theory has frequently been misread as both crudely economically determinist and simply a theory of Fordism, it is true that most regulationists implicitly adhered to a belief that economic processes should ultimately subordinate others (Gibson-Graham, 1996). Furthermore, analyses based upon identifying coherence over years and decades run the risk of crudely imposing form where there is none and emphasizing difference rather than continuity (Thrift, 1989). And yet, just because a venture is risky and potentially problematic does not mean that it should not be attempted. While ultimately extant regulationist accounts of financialization are partial, as with their earlier accounts of Fordism they do provide a useful reminder that the economic, the social and the cultural are intertwined. For example, a regulationist approach may note the widespread and growing tendency in western capitalist countries for the logics of finance to penetrate social life and subject it to critical interrogation as to the economic sustainability of the process. Yet, for this to develop force, it would also need to explore the cultural specificity of the process (the form of financialization is different in different geographical contexts, and there are varying degrees of resistance to the supremacy of finance); the ways in which the framing of the rules at the micro level cascades through the economy; and the ways that narratives construct and naturalize the place of finance. Therefore, this should not simply be a regulation theory which visits the twenty-first century through the lens of Fordism, but one which attempts a coherent integration of cultural and economic analysis, drawing upon Boltanski and Chiapello (1999) whose explanation of cultural and economic change argues that inherent in the creative, Schumpeterian, nature of capitalism is a capacity to feed off its critics: 'The main agent in the creation and transformation of the spirit of capitalism is its critique' (1999: 555; see also Guilhot, 2000; and Tilly, 1999).

NOTES

I would like to thank my section editor, Trevor Barnes, for his forbearance at the extreme delay in finishing this chapter and the adept way that he dealt with the first draft. Thanks too to ESRC for their support for some of my work on finance through the research grants 'Regulating finance: the political geography of financial services' and 'Hard borders and soft geographies: reregulating global finance' (with Gordon Clark). The contents, of course, remain my responsibility.

1　This is written with a degree of envy by someone who tried – and failed – to do the same.

2　In some cruder political economic accounts, global financial institutions directly correspond to Marx's dictum that the modern state is 'but the executive committee of the bourgeoisie'. Van der Pijl (1998), for example, sees the IMF and the OECD simply and largely unproblematically as tools of international financiers.

3　Derivatives have existed for agricultural products for over 400 years, although they became formalized in Chicago in 1848.

4　Indeed, in spring 2001 I interviewed one of the authors of *Dow 36,000* who remained convinced of the utility of the analysis and insisted that the 'correction' during 2000 and 2001 was an aberration along the way to higher share values.

5　Like all good economists, Thaler and his colleagues have set up a fund management company which incorporates their theories of investor behaviour (see www.fullerthaler.com).

6　This has strong resonances with the insights of Fligstein (1990; 2001), who argues that social relations within and between firms and their relationships with the state are pivotal to understanding how stable markets emerge.

7 Few people even refer to the Bank of England as the 'Old Lady' any longer.

8 This is a two-way process. The language of 'moral hazard', a theory particularly beloved of neoliberal economists and politicians which posits that the effect of government action to rescue bankrupt institutions is to encourage irresponsible behaviour on the part of individuals, developed in the USA in the nineteenth century in response to a perception that new forms of insurance threatened public morality: 'The rhetoric of moral hazard permitted the insurance men to deny that insurance broke with conventional morality, and to believe their denial, even as the enterprise they built traveled down the road toward the abandonment of morality in favor of a populational, actuarial understanding of that world' (Baker, 1996: 260; although subsequently Baker, 2000, modifies his claim that conventional morality was abandoned).

9 Nigel Thrift has similarly subjected the notion of the 'new economy' to critical interrogation, and argues that it is a constructed entity rather than something which has emerged as a result of some overarching economic rationality: 'by the mid 1990s, the new economy had already become a stable rhetorical form, in common usage in business and government, and seeping into popular culture. In effect the new economy had become a kind of *brand*, compounding in one phrase the attractions and rewards of a new version of capitalism' (2001: 415). The new economy is viewed as a form of ramp created by five sets of stakeholders: the cultural circuit of capital (the machine for producing and disseminating knowledge to business elites); governments; non-business-school academics ('Economists, in other words, began to produce a formal body of knowledge which could act as serious confirmation of more general (and rather flighty) business knowledge', 2001: 415); managers; and information and communications technology which 'has now reached the point where it can be counted as having its own agency, of a sort'.

REFERENCES

Abolafia, M. (1996) *Making Markets: Opportunism and Restraint on Wall Street.* Cambridge, MA: Harvard University Press.

Aglietta, M. (1979) *A Theory of Capitalist Regulation.* London: Verso.

Aglietta, M. (1988) 'L'ambivalence de l'argent', *Revue Français d'Economie* 3 (3): 87–133.

Aglietta, M. (1998) 'Capitalism at the turn of the century: regulation theory and the challenge of social change', *New Left Review* 232: 41–90.

Aglietta, M. and Orléan, A. (1982) *La Violence de la monnaie.* Paris: PUF.

Aglietta, M. and Orléan, A. (eds) (1998) *La Monnaie souveraine.* Paris: Odile Jacob.

Altvater, E. (1993) *The Future of the Market: an Essay on the Regulation of Money and Nature After the Collapse of 'Actually Existing Socialism'.* London: Verso.

Amin, A. and Thrift, N.J. (1992) 'Neo-Marshallian nodes in global networks', *International Journal of Urban and Regional Research* 16: 571–87.

Baker, T. (1996) 'On the genealogy of moral hazard', *Texas Law Review* 75: 237–92.

Baker, T. (2000) 'Insuring morality', *Economy and Society* 29: 559–77.

Bernstein, P. (1993) *Capital Ideas: The Improbable Origins of Wall Street.* New York; Free Press.

Boden, D. (1993) *The Business of Talk.* Cambridge: Polity.

Boden, D. and Molotch, H. (1993) 'The compulsion of proximity', in R. Friedland and D. Boden (eds) *Now/Here: Time, Space and Modernity.* Berkeley: University of California Press.

Boltanski, L. and Chiapello, E. (1999) *Le Nouvel Esprit du capitalisme.* Paris: Gallimard.

Bordo, M.D. and Eichengreen, B. (eds) (1993) *A Retrospective on the Bretton Woods System: Lessons for International Monetary Reform.* Chicago: Chicago University Press.

Boyer, R. (2000a) 'Is a finance led growth regime a viable alternative to Fordism? A preliminary analysis', *Economy and Society* 29: 111–45.

Boyer, R. (2000b) 'The political in the era of globalization and finance: focus on some Régulation School research', *International Journal of Urban and Regional Research* 24: 274–322.

Butler, J. (1987) *Bodies that Matter.* London: Routledge.

Callon, M. (1998) 'Introduction: the embeddedness of economic markets in economics', in M. Callon (ed.) *The Laws of the Markets.* Oxford: Blackwell/ Sociological Review. pp. 1–57.

Clark, G.L. (1998) 'Rogues and regulation in global finance: Maxwell, Leeson and the City of London', *Regional Studies* 31: 221–36.

Clark, G.L. (2000) *Pension Fund Capitalism.* Oxford: Oxford University Press.

Clark, G. and O'Connor, K. (1997) 'The informational content of financial products and the spatial structure of the global finance industry', in K. Cox (ed.) *Spaces of Globalization.* New York: Guilford.

Clark, G.L., Mansfield, D. and Tickell, A. (2001) 'Emergent frameworks in global finance: accounting standards and German supplementary pensions', *Economic Geography*, 77 (3): 250–71.

Connell, R.W. (1987) *Gender and Power: Society, the Person and Sexual Politics.* Cambridge: Polity.

Courtney, C. and Thompson, P. (1995) *City Lives: The Changing Voices of British Finance.* London: Methuen.

Davis, G.F. and Greve, H.R. (1997) 'Corporate elite networks and governance changes in the 1990s', *American Journal of Sociology* 103: 1–37.

de Goede, M. (2001) 'Discourses of scientific finance and the failure of Long Term Capital Management', *New Political Economy* 6: 149–70.

Dodd, N. (1995) *The Sociology of Money.* Cambridge: Polity.

Fay, S. (1996) *The Collapse of Barings*. London: Richard Cohen Books.

Fine, B. and Lapavistas, C. (2000) 'Markets and money in social theory: what role for economics?', *Economy and Society* 29: 357–82.

Fligstein, N. (1990) *The Transformation of Corporate Control*. Cambridge, MA: Harvard University Press.

Fligstein, N. (1996) 'Markets as politics: a politico-cultural approach to market institutions', *American Sociological Review* 61: 656–73.

Fligstein, N. (2001) *The Architecture of Markets*. Princeton: Princeton University Press.

French, S. (2000) 'Re-scaling the economic geography of knowledge and information: constructing life assurance markets', *Geoforum* 31: 101–19.

Froud, J., Haslam, C., Johal, S., Leaver, A., Williams, J. and Williams, K. (2000) 'Shareholder value and finan-cialisation: consultancy promises, management moves', *Economy and Society* 29: 13–27.

Froud, J., Johal, S., Haslam, C. and Williams, K. (2001) 'Accumulation under conditions of inequality', *Review of International Political Economy* 8: 66–95.

Galbraith, J.K. (1975) *The Great Crash of 1929* London: Penguin.

Ganssmann, H. (1988) 'Money – a symbolically general-ized medium of communication? On the concept of money in modern sociology', *Economy and Society* 17: 285–316.

Gapper, J. and Denton, N. (1996) *All That Glitters: The Fall of Barings*. London: Hamilton.

Gibson-Graham, J.K. (1996) *The End of Capitalism (As We Knew It): A Feminist Critique of Political Economy*. Oxford: Blackwell.

Gilbert, E. (1998) '"Ornamenting the façade of hell": icono-graphies of 19th-century Canadian paper money', *Envi-ronment and Planning D: Society and Space* 16: 57–80.

Glassman, J. and Hassett, K. (1999) *Dow 36,000*. New York: Random House.

Grahl, J. (2000) 'Money as sovereignty: the economics of Michel Aglietta', *New Political Economy* 5: 291–316.

Granovetter, M. (1985) 'Economic action and social structure: the problem of embeddedness', *American Journal of Sociology* 91: 481–510.

Guilhot, N. (2000) 'Review of L. Boltanski and E. Chiapello *Le Nouvel Esprit du capitalisme*', *European Journal of Social Theory* 3: 355–64.

Halford, S. and Savage, M. (1995) 'Restructuring organi-sations, changing people: gender and restructuring in banking and local government', *Work, Employment and Society* 9: 97–122.

Halford, S., Savage, M. and Witz, A. (1997) *Gender, Careers and Organisations: Current Developments in Banking, Nursing and Local Government*. Basingstoke: Macmillan.

Harvey, D. (1982) *The Limits to Capital*. Oxford: Blackwell.

Helleiner, E. (1999) 'Historicising territorial currencies: monetary space and the nation-state in North America', *Political Geography* 18: 309–39.

Jessop, B. (1995) 'The regulation approach and gover-nance theory: alternative perspectives on economic and political change', *Economy and Society* 24: 307–33.

Jones, A. (1998) '(Re)producing gender cultures: theorizing gender in investment banking recruitment', *Geoforum* 29: 451–74.

Kindleberger, C. (1978) *Manias, Panics and Crashes* New York: Basic.

Lee, R. (2000) 'Radical and postmodern? Power, social relations and regimes of truth in the social construction of alternative economic geographies', *Environment and Planning A* 32: 991–1009.

Leeson, N. (1996) *Rogue Trader* London: Little Brown.

Lewis, M. (1989) *Liar's Poker*. London: Coronet.

Leyshon, A. (1996) 'Dissolving difference? Money, dis-embedding and the creation of "global financial space"', in W. Lever and P.W. Daniels (eds) *The Global Economy in Transition*. London: Longman. pp. 62–82.

Leyshon, A. and Tickell, A. (1994) 'Money order? The discursive construction of Bretton Woods and the making and breaking of regulatory space', *Environment and Planning A* 26: 1861–90.

Leyshon, A. and Thrift, N.J. (1997) *Money/Space: Geographies of Monetary Transformation*. London: Routledge.

Lipietz, A. (1983) *Le Monde enchanté*. Paris: La Découverte.

MacKenzie, D. (2000) 'Long Term Capital Management and the sociology of finance', *London Review of Books* 13 April: 1–5 (http://www.lrb.co.uk/v22/n08/mack 2208.htm).

Martin, R. (1994) 'Stateless monies, global financial inte-gration and national economic autonomy: the end of geography?', in S. Corbridge, N.J. Thrift and R. Martin (eds) *Money, Power and Space*. Oxford: Blackwell. pp. 253–78.

Martin, R. (1999) 'The new economic geography of money' in R. Martin (ed.) *Money and the Space Economy* London: Wiley. pp. 3–28.

Maurer, B. (2000) 'The general equivalent and alternative currencies: money, math and the failure of finance'. Paper presented at the American Anthropological Asso-ciation, San Francisco, 15–19 November and available from the author at University of California, Irvine.

McAndrews, J. and Stefanadis, C. (2000) 'The emergence of electronic communications networks in the US equity markets', *Current Issues in Economics and Finance* 6 (12): 1–4.

McCloskey, D.N. (1985) *The Rhetoric of Economics*. Madison: University of Wisconsin Press.

McDowell, L. (1997) *Capital Culture: Gender at Work in the City*. Oxford: Blackwell.

McDowell, L. (2001) 'Men, management and multiple masculinities in organisations', *Geoforum* 32: 181–98.

McDowell, L. and Court, G. (1994a) 'Gender divisions of labour in the post-Fordism economy: the maintenance of occupational sex segregation in the financial services sector', *Environment and Planning A* 26: 1397–418.

McDowell, L. and Court, G. (1994b) 'Performing work: bodily representations in merchant banking' *Environ-ment and Planning D Society and Space* 12: 253–78.

McDowell, L. and Court, G. (1994c) 'The missing subject in economic geography' *Economic Geography*, 70 (3): 229–51.

McMurtry, J. (1999) *The Cancer Stage of Capitalism.* London: Pluto.

Mitchell, D. (2000) *Cultural Geography: A Critical Introduction.* Oxford: Blackwell.

Moran, M. (1991) *The Politics of the Financial Services Revolution.* London: Macmillan.

Muniesa, F. (2000) 'Un robot walrasien: cotation électronique et justesse de la découverte des prix', *Politix* 13: 121–54.

Muniesa, F. (2001) 'Performing prices: the case of price discovery automation in financial markets', in H. Kalthoff, R. Rottenburg and H.-J. Wagener (eds) *Facts and Figures: Economic Representations and their Rhetorical Forms.* Marburg: Metropolis. pp. 289–312.

O'Barr, W.M. and Conley, J.M. (1992) *Fortune and Folly: The Wealth and Power of Institutional Investing.* Homewood: Irwin.

O'Brien, R. (1991) *Global Financial Integration: The End of Geography?* London: Pinter.

Pacione, M. (1999) 'The other side of the coin: local currency as a response to the globalisation of capital', *Regional Studies* 33: 63–72.

Passos, N. (1996) 'The mirror of global evils: a review essay on the BCCI affair', *Justice Quarterly* 12: 801–29.

Preda, A. (2001) 'Sense and sensibility: or, how social studies of finance should be(have)', *Economic Sociology: European Economic Newsletter* 2 (2): 15–18 (http://www.siswo.uva.nl/ES).

President's Working Group on Financial Markets (1999) *Hedge Funds, Leverage, and the Lessons of Long Term Capital Management.* Washington: President's Working Group on Financial Markets.

Previts, G.J. and Merino, B.D. (1999) *A History of Accountancy in the United States: The Cultural Significance of Accounting.* Columbus: Ohio State University Press.

Pryke, M. and Allen, J. (2000) 'Monetized time–space: derivatives – money's "new imaginary"?', *Economy and Society* 29: 264–84.

Roper, M. (1994) *Masculinity and the British Organization Man.* Oxford: Oxford University Press.

Rotman, B. (1987) *Signifying Nothing: The Semiotics of Zero.* Basingstoke: Macmillan.

Rubinstein, M. (2001) 'Rational markets: Yes or no? the affirmative case', *Financial Analysts Journal* May/June: 6–10.

Russell, T. and Thaler, R.H. (1985) 'The relevance of quasi-rationality in competitive markets', *American Economic Review* 75: 1071–82.

Sayer, A. (1999) 'Valuing culture and economy' in L. Ray and A. Sayer (eds) *Culture and Economy after the Cultural Turn.* London: Sage. pp. 53–75.

Shiller, R.J. (1997) 'Human behaviour and the efficiency of the financial system' in J.B. Taylor and M. Woodford (eds) *Handbook of Macroeconomics.* Princeton: Princeton University Press.

Shiller, R.J. (2000) *Irrational Exuberance.* Princeton: Princeton University Press.

Simmel, G. (1990) *Philosophy of Money.* London: Routledge.

Stanley, C. (1996) *Urban Excess and the Law: Capital, Culture and Desire.* London: Cavendish.

Swyngedouw, E.A. (1996) 'Producing futures: global finance as a geographical project', in W. Lever and P.W. Daniels (eds) *The Global Economy in Transition.* London: Longman. pp. 135–64.

Thaler, R.H. (ed.) (1993) *Advances in Behavioral Finance.* New York: Russell Sage Foundation.

Thorne, L. (1996) 'Local exchange trading systems in the United Kingdom: a case of reembedding?', *Environment and Planning A* 28: 1361–76.

Thrift, N.J. (1987) 'The fixers: the urban geography of international capital', in M. Castells and J. Henderson (eds) *Global Restructuring and Territorial Development.* London: Sage. pp. 219–54.

Thrift, N.J. (1989) 'New times and spaces: the perils of transition models', *Environment and Planning D: Society and Space* 7: 127–9.

Thrift, N.J. (1994) 'On the social and cultural determinants of international financial centres: the case of the City of London', in S. Corbridge, N.J. Thrift and R. Martin (eds) *Money, Power and Space.* Oxford: Blackwell. pp. 327–55.

Thrift, N.J. (1999) 'Capitalism's cultural turn', in L. Ray and A. Sayer (eds) *Culture and Economy after the Cultural Turn.* London: Sage. pp. 135–62.

Thrift, N.J. (2001) 'It's the romance, not the finance, that makes the business worth pursuing: disclosing a new market culture', *Economy and Society,* 30 (4): 412–32.

Thrift, N.J. and Leyshon, A. (1992) 'In the wake of money: the City of London and the accumulation of value', in R.J. Johnston and G.A. Hoekveld (eds) *Regional Geography: Current Developments and Future Prospects.* London: Routledge. pp. 282–311.

Tickell, A. (1996) 'Making a melodrama out of a crisis: reinterpreting the collapse of Barings Bank', *Environment and Planning D: Society and Space* 14: 5–33.

Tickell, A. (1998) 'Creative finance and the local state: the Hammersmith and Fulham swaps affair', *Political Geography* 17: 865–81.

Tickell, A. (2000a) 'Global rhetorics, national politics: pursuing bank mergers in Canada' *Antipode* 32: 152–75.

Tickell, A. (2000b) 'Unstable futures: creating risk in international finance' *Geoforum* 31: 87–99.

Tickell, A. (2001) 'The transformation of financial regulation in the United Kingdom', in M. Bovens, P. Hart and G. Peters (eds) *Success and Failure in Public Governance.* Cheltenham: Elgar. pp. 419–36.

Tickell, A. and Peck, J.A. (1992) 'Accumulation, regulation and the geographies of post-Fordism: missing links in regulation theory', *Progress in Human Geography,* 16: 190–218.

Tilly, C. (1999) 'Epilogue: where now?', in G. Steinmetz (ed.) *State/Culture: State Formation after the Cultural Turn.* Ithaca: Cornell University Press. pp. 407–20.

Tsing, A. (1999) 'Inside the economy of appearances', *Public Culture* 12: 115–44.

Tversky, A. and Kahneman, D. (1981) 'The framing of decisions and the psychology of choice', *Science* 211: 453–8.

van der Pijl, K. (1998) *Transnational Classes and International Relations.* London: Routledge.

Walter, A. (1992) *World Power and World Money.* Brighton: Harvester Wheatsheaf.

Williams, C.C. (1996) 'Local exchange trading systems: a new source of work and credit for the poor and unemployed' *Environment and Planning A* 1395–415.

Williams, C.C. and Windebank, J. (1998) *Informal Employment in the Advanced Economies: Implications for Work and Welfare.* London: Routledge.

Zeckhauser, R., Patel, J. and Hendricks, D. (1991) 'Nonrational actors and financial market behavior', *Theory and Decision* 31: 257–87.

Zelizer, V. (1994) *The Social Meaning of Money.* New York: Basic.

Zelizer, V. (1998) 'The proliferation of social currencies', in M. Callon (ed.) *The Laws of the Markets.* Oxford: Blackwell/Sociological Review. pp. 58–68.

Zelizer, V. (2000) 'Fine tuning the Zelizer view', *Economy and Society* 29: 383–9.

Zukin, S. and DiMaggio, P. (eds) (1990) *Structures of Capital: The Social Organization of the Economy.* Cambridge: Cambridge University Press.

6

A Cultural Economic Geography of Production

Meric S. Gertler

THE CULTURAL TURN IN ECONOMIC GEOGRAPHY: IS THAT ALL THERE IS?

Culture is now solidly on the agenda of economic geographers, and nowhere is this more evident than in their study of production processes and dynamics. There are few who would not regard this as a very good thing. After all, the classical intellectual foundations of economic geography, whether the minimalist industrial location calculus of Alfred Weber, the hydraulic engineering models of market equilibration devised by Léon Walras, or the input–output systems of Wassily Leontief, were all based on mechanical metaphors (Barnes, 1996). They were also thoroughly marginalist in their approach. If one *held constant* a variety of considerations, factors and processes known to be in more or less constant flux (such as product or process technologies, market demand, the number and strength of competitors, sources of supply for goods and services), one could designate the optimal location for a production facility or predict how a perturbation in one part of a regional or national production system (say, a rapid increase in demand for a particular product, or the creation of a major new production facility) would affect all other parts of the economy. Of course, these approaches either held constant or assumed away all of the most interesting and important aspects of economic life. Looking back, perhaps the most remarkable aspect of this era in the life of the 'economic sciences' was that so many claimed to find something of value in this work for so long (Thrift, 2000a).

The growing popularity of Marxian political economy in the 1970s and 1980s was in many ways a breath of fresh air. It brought a new appreciation of the role of power to economic geographers' analyses of workplace change and the 'inconstant geography' of capitalist production systems (Storper and Walker, 1989). Ultimately, however, it too came to be regarded by many as too reductionist to capture the most beguiling and elusive features of economic dynamism. Its heavy reliance on the concept of class as the principal determinant of interests, identity and behaviour proved to be too limiting (Gibson-Graham, 1996). At a time when 'the emancipatory politics of class struggle' had largely been rejected in favour of 'the representational politics of political, cultural and environmental recognition', political economic approaches within economic geography had clearly lost steam (Crang, 1997: 3).

Therefore, the move towards a cultural view of the geographical economy beginning sometime in the mid 1980s signalled a growing interest in fundamental questions pertaining to economic change which had not, until then, received their proper due. What are the social dimensions of technological change in production systems, and how are these situated in localities and regions? How do local production practices interact and intersect with the global economy, and must the process of globalization obliterate all differences and distinctive characteristics of regional and national production systems? How are economic processes structured and shaped by social institutions, and at what geographical scales do these institutions exert their influence?

These are undoubtedly important questions, and their centrality within economic geography in recent times has helped enliven the field considerably (Barnes and Gertler, 1999; Clark et al., 2000a; Lee and Wills, 1997; Sheppard and Barnes, 2000).[1] It has also served to build valuable bridges to scholarship in other related fields, such as economic sociology, industrial economics and the study of material culture, where interesting things have been happening recently. However, the cultural turn in economic geographers' study of production is already, in some ways, old news.[2] The question now to be asked is: what has it delivered? The answer I would venture at this stage is: not as much as had been promised or hoped for originally. For every advance towards enlightenment there has been an equal measure of confusion. While the introduction of cultural arguments has greatly enriched and enlivened the study of production questions, it has also raised pressing new questions without answering them, and has left many important older questions unanswered or neglected. Because of this, some constructive criticism concerning this larger project and its progress thus far is appropriate.

In this chapter, my aim is to address these developments by taking a sceptical view of what has been achieved to this point. My purpose is *not* to belittle or denigrate the important and useful work that has been completed thus far, for I agree wholeheartedly with Thrift (2000a) that, thanks to the cultural turn, we can never again look at 'the economic' in quite the same way – nor should we. My goal is to challenge the sense of self-satisfaction that has, in my view, permeated the subdiscipline rather prematurely, by posing some unsettling questions. I begin by attempting to distil the recent ferment around the cultural and its relationship to the economic (at least in the realm of production)[3] into three big ideas: the rediscovery of 'the social' in production systems, the rise of the learning paradigm and the idea of regional culture, and the evolutionary dynamics of local production systems. I then offer up a critical commentary on this newish cultural economic geography, focusing on several important areas of controversy. In particular, I argue that we have made surprisingly little progress in advancing a plausible theoretical understanding of one of the most central entities within capitalist production systems – the firm. As a consequence, our ability to provide useful answers to some very fundamental questions – like 'When and why is the local important in production and innovation processes?' or 'Under what circumstances does production knowledge travel between places?' – remains

very poorly developed. I conclude by enunciating some other important questions that have been neglected or left for dead, but which are still pressing and in need of our serious attention.

A CULTURAL ECONOMIC GEOGRAPHY OF PRODUCTION: BIG IDEAS

Just where did this renewed interest by economic geographers in the culture of production come from? Why is the cultural seen as being so important now? I think it is possible to answer this by singling out key debates and questions at several different scales of analysis. At the macro scale, the tremendous interest and attention given to the success of 'Japan, Inc.' during the 1980s (Oliver and Wilkinson, 1988) and Germany's 'Rhineland model' of stakeholder capitalism in the 1990s (Hutton, 1995) has foregrounded what many have inferred to be the central role of national cultures in determining economic success (Sayer and Walker, 1992). After all, though it seems somewhat unlikely given the tenor of commentary in today's mainstream business press, these two models were not all that long ago much admired by academics and business writers alike. Their successes were most commonly interpreted in terms of national business cultures, not only shared by all national firms, but also deeply engrained in the psyches and mentalities of workers and the general public. In retrospect, a lot of this writing strikes one now as more than a little crude and naive, for reasons I shall enumerate below.

At the micro scale of the individual firm, the concept of 'corporate culture' came into vogue in the 1990s as a central variable capable of explaining both spectacular successes – such as the rise of Dell and Microsoft – and colossal failures – the chastening of Xerox, DEC or Lockheed in the world of big business (Schoenberger, 1997; 2000). More recently, the idea has also been invoked to understand the problems emerging in the wake of highly publicized mergers between corporate giants (for example, the fundamental incompatibilities between Daimler and Chrysler, or between BMW and Rover, only a few years after their much celebrated unions). The prevailing narratives have laid virtually all the blame upon incommensurate corporate (and/or national) cultures that have produced worldviews and behaviours so fundamentally divergent that they cannot be overcome.

*Big idea number one: the rediscovery
of the social in production*

However, from the perspective of economic geography, it is a set of meso-scale developments that have probably held the greatest significance. Beginning in the mid 1980s, economic geographers and other social scientists began to discern some important changes in the nature of capitalist competition and production systems. One line of thought emphasized the transition from mass production and competition based on lowering average costs and prices, toward batch or customized production and competition based on quality, performance and distinctiveness (Piore and Sabel, 1984; Scott, 1988). In response to the macro-economic stagnation of the 1970s, which had caused the aggregate purchasing power of national economies to stop rising or actually decline, firms appeared to be seeking new ways to gain market share based on the identification and servicing of smaller, qualitatively distinctive market 'niches'. In order to succeed at this game, they required new skills and production practices, both private and social.

Inside the firm, the transition from producing large runs of more or less standardized goods to small-batch or custom (one-off) production required firms to implement new, more flexible process technologies and practices – versatile, computerized machines, multiskilled and multitasked workers, and novel approaches to workplace organization that enabled quality improvements to be identified and implemented on a continuous basis. Closely related to these internal changes was a restructuring in the wider social division of labour between firms. Most important here was a process of vertical disintegration. Under an older mode of organization, vertically integrated firms had performed a long sequence of production operations themselves, sometimes extending from the processing of raw material inputs to the production and distribution of finished goods. As firms reorganized production, they chose to perform fewer functions for themselves, turning instead to external specialist suppliers of goods and services. Hence, discrete elements of the production process that were once provided within the bounds of the legal entity known as the firm were now being acquired through a market transaction between the firm and its suppliers. The same principles of specialization and division of labour operating inside the firm were now being exploited at a social scale of organization *between* firms.

The principal virtue of this new social organization of production was that it too enhanced the overall flexibility of producers, both individually and collectively. As each firm's production needs changed (according to rapidly changing and increasingly fickle market demand), so too would its input needs. Under such conditions, it proved to be faster and/or more efficient to draw upon the specialized offerings of external suppliers – mixing, matching and changing inputs (and suppliers) at will and on short notice.

As production systems transformed themselves toward this increasingly social basis of organization, so the story goes, the importance of proximity – i.e. geography – was also greatly enhanced. Given that the market exchanges or transactions amongst this multitude of now vertically disintegrated firms would necessarily become more frequent, less predictable and rapidly changing, there were real cost advantages arising from spatial concentration. The closer they were to one another, the lower the transaction costs – that is, all costs associated with achieving successful market exchange of goods and services. However, there is a second and more fundamental advantage produced by spatial proximity, related more to the growing importance of innovation, learning and culture in production systems. As we shall see below, it is these aspects of the new mode of production organization that embody the truly social nature of socially organized production systems, since they implicate forms of interaction between firms that go far beyond simple market transactions.

*Big idea number two: learning
and regional culture*

In order to meet or anticipate the demands of rapidly shifting markets, and in a competitive environment in which product life cycles had become dramatically shorter, the onus on firms to achieve successful innovations in products and processes had become paramount. In some accounts, couched more in the language of long waves, these conditions were seen to stem from an epochal transition of capitalist societies towards a new 'techno-economic paradigm' based on microelectronic and information technologies (Freeman and Perez, 1988). The early stages of such a transition are typically characterized by rampant 'creative destruction' as products and processes based on the new paradigm are generated to replace old ones. Hence, an alternative approach views this heightened importance of innovation as being driven less by market conditions, and more by fundamental changes on the supply side of the economy.

Whatever the driving force, as production processes had become increasingly socialized, so too had the process of innovating. In place of an old 'linear' model of innovation, in which new ideas were developed in the isolation of research and development labs, then 'pushed' onto the market by firms, the new model of innovation was interactive and recursive. It was said to depend on close, repeated interaction between firms (technology producers) and their customers (technology users) over extended periods of time (Lundvall, 1988). This interaction was underpinned by the frequent sharing of proprietary technical and market information and – here's where cultural processes finally enter the picture – it became clear that this kind of knowledge flow (or inter-firm learning) was best supported by *closeness* (Gertler, 1995).

Why? To begin with, much of the knowledge passing between innovation partners is said to be highly confidential and crucial to the competitive advantage of the firms involved. In such cases, it has been argued, repeated interaction over long periods of time – as well as cultural commonality and personal relationships – serve to build up trust or 'social capital' between transacting parties (Putnam, 1993), discouraging opportunistic use of the knowledge exchanged and thereby facilitating its flow. Second, because this knowledge is frequently finely nuanced, tacit and context-specific, this form of learning is said to be most effective when the partners achieve the deeper understanding that is only possible when they share a basic linguistic and cultural commonality. In Storper's (1997) characterization, the central idea is that the interrelationships or dependencies that develop between firms through the market exchange of goods and services come to be supplemented – if not eclipsed – by extra-market bonds, linkages and commonalities: what he and others have called untraded interdependencies.

The ideas of Putnam, Storper and others owe more than a little debt to the much earlier work of Karl Polanyi (1944), who demonstrated the extent to which economic processes and activities are always shaped by, or 'embedded' within, a social or cultural context. According to Polanyi, and to Veblen (1919), this context inheres in the formal and informal institutions that produce and reproduce norms, conventions, customs and habits, shared by a group of economic actors, which define the grooves along which economic behaviour runs (see also Granovetter, 1985, who has been most responsible for the resurgence of interest in the concept of embeddedness). This is where the notion of 'regional culture' enters the picture.[4] The idea put forth in the literature is that the shared social attributes (conventions, routines, habits, customs

and understandings) facilitating the kind of inter-firm learning and embeddedness described above are regionally defined. That is, economic behaviour is *embedded in regional cultures*.

Moreover, these cultures will vary substantially from region to region, and not always in a happy way. In fact, within the literature on industrial districts and learning regions the world came to be divided into a handful of essentially ideal-typical regional formations. There was the 'Holy Trinity' of charmed places that were blessed with favourable regional cultures: the industrial districts of central Italy, where direct inter-firm collaboration of all types was said to be rampant and deeply ingrained in the regional culture (Piore and Sabel, 1984); the machinery and automotive districts of Baden-Württemberg in south-western Germany, where the institutions of the state government encouraged and accommodated indirect horizontal cooperation (between competitors) but direct vertical collaboration (between buyers and suppliers) (Herrigel, 1996); and Silicon Valley, the grand-daddy of them all, where a common regional culture was produced (and inter-firm learning supported) by a prodigious ability to spin off new firms from old, as well as astounding rates of labour market promiscuity (serial employment relations?) by which key information circulated rapidly throughout the region (Saxenian, 1994).

There were also the hard-luck cases: once-successful places where local cultures fostered ties so strong, structures so rigid, and attitudes so unbending that newcomers and new ways of doing things encountered insurmountable barriers to entry – think Germany's Ruhr Valley (Grabher, 1993), Massachusetts' Route 128 (Saxenian, 1994) or the Swiss Jura (Glasmeier, 2001).

Then there were the reclamation projects: places like South Wales or the Basque country, diagnosed as suffering from too little (rather than too much) embeddedness and a weakly developed collaboration culture, that nevertheless sought to mend their errant ways through a variety of locally orchestrated, concerted actions aimed at matchmaking or social engineering of inter-firm relations (Cooke and Morgan, 1998).

Big idea number three: the evolutionary dynamics of local production systems

To complete the trilogy of big ideas, we turn to one imported from the recently emerging field of evolutionary economics. Barely 20 years old,[5] it remains very much on the fringe of the mainstream, since it concerns itself with the epiphenomena of capitalist economies: trivial things such as real industries, history, institutions and places. A few key concepts have been especially influential to economic geographers' study of production systems (Barnes, 1997). First, economic systems change over time, but they do so in ways that are to some extent shaped and constrained by past decisions, random events and accidents of history. Current decisions and events are not determined by past ones, but they are conditioned by them. As a result of past events and choices, certain choices today are easier to pursue, others less so. This is the key idea of path dependency. Walker captures the idea succinctly:

> One of the most exciting ideas in contemporary economic geography is that industrial history is literally embodied in the present. That is, choices made in the past – technologies embodied in machinery and product design, firm assets gained as patents or specific competencies, or labor skills acquired through learning – influence subsequent choices of methods, designs, and practices. (2000: 126)

The past may be embodied within material objects such as machinery, buildings and physical infrastructure, or through the experiences of individuals (alone or in groups). Part of this past is also embodied in institutions – social structures that shape the attitudes, norms, expectations and practices of individuals and firms through formal or informal means of regulation – meaning that path dependency has a strong social dimension. In essence then, what we have already described above as 'regional culture' or 'embeddedness' can be thought of as a real and significant component of 'industrial history … literally embodied in the present'. Culture becomes a part of the historical baggage (sometimes useful, sometimes a liability) associated with particular regions.

There is another potent idea within the evolutionary approach, closely associated with the path dependency notion – increasing returns – which has actually been appropriated from the much earlier work of economists on the fringe of the discipline (Myrdal, 1957; Young, 1928). It refers to the process in which, once a particular economic change occurs, it becomes self-reinforcing. A particular technological design, once adopted by a critical mass of early users, becomes a standard. After this happens, the market for this design will expand even further. Initial growth begets further growth. Moreover, even though a particular technology has become a standard in its field, this will not necessarily indicate that it is unequivocally superior to available alternatives. Its dominance may instead be based on the fact that it was the first viable

technology in the marketplace, that many supplying businesses, complementary technologies and institutions, plus a large community of developers and users, have been created to support its use. Once this unfolds to the point where perfectly viable – if not superior – alternatives cannot easily be adopted, a situation of 'lock-in' is said to have been reached (Arthur, 1989; David, 1985).

The twin concepts of path dependency and increasing returns have obvious relevance in understanding the historical paths taken by production regions. Once a region establishes itself as an early success in a particular set of production activities, its chances for continued growth are very good indeed. While this may be to some extent reducible to the success of dominant 'lead' firms in the region (e.g. Microsoft in Seattle), the really interesting aspects of this process have more to do with the collective processes and forces at work: local social and economic institutions and, yes, culture. By the same token, as cases such as the Ruhr or Route 128 suggest, ailing places may also be difficult to turn around for the same reasons. Once a path-dependent trajectory of decline becomes firmly established, institutional and cultural lock-in will make deviation from this path a serious challenge.

CULTURE CONTROVERSIES

I am certain I have done a disservice to the literature on the cultural economic geography of production by reducing the hefty weight of its intellectual output to three big ideas. But if you were to sit me down and force me to enunciate what is/was new, different, distinctive and most significant in this work – reduced to its bare essence – this is pretty much the list I would produce. Nevertheless, the danger of engaging in such a summary exercise is that it creates the mistaken impression of consensus and completion where none actually exists. In fact, there is plenty of disagreement and lack of convergence, as well as much unfinished business. Perhaps surprisingly, one of the most troublesome areas concerns the economic-geographical theory of the firm – a theory that ought to be able to provide answers to fundamental questions like: why do firms in particular places adopt particular production and innovation practices, and not others? What forces and processes determine what a firm 'knows', and how easy is it to transfer this knowledge from one place to another? Or to use the language of the previous section, must learning by firms really conform to the geographies of local cultures?

The knowledge foundations of regional cultures of production and innovation

As outlined in the previous section, the new cultural economic geography of production asserts a clear role for the local – a generally counterintuitive result in the supposed age of the global information economy. This idea has become something of a badge of honour amongst economic geographers, who now regularly delight in disabusing their spatially challenged colleagues in other disciplines of simplistic ideas about 'the end of geography' (O'Brien, 1992), 'the death of distance' (Cairncross, 1997) or 'the end of the city as we knew it' (Mitchell, 1995). Besides the satisfaction of scoring points with one's economist friends, it is comforting to be reassured that 'geography matters'. Moreover, the political possibilities associated with this position defy the stifling determinism of the 'global economy' view of the world, in that local politics are at least open to contestation by local stakeholders.[6] And yet, just when it appears that all is right with the world, along comes some unsettling little complications. It turns out that the conceptual foundations for arguments promoting the primacy of the local – particularly as they concern the knowledge and behaviour of the firm, arguably the principal actor within capitalist production systems – may not be as solid or clearly articulated as they should be.

The problems begin with the concept of tacit knowledge and its relationship to the local. In much of the recent literature on regional cultures of production, the distinction between tacit ('know how') and codified ('know that') knowledge emerges as central. The touchstone for this distinction is Michael Polanyi's oft-quoted phrase that 'we can know more than we can tell' (1966: 4) – alluding to a form of knowledge that is 'imperfectly accessible to conscious thought' (Nelson and Winter, 1982: 79). Underlying this is a deeper distinction between forms of knowledge (explicit or codified) that can be effectively expressed and shared using symbolic forms of representation and other forms of knowledge (tacit) that defy such representation. Hence, the tacit component of the knowledge required for successful performance of a skill defies codification or articulation either because the performer herself is not fully conscious of all the 'secrets' of successful performance, or because the codes of language are not well enough developed or sufficiently universal to permit explication and translation.

To show why the concept of tacit knowledge is so important to the question of local cultures of production requires a bit of a digression into the theory of the firm. Recently fashionable theories of the firm draw their inspiration from the pioneering (and long-neglected) work of Edith Penrose (1959). This approach has been labelled the competence- or resource-based theory of the firm because it emphasizes the importance of the competences, capabilities or resources of individual workers and managers within the firm, as well as the larger collectivity of customs, procedures, routines and practices that define the firm. It is the collective or social aspect of this capability which is most interesting, but also the most difficult to grasp fully when trying to understand why firms behave the way they do. According to this view, individuals bring to the firm a set of attributes and competences based on their formal education and broader socio-economic background, as well as their prior work experiences. But the firm's capabilities are constituted by more than the simple sum of its individual employees' capabilities. *Every firm* has a unique set of organizational structures, relationships, rules and routines (explicit and implicit) that help coordinate the actions of individuals inside the firm in order to achieve purposive goals – both the day-to-day objectives entailed in successful production and the longer-term goals set forth in a firm's broader strategy (Kay, 1993).

These features, together with the common learning experiences of workers within the firm, also serve as a means for enabling the social production and sharing of knowledge, through the creation of a distinctive 'knowledge context' or framework. As Howells puts it, 'What individuals within the firm know and can achieve depends in part on the nature and characteristics of the firm's context' (2000: 55). Common codes of communication are essential to this process, since one of the principal challenges confronting the firm is how to absorb, share and transform both codified and tacit knowledge throughout the organization. In the absence of shared codes of communication, the transmission of this knowledge – that is, learning processes within the firm – will be impeded. Once again quoting from Howells, who acknowledges the work of Metcalfe and De Liso (1998):

> The tacitness of much knowledge, its indivisibility in use, the uncertainty of its values in different contexts, its proprietary nature, and the fact that much of what is known is jointly produced by the firm's activities … means that the firm provides a central contextual role in harnessing knowledge to produce new innovative capabilities. (2000: 56)

Hence we can now see that, according to the resource- or competence-based view, the firm's primary purpose is to absorb, share and transform (that is, process) knowledge, thereby leading to new or better products and processes. Moreover, the firm's competitive edge is defined by its success in producing distinctive competences through the management of its knowledge (Thrift, 2000b). And the emphasis on knowledge processing fits well with the greater attention recently given to the 'knowledge-based' economy (OECD, 1996). But notice as well that the principal emphasis here is very much on *the firm itself* and its ability to define its own context for knowledge production and sharing.[7] This is the point where the argument veers into highly contested terrain.

As noted above, a key argument emerging from this competence-based perspective holds that the success of the firm has become increasingly dependent on its ability to gain access to tacit knowledge. As Maskell and Malmberg have recently put it, when all firms have unimpeded access to explicit/codified knowledge, the creation of unique capabilities and products depends on the production and use of tacit knowledge:

> Though often overlooked, a logical and interesting consequence of the present development towards a global economy is that the more easily codifiable (tradable) knowledge can be accessed, the more crucial does tacit knowledge become for sustaining or enhancing the competitive position of the firm. (1999: 172)

And recalling from our earlier discussion of learning, culture and the context-specific nature of tacit knowledge, it is the growing importance of this form of knowledge to the firm that has, according to the now widely accepted argument, been responsible for asserting the resurgence of the local in an era of the globalizing economy. In other words, tacit knowledge is most easily shared locally, while knowledge must be codified in order for it to 'travel' globally. This is because (1) local proximity enables and promotes the direct, face-to-face interaction necessary to support tacit knowledge transmission, and (2) common local origins also generate the shared understandings (equals local culture) necessary to support the easy translation of tacit ideas between two interacting economic agents.

Recently, this tendency to link the tacit with the local scale and the codified with the global scale has been subjected to critical scrutiny, exposing an interesting new twist in economic geography's cultural turn (Allen, 2000; French, 2000). While this critique of the localist approach has helped focus our attention on some critical issues and weaknesses in the (now) mainstream cultural economic geography of production, it ultimately disappoints as well by failing to offer up a conceptually robust alternative.

For example, Allen represents this tacit/local, codified/global dualism as something more akin to the learning region literature's big lie[8] rather than its big idea: 'largely a flawed, if not spurious, exercise' (2000: 30). He blames the propagation of this oversimplified dualism on what might be described as an unfortunate family reunion, featuring the Polanyi brothers. He complains that Michael Polanyi's (1958; 1966) essential insights into the context-specific nature of tacit knowledge have been combined with, and corrupted by, brother Karl's (Polanyi, 1944) seminal ideas on the social and institutional embeddedness of economic action – a kind of conceptual sleight-of-hand attributed to a 'powerful set of discourses' with a disturbing 'inability to keep apart the ideas of the two Polanyi brothers', resulting in this 'conflation' (Allen, 2000: 27).

Allen openly challenges the idea that tacit knowledge is 'solely the creation of territorially specific actions and assets' (2000: 27), or that 'face-to-face presence and proximity are paramount' (2000: 28). Instead, he offers up alternative assertions about the importance of 'distanciated contacts … [and] "thick" relationships [that] may span organizational and industry boundaries', realized through 'people moving to and through "local" contexts, to which they bring their own blend of tacit and codified knowledges'. He continues:

> What matters in such situations is not the fact of local embeddedness, but the existence of relationships in which people are able to internalize shared understandings or are able to translate particular performances on the basis of their own tacit and codified understandings. (2000: 28)

Communities of practice: is all knowledge local?

In highlighting distanciated, thick relationships, Allen implies that tacit knowledge is not the prisoner of local culture, but can in fact flow across long distances so long as the relationships between actors involved in this flow are strong enough. In making this claim, Allen situates himself implicitly within the recent literature on 'communities of practice' emerging from organizational studies (Amin, 2000; Wenger, 1998), which emphasizes the possibilities for non-local learning crossing geographical, organizational and cultural divides.[9] Received wisdom is always worth interrogating, and Allen's unsettling analysis draws our attention to some very fundamental questions. *Can* we be so sure that tacit knowledge really is embedded in local culture to the extent that is widely implied or assumed in the new cultural economic geography literature? Are proximity and face-to-face contact *really* so important? Isn't this what the information technology revolution was supposed to address?

Good questions all, and the message of Allen, Amin and others is that we should not presume to know the answers simply because of the 'accepted canon' within the new church. Fair enough. But before we wax too enthusiastic about the possibilities of theological revolution, we need to address two weaknesses in the critique of the localist position. First, we need to re-examine the conceptual foundations underlying the localist argument. While Allen is undoubtedly correct in placing the emphasis on 'thick relationships' and 'shared understandings', pleasantly enigmatic terms like 'distanciated' may scan well, but offer us little more than fleeting insight since they are so open to retranslation and interpretation themselves. Allen's critique of the now standard take on tacit knowledge exposes the fact that its conceptual foundations are indeed far from solid. Yet, even with this critique, we still have no clear idea of the forces and processes stimulating the emergence of thick relationships and shared understandings between firms.

Second, it is hard to see how merely asserting that 'the local does not really matter' advances the state of our understanding in any appreciable way, so long as statements such as this remain unsubstantiated empirically. Instead, we need to get beyond bald assertions to consider the evidence at hand – something, alas, for which neither the proponents nor the critics have shown overwhelming concern, up to this point. In other words, we need to engage in more 'learning by doing' ourselves since there is apparently still much hard work to be done, both conceptually *and* empirically, to unravel these processes and relationships more carefully.

Two recent studies by economic geographers – in sectors as diverse as motor sport engineering and life insurance – begin to illustrate the true complexities of knowledge formation and translation over space. Both of these cases, in their own way, deepen our understanding of the precise mechanisms by which local and non-local processes of knowledge formation intersect *at particular sites of production*. In the case of life insurance, key 'learning intermediaries' – each with their own distinct operative scales – interact and combine to produce knowledge locally (French, 2000: 110). When it comes to producing and applying knowledge in the motor sport industry, the longitudinal study of top designers' work histories reveals a great deal about the

geographical structure and channels of knowledge flows (Henry and Pinch, 2000). At the same time, spatial propinquity is undeniably important in creating a discursive dynamics of innovation as performed through 'gossip, rumour and observation' (2000: 136).

Another helpful source of insight is the new literature on 'knowledge management' emerging from the business schools and international consulting community. To be sure, this literature is filled with the typical puffery about 'harnessing the organization's knowledge assets' to 'unlock the firm's innovative potential'.[10] But when one strips away all of the marketing hyperbole, one is confronted with a striking realization: while large corporations recognize the economic value of tacit knowledge in producing distinctive competitive advantages, they also know *how difficult it is to transpose and translate it from one local context to another*. Even the 'success stories' in this literature (von Krogh et al., 2000) are not very convincing, and betray a troubling lack of sophistication in their understanding of how 'shared understandings' develop, or how tacit knowledge and local cultures are produced (Gertler, 2001b).[11]

Far more enlightening is the recent work of Brown and Duguid whose intriguing book *The Social Life of Information* (2000) tackles these issues head-on. Brown and Duguid situate their work very much within the 'communities of practice' approach. However, in contrast to Wenger and others, they make a key distinction between this construct and the related idea of 'networks of practice' – for example, the 25,000 service reps working worldwide for Xerox Corporation. Their detailed research of knowledge flows in large organizations leads them to the following conclusion:

> Networks of this sort are notable for their reach – a reach now extended and fortified by information technology. Information can travel across vast networks with great speed and to large numbers but nonetheless be assimilated in much the same way by whomever receives it. By contrast, there is relatively little reciprocity across such networks: that is, network members don't interact with one another directly to any significant degree. When reach dominates reciprocity like this, it produces very loosely coupled systems. Collectively, such social systems *don't take action and produce little knowledge*. (2000: 142, emphasis added)

In contrast to this, Brown and Duguid see communities of practice as 'relatively tight-knit groups of people who know each other and work together directly … face-to-face communities' in which reach is limited but 'reciprocity is strong' (2000: 143). It is only these small groups that 'allow for highly productive and creative work to develop collaboratively'.

So perhaps, after having considered some recent empirical evidence on the matter, the non-local flow of tacit knowledge is (still) more difficult to achieve than the sceptics might think. Thus, we appear to have come full circle, back to the starting argument that local culture matters. But before we buy in, what about the global reach of large, wealthy and powerful corporations? Aren't their resources, organizational structures, standardized practices and corporate cultures sufficiently well developed to overcome the simple friction of distance or the stickiness of local and regional cultures? Surely corporate culture and practice have the ability to trump physical and cultural divides, bringing about a convergence of practices across national and regional boundaries, right? Certainly, one implication of the competence-based theory of the firm reviewed above is that firms can be 'masters of their own destiny' rather than 'slaves to geography'.[12] Recall that this approach argued that the manner in which firms absorb, share and transform (or process) knowledge is determined by each firm's unique knowledge context or framework, which is constituted as a set of *firm-specific* structures, relationships and routines. In other words, in this view the characteristic practices of firms are shaped or determined by a socially constructed process *internal to* the firm itself – something very much akin to the idea of corporate culture.

Corporate culture

Although this concept continues to be frequently invoked without due regard to its specific meaning, more careful analysts have sought to define it in a variety of ways. Howells, for example, defines it as 'shared routines and patterns of working ... common decision-making procedures ... a common organisational information and routine base ... a firm's distinctive set of decision rules or routines ... [which] in turn help shape a tradition of practice within the firm' (2000: 55). McDowell writes of organizational culture as 'explicit and implicit rules of conduct' responsible for 'inculcating the desirable embodied attributes of workers, as well as establishing the values and norms of organisational practices' (1997: 121). Hudson takes a slightly different tack by observing that 'Firms, governments and other organizations have a collective memory *beyond* that of any individual or group of individuals' (2001: 32). Glasmeier emphasizes the role of collective identity and belonging in culture; she singles out

'shared understandings ... the prevailing belief system' as well as 'the rules and practices, identities and aspirations at the most intimate level' which make possible 'collective and purposeful action' (2001: 58).

Taking perhaps the broadest view of all (and one embraced by Glasmeier), Schoenberger argues that:

> culture is inherently and deeply implicated in what we do and under what social and historical circumstances, in how we think about or understand what we do, and how we think about ourselves in that context. It embraces material practices, social relations, and ways of thinking. Culture both produces these things and is a product of them in a complicated and highly contested historical process. (1997: 120)

She adds that culture and human action are reflexively related: 'humans, through their actions and social relations, produce culture at the same time that culture produces them' (1997: 121). Concerning corporate culture specifically, Schoenberger makes two further key points: first, that it is ultimately about power, and hence it is appropriate to speak of a 'dominant' corporate culture (1997: 121), produced by top management; and second, that the most fundamental form of this is 'the power to define a very particular social order – the firm – and its relationship to its environment' (1997: 122), in essence, the very identity of the firm itself.

Thus, a closer reading of the meaning and origins of corporate culture reveals it to be as much a product of its social-cultural-institutional environment as it is a reflection of the firm's own volition (or the vanity of its senior managers). In other words, while corporate culture certainly cannot be 'read off' from the larger institutional environment around it, it will surely be shaped, influenced and constrained by this broader framework.

Schoenberger's fascinating case studies of the frailties and myopia of large American companies such as Xerox, DEC and Lockheed underscore the limits to the reach and potency of corporate culture in the face of significant international and interregional institutional discontinuities, shortcomings that amount to a 'cultural crisis of the firm'. Her approach resonates strongly with the important findings of two other recent literatures from outside geography: the work on national business systems and 'varieties of capitalism' (Berger and Dore, 1996; Boyer and Hollingsworth, 1999; Soskice, 1999; Whitley, 1999) and a closely related literature on corporate practices, governance and national origins (Doremus et al., 1998; O'Sullivan, 2000; Yeung, 2000). Based on both conceptual argument and empirical evidence, this work demonstrates how the strategies and practices of even large firms are shaped by (and embedded within) the various divergent national institutional frameworks in which they operate.

How do these insights help us navigate our way through the recent debate on the role of the local and the ability of knowledge to 'travel'? According to this view, institutional context – and *not* geographical proximity *per se* or the availability of information technology – acts to define the conditions under which knowledge can actually be translated and transferred from one locale to another. In other words, physical proximity is not the real source of 'thickness' in relationships, nor can this strength be guaranteed by the availability of information technologies designed to substitute for actual collocation. Instead, the bedrock preconditions on which strong relationships and shared understandings develop is institutional affinity or similarity.

When firms (or individual workers) operate according to a common set of norms, routines, conventions and assumptions about the way the economic world works (what we might call a common culture), they are more likely to share those understandings that promote knowledge flow effectively between them, no matter what the intervening physical distance. But these very routines, norms, conventions and assumptions are themselves strongly influenced by the institutional structures that delimit how labour and financial markets operate, how corporate governance, employment relations and work are organized, and how competition and inter-firm relations develop.[13] In other words, a more complete theory of the firm and the role of regional culture needs to be set within a wider matrix of institutional forces and processes that influence (but do not fully determine) their actions. In my view, neither the new cultural economic geography literature *nor* its critics have grasped the significance of this basic insight. For this reason, the debate remains pitched at a distressingly superficial level of analysis.

LIFE AFTER CULTURE: WHAT COMES NEXT?

The preceding discussion has, I believe, demonstrated that economic geographers live in interesting times, even if they can't quite reach consensus on some of the most fundamental questions confronting them today. On questions like 'When is proximity important, and why?', 'How easily does tacit knowledge flow between places, and how is it produced in the first place?', a culturally informed analysis has taken us into some interesting new terrain, even as it raises many unanswered questions. We now understand production to be imbued with cultural processes and forces emanating from the region, the nation-state and the firm.

One may well ask, however, whether the gains achieved from taking this turn have fulfilled the promise expected. In particular, the interest in cultural aspects of production in the economy may have diverted attention away from some of the older, but still pressing, 'core questions' within the economic realm. While it may be true, as Thrift has argued, that earlier work on these topics 'had lost any sparkle of innovation' (2000a: 692), others contend that there remain 'immensely real substantive issues and purposive human practices that have always been and still are fundamentally at stake' (Scott, 2000: 34). Where do we go from here? In this concluding section, I comment on the continuing relevance of some of these 'real issues' and suggest some possibly fruitful ways forward.

Cooperation or conflict?

To begin with, the overriding interest in innovation, knowledge and learning in today's cultural economic geography is undoubtedly well justified, for reasons that are evident in the preceding discussion.[14] However, we seem to have introduced some topical blinders and biases in the process, leaving behind some questions of immense importance that once occupied us all quite a bit.[15] For instance, within the field of innovation studies itself, we have probably all been guilty of overemphasizing cooperation and collaboration between firms, and downplaying older themes like competition and conflict. In a recent paper, Malmberg and Maskell (2002) point out that the true essence distinguishing many local and regional clusters may not in fact be cooperation and culturally grounded harmony but, rather, intense rivalry, competition, observation, spying and other forms of anti-social behaviour. Their point is that this kind of activity is also fostered by spatial concentration, and is capable of conferring real benefits on the firms involved.

This point may be more profound than it appears at first blush. Consider the case of one of the world's paradigmatic clusters – Silicon Valley. In her landmark study, Saxenian (1994) was at pains to challenge the then-prevailing representation of the Valley as the creation of gifted and daring entrepreneurs working more or less in isolation from one another. Her take was different: that the rise (and especially the early 1990s recovery) of the Valley was really a story about a finely articulated local social division of labour, social cooperation, and a widely shared regional culture of openness and mobility that promoted the free flow of knowledge. This does not mean that it remained closed to outsiders: indeed, to the contrary, it attracted large numbers of workers and entrepreneurs from other regions and countries (Saxenian, 1999). Nor does she exclude competition from her analysis.[16] Nevertheless, according to Saxenian, the Valley owes much of its continuing success to a regional culture of cooperation and trust, and its ability to integrate talented newcomers into its socially organized production system.

However, this view has recently been contested by Cohen and Fields (1999) who interpret Silicon Valley's success in rather different terms. They emphasize competition and secrecy to a much greater extent than Saxenian did. Moreover, they argue that the kind of trust prevalent in the Valley is based far less on deep knowledge acquired through direct, close interaction over time, and far more on the circulation of information – through indirect contact – concerning the performance-based reputation of potential suppliers or customers. While they too emphasize the importance of a highly articulated social division of labour (which has produced, among other things, a well-developed array of specialized legal and other professional services), their characterization of the region's economic organization seems much closer to the traditional urban economist's notion of agglomeration economies realized through market exchange.

Power and (work) place

Going farther afield, one is struck by how infrequently questions of power – in the workplace, in inter-firm relations or in communities – appear in today's cultural economic geography of production literature.[17] It is as if the infatuation with innovation and 'the new' in much of this literature has diverted attention away from questions concerning 'social relations of production' in the older (less sparkly) sense of the phrase. These two fields of enquiry need not be seen as mutually exclusive. For example, inside the firm these social relations still play a major part in determining workers' roles in implementing new technologies and modes of workplace organization (Kochan and Osterman, 1994; Kumar and Holmes, 1997), and in producing and transmitting tacit knowledge for the firm to exploit for commercial advantage.[18] Social relations and the

distribution of power between workers and managers are also implicated in the manner in which the spoils from innovation are distributed, which may in turn influence the commitment of workers to the firm's corporate strategy and culture. In turn, these workplace social relations do not develop in a vacuum, but are influenced and enabled by the regulatory climate (state/provincial, national and even supranational) in which the firm is situated (Gertler, 1997; 2001a; Peck, 1996; Rutherford and Gertler, 2001; Wills, 2000).

Interestingly, an alternative view of social relations of production has begun to emerge in recent work on the 'new economy', in which the firm has receded into the background and the worker has come to the fore. Of course, we are not talking about just any old workers here. This is a literature that has become enraptured with what Richard Florida (2001) and others refer to as 'talent' – highly skilled, much sought-after new economy workers.[19] Strongly implicit in this work is the argument that the central relationship within the capitalist knowledge economy is no longer the employment relation between worker and manager/owner. Instead, in a seller's market in which talented labour is in short supply and able to choose between employers (and communities), the principal relationship is now between workers and local labour markets. Talented labour chooses where to live and work based on the local density of employment opportunities as well as broader social and quality-of-life characteristics of local communities. These include a range of 'soft' cultural attributes such as social variety, openness to difference, tolerance and low barriers to entry by newcomers – in addition to the normal list of desirable place attributes (cultural and entertainment amenities, appealing natural environment and built form, and so forth). Under such conditions, occupational groupings and affinities constitute the principal source of social interaction providing labour market information and security (Markusen, 2001). Clearly, this kind of 'fast company' scenario applies only to a special class of worker and perhaps only under certain phases of the business cycle when demand for such labour is especially strong (Thrift, 2000b). It remains to be seen just how lasting and widespread this transition away from the firm and the employment relation and toward the community turns out to be. With recent evidence indicating that even the 'new' economy cannot escape the vagaries of the business cycle, I would venture that at best the jury is still out.

In other instances where power relations are implicit in the work of economic geographers, the treatment is still quite unsophisticated. I refer here to the work on multinational corporations and the global economy, where most of us still labour under ridiculously simple-minded assumptions about the ease and power with which these corporate behemoths can transfer their products and practices from their home turf to new markets and production spaces, erasing local cultures of consumption and production (and national borders) along the way. I have already reviewed the arguments to the contrary, coming out of the institutionalist literature on national business systems and varieties of capitalism. However, it is possible to read too much into this critique. A recently fashionable view pursues an equally extreme alternative, by claiming that the global corporation has become re-embedded in local clusters of knowledge production, *wherever* these may be, thereby freeing themselves from the geographical legacies of their local and national origins. Appealing as this vision may be to some, the accumulated evidence thus far suggests that there remains a substantial gap between the rhetoric of local embedding and the reality of corporate spatial organization and practices (Doremus et al., 1998; Gertler et al., 2000). If embedding takes place anywhere, this is still largely in one's original home market.

Continuing this theme, the power of local, regional and national communities to shape the contours of production systems represents another somewhat neglected issue in today's economic geography.[20] A culturally inflected approach to this question has emphasized the importance of public investments in education and scientific research. The state has also been called upon to help modify the behaviour of economic agents by inculcating cultures of individual enterprise and entrepreneurship, and especially cultures of cooperation, where none existed before – what Jessop (2000: 76) refers to as a fundamental shift from government to governance and *meta-governance*.[21] Taking the longer view, the presence of the public sphere in the realm of production has historically been considerably larger than this, suggesting that we need to resist the urge to reduce its role to that of educator, coach and matchmaker.

Through its control of key economic and social institutions, the state's continuing role in shaping the behaviour and practices of firms, and in imposing order on an otherwise unruly world of markets, remains vital to our understanding of how production works and how production cultures are themselves produced at a variety of spatial scales. This role for the state as a producer and regulator of the 'soft infrastructure' of the economy is as old as capitalism itself (Polanyi, 1944). Of course these institutions, while durable, also evolve through time and are themselves

shaped by local, national and international politics and power struggles (Clark et al., 2000b). It is here that we need to reconsider explicitly the third big idea from the cultural economic geography of production – path dependency. Whether at the national or the regional scale, the influence of human action can and does bring about change within institutional structures. Our present level of understanding of the evolutionary dynamics of lock-in versus change still remains woefully underdeveloped, with seemingly little or no room for local or national politics.

In summing up, there is little doubt that a cultural economic geography of production has opened up new conceptual and empirical terrain previously considered off-limits to economic geographers, and the field is richer for it. Certain truisms – once unthinkable – have now become widely accepted. Culture and economy are now accepted as two sides of the same coin. Seemingly natural structures such as markets are now understood to be deeply embedded. One simply cannot understand how economies in particular places work (or how production processes unfold) without considering the broader social matrix within which they are situated. While this has always been true, our awareness of this relationship, and of the importance of local cultures of production, has been heightened by the growing importance of two major phenomena: the social division of labour in production systems and the role of knowledge production and sharing (learning). All the same, with these basic insights now effectively assimilated within the canon of economic-geographical thought, it is now time to take our analysis to another level of sophistication. In the heady early days of economic geography's cultural turn, the choice posed by many – to which I alluded at the beginning of this chapter – was between cultural economy and political economy. As we enter a more mature phase, perhaps we shall see greater value in reworking earlier themes and questions concerning the geography of production. How about a cultural political economy of geography?

NOTES

The author wishes to acknowledge the Social Sciences and Humanities Research Council of Canada and the Goldring Chair in Canadian Studies at the University of Toronto for their continuing support. He would also like to thank Trevor Barnes, Gordon Clark, Norma Rantisi and the editors for their very helpful comments and insights.

1 Thrift describes economic geography's cultural turn as 'probably the most important event to impact on the

sub-discipline in the last ten or fifteen years' (2000a: 689). He adds, 'there is no doubt that the cultural turn saved economic geography from what might otherwise have been a musty oblivion. By the 1980s, economic geography was in a pretty moribund state, at risk of boring its audience to death' (2000a: 692).

2   Just as in the economy itself, the 'product life cycle' of academic ideas in human geography seems to have become dramatically shorter than ever before. Crang attributes this 'powerful pressure for subdisciplinary reinvention' to 'the institutional dynamics of a field of enquiry in which academic capital was, and is, produced primarily through intellectual innovation rather than loyalty to existing luminaries' (1997: 3). Thrift apparently agrees: 'human geography is a profoundly trendy subject' (2000a: 692).

3   As Walker (2000) has so eloquently reminded us, 'production' is not confined to factories or the making of physically tangible products. Large and growing proportions of most national and regional economies – whether measured in terms of employment or the value of output – are dedicated to the production of services and intangible commodities. Some of these are directly or indirectly related to the production and consumption of tangible products; others are produced and consumed independently. As Walker puts it: 'Production is making something. That something may be as rock solid as a car, as passing as a meal, or as shadowy as a program flickering on a TV screen … But production is, in all cases, an act of human labor; it involves work, plain and simple' (2000: 113–14).

4   For a fuller, critical examination of this theme, see Gertler (1997).

5   Most would identify the founding of this school of thought with the publication in 1982 of Nelson and Winter's seminal work *An Evolutionary Theory of Economic Change*. For another classic statement, see Nelson (1995).

6   While some see a new kind of Schumpeterian neoliberalism emerging from the literature on local production culture and learning regions (Jessop, 2000; Lovering, 1999), others emphasize that a local 'productivity' or 'competitiveness agenda' can be defined inclusively, so that rising real wages and richer employment prospects become central goals for local development policy (Cooke and Morgan, 1998).

7   It is this emphasis on the firm as the primary producer of competence and capability that distinguishes the resource-based approach from a broader institutionalist perspective.

8   To be perfectly explicit, Allen refers to this idea as ' "the truth" … [an] account of economic knowledge [that] has the power to render itself true' (2000: 27).

9   For a more detailed consideration of this literature, see Gertler (2001a; 2001b).

10   See Gertler (2001b) for further commentary on this.

11   In contrast to Allen (2000), Gertler (2001b) argues that the real problem lies not in 'conflating' the ideas of the two Polanyi brothers, but in failing to understand properly Karl Polanyi's arguments about the institutional foundations of economic action. Many of the practices of managers and workers in firms are, in fact, 'embedded' in (that is, influenced by) national macro-institutional structures (shaping behaviour in capital and labour markets, corporate governance and inter-firm relations) of which individual actors may remain partly or completely unaware. Seen in this light, the notion of 'embedding' is problematic only if it is taken to imply necessarily *local and regional* embedding. At the same time, Michael Polanyi's analysis of the role of 'context' in shaping tacit knowledge would have benefited considerably from a closer reading of his older sibling's (pre-existing) work on institutions and their ability to define the social context for human action. Without any kind of serious institutional analysis, Michael is compelled to develop a more truncated, cognitive-experiential understanding of tacitness, based on a performer's lack of self-awareness of tacitly held know-how and/or an inability to describe or explain to others how a particular skill is performed.

12   In a recent paper, I provide a critical review of ongoing debates on international and interregional convergence of industrial practices (Gertler, 2001a).

13   There is now a very lively debate taking shape within the field of international political economy over the extent to which national institutional structures cohere to achieve system-wide 'complementarity' within distinctive national models (Amable, 2000). Clark et al. (2001) raise some important questions about the ease with which this coherence can be achieved and sustained over time, while also arguing that the agency of the firm needs to be more prominently represented to achieve better balance in the next generation of conceptual models.

14   *Mea culpa*. I have participated in this venture myself – see Gertler and Wolfe (2002).

15   For another, much more carefully considered expression of this sentiment, see Hudson (2001).

16   This is evident in the subtitle of her book: *Culture and Competition in Silicon Valley and Route 128*.

17   The recent works of Schoenberger (1997), Allen (2000) and French (2000) represent important exceptions to this general state of affairs. As noted above, in Schoenberger's analysis, corporate culture is all about the power to define the fundamental identity of the firm. More recently, Johnson and Lundvall (2001) have come round to the view that power relations may well play a role in the production, distribution and ownership of tacit knowledge.

18   Recalling the classic problems arising from the separation of conception and execution in the firm's division of labour, one might say that the employer's task of identifying and appropriating the tacit knowledge produced by their workers in doing their daily jobs remains a principal objective and challenge in firms' pursuit of high-quality, innovative production – what I have elsewhere referred to as 'the original tacit knowledge problem' (see Gertler, 2001b).

19 While Florida's use of a performative metaphor such as 'talent' may be unconscious, others have more consciously adopted a perspective in which the performance aspects of work are front and centre (Thrift, 2000b).

20 The recent work of labour geographers such as Herod, Wills and others stands out as a notable exception to this general condition. For a summary of much of this work, see Rutherford and Gertler (2002).

21 In a lovely, subtle allusion to Karl Polanyi who famously stated that 'Laissez-faire was planned; planning was not' (1944: 141), Jessop discusses meta-governance in the following terms: 'Governments on various scales are becoming more involved in organising the self-organisation of partnerships, networks, and governance regimes' (2000: 76) .

REFERENCES

Allen, J. (2000) 'Power/economic knowledge: symbolic and spatial formations', in J.R. Bryson, P.W. Daniels, N. Henry and J. Pollard (eds) *Knowledge, Space, Economy*. London: Routledge. pp. 15–33.

Amable, B. (2000) 'Institutional complementarity and diversity of social systems of innovation and production', *Review of International Political Economy* 7: 645–87.

Amin, A. (2000) 'Organisational learning through communities of practice'. Paper presented at the Workshop on The Firm in Economic Geography, University of Portsmouth, UK, 9–11 March.

Arthur, B. (1989) 'Competing technologies, increasing returns and lock-in by historical events', *Economic Journal* 99: 116–31.

Barnes, T.J. (1996) *Logics of Dislocation*. New York: Guilford.

Barnes, T.J. (1997) 'Theories of accumulation and regulation: bringing life back into economic geography – introduction to section three', in R. Lee and J. Wills (eds) *Geographies of Economies*. London: Arnold. pp. 231–47.

Barnes, T.J. and Gertler, M.S. (eds) (1999) *The New Industrial Geography: Regions, Regulation and Institutions*. London: Routledge.

Berger, S. and Dore, R. (eds) (1996) *National Diversity and Global Capitalism*. Ithaca: Cornell University Press.

Boyer, R. and Hollingsworth, J.R. (1999) *Contemporary Capitalism: The Embeddedness of Institutions*. Cambridge: Cambridge University Press.

Brown, J.S. and Duguid, P. (2000) *The Social Life of Information*. Boston: Harvard Business School Press.

Cairncross, F. (1997) *The Death of Distance*. Boston: Harvard Business School Press.

Clark, G.L., Feldman, M.P. and Gertler, M.S. (eds) (2000a) *The Oxford Handbook of Economic Geography*. Oxford: Oxford University Press.

Clark, G.L., Mansfield, D. and Tickell, A. (2000b) 'The German social market in the world of global finance: pension investment management and the limits of consensual decision making'. Working Paper 00-06, School of Geography and the Environment, University of Oxford.

Clark, G.L., Tracey, P. and Lawton Smith, H. (2001) 'Rethinking comparative studies: an agent-centred perspective'. Paper presented at the Annual Meeting of the Association of American Geographers, New York City, February.

Cohen, S. and Fields, G. (1999) 'Social capital and capital gains in Silicon Valley', *California Management Review* 41: 108–30.

Cooke, P. and Morgan, K. (1998) *The Associational Economy: Firms, Regions, and Innovation*. Oxford: Oxford University Press.

Crang, P. (1997) 'Cultural turns and the (re)constitution of economic geography', in R. Lee and J. Wills (eds) *Geographies of Economies*. London: Arnold. pp. 3–15.

David, P. (1985) 'Understanding the economics of QWERTY', *American Economic Review* 75: 332–7.

Doremus, P., Keller, W., Pauly, L. and Reich, S. (1998) *The Myth of the Global Corporation*. Princeton: Princeton University Press.

Florida, R. (2001) 'The economic geography of talent'. Paper presented at the Annual Meetings of the Association of American Geographers, New York City, 1 March (http: //info.heinz.cmu.edu/~florida/).

Freeman, C. and Perez, C. (1988) 'Structural crises of adjustment: business cycles and investment behaviour', in G. Dosi, C. Freeman, G. Silverberg and L. Soete (eds) *Technical Change and Economic Theory*. London: Pinter. pp. 38–66.

French, S. (2000) 'Re-scaling the economic geography of knowledge and information: constructing life assurance markets', *Geoforum* 31: 101–19.

Gertler, M.S. (1995) '"Being there": proximity, organization and culture in the production and adoption of advanced manufacturing technologies', *Economic Geography* 71: 1–26.

Gertler, M.S. (1997) 'The invention of regional culture', in R. Lee and J. Wills (eds) *Geographies of Economies*. London: Arnold. pp. 47–58.

Gertler, M.S. (2001a) 'Best practice? Geography, learning and the institutional limits to strong convergence', *Journal of Economic Geography* 1: 5–26.

Gertler, M.S. (2001b) 'Tacit knowledge and the economic geography of context, or the undefinable tacitness of being (there)'. Paper presented at the Nelson and Winter DRUID Summer Conference, Aalborg, Denmark, 12–15 June.

Gertler, M.S. and Wolfe, D.A. (eds) (2002) *Innovation and Social Learning*. Basingstoke: Macmillan/Palgrave.

Gertler, M.S., Wolfe, D.A. and Garkut, D. (2000) 'No place like home? The embeddedness of innovation in a regional economy', *Review of International Political Economy* 7: 688–718.

Gibson-Graham, J.K. (1996) *The End of Capitalism (As We Knew It)*. Oxford: Blackwell.

Glasmeier, A.K. (2001) *Manufacturing Time: Global Competition in the Watch Industry, 1795–2000*. New York: Guilford.

Grabher, G. (1993) 'The weakness of strong ties: the lock-in of regional development in the Ruhr area', in G. Grabher (ed.) *The Embedded Firm: On the Socio-Economics of Interfirm Relations*. London: Routledge. pp. 255–78.

Granovetter, M. (1985) 'Economic action and social structure: the problem of embeddedness', *American Journal of Sociology* 91: 481–510.

Henry, N. and Pinch, S. (2000) '(The) industrial agglomeration (of Motor Sport Valley): a knowledge, space, economy approach', in J.R. Bryson, P.W. Daniels, N. Henry and J. Pollard (eds) *Knowledge, Space, Economy*. London: Routledge. pp. 120–41.

Herrigel, G. (1996) *Industrial Constructions: The Sources of German Industrial Power*. Cambridge: Cambridge University Press.

Howells, J. (2000) 'Knowledge, innovation and location', in J.R. Bryson, P.W. Daniels, N. Henry and J. Pollard (eds) *Knowledge, Space, Economy*. London: Routledge. pp. 50–62.

Hudson, R. (2001) *Producing Places*. New York: Guilford.

Hutton, W. (1995) *The State We're In*. London: Cape.

Jessop, B. (2000) 'The state and the contradictions of the knowledge-driven economy', in J.R. Bryson, P.W. Daniels, N. Henry and J. Pollard (eds) *Knowledge, Space, Economy*. London: Routledge. pp. 63–78.

Johnson, B, and Lundvall, B.-Å. (2001) 'Why all this fuss about codified and tacit knowledge?' Paper presented at the DRUID Winter Conference, Korsør, Denmark, 18–20 January.

Kay, J. (1993) *Foundations of Corporate Success*. Oxford: Oxford University Press.

Kochan, T.A. and Osterman, P. (1994) *The Mutual Gains Enterprise*. Boston: Harvard Business School Press.

Kumar, P. and Holmes, J. (1997) 'Canada: continuity and change', in T.A. Kochan, R.D. Lansbury and J.P. MacDuffie (eds) *After Lean Production: Evolving Employment Practice in the World Auto Industry*. Ithaca: Cornell University Press. pp. 85–108.

Lee, R. and Wills, J. (eds) (1997) *Geographies of Economies*. London: Arnold.

Lovering, J. (1999) 'Theory led by policy? The inadequacies of "the new regionalism" illustrated from the case of Wales', *International Journal of Urban and Regional Research* 23: 379–95.

Lundvall, B.-Å. (1988) 'Innovation as an interactive process: from user–producer interaction to the national system of innovation', in G. Dosi, C. Freeman, G. Silverberg and L. Soete (eds) *Technical Change and Economic Theory*. London: Pinter. pp. 349–69.

Malmberg, A. and Maskell, P. (2002) 'The elusive concept of agglomeration economies: towards a learning-based theory of spatial clustering', *Environment and Planning A* 34: 429–49.

Markusen, A. (2001) 'Targeting occupations rather than industries in regional and community economic development'. Paper presented at the Annual Meetings of the Association of American Geographers, New York City, 28 February to 4 March.

Maskell, P. and Malmberg, A. (1999) 'Localised learning and industrial competitiveness', *Cambridge Journal of Economics* 23: 167–86.

McDowell, L. (1997) 'A tale of two cities? Embedded organisations and embodied workers in the City of London', in R. Lee and J. Wills (eds) *Geographies of Economies*. London: Arnold. pp. 118–29.

Metcalfe, J.S. and De Liso, N. (1998) 'Innovation, capabilities and knowledge: the epistemic connection', in R. Coombs, K. Green, A. Richards and V. Walsh (eds) *Technological Change and Organization*. Cheltenham: Elgar. pp. 8–27.

Mitchell, W. (1995) *City of Bits: Space, Place, and the Infobahn*. Cambridge, MA: MIT Press.

Myrdal, G. (1957) *Economic Theory and Under-Developed Regions*. New York: Harper and Row.

Nelson, R.R. (1995) 'Recent evolutionary theorizing about economic change', *Journal of Economic Literature* 33: 48–90.

Nelson, R.R. and Winter, S.G. (1982) *An Evolutionary Theory of Economic Change*. Cambridge, MA: Harvard University Press.

O'Brien, R. (1992) *Global Financial Integration: The End of Geography*. London: Pinter.

OECD (1996) 'The knowledge–based economy', *Science, Technology and Industry Outlook* 229–56.

Oliver, N. and Wilkinson, B. (1988) *The Japanization of British Industry*. Oxford: Blackwell.

O'Sullivan, M. (2000) *Contests for Corporate Control: Corporate Governance in the United States and Germany*. Oxford: Oxford University Press.

Peck, J. (1996) *Work-Place: The Social Regulation of Local Labour Markets*. New York: Guilford.

Penrose, E. (1959) *The Theory of the Growth of the Firm*. Oxford: Blackwell.

Piore, M. and Sabel, C. (1984) *The Second Industrial Divide*. New York: Basic.

Polanyi, K. (1944) *The Great Transformation*. New York: Rinehart.

Polanyi, M. (1958) *Personal Knowledge: Towards a Post-Critical Philosophy*. London: Routledge and Kegan Paul.

Polanyi, M. (1966) *The Tacit Dimension*. New York: Doubleday.

Putnam, R. (1993) *Making Democracy Work*. Princeton: Princeton University Press.

Rutherford, T. and Gertler, M.S. (2002) 'Labour in "lean" times: geography, scale and the national trajectories of workplace change', *Transactions, Institute of British Geographers* 8: 1–18.

Saxenian, A. (1994) *Regional Advantage: Culture and Competition in Silicon Valley and Route 128*. Cambridge, MA: Harvard University Press.

Saxenian, A. (1999) *Silicon Valley's New Immigrant Entrepreneurs*. San Francisco: Public Policy Institute of California.

Sayer, A. and Walker, R.A. (1992) *The New Social Economy: Reworking the Division of Labor*. Oxford: Blackwell.

Schoenberger, E. (1997) *The Cultural Crisis of the Firm*. Oxford: Blackwell.

Schoenberger, E. (2000) 'The management of time and space', in G.L. Clark, M.P. Feldman and M.S. Gertler (eds) *The Oxford Handbook of Economic Geography*. Oxford: Oxford University Press. pp. 317–32.

Scott, A.J. (1988) *Metropolis*. Berkeley: University of California Press.

Scott, A.J. (2000) 'Economic geography: the great half-century', in G.L. Clark, M.P. Feldman and M.S. Gertler (eds) *The Oxford Handbook of Economic Geography*. Oxford: Oxford University Press. pp. 18–44.

Sheppard, E. and Barnes, T.J. (eds) (2000) *A Companion to Economic Geography*. Oxford: Blackwell.

Soskice, D. (1999) 'Divergent production regimes: coordinated and uncoordinated market economies in the 1980s and 1990s', in H. Kitschelt, P. Lange, G. Marks and J.D. Stephens (eds) *Continuity and Change in Contemporary Capitalism*. Cambridge: Cambridge University Press. pp. 101–34.

Storper, M. (1997) *The Regional World*. New York: Guilford Press.

Storper, M. and Walker, R.A. (1989) *The Capitalist Imperative*. Oxford: Blackwell.

Thrift, N.J. (2000a) 'Pandora's box? Cultural geographies of economies', in G.L. Clark, M.P. Feldman and M.S. Gertler (eds) *The Oxford Handbook of Economic Geography*. Oxford: Oxford University Press. pp. 689–704.

Thrift, N.J. (2000b) 'Performing cultures in the new economy', *Annals of the Association of American Geographers* 90: 674–92.

Veblen, T. (1919) *The Place of Science in Modern Civilization and Other Essays*. New York: Huebsch.

von Krogh, G., Ichijo, K. and Nonaka, I. (2000) *Enabling Knowledge Creation: How to Unlock the Mystery of Tacit Knowledge and Release the Power of Innovation*. Oxford: Oxford University Press.

Walker, R.A. (2000) 'The geography of production', in E. Sheppard and T.J. Barnes (eds) *A Companion to Economic Geography*. Oxford: Blackwell. pp. 113–32.

Wenger, E. (1998) *Communities of Practice: Learning, Meaning and Identity*. Cambridge: Cambridge University Press.

Whitley, R. (1999) *Divergent Capitalisms: The Social Structuring and Change of Business Systems*. Oxford: Oxford University Press.

Wills, J. (2000) 'Being told and answering back: knowledge, power and the new world of work', in J.R. Bryson, P.W. Daniels, N. Henry and J. Pollard (eds) *Knowledge, Space, Economy*. London: Routledge. pp. 261–76.

Yeung, H.W. (2000) 'The dynamics of Asian business systems in a globalizing era', *Review of International Political Economy* 7: 399–433.

Young, A.A. (1928) 'Increasing returns and economic progress', *The Economic Journal* 38: 527–42.

7

Cultures of Consumption

Don Slater

The literature on consumption has grown enormously over the past 15 years, now constituting a recognized subdiscipline within many social sciences and humanities (Miller, 1995). While consumption has featured significantly as an issue in modern western thought since at least the eighteenth century, it was rarely regarded as a socially consequential object of study in its own right. Consumption was widely considered both too trivial and too eccentrically individual to figure largely in social analysis. It appeared rather as an object of moral-political judgement, an index of either the growth of liberal freedoms or the moral and cultural degeneration within modern commercial society.

By contrast, the huge contemporary interest in consumption rests on three broad premises, each of which places culture at the centre of social processes, and in ways that have made consumption studies almost paradigmatic of the 'cultural turn' in social thought. Firstly, notions such as 'material culture' or 'common culture' stress that consumption is central to social and cultural reproduction. All acts of consumption are profoundly cultural. Even ostensibly 'natural' and mundane processes such as eating invoke, mediate and reproduce those structures of meaning and practice through which social identities are formed and through which social relations and institutions are maintained and changed over time. The consumption of a family meal requires complex frameworks of meaning that adjudicate just what counts as food, how it is properly prepared and presented, and what is good or bad in terms of such disparate issues as health, gender roles and powers, ethical relations of care, the identity of the family and its religion, social status and so on. In the extended process of consumption – shopping, buying, using – people raise and negotiate the most central questions as to who they are and what they need.

The second premise has been a concern with 'consumer culture' as a characterization of modern market society (Slater, 1997a; Slater and Tonkiss, 2001), and more specifically as an increasingly central feature of what came to be known as the postmodern (Featherstone, 1991). Consumption as cultural process may be central to all human society, but only the modern west came to define itself as a consumer culture or consumer society. The underlying claim here is that as a result of modernization processes such as marketization, the decline of traditional status systems and the rise of cultural and political pluralism, private, market-based choice has become increasingly central to social life. At the extreme point, the neoliberal projects of the 1980s recognized and promoted this by seeking to redefine all social processes (e.g. education, health provision, democracy) according to the paradigm of consumption such that in the field of education, for example, students become 'consumers' and their 'demand' sovereign. In a consumer culture, then, key social values, identities and processes are negotiated through the figure of 'the consumer' (as opposed to, say, the worker, the citizen or the devotee); central modern values such as freedom, rationality and progress are enacted and assessed through consumerist criteria (range of choice, price calculations and rising affluence, respectively); and the cultural landscape seems to be dominated by commercial signs (advertising, portrayals of 'lifestyle' choices through the media, obsessive concern with the changing meanings of things).

Finally, it is partly through the study of consumption that we have come to better understand the role of culture in the constitution of economic processes and institutions. Consumption is not a cultural endpoint or addition to 'truly' economic processes of production or formally modelled market exchange, nor can it be reduced to quantitative measures of 'demand'. To the contrary, the study of consumption cultures leads us to examine the construction of objects, exchanges and relationships across a wide range of interconnected sites and processes.

The specific conjuncture of cultures of consumption with geographical perspectives moves in two directions, each modifying the other in exciting ways. On the one hand, geography has been both influenced by and a major contributor to several major themes that cut across the entire consumption field. Firstly, a cultures of consumption perspective regards consumption as an active process of making and using meanings and objects, and the consumer as a subject active in the constitution of its own subjectivity and world. For example, Miller, et al. (1998), as discussed further below, look at shopping spaces as sites which refract gender, class and ethnicity through the consumer's active understanding, use and negotiation of consumption landscapes.

Secondly, geographers along with other scholars have largely come to reject the presumption that production simply determines consumption or that consumer choice transparently directs production. The focus has instead been on the complex and contradictory connections between different moments in the making of material cultures. Hence, for example, the idea of the active consumer leads directly to the possibility that consumers are themselves productive in their appropriation of things – making new meanings, uses and relationships – and that production has therefore to be understood as a distributed process, one that inhabits multiple sites (Suchman, 1999). A profoundly consequential result of this is a concern to reconnect political economy with cultural analysis in new and more complex ways (for example, du Gay and Pryke, 2001; Wrigley and Lowe, 1996).

Thirdly, consumption studies has promoted new methodological concerns. Above all, there has been a major 'ethnographic turn' that owes a great deal to issues raised by cultures of consumption: the focus on both the culturally active consumer and the distributed nature of economic-cultural processes requires us to probe deeply into the detailed and particular conjunctures that make up any act or process of consumption and that relate it to broader social contexts (Jackson, 1995–96). This ethnographic turn was taken not only as a corrective to older political economies that pretended to derive consumption unproblematically from structural determinants, but also in response to semiotic and postmodern currents for whom consumption could be derived from readings of objects and spaces without examination of actual and particular consumers and consumption practices.

At the same time, it is important to consider geography's specific contribution to this field in attending to the connections between consumption and space, a set of issues that have become central to all studies of consumption cultures. Crudely, we might think about the relationship between these terms in two directions: on the one hand, consumption is spatially constructed and distributed; on the other hand, important social spaces are constructed in relation to consumption. The cases of retailing and globalization, discussed below, indicate how intertwined these two relationships can be, but it might be useful to think about them separately for a moment. On the one hand, modern consumption emerges from a division between production and consumption that is partly spatial: the spatial segregation of labour and leisure, work and home, public and private. Indeed, the central commodities of modern life – home and automobile – are premised on this spatiality and the need to move between production and consumption spaces (Aglietta, 1979), while key social spaces such as city and suburb are marked out accordingly. One could also think about such notions as commodity fetishism and the split between the politics of production and consumption in spatial terms. The market as a mediation seems to purify consumer goods of any traces of their conditions of production, which are only visible somewhere else (see, for example, Ross' 1997 account of the problems faced by consumer campaigns against commodities such as Nike that involve sweated labour). Frameworks such as commodity chain analysis aim at making these connections visible again (Fine and Leopold, 1993), while much discussion of the so-called 'new economy' is concerned with the possible breakdown of these older divisions in the confused spatiality of the internet and the confused materiality of information goods (Poster, 2001).

On the other hand, space is not an objective container or structure that moulds consumption; consumption is crucial (and perhaps increasingly crucial) in constituting social spaces. This is a long-term theme within much cultural anthropology in which the sharing of goods and meanings in consumption is fundamental to moulding the material and cultural form of households (for example, Bourdieu's 1973 classic account of the Berber house), nation and ethnicity (including the media consumption that helps constitute 'imagined communities': Anderson, 1986). Consumption may also be crucial in our construction of spaces of which we have no direct experience. Consider, for example, the geographical knowledges that are constructed out of our understandings of the origins of goods, such as the ethnicity of spices and 'exotic' or cosmopolitan cuisines (Cook and Crang, 1996; Crang, 1996), or consider tourism as a construction of space through the consumption of place (Urry, 1990). As Crang argues: 'cultural lives and economic processes are characterized not only by the points in space where they take and make place, but by the movements to, from and between those points' (1996: 47). In their mobility, goods make new spatial connections and spatial knowledges. There is also a more specific set of arguments that, under conditions of post-Fordism or postmodernism or 'new economy', consumption has become increasingly central to the constitution of social spaces. This might be specifically analysed in terms of the centrality of retail spaces to the fate of cities (as discussed below), either transforming them into centres of consumption and leisure or exporting these functions to exurban areas which now compete with city spaces (for example, Soja, 1989; 1996; 2000).

Finally, we might add to these themes another contribution from cultural geography to the study of cultures of consumption: that of 'scale' and 'scaling', a very fruitful concept that was completely missed by consumption scholars from other disciplines. Bell and Valentine (1997: 12), for example, drawing on Smith (1993), structure their discussion of food consumption according to the various scales of body, home, community, city, region, nation and globe. Each level involves different aspects, conditions and processes of consumption; but equally each level is partially constituted through different consumption processes. It is easy to see both the differences and connections between the construction of specific family relationships through different food practices, and vice versa; and the construction of national or regional diets (and of national cultures and identities through different diets). We can also see that the production of a certain kind of body through diet may be scaled up to the national or global level (the production

of a Californian-style 'hard-body look' is one way of imagining a global culture), while the global organization of food chains equally scales down to the structures within which everyday body practices are carried out.

In the rest of this chapter we will try to draw out some of these themes as they have developed in the study of cultures of consumption. I will first look at the relationships drawn between culture and economy in older traditions of thought on consumption, proceeding then to consider the perspectives that underlie more recent culturalist approaches to consumption. The final two sections, on shopping and on globalization, look at these themes in terms of the two most dynamic and consequential conjunctures of thinking about consumption: space and economy.

ECONOMY AND CULTURE

Today, consumption has come to represent the site on which culture and economy most dramatically converge. Historically, it has marked a central point of division between them. For conventional economics, for example, consumption has always represented a process that takes place outside the economy, for two reasons. Firstly, economics is largely associated with the production and distribution of goods, whereas consumption is defined as the mere 'using up' of things, their destruction in use. Secondly, in conventional economics, actors enter the marketplace with their needs and wants already formed outside it, through cultural or biological or 'subjective' processes of taste formation that are not considered part of the economists' remit. Once inside the market, consumers supposedly then make price-rational calculations in relation to 'utility', which is not culture, but an abstraction from culture which is manifested in the form of variable quantities of demand at different prices. Cultures of consumption are therefore the backdrop to economic life, but play no role within it, or in analysing it. Conversely, if consumption is a cultural process that should take no part within the economy, it is also the case that economic processes should play no role within culture: conventional economics relies on the autonomy of supply and demand. For example, Galbraith (1972) characterized the marketing mix, comprising cultural interventions such as advertising and design, as a 'revised sequence' which destroyed markets by allowing the cultural control of corporations over the taste forming processes which generated demand for their goods.

Indeed, the critical stalemate that stymied thinking about consumption until quite recently was structured by an opposition between economy and culture. On the one hand, liberal traditions (including neoclassical economics) assumed, as we have just noted, the autonomy of consumption processes from economic ones, and saw this as central to both the autonomy and the 'authority' (Keat et al., 1994) of the consumer: economic processes should respond to cultural (or biological or simply 'subjective' and individual) determinations of needs and wants that occur elsewhere. This view is particularly inscribed in the notion of 'consumer sovereignty'. The broad liberal tradition, from Hobbes onwards, has privileged the liberty of the self-determined individual, possessed of self-defined desires and interests, and placed this figure at the centre of moral, political and economic good. The market is crucial to this conception as a space in which individuals are ideally freed from external social regulation. At the same time, the individualist premises of liberal thought are methodologically inimical to a cultural approach to consumption, in so far as 'culture' assumes meanings shared within collectivities with consequential dynamics and identities. In Margaret Thatcher's immortal formulation, for a neoliberal 'There is no such thing as society, there are individual men and women and there are families' (quoted in Heelas and Morris, 1992: 2).

On the other hand, critical traditions – of both the right and the left – have tended to regard consumption as the site of major incursions of economic processes into culture and everyday life. For them, modern consumer culture did not register the triumph of individual freedom but rather expressed the dominance of market exchange and industrial process over human life and meanings, apparently rendering them inauthentic or otherwise debased. Both conservative and progressive critics have tended to start from a somewhat nostalgic view of premodern life as an organic community characterized by a direct and largely transparent relationship between production and consumption: most goods were produced by people who were also final consumers, or in direct contact with final consumers. In this idealized world prior to capitalism, only a small fraction of consumption was mediated by markets and commodities. Culture therefore evolved (or more often was held stable) through the internal rhythms of collective life rather than through the pursuit of commercial interests or the impersonal structures of commodity exchange.

In such perspectives, the market drives a wedge through the previously organic relation between production and consumption, and monetary values become the only ones that now adjudicate social worth and distribute social goods. For conservatives this has meant that social status and cultural goods are now opened up to anyone who has the money to buy them, hence threatening those social traditions and hierarchies which – in premodern societies – ensured the transmission of 'authentic' values. For progressives, it has meant that all social and cultural values are bound up with commodity exchange, hence subordinate to the logics of profit and exploitation. In either case, consumption tends to mark the process through which culture is colonized by economic forces, and cultural critique puts forward a model of culture as an ideal realm purified of commercial interests. It is important to note that the very word 'culture' developed its modern meaning in the eighteenth century in relation to the rise of commercial society. Raymond Williams (1976; 1985), for example, argued that a 'culture and society' tradition emerged which sought to define values that it believed were previously embedded in traditional ways of life but which were now under assault from industrial civilization and the 'cash nexus'. Both conservative and progressive intellectuals sought to map a terrain of authentic culture that could be defended from capitalist modernization on the basis of values that could not be reduced to market prices and individual choices. In relation to consumer culture, this largely took the form of attacks on the commercial debasement and industrial management of public taste, leisure and consciousness in perspectives as diverse as cultural criticism and critical theory.

The most commanding formulation of market-mediated culture as alienation is undoubtedly that of Marx. Marx's understanding of pre-capitalist social order is largely romantic. However brutal the old world might be, it is characterized by a transparency and directness of the relationship between production and consumption, summarized in the model of production of use values rather than exchange value. The commodity form – production of exchange value for the market; labour as commodity – means that workers sell their labour power in one market for the cash with which they might purchase the means of consumption in quite other markets. This means, firstly, that their (concrete) labour is disconnected from its own product: we do not work directly to satisfy our needs. Capitalist consumption is utterly warped by alienation. Secondly, technically, labour is denied ownership of the means of production; under conditions of exploitation (the extraction of surplus value) labour as a whole receives only a portion of the value it produces and is therefore *quantitatively* unable to purchase as means of consumption all that it in fact produced. It is significant that for Marx (as for Keynes) this has meant not only the relative poverty of workers but (more importantly for all of them) technical crisis tendencies within capitalism, which suffers periodic catastrophes as a result of endemic underconsumption or overproduction.

The split between production and consumption occasioned by market mediation is a temporal and spatial displacement that Marx identified through the term 'commodity fetishism'. This is Marx's key statement of the structural separation of production and consumption, a disconnectedness that is mediated and at the same time obscured by the market. As is the case throughout Marx's work there is a fusion of the ethical and the technical: market mediation not only mystifies the social order and constitutes the condition for alienation, but it is also economically unstable and crisis prone. Capitalists – who are driven by competitive forces to increase the scale of their production – cannot know in advance what expenditures of labour will later be deemed 'socially necessary' by effective consumer demand in the market; they therefore constantly court individual bankruptcy and collective catastrophe in the form of the trade cycle.

At the same time, market mediation allows for a relatively autonomous space of commodity representations, the elaboration of packaging, branding, advertising and so on, carried out by functionally differentiated firms or departments (Haug, 1986; Richards, 1991). Much work on consumption has been either a critique or a phenomenology of commodity fetishism. This is obvious in the case of the theme of reification in western Marxism (Lukacs, Adorno, Habermas). This considers not only the cultural consequences of production for the market, but also the political consequences of a social order which appears as the product not of human labour but rather of the (quantitative) relation between atomized 'things'. For Lukacs and Adorno, the entire social landscape appears to individuals as a consumable spectacle – a literally natural landscape, dominated by natural forces – rather than as a historical product of human action and the historical site of active social intervention. Less obviously related to commodity fetishism are more recent postmodern approaches such as that of Baudrillard, discussed below, in which consumption appears as a spectacle of signs completely detached from other social relations and processes. The link runs backwards through the Situationists' 'society of the spectacle' (Debord, 1991; Plant, 1992) and Lefebvre's critique of everyday life as alienation (1947; 1971), both of which are firmly rooted in the framework of early Marx. Society is experienced entirely through the detached signs that it produces through market mediation.

It is important to note that critical perspectives on consumption have been largely characterized by a 'productivist bias' in which consumption is derived from characterizations of modes of production or the industrial order. This is frequently based on equating the production/consumption dichotomy with the economy/culture dichotomy and giving analytical priority to the first term in each case. Hence the focus is on how forms of consumption are structurally determined by processes and institutions such as advertising and marketing or the changing forms of mass production. The problem is not simply that this produces an image of the consumer as 'cultural dope' or 'dupe', as passive victim, but also that the effectivity of production systems in securing cultural ends is too often assumed without proper (and probably ethnographic) investigation of the actual consumption practices that consumers engage in: they are simply 'read off' of industrial processes. As has been well understood in studies of media consumption, for example, the issue of power over consumption cannot be resolved into either structures (however mighty they may be) or self-determining agents (for example, Morley, 1992). Moreover, in investigating actual consumer and producer practices we find a complexity of interconnections that cut across both the production/consumption and economy/culture divides.

This issue comes to the fore again in the most important recent productivist framework for understanding contemporary transformations within consumption cultures. Historically, modern consumer culture is often associated with the rise of mass production and corresponding mass consumption at the turn of the twentieth century, developing into what is often characterized as a Fordist system during the post-war period. As most thoroughly analysed by the French regulationist school (Aglietta, 1979; Lee, 1993; Lipietz, 1992; see also Lee, 2000), the general course of industrialization and marketization had little effect on the broad mass of the population – which continued to cater for most of its needs through non-commodities – until the rise of mass production, exemplified by the Fordist flow-past assembly line with its intensive technical division of labour, high productivity, aesthetic standardization of goods and decreasing unit costs. The double need to ensure workplace discipline in increasingly alienated production processes, and to ensure sufficient effective demand to sell the huge volume of output, were both to be solved by promising workers (who are also consumers) a steadily rising standard of living defined through consumption norms. These were institutionalized through such mechanisms as national industrial relations agreements, underwritten by the state, and Keynesian demand management. Therefore Fordism is a means of institutionalizing and stabilizing the split between production and consumption, identifying private consumption as the sphere in which modern citizens might experience progress, freedom and self-determination, and establish culturally meaningful ways of life within their private sphere of consumption.

Post-Fordism – usually dated from the early 1970s – represents a new mode of articulating and stabilizing the relation between economy and culture, production and consumption. It is a response to both the perceived limits of Fordism (e.g. insupportable risks of investment in inflexible mass production facilities in a context of saturated consumer markets; increased workplace alienation; the ability of the state to avert crisis tendencies) and the emergence of new technical and organizational opportunities. The latter tend to promote structures of consumption that are not 'mass' but segmented and specialized, flexible, 'small batch'. For example, the increased role of knowledge and information in production (computer-aided design and robotization) allows for production lines to be changed by cheaply reprogramming rather than by scrapping expensive plant; increasingly fragmented media (non-broadcast television, internet) allow targeting of smaller, more specialized market niches; marketing and advertising – the conceptual and symbolic definition of goods and services – take on commanding and coordinating positions within firms. Moreover, the idea of post-Fordism converges with broader characterizations of socioeconomic change in the direction of increasing 'dematerialization' or 'informatization' (discussed below) in which commodities are defined, produced and distributed in relation more to their signification than to their materiality. The upshot is the increasing centrality of cultural processes and logics within both production and consumption and their articulation.

The various brands of post-Fordist theory, and formulations of 'new economy', are certainly open to considerable debate. However, their influence has been enormous, particularly in forming and more latterly in convergences between consumption studies and economic sociology (Callon, 1998; Slater and Tonkiss, 2001). The Fordist/post-Fordist framework paradoxically derives its pictures of consumption entirely from transformations in production and economy (narrowly conceived) while at the same time points us towards the absolute and increasing centrality of consumption, and indeed culture, in the reproduction of the economic order. Hence it has moved analysis, almost in spite of itself, towards a concern with the interconnections between economy and culture to the extent that it has provided the core set of presumptions behind the most culturalist accounts of consumption to date: theories of postmodernity (Harvey, 1989; Jameson, 1984; Lee, 1993), discussed below.

CONSUMPTION AND CULTURE

It could be said that neither liberal nor critical traditions seriously examined consumption as culture. For liberals, consumption lay within the private domain of the individual and only reached visibility in the form of demand, the result of rational abstraction from a culture of needs and wants that was itself unexamined. For critics, contemporary consumer culture was the inauthentic and manipulated result of productive forces which were the only important focus of investigation: actual contemporary consumption was simply an index of debasement; the only alternative form of consumption was the utopian or nostalgic relationship to needs that came before or after capitalism.

We could therefore argue that the emergence of a research agenda that is explicitly concerned with cultures of consumption is relatively recent (perhaps two decades old), and has drawn on two kinds of resources: firstly, traditions and methodologies for thinking about the way in which meaningful goods play a part in the reproduction of everyday life; and secondly, accounts of those specifically modern conditions which have given consumption a strategic place in negotiating status and identity.

Different lineages could be claimed but we might point to three major traditions that place the 'meanings of things' on the centre stage of analyses of consumption as an aspect of cultural reproduction. Firstly, the various schools of semiotics, drawing on the model of structural linguistics, provided a methodology for treating all objects as signs within a social circulation of meaning, and ones capable of bearing significations that were irreducible to the functionality of, or instrumental orientation towards, goods. The exemplary analysis is still Barthes' *Mythologies* (1986), which involved a virtuoso reading of mundane objects and events (French wine, landscapes or wrestling matches) in relation to ideological structures of meaning. Objects both bore and reproduced deep structural ways of seeing the world. Objects and their representation (for example in advertising) are able to take on second-order meanings – connotations – which are fundamentally ideological and therefore mystify the consumer's identity and position within social relations. In a famous example, a pasta product can come to signify nationality (Italianness) within a system of ethnic significations that have no proper grounding in the materiality or use of the object (Barthes, 1977). This approach has had huge influence in cultural studies of consumption, becoming one of its two most conventional methodologies (the other being ethnography) (Cook, 1992; Dyer, 1982; Leiss et al., 1986; Myers, 1999; Williamson, 1978). Consumer culture can be read as a complex text and site of ideological work. Later developments of this approach – roughly poststructuralist – have emphasized the fluidity, ambivalence and unpredictability of these structures of meaning and of the values given to objects within them, hence foregrounding both creativity and contestation (for example, de Certeau, 1984; Fiske, 1989) in the way consumers deploy consumption acts and objects.

Secondly, in the tradition of material culture studies within anthropology, function is only one aspect of the meaning of goods (and indeed one that is only really analytically separated out by western observers). Goods and their uses reflect, communicate and are instrumental in reproducing cosmologies. As Mary Douglas writes: 'Forget that commodities are good for eating, clothing and shelter; forget their usefulness and try instead the idea that commodities are good for thinking; treat them as a nonverbal medium for the human creative faculty' (Douglas and Isherwood, 1979). In Douglas' work, consumption goods and rituals make up a social information system through which schemes of social classification are deployed and controlled. Douglas is particularly concerned to demonstrate

that consumption systems are, in effect, complete 'cosmologies', they order an entire moral universe: 'The choice between pounding and grinding [coffee] is … a choice between two different views of the human condition' (1979: 74). Such a perspective also makes perfect sense to any author working within a framework of objectification derived from Marx or Simmel (see, above all, Miller, 1987). Objectification suggests that the relation of need between the individual and the object world is an essentially dialectical one of constant mutual transformation through praxis. Modern consumer culture is one aspect of the monumental development of productive, transformative forces under capitalism, which is simultaneously a world-historical transformation and development of human need, or – to use Simmel's terminology – a massive development of objective culture, which subjective culture struggles hard to assimilate.

Finally, we might point to the tradition of cultural studies, in many respects a development of both semiotics and of the anthropological notion of culture as the meaningful patterning of a whole way of life. However, cultural studies has always had a populist and spectacular dimension – exemplified in studies of subculture and popular expressive forms – that regards consumer goods as sites for the articulation of contradiction and opposition: for example, the punk's transformation of black bin-liners into enactments of working-class, urban nihilism. Cultural studies emerged from a heavily structuralist phase (emphasizing ideological determination of meaning), as well as a fixation on the spectacular and oppositional (rather than on mundane or conformist) consumption. However, over the past 15 years, it has increasingly recognized that *all* consumption involves creative symbolic labour. Willis (1990), for example, focuses on how people make sense of – and therefore make *different* sense of – objects in the act of assimilating them. Consumption is therefore always an active cultural process; at the same time, it is clear that capitalism has delivered into the hands of ordinary people a massive cultural resource for the making of meaning, a huge site of 'common culture'.

Status, identity and meaning

If consumption is always cultural, what – if anything – has changed in the contemporary social landscape? In what sense might consumption and commercial culture have become more socially central? All three of the previously mentioned approaches would point to quantitative and qualitative shifts over the modern period. There is massively *more* material culture; and, at the same time, the elaboration of that material culture follows a hectic rhythm dictated by the drive to increased sales and profits. Modern consumer culture is associated with a destabilization of meanings within consumption and with the instrumentalization of these meanings through functionally differentiated market institutions such as advertising and marketing, mass media and design.

The most important contemporary accounts of this transformation address it by way of a contrast between traditional and post-traditional social orders. As noted above, traditional consumption is associated with a stability due to regulation by tradition and a fixed status order, often formalized in explicit sumptuary laws. Key aspects of consumption such as food, housing and dress are determined not by individual choice but by custom and ascribed status. Modernity is then associated with something like an institutionalized identity crisis in status orders; people's positions within them and ways of signifying those positions through lifestyles are all rendered unstable. Giddens (1991), for example, points to features such as methodical doubt of all authority and knowledge, the plurality of life-worlds that individuals must negotiate in their daily lives, the increasing mediation of possible lifestyles as conveyed through public representations, and the absence of fixed and ascribed identities. In such conditions, as Giddens puts it, 'We have no choice but to choose' (1991: 81). Indeed, as a requirement of modern social life we have to forge identities through the production of 'reflexive narratives of the self', the constitution of coherent identities by all means available. These include the patterns of consumption that are provisionally fixed into relatively stable 'lifestyles' and public representations of lifestyles (by individuals, public authorities and media representations, including advertising).

Unsurprisingly, this instability of meaning, identity and consumption is associated with heightened anxiety over consumption choices which – as choices – are both problematic and yet read as profoundly expressive of a choosing self: we do not know what choices are 'right', but we know that any choice will be interpreted as a moral comment on who we think we are (but see also Warde, 1994a; 1994b). For example, a great deal of research focuses on what Featherstone (1991) describes as the production of an 'outer body' or appearance through bodily regimes such as dieting, exercise and cosmetic transformation, including surgery (see also Finkelstein, 1991). Failure in such disciplinary regimes of consumption (e.g. being overweight) deeply implicates the moral and social worth of the self; and yet diet regimes and ideal body shapes change rapidly, and conflicting imperatives and advice coexist. The depth of this identity crisis and consumption anxiety has often been associated with deeper social pathologies of the modern personality 'type'. Riesman's (1961) account of the other-directed self, Lasch's (1979) critique of the narcissistic personality, and Sennett's (1977) critique of the modern injunction to 'be authentic' under conditions of constant performance, are all diagnoses of the modern attachment of the truth of the self to the consumerist surface of its body, appearance and style of life. In a related vein, Daniel Bell (1979) summed up a tradition of reading consumption as the focal point of a 'hedonistic ethic' that undermines a more traditional and early modern ethics of the self grounded in character, religion and work.

Status, semiotics and postmodernism

This version of modernity emphasizes the contrast with traditional order, particularly around the fixity of individual identity and status. This is an old theme, often captured by notions such as 'status symbol' and 'conspicuous consumption', both terms from Veblen (1899) who was pointing out the strategic role of consumption and leisure practices in establishing social distinction under conditions of social mobility, mainly within the context of an ever-rising middle class. For Veblen, the entire point of a status symbol was that it was a pure sign: such things as immaculate etiquette or exquisite taste in antiques served no function whatsoever but merely indicated that one had the wealth, and therefore leisure, to do no useful work and to devote oneself to being well bred. Consumption was the site on which to *signal* this, and was therefore a marker of pure difference. The dynamic of consumption was given by a process of emulation and devaluation: rising middle classes attempted to ape these consumption symbols, which consequently lost their value – and had to be replaced by new status symbols – as they no longer signified distinction.

As often noted (for example, Miller, 1987), this argument is not a million miles away from much of Bourdieu's (1984), in which battles to legitimate particular criteria and hierarchies of cultural value and taste are central to the exercise of power, not only in culture but also in economy. While the complexity of this argument, in particular Bourdieu's account of transactions between different systems for signifying status (e.g. cultural, social and economic capital), obviously goes way beyond Veblen, he still pays little attention to the content of consumption, which is valued not in its own right but, again, only as a token of status difference. There are two aspects of Bourdieu's work, however, that transform this way of looking at consumption. Firstly, in contrast to Veblen (as well as Durkheimian-derived analysts such as Douglas), cultural consumption is treated as part of the *constitution* of class and power difference, not merely as reflecting or reproducing existing class structures that are rooted entirely in economic structures. Secondly, Bourdieu's notion of habitus attempts to map out the interface between structure and agency: instead of actors as conscious manipulators of signs, or as manipulated by them, habitus addresses the way in which actors internalize – bodily and through experience – patterns of acting within their objective social position. In both respects, consumption is treated as a serious and relatively autonomous aspect of social reproduction.

A far more extreme version of this line of thought is to be found in Baudrillard (1970; 1981). For Baudrillard, as for Veblen and much of Bourdieu, the crucial aspect of consumption is the object as sign and hence as a marker of social distinction. In Baudrillard's (1981) critique of Veblen (which applies to Bourdieu), 'function' itself becomes just another sign rather than an external reference point, the location of the object's authenticity. We might want to signify functionality through the design of, say, the kitchen appliances we choose, but this itself is a mark of a 'modern style' (or perhaps of an anti-consumerist politics): it distinguishes us from others through our choices within a system of signs. Ultimately what we are really buying into in any act of consumption is not the object and its uses but rather the overall system of representations and our position in the matrix of differences it maps out and signals to others. However what is radical in Baudrillard is that along with function he discards any objectivity to which the system of signs might refer, including the structures of social distinction themselves. The triumph of the sign through consumer capitalism is a triumph over all reality: the code dominates production and generates contemporary material reality, and it overwhelms all social status. Hence, it produces a 'hyperreal', a domain of exhaustive experience and meaning that substitutes for what has previously been identified as 'the social' and indeed accounts for the 'death of the social' itself (Baudrillard, 1983; 1994).

Baudrillard moves along this route by translating the notion of social distinction into the language of semiotics, discussed above. Baudrillard takes on the semiotic methodology and takes it very literally: goods as linguistic terms are completely detached from their referents, their value being determined internally to the code. At the same time, Baudrillard understands this approach not only as a methodological move but also as a historical development, what increasingly comes to be thought of as 'the postmodern'. He himself produces a very grand narrative of the progressive eclipse first of reference (in the form of the object's use value), then of sociality (exchange value), finally resulting in the dominance of sign value over social reality, such as it is. Baudrillard's own stance can be interpreted to fit well within older traditions of mass culture critique (to which he was directly related through the Marxism of Lefebvre and the Situationists). His work points to the complete dominance of a totalistic 'spectacle' which can only be countered through a nihilistic embrace on the part of 'the masses'.

In fact, the overall development of consumption studies has been in completely the opposite direction from Baudrillard, whatever it might owe him rhetorically or methodologically, towards an optimistic postmodernism (Hebdige, 1988) which treats the increasing commodity 'aestheticization of everyday life' (Featherstone, 1991), fragmentation of identity and apparently decreasing relevance of older social divisions as the opportunity to treat consumer culture as a kind of ironic and hedonistic playground. Bauman (1990) and Maffesoli (1996), for example, emphasize the 'neotribalism' of a consumer culture in which densely meaningful goods are like costumes in which people dress up in order to enact their current elective, but flexible, social memberships and allegiances. The very profusion and motility of signs – which in Baudrillard points to nihilism – has more generally been taken to suggest the opening of a space for consumer creativity (Willis, 1990) or resistance and rebellion (de Certeau, 1984; Fiske, 1989). Consumption is an always active process of assimilation, hence also one that is unpredictable and undetermined.

GOING SHOPPING

Shopping and retailing both exemplify the different variations of postmodern thought and have constituted the most decisive site for the conjunction of cultural geography and consumption studies. Indeed, the beginning of geography's major contribution to contemporary consumption studies was probably marked by two mid 1990s publications: a special issue of *Environment and Planning A* on *Changing Geographies of Consumption*, edited by Peter Jackson (1995–96); and the 'new retail geography' heralded by Wrigley and Lowe (1996; see also sections of Jackson et al., 2000). This is probably unsurprising since places of purchase concretely spatialize people's encounters with commodities, while – conversely – so much contemporary social space seems structured in relation to consumption. Shopping became a central and evocative issue with the rise of postmodern theory in the 1980s (the huge literature includes Bowlby, 1985; Chaney, 1991; Falk and Campbell, 1997; Ferguson, 1992; Gottdiener, 1997; Laermans, 1993; Langman, 1992; Nava, 1987; Nixon, 1992; Ritzer, 1999; Shields, 1992a; Slater, 1993), in which it came to stand for a central site through which the postmodern triumph of the sign could be studied and was enacted. As Glennie and Thrift (1996) point out, this research focus could take at least two quite opposed forms. On the one hand, a largely productivist and pessimistic line of thought looked at the new centrality of consumption and of shopping sites as a function of transformations in capital and the increased velocity and fluidity of circulation, itself partly a consequence of the ever-greater role of signifying processes in capitalist accumulation (for example, Harvey, 1985; 1989; Jameson, 1984). This involved new forms of rationalization of retail, including a move away from the more Fordist organization of the supermarket to the construction of more complex cultural spaces that provided a range of experiences, treated shopping as part of a total leisure experience (rather than the functional satisfaction of consumer needs through goods), and resulted in the production of spaces that had the character of 'dreamworlds' (Williams, 1982) – the self-enclosed, 'hyperreal' 'no-space' of the out-of town shopping mall or downtown retail development. The former was emblematized in developments such as Edmonton mall in Canada or MetroCentre in the UK (Chaney, 1991), which shifted central city retail functions out of town to new spaces

entirely constructed in relation to consumption practices; the latter by modern complexes such as the Bonaventura in Los Angeles (Jameson, 1984) or heritage recoveries of older, pre-industrial marketplaces such as Quincy market in Boston or Covent Garden in London. All of these developments seemed to reach backwards to simulate historical models of retail space – the arcade, the department store or the marketplace itself (M. Miller, 1981; Slater, 1993; Williams, 1982). At the same time, authors such as Zukin and DiMaggio (1990; Zukin, 1991), Soja (1989; 2000) and Harvey (1985; 1989) argued that these developments need to be seen – however they may present themselves – as a conflict of power between new forms of centralizing capital and the previously more diverse and chaotic spaces formed by organic city development. In Zukin's terminology, there was a 'battle for downtown' between 'landscapes of power' (the reformatting of urban space by new retail capital) and the 'vernacular' city life that previously inhabited these spaces (or which were sidelined by a move out of town).

The more optimistic reading of these developments focused on the emergence of new forms of subjectivity that seemed well adapted to these spaces and which also seemed to emblematize the postmodern. Firstly, consumption sites were recognized as providing new locations of social centrality (Shields, 1992a; 1992b). Just like the town centres they so often replaced or shifted to new locations, these sites congregated and focused the activities and signs through which people enact and experience civic identity and civic life. It is a matter of a lot more than shopping even if shopping is the central occasion for congregating. Visibility – of people, goods and settings – plays a central role here in acting out the social (which has led to a renewed focus on Benjamin's elaboration of the figure of the flaneur). Issues of policing entry (Davis, 1990) are crucial in regulating entry into sociality (and exclusion from these spaces is a real social exclusion) and in producing (commercially) desired images of the social (for example, the exclusion of unruly youth, or the poor or ethnic groups). Secondly, the new retailing practices and subjectivities were associated with consumers who were both highly reflexive and fluid in their relationship to the myriad signs on offer. Going shopping was given the character of a preparation for a costume party, in which we try on or play at multiple identities and desires through various imaginative encounters with goods and their significations: not just in buying and owning, but also in looking and browsing, watching other consumers and moving through sign worlds, we imaginatively try on identities. Both the reflexivity of the consumer – their 'knowingness' and semiotic skills – and their supposed playfulness or ironic, flaneur-style distance from commitment are also associated with a new fluidity in social identities and memberships, as we have noted in Bauman's and Mafessoli's ideas of neotribes and elective memberships: new spaces of consumption both enable and arise from a condition in which people can elect (imaginatively or really) their cultural and subcultural allegiances.

What came to be known in the mid 1990s as the 'new retail geography' – an important corrective to the postmodern excesses current in other disciplines – sought to evade the problems of both these positions, both the productivist and culturalist extremes as well as their overly facile pessimism or optimism. Firstly, it correctly recognized retail as a primary site on which one could and indeed had to connect political economy and cultural processes rather than reductively to assign a dominant position to either of them. For example, Lowe (2000) demonstrates that new retail megastructures, planned by global capital, can be transformed into true 'places' by local authorities, consumers and users. Secondly, it cleared the path for new strategies of empirical engagement, eschewing both macro-analysis and simple semiotic readings of spaces and discourses. The primary need was, and is, for ethnographic investigations that bring to light the ways in which people actually use and experience these retail spaces, and how they are linked into longer chains of provision, both 'downwards' to the consumers' lifeworld and 'upwards' into commercial and industrial organization and social regulation. Moreover, the ethnographic approach gives a more concrete sense of how more durable identities such as gender, ethnicity and age mediate these retail practices but are also partly constituted through them. The study of London shopping centres and high streets by Miller et al. (1998), for example, provided a particularly rich account of the relation of these social spaces to complex social and local histories, rather than to a postmodern play of styles. Thirdly, as particularly emphasized by Miller (1998; also Slater, 1997a), postmodern readings of new consumption spaces were perversely informed by a highly individualistic orientation. These consumers, unlike the sovereign ones of economic theory, might be fragmented and motile

subjects, but they were nonetheless depicted as individual ones. Miller's work focused on the consumer's connectedness to significant others: going shopping is not so much the act of identity-seeking subjects entering a supermarket of style as the process by which people (and largely women) provision the lives in which they are embedded, and hence in which they must construct the needs of their children and partners as much as they may imaginatively play with their own. Shopping is, as Miller writes, an act of love.

MASS CONSUMPTION AND GLOBAL CULTURE

Similar issues also arise in relation to increasing scales of both production and consumption, which is also a central concern for cultural geography. Globalization is hardly a new concern in that capitalism has always been associated with an internationalization of trade and production relations. Early liberal arguments for capitalism emphasized both increased awareness of interdependence and the stance of rational calculation that attended the development of commerce (Hirschman, 1977). By contrast, Marx provided some of the most vivid images of capitalism as a force that is driven to explore the world for new 'use values', hence bringing formerly isolated populations into competitive market conditions for both labour and consumer goods. In the process, Marx argued, non-capitalist social relations and cultures are dissolved. Significantly, consumption has long been equated with mass consumption, a central means through which concepts of mass culture and mass society were understood. Again, the central image is the ineluctable dissolution of previous material cultures in the face of globalizing commodity production. Early arguments about a global consumer culture echoed the structure of mass consumption and mass culture theories, often in the form of 'Americanization' theses.

There are in fact several different claims embedded in such formulations: firstly, a claim as to the homogeneity of consumption under regimes of massification or globalization; secondly, a claim as to the inevitability and smoothness of the successful spread of consumer culture; and thirdly, a set of value claims usually centring on either the quality or the authenticity of life under consumer culture (Miller, 2001; Wilk, 2001). Concern about both mass consumption and homogenized global culture took the form of debates over Americanization during much of the twentieth century. America seemed to be the point of origin and power for specifically enticing goods, for a system of production, marketing and consumption, and for a generally materialistic value system that equated freedom and progress with increasing satisfaction of private wants. Successful export of American consumerism seemed based not only on the inherent dynamism of the system but also on the political, military and media power that ensured a global reach and a globe dominated by Coca-Cola and McDonald's. However, the presumption that the export of American goods, services and media representation directly translates into homogeneous global culture rests at least partly on the same image of the passive consumer and the automatic determination of consumption by production that also underwrote mass consumption perspectives. Media studies of the different sense made of ostensibly hegemonic global culture products (for example, Ang, 1985; Liebes and Katz, 1990; Silj, 1988) pioneered a concern with the local mediation of goods that clearly extended to consumption in general (Howes, 1996; Miller, 1994). Consumption is an active process of assimilation at the global level, given the overall structure of unequal power, a process captured by the unlovely word 'glocalization'. Moreover, it is argued that multinational companies are as clear about the lack of global homogeneity as academics are learning to be, moving from older models of international marketing to their own versions of glocalization (for example, Kline, 1993; 1995).

Another important perspective that has served to introduce a more complex spatial sense of local–global connections into consumption studies has been commodity chain analysis or systems of provision approaches. These labels embrace quite some divergence but we might take as paradigm cases two non-geographers. Mintz's (1986) *Sweetness and Power* demonstrates how production and consumption of a single commodity – sugar – brings together spatially dispersed histories. Fine and Leopold (1993) advocate a systems of provision approach in which relations internal to a commodity sector are shown to structure each other. Examples from the food and clothing industries are used to argue that this kind of analysis throws up findings that would be counter-intuitive on the basis of analysis of consumption as a separate social moment. Work within this perspective generally points to the multiple lines of mediation and connection, in

which consumption structures production as much as the other way around, and cultural and financial intermediaries – above all, marketing and retailing – take on decisive strategic roles. At the same time, this approach places the spatial distribution of these connections to the fore.

Globalization of consumer culture is also not a particularly even process. The older image of American dominance has given way to a concern with competition between regional blocks (for example, the power of Asian production and consumption), and with conflict directly provoked by consumerism as a value system (e.g. Castells, 1997, on the resurrection of traditionalist identities and politics). Appadurai (1990) offers a particularly complex attempt to map the different economic, social, political and cultural flows that generate this unevenness. Moreover, as developed in his earlier work (Appadurai, 1986), the idea that the rise of consumer and commodity relations is inexorable rests on a mistaken assumption that these processes are irreversible within any system of consumption (see also Carrier, 1994). In fact, within consumption objects move into and out of commodity status, from consumerist frameworks to many others.

Finally, globalization arguments, like the mass consumption ones before them, generally assume an opposition between pristine indigenous cultures existing before the intrusion of consumer culture, and their afterlife as commodity cultures – a fall from grace. Even where gains such as wealth and standard of living are conceded, there is a sense that consumer culture is neither as good nor as authentic as what came before. Names such as McDonald's or Nike are identified with global culture and the evils of production (environment, labour relations and exploitation) (Klein, 2000). Anthropological research (most notably Thomas, 1991) has pointed out both the extent and the complexity of trade relations and non-immediate consumption in non-modern societies, as well as the 'entanglement' of supposedly pristine consumption cultures in wider and negotiated social networks.

Romanticization of the premodern is one problem here, as is the idea that autonomy and isolation could ever be a proper standard for assessing cultures. To do so is to reify them outside all the history of contact and communication. It is also to assume the kind of cultural absolutism that underlies all relativism: the assumption that whatever a people value is unquestionable so long as it has been self-defined (Slater, 1997b). Rather more interesting is the ethical framework evolved by Sen and Nussbaum (Sen, 1985; 1987) which seeks to understand development politics in terms of 'empowerment' and citizenship: that minimum levels of consumption (whose content is defined specifically for different communities) are required to achieve such values as self-determination and democracy (see also Doyal and Gough, 1991; Soper, 1990).

Finally, in addition to the emphasis on heterogeneity, unevenness and the issue of authenticity, contemporary approaches to the globalization of consumption have been heavily marked by a more general stress on enculturation of the economy and on notions of information or network society. For example, Malcolm Waters offers the slogan: 'material exchanges localize; political exchanges internationalize; and symbolic exchanges globalize' (1995: 9). This is to argue that the increasingly dematerialized form of goods has an inherent tendency to global scales of operation. Similarly, Appadurai (1986; 1990; 1995; like Castells, 1996) uses a language of 'flows' and 'scapes' to capture the way in which global movements of goods, people, signs and so on are increasingly overlapping both in terms of geography and across social moments (culture, society, economy, politics). Although Waters' position does not assume homogenization or global culture, it is certainly in tune with critics such as Klein (2000) or Goldman and Papson (1998) for whom the modern form of the multinational corporation is exemplified by Nike: it owns no factories or other industrial apparatus and yet is able to coordinate worldwide production as well as a seemingly international cultural allegiance under the aegis of a symbol, the brand name and its 'swoosh' logo.

The emergence of the internet and e-commerce has come increasingly to symbolize and perform a new geography of consumption in which the circulation and exchange of goods are dematerialized and hence rendered 'frictionless' and 'disintermediated'. While it is evident that the precise relationship between online and offline commerce is still being worked through by both producers and consumers, the internet seems at least capable of reforming markets through global competition, through the identification or organization of consumer groups independently of physical location, and through new processes of both commodification and decommodification which involve challenges to the very idea of 'a product' (Coombe, 1998; Lury, 1993; 1996; Miller and Slater, 2000; Slater, 2001).

NEW DIRECTIONS

Consumption studies has become a fairly well-defined and well-established field within a number of disciplines. It has opened up a range of research agendas that are now being pursued as 'normal science'. It is well accepted that consumption is a significant issue of cultural, social and economic reproduction, not to be treated as private, natural or trivial. We might therefore want to read some of the tea-leaves of our intellectual situation to see where things might go next. Several tendencies stand out from the current state of the field.

Firstly, in consumption studies as elsewhere, the high-water mark of debates over postmodernity has long passed. For this field, arguably more structured by (and structuring of) these debates than many others, this means a move away from an obsessive concern with the relation between identity and culture, and away from encountering consumption through processes of signification rather than broader constructions of social relationships and practices. Symptomatic of this shift is a new concern with mundane rather than spectacular and expressive consumption (for example, Gronow and Warde, 2001; Warde and Martens, 2000), which includes a concern with consumption as habitual, routine and embedded in the practical reproduction of everyday life rather than as directly consequential for self, identity and status (Ilmonen, 1997). Finally, the postmodern roots of the first wave of consumption studies were marked, as previously noted, by a bias towards individual hedonism that peculiarly mirrored liberal traditions. Partly under the impact of both ethnographic and feminist studies, there is a greater concern with mundane consumption as social and interpersonal, and concerned with the needs of others as much as of self (Miller, 1998).

Secondly, the shift away from the postmodern agenda involves a renewed concern with the relation between consumption and persistent social structures of power and inequality (for example, Edwards, 2000). Debates within cultural studies have been particularly important in pointing up the movement from highly optimistic accounts of consumption as liberating and empowering (Fiske, 1989; Nava, 1992; Nava et al., 1996) towards understanding it as structured by the same constraints that long exercised an older political economy. Figures such as Bauman (1998) and McRobbie (1998; 1999) have been exemplary in trying to move back onto this terrain without losing the insights gained from the long 'cultural turn'. Central also to this development is a revived concern with ecological and environmental issues and the problematization of consumption through an experience of its social limits, costs and risks. Indeed, the issue of 'risk' has to some extent replaced the earlier, more identity-oriented, notion of 'anxiety' within consumption studies (Beck, 1992; Beck et al., 1994; Halkier, 2001; Warde, 1994a; 1994b).

Thirdly, having asserted consumption as a significant social instance in its own right, particularly against the 'productivist bias' in much previous social thought, the research tendency is now to reconnect consumption and production, focusing on continuities and interconnections, not least through more integrated accounts of markets and market behaviours (for example, du Gay, 1996; 1997; Slater, 2001; Slater and Tonkiss, 2001). This tendency has been given a considerable impetus by the rise of the internet and e-commerce which evidences blurred boundaries between production and consumption as well as an ever more globalized reach for both.

REFERENCES

Aglietta, M. (1979) *A Theory of Capitalist Regulation: The US Experience*. London: Verso.

Anderson, B. (1986) *Imagined Communities*. London: Verso.

Ang, I. (1985) *Watching 'Dallas': Soap Opera and the Melodramatic Imagination*. London: Methuen.

Appadurai, A. (1986) *The Social Life of Things: Commodities in Cultural Perspective*. Cambridge: Cambridge University Press.

Appadurai, A. (1990) 'Disjuncture and difference in the global cultural economy', *Theory, Culture and Society* 7: 295–310.

Appadurai, A. (1995) *Modernity at Large: Cultural Dimensions of Globalization*. Minneapolis: University of Minnesota Press.

Barthes, R. (1977) *Image–Music–Text*. London: Fontana.

Barthes, R. (1986) *Mythologies*. London: Paladin.

Baudrillard, J. (1970) *The Consumer Society: Myths and Structures*. London: Sage, 1998.

Baudrillard, J. (1981) *For a Critique of the Political Economy of the Sign*. St Louis: Telos.

Baudrillard, J. (1983) *Simulations*. New York: Semiotext(e).

Baudrillard, J. (1994) *In the Shadow of the Silent Majorities*. New York: AK.

Bauman, Z. (1990) *Thinking Sociologically*. Oxford: Basil Blackwell.

Bauman, Z. (1998) *Work, Consumerism and the New Poor*. Cambridge: Polity.

Beck, U. (1992) *Risk Society: Towards a New Modernity*. London: Sage.

Beck, U., Giddens, A. and Lash, S. (1994) *Reflexive Modernization: Politics, Tradition and Aesthetics in the Modern Social Order*. Cambridge: Polity Press.

Bell, D. (1979) *The Cultural Contradictions of Capitalism*. London: Heinemann.

Bell, D. and Valentine, G. (1997) *Consuming Geographies: We Are Where We Eat*. London: Routledge.

Bourdieu, P. (1973) 'The Berber house or the world reversed', in M. Douglas (ed.) *Rules and Meanings*. Harmondsworth: Penguin. pp. 98–110.

Bourdieu, P. (1984) *Distinction: A Social Critique of the Judgement of Taste*. Cambridge, MA: Harvard University Press.

Bowlby, R. (1985) *Just Looking: Consumer Culture in Dreiser, Gissing and Zola*. Andover: Methuen.

Callon, M. (ed.) (1998) *The Laws of the Market*. Oxford: Blackwell Sociological review.

Carrier, J.G. (1994) *Gifts and Commodities: Exchange and Western Capitalism since 1700*. London: Routledge.

Castells, M. (1996) *The Rise of Network Society*. Oxford: Blackwell.

Castells, M. (1997) *The Power of Identity*. Oxford: Blackwell.

Chaney, D. (1991) 'Subtopia in Gateshead: the Metro-Centre as a cultural form', *Theory, Culture and Society* 7: 649–68.

Cook, G. (1992) *The Discourse of Advertising*. London: Routledge.

Cook, I. and Crang, P. (1996) 'The world on a plate', *Journal of Material Culture* 1 (2): 131–53.

Coombe, R.J. (1998) *The Cultural Life of Intellectual Properties: Authorshop, Appropriation, and the Law*. London: Duke University Press.

Crang, P. (1996) 'Displacement, consumption and identity', *Environment and Planning A* 28: 47–67.

Davis, M. (1990) *City of Quartz: Excavating the Future in Los Angeles*. London: Verso.

Debord, G. (1991) *Comments on the Society of the Spectacle*. London: Verso.

de Certeau, M. (1984) *The Practice of Everyday Life*. Berkeley: University of California Press.

Douglas, M. and Isherwood, B. (1979) *The World of Goods: Towards an Anthropology of Consumption*. Harmondsworth: Penguin.

Doyal, L. and Gough, I. (1991) *A Theory of Human Needs*. London: Macmillan.

du Gay, P. (1996) *Consumption and Identity at Work*. London: Sage.

du Gay, P. (ed.) (1997) *Production of Culture, Cultures of Production*. London: Sage.

du Gay, P. and Pryke, M. (2001) *Cultural Economy: Cultural Analysis and Commercial Life*. London: Sage.

Dyer, G. (1982) *Advertising as Communication*. London: Methuen.

Edwards, T. (2000) *Contradictions of Consumption*. Milton Keynes: Open University Press.

Falk, P. and Campbell, C. (eds) (1997) *The Shopping Experience*. London: Sage.

Featherstone, M. (1991) *Consumer Culture and Postmodernism*. London: Sage.

Ferguson, H. (1992) 'Watching the world go round: atrium culture and the psychology of shopping', in R. Shields (ed.) *Lifestyle Shopping: The Subject of Consumption*. London: Routledge. pp. 21–39.

Fine, B. and Leopold, E. (1993) *The World of Consumption*. London: Routledge.

Finkelstein, J. (1991) *The Fashioned Self*. Cambridge: Polity.

Fiske, J. (1989) *Reading the Popular*. Boston: Unwin Hyman.

Galbraith, J.K. (1972) *The New Industrial Estate*. Harmondsworth: Penguin.

Giddens, A. (1991) *Modernity and Self-Identity: Self and Society in the Late Modern Age*. Cambridge: Polity.

Glennie, P. and Thrift, N. (1996) 'Consumption, shopping and gender', in E. Wrigley and M. Lowe (eds) *Retailing, Consumption and Capital: Towards the New Retail Geography*. London: Longman. pp. 221–37.

Goldman, R. and Papson, S. (1998) *Nike Culture: The Sign of the Swoosh*. London: Sage.

Gottdiener, M. (1997) *The Theming of America: Dreams, Visions and Commercial Spaces*. Oxford: Westview Press.

Gronow, J. and Warde, A. (eds) (2001) *Ordinary Consumption*: Harwood Academic.

Halkier, B. (2001) 'Consuming ambivalences: consumer handling of environmentally related risks in food', *Journal of Consumer Culture* 1 (2): 205–24.

Harvey, D. (1985) *The Urbanisation of Capital*. Oxford: Blackwell.

Harvey, D. (1989) *The Condition of Postmodernity: An Enquiry into the Origins of Culture*. Oxford: Blackwell.

Haug, W.F. (1986) *Critique of Commodity Aesthetics: Appearance, Sexuality and Advertising*. Cambridge: Polity.

Hebdige, D. (1988) *Hiding in the Light: On Images and Things*. London: Comedia.

Heelas, P. and Morris, P. (eds) (1992) *The Values of Enterprise Culture: The Moral Debate*. London: Routledge.

Hirschman, A. (1977) *The Passions and the Interests*. Princeton: Princeton University Press.

Howes, D. (ed.) (1996) *Cross-Cultural Consumption: Global Markets, Local Realities*. London: Routledge.

Ilmonen, K. (1997) 'Consumption and routine'. Paper presented at the European Sociological Association Conference, Essex.

Jackson, P. (ed.) (1995–1996) *Changing Geographies of Consumption;* special issue of *Environment and Planning A* (27/28).

Jackson, P., Lowe, M., Miller, D. and Mort, F. (eds) (2000) *Commercial Cultures: Economies, Practices, Spaces*. London: Berg.

Jameson, F. (1984) 'Postmodernism, or the cultural logic of late capitalism', *New Left Review* 146: 53–92.

Keat, R., Whiteley, N. and Abercrombie, N. (1994) *The Authority of the Consumer*. London: Routledge.

Klein, N. (2000) *No Logo*. London: Flamingo.

Kline, S. (1993) *Out of the Garden: Toys, TV and Children's Culture in the Age of Marketing*. London: Verso.

Kline, S. (1995) 'The play of the market: on the internationalization of children's culture', *Theory, Culture and Society* 12 (2): 103–29.

Laermans, R. (1993) 'Learning to consume: early department stores and the shaping of the modern consumer culture (1860–1914)', *Theory, Culture and Society* 10 (4): 79–102.

Langman, L. (1992) 'Neon cages: shopping for subjectivity', in R. Shields (ed.) *Lifestyle Shopping: The Subject of Consumption*. London: Routledge. pp. 40–82.

Lasch, C. (1979) *The Culture of Narcissism*. London: Abacus.

Lee, M. (1993) *Consumer Culture Reborn: The Cultural Politics of Consumption*. London: Routledge.

Lee, M. (ed.) (2000) *The Consumer Society Reader*. Oxford: Blackwell.

Lefebvre, H. (1947) *Critique of Everyday Life*. London: Verso, 1991.

Lefebvre, H. (1971) *Everyday Life in the Modern World*. New York: Harper and Row.

Leiss, W., Kline, S. and Jhally, S. (1986) *Social Communication in Advertising: Persons, Products and Images of Well-Being*. London: Methuen.

Liebes, T. and Katz, E. (1990) *The Export of Meaning: Cross-Cultural Readings of Dallas*. London: Macmillan.

Lipietz, A. (1992) *Towards a New Economic Order: Post-Fordism, Ecology and Democracy*. Cambridge: Polity.

Lowe, M. (2000) 'From Victor Gruen to Merry Hill: reflections on regional shopping centres and urban development in the US and UK', in P. Jackson, M. Lowe, D. Miller and F. Mort (eds) *Commercial Cultures: Economies, Practices, Spaces*. London: Berg. pp. 123–42.

Lury, C. (1993) *Cultural Rights: Technology, Legality and Personality*. London: Routledge.

Lury, C. (1996) *Consumer Culture*. Cambridge: Polity.

Maffesoli, M. (1996) *The Time of the Tribes*. London: Sage.

McRobbie, A. (1998) *British Fashion Design: Rag Trade or Image Industry?* London: Routledge.

McRobbie, A. (1999) *In the Culture Society: Art, Fashion and Popular Music*. London: Routledge.

Miller, D. (1987) *Material Culture and Mass Consumption*. Oxford: Basil Blackwell.

Miller, D. (1994) *Modernity – An Ethnographic Approach: Dualism and Mass Consumption in Trinidad*. Oxford: Berg.

Miller, D. (ed.) (1995) *Acknowledging Consumption: A Review of New Studies*. London: Routledge.

Miller, D. (1998) *A Theory of Shopping*. Cambridge: Polity.

Miller, D. (2001) 'The poverty of morality', *Journal of Consumer Culture* 1 (2): 225–44.

Miller, D. and Slater, D. (2000) *The Internet: An Ethnographic Approach*. London: Berg.

Miller, D., Jackson, P. and Thrift, N. (eds) (1998) *Shopping, Place and Identity*. London: Routledge.

Miller, M. (1981) *The Bon Marche: Bourgeois Culture and the Department Store*. London: Allen & Unwin.

Mintz, S. (1986) *Sweetness and Power: The Place of Sugar in Modern History*. Harmondsworth: Penguin.

Morley, D. (1992) *Television, Audiences and Cultural Studies*. London: Routledge.

Myers, G. (1999) *Ad Worlds: Brands, Media, Audiences*. London: Arnold.

Nava, M. (1987) 'Consumerism and its contradictions', *Cultural Studies* 1 (2): 204–10.

Nava, M. (1992) *Changing Cultures: Feminism, Youth and Consumerism*. London: Sage.

Nava, M., Blake, A., MacRury, I. and Richards, B. (eds) (1996) *Buy This Book: Contemporary Issues in Advertising and Consumption*. London: Routledge.

Nixon, S. (1992) 'Have you got the look? Masculinities and shopping spectacle', in R. Shields (ed.) *Lifestyle Shopping: The Subject of Consumption*. London: Routledge. pp. 149–69.

Plant, S. (1992) *The Most Radical Gesture: The Situationist International in a Postmodern Age*. London: Routledge.

Poster, M. (2001) *What's the Matter with the Internet?* Minneapolis: University of Minnesota Press.

Richards, T. (1991) *The Commodity Culture of Victorian England: Advertising and Spectacle, 1851–1914*. London: Verso.

Riesman, D. (1961) *The Lonely Crowd: A Study of the Changing American Character*. New Haven: Yale University Press.

Ritzer, G. (1999) *Enchanting a Disenchanted World: Revolutionizing the Means of Consumption*. London: Sage.

Ross, A. (ed.) (1997) *No Sweat: Fashion, Free Trade and the Rights of Garment Workers*. London: Verso.

Sen, A. (1985) *Commodities and Capabilities*. Amsterdam: Elsevier.

Sen, A.K. (1987) *On Ethics and Economics*. New York: Blackwell.

Sennett, R. (1977) *The Fall of Public Man*. New York: Faber and Faber.

Shields, R. (ed.) (1992a) *Lifestyle Shopping: The Subject of Consumption*. London: Routledge.

Shields, R. (1992b) 'Spaces for the subject of consumption', in R. Shields (ed.) *Lifestyle Shopping: The Subject of Consumption*. London: Routledge. pp. 99–113.

Silj, A. (1988) *East of Dallas: The European Challenge to American Television*. London: BFI.

Slater, D.R. (1993) 'Going shopping: markets, crowds and consumption', in C. Jenks (ed.) *Cultural Reproduction*. London: Routledge. pp. 188–209.

Slater, D.R. (1997a) *Consumer Culture and Modernity*. Cambridge: Polity.

Slater, D.R. (1997b) 'Consumer culture and the politics of need', in M. Nava, A. Blake, I. MacRury and B. Richards (eds) *Buy This Book: Contemporary Issues in Advertising and Consumption*. London: Routledge. pp. 51–63.

Slater, D.R. (2001) 'Capturing markets from the economists', in P. du Gay and M. Pryke (eds) *Cultural Economy: Cultural Analysis and Commercial Life*. London: Sage.

Slater, D. and Tonkiss, F. (2001) *Market Society: Markets and Modern Social Thought*. Cambridge: Polity.

Smith, N. (1993) 'Homeless/global: scaling places', in J. Bird, B. Curtis, T. Putman, G. Robertson and L. Tickner (eds) *Mapping the Future: Local Cultures, Global Change*. London: Routledge. pp. 87–119.

Soja, E. (1989) *Postmodern Geographies*. London: Verso.

Soja, E. (1996) *Thirdspace: Journeys to Los Angeles and other Real-and-Imagined Spaces*. London: Blackwell.

Soja, E. (2000) *Postmetropolis: Critical Studies of Cities and Regions*. London: Blackwell.

Soper, K. (1990) *Troubled Pleasures: Writings on Politics, Gender and Hedonism*. London: Verso.

Suchman, L. (1999) 'Working relations of technology production and use', in D. Mackenzie and J. Wajcman (eds) *The Social Shaping of Technology*. Buckingham: Open University Press. pp. 258–65.

Thomas, N. (1991) *Entangled Objects: Exchange, Material Culture and Colonialism*. Cambridge, MA: Harvard University Press.

Urry, J. (1990) *The Tourist Gaze: Leisure and Travel in Contemporary Societies*. London: Sage.

Veblen, T. (1899) *The Theory of the Leisure Class: An Economic Study of Institutions*. New York: Mentor, 1953.

Warde, A. (1994a) 'Consumers, identity and belonging: reflections on some theses of Zygmunt Bauman', in R. Keat, N. Whiteley and N. Abercrombie (eds) *The Authority of the Consumer*. London: Routledge. pp. 58–74.

Warde, A. (1994b) 'Consumption, identity-formation and uncertainty', *Sociology* 28 (4): 877–98.

Warde, A. and Martens, L. (2000) *Eating Out: Social Differentiation, Consumption and Pleasure*. Cambridge: Cambridge University Press.

Waters, M. (1995) *Globalization*. London: Routledge.

Wilk, R. (2001) 'Consuming morality', *Journal of Consumer Culture* 1 (2): 225–44.

Williams, R. (1976) *Keywords: A Vocabulary of Culture and Society*. Glasgow: Fontana.

Williams, R. (1982) *Dream Worlds: Mass Consumption in late 19th Century France*. Berkeley: University of California.

Williams, R. (1985) *Culture and Society: 1780–1950*. Harmondsworth: Penguin.

Williamson, J. (1978) *Decoding Advertisements: Ideology and Meaning in Advertising*. London: Boyars.

Willis, P. (1990) *Common Culture: Symbolic Work at Play in the Everyday Cultures of the Young*. Milton Keynes: Open University Press.

Wrigley, N. and Lowe, M. (1996) *Retailing, Consumption and Capital: Towards a New Retail Geography*. Harlow, Longman.

Zukin, L.A. (1991) *Landscapes of Power: From Detroit to Disney World*. Berkeley: University of California Press.

Zukin, S. and DiMaggio, P. (eds) (1990) *Structures of Capital: The Social Organization of the Economy*. New York: Cambridge University Press.

Section 3

CULTURENATURES Edited by Sarah Whatmore

Introduction: More than Human Geographies

Sarah Whatmore

As every undergraduate knows, geography stakes its disciplinary identity on being uniquely concerned with the interface between human culture and natural environment. Nowhere is this better epitomized than in the early work of Carl Sauer (1925) and the legacy of the Berkeley School with its emphasis on 'cultural landscape' in which 'culture is the agent [and] the natural area the medium'. However, even here, it is evident that this definitive geographical concern assumes that everything we encounter in the world already belongs either to 'culture' or to 'nature', a division entrenched in the very fabric of the discipline and reinforced by the faltering conversation between 'human' and 'physical' geography. In consequence, as human geographers set about trafficking between culture and nature, a fundamental asymmetry in the treatment of the things assigned to these categories has been smuggled into the enterprise. Geographies, like histories, become stories of exclusively human activity and invention played out over, and through, an inert bedrock of matter and objects made up of everything else. It is a story that is writ large in disciplinary texts like *Man's Role in Changing the Face of Earth* (Thomas et al., 1956) and that percolates through diverse currents of cultural geography, whether in the guise of Marxist concerns with the 'production of nature' (for example, Mitchell, 1996;

Smith, 1984) or representational concerns with the cultural politics of landscape (e.g. Barnes and Duncan, 1992; Cosgrove and Daniels, 1988). But it is those geographies associated with the so-called 'cultural turn' that have most intensified these divisions between the natural and the cultural, championing the 'agent' over the 'medium' to such an extent that the world is rendered an exclusively human achievement in which 'nature' is swallowed up in the hubris of social constructionism (Demeritt, 1998). This section of the *Handbook* deals with culture's 'outside' and the fate of this pervasive exteriorization for the things ascribed to 'nature' (Wolfe, 1998).

The three chapters in this section endeavour to retrace some of the ways in which 'nature' has been evacuated in cultural geography and, more importantly, some of the more imaginative responses to the culture–nature antinomy that this evacuation has occasioned in geography and elsewhere (see Latour, 1993). The guiding ethos of these chapters is their dissatisfaction with the binary terms in which the question of nature has been posed and their engagement with currents in critical thinking that unsettle the *humanist* assumptions underlying their perpetuation in cultural geographies of various kinds (see Hayles, 1999). I would highlight four sets of conversations as the most influential amidst these efforts to fashion more-than-human geographies in which, as Michel Serres puts it, 'there is sense in space before the sense that signifies' (1991: 13). The first of these is with environmental history (for example, Bird, 1987; Cronon, 1995), exploring shared concerns with recovering the forceful socio-materialities of earth and water – with redeeming plants and creatures in the assemblage and durability of particular cultural practices from the narrative monopoly of human designs. The second conversational tack is with the vein of anthropological interest in material culture (for example, Appadurai, 1986; Ingold, 2000) that emphasizes the 'mutualism' of social agency such that, for example, a tree stump 'affords' sitting as a product of the relationship between the form and scale of the human body and that of the material object (Graves-Brown, 2000: 4). These two conversational threads are of long standing and refresh unfashionable currents in cultural geography associated with natural history, such as domestication (see Anderson, 1997) as well as more fashionable ones like postcolonialism (see Willems-Braun, 1997).

The third and fourth of these conversational influences are of rather more recent genesis, and have seen cultural geography engaging with the proliferation of 'extra-disciplinary' bodies of intellectual work, particularly those associated with feminism and science studies. In the case of science studies, the rich flowering of interest in the knowledge practices of science (including geography) as important analytical subjects in their own right, particularly in its feminist and actant network theory (ANT) variants (e.g. Haraway, 1997; Latour, 1999), have done most to redistribute the 'agency' of their socio-material fabrications through the associational performance of all manner of bodies, devices, documents and codes. Here, the spatial formations of these knowledge practices have been rendered significant in ways previously reserved for their historicity. Finally, feminist work on embodiment in terms both of situating knowledge practices and claims and of attending to the intercorporeal affects of differently embodied kinds (for example, Kirkby, 1997; Weiss, 1999) has enjoined a lively re-examination of the categorical distinction between the human and the non-human that has marked the geographical enterprise so deeply, not

least its cultural variants. Pulling these various conversations together to identify a common theme in the work critically reviewed in the chapters in this section, it is the shift from representational, or discursive, to what Nigel Thrift (1996) has called non-representational, or performative, social theories that most stands out. This shift in the vocabularies of cultural geography promises to be particularly important for the terms on which the human and the non-human are admitted into geographical analysis, permitting more promiscuous and volatile configurations of the social and the material that complicate the laboured divisions and *rapprochements* between culture and nature (Whatmore, 2002).

The three chapters in this section take up and explore these currents differently, and some others besides. The first chapter, by Noel Castree, surveys the ways in which the 'question of nature' has been framed in human geography and reviews the enduring concerns and dilemmas of its Marxist heritage. The second chapter, by Jennifer Wolch, Jody Emel and Chris Wilbert, explores one of the most radical efforts in cultural geography at the millennium to take up questions of non-human agency by making space for animals (see Wolch and Emel, 1998). Here, the theoretical and methodological impulses that have shaped the 'new' cultural geography are charged with privileging of cognition and language as the markers of a decidedly 'human' geography that takes no account of our ethical kinship with other animal kinds. The third chapter, by Steve Hinchliffe, examines the idea of 'landscape' and reassesses the legacy of 'building' approaches to the question of nature in cultural geography as against various efforts to articulate a 'dwelling' perspective in which the human is situated within, rather than divorced from, the fabric of heterogeneous worldly habitations.

REFERENCES

Anderson, K. (1997) 'A walk on the wildside: a critical geography of domestication', *Progress in Human Geography* 21(4): 463–85.

Appadurai, A. (ed.) (1986) *The Social Life of Things: Commodities in Cultural Perspective*. Cambridge: Cambridge University Press.

Barnes, T. and Duncan, J. (eds) (1992) *Writing Worlds: Discourse, Text and Metaphor in the Representation of Landscape*. London: Routledge.

Bird, E. (1987) 'The social construction of nature: theoretical approaches to the history of environmental problems', *Environmental History Review* 11: 255–64.

Cosgrove, D. and Daniels, S. (1988) *The Iconography of Landscape: Essays on Symbolic Representation, Design and Use of Past Environments*. Cambridge: Cambridge University Press.

Cronon, W. (1995) 'The trouble with wilderness; or, getting back to the wrong nature', in W. Cronon (ed.) *Uncommon Ground: Toward Reinventing Nature*. New York: Norton. pp. 69–90.

Demeritt, D. (1998) 'Science, social constructivism and nature', in B. Braun and N. Castree (eds) *Remaking Reality: Nature at the Millennium*. London: Routledge. pp. 173–83.

Graves-Brown, P. (ed.) (2000) *Matter, Materiality and Modern Culture*. London: Routledge.

Haraway, D. (1997) *Modest_Witness @ Second_Millennium: FemaleMan©_Meets_Oncomouse™*. London: Routledge.

Hayles, K. (1999) *How We Became Post-Human*. Chicago: University of Chicago Press.

Ingold, T. (2000) *The Perception of the Environment*. London: Routledge.

Kirkby, V. (1997) *Telling Flesh: The Substance of the Corporeal*. London: Routledge.

Latour, B. (1993) *We Have Never Been Modern*. Hemel Hempstead: Harvester Wheatsheaf.

Latour, B. (1999) *Pandora's Hope: Essays on the Reality of Science Studies*. Cambridge, MA: Harvard University Press.

Mitchell, D. (1996) *The Lie of the Land: Migrant Workers and the California Landscape*. Minneapolis: Minnesota University Press.

Sauer, C. (1925) 'The morphology of landscape', reprinted in J. Leighly (ed.) *Land and Life: Selections from the Writings of Carl Ortwin Sauer*. Berkeley: University of California Press, 1963. pp. 315–50.

Serres, M. (1991) *Rome: The Book of Foundations*. Stanford: Stanford University Press.

Smith, N. (1984) *Uneven Development*. Oxford: Blackwell, 1990.

Thomas, W., Sauer, C., Bates, M. and Mumford, M. (eds) (1956) *Man's Role in Changing the Face of the Earth*. Chicago: University of Chicago Press.

Thrift, N. (1996) *Spatial Formations*. London: Sage.

Weiss, G. (1999) *Body Images: Embodiment as Intercorporeality*. London: Routledge.

Whatmore, S. (2002) *Hybrid Geographies*. London: Sage.

Willems-Braun, B. (1997) 'Buried epistemologies: the politics of nature in (post)colonial British Columbia', *Annals of the Association of American Geographers* 87 (1): 3–31.

Wolch, J. and Emel, J. (eds) (1998) *Animal Geographies*. London: Verso.

Wolfe, C. (1998) *Critical Environments: Postmodern Theory and the Pragmatics of the 'Outside'*. Minneapolis: Minnesota University Press.

8

Geographies of Nature in the Making

Noel Castree

We need different ideas because we need different relationships. (Raymond Williams, 1980: 85)

My task is an unenviable one: to think the history of thinking about the relations between 'culture' and 'nature', all the while keeping sight of the latter's elusive materiality. The difficulties of undertaking such a task are legion. To begin, there's the legendary problem of fixing the meaning of my two 'keywords', which is why they hang so precariously between scare quotes. As Raymond Williams famously noted, the terms 'culture' and 'nature' are formidably polysemic or, if you prefer, semantically promiscuous. This immediately makes any discussion of the relation between the two unstable. Unable to pin down their cognitive content, one is forced to use signifiers whose meaning is constantly shifting.[1] The result, potentially at least, is a form of argumentative sea-sickness.

Added to this, things get no easier when we try to think about the materiality of culture and nature.

It's become habitual in post-Enlightenment thought to oppose the material and the ideal, that is, 'hard' facticity and 'soft' discursivity. In geography, anthropology, history and beyond, there's a whole tradition of research into the material making and remaking of nature/s by and through culture/s. In recent years, the 'matter of nature' – to borrow Fitzsimmons' (1989) felicitous phrase – has taken on a heightened importance across the disciplines and in the wider world. Greenhouse warming, species extinction, acid deposition, ozone thinning, resource exhaustion, genetic modification: these and other purportedly epochal transformations of the natural have alarmed and preoccupied academics, policy makers and publics worldwide. And what is at stake here is nature's materiality: that is, its ontological existence, capacities, powers and consequences. From the plaintive cry that nature has ended to the brave new world promises of transgenic technologies, the focus of attention is the physicality of those things we conventionally call natural. But, as I hope to show in this chapter, the materiality of nature is far from self-evident. This is not just because the common equation of the material and the natural is itself problematic – after all, isn't culture every bit as material as nature? (Graves-Brown, 2000) – but also because the closer we get to knowing nature's materiality, the more that materiality is deferred. Herein, I will argue, lies a persistent dilemma for research into the culture–nature nexus. The conclusion I will ultimately drive towards is that attempts to think about the matter of nature take us to the very limits of thinking itself.[2]

The root of the problem, I want to argue, is the analytical cast of so much post-Enlightenment thinking about culture and nature. With Descartes and Kant as its philosophical flag-bearers, the analytical mindset works with self-sufficient abstractions and seeks out binarisms. Thus the exclusive categories of culture and nature line up with a host of other suspect dualisms: human/non-human, mind/world, representation/reality, epistemology/ontology, ideality/materiality. The critique of the analytical imaginary is long standing and familiar and I make no claims to originality in repeating it here. However, I also want to argue that even its apparent opposite – a relational imagination – can suffer the same cognitive maladies. With Hegel, Leibniz and Spinoza as its philosophical flag-bearers, the relational worldview tries to stitch back together that which is rent asunder by the analytical lexicon. As we'll see, by expanding the meaning of materiality successive authors have sought to breach the culture–nature divide. But by crossing one Maginot line, these authors still run up against another: that which separates nature's materiality from thought about it. In short, relationalism (or certain versions of it) has its own analytical limits.

In order to give the discussion some focus I propose to review a history of Marxist work into the culture–nature nexus. This may seem an odd focus for two reasons. First, Marxism (historically at least) is often seen as a productivist, economistic tradition of thought with relatively little to say about culture and even less about nature. Second, in geography Marxism has predominantly been deployed in urban and regional economic research, not in work on culture–nature. However, my decision to examine Marxian thinking is not, in fact, as peculiar as it seems. To begin, I use the term 'Marxism' in the broadest sense to designate a variegated tradition of thought with multiple branches (Berman, 2000; McLellan, 1999). Marxism is anything but a unified thought-field; if it has any coherence then it's loose indeed, hinging on Marx's political economy as a common reference point (however critically and vestigially). Thus, when I use the word 'Marxism' in this chapter I refer to everything from the 'classical' work of Marx through to 'neo-' or 'post-Marxist' writing.[3]

Equipped with this ecumenical definition, a whole world of Marxian work into culture and nature hoves into view.[4] In addition to this, much of this work spans the disciplines. Geography hardly has exclusive rights to the study of culture and nature. Though much of my discussion will refer to the research of geographers, it would be parochial indeed to disbar any wider discussion of how Marxists have grappled with the nature–culture nexus. Indeed, one of my points will be that the dilemmas faced by Marxist geographers are by no means exclusive to them. Finally, my focus on Marxism is, arguably, particularly apposite for any discussion of the materiality of nature and culture. For the Marxian tradition, of course, has always boasted about its hard-headed materialism since Marx and Engels famously declared that existence precedes consciousness. More than most approaches to culture and nature, therefore, one might expect Marxism to offer real insights into materialism (Williams, 1978). The insights, I hope to show, are real enough. But, despite themselves, they only serve to demonstrate how difficult it is to understand the materiality of nature (and culture) at all.[5]

The chapter is structured as follows. In the next section I show briefly how several Marxists, following Marx, have held the problematics of culture and nature apart and why this is not a defensible option. I then explore ten ways in which other Marxists have tried to rethink the culture–nature connection. These ways can be captured in the motifs of determinism/determination, articulation/dialectic, coupling/conjunction, embedding, process/construction and materialization. As I move from the first to the last of these we see how Marxists have increasingly challenged analytical approaches to culture and nature. This challenge has entailed a stretching of the concept of materiality to encompass more and more things. As I proceed, I will explore the strengths and weaknesses of these ten attempts to show how culture enters into the constitution of nature. However, apropos of my earlier comment about how slippery the terms 'culture' and 'nature' are, we see the meaning of each shifting somewhat as I explore the different options. Though this definitional variance does make the argument rather fluid, my main point will, I hope, be clear enough: namely, that even the most relational of Marxist imaginaries runs up against the problem of how to grasp nature's materiality. This is, I submit, a problem for *all* attempts to understand how culture/s make nature/s. As such, the dilemmas of Marxism have a wider relevance. What is required is nothing less than a rethinking of how we think about the matter of culture and nature.

HOLDING CULTURE AND NATURE APART

Until relatively recently the Marxian tradition tended to hold the problematics of culture and nature apart, reflecting a more deep-seated division of thought in the western habitus. Whether or not Marx intended it to be so, his work has been read[6] as treating issues of culture and nature in relative isolation and for much of the twentieth century his epigones mimicked this bifurcated worldview.[7] It's not hard to understand why. On the one side, Marx famously said very little about nature in his major writings. What he did say – as Schmidt's (1971) careful act of historical recovery showed – tended to emphasize the relations between nature (taken as a non-human domain) and mode of production. On the other side, though Marx said much more about culture – from his writings on art to his more general claims about the relationships between 'base' and 'superstructure' – the implication was that it was at a considerable ontological and causative remove from nature.

If culture, as Marx at times insisted, was but a reflex of the economic, and if nature's brute facticity underpinned, enabled and was sometimes threatened by the gyrations of economic life, then culture and nature existed in different registers altogether. It was all very well for the Marx of the *Grundrisse* to insist famously that all reality is a complex 'unity of the diverse', but his own writings tended to dissociate two key moments of that unity as if 'culture' and 'nature' were somehow held apart by an 'economic' middleman. Added to this was his well-known ambivalence about science. Though Marx was frequently critical of certain types of scientific thinking, several of his writings instantiate a remarkably conventional distinction between 'natural' and 'social' science (Aronowitz, 1988: Part I; Wilson, 1991). Armed with this distinction, Marx could claim that his own 'critical science' of society would explain culture as the ideological clothing on the body of capitalism, while leaving nature to the biologists, physicists and chemists who, during the late nineteenth century, were fast achieving disciplinary legitimacy in the new western universities (Vogel, 1996). Again, the result was to hold nature and culture at a distance. After all, the former was the domain of a trenchant materialism which could be known 'objectively' – and after Marx's death Engels (1956) controversially went on to specify nature's 'dialectical laws' – while culture was the paper house erected duplicitously on the foundations of production.

Depending on how Marx is read, then, his various writings on culture and nature seem to instantiate two materialities and one ideality. While nature undoubtedly 'mattered' to Marx, he seemed to take its materiality for granted as both condition and resource for human labour[8] – hence the lack of a full explication of the natural in his work. The materiality that fascinated Marx much more, of course, was the domain of practices and relations captured by the abstraction 'the economy': hence the thousands of pages Marx devoted to developing a theoretical vocabulary to anatomize capitalism. It remains unclear which of these two materialities mattered most for Marx – Schmidt, for example, argued that Marx gave priority to the motive force of modes of production,[9] while Italian Marxist Sebastiano Timparano (1975) insisted on nature's primacy – but they were, it seems, posited as ontologically different and distinct.[10]

This 'two materialities' approach was to become particularly clear in the agrarian political economy of Mann and Dickinson (1978) who, revisiting Marx, famously identified the problem that nature's stubborn physicality posed for capital. Taking the case of commercial agriculture, Mann and Dickinson showed that the physical properties of soils and plants posed a 'barrier to accumulation' for rural capital – in particular, by slowing down capital turnover time, which directly affects profitability (see Henderson, 1998). A similar, though more abstract, focus on the intractable physicality of nature also arguably animates the new 'eco-Marxist' theorizing of Elmar Altvater (1993), Ted Benton (1989), James O'Connor (1998) and Paul Burkett (1999). Finally, boasting about his 'tough-minded materialism' as a counterpoint to German philosophical idealism and bourgeois ideology, culture was figured by Marx as an ontological poor relation: real enough, to be sure, but somehow phantom-like and insubstantial. It was, in short, a domain of ideas that could have no existence if nature and economy did not together materially undergird it.

The problems here are plain to see. On the one hand, it's not unreasonable to insist that nature, economy and culture are irreducible to one another (and this is the kind of argument that critical realists like Andrew Sayer, 1999, would make today). After all, if nature is seen as a realm of physical processes, the economy as a domain of instrumental-technical practices and culture as a sphere of symbolic interaction then they deserve, quite properly, to be theorized in relative isolation prior to any attempt to grasp their mutual interactions.[11] But Marx's approach is implausible in other key ways. First, the ontological hierarchy installed between economy and culture – and by implication nature and culture – these days looks decidedly shaky. Quite how one can claim that culture is somehow 'less material' than other moments of socio-environmental life is difficult to fathom. Moreover, the attempt to deny culture any real autonomy outside the marionette movements dictated by the economic base (itself erected upon natural foundations) seems today to be theoretically crude. In short, Marx's severance of the natural and the cultural is operationalized through a set of dubious distinctions and causal hierarchies.

Of course, if one reads Marx's work as implying that capitalism is a culturally specific and expansive economic system, originating in western Europe, then the argument that he separated nature and culture falls apart. From this perspective – which sees modes of production as distinctive 'ways of life' – Marx's scattered remarks about nature can be seen as a tantalizing commentary on how capitalism materially remakes nature in its image. As such, the minor cottage industry on what Marx 'really' said about nature – from Schmidt through to the recent work of the eco-Marxists mentioned above – can also be viewed as an ecological critique of the 'culture of capitalism': a 'culture' that makes nature but a means to the soulless end of accumulation.[12] However, while fair enough in its own right, this attempt to equate the economic and the cultural inevitably misses far too much about those various practices and representations we have now come to associate with 'culture'. Additionally, it's probably true to say that neither Marx nor Schmidt et al. intended their writings on nature to be read primarily as a critique of capitalism's peculiar 'cultural economy'.

In any case, several of Marx's twentieth-century epigones explicitly dissented from Marx by seeking to bind together the problematics of culture and nature in a more formal way, in part by redefining the meaning of the two terms. Key here were those 'western Marxists' Adorno, Horkheimer and Marcuse. Through a creative fusion of Marx, Weber, Freud and other theorists, their melancholic diagnosis of 'instrumental reason' posited nature as capitalist modernity's utopian 'other': at once an unattainable source of redemption and yet something, tragically, which was increasingly dominated by western societies (Vogel, 1996). Then, to offer a second, very different, less theoreticist example, there were the magisterial writings of French historian Fernand Braudel. Never a Marxist in the orthodox sense, Braudel dissolved the false partitions dividing nature, economy and culture in a seamless story of human–environment translations and transformations.

I will say more about the Frankfurt School, the kind of 'embedded' Marxism that Braudel articulated, and a host of other Marxian attempts to link nature and culture later, but what's notable is that these and other critical reworkings of Marx's materialism were largely ignored during human geography's early engagements with historical materialism. Overlooking virtually a century of Marxist work after Marx, the peculiarity of Marxism's importation into the discipline was just how orthodox it was, with David Harvey as the principal advocate of a 'classical' version. This, I think, helps explain why Marxist geographers – like Marx himself, only 100 years later – for a long time tended to treat culture and nature separately, if at all. Virtually ignored through the 1970s and 1980s, when Marxist geographers were preoccupied with studying the capitalist space economy, they received initial attention only in isolation.

Thus, to take two representative examples, there was Denis Cosgrove's (1985) cultural materialism on the one hand – a new, innovative approach to cultural landscapes – and Neil Smith's (1984) insightful theorization of nature on the other.[13] True, Cosgrove's work did impinge directly on the culture–nature question, in part because it drew upon the synthetic materialism of Raymond Williams (of which more anon) (Daniels, 1989). But by and large Cosgrove and Smith were speaking different languages, albeit within the broad parameters of the Marxist tradition. And yet all along, of course, human geography had had its own tradition of thought – and a materialist one at that – which placed the culture–nature nexus centre-stage: namely, Sauer's brand of cultural landscape study, which was at once intellectually ambitious and empirically exacting.[14] Far too 'bourgeois' and conventional for the early Marxist geographers, this Sauerian attempt to synthesize cultural process and environmental change was largely bypassed; even as an object of critique it barely figured in the first writings on nature by geographical Marxists like Neil Smith, Phil O'Keefe (Smith and O'Keefe, 1980) and David Harvey (1974). So it was that, until recently at least, human geographers working within, or responding critically to, the Marxist tradition illicitly kept nature and culture in separate intellectual boxes.

Of course, this has all changed over the last decade or so, in part because human geography's 'cultural turn' has been achieved by extending the compass of culture and breaking down those analytical divisions that have for too long imprisoned it in a discrete ontological space. But, as I noted above with the examples of Adorno et al. and Braudel, Marxists outside geography have for decades sought to span the culture–nature divide. These spanning operations, within geography and without, have entailed a challenge to the 'common-sense' Cartesian-Kantian mindset that many Marxists since Marx have flirted with, even as they have sought to accent the relational worldview Marx critically adopted from Hegel and his forebears. I now want to explore these different attempts to think the culture–nature relation from a materialist perspective. My review is illustrative rather than exhaustive and in no sense chronological. As we'll see, the closer Marxists and their interlocutors seem to get to the matter of culture–nature, the more elusive that materiality becomes.

DETERMINISM AND DETERMINATION

One solution to the dichotomization of culture and nature has been to think about the causal relations between the two domains, however the latter are defined. This is not quite the transgression of categorial boundaries that it seems because, as I'll now illustrate, resort to the motifs of 'determinism' and 'determination' has still left Marxists saddled with the pitfalls of more analytical approaches to culture and nature. Moreover, these motifs fail to resolve the tricky issue of where the materialism that Marxism vouchsafes begins and ends.

Of the several examples I could offer, I want to focus on just three cases of the 'determinist' solution to theorizing culture–nature relations. The first concerns the maverick Marxism of Karl Wittfogel who inverted Marx's familiar theses about the primacy of material production to stress the 'deeper' materiality of nature. His *Oriental Despotism* (1964) advocated a peculiar form of Marxist environmental determinism (what he called 'geographical materalism') which argued that social formations – a mesh of economic, culture, politics and more – are decisively controlled by their natural resource base (Peet, 1985), an argument later echoed independently by Timpanaro (1975) and taken up explicitly by environmental historian Donald Worster (1985) in his exploration of 'hydraulic capitalism' in the western USA. The second and third examples, by contrast, emphasize the effectivity of 'non-economic' forces in concealing social truths by resort to dubious conceptions of nature. Here it is ideas deriving from, and functional to, the economic 'foundations' of capitalist society which make the notion of nature a cultural construct. Thus Neil Smith (1984) spoke of general and enduring 'ideologies of nature' in the west which were designed to conceal the reality of nature's creative destruction at the hands of capital. Similarly, David Harvey (1974) talked about the 'ideology of science' operative in the neo-Malthusian debates of the early 1970s on the global 'population bomb'. Whether Smith and Harvey would label these ideologies as belonging to the cultural domain is moot, but they certainly seemed to arise, directly and functionally, from economic relations. For both writers, then, the 'culture of nature' under capitalism is a duplicitous one that serves dominant class interests.

The limits of a deterministic marrying of culture and nature are not hard to grasp. This is why other Marxists have preferred the more flexible and open language of determination. Some of the best examples come from Marxist analysts of the practice of natural science. From the mid 1950s, inspired by Merton's new sociology of science, provoked by J.D. Bernal's writing and dismayed by the Frankfurt School's bleak reading of scientific-technical reason, several Marxists sought to offer a more careful account of western capitalist science in action. Early critics – like the Radical Science Journal Collective (1974), Rose and Lewontin (1984) and Robert Young (1979) – tended to focus on laboratory science and scientific theories, while more recent Marxists have turned their gaze to field science. Good examples of the latter are Goodman et al.'s *From Farming to Biotechnology* (1987), Lawrence Busch et al.'s *Plants, Profits and Power* (1991) and Jack Kloppenburg's *First the Seed* (1988), where plant biology is set in its specific socio-economic context. In each case science was/is seen as infused with class interests, in terms of foci of research, the norms of the scientific community, scientific funding and so forth.

The dominant class agendas and economic imperatives of capitalist production were thus seen not as *controlling* scientific practice but as exerting definite pressures with definite outcomes in terms of what 'truths' about nature were discovered and legitimated. As in the case of Wittfogel, Smith and Harvey, these are only analyses of culture and nature if one stretches the meaning of the former term to include scientific communities, themselves embedded in a mesh of economic relations, institutional sites and state policies. But this is, of course, precisely what contemporary post-Marxist analysts of science like Donna Haraway and Hilary Rose have subsequently done in an effort to argue that it is not immune from the 'subjective' forces of culture.

The motifs of determinism and determination have helped take Marxists beyond the unhelpful separation of culture and nature. But they are freighted with problems. Though the determinist approach stages an encounter between culture and nature, it does so only by way of spurious ontological hierarchies and distinctions secured by theoretical fiat. Thus in Wittfogel's case nature reigns causally supreme, while Smith and Harvey stress economic prime causes while seeing ideology as a sort of cultural negative. The causal hierarchy is much softer when the motif of determination is invoked, but again some overly tidy abstractions are posited such that – to revisit the examples used above – the 'culture' of science is related to, but discrete from, 'economic interests'. Rather than taking these distinctions as givens it is important to reflect on their origins and epistemic status. Likewise, it's worth asking why some Marxists have routinely considered that nature (for example, Wittfogel) or economy (for example, Smith and Harvey) are somehow 'more material' than culture. On top of all this, and finally, the determinist/determination approach to culture and nature leaves one crucial issue unresolved. However its materiality is figured, how can Marxists come to *know* the nature they speak of? If, after Harvey and Smith, culture 'conceals' the realities of nature, or if, after Goodman et al., Busch et al. and Kloppenburg, science as class culture constructs its own truths, then how can Marxists claim an 'ideology-free' understanding of the natural? This, as I will go on to show, is a fundamental problem for the Marxist tradition in geography and without, and one that is inadequately addressed by successive redefinitions of the materiality of culture and nature.

ARTICULATION AND DIALECTIC

One way that Marxists have challenged the conceptual policing and dilemmas intrinsic to determinative approaches is to think of culture and nature as articulated dialectically. As with almost all Marxian work in this area, a key role for the economic is retained but, instead of the ontological triad being conceived vertically (nature underpins economy which, in turn, shapes culture), the metaphors used are more horizontal (no more base and superstructure). This makes a lot of sense. As Maurice Godelier aptly put it, 'since a society has neither a top nor a bottom … the distinction between infrastructure and superstructure … cannot be taken as a distinction between levels or instances' (1986: 18–19). As we'll see momentarily, this is not to 'demote' economic relations but rather to resituate them in a more complex field of relations that includes organic and inorganic entities, mental representations, symbolic codes and communicative interaction. Here, then, the cultural and the natural are brought closer together by way of a more fluid theoretical imagination.

To illustrate what I mean, let me look briefly at the work of geographers Donald Moore (1996) and Michael Watts (Pred and Watts, 1992; Watts, 1998), whose thinking is informed by Antonio Gramsci and Althusser/Benjamin respectively. Moore's Marxian political ecology attempts to make sense of conflicts over access to resources among men, women, peasants and state actors in eastern Zimbabwe during the 1980s. As a way of moving beyond Marxism's 'tired orthodoxies' (Moore, 1996: 126) Moore turns to Gramsci in order to 'underscore … how symbolic struggles effect material transformation … [C]ultural meanings are *constitutive* forces … not simply reflections of a material base' (1996: 127). Like some of the best work in so-called 'Third World political ecology', Moore's approach is synthetic, attuned to the

contingent unfolding of events in a specific time and place. There are two things to note here. First, for Moore culture and nature remain relatively distinct, but culture becomes the means *by and through which* struggles over nature occur. Second, culture here is granted a positivity – it is not merely a domain of ideological deception. As such, it becomes something like a full partner in the nature–economy–culture triad, invested with its own material effectivity. This is true for both those in positions of power – for example, the managers of the state nature park in Moore's case study – and those who seek to resist the operations of power – in Moore's case study, indigenous groups who've been on the wrong end of European colonialism for over a century. In short, culture is the vital place where diverse actors give voice to, and contest, specific economy–environment interactions.[15]

Similar themes animate Michael Watts' brilliant interpretation of the 'petrocapitalism' fuelling the dramatic economic, cultural and political history of modern Nigeria. Using a creative synthesis of Althusserian and Benjaminian ideas, Watts examines symbolic discontent among followers of the Muslim Maitatsine movement centred on the city of Kano. In the context of rapid industrialization, a product of Nigeria's oil exports which linked it to the hidden hand of global markets, Watts shows how the Maitatsine drew upon fantastical religious wish images to contest the secular moral economy fast becoming hegemonic in 1980s Nigeria. As with Moore's study, Watts' point is that while nature, economy and culture are irreducible, it's unproductive to see them as separate or as hierarchically positioned in ontological terms. Each domain has a 'relative autonomy', if you will, which ensures that the materiality of each is recognized rather than diminished. To be sure, Nigeria's 'fast capitalism' lies at the centre of the drama for Watts, but nature is not a passive surface, nor is culture economy's pale florescence (Watts, 1998). Rather, each mutually conditions the other; they articulate in changing and contingent ways that amount to a complex, one might even say 'overdetermined', dialectic that is not about famous last instances.[16]

With Moore and Watts we see a more relational and less analytical approach to the culture – nature nexus, a nexus infused by economic relations. Because both writers examine empirical complexities, they avoid the rather rigid categorial and causative claims of more theoretical forms of Marxism. True, their relational imaginaries have their limits: to wit, the culture–nature distinction is, in the final instance, maintained throughout. But the point is that the operations of culture and nature, mediated as they are by political economic relations, cannot be understood discretely. With culture taken to be every bit as material as the 'hard' economic and ecological realities it seeks to make sense of, the motifs of articulation and dialectic suggest an approach which *relativizes* the limits and powers of culture/nature/economy in any given context. In short, these domains mutually and materially determine one another, echoing Marx's plea that any proper (*sic*) analysis of concrete situations should attend to 'diverse determinations'.

Of course, knowing which determinations are at work requires that the analyst comes equipped with appropriate abstractions. And here again, even with the 'matter' of nature and culture on a more even footing, we find Marxists like Moore and Watts resorting to an inherited language of difference which specifies, albeit in broad terms, that the two domains are ontologically distinct though related. What are needed, I would argue, are Marxian metaphors that are rather more relational than those of articulation and dialectic – ones which can push Cartesian-Kantian binaries even further to the margins. Moreover, the epistemic problem of how Marxists can know the materiality of culture–nature–economy remains relatively unproblematized in Moore's and Watts' work. Indeed, one can detect an almost traditional attachment to epistemic realism in the writings of both men, since one is left to assume that the cognitive claims made are more-or-less accurate 'reflections' of the material world being discussed. As Bartram and Shobrook put it, echoing Marcus Doel (2000), 'within the "enclosed" epistemology of nature–society dialectics there is an assumption that nature is still out there to be reclaimed, protected or remade' (2000: 373).

COUPLING AND CONJOINING

This brings me to the notions of nature–culture couplings and conjunctions. Here culture is seen as being 'in' the natural, either directly or at a distance. In other words, nature and culture are seen here as necessarily entangled (see Thomas, 1991). Some of the most productive Marxian work in this area has been based on a creative rereading of Marx's account of commodity fetishism; other work has attended more to the specific form of 'value' dominant in capitalist societies. In both cases the materiality of culture and nature is not only redefined; the epistemic status of knowledge of that materiality also begins to be problematized. Let me take each in turn.

In rather different but equally illuminating ways, anthropologist Michael Taussig (1980) and geographer Allan Pred (1998) have sought to stress the physical imbrications of culture and nature. They achieve this by way of unconventional readings of what counts as the 'economic' and, more particularly, by way of a respecification of commodity fetishism. In *The Devil and Commodity Fetishism in South America* (1980), Taussig's well-known study of how indigenous peasant communities involved in mining and plantation agriculture in South America responded to the sudden intrusion of market relations, he effectively writes an anthropology of capitalism in order to thoroughly denaturalize it. He does this not by treating capitalism as a culturally specific economic system that has spread out historically from Europe – the appealing but rather urbane attempt to collapse the economy–culture division that I mentioned earlier – but in a more interesting way. Echoing Lukacs' (1971) brilliant account of reification, Taussig seeks to 'examine the connections between the deepest categories of thought and the social practices (and social contradictions) within which that thought arises' (Vogel, 1996: 20–1).

In other words, social ideas and thoughts – what Marx once called 'material forces' in their own right – are not mere superstructural ephemera but the very 'cultural' categories that enable daily economic (and other) practices to occur. In capitalism, for Taussig as for Lukacs, the commodity is emblematic of a peculiarly estranged worldview, where 'things' are taken as ontological givens that can be separated from social relations, personal identities and ecological contexts. As Taussig put it, 'things stand in some way ... for social relations. But unless we realize that ... social relations ... are *themselves* signs ... defined by categories of thought that are also the product of society ... we remain victims of ... the semiotic we are seeking to understand'. (1980: 9, emphasis added). What his book does, in thrilling anthropological detail, is to show how peasant communities confronted with commodity culture seek to make sense of, and contest, the mining/farming of natural landscapes. The result is to show the fetishism and cognitive category mistakes intrinsic in the capitalist 'worldview', a worldview which fails to see the inextricable relations between social labour, worker exploitation and natural resource extraction.[17]

Approaching commodities from a rather different direction, Allan Pred has recently tried to expand Marx's sense of them as social relations to think of them as natural-social-cultural relations. In this his thinking resonates with that of anthropologist Arjun Appadurai (1986). His argument works by attending to the two 'ends' of the commodity chain that Marx, in his late political economy, artificially closed off for discussion. On the one side, along with many contemporary cultural geographers, Pred insists that we attend to the moment of commodity consumption as an active partner in the reproduction of economic life. On the other hand, he argues that commodity fetishism – that structural misrecognition of both the origins and social content of capitalist goods – has a hitherto little-explored link to nature. For, though value in capitalism is for Marxists defined in terms of social labour, the material wealth (use value) embodied in commodities ultimately derives from the extraction and remaking of natural entities. By 'denaturalizing consumption' Pred thus seeks to make visible a whole world of hidden connections between, say, the purchase and use of elaborately marketed cashmere garments and the socio-ecological exploitation attendant upon their production. For Pred, we must, in sum, trace the whole chain of connections from consumption (culture) through economy (social labour) to nature (use values) – and back again – in an act of profane illumination.[18] Similar arguments have been made by Elaine Hartwick (1998) in her materialist deconstruction of the semiotics of diamond consumption in the west.

Pred's analysis links well with the second body of Marxian work on culture–nature couplings and conjunctions I want to consider. This work focuses more upon the form of value dominant in capitalist societies and the way it affects human–environmental relations. Though pitched at different levels of abstraction, the arguments articulated in Chapter 7 of David Harvey's *Justice, Nature and the Geography of Difference* (1996) and in John Foster's edited book *Valuing Nature* (1997) have an elective affinity. Harvey anatomizes the 'value form' which, in capitalist societies, environments, cultural artifacts and social relations assume. Foster looks at the hegemonic languages used to measure and act upon the environment in a world where capitalist money (the representation of labour value) serves to commensurate diverse systems of cultural evaluation. In both cases labour value is treated not simply as an 'economic' measure but as one of several possible value systems in western and non-western cultures. Harvey's and Foster's point is not just that labour value is an increasingly dominant value sphere worldwide – thereby displacing other ways of valuing nature – but that it envelops nature both discursively (as, for example, in the language of 'free market environmentalism') and physically (as when natural entities become means to the end of value

expansion). Here, then, we see labour value as the chrematistic thread that brings nature into the ghostly universe of capitalist social relations.[19]

The strengths of the motifs of coupling and conjunction are threefold. First, culture, nature and economy almost dissolve one into the other; though ontologically different they are practically interfused. Second, in the studies reviewed above, little or no attempt is made to install causal hierarchies; the matter of culture and nature is relatively coequal. Finally, the criticism of fetishism and the critique of value both call into question conventional or received ways of knowing nature – thus relativizing the question of how the matter of socio-environmental life is to be understood 'correctly'. Depending on where one is situated historico-geographically – Taussig's peasants, Pred's consumer, Harvey's and Foster's unthinking appropriator of capitalist value definitions – one's knowledge of nature varies dramatically. And yet, this said, in Taussig, Pred, Harvey and Foster there's arguably a hint of residual naturalism. That is, one detects a yearning to 'uncover' 'hidden realities' with the implication that what is revealed is somehow of 'deeper' importance than other moments of, or perspectives on, the socio-ecological world in which we exist. With Taussig and Pred, the motif of fetishism threatens to revert to a depth metaphysics where some material 'essence' lies behind duplicitous 'appearances' (Castree, 2001; Cook and Crang, 1996). Similarly, with Harvey and Foster, there's an epistemic doubleplay, wherein the Marxist analyst lays claim to corrective vision for the ideologically near sighted. In other words, at some level all four authors claim – implicitly and unreflexively – to stand above or outside that which they critique: a 'critical distance' that precisely allows them to make the cognitive claims they make.

EMBEDDING

This problem is not resolved in the fourth of the Marxian motifs for conceptualizing culture–nature relations that I want to review here: that of embedding. This motif directs us to work that is less theoretical and more empirically grounded than much of that I've discussed so far. Though they never formally described their work using the term, the concern to disclose the 'embeddedness' of economic activities in specific ecologies and cultures arguably animated the very different writings of Fernand Braudel and Karl Kautsky. Abjuring the seemingly arid abstractions of Marxian modes of production theory, both men sought to trace the intimate, organic ties binding specific social formations to specific environments. True, Braudel was concerned with the *longue durée* of regional change, while Kautsky (1988) was concerned with the 'problem' of agriculture in modern capitalism. For Braudel (1972; 1973), the issue was how diverse places, ecologies and cultures in southern Europe, from the Middle Ages onwards, became connected together in extended mercantile trading networks and, later, a nascent capitalist economy. For Kautsky, in *The Agrarian Question*, the issue was more specific: how and why were 'pre-capitalist' class relations surviving in an economic sector (agriculture) at once directly dependent on nature and vital to sustaining the wider expansion of a capitalist way of life in the nineteenth century? But these differences notwithstanding, in both cases an attempt was made to respect the routinized flows of materials and practices wherein the labels 'cultural' and 'natural' became fuzzy to the point of being slightly irrelevant. Culture, nature and economy were each depicted as medium and outcome of, or condition and cause for, the others.

In more recent times we owe our understanding of the term 'embeddedness' to Karl Polanyi (1944) and Mark Granovetter (1985). In spirit at least, something of what Polanyi and Granovetter were trying to communicate about 'the economy' has been achieved *vis-à-vis* nature–culture by a new generation of Third World political ecologists and by a cohort of agrarian political economists. I can only consider one of many possible examples here: Miriam Wells' wonderful *Strawberry Fields* (1996), a detailed study of class and work in rural California. Going against the grain of Marx's thinking, Wells seeks to explain the 'reversion' from wage labour to share-cropping in strawberry production. What emerges is a nuanced story of how ecology, class, 'moral discourse', cultural norms and the law are mutually shaped in a particular time and place with very specific effects. Refusing Marx's apparently presumptive materialism, Wells' analysis is, in other words, responsive to local detail not theoretical diktat – a move that has clearly bothered some (Walker, 1997). Consequently, *Strawberry Fields* identifies no 'essential' processes and directs our attention instead to what Williams (1978: 7) aptly called the 'constituted materiality' that is distributed throughout the nature–economy–culture nexus.

It was Williams himself, of course, who in the 1970s and 1980s gave theoretical voice to the sensitivity that authors like Wells have since sought to express.[20] In *Problems in Materialism and Culture* (1980) he challenged the balkanization of

knowledge in Marxism that confined materialism to a privileged ontological zone. As such, his 'cultural materialism' was a creative attempt to rethink nothing less than the materiality of historical materialism. More than other cultural Marxists of his generation – such as E.P. Thompson – Williams was deeply interested not only in how culture and nature were connected but in how these categories themselves emerged and functioned in capitalist societies. Though these interests were crisply articulated in his more theoretical essays, it was in Williams' novels – such as *Border Country* – that the sedimentation of culture, environment and production were most sensitively explored (Harvey, 1995). In a sense what Williams shows is that we live in an a world of interconnected 'local capitalisms' whose common capitalist character is deeply modified by lived cultures and animate ecologies. This said, Williams' thinking about culture and nature cannot easily be pigeonholed under the category 'embedding'. But since I don't want this essay to become an exploration of Williams' polymorphous *oeuvre*, I will simply state that some of the other work reviewed in this essay owes a real debt to his thinking in this area.

Like the work considered in the previous section there is much to recommend these attempts to embed culture and nature in each other. But a persistent problem that I have highlighted throughout this chapter remains. Even if culture and nature are placed on an equal onto-explanatory footing, with economic relations hovering somewhere nearby, this does not answer a key question raised earlier: how can one come to know the matter of the nature that is being remade by, and yet helping to constitute, specific social formations? For if, as all of the authors I've discussed so far indicate in different ways, nature *as such* scarcely exists outside cultural, economic and other relations, then presumably we still need a way to authorize any claims about it. This is particularly so if we are to respect the fact that while those things we call natural do not exist *tout court*, they nonetheless have material capacities that are not *the same* as those of culture or economy (however we define the latter two terms). As I want to illustrate in the next two sections, several contemporary Marxists (orthodox, neo- and post-) have only really addressed this question by displacing it.

PROCESS AND CONSTRUCTION

One solution to the question posed above is to reject it. After all, the question rests on the hope that the materiality of nature/s in the making can be grasped with some degree of accuracy (by either people 'on the ground' or academic Marxists analysing them) – and that implies a distinction between the knower (a 'cultural', 'economic' and 'social' being) and the known ('nature'), even as they're practically related and united. In other words, the question seems to revert to a dualism of the very kind that, as I've shown, several Marxists studying culture, nature and economy have sought to overcome. What is required, it seems, is an *even more* relational approach than the motifs of coupling, conjunction and embedding offer us. In recent years, two versions of this more relational approach stand out.

The first is a process-based approach that insists on the absolutely seamless intertwining of, and perpetual exchange between, the economic, cultural, natural and all other moments of socio-ecological life. Drawing upon Bertell Ollman's (1993) interpretation of Marx, Smith's *Uneven Development* and the 'amodern' arguments of Bruno Latour (1993), it's an approach advocated by Erik Swyngedouw (1999) in his inquiries into the waterscapes of Spain and Ecuador. The watchword of Swyngedouw's anti-analyical imagination is 'metabolism': the ceaseless 'glocal' interleaving of people, things, materials and artifacts under the aegis of capital. Capitalism is thus not above or outside the rhythms of daily life. Rather, economy, culture, nature and much more all get churned up in the driverless juggernaut that is capitalism, such that ontological 'insides' and 'outsides' are difficult to fathom. This holistic view of 'socio-nature' is advocated at a more philosophical level by David Harvey (1996; Chapter 2) and dovetails with his views on the peculiarity of capitalist value (mentioned above). For Harvey, socio-environmental life is an indissoluble material process of multiple 'moments' and material 'permanences', and the study of them is part of this process, meaning that all knowledge of them is necessarily situated and provisional.

Where Swyngedouw and Harvey arguably accent the practical, physical aspects of socio-natural change, other authors have looked more to the power of discourse. Here the question as to how one might come to know the materiality of the nature that is so intimately inscribed in the materiality of culture and economy is subject to a linguistic *coup de grâce*. For the argument is this: that since the matter of nature can only be known by beings who are ineluctably cultural it is pointless to inquire about materiality as such. In other words, the material 'power' of linguistic conventions is so intractable that knowing nature becomes an issue *internal* to culture (including,

presumably, the 'culture' of the academic analyst too). The relationality of nature and culture here becomes asymmetrical, though their materiality remains nominally equal. There are 'weak' and 'strong' versions of this argument. Gavin Bridge (1997) accents this cultural-linguistic constructionism in his research into the contemporary mining industry. He focuses on the industry's 'environmental narratives' to show how culturally specific and carefully choreographed representations of nature circulate through the media and society. His point is not that metals like gold and copper don't exist. Rather, he is at pains to show how 'representation' can become 'reality' because it is inherently performative. Discourse is not, *contra* Smith (1984), mere 'ideology' for Bridge. Rather, it functions for mining companies as a tool that allows them to tell a very specific story about the economically driven excavation of pits, quarries and hillsides: it is productive, proactive, material.

A stronger version of this discursive constructionism has been advocated by anthropologist Arturo Escobar (1996; 1999). More indebted to poststructuralism than Bridge, Escobar's synoptic account of contemporary economy–environment relations links different phases of capitalism with distinctive 'discursive regimes'. In its current 'ecological phase', global capitalism, so Escobar argues, redescribes natural entities in such a way that they become available for a new, 'greener' round of exploitation than heretofore. No longer simply a 'resource' to be exploited, nature has become something to be 'banked', collected and partitioned in the brave new world of 'sustainable development', biomedicine and genetic modification (Katz, 1998). More than Bridge, Escobar's point is that the power of language goes 'all the way down': there is simply *no way* to access nature's materiality outside the materiality of signification. In a capitalist world, therefore, discourse becomes as much of a 'productive force' as any harvester's plough, manufacturer's machine or scientist's pipette. It is a grid that both enables and disables those things we come to call 'economic' activities and 'natural' entities (see also Darier, 1999). It is, in short, both world disclosing and world constituting.

Persuasive as they might seem, the processual and constructivist approaches to nature and culture within Marxism are not wholly consistent. To begin with, while the processual view seems to dissolve all distinctions (which now become 'moments') by envisioning a world of 'distributed materiality', there's no doubt that for Swyngedouw and Harvey it is 'economic' forces that are prime. That is, human labour becomes the mediating moment that glues all the others together. Somewhat differently, short of environmental discourses emerging *sui generis*, both Bridge and Escobar imply that they are functional to capital. This argument takes us back, ironically, to those rather simplistic motifs of determinism and determination that I considered near the start of this chapter. For it is otherwise difficult to explain where environmental discourses come from and to what end they exist. Moreover, by implying that discourse (culture) is founded upon production (economy), Bridge and Escobar paradoxically take as a given the kind of distinction they're at pains to show is a product of discourse itself. But more serious than this 'performative contradiction' is another. For as with pretty much all of the work I've reviewed so far, Swyngedouw, Harvey, Bridge and Escobar want, at some level, to speak for nature – to ascertain its materiality – even as they show how inapproachable that materiality might be. Politically, Swyngedouw and Harvey want to contest the 'creative destruction' of environment in the interests of a 'greener' Marxism. Likewise, Bridge wants to contest the practical agenda written into mining discourse, while Escobar touts those new social movements which are seeking to resist the appropriation of nature by capitalist interests. In each case, a paradoxical attempt to know nature *qua* nature (albeit an always already 'impure' nature) runs up against the impossibility of exiting the practical (labour) or representational (discourse) moment.

MATERIALIZATION

This problem brings me to the final cluster of Marxian work on the culture–nature question. It's arguably the most ingenious cluster of all and one where traditional Marxist concepts have a rather vestigial presence. Here the matter of nature is brought so deeply within the realm of culture that there's simply no epistemic or ontological space left to conceive of grasping the former's physicality at all. That is, nature's 'truths' are, quite literally, disclosed in and through webs of words that, iteratively, inform social and economic practices. Nature may indeed have a consequential materiality – whether its genes, rocks or weather – but the argument is that we can never know it, only specific, constructed *versions* it. To use a very cumbersome phrase: the materiality of nature is 'materialized' for us discursively. There is no conceivable 'outside' to linguistic practices. Indeed, the very distinction between inside and outside, representation and

reality, language and object *is itself an emergent effect*. A truly relational approach to nature and culture would thus concede that there is only a world of material entities and connections that are necessarily understood by way of discursive practices (see Doel, 2000). The things we call 'natural' assuredly exist, and we engage with them practically, but we can know them only within a linguistic universe that duplicitously assures us there's an ultimate referent upon which we can hang our signifiers.

This 'hyper-constructionist' or 'deep discursivity' argument, which may seem highly 'un-Marxist', can be traced to at least two sources: the Frankfurt School (and specifically Adorno and Horkheimer), which I mentioned in passing earlier, and more recent attempts to marry Marx with Heidegger and Derrida. These more recent attempts, as I'll show momentarily, address a key weakness in Adorno and Horkheimer's approach but also take us to the abyss of another cognitive dilemma. The Frankfurt School was famously critical of the 'productivist' cast of Marx's political economy and, partly inspired by Lukacs' example, developed a wider critique of capitalist modernity. Horkheimer and Adorno's notoriously difficult *Dialectic of Enlightenment* (1972) and Adorno's later work on 'non-identity' together delivered a shattering indictment of the modern condition. Arguing that Enlightenment had turned back upon itself, they diagnosed a dismal world where a regnant instrumental reason – at once a state of mind and a practical activity – dominated both people and nature. With the economy one of several 'spheres' governed by this monological 'culture' of reason, there was, for Adorno and Horkheimer, no way to exit from the iron cage. Since, for them, all other ways of knowing had been squeezed to the margins, they could only turn to nature as a quasi-mythical, redemptive realm that was utterly unapproachable. Smith (1984: Chapter 1) argues that this entails a residual dualism in Horkheimer and Adorno's work. This is not incorrect. However, I think it's more accurate to say that their inability to avoid direct appeals to nature issues in a paradox. As Stephen Vogel puts it, 'it is not clear how … an appeal to … immediacy underlying the subject's acts of mediation to which those acts might be said to contribute … could possibly be justified' (1986: 79).

So if Adorno and Horkheimer were unable to avoid a residual, yet impossible, naturalism/realism, are there any alternatives? Writing in the wake of Judith Butler's (1993) adaptation of Derrida and Heidegger to the question of the body, Bruce Braun (2000) has recently sought to rework Marxian ideas so as to avoid any hint of cognitive dualism or ontological 'outsides'. In a study of discourses of geology in Victorian British Columbia, Braun's concern is to understand how the province's resources were made visible and legible as objects for calculable economic action. Criticizing Marxist work that either seeks or assumes a material nature somehow knowable outside discursive-material practices, he focuses on how colonial geologists like George Dawson 'ordered' and 'enframed' the 'natural landscape' and has thereby 'stresse[ed] the "implosion" of the epistemological and the ontological ... achieved continuously in the mundane practices of daily life' (2000: 14). Note that in bringing 'science' – in this case a field science – within the orbit of culture, Braun leaves absolutely no room for a 'non-cultural' knowledge of, or access to, nature. He thus combines an activist view of knowledge (including, presumably, his own) and a materialist view of nature, while insisting that the latter can *never* be approached with any immediacy. As Massey puts it in a different context, there's 'an ever-changing and causal relationship between intelligibility and materiality' (2000: 10).

Though Braun never phrases his work in these terms, his approach echoes a philosophical tradition, stretching back to Hegel, Leibniz and Spinoza, that emphasizes the active and social character of knowledge. *Contra* the categorical distinctions of the Cartesian–Kantian tradition, Braun sees all such distinctions as discursive products. As such, he arguably pushes beyond the limit case of binary thinking found in Adorno and Horkheimer while avoiding the monistic position that 'discourse is all there is'. Because discourse is for Braun a culturally fashioned form of *practice*, he is able to approach nature's materiality without laying claim to knowing it 'as it is'. What he does, in effect, is ask the following question: which natures matter, how and with what consequences? His answer, vitally, is that there is no single answer since natures are materialized in historico-geographically contingent and variant ways.

CONCLUSION: CULTURES/ NATURES/MARXISMS

In this chapter I've sought to review a diverse tradition of Marxist work on nature and culture (and economy). I've sought to avoid a parochial focus on geography by situating the discipline's Marxists within a more expansive intellectual terrain. Moreover, I've sought to avoid teleology in my account. As we've seen, the journey from determinism to materialization has not been a linear one. It's important not to fall into the presentist trap of believing that the most recent Marxian work on culture and nature somehow transcends all that has gone before. By way of a conclusion, I want to reflect on where Marxists might go next in their attempts to understand the materiality of those things we call natural.

Perhaps the place to begin is with work like that of Braun, which seems to address the conceptual divisions and dilemmas of the Marxist work reviewed in this chapter. One of the objections to Braun's focus on 'materialization' is that it overemphasizes the power of language and culture. This objection could as easily come from 'green' Marxists – like James O'Connor or David Harvey – as it could from deep ecologists and others with an environmental sensibility. It amounts to the claim that if it's impossible to say with any certainty what 'materially' is happening to nature today, then we lack the political means to judge 'better' and 'worse' imbrications of culture–nature–economy. This lack seems particularly problematic in light of the perceived environmental problems of our time. Braun's response might reasonably be as follows: since discourse is performative, and since nature cannot be known as such, it's politically possible – indeed necessary – to engage in 'strategic essentialism'. That is, Braun might argue that it's right and proper to speak in the name of nature *tout court* in the interests of a more socially and ecologically just world – even as we acknowledge the hazards/impossibility of claiming to 'know nature'. Here, then, a Marxian diagnosis of current socio-environmental 'wrongs' and the quest for future 'goods' would entail making reflexive 'truth claims' about the state of culture–nature in a capitalist world without supposing that those truths are singular or value-free.

So far so good. But my worry is that the focus on 'materialization' as a means of getting beyond analytical ways of grappling with culture and nature is far too lexical and anthropocentric. It is lexical because it is primarily concerned with the cultural politics of representation. Though language is, quite rightly, granted an effectivity, the implication is that *words* (written and spoken) are the primary ways we engage with the world as capable agents. They, *pace* Braun, apparently enable and disable other forms of practice. This semio/phonocentrism links to the problem of anthropocentrism. Once it is conceded that the possession of language (however expansively defined) is the starting point for inquiry, it becomes very hard to acknowledge other ways in which the materiality of nature might make itself felt. Indeed, for all its subtlety, Cheah (1996) regards the materialization approach adopted by Braun and Butler to be essentially neo-Kantian: for all its claims to relationally overcome the representation–reality distinction, Cheah argues, it ultimately only ever defers it in endless regress. But, if this is so, how might Marxists – and others – think about the materiality of nature without becoming impaled on the horns of dualist/analytical dilemmas? For doesn't my question about language itself depend on a dubious notion of a non-linguistic 'outside'?

One possibility is to explore the relatively uncharted terrain of 'non-representational theory' (Thrift, 1996). What's interesting about this is not that it argues that 'language does not simply re-present' or even that 'there's more to the world than language'. Rather, in Thrift's view, it entails a recognition that our collective efforts to understand the world – as analysts and ordinary people – entail continuous, mundane physico-linguistic *interventions in* the world such that practice and utterance are iteratively united and changing. Knowing is doing, and vice versa. Another, related possibility is to pursue the 'materialist semiotics' advocated by Bruno Latour, John Law and Michel Serres which 'flattens' the ontological universe and treats language as just one of several sensible means by which humans relate to nature (Whatmore, 1999). In geography this second approach to culture–nature is having a considerable impact and, as I noted, has been appropriated in limited ways by Marxists like Swyngedouw. Third, we might want to consider Haraway's (1991; 1997) ongoing attempts to escape the onto-epistemic straitjacket of Enlightenment thinking without altogether abandoning the commitment to the idea that the world is structured and can be systemically explained (see Haraway and Harvey, 1995).

But if the dominant mindsets in geography and beyond – whether analytical or relational, Marxist or otherwise – are to truly grasp nature's materiality they are going to have to more forcefully explore and acknowledge truly tactile, sensual and embodied ways of figuring the culture–nature nexus. This would entail a relational approach to culture–nature but one where the material capacities are myriad, variable, lively and shared (see Ingold, 2000). This approach would be sensitive to historical-geographical difference but would also look for the ways in which relations to nature are nonetheless consequentially ordered in a capitalist world (Castree, 2002; Goodman, 1999; 2001). In short, it would not be a shapeless approach in which all manner of non-human and non-linguistic actors randomly figure. Rather, it would critically scrutinize the material ties between those things we call

'cultural', 'natural' and 'economic' while acknowledging that these categories cannot be taken as either given or adequate. It would, in short, retain the explanatory-diagnostic tenor of previous Marxisms but move beyond their lexical, anthropomorphic cast. As such, it would also need to find a language (not necessarily verbal) to express non-cognitive aspects of those material things we call nature, economy and culture.

This is, of course, easier said than done, but crucial nonetheless in a world where 'nature's' materiality is being remade at macro and micro scales (Castree and Braun, 1998). As I've sought to show here, it remains enormously difficult to escape our collective intellectual heritage. If the Marxian tradition is anything to go by, it remains an open question whether we can genuinely 'unthink' our approaches to culture and nature. If Marxism is still to matter as a critical voice in debates on our current and future socio-ecological condition, it must challenge – not merely work within – inherited understandings of the materiality of nature and culture. Otherwise, we may, ironically, find ourselves repeating the very same representational practices that have led Marxists to question the sanity of capitalism in the first place – only now framed in the comforting idiom of an 'emancipatory geography'.

NOTES

So promiscuous are the categories 'culture' and 'nature' that writing this chapter was an arduous task, one compounded by the fact that Marxists, where they haven't relegated culture to economy's poor relation, have said relatively little over the decades about nature. My thanks go to Sarah Whatmore and Steve Hinchliffe for helping to improve what's still a rather ramshackle argument.

1 For a good overview of the polymorphous terms 'nature' and 'culture' see, respectively, Kate Soper's *What Is Nature?* (1995) and Chris Jenks' *Culture* (1993).
2 In effect, I want to revisit here the problematic status of 'materiality' that I broached in a very particular way in Castree (1995).
3 Moreover, in thematic terms my discussion will range from commodities and consumption to the culture of science to discourses of nature and beyond.
4 Though this work is complicated by the fact that culture and nature must also be linked with one other equally complex concept: the economy. As is well known, recent years have seen many human geographers attempting to rethink the culture–economy nexus (e.g. Lee and Wills, 1997; Thrift and Olds, 1996). It's not my intention to revisit the debates here, though the same impulse to question received categorial distinctions does inform my account.

5 Note that in this chapter I am interested less in 'nature' and 'culture' as ideas having a material existence – though this existence is undoubtedly real and important – and more in the 'materiality' of those myraid things we choose to call natural and cultural.

6 And 'reading' – in the active, interpretive sense of the word – really is the point here. There is, in my view, little point in trying to grasp what Marx 'really' said about any subject. His diverse corpus has now been used in so many ways and contexts that the 'real Marx' scarcely exists any more. Accordingly, in this section I work with a strategic fiction: that it's possible to say what Marx 'did' and 'did not' say in his writings about nature and culture.

7 See, for instance, Gamble (2000) where the chapters on nature and culture barely overlap.

8 And David Goodman (2001) has shown how the concept of the labour process has functioned within Marxism – notably in agro-food studies – as a means of privileging economic over natural matters.

9 Neil Smith (1984) has taken this argument forward with his thesis about the 'production of nature'.

10 It's interesting that the recent efforts to rethink the culture–economy distinction in geography have not been repeated for nature–economy. To my mind, much of the new political economy of environment works *within* this distinction rather than challenging it.

11 Though the point, of course, is that not everyone would define nature, economy and culture in these ways.

12 And it was Weber, of course, who claimed that capitalism was, quite literally, a culturally specific economic system in his work on the Protestant ethic.

13 Indeed, what is striking is how two Marxist geographical books – both landmark texts – could be published at the same time and yet say so little to the concerns of each other. Smith's *Uneven Development* effectively bracketed out questions of culture, while Cosgrove's *Social Formation and Symbolic Landscape* gives the concept of nature none of the careful scrutiny and reformulation offered by Smith.

14 In addition to this, many geographers working on human–environment relations drew upon 'cultural ecology' and looked to anthropology for ways of approaching 'material culture'.

15 See also George Henderson's (1999) marvellously subtle analysis of the material transformation of rural, late-nineteenth-century California and its representation in popular literature.

16 For another Marxist study that is similarly attuned to the overdetermination of nature/culture/economy see Alicja's Muszynski's *Cheap Wage Labour: Race and Gender in the Fisheries of British Columbia* (1996).

17 The thinking of Peter Dickens (1996) on the alienation from nature intrinsic to capitalist societies echoes some of Taussig's ideas in this regard.

18 In the work of Fine et al. (1996) this commodity chain approach has been given an even greater ecological emphasis through a concentration on food, a 'cultural commodity' *par excellence* but one whose organic character affects both the production and consumption ends of the chain; see also Goodman and Redclift (1991).

19 Though not exclusively or even largely Marxian in tone, the journal *Environmental Values* is doing much to critique the hegemonic valuation of nature in capitalist societies.

20 Though it's in his novels that Williams' understanding of culture and nature is arguably best expressed.

REFERENCES

Altvater, E. (1993) *The Future of the Market*. London: Verso.

Appadurai, A. (1986) *The Social Life of Things*. Cambridge: Cambridge University Press.

Aronowitz, S. (1988) *Science As Power*. Minneapolis: University of Minnesota Press.

Bartram, R. and Shobrook, S. (2000) 'Endless/end-less natures: environmental futures at the fin-de-millennium', *Annals of the Association of American Geographers* 90: 370–80.

Benton, T. (1989) 'Marxism and natural limits', *New Left Review* 187: 51–81.

Berman, M. (2000) *Adventures in Marxism*. London: Verso.

Braudel, F. (1972) *The Mediterranean and the Mediterranean World in the Age of Philip II*. London: Collins.

Braudel, F. (1973) *Capitalism and Material Life, 1400–1800*. London: Weidenfeld and Nicolson.

Braun, B. (2000) 'Producing vertical territory', *Ecumene* 7: 7–46.

Bridge, G. (1997) 'Excavating nature', in A. Herod, G. Tuathail and S. Roberts (eds) *An Unruly World?* London: Routledge. pp. 219–44.

Burkett, P. (1999) *Marx and Nature*. New York: St Martin's.

Busch, L., Lacy, W., Burkhand, J. and Lacy, L. (1991) *Plants, Profits and Power*. Oxford: Blackwell.

Butler, J. (1993) *Bodies That Matter*. New York: Routledge.

Castree, N. (1995) 'The nature of produced nature', *Antipode* 27: 12–48.

Castree, N. (2001) 'Commodity fetishism, geographical imaginations and imaginative geographies', *Environment and Planning A*, 33, 10: 1519–25.

Castree, N. (2002) 'False antitheses? Marxism, nature and actor-networks', *Antipode*, 34, 1: 114–36.

Castree, N. and Braun, B. (1998) 'The construction of nature and the nature of construction', in B. Braun and N. Castree (eds), *Remaking Reality*. London: Routledge. pp. 1–36.

Cheah, P. (1996) 'Mattering', *Diacritics* 26: 108–39.

Cook, I. and Crang, P. (1996) 'The world on a plate', *Journal of Material Culture* 1: 131–53.

Cosgrove, D. (1985) *Social Formation and Symbolic Landscape*. London: Croom Helm.

Daniels, S. (1989) 'Marxism, culture and the duplicity of landscape', in R. Peet and N. Thrift (eds) *New Models in Geography*, vol. II. London: Unwin Hyman, pp. 196–220.

Darier, E. (ed.) (1999) *Discourses of the Environment*. Oxford: Blackwell.

Dickens, P. (1996) *Reconstructing Nature*. London: Routledge.

Doel, M. (2000) *Postructural Geographies*. Edinburgh: Edinburgh University Press.

Engels, F. (1956) *The Dialectics of Nature*. Moscow: Progress.

Escobar, A. (1996) 'Constructing nature', in R. Peet and M. Watts (eds) *Liberation Ecologies*. London: Routledge. pp. 46–68.

Escobar, C. (1999) 'After nature', *Current Anthropology* 40, 1: 1–16.

Fine, B. and Wright, J. (1996) *Consumption in the Age of Affluence*. London: Routledge.

Fitzsimmons, M. (1989) 'The matter of nature', *Antipode* 21: 106–20.

Foster, J. (ed.) (1997) *Valuing Nature*. London: Routledge.

Gamble, A. Marsh, D. and Taut, T. (eds) (1999) *Marxism and Social Science*. London: Macmillan.

Godelier, M. (1986) *The Mental and the Material*. London: Verso.

Goodman, D. (1999) 'Agro-food studies in the "Age of Ecology"', *Sociologia Ruralis* 39: 17–38.

Goodman, D. (2001) 'Ontology matters', *Sociologia Ruralis* 41: 182–200.

Goodman, D. and Redclift, M. (1991) *Refashioning Nature*. London: Routledge.

Goodman, D., Sorj, B. and Wilkinson, J. (1987) *From Farming to Biotechnology*. Oxford: Blackwell.

Granovetter, M. (1985) 'Economic action and social structure: the problem of embeddedness', *American Journal of Sociology* 91: 418–510.

Graves-Brown, P.-M. (2000) *Matter, Materiality and Modern Culture*. London: Routledge.

Haraway, D. (1991) *Simians, Cyborgs and Women*. London: Routledge.

Haraway, D. (1997) *Modest_Witness@Second_Millenium: FemaleMan©_Meets_Oncomouse™*. London: Routledge.

Haraway, D. and Harvey, D. (1995) 'Nature, politics and possibilities', *Society and Space* 13: 507–27.

Hartwick, E. (1998) 'Geographies of consumption', *Society and Space* 16: 423–37.

Harvey, D. (1974) 'Population, resources and the ideology of science', *Economic Geography* 50: 256–77.

Harvey, D. (1995) 'Militant particularism and global ambition', *Social Text* 42: 69–98.

Harvey, D. (1996) *Justice, Nature and the Geography of Difference*. Oxford: Blackwell.

Henderson, G. (1998) 'Nature and fictitious capital', *Antipode* 30: 73–118.

Henderson, G. (1999) *California and the Fictions of Capital*. Oxford: Oxford University Press.

Horkheimer, M. and Adorno, T. (1972) *Dialectic of Enlightenment*. New York: Continuum.

Ingold, T. (2000) *The Perception of the Environment*. London: Routledge.

Jenks, C. (1993) *Culture*. London: Routledge.

Katz, C. (1998) 'Whose culture, whose nature?', in B. Braun and N. Castree (eds) *Remaking Reality*. London: Routledge. pp. 46–63.

Kautsky, K. (1988) *The Agrarian Question*, 2 vols. London: Zwan.

Kloppenburg, J. (1998) *First the Seed*. Cambridge: Cambridge University Press.

Latour, B. (1993) *We Have Never Been Modern*. Cambridge: Harvard University Press.

Lee, R. and Wills, J. (eds) (1997) *Geographies of Economies*. London: Arnold.

Lukacs, G. (1971) *History and Class Consciousness*. Cambridge, MA: Harvard University Press.

Mann, S. and Dickinson, J. (1978) 'Obstacles to the development of capitalist agriculture', *Journal of Peasant Studies* 5: 466–81.

Massey, D. (2000) 'Geography on the agenda', *Progress in Human Geography* 25: 5–17.

McLellan, D. (1999) 'Then and now: Marx and Marxism', *Political Studies* 47 (5): 955–66.

Moore, D. (1996) 'Marxism, culture and political ecology', in R. Peet and M. Watts (eds) *Liberation Ecologies*. London: Routledge.

Muszynski, A. (1996) *Cheap Wage Labour: Race and Gender in the Fisheries of British Columbia*. Kingston: Queens–McGill University Press.

O'Connor, J. (1998) *Natural Causes*. New York: Guilford.

Ollman, B. (1993) *Dialectical Investigations*. New York: Routledge.

Peet, R. (1985) 'Introduction to the life and thought of Karl Wittfogel', *Antipode* 17: 3–20.

Polanyi, K. (1944) *The Great Transformation*. Boston: Beacon.

Pred, A. (1998) 'The nature of de-naturalised consumption', in B. Braun and N. Castree (eds) *Remaking Reality*. London: Routledge, pp. 150–68.

Pred, A. and Watts, M. (1992) *Reworking Modernity*. New Brunswick: Rutgers University Press.

Radical Science Journal Collective (1974) 'Editorial statement', *Radical Science Journal* 1: 1–3.

Rose, S. and Lewontin, R. (1984) *Not in Our Genes*. Harmondsworth: Penguin.

Sayer, A. (1999) 'Valuing culture and economy', in L. Ray and A. Sayer (eds) *Culture and Economy after the Cultural Turn*. London: Sage. pp. 53–75.

Schmidt, A. (1971) *The Concept of Nature in Marx*. London: New Left Books.

Smith, N. (1984) *Uneven Development*. Oxford: Blackwell.

Smith, N. and O'Keefe, P. (1980) 'Geography, Marx and the concept of nature', *Antipode* 12: 30–9.

Soper, K. (1995) *What is Nature?* Oxford: Blackwell.

Swyngedouw, E. (1999) 'Modernity and hybridity', *Annals of the Association of American Geographers* 89: 443–65.

Taussig, M. (1980) *The Devil and Commodity Fetishism in South America*. Chapel Hill: University of North Carolina Press.

Thomas, N. (1991) *Entangled Objects*. Cambridge, MA: Harvard University Press.

Thrift, N. (1996) *Spatial Formations*. London: Sage.

Thrift, N. and Olds, K. (1996) 'Refiguring the economic in economic geography'. *Progress in Human Geography* 20: 311–37.

Timparano, S. (1975) *On Materialism*. London: New Left.

Vogel, S. (1996) *Against Nature*. Buffalo: State University of New York.

Walker, R. (1997) 'Fields of dreams', in D. Goodman and M. Watts (eds) *Globalising Food*. London and New York: Routledge. pp. 273–86.

Watts, M. (1998) 'Nature as artifice and artefact', in B. Braun and N. Castree (eds) *Remaking Reality*. London: Routledge, pp. 243–68.

Wells, M. (1996) *Strawberry Fields*. Ithaca: Cornell University Press.

Whatmore, S. (1999) 'Hybrid geographies', in D. Massey, J. Allen and P. Sarre (eds) *Human Geography Today*. Oxford: Polity. pp. 22–40.

Williams, R. (1978) 'Problems of materialism', *New Left Review* 109: 3–17.

Williams, R. (1980) *Problems in Materialism and Culture*. London: Verso.

Wilson, H.T. (1991) *Marx's Critical/Dialectical Procedure*. New York: Routledge.

Wittfogel, K. (1964) *Oriental Despotism*. New York: Basic Books.

Worster, D. (1985) *Rivers of Empire*. New York: Pantheon.

Young, R. (1979) 'Science is social relations', *Radical Science Journal* 5: 65–129.

9

Reanimating Cultural Geography

Jennifer Wolch, Jody Emel and Chris Wilbert

They are first and foremost themselves, despite the many meanings we discover in them. We may move them around and impose our designs upon them. We may do our best to make them bend to our wills. But in the end they remain inscrutable, artifacts of a world we did not make whose meaning for themselves we can never finally know. (Cronon, 1995: 55)

Nature is replete with animate, sentient beings possessed of agency as well as instinct – namely, animals. Many animals are essential to human survival, are foundational to our ontology and epistemology, and form the basis of countless cultural norms and practices. Curiously, although explaining relatious between nature and human society has long been a central goal for geographical research, animals have largely appeared as biotic elements of ecological systems, available for human use, or forms of symbolic capital. Have animals simply been too insignificant, too irrelevant to human affairs, and thus unworthy of serious contemplation as subjects or actors? Or are animals so indispensable to people and tied to visions of progress and the good life, that even geographers, trained to analyse nature–society relationships, have been unable to see them?

As the twenty-first century opens, geographers in the United States and Britain are bringing the 'animal question' to the forefront of geographic debates, and in ways different from in the past. Why animals, and why now? The reasons are complex, but several forces are at work. One is that the plight of animals has never been more serious than it is today. Billions of animals annually are killed in factory farms, poisoned by toxic pollutants and waste, and driven from their homes by logging, mining,

agriculture and urbanization. Dissected, re-engineered and used for spare body parts, animals are kept in captivity and servitude, discarded when their utility to people wanes. Another factor is that these multiple assaults have stimulated a vocal politics of resistance and fears about the 'end of nature'. In addition, stimulated in part by these environmental and socio-ethical dynamics, intellectual currents within the academy have blurred boundaries between humans and animals, and encouraged recognition of human–animal bonds at multiple levels of analysis and spatial scales.

This chapter traces relationships between geography and animals over the past century. In the remainder of this introduction, we examine the historical origins of zoögeography, a subfield of physical geography, as well as the parallel rise of a culturally oriented geography of animals. Both approaches had receded from view by the mid twentieth century. After several decades, however, another vision of animal geography – sharing little with either zoögeography or cultural animal geography – emerged. And so we turn to consider this new approach and its implications.

Although never prominent in the discipline, geographers have long shown some interest in animals. Indeed, an identifiable branch known as 'animal geography' was actively researched, at least since Marion Newbigin's *Animal Geography* (1913) (Figure 9.1) and her posthumously published *Plant and Animal Geography* (1936). Newbigin pointed to the need for distributional studies of animal populations, examining floral and faunal regions and their relations (Maddrell, 1997). Animal geography was also accorded a role in Hartshorne's grand statement of what geography 'is and ought to be' in *The Nature of Geography* (1939). Hartshorne regarded the apex

Figure 9.1   *New World monkeys* (from M.I. Newbigin, *Animal Geography*, 1913)

of geographical inquiry as the study of 'areal differentiation', requiring attention to unique assemblages of phenomena (natural and human) that seemingly create distinctions between different regions of the earth. This apex – 'regional geography' – was reached through integrating findings of 'systematic geographies', synthesized with reference to particular areas. Hartshorne portrayed animal geography, allied to zoology, as one such systematic subfield of geography.

Thus through this first half of the twentieth century animal geography was an active if small portion of the discipline. Two approaches to the field were clearly articulated, reflecting the widening divide between physical and human geography. First was zoögeography, focused mainly on animal distributions, and rooted in physical geography, zoology and the emerging science of ecology. Second was a culturally oriented geography of animals, focused on animal domestications, grounded in human geography and social sciences.

Zoögeographers, typically affiliated with physical geography, focused on geographic distributions of animals, and determinants of distributional patterns at both small and large scales, incorporating notions of *space, spatial patterns* and *spatial relations* into their research.[1] Core questions related to where different types of animals live – breed, forage, migrate, give birth – and determinants of animal distributions such as climate, topography, hydrology, soils and vegetation, other fauna, and species-specific habitat preferences. The approach was conventionally scientific, utilizing field techniques, mapping procedures, quantitative analyses, hypothesis testing and model specification. The ambition was to establish general zoögeographical laws of how animals (populations and species) arranged themselves across the earth's surface, or at smaller scales, to establish patterns of spatial covariation between animals and other environmental factors.

Zoögeography rarely contributed directly to debates about animal interactions with human society. But zoögeographers were not unconcerned about anthropogenic influences on animals, and in turn what animals might mean to people. Cansdale thoughtfully explored the relations between animals and humans in terms of 'competition, conflict, domestication and biological control' (Cansdale, 1950; 1951a; 1951b; 1951c; 1951d; see also Cansdale, 1952). Even earlier, Eagle Clark (1896) had recited human influences on bird migration, including the frequent deaths of migrating birds dashing themselves against the lanterns of light stations in foggy weather, while Moebius (1894) despaired about the 'violent encroachment' of humans chasing the whale. The pioneering work of Fitter on *London's Natural History* (1945) noted that human–animal interactions often intensified around the city, and that many animals were adapting to the environs of the metropolis (Matless, 1998). Nonetheless, attention to society–animal relations was patchy and untheorized. Animals were regarded with a detached scientific eye as natural objects to be tracked, trapped, counted, mapped and modelled – and devoid of any 'inner life', consciousness or agency.

Linkages between zoögeography and the cognate scientific disciplines of zoology, ecology and biology were strong (Clark, 1927: 102), owing to the scientific training of many zoögeographers. Over time, these fields came to dominate the study of animal distributions; geographers increasingly focused on plant distributions instead. By mid-century, Davies would claim that scientific 'animal geography is perhaps the branch of geography least practised by geographers, and the one they most cheerfully abandon to the systematic sciences' (1961: 412).

At just this time, Bennett proposed a *cultural* animal geography, to return animal geography 'into the geographical fold' from which he thought it had become disengaged (1960: 13–14; see also Bennett, 1961). He called for research on human–animal interactions, involving studies of how humans influence animal 'numbers and distributions', echoing zoögeography's emphasis on space and spatial distributions. Bennett also urged examinations of animals' response to domestication, subsistence hunting and fishing, and more indirect human impacts such as fire, war and travel, and further suggested studies of animal influences on human life (through destroying crops or carrying and causing disease). Here, animals were a key element of the natural environment that 'determined' the human geography (of settlement, agriculture and industry) in places and world regions (see also Anderson, 1951).

Bennett's proposals dovetailed with cultural ecology, already emerging in both North America and Europe, which focused on the origins of animal domestication, and while concerned with distributions and diffusions of domesticates, was characterized by attention to *place, region* and above all *landscape*. These emphases grew out of

historical approaches to biogeography as influenced by humans, as well as anthropology and archaeology, in which animals featured in discussions about origins of agriculture.[2] Perhaps the central figure in the US studies of animal domestication was Carl Sauer. Reacting against a crude environmental determinism characteristic of turn-of-the-century geography, Sauer was influenced by German geographers such as Hahn, Ritter and Hettner (Dickinson, 1976: 319; Livingstone, 1992: 297). His method was to analyse 'sequential landscape cross sections' to reveal the coevolution of environments and culture in places and regions and understand the 'morphology' of what he termed 'the cultural landscape'. Presaging later environmentalists, the politico-moral concerns of Sauer and the Berkeley School revolved around environmental degradation (Sauer, 1925; 1956; Price and Lewis, 1993).[3]

Animals were not primary subjects in Sauerian geographies. However, studies of domestication and diffusion of animal husbandry, cultural and economic roles of animals in agrarian societies, and environmental changes attendant upon agriculture- and livestock-based lifeways, revealed the importance of certain animals to cultural practices and environmental conditions. Sauer's pioneering text *Seeds Spades, Hearths and Herds* (1952) documented the role of animal domestication in the conversion of 'natural landscapes' into 'cultural landscapes'. Notably Sauer's approach resisted simplistic economic explanations of society–animal relations, and refused to regard domestication as simply an adaptive response to human needs for animal products. Donkin, a British geographer emulating Sauer's example, argued that economic benefits of domestication accrued only after the fact; domestication was motivated more by religious and ceremonial reasons. Moreover, other practices such as pet keeping depended partly 'on the physical and behavioural characteristics of particular species' (Donkin, 1991: ix; see also Donkin, 1985; 1989).

British and continental developments in this area were shaped by Sauerian traditions as well as their common European intellectual heritage. In the UK animals appeared in the more anthropologically oriented writings of H.J. Fleure[4] (1919; 1937) and C. Daryll Forde.[5] Both made significant contributions to anthropology and archaeology and cross-fertilization between all three disciplines continued to enrich the cultural-ecology approach. Linkages were reinforced by the shared reliance upon archaeological evidence on domestication, common methodologies (Renfrew, 1983; Wagstaff, 1983), use of ecological concepts (though this had less appeal in Europe: Harris, 1996: 442; Simmons, 1980: 150) and a focus on material culture.

Later British geographers such as Harris and Donkin were influenced by Sauer. Harris initially focused on goat domestication, and devel-oped evolutionary models for domestication stressing the ecosystems in which early domestications were embedded (Harris, 1962; 1977; 1996; Simmons, 1980: 150). Donkin produced meticulous work on domesticated, or transitional, animals in the Americas, such as the peccary, the muscovy duck and the cochineal insect, incorporating the study of language, based on historical records, indigenous folklore and analysis of local dialects (Donkin, 1977; 1985; 1997). Yet, much 'traditional' cultural geography (or cultural ecology) treated domestic animals simply as cultural artifacts, an evolutionary technological development, little more than a medium of further environmental transformation (Fraser-Darling, 1956). McKnight, for example, suggested that Australia's feral livestock 'represents an immense amount of wandering protein on the hoof' constituting a 'biomass approaching 800,000,000 pounds' but with deleterious ecological (as well as economic) effects on areas in which they were concentrated (1976: 93–4). While stressing human powers and agency, as well as ecological processes, possible roles of non-human animals were minimized, limiting human–animal relations to dominatory or symbiotic forms.[6]

By the 1960s, cultural geography was under heavy attack, partly owing to the Berkeley School's organic approach to 'culture' and its treatment of culture–economy relations. For some cultural ecologists such as Clutton-Brock, culture was 'a way of life imposed over successive generations on a society of humans or animals by its elders' (1994: 29).[7] Such a view seemed too simple. Moreover, with only two interacting realms (cultural and biological), analysis of how activities and meanings are contested and imprecated in power relations was precluded. Sauerian ideas were also criticized based on re-examinations of archaeological evidence (Bender, 1975: 21; Blumler, 1993; Rodrigue, 1992). In particular, Rodrigue (1992: 336) criticized Sauerian beliefs in the non-economic origins of animal domestication, and argued on the basis of archaeological research for a more materialist lens. Thus cultural geography arguably became a backwater, and its central questions about human–environment relations receded from view as well.

RETHINKING CULTURE, NATURE AND SUBJECTIVITY

By the last quarter of the twentieth century, the term 'animal geography' had vanished from geographic discourse. In the 1990s, however, interest revived, inspired by the encounter between human geography and social theory, cultural studies, selected natural sciences and environmental ethics. In the US, efforts to 'bring the animals back in' were taken up by Wolch and Emel, who discovered some isolated albeit prescient and intriguing attempts to address the animal question. Tuan's *Dominance and Affection* (1984), which traced inherently unequal and 'paternalist' power relations entailed in pet keeping, was perhaps the best example. Trying to stimulate a broader discourse about animals and society, Wolch and Emel organized a 1995 thematic issue of *Society and Space* and the edited book *Animal Geographies* (1998). Here, Philo wondered what might develop if concepts of the 'new' cultural geography were applied to human–animal relations. Together with Wilbert, Philo posed the 'animal question' within the context of British cultural geography, resulting in the collection *Animal Spaces Beastly Places* (2000).

These efforts touched off the flowering of discourse that, harking back to Newbigin, we simply term *animal geography*.[8] What stimulated this re-emergence? Figure 9.2 is an attempt to untangle the relevant context and intellectual currents. Certainly, growing concern about environmental degradation, habitat loss and species endangerment, and the plight of animals relegated to the killing zones of shelter, lab, factory farm and slaughterhouse, served as central contexts motivating a renewed focus on animals. During the 1970s and 1980s, hundreds of new organizations sprang up to lead social movements in defence of the environment and animals (pets, farm animals, wildlife, lab animals). The most radical, such as PETA and the Animal Liberation Front, challenged people to rethink their relationships with animals, drawing explicit linkages, for example, between racism, sexism and 'speciesism'; slavery and animal captivity; and the Holocaust and factory farms, fur farms and research labs.

Against this turbulent backdrop, the emergence of new research in social theory and cultural studies led to a profound *rethinking of culture* and especially *a rethinking of subjectivity*. Simultaneously, however, natural science disciplines concerned with the environment and animals *per se* (including cognitive psychology,

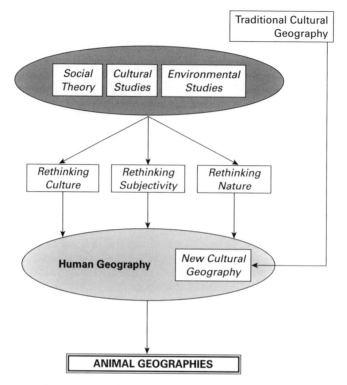

Figure 9.2   *Emergence of animal geographies*

ethology, landscape ecology and conservation biology), as well as the 'new' environmental history, collided with social theoretic and cultural studies perspectives. This engendered a *rethinking of nature* and also shed different light on the subjectivity of 'natural aliens' – to use Neil Evernden's (1993) phrase. Along with many natural scientists, geographers from various intellectual traditions – political economy (especially neo-Marxism), poststructuralism, feminism and science studies – began arguing for animal subjectivity and the need to unpack the 'black box' of nature to enliven understandings of the world.

The reconceptualizations of culture, nature and subjectivity allowed for the emergence of arguments concerning *human identity and animal subjectivity*. In particular, the focus was on animals' role in the social construction of culture and individual human subjects, and the nature of animal subjectivity and agency itself. Topics for animal geographers included the human–animal divide, especially how and why this line shifts over time and space, and links between animals and human identities, namely, the ways in which ideas and representations of animals shape personal and collective identity. The recognition of animal subjectivity led to questions of animal agency *per se* and what it might mean for both everyday human and animal lives.

Debates about the social construction of landscapes and places led animal geographers to explore how animals and the networks in which they are enmeshed leave imprints on particular places, regions and landscapes over time, prompting studies of *animals and place*. The places considered include specific sites, such as zoos, 'borderland' communities in which humans and free animals share space, and places in the grip of powerful forces of economic or social change affecting both people and animals, especially those caught up in the worldwide trade in captive or domesticated animals.

Lastly, arguments about animal subjectivity led some geographers toward environmental ethics and especially a rethinking of *animals in the moral landscape*. This focus raised ethical questions associated with human–animal relations. How are notions of animal subjectivity and animal rights incorporated into quotidian ethical frameworks operative in the non-western world? How can animal wellbeing and species existence be protected under conditions of rapid urbanization, economic globalization and free trade, in which they are so integrally enmeshed – even when they are 'charismatic' and/or protected under worldwide conventions? If animals are inherently linked to place-making and place-specific moral landscapes, how can they be integrated within broader notions of geoethics?

In what follows, we track these three basic themes in animal geography: human identity and animal subjectivity, animals and the making of place, and the role of animals in the moral landscape.

Human identity and animal subjectivity

Questions about human–animal differences and the nature of the human–animal divide have proliferated recently. Beginning in the 1970s, new findings on animal thinking, culture and politics from comparative psychology, primatology and cognitive ethology underscored animal consciousness and capacities to think and act (Goodall, 1986; Griffin, 1976; 1984). Highly publicized research concerned linguistic capabilities of primates and their social and political behaviour in the wild (de Waal, 1982; Fossey, 1983; Galdikas, 1995; Goodall, 1986). Studies of other species revealed complex communications systems and social organizations (Cheney and Seyfarth, 1990; Morton and Page, 1992). Behaviour-specific studies, such as those on animal play, revealed the continuity of vertebrate behavioural repertoires and their complex social functions: kangaroos, pronghorn antelope, ravens and even reptiles engage in play (Bekoff and Byers, 1998). Simultaneously, genetic engineering, cloning and xenotransplants increasingly called into question boundaries between machines, animals and humans (see Haraway, 1991; Sheehan and Sosna, 1991; Taylor et al., 1997). Such complex machine–human–animal interdependencies raised possibilities of a 'cyborg' world. Beyond speculation, however, scientists created computer implants designed to correct human medical problems, and an array of chimeras, for example a mouse with a human ear (Toufexis, 1995) and pigs with human organs to be 'harvested' for transplantation to people.

Poststructural and feminist refusals of 'man, the subject' shifted the animal out from the cultural margins 'by destabilizing that familiar clutch of entrenched stereotypes which works to maintain the illusion of human identity, centrality and superiority' (Baker, 1993: 26).[9] Exploiting this move, Wolch (1996; 1998) argued for reconceptualizing the human–animal divide. Humans may not be able to literally 'think like a bat' but, rather than nihilistic relativism or denial of human–animal differences, the human response

could recognize that both people and animals are embedded in social relations and networks with others (both human and non-human) upon which their social welfare depends. This realization allows for the recognition of kinship but also difference since identities are defined by the ways in which we are similar to, as well as different from, related others. Recalling Haraway's 'cyborg vision' Wolch argued that people should come to know – however partially – the animals with whom they coexist, thereby sustaining webs of connection and an ethic of respect, mutuality, caring and friendship (Wolch, 1998: 122). The human–animal divide is thus transformed from an oppositional dualism to a network of intricate interdependencies.

For other animal geographers, challenges to unitary subjectivity and their disciplinary predisposition to acknowledge time–space contingency in nature–society relations led to empirical conceptions of the human–animal divide. Anderson (1997) reconsidered the story of domestication, situating certain chapters in European political discourses about human uniqueness. Arguing that domestication could not be understood by seeing animals as natural resources or as filling a narrow religio-cultural role, she drew upon theories of power and identity, contending that domestication underlay a dualistic model of human and animal in western cultures, with implications for both sides of the divide. Moreover, a legacy of the domesticatory project was colonial oppression and racialization of indigenous peoples who were treated 'like animals' and ideologies of human improvement that resulted in 'civilizing' efforts, eugenics and assimilationist policies.

Also recognizing how boundary definitions are in part determined by the normalization of human–animal relations in place, Elder et al. (1998) argued that while concepts of human and animal are universally understood, the boundary is in flux and varies geographically. Such repertoires are partly environmentally determined: the diversity of animals available to use is shaped by environmental factors, as are animal-based modes of subsistence (for example, pastoralism). In addition, however, ideas about animals (like other aspects of culture) evolve in place over time owing to social or technological change generated within a society, or externally by migrations or invasions. Thus values and practices concerning cosmological, totemic or companionate relations between people and animals, and material uses of animals, shift owing to social dynamics, technological change or culture contact. The result is a dynamic but place-specific assemblage of animals, valued

Figure 9.3 *Live animal market in LA's Chinatown*
(photo credit: Marcie Griffith)

and used according to particular, legitimized codes.

As the 'empire comes home' in the form of international migration to the west, 'out-of-place' animal practices risk being interpreted as trangressions of the human–animal divide. Wolch and her co-workers investigated attitudes toward animals among women of diverse racial/ethnic and cultural background living in central Los Angeles, and particularly explored racialization based on animal practice (such as shopping at live animal markets; see Figure 9.3). For example, African American study participants segmented the animal world into three categories: 'food', 'pet' and 'wildlife' (Wolch et al., 2000). 'Food' animals were simply necessary for survival; people had to distance themselves from their unfortunate fate. Pets and wild animals, in contrast, demanded compassion. Participants argued that people should help wildlife in distress, just as people should help each other regardless of colour, hinting at their solidarity with animals as brethren due to their outsider status. A willingness to tolerate dog eating among South East Asian immigrants, however, reflected their own experience as a marginalized group in American society, and their sensitivity

to racialization based on colour and culture. Dog eating in the Philippines is tied to both tradition and economics, its prevalence varying by class and region (Griffith et al., 2003). Although dog eating is not common among US Filipinos, associations of dog eating with Filipinos appeared to exacerbate Anglo racial intolerance. Like the African American women, the Filipinas hesitated to condemn other groups whose animal practices, while alien or distasteful, were rooted in their particular culture, instead adopting a position of cultural relativism.

Stimulating new considerations of human as well as animal representations and identities, critical race and postcolonial theorists highlighted connections between race and representations of 'animality' while feminists and others working on sexuality and the body emphasized the importance of animals in body part coding. Stallybrass and White's (1986) discussion of Jonson's *Bartholomew Fair*, for example, revealed how non-human and human animal bodies were mutually inscribed. Similarly, animals were identified as the 'absent referent' in the racialization and brutalization of others, borrowing from and contributing to postcolonial theory (Adams, 1990; Spiegel, 1988). Animal geographers expanded on these insights, focusing on animals' role in the formation of heterogeneous identities – individual and collective – that people adopt or have ascribed to them. These identities may be linked to particular eras, places and nations, and to racial/ethnic, cultural or gendered identities.

Anderson (1995), for example, developed a cultural critique of the zoo as an institution that inscribes various human strategies for domesticating, mythologizing and aestheticizing the animal universe. Using the case of Adelaide, South Australia, she charted the mutable discursive frames and practices through which animals were fashioned and delivered to the South Australian public by the Royal Zoological Society of South Australia. Through their constructions of nature and animals, zoo practices consolidated and legitimated Australian colonial identity, naturalized colonial rule and oppression of indigenous peoples, and reinforced gendered and racialized underpinnings of 'human' boundary-making practices in relation to 'non-human' animals. Ryan (2000) similarly showed how big game hunting and photography in Africa by British colonialists reinforced and consolidated British imperialism, empire and masculinity. Also connecting animal representations to constructions of human masculinity in processes of place domination, Emel (1995) wrote about nineteenth – and twentieth – century wolf eradication efforts in the American West (Figure 9.4).

Figure 9.4 *Trapping 'Old Three-Toes'* (from S. Young and E. Goldman, *The Wolves of North America*, 1944)

Wolf representations emphasized so-called savagery, lack of mercy, unfair habits of pack hunting, and cowardice – all contravening norms of masculinity in the American frontier centred around virility and prowess, sporting honour, and willingness to kill in the name of chivalry, morality, progress and civilization. Emel showed how these sorts of images not only devastated wolves and other animals, but were analogous to racist and sadistic treatment of people falling below European-American males on the hierarchy of beings.

Animals were imbricated as well in the constructions of urban identity in Victorian Britain. Philo (1995) explored the emergence of a distinctive Victorian, urban identity associated with standards of civility, public decency, sexual licence and norms of compassion, in contrast to rural stereotypes. Live meat markets and in-town slaughterhouses, such as Smithfield in London, violated these standards and norms, obliging civilized city dwellers to witness sexual intercourse among animals on their way to market, exposing their delicate senses to the violence of auction and slaughterhouse, and risking their moral decay by forcing them to mingle with drovers perceived by bourgeois reformers as inclined to drink and sexual excess. To distance the emerging urban order from its rural and animal origins (as well as for sanitary and organizational reasons), meat markets and slaughterhouses were excised from the city, reinforcing urban identities defined in opposition to a countryside populated by beastly people and animals.

Howell's (2000) study of dog-stealing in Victorian London shows both dogs and bourgeois women as victims of a patriarchal society, confined to domestic captivity but vulnerable to the actions of lower-class men, the venal public world of cash, commerce and calculation, and the dangers and degradations lurking in the city's poorest districts. Borrowing from Virginia Woolf's satirical tale of the theft of Elizabeth Barrett's dog written from the dog's point of view, he articulates a political geography of dog-stealing characterized by class antagonisms and exploitation of rich by poor, and deeply ingrained practices of 'domestication' itself – of both dogs and women confined by Victorian ideals of femininity, obedienced to male authority and middle-class domesticity.

Critiques of science provided another avenue to rethinking the animal–human nexus, particularly the question of animal subjectivity. Scholars in science studies, as well as some self-reflexive scientists (see Birke and Hubbard, 1995), promoted a critical reading of scientific results and emphasized how social interactions shape their content. Haraway (1989) and others argued that foundational primatology texts were ethnocentric and androcentric, and questioned whether knowledge of the 'natural' world and its non-human inhabitants could be anything other than partial and subjective (Braun and Castree, 1998; Demeritt, 1994; Willems-Braun, 1997).

Such arguments about partial perspective and subjectivity in our knowledge of the natural world, the social construction of nature and animals, and shifting networks linking humans, nature and quasi-objects, prompted responses ranging from cautionary to hostile among natural scientists (and some social scientists: see Murdoch, 1997). Natural scientists worried about potential practical implications for animals and nature more generally, arguing that science critics addressed inadequately, or failed to address altogether, the dilemma that they themselves posed. If nature and human knowledge of it is fragmented, partial and socially constructed, could people take responsibility for those sentient life forms historically relegated to the 'natural' realm – whose habitats were being so deeply degraded by anthropogenic change? By stripping scientists of any authority or special knowledge, critics of science effectively denied a powerful voice endeavouring to speak thoroughly, thoughtfully and with some accuracy for non-human life. A real prospect was further marginalization and silencing of these already oppressed animal 'others' (Cronon, 1995; Shepard, 1995; Soule and Lease, 1995).

This response sparked a series of 'science wars' noted for vicious caricatures of both realism and relativism/constructivism. Some in science studies sought to move beyond this binary (Rouse, 1992), by showing how knowledge is neither just socially constructed nor simply real, and developing new vocabularies to incorporate non-human agency into social analysis (Whatmore, 1999: 27). Within such performative approaches, both subjectivity and agency

are relational effects generated by a network of heterogeneous, interacting, materials rather than inherently innate properties and powers. But debates continue, clearly sensitizing animal geographers to the centrality of knowledge produced by the environmental sciences, whilst leading them to understand the limits to such knowledge. Especially for human geographers trying to take animal subjectivity seriously, simply ignoring scientific findings was not an option. Rather, it became imperative to investigate those scientific disciplines dedicated to understanding animal behaviour, animal–habitat relations, and human influences on animal life chances – such as ethology, cognitive psychology, conservation biology and landscape ecology (Whatmore and Thorne, 1998: 451).

Actor network theory (ANT), derived from poststructuralism, provided new analytics for studying society, nature and technology together, whilst avoiding the denial of differences between, and within, such categories (Law, 1992). ANT proponents argued that analytically there was no *a priori* distinction to be made between humans and non-humans, that dividing lines between people, machines and animals are subject to change and negotiation. Machines and animals may gain or lose attributes, such as intelligence or independence, while people too may take on and lose attributes of machines and animals (Law, 1992: 383). Since humans cannot be disentangled from non-humans, non-humans – including animals – are partly constitutive of society. The analytic treatment of animal subjectivity by animal geographers has thus flowed both from scientific information about their behaviour and cognition, and from considering them as groups socially constructed as minorities or 'outsiders'. Whatmore and Thorne (1998), for example, arguing for consideration of animal agency, followed Ingold (1992) in suggesting that animals were best seen as 'strange persons' to be treated analytically in the same way as human groups. Similarly, Philo (1995) sought to view animals rather as marginalized, socially excluded people; he speculates on the terror of cattle at the hands of drovers and the possibility of their transgression of human-set limits, as they jump through shop windows and engage in 'beastly' sexual conduct.

Neither Whatmore and Thorne nor Philo explored animal thinking and behaviour to better understand the subjectivity of the creatures they were trying to understand. In contrast, Gullo et al. (1998) excavated scientific literature on cougar ecology and behaviour, to explicitly assess lion attitudes toward people and how, if at all, they might have changed with the encroachment of urbanization into their habitats. This strategy is fraught with difficulties, including lack of sufficient scientific knowledge (itself often influenced by behaviourist or evolutionary psychological views, or constructed to minimize animal subjectivity and agency) and the temptation to indulge in excessive anthropomorphism. But such work attempts to go beyond speculation and engage animals on their own terms to try to better understand their interactions with people.

The specific question of animal agency has provoked consternation amongst those committed to conceptions of agency that require conscious intentionality that can result in resistance to, or transgression of, material or socially constructed boundaries – a form of agency that many argue only humans possess and is thus constitutive of the human-animal divide. Wilbert (2000) sought to destabilize anthropocentric conceptions of agency, pointing to anthropomorphic radical ecological discourses of human and animal resistance to 'civilization'. Rather than simply rejecting these ideas, he used them to question whether conscious intentionality was necessary for acknowledging the agency of non-humans, framing non-human agency as a relational effect similar to ANT to argue that animals may indeed 'resist'. However, as Thrift (2000) argued, representing actions in terms of resistance may also be too restrictive and negative a conception. Other animal geographers have utilized Latour's notion of 'agency-as-effect' to counter the notion that only humans possess agency. Woods (2000), for example, in his study of the British hunting debate, argues that the 'ghostly representations' of deer and fox constitute 'agency as effect' – unintended yet powerful in terms of disrupting the spatial imaginaries of both city and countryside. Another example is Whatmore and Thorne's (1998) analysis of wildlife trading networks, which relied on performative conceptions termed *topologies of wildlife* that reveal how definitions of wildlife emerge out of complex interrelations within heterogeneous social networks, and also how wildlife matter as active agents, sensible creatures who are subjects as well as objects present in multidimensional forms (Whatmore and Thorne, 1998: 451).

For some, this strategy may seem insufficient to capture animal decision-making and how resulting decisions, predicated on both their social and biological subjectivity, shape and are shaped by human action. Thus a challenge facing animal geography is the design of research that considers both human and animal agency simultaneously. The most productive model for such work may be an ethnography informed by scientific evidence on animal minds, behaviour and ecology. Elizabeth Marshall Thomas' (1994) narrative about relations between the Ju/wasi and lions of the Kalahari provides a tantalizing example of what such work can reveal. She recounts the relations between Ju/wa hunters and lions during the 1950s, when she and her family lived amongst the Ju/wasi. The Ju/wa hunters respected lions and studied lion hunting tactics to inform their own efforts. Lions also accorded the Ju/wa respect, however. After a long chase and kill, for example, lions would sometimes beat the men to the carcass. The men would speak to the lions gently, toss a stone at them, and tell them softly to be gone. The lions would grumble but leave peacefully. But sometimes the humans robbed the lions of their kill, without challenge.

Since the 1950s, however, the dramatic growth in human population, the emergence of large-scale cattle ranches, wildlife tourism and hunting, and the establishment of a nature reserve (from which the Ju/wasi were removed), meant increasing pressure on lions. The lion nation was effectively divided into subgroups whose culture and interactions with humans had diverged by the 1980s. In one subpopulation, lions altered traditional hunting habits; emerging only at night, they took livestock since their prey population was badly depleted, and were sometimes shot for this by cattle-keepers. They also stopped roaring, seeking to live discretely and minimize contact with people. Another group, however, living in the reserve, had lost contact with people (except those who hunted them with gun or camera), and within two lion generations their tradition of peaceable relations with humans had been lost; they were openly aggressive, and would attack people. A third group had created still a third cultural form, characterized by different hunting patterns, and a new brazenness but no violence toward people; some residual respect seemed to remain. Humans had, over time, gone from fellow hunters to lion enemies (or potential enemies) who no longer treated lions with respect and thus were often no longer accorded respect by lions either. Thomas' tale fully captures vital aspects of human identity as shaped by animals, animal agency as shaped by human action, and dynamics of change in human-animal relations.

Animals and place

Currents in social theory and cultural studies stimulated a rethinking of how nature, culture and subjectivity were embodied in landscape, led by 'new' cultural geographers. The 'new' cultural geography emphasized symbolic qualities of landscape and avoided the morphology and metaphors of natural agency characteristic of traditional cultural geography. Key questions were how nature and landscape were culturally constructed and produced, how they could be read textually, and how landscape representations reproduced and transformed societies (Barnes and Duncan, 1992; Cosgrove, 1983; Cosgrove and Daniels, 1988; Price and Lewis, 1993).

Critical of Sauer's superorganicism, 'new' cultural geographers used poststructuralism, discourse theory and deconstruction to identify landscape not as the product of a realist 'nature' or an enigmatic 'culture group' but as a 'text' composed of signs and symbols whose hegemonic reading both represented and reproduced power relations, knowledge claims and discourses that initially inscribed them (Cosgrove and Jackson, 1987; Duncan, 1980; Jackson, 1989). Simultaneously, they promoted alternative readings of landscape by those marginalized and/or exploited by virtue of their gender, race, class and sexual orientation (categories themselves often 'naturalized' or 'animalized' by patriarchal or colonial powers: Anderson, 2000, Davis, 1998). Such approaches were in the contradictory position of making discursive space for animal 'others' while criticizing any notion of an extra-discursive or external nature of which animals might be a part or from which they might act.

For many political ecologists, animal rights theorists, ecofeminists and activists, such views also seemed to deny the very liveliness of the world. Moreover, denaturalizing nature and treating geographic places as cultural productions denied the agency of nature and especially animals. As Demeritt argued, 'the metaphor of landscape as text ... suppresses any trace of other, nonhuman actors from the production of landscape' (1994:163).

This 'writing out' of nature catalysed lively debate with 'new' environmental historians such as Worster and Cronon, trying to demonstrate nature's agency. Their project was to understand how culture–nature boundaries were (re)drawn owing to cultural process, ecological features and agency, and their interactions. Increasingly, animals were included within their understanding of ecological agency. Merchant (1989) and Cronon (1983), both writing on colonial New England, exposed the significance of the region's indigenous animals. Commodification and consumption of beaver especially led to its eventual demise, but was also fundamental to the evolution of local, regional and transatlantic ecologies, economies, institutions and social relations.

Crosby (1986) identified the instrumental roles played by Old World animals in European global expansion and socio-ecological domination. And in *Nature's Metropolis* (1991) Cronon explicated the roles of animals in organization and production of space and spatial relations in nineteenth-century Chicago, deftly exposing the spatial implications of Chicago's evolution as supplier of beef to a growing nation in which cattle – in both embodied and disassembled forms – were primary actors linking economic and political geographies of western producers with East Coast consumers. Cattle were thus agents embodying the complex history of urban–rural, east–west and class-based relations, serving to shape urban spatial structures, institutions and social relations in Chicago.

The 'new' cultural geography versus environmental history debate stimulated a reconsideration of the role of animals in the making of place, region and landscape. Matless' (1994; 1996; 2000) research on the British Broadlands, for instance, examined two ways in which animals of the Broads defined a relational human identity and how differing expert practices and technologies – looking, touching, testing, killing, listening, tasting – each produced claims to authority concerning local conservation and defined acceptable behaviour toward local animals and ecology. Similarly, Proctor's (1998: see also Proctor and Pincetl, 1996) work showed that controversies over old growth forest were rooted not only in actual Pacific Northwest landscapes replete with declining timber towns, but also the region's 'moral geography' symbolized by the spotted owl. Exploring debates over listing the owl as a threatened species, Proctor found that the owl's political force emanated from its multidimensional symbolism of Pacific Northwest landscapes. Under pressures of timber industry restructuring and flight, Northwest residents linked owl protection to loss of timber jobs, death of timber-dependent communities, and elimination of a traditional rural way of life; environmental groups saw the owl as a symbol of nature's wisdom, harmony and balance. Thus place and region were discursively created around the owl itself, symbolizing sagacity of nature for some but community collapse for others.

Domesticated animals are also powerful symbols of places and ways of life and livelihood. Place-specific breeds are intimately connected to the histories and cultures of places and regions. In Britain, for example, Cotswold Sheep, Gloucester Old Spot Pigs, and Leicester Longwool Sheep arose in particular places and were adapted to place-specific features (climate, forage, etc.). Globalization of livestock production has led to widespread rural decline, however, resulting in the disappearance of many traditional breeds. Yarwood and Evans (2000) considered how recent shifts in capitalist agriculture stimulated both rural decline and efforts to reinvigorate the countryside through agrotourism and alter the rural landscape to preserve its rural character. Thus family farms became theme parks starring old, rare and endangered livestock breeds, now powerful – and fungible – symbols of cultural heritage. Strategies for rare breed protection, along with agrotourism, altered spatial distributions of rare breeds and transformed the links between breeds, place and culture.

Others, such as Ufkes (1995), investigated how globalization-driven livestock sector restructuring and polarized consumer demands reshaped the US 'heartland'. Here, large-scale intensive farming emerged to meet worldwide demand for cheap meat, along with boutique producers targeting affluent consumers worried about health and food safety. For pigs, new meatpacking practices centred around lean meat production and involved new genetics, feeding regimes, facilities construction and management practices 'down on the farm'. Leaner hogs spurred demand for an array of new, commercial inputs designed to 'build a better pig' – with devastating implications for animal welfare. These shifts also profoundly altered rural Midwestern places, as enormous packing plants were sited close to large-scale hog raising facilities and their smaller-scale contract producers, transforming nearby towns as workers – mostly immigrants – arrived to fill packing plant jobs.

Animal geographers also studied the role of animals in the creation of smaller-scale places, the zoo being a prime example. Stating that 'If the zoo is a "space", Adelaide Zoo is a "place"', Anderson (1995: 282) advanced general claims about zoos as pivotal sites in the cultural construction of nature, and also particular claims about colonial and national identity in the Adelaide Zoo as a landscape. With greater attention to the architectural features of place, Gruffudd (2000) also tackled the London Zoo, by examining early-twentieth-century debates around Lubetkin's modernist penguin pool and gorilla house. Jumping little more than a century ahead, Davies (2000) considered the creation of the electronic or cyberzoo, through an ANT-based analysis. Unlike traditional zoos that trade in animal bodies, the electronic zoo – global in reach – appropriates digital animal images, circulates them for production of movies and TV, and thus derives an accelerated accumulation of value from the recirculating images. Meanwhile

responsibility for animals as embodied subjects is opaque and dispersed to nature parks and reserves. Not insignificantly, the concentration purely upon visual consumption of animals in the electronic zoo also curtails both (the already limited) human experience with animals and the possibilities of animal agency. Animals, it would seem, are perfectly caged in cyberspace zoos.

Animal geographers have also focused on places characterized by the presence/absence of animals, and on how human–animal interactions can create distinctive landscapes and landscape imagery. Emel's (1995) work, discussed above, documents how landscapes of the nineteenth-century American West, once mythic and majestic but also malevolent and threatening the continent's very civilization, were symbolized and embodied by the active presence of wolves. Philo (1995) and Howell (2000), in contrast, focus on nineteenth-century London, revealing the city as a place of crazily congested Victorian commercial capital laced by the unsettling sights and sounds of live meat markets, but also a patchwork of stately bourgeois neighbourhoods filled with urban domesticates (lap dogs and virtuous women) standing cheek by jowl with rabbit warrens of the deviant and dangerous classes symbolized by depraved dog-stealers. Philo's (1995) inquiry into the plight of livestock animals sent to live meat markets and slaughterhouses also showed how particular traditional nature/culture dualisms led London to become a landscape of exclusion for animals labelled as 'wild', 'unclean', 'immoral', 'unhygienic'. Only animals regarded as 'tame', 'clean', 'moral' or 'aesthetic' were allowed into the cultural fold, and as Howell points out, then only for certain classes and in some districts. Such dualisms, and the interactions they engendered, promoted mutually exclusive spaces and places for humans and most animals.

The urban–wildlands border zones of metropolitan regions remain stubbornly permeable to both people and animals and, despite routine exterminations, even inner cities host 'a shadow population of non-humans spanning the phylogenetic scale', leading Wolch, et al. (1995: 736) to propose a 'transpecies urban theory' alert to such realities. Picking up on this, Gaynor (1999) looked at productive animal keeping in Perth, reviewing attempts by urban managers to exclude chickens, goats and pigs – once common in Australia's urban backyards – from residential areas in order to sanitize the city. Working- and middle-class householders especially regarded animal keeping as legitimate; despite public assertions of health and amenity related negative externalities, they routinely subverted regulations. Gaynor (1999: 13) argued that productive animal exclusion constituted an ideological attack on working-class practices, privileging consumptive uses of animals (for example, pets) favoured by the more affluent, and advocated 're-animating the suburbs' through promotion of backyard animal keeping, community gardens and city farms.

In Hull, Griffiths et al. (2000) explored human responses to feral cats in relation to ideas about the proper order of urban places. Here feral cats were a marginalized social grouping existing within human society. Drawing on psychoanalytical studies, Griffiths et al. drew attention to how wilder spaces, and the place of animals in the city are uncertain and often contested. Whereas some people saw wild places associated with feral cats as sites of anxiety and aversion, others viewed them as refuges for an otherwise lost wild nature. Responses to feral cat colonies among local residents were affected by their constructions of the built environment, rendering cat spaces either discrepant or acceptable urban features, and by ideas of feral cats as either legitimately wild or domestic 'convicts on the loose'. Feral colonies served both to fracture and to cement social relationships between people; allotment plot holders were suspicious of any local council interest in the feral cat colony that they supported, while other colonies engendered conflict between cat protectors and those arguing for their removal. Such studies show the complexity of human–animal spatial orderings in the city and the ambiguity of resident attitudes toward nature and civilization as manifest in place.

Tackling much larger felines, Gullo et al. (1998) considered the changing relations between people and mountain lions in California, where cities encroach on cougar, coyote, bear and golden eagle habitats. During the late twentieth century, urbanization-driven increases in human–cougar interactions along with scientific discord over cougar ecology stimulated a renegotiation of cougar population management. A polarized public discourse around cougars emerged, characterized by renewed advocacy of trophy hunting by gun/hunting lobbyists, and proposals from ecologists for wildlife reserves, movement corridors and buffer zoning to protect both people and cougars in the rapidly changing metropolitan fringe. Media coverage revealed and reinforced changes in public attitudes toward cougars, whose character was reconstructed from symbol of wilderness heritage to cold-blooded serial killer, as fringe habitats become human-dominated places. Seeking a more symmetrical perspective, Gullo et al. also considered attitudes towards people among cougars, highlighting cougar capacity for learning and behavioural change. Both human and cougar attitudes present difficulties for predator management but also suggest potentials for coexistence through mutual learning and behavioural modification – an approach that could increasingly characterize nature–culture relations in border landscapes.

Michel (1998) also considered the role of animals in place formation, in the context of southern California golden eagle endangerment and conservation/rehabilitation. Urban growth rhetoric, real estate interests and scientific environmentalism shape eagle conservation politics but so do community activists who contest scientific eagle habitat conservation planning. Activism among eagle rehabilitators and wildlife educators, grounded in struggles to save injured eagles and starving eagle chicks, and to nurture responsibility and consideration for animals among children, was thus imbued with gendered notions of motherhood and family and an ethic of care. Michel argued that eagle rehabilitation and wildlife education for children constitute a personal politics of both animal and human social reproduction that asserts the agency of wildlife in defining pathways to human–animal coexistence and shared places.

Places are not only shaped by human interactions with charismatic megafauna. Waley (2000) traces the development of new ecological thinking and how it led to new riparian landscapes in urban and suburban Japan.[10] European ecological landscaping theory, adapted to the Japanese context to design restored urban river habitats, often prioritized animals linked to threatened ways of life, such as rice farming, and harnessed the rich symbolism of fireflies, dragonflies and fish. Yet such ecological landscaping was feared as privileging animals over people, or countering 'traditional' human–animal relations. But the most vigorous opposition to river landscapes that encourage coexistence came from powerful engineering and building construction interests. The result, predicted Waley, would be humanized urban streamscapes unable to support riparian wildlife.

Trying to think prospectively about animals and urban places, Wolch (1996) argued that seeing animals as subjects suggests that creation of a 'zoöpolis' – a place in which people and animals coexist – might help re-establish networks of care between people and animals. Despite perceptions of city and country as binary opposites, cities host a wide range of human–animal interactions (Figure 9.5). By extending understandings of human relations with animals in the city, Wolch intimated that the ideal of zoöpolis

Figure 9.5 *Atlanta's long-horned steers*

might create not only physical places for animals in the city, but also political space for social movements intent on peace between the species.

Animals and the moral landscape

Environmental movements of the 1970s were expressly normative, concerned with the question of how to fashion more ethical ways of living with and in nature. They stimulated a flowering of work on environmental ethics, which ultimately produced polarized debates about animal subjectivity and moral status. Traditional ethical frameworks were used to argue in favour of the rights of nature (Nash, 1989) and the standing of natural objects in legal contexts (Stone, 1984). Much of the effort, grounded in the work of Aldo Leopold, emphasized 'land ethics' and environmental stewardship (Callicott, 1989); like the larger environmental movement, this work was concerned with species and habitats, and their protection from anthropogenic impacts such as development and pollution. Informed by wildlife biology, forestry and the other environmental management sciences, the lives of individual animals *per se* were not a consideration, nor was their subjectivity problematized.

The deep ecology movement also sprang up within environmental ethics (Devall and Sessions, 1985; Fox, 1990). Less concerned with practicalities of legal arguments, and emphasizing human unity with nature and ethical duties to earth, as well as rejecting anthropocentrism, deep ecology was attacked especially on the basis of its weak political critique (Salleh, 1993). Moreover, its tendency to dissolve human–animal differences obscured the uniqueness of specific animals (and their ecological requirements) and thus glossed over the nature of their subjectivity, leaving little to say about how individual animals should be treated (Plumwood, 1993).

Stimulated by a very different arm of the environmental movement – animal rights – and with different philosophical origins were philosophers such as Peter Singer (1975) and Tom Regan (1983), who attempted to utilize standard ethical frameworks to argue that animals fell within the purview of human moral consideration. An enormous literature on the moral status of animals ensued, concerned with the inherent value of and/or consideration due to individual animals (not aggregates such as populations or species). This work drew on findings from ethology and cognitive psychology to argue that human–animal differences were in degree, not in kind, and thus to reject the anthropocentric ethics of western philosophy.

These streams within environmental ethics ran in parallel course but ultimately converged, sparking intense clashes. Over time, however, environmental ethics incorporated insights from social theory, especially feminism and postmodernism, generating a critique both of traditional rights approaches to animals and of utilitarian ethics, ultimately leading to a postmodern environmental ethics (Oelschlaeger, 1995) and a feminist ethics of care (Plumwood, 1993). The latter was especially influential in thinking about nature and its subjectivity, since it takes a contextual approach to ethical choice, stresses the situatedness and partialness of knowledge, and emphasizes the interconnectedness of living creatures and environments. In particular, the subjectivity of animals was increasingly emphasized (Adams and Donovan, 1995).

Evolving in conjunction with these debates, justice for both people and animals became paramount for many animal geographers. Issues of justice and ethics in human–animal relations have been approached in distinctive manners, however. Some animal geographers have considered specific issues of animal rights to sustenance. Wescoat (1995), for instance, maintained that little attention had been given to animals' access to water in different cultural and legal contexts, with the 'right of thirst' in Islamic law constituting an important exception. He outlined the doctrinal bases for the right of thirst and clarified the sense in which it is a 'right' and is 'Islamic', before proceeding to assess the relevance of this water law in two geographic contexts, the Islamic Republic of Pakistan and the American West. The comparison indicated that direct relevance for Pakistan is more complex than expected, since the right of thirst is not legally required; yet this religious imperative stimulated the emergence of animal welfare and rights activities. In the US the right of thirst is indirectly relevant, its moral imperative revealing western water law as morally inadequate, and provides a cross-cultural example of how duties to provide water for animals might be expanded.

Religious perspectives have also influenced ideas about the ethics of meat eating in India. Robbins (1998), for example, argued that India's internationalizing livestock sector intensified long-standing political struggles between ethnoreligious groups for whom animals have historically served as goddesses, high-interest capital and/or food. He showed how, in Rajasthan, transformations in international meat demand catalysed shifts in agricultural production and changes in caste identity and notions of property, altering bargains between people and domesticated animal species. Moreover, conservative Hindu politicians blamed meat eating for declining morality, calling for economic isolationism and a return to the (mythical) past of pure vegetarianism, and fuelling anti-Muslim sentiment; Muslim livestock farmers and butchers are caught in this Hindu–Muslim power struggle.

Animal geographers have also focused on animal protection movements, and how an animal's position within the movement's hierarchy of effort influences its fate. Whatmore and Thorne (1998), for instance, considered how broadnosed crocodiles were initially listed and then downlisted under the Convention on the International Trade in Endangered Species (CITES), a global conservation policy privileging existence rights for species versus individual animals. Downlisting and 'sustainable use' for these crocodiles may relegate them to factory farm lives: birth in a concrete basin, permanent enclosure for rapid fattening, slaughter, and reincarnation as purses, boots and belts for global consumers. Thorne (1998) also considered how a perceived abundance of kangaroos eliminated them from ethical consideration by conservationists who place value only on the rare or endangered. Considered 'pests' by farmers, abundant by wildlife managers, and roaming a vast, empty territory, tens of thousands of kangaroos are shot annually despite increasing loss and fragmentation of kangaroo habitat, and the animal's status as a powerful tourist attraction and symbol of Australian nationhood.

Others have developed more general ethic frameworks for thinking about human–animal interactions. Lynn's (1998) comparison of geographical ethics – or geoethics – with other ethical traditions revealed the value in embracing the concept of 'geographical community' to encompass ethical questions involving people, animals and nature. Lynn made a case for including animals in the moral community, and proposed four normative principles to guide human–animal relations and resolve moral dilemmas inherent in sharing space with animals.

Jones (2000) also called for an ethics that accounts for differing spatial contexts and practices. Seeking to adapt Levinas' ethics of the encounter to human–animal interactions, Jones argued that all encounters between humans and animals are ethically charged. Treatment of animals in one form or situation or space may be deemed unethical in another, and the individuality of non-human others is often erased, except in the cases of pets or charismatic creatures. Even most conservation initiatives are driven by notions of scarcity, which favours ethical consideration. Scarcity may transform animals in places like zoos into gene banks representing an animal population, or they may be individualized only as 'attractions' or commodities. Instead Jones calls for ethical consideration of individual non-humans, and for bringing the geoethical gaze to bear on often forgotten spaces – the seas or fish farms. Jones suggests that the face-to-face ethics of encounter require us to see the individuality of non-humans. However, there are many problems with such a challenging ethical view; feminists argue that it may result in the denial of ethical responsibility to women, as well as what Irigaray (1991) terms 'the face of the natural universe' – which in some ways is what Jones seeks to redress.

Several animal geographers have considered how animals have shaped the 'moral landscapes' of particular places and regions. As mentioned earlier, Matless considered the role of animals in creating the 'moral geographies' of particular places and regions. Specifically, he analysed Broadland's 'moral geography' or the constellations of ideas about how human life should be lived in relation to given environments, some predicated on a violent, hunting-based approach to wildlife, others on a preservationist approach 'warranting quiet observation rather than loud killing' (1994: 141; 2000). Such conflicting cultures of nature decisively shaped local society–animal relations and ideas of how to interact with nature within this place. Proctor (1998), too, argued that the spotted owl conflict was part of a long-standing debate over the Pacific Northwest's moral landscape, as revealed in relations between people and forests. Environmentalists argued that old growth forests and wildlife predated and existed apart from people, and thus people had a moral obligation not to destroy them, while pro-timber advocates saw logging as a way to better manage and sustain the forest, with human welfare dependent upon it. Caught between these two visions of the region's moral landscape was the spotted owl.

If animals are 'granted' subjectivity, agency and maybe culture (see Clutton-Brock, 1994; Whatmore and Thorne, 1998), but their survival opportunities are 'produced' by humans, how do human groups decide what those opportunities will be? And *who* gets to decide? Here many animal geographers depart from those writing from the same theoretical positioning in the nature/culture debates (for example, Braun and Castree, 1998). Granting some subjectivity and culture to animals, animal geography (in most manifestations) requires not only an emphasis on human 'survivable futures' (Katz, 1998) but also an emphasis on *animal* survivable futures. Social and environmental justice as currently understood is broadened to include animal justice as well. In response, Elder et al. (1998) recommended a 'pratique sauvage' or radical democracy encompassing not only subaltern people but animals too.

THE FUTURE OF ANIMAL GEOGRAPHY

Geography, as a discipline, has provided significant leadership in explicating the history and cultural construction of human and non-human animal relations, as well as their gendered and racialized character and their economic embeddedness. This work must continue. There are wide areas of barely touched terrain in comparative cultural analyses, economies of animal bodies, and the geographical history of human–animal relations that need articulation and examination.

Animal geographies also offer a fecund opportunity for integrating social and physical streams within geography, and to harness geographic information systems (GIS) in the effort to better understand human–animal relations. Biogeographers focus on plants; few geographers contribute to the *Journal of Biogeography's* occasional 'aspects of zoögeography' section compared with contributions from other scholars, and there are only a handful of geographers in the journal's editorial team (although this handful does include the current editor). This situation may, in large part, be driven by the increasingly widespread availability of remotely sensed data on the biosphere, most of which relate to plants rather than to inconveniently mobile animals, and which must be translated into complex habitat suitability models in order to understand implications for animals.

Consequently, empirical works on animals are scarce, particularly in urban areas. Except for rare animals in 'wildernes' or rural settings, data on animals tends to be scant and inadequate for most sorts of analyses. Existing data, usually developed by ecologists or conservation biologists, generally ignore human–animal relations, except anthropogenic habitat degradation (however, see Carter, 1997). New developments in GIS, however, hold some potential to creatively combine quantitative as well as qualitative data on habitat with information on human populations and land use practices, in order to facilitate analysis of how human activities affect animals as groups and individuals. Such studies could be theoretically framed in various ways, as the tools of GIS become more flexible. Even an approach such as ANT could conceivably be used to structure a GIS-based analysis of how people, things and places engaged in an animal-centred network are located, interact and impact various actors – especially animals.

Conversely, blossoming areas of scholarship in social geography also need to recognize the presence of animals in place, and the agency of animals in creating those places. Both urban ecology and rural studies are growing fields that cannot escape people–animal interactions. Following actor network theory, we find a myriad of heterogeneous materials also making up the social, some of which are animals. This encourages us to rethink the possibility of an animal geography, never mind a human geography. For if actors are networks of heterogeneous materials, it would seem to make little sense to delineate an animal geography when humans and non-humans so promiscuously mix and exchange properties (Murdoch, 1997: 332). This, then, would lead to studies in which human and physical geography are necessarily closely integrated.

More importantly, perhaps, geographers need to communicate their findings to a broader public so that the politics of animal geographies can be better informed. If, in fact, nature and animal lives are socially sanctioned and produced, decisions regarding those futures must be made with as much illumination as possible. Biotechnology is a fast-moving industry that challenges extant animal lives on a daily basis. So too, resource development, urbanization and agricultural expansion continually threaten existing animal ecologies and geographies. But as geographers and others writing on human–animal interactions have shown, narratives or representations of animals are culturally and economically scripted. The struggles between groups to create their 'places', livelihoods, and future visions will also be struggles to impose particular narratives and representations as the correct interpretation. The historical and everyday construction of these disparate narratives and representations needs considerably more attention from scholars in order for people to 'see' that they do not derive from natural law, a deistic nationalism or traditionalism, or some other source of mysticism.

Just as compelling is the need to outline the way in which the politics of animal geographies might be addressed in democratic fashion. If 'rights talk' is to be rejected by critical geographers and others as a capitalist and patriarchal artifact, and if the nature/culture dualism is to be dismantled as a modernist mistake or sham, then some pragmatics of animal futures must be adopted to take the place of such condemned referents. Such a pragmatics requires an immense educational project and a new public arena for the agonistic politics that must ensue after the falling away of old moral pillars and culturally and economically determined categories. The constant emphasis on democratic decision-making and the need to appreciate 'militant particularisms' (Williams, 1989) by those on the left who have contributed most to the new cultural geography (and animal geographies) leaves little room for the bureaucrat or expert (in most cases scientists of one ilk or another). Even environmentalists and animal rights activists – the 'civil society' monitors and advocates pushing and shoving the bureaucrats into action on animal issues, quite often using the expertise of the scientist – have been vilified or called into question. Who does this leave to make decisions about animal lives, to speak for those who do not speak for themselves, to imagine animal futures? Are there to be any requirements for participation? Over what territories are decision-makers to have sovereignty: Backyards, parks, towns, villages, nations, the globe? Clearly, we need voices with authority to speak on behalf of animals, over specified domains; such voices should include lay people, non-experts and indigenous peoples, as well as scientists willing to learn from others with more situated knowledge and perspectives.

Finally, recognizing animal agency and subjectivity means that moral choices must be made. Animals do survive at the leisure or predisposition of humans. There is an implicit set of assumptions or moral choices embedded within much of the new cultural geography writing dealing with animals and nature. These assumptions and choices need excavation and honest enunciation. In fact, an area of scholarship that has gone far in critiquing the lack of self-reflexivity in natural and physical science and in economics has shown little evidence of willingness to follow suit itself, to confront the anthropocentric character emanating from the discourse as a whole.

One has only to look at the environmental residual of Marxist-led development in Russia and China to see why many greens (and animal advocates) are as uncomfortable with a 'no limits' Marxist-informed nature/culture discourse as they are with a western-capitalist-informed discourse and structure. Are there no limits with which to be concerned? No problematic scarcities? Is cloning merely an extension of domestication and breeding practices and the creations of chimeras simply another step forward in an already hybridized nature? Those who would argue that social science, and more specifically geography, should not address such questions in the domain of morality or politics, ignore the great strides that feminist, Marxist, poststructuralist and other theorists have made in the past few decades. At the peril of trivializing their discipline, geographers ignore the immense political import of their scholarship in this heady field of nature/culture discourse and animal–society relations. Precisely because there is no foundational 'truth' or 'nature', talk about morality is not out of bounds. Without morality talk, animal geographers risk adopting a project much like Habermas' wherein democracy is grounded in 'domination-free communication', an ideal most of us would recognize as unrealizable.

Animal geography – undergirded by a critique of capitalism, an awareness of positionality and a deployment of poststructuralist methods of analysis, and manifesting a cautionary position regarding 'science' – offers an important ethical perspective on these questions. It highlights the importance of a politics of care and the admission of animals as legitimate beings 'first and foremost themselves'. Such a stance may help overcome some tensions between radical social theorists and more conventional environmentalists, both of whom seek to find more humane futures for people and other sentient beings. More specifically, morality talk in animal geography aims at preventing suffering and cruelty, and fostering happiness and comfort. As 'suffering', 'cruelty', happiness' and 'comfort' mean different things to different people and other animals, therein lies the rub and the importance of political process. Although crucial, it is not enough to illuminate the multiplicity of meanings; we must also imagine how these meanings are created, changed and coexist (or not), and the nature of new institutions that can foster comfort and happiness for animals as well as people. Those undertaking this latter task must have cultivated moral selves, lest these social institutions become too disciplinary and coercive. This is a complex challenge. Ethical/moral dilemmas are more complicated than ever (for example, bio-engineered animals and viruses), and most societies, under the influence of a post-Keynesian neoliberal ideology, seem headed toward more individualization, less education for citizenship, and less moral/ethical training or sensitization. It is nonetheless incumbent upon intellectuals to help reconstruct modern organizations so that, to use Bauman's (1995: 148–9) words, 'massive participation in cruel deeds' becomes impossible rather than possible, and without witness.

NOTES

We are grateful to Chris Philo for his comments about the historical development of animal geography, and to Sarah Whatmore for her fine editorial advice and recommendations. The support of the US National Science Foundation (Program in Geography and Regional Science) and the US Sea Grant Program, which was critical to research by Wolch and her colleagues that is reported in the chapter, is kindly acknowledged. Please direct correspondence to Jennifer Wolch, Department of Geography, University of Southern California, Los Angeles, California 90089–0255, USA.

1 Some zoögeographers also used concepts of place and region: see Cansdale (1949), Davies (1961) and Thompson (1905).
2 German geographers, such as Hahn, were prominent influences, as were Vidal de la Blache, who discussed animal domestication historically in relation to transportation, and other French regionalists (James, 1972: 252–3), such as Vidal's student Brunhes (1920), who argued that animal 'conquest' and animal-raising, as well as the 'devastations' they caused were 'essential facts' of human geography (James, 1972: 250; Vidal, 1926: 354ff).

3 Sauer co-organized a Yale University conference on the destructive environmental impacts of development, published as *Man's Role in Changing the Face of the Earth* (Thomas, 1956), that served as a wake-up call to social and natural scientists.

4 Particularly those undertaken with Harold Peake such as Peake and Fleure (1927). See Campbell (1972) and Dickinson (1976) for analyses of Fleure's work.

5 It may be argued that Forde's similarities emerge partly from his time with Kroeber in California in the 1920s (Kuper, 1996: 120), if we accept that Sauer's notion of culture was strongly influenced by Kroeber. Peter Jackson (1979) also noted this similarity between Sauer's approach and that of Fleure and Forde.

6 Though Donkin's work was an exception. Donkin asserted that intrinsic qualities of some animals played a major role in determining which were domesticated and that animals had some capacity to rebel against the imposition of human requirements, raising notions of non-human agency and anticipating the enlarged 'space' recently given to animals (Philo, 1995).

7 This definition, common to many cultural ecologists, is heavily influenced by Julian Steward (1955).

8 We use this term rather than 'cultural animal geography' – or for that matter zoögeography – because while UK animal geography is rooted in the 'new' cultural geography, intellectual origins in the US are far more eclectic. To retain this openness and encourage dialogue not only within human geography but also across the physical–human divide, we embrace the more inclusive referent.

9 But decentring the human subject also presents problems for animal subjectivity – similar to those recognized for 'women' by feminists and for the 'subaltern' by postcolonial theorists. Entitlements in law, ethics and elsewhere have been constructed in reference to essential qualities of 'subjectivity'. The loss of legitimacy for those subject positions and their referential 'rights' undoes the persuasive utilitarian and natural rights arguments made for animals by philosophers Peter Singer (1975) and Tom Regan (1983), among others. How to negotiate the oppression of animals (and people) in the absence of subjectivity has been as important to animal activists and theorists as to those feminists, postmodernists and multicultural theorists for whom the whole idea of 'rights' has been discredited.

10 Though, as mentioned earlier, Donkin's (1977) study of the cochineal insect is one of the finest geographical studies of animals.

REFERENCES

Adams, C. (1990) *The Sexual Politics of Meat*. New York: Continuum.

Adams, C. and Donovan, J. (eds) (1995) *Animals and Women: Feminist Theoretical Explorations*. Durham and London: Duke University Press.

Anderson, K. (1995) 'Culture and nature at the Adelaide Zoo: at the frontiers of "human" geography', *Transactions of the Institute of British Geographers* 20 (NS): 275–94.

Anderson, K. (1997) 'A walk on the wild side: a critical geography of domestication', *Progress in Human Geography* 21: 463–85.

Anderson, K. (2000) '"The beast within": race, humanity, and animality', *Environment and Planning D: Society and Space* 18: 301–20.

Anderson, M.S. (1951) *The Geography of Living Things*. London: English Universities Press.

Baker, S. (1993) *Picturing the Beast*. Manchester: Manchester University Press.

Barnes, T.J. and Duncan, J.S. (1992) 'Introduction', in T.J. Barnes and J.S. Duncan (eds) *Writing Worlds: Discourse, Text and Metaphor in the Representation of Landscape*. New York: Routledge. pp. 1–17.

Bauman, Z. (1995) *Life in Fragments: Essays in Postmodern Morality*. Oxford: Blackwell.

Bekoff, M. and Byers, J.A. (1998) *Animal Play: Evolutionary, Comparative, and Ecological Perspectives*. Cambridge: Cambridge University Press.

Bender, B. (1975) *Farming in Prehistory: From Hunter-Gatherer to Food Producer*. London: Baker.

Bennett, C.F. (1960) 'Cultural animal geography: an inviting field of research', *Professional Geographer* 12 (5): 12–14.

Bennett, C.F. (1961) 'Animal geography in geography textbooks: a critical analysis', *Professional Geographer* 13: 13–16.

Birke, Lynda and Hubbard, Ruth (1995) *Reinventing Biology*. Bloomington: Indiana University Press.

Blumler, M. (1993) 'On the tension between cultural geography and anthropology: commentary on Rodrigue's early animal domestications', *Professional Geographers* 45: 359–63.

Braun, B. and Castree, N. (eds) (1998) *Remaking Reality: Nature at the Millennium*. London: Routledge.

Brunhes, J. (1920) *Human Geography: An Attempt at a Positive Classification, Principles and Examples*. London: Harrap.

Callicott, J. Baird (1989) *In Defense of the Land Ethic: Essays in Environmental Philosophy*. Albany, NY: State University Press of New York.

Campbell, J.A. (1972) 'Some sources of the humanism of H.J. Fleure'. School of Geography Research Papers no. 2, University of Oxford.

Cansdale, G.S. (1949). 'Some problems in animal geography', *Geographical Magazine* 22: 108–9.

Cansdale, G.S. (1950). 'Animals and man I: the results of competition and conflict', *Geographical Magazine* 23: 74–84.

Cansdale, G.S. (1951a) 'Animals and man II: domestication – providers of food and clothing', *Geographical Magazine* 23: 390–9.

Cansdale, G.S. (1951b). 'Animals and man III: domestication – transport animals, pets and allies, *Geographical Magazine* 23: 524–34.

Cansdale, G.S. (1951c). 'Animals and man IV: chance and deliberate introductions', *Geographical Magazine* 24: 155–62.

Cansdale, G.S. (1951d) 'Animals and man V: biological control', *Geographical Magazine* 24: 211–17.

Cansdale, G.S. (1952) *Animals and Man*. London: Hutchison.

Carter, J. (1997) 'Nest-site selection and breeding success of wedge-tailed shearwaters *Puffinus pacificus* at Heron Island', *Australian Geographical Studies* 35: 153–67.

Cheney, D.L. and Seyfarth, R.M. (1990) *How Monkeys See the World*. Chicago: University of Chicago Press.

Clark, A.H. (1927) 'Geography and zoology', *Annals of the Association of American Geographers* 17: 101–45.

Clutton-Brock, J. (1994) 'The unnatural world: behavioural aspects of humans and animals in the process of domestication', In A. Manning and J. Serpell (eds) *Animals and Human Society: Changing Perspectives*. London: Routledge.

Cosgrove, D. (1983) 'Towards a radical cultural geography: problems of theory', *Antipode* 15: 1–11.

Cosgrove, D. and Daniels, S. (1988) *The Iconography of Landscape*. Cambridge: Cambridge University Press.

Cosgrove, D. and Jackson, P. (1987) 'New directions in cultural geography', *Area* 19: 95–101.

Cronon, W. (1983) *Changes in the Land: Indians, Colonists, and the Ecology of New England*. New York: Hill and Wang.

Cronon, W. (1991) *Nature's Metropolis*. Chicago: University of Chicago Press.

Cronon, W. (1995) 'In search of nature', in W. Cronon (ed.) *Uncommon Ground*. New York: Norton.

Crosby, A. (1986) *Ecological Imperialism*. New York: Cambridge University Press.

Davies, G. (2000) 'Virtual animals in electronic zoos: the changing geographies of animal capture and display', in C. Philo and C. Wilbert (eds) *Animal Spaces, Beastly Places: New Geographies of Human-Animal Relations*. London: Routledge pp. 243–56.

Davies, J.L. (1961) 'Aim and method in zöogeography', *Geographical Review* 51: 412–17.

Davis, M. (1998) *Ecology of fear: Los Angeles and the Imagination of Disaster*. New York: Metropolitan Books.

Devall, B. and Sessions, G. (1985) *Deep Ecology: Living as if Nature Mattered*. Salt Lake City, UT: Peregrine Smith.

Demeritt, D. (1994) 'The nature of metaphors in cultural geography: geography and environmental history', *Progress in Human Geography* 12: 163–85.

De Waal, F. (1982) *Chimpanzee Politics*. New York: Harper and Row.

Dickinson, R.E. (1976) *The Regional Concept: The Anglo-American Leaders*. London: RKP.

Donkin, R.A. (1977) 'Spanish Red: an ethnographical study of cochineal and the opuntia cactus', *Transactions of the American Philosophical Society* 67: 5–84.

Donkin, R.A. (1985) 'The peccary: with observations on the introduction of pigs to the new world', *Transactions of the American Philosophy Society* 75: 1–145.

Donkin, R.A. (1989) *The Muscovy Duck, Cairina Moschata Domestica: Origins, Dispersal and Associated Aspects of the Geography of Domestication*. Rotterdam: Balkema.

Donkin, R.A. (1991) *Meleagrides: An Historical and Ethnogeographical Study of the Guinea Fowl*. London: Ethnographica.

Donkin, R.A. (1997) 'A servant of two masters?', *Journal of Historical Geography* 23: 247–66.

Duncan, J. (1980). 'The superorganic in American cultural geography', *Association of American Geographers* 70: 181–98.

Eagle Clark, W. (1896) 'Bird migration in the British Isles: its geographical and meteorological aspects', *Scottish Geographical Magazine* 12: 616–26.

Elder, G., Wolch, J. and Emel, J. (1998) 'Le pratique sauvage: race, place, and the human–animal divide', in J. Wolch, and J. Emel, (eds) *Animal Geographies: Place, politics, and Identity in the Nature–Culture Borderlands*. London: Verso. pp. 72–90.

Emel, J. (1995) 'Are you man enough, big and bad enough? Ecofeminism and wolf eradication in the USA', *Environment and Planning D: Society and Space* 13: 707–34.

Evernden, N. (1993) *The Natural Alien*. Toronto: University of Toronto Press.

Fitter, R. (1945) *London's Natural History*. London: Collins.

Fleure, H.J. (1919) 'Human regions', *Scottish Geographical Magazine* 35: 94–105.

Fleure, H.J. (1937) 'Geography and the scientific movement', *Geography* 22.

Fossey, D. (1983) *Gorillas in the Mist*. Boston: Houghton-Mifflin.

Fox, W. (1990) *Toward a Transpersonal Ecology: Developing New Foundations for Environmentalism*. Boston, MA: Shambhala.

Fraser-Darling, F. (1956) 'Man's ecological dominance through domesticated animals on wild lands', in W.L. Thomas, *Man's Role in Changing the Face of the Earth*. Chicago: University of Chicago Press.

Galdikas, Biruté (1995) *Reflections of Eden: My Years with the Orangutans of Borneo*. Boston: Little, Brown.

Gaynor, Andrea (1999) 'Regulation, resistance and the residential area: the keeping of productive animals in twentieth-century Perth, Western Australia', *Urban Policy and Research* 17: 7–16.

Goodall, J. (1986) *The Chimpanzees of Gombe: Patterns of Behaviour*. Cambridge, MA: Belknap Press.

Griffin, D. (1976) *The Question of Animal Awareness: Evolutionary Continuity of Mental Experience*. New York: Rockefeller University Press.

Grifffin, D. (1984) *Animal Thinking*. Cambridge, MA: Harvard University Press.

Griffith, M., Wolch, J. and Lassiter, U. (forthcoming) 'Animal practices and the racialization of Filipinas in Los Angeles', *Society & Animals*.

Gruffudd, P. (2000) 'Biological cultivation: Lubetkin's modernism at London Zoo in the 1930s', in C. Philo and C. Wilbert (eds) *Animal Spaces, Beastly Places: New Geographies of Human–Animal Relations*. London: Routledge. pp. 222–42.

Gullo, A., Lassiter, U. and Wolch, J. (1998) 'The cougar's tale', in J. Wolch and J. Emel (eds) *Animal Geographies: Place, Politics, and Identity in the Nature-Culture Borderlands*. London: Verso. pp. 139–61.

Haraway, D. (1989) *Primate Visions: Gender, Race, and Nature in the World of Modern Science*. New York: Routledge.

Haraway, D. (1991) *Simians, Cyborgs and Women*. New York: Routledge.

Harris, D.R. (1962) 'The distribution and ancestry of the domestic goat', *Proceedings of the Linnaen Society of London* 173: 79–91.

Harris, D.R. (1977) 'Alternative pathways to agriculture', in C.A. Reed (ed.) *Origins of Agriculture*. The Hague: Mouton. , pp. 179–244.

Harris, D.R. (1996) 'Domesticatory relationships of people, plants and animals', in R. Ellen and K. Fukui (eds) *Redefining Nature*. Oxford: Berg.

Hartshorne, R. (1939) *The Nature of Geography: A Critical Survey of Current Thought in the Light of the Past*. Lancaster, PA: Association of American Geographers.

Howell, P. (2000) 'Flush and the *banditti*: dog-stealing in Victorian London', in C. Philo and C. Wilbert (eds) *Animal Spaces, Beastly Places: New Geographies of Human–Animal Relations*. London: Routledge. pp. 35–55.

Ingold, T. (1992) 'Culture and the perception of the environment', in E. Croll and D Parkin (eds) *Bush Base, Forest Farm: Culture, Environment and Development* London: Routledge. pp 39–56.

Irigaray, L. (1991) 'Questions to Emmanuel Levinas', in R. Bernasconi and S. Critchley (eds) *Re-Reading Levinas*. London: Athlone.

Jackson, P. (1979) 'A plea for human geography', *Area* 12: 110–13.

Jackson, P. (1989) *Maps of Meaning: An Introduction to Cultural Geography*. London: Unwin Hyman.

James, P.E. (1972) *All Possible Worlds: A History of Geographical Ideas*. Indianapolis: Odyssey.

Jones, O. (2000) '(Un)ethical geographies of human non-human relations: encounters, collectives and spaces', in C. Philo and C. Wilbert (eds) *Animal Spaces, Beastly Places: New Geographies of Human–Animal Relations*. London: Routledge. pp. 268–91.

Katz, C. (1998) 'Whose nature, whose culture? Private productions of space and the "preservation of nature"', in B. Braun and N. Castree (eds) *Remaking Reality: Nature at the Millenium*. London: Routledge.

Kuper, A. (1996) *Anthropology and Anthropologists: The Modern British School'* (3rd edn.). London: Routledge.

Law, J. (1992) 'Notes on the theory of the actor-network: ordering, strategy and heterogeneity', *Systems Practice* 5: 379–93.

Livingstone, D.N. (1992) *The Geographical Tradition: Episodes in the History of a Contested Enterprise*. Oxford: Blackwell.

Lynn, W.S. (1998) 'Animals, ethics and geography', in J. Wolch and J. Emel (eds) *Animal Geographies: Place, Politics, and Identity in the Nature–Culture Borderlands*. London: Verso. pp. 280–97.

Maddrell, A.M.C. (1997) 'Scientific discourse and the geographical work of Marion Newbigin', *Scottish Geographical Magazine* 113: 33–41.

Matless, D. (1994) 'Moral geography in Broadland', *Ecumene* 1: 127–56.

Matless, D. (1996) 'Visual culture and geographical citizenship: England in the 1940s', *Journal of Historical Geography* 22: 424–39.

Matless, D. (1998) *Landscape and Englishness*. London: Reaktion.

Matless, D. (2000) 'Versions of animal–human: Broadland, *c.* 1945–1970', in C. Philo and C. Wilbert (eds) *Animal Spaces, Beastly Places: New Geographies of Human–Animal Relations*. London: Routledge. pp. 115–40.

McKnight, T.L. (1976) *Friendly Vermin: A Survey of Feral Livestock in Australia*. Berkeley: University of California Press.

Merchant, C. (1989) *Ecological Revolutions: Nature, Gender and Science in New England*. Chapel Hill, NC: University of North Carolina Press.

Michel, S. (1998) 'Golden eagles and the environmental politics of care', in J. Wolch and J. Emel (eds) *Animal Geographies: Place, Politics, and Identity in the Nature– Culture Borderlands*. London: Verso. pp. 162–90.

Moebius, C. (1894) 'The geographical distribution and habits of whales', *Geographical Journal* 4: 266–8.

Morton, E.S. and Page, J. (1992) *Animal Talk*. New York: Random House.

Murdoch, J. (1997) 'Inhuman/nonhuman/human: actor-network theory and the prospect for a nondualistic and symmetrical perspective on nature and society', *Environment and Planning D: Society and Space* 15: 731–56.

Nash, R. (1989) *The Right of Nature: A History of Environmental Ethics*. Madison, WI: University of Wisconsin Press.

Newbigin, M.I. (1913) *Animal Geography: The Faunas of the Natural Regions of the Globe*. Oxford: Clarendon.

Newbigin, M.I. (1936) *Plant and Animal Geography*. London: Methuen.

Oelschlaeger, M. (ed.) (1995) *Postmodern Environmental Ethics*. New York: SUNY Press.

Peake, H. and Fleure, H.J. (1927) *Peasants and Potters: The Corridors of Time, vol. 3*. Oxford: Clarendon.

Philo, C. (1995) 'Animals, geography and the city: notes on inclusions and exclusions', *Enviroment and Planning D: Society and Space* 13: 655–81.

Philo, C. and Wilbert, C. (eds) (2000) *Animal Spaces, Beastly Places: New Geographies of Human–Animal Relations*. London: Routledge.

Plumwood, V. (1993) *Feminism and the Mastery of Nature*. London: Routledge.

Price, M. and Lewis, M. (1993) 'The reinvention of cultural geography', *Annals of the Association of American Geographers* 83: 1–17.

Proctor, J. (1998) 'The spotted owl and the contested moral landscape of the Pacific Northwest', in J. Wolch and J. Emel (eds) *Animal Geographies: Place, Politics, and Identity in the Nature–Culture Borderlands*. London: Verso. pp. 191–217.

Proctor, J.D. and Pincetl, S. (1996) 'Nature and the reproduction of endangered species: the spotted owl in the Pacific Northwest and southern California', *Environment and Planning D: Society and Space* 14: 683–708.

Regan, T. (1983) *The Case for Animal Rights*. Berkeley: University of California Press.

Renfrew, C. (1983) 'Geography, archaeology and environment 1: archaeology', *The Geographical Journal* 149: 316–23.

Robbins, P. (1998) 'Shrines and butchers: animals as deities, capital, and meat in contemporary North India', in J. Wolch and J. Emel (eds) *Animal Geographies: Place, Politics, and Identity in the Nature–Culture Borderlands*. London: Verso. pp. 218–40.

Rodrigue, C. (1992) 'Can religion account for early domestication?', *Professional Geographer* 44 (4): 417–30.

Rouse, J. (1992) 'What are cultural studies of scientific knowledge', *Configurations* 1: 1–22.

Ryan, J. (2000) '"Hunting with the camera": photography, wildlife and colonialism in Africa', in C. Philo and C. Wilbert (eds) *Animal Spaces, Beastly Places: New Geographies of Human–Animal Relations*. London: Routledge. pp. 203–21.

Salleh, A. (1993) 'Class, race, and gender discourse in the ecofeminist/deep ecology debate', *Environmental Ethics* 15 (3): 225–44.

Sauer, C. (1925) 'The morphology of landscape', University of California Publications in *Geography* 2: 19–54.

Sauer, C. (1956) 'The agency of man on earth', in W. Thomas (ed.), *Man's Role in Changing the Face of the Earth*. Chicago: University of Chicago.

Sauer, C.O. (1952) *Seeds, Spades, Hearths and Herds*. New York: American Geographical Society.

Sheehan, J. and Sosna, M. (eds) (1991) *The Boundaries of Humanity: Humans, Animals, Machines*. Berkeley: University of California Press.

Shepard, P. (1995) *The Others: How Animals Made Us Human*. Washington: Island.

Simmons, I.G. (1980) 'Biogeography', in E.H. Brown (ed.) *Geography, Yesterday and Tomorrow*. Oxford: Oxford University Press.

Singer, Peter (1975) *Animal Liberation*. New York: Avon.

Soule, M. and Lease, G. (eds) (1995) *Reinventing Nature? Responses to Postmodern Deconstruction*. Washington, DC: Island Press.

Spiegel, M. (1988) *The Dreaded Comparison: Human and Animal Slavery*. Denmark: Heretic.

Stallybrass, J. and White, A. (1986) *The Politics and Poetics of Transgression*. London: Methuen.

Steward, J.H. (1955) *Theory of Cultural Change*. Urbana: University of Illinois Press.

Stone, C. (1974) *Should Trees have Standing? Toward Legal Rights for Natural Objects*. Los Altos, CA: W. Kaufmann.

Taylor, P., Halfon, S.E. and Edwards, P. (eds) (1997) *Changing Life: Genomes, Ecologies, Bodies, Commodities*. Minneapolis: University of Minnesota Press.

Thomas, E.M. (1994) *The Tribe of Tiger: Cats and their Culture*. New York: Simon & Schuster.

Thomas, W.L. (1956) *Man's Role in Changing the Face of the Earth*. Chicago: University of Chicago Press.

Thompson, J.A. (1905) 'How to teach the geographical distribution of animals', *Geographical Teacher* 3: 116–19.

Thrift, N. (2000) 'Entanglements of power: shadows', in J. Sharp, P. Routledge, C. Philo and R. Paddison (eds) *Entanglements of Power: Geographies of Domination and Resistance*. London: Routledge. pp. 269–78.

Thorne, K. (1998) 'Kangaroos: the non-issue', *Society and Animals* 6: 167–82.

Toufexis, A. (1995) 'An eary tale', *Time Magazine*, 6 November, p. 60.

Tuan, Y.-F. (1984) *Dominance and Affection: The Making of Pets*. New Haven: Yale University Press.

Ufkes, F.M. (1995) 'Lean and mean: US meat-packing in an era of agro-industrial restructuring', *Environment and Planning D: Society and Space* 13: 683–706.

Vidal de la Blache, P. (1926) *Principles of Human Geography*. London: Constable.

Wagstaff, J.M. (1983) 'Geography, archaeology and environment 2: a human geographer's view', *The Geographical Journal* 149: 323–5.

Waley, P. (2000) 'What's a river without fish? Symbol, space and ecosystem in the waterways of Japan', in C. Philo and C. Wilbert (eds) *Animal Spaces, Beastly Places: New Geographies of Human–Animal Relations*. London: Routledge. pp. 159–81.

Wescoat, J. (1995) 'The "right of thirst" for animals in Islamic law: a comparative approach', *Environment and Planning D: Society and Space* 13: 637–54.

Whatmore, S. (1999) 'Hybrid geographies: rethinking the human in human geography', in D. Massey, J. Allen and P. Sarre (eds) *Human Geography Today*. Cambridge: Polity.

Whatmore, S. and Thorne, L.B. (1998) 'Wild(er)ness: reconfiguring the geographies of wildlife', *Transactions of the Institute of British Geographers* 23: 435–54.

Wilbert, C. (2000) 'Anti-this-against-that: resistances along a human/non-human axis', in J. Sharp, P. Routledge, C. Philo and R. Paddison (eds) *Entanglements of Power: Geographies of Domination and Resistance*. London: Routledge. pp. 238–55.

Willems-Braun, B. (1997)' Buried epistemologies: the politics of nature in (post) colonial British Columbia', *Annals of the Association of American Geographers* 87: 3–31.

Williams, R. (1989) *Resources of Hope: Culture, Democracy, Socialism*. London: Verso.

Wolch, J. (1996) 'Zoöpolis', *Capitalism Nature Socialism* 7: 21–48.

Wolch, J. (1998) 'Zoopolis', in J. Wolch and J. Emel (eds), *Animal Geographies*. London: Verso.

Wolch, J. and Emel, J. (eds) (1995) Theme issue on `Bringing the Animals Back In', *Environment and Planning D: Society and Space* 13: 631–760.

Wolch, J. and Emel, J. (eds) (1998) *Animal Geographies: Place, Politics and Identity in the Nature–Culture Borderlands*. London: Verso.

Wolch, J., West, K. and Gaines, T.E. (1995) 'Transspecies urban theory', *Environment and Planning D: Society and Space* 13: 735–60.

Wolch, J., Gullo, A. and Lassiter, U. (1997) 'Changing attitudes toward California's cougars', *Society and Animals* 5, 2: 95–116.

Wolch, J., Brownlow, A. and Lassiter, U. (2000) 'Constructing the animal worlds of inner-city Los Angeles', in C. Philo and C. Wilbert (eds) *Animal Spaces, Beastly Places: New Geographies of Human–Animal Relations*. London: Routledge. pp. 71–97.

Woods, M. (2000) 'Fantastic Mr Fox? Representing animals in the hunting debate', in C. Philo and C. Wilbert (eds) *Animal Spaces, Beastly Places: New Geographies of Human–Animal Relations*. London: Routledge. pp. 181–202.

Yarwood, R. and Evans, E. (2000) 'Taking stock of farm animals and rurality', in C. Philo and C. Wilbert (eds) *Animal Spaces, Beastly Places: New Geographies of Human–Animal Relations*. London: Routledge. pp. 98–114.

10

'Inhabiting' – Landscapes and Natures

Steve Hinchliffe

This chapter opens up a number of questions regarding human and non-human relations. The focus is on landscape practices, which shape and are shaped by those relations. In the first half, I review some of the main ways in which geographers have dealt with the relationships between landscapes and nature. Simplifying, I divide these approaches into landscape tectonics and landscape semiotics. Finding resources in both 'traditions', I argue in the second half that there are ways of engaging with landscapes and natures that refuse to see either as pure culture (the nature of no nature) or as raw matter (the nature of nature). The intention is to avoid any understanding of nature that reduces 'it' to primary (or for that matter secondary) properties (a tactic I will refer to pejoratively as a first nature politics) and yet, at the same time, to refuse to obliterate spaces of nature by reading all instances of human/non-human relations as somehow culturally determined. In some ways following Castree (Chapter 8, this volume), I argue that avoiding the classic pitfalls of natural and/or cultural determinism requires something more than an analytical imagination. Therefore, in the latter parts of the chapter, I review a number of approaches which attempt to inhabit landscapes as living relations, with all their differences, continuities, discontinuities and entangled formations. I look for various possibilities in cultural geography and its surroundings, including science studies, feminist theory and poststructuralism, for developing a sensitized geography of landscapes and natures. Rather than fixing the terms, the goal becomes one of finding ways of understanding landscape, nature and inhabitation that are experimental and potentially creative.

By way of background, it is useful to dwell upon the importance of arguing for change in the ways in which landscapes and natures are understood. To caricature a conventional argument, as things currently stand, people, landscapes and natures are 'out of joint'. And in conventional environmental politics, this tends to mean that somewhere and at some point in the dark past of urban-industrial society, the joins between people and their environments have been ruptured. The implicit and sometimes explicit aim is to rejoin the worlds of culture, economy and humans with the already constituted worlds of nature, ecology and non-humans. Such views can be found in certain versions of bioregionalism (although see McGinnis, 1999, for a range of bioregional writing) and in a variety of environmentalisms (for a review see Dobson, 1990). I want to avoid such a judgement in this chapter, and steer clear of a politics and an ethics which found themselves on a universal first nature (or even imagine a universalized second nature upon which to build an unchanging ethical system). Yet, at the same time, there is something about being out of joint which can present the possibility for new forms of environmental politics.

Indeed, the sense of being out of joint that I want to pursue in this chapter is one which invites attempts to make new articulations, to experiment with connections. That these attempts cannot be made solely on the basis of human volition starts to open up what a politics of inhabitation might involve. Inhabiting is a more than human affair. Equally, inhabitation is not simply a matter of adding in non-humans. Indeed, this is not about 'social interactions between already constituted objects' (Rajchman, 2000: 12), be they human, non-human or any other segmented identity. As such this is not simply about representing landscaping elements or speaking of and for others. A politics of representation can only be, if anything, the imperfect start to a

politics of inhabitation. More schematically, the
argument pursued here relies on two forms of
politics.

First, there is a *politics of representation*: there
is a politics of recovery with which to be
engaged. There are cultural geographies of landscaping to be written which engage in a different
politics of representation – a politics that takes
the presences and performances of humans and
non-humans of all kinds, shapes and sizes as
matters of potential importance. Second, there is
a *politics of inhabitation*: more than an attempt to
restock the pages of cultural geography with the
missing masses, there is another, less obviously
liberal democratic, motive at work. This is not
simply a matter of a liberation of the oppressed
(or even return of the repressed), although such a
politics is far from redundant. It is also a matter
of experimenting with styles of inhabiting, styles
that manage to re-cover and re-cognize without
covering over everything (inventing itself as a
final vocabulary), or imagining that cognition
is a matter only for human minds and human
minds alone.

The politics of representation and of inhabitation that inform this chapter are, then, neither
mutually exclusive nor in competition. Nevertheless, one of the main aims is to open up a landscape and nature geography that is aware of the
limits to representation, and is thereby sensitized
to the orders and indeterminacies that are
involved in inhabiting. I start with the practices
and meanings of 'landscape' as they have
worked themselves through in cultural geography, relating these specifically to the representation of, and styles of inhabitation with,
non-human natures.

GEOGRAPHIES OF LANDSCAPE AND NATURE: TECTONICS AND SEMIOTICS

*Tectonics: the study of the building of
landscape form*

Landscape tectonics, in the sense I use it here, is
crudely summarized as the material building of
landscape. One particularly influential strand to
this approach has been Carl Sauer's (1925; 1966)
writing on the cultural and material shaping of
landform and landscape. At its best, this work
foregrounds the ways in which landscaping is
always a coproduction – involving humans
(cultures) and non-humans (natures). Whilst
Sauer's earlier work tended to presuppose a

pre-human landscape which was processed by
various waves of human occupants, his later
work on cultural landscapes managed to successfully dispel those representations of American
wilderness as devoid of cultural production (the
narrative basis for various forms of colonization,
including, more lately, colonization performed
through the exclusionary practices of natural
landscape conservation: see Escobar, 1995;
Wilson, 1992). Indeed, Sauer and his followers
have been particularly successful in providing a
counterpoint to those understandings of landscape that seek to derive normative value from a
myth of pre-human natural purity. Nevertheless,
a strong nostalgia for the 'natural', maintained
for example through the Sauer-influenced volume
Man's Role in Changing the Face of the Earth
(Thomas, 1956), seemed to underscore a fairly
robust sense of a division between cultures and
nature, or, at the very least, between naturalized
(pure and unified) versions of culture and those
problematic modern versions of industrialized,
urbanized society. The tendency has been, therefore, to treat the process of landscape formation
as the result of *interactions* between natural and
cultural processes – both of which tended to be
portrayed as somehow definable in, and through,
the absence of the other (Figure 10.1).

As Demeritt (1994a) demonstrates, the inheri-
tors of this 'interactive' version of landscape
tectonics include the wave of landscape and
environmental historians writing in the 1980s
and 1990s. These writers, partly provoked by a
growing environmentalist critique of modern
society, sought to recover 'the earth itself'
(Worster, 1988: 289) as a vital and autonomous
component of landscape evolution. For Worster,
all landscapes are the result of interactions
between nature and culture (1990: 1144), and
any account that denies one or the other will fail
to represent the full tectonics of landscape. In
one sense, Worster is surely right to unsettle
assumptions of humans as sole agents in the
making of landscape and environmental histories.
The main danger is, however, that the physical
world he evokes resembles a universal, timeless
and spaceless nature whose primary properties
can be revealed or derived (Demeritt, 1994a; see
also Demeritt, 1994b; and the reply from
Cronon, 1994).

The result is politically fragile. The combination of a concern for non-human nature which at
one and the same time is included in landscape
accounts but also is 'naturalized' (that is,
extracted from the histories and geographies of
worldly affairs) is a strategy that buys political
time but at a cost which is more than a matter
for academic pedantry alone. For example, the

Figure 10.1   *Alan Sonfist, 'Time Landscape', New York, 1965. A fenced area of Manhattan is meant to evoke a sense of what the island might have looked like before colonization. Good fencing keeps out the rubbish and most street people (photo and caption text in Napier 1992: 46, copyright of the author)*

prehistorical first nature that inhabits Cronon's (1991) writing of Chicago (if not his later collective work, see Cronon, 1996, or even his introduction to *Nature's Metropolis*) is informed by bio-energetics and a trophic-dynamic model of ecosystems – which are themselves wrapped up in a broad political nexus (see Demeritt, 1994a). So much so that Demeritt quite rightly questions the degree to which this first nature exists as ontologically prior to human history:

> Ecosystem ecology got its start as radiation ecology, but the insistent press of the outside world upon the modern science of ecology hardly stops there. Integral to the metaphor of ecosystem are cybernetics, the

mathematics of command and control, first developed to control automatic anti-aircraft guns and now used to guide the US Navy's Cruise missiles and the automatic trading program of institutional commodities bankers. (1994a: 177)

This is, it is important to add, not to say that this science is necessarily flawed, nor is it to say that another acultural position or god-like viewing platform is possible (see Haraway, 1991). Rather, it is to say that the natures that we (possibly rightly) want to include in landscape histories and geographies are unlikely to be innocent. Nor are they likely to be accessible as a set of unmediated (or even mediated) primary properties (a matter to which I will return). To be sure,

Cronon is well aware of the need to avoid this universal nature at the same time as wanting to hold onto the political project of environmental history. For example, Cronon is well aware of the limitations of Clementsian climax ecology, with its projection of balance and harmony onto the external, natural world (see Cronon, 1996, and the chapter by Barbour, 1996, in the same volume). Nevertheless, and despite some exemplary writings in this field, there remains a nagging doubt that environmental and landscape historians have not yet provided a means of engaging with the natures of landscape which avoids either smothering them in cultural processes or allocating to them first-order, timeless and spaceless properties. For the most part, and despite the best of intentions, their accounts tend only to delay the moment when a naturalized universal nature reenters the story.

The practical and political reasons for rejecting any recourse to first nature are amply demonstrated in some of the more recent work on environmental and landscape histories that is emerging from studies in political ecology and new biogeographies (see Fairhead and Leach, 1998; Moore, 1996; Zimmerer, 2000). For example, through a wonderfully detailed and thorough study of forestry practices in several West African nation-states, Fairhead and Leach argue that:

> Not only did the development of scientific ideas about West African forests have its own complex intellectual history and sociology, in which certain theories or debates were able to rise to the exclusion of others. But also, and crucially, these views dovetailed with the administrative and political concerns of the institutions with which they co-evolved in a process of mutual shaping. Ideas about forest-climate equilibria, or the functioning of relatively stable forest ecosystems, for instance, fed directly into a conceptual framework and set of scientific practices for conservation, which was about external control. (1998: 189)

The result of this 'ecology of understanding' produced what Fairhead and Leach identify as a simplification and homogenization of forestry knowledge. It led to the valorization of a first nature (an ecological bottom line) which itself contributed to an oversimplified account of deforestation. Critically, and as a result, 'the complex, unexpected social and ecological dynamic' (1998: 190) of living forests remained outside authorized understandings. A rather purified, natural systems model of forest dynamics formed the yardstick against which social systems of forestry practice were, normally unfavourably, measured. This exclusion of human/non-human relations from ecological understanding of deforestation and aforestation practices resulted in a

tendency to treat people living in forest zones as strangers (see also Hecht and Cockburn, 1989). Not only that, they were also to become unwelcome strangers in a land where '"nature" and its national and international guardians have come to claim a right' (Fairhead and Leach, 1998: 192). It should be noted that denouncing this external authority of modern environmental conservation discourse and instead celebrating flux and dynamism is a risky venture. Whilst Fairhead and Leach point towards a form of participatory pluralism which recognizes the importance of power relations in the making of landscapes, others adopt a language that seems to echo the hyperbole of laissez-faire market capitalists (see Stott, 1998).

Even the growing level of awareness regarding the political and ecological importance of participatory forms of landscape management has, by and large, failed to dislodge the basic foundations and authority of this first nature politics. Participation becomes, in many cases, a means of meeting what are preset expert goals and objectives (for a parallel example in UK nature conservation practice, see Goodwin, 1998). Or else, it becomes a means to order people and practices in terms of their naturalness, their conservation compatibility (Zimmerer, 2000: 357), or their suitability as timeless guardians of a timeless first or second nature (see Ingold and Kurttila, 2000, for a more developed argument concerning the dynamics of knowledge and practice, focusing in their case on Finnish landscapes).

Fairhead and Leach start to develop an unsettling of the natures and cultures that make up forest and savanna zone landscapes in West Africa. Following earlier work by Hawthorne (1996), forest landscapes are understood no longer as intricately balanced and likely to fall apart at the slightest disruption, but as 'an ad hoc assemblage of species thriving after millennia of disturbance' (Fairhead and Leach, 1998: 185). What starts to emerge is a sense of living landscapes which cannot be reduced to either a pre-existing culture or a pre-exiting nature. These landscapes are not solely about interactions of an already constituted nature and a culture which somehow can be defined in the absence of its human/non-human relations. Rather, Fairhead and Leach start to point to a coproduction of landscapes, cultures and natures. To be sure, the theoretical delicacies of this achievement are of little concern to Fairhead and Leach. But, in order for them to be able to imagine alternative accounts, the authors reach beyond Sauerian landscape history (a tradition to which they nevertheless explicitly see themselves as belonging) to a second set of general responses to the landscape question. As the title of their book, *Reframing Deforestation*, suggests, landscape is not only a tectonic affair, it is also a way of seeing and a matter for semiotics. It is to these approaches that I will now turn.

Semiotics: the study of the building of landscape meanings

If 'nature', and for that matter 'culture', tended to be treated as unproblematic matters in landscape history and cultural ecology, then no such comfort was available to geographers in this second tradition. Landscape, as object or form, was turned into one, among many, 'ways of seeing' (see Berger, 1972) and, in other cases, as one among a number of possible textual inscriptions and descriptions of meaning. In what is now a familiar and well-worn critique of the Sauerian approach to landscape form, geographers who have sought to denaturalize landscape as a way of seeing have highlighted a tendency to uncritically adopt what was a historically and geographically specific approach to the study of landscape.[1] Meanwhile, those geographers who have explored landscape-as-text metaphors have similarly highlighted the historical, political and cultural means through which landscapes are written and read (see Duncan, 1995). In the following I treat both traditions as examples of attempts to *interpret* landscapes, their production and their reproduction. The focus, again, is on the spaces of nature that these analyses allow for or produce.

Mitchell's identification of what he terms an 'encounter' way of seeing in the landscape and environmental narratives of the 'new western history' is a recent example of an approach to landscape interpretation which reflects upon the role of visual subjects:

> [R]epresentations of landscape are bound with a particular 'way of seeing' the landscape that understands it to be something always already there, something simply to be *encountered* (rather than actively constructed). (1998: 9, emphasis added)

Given the conflicts that mark North American colonial history and geography, it is possibly even more surprising that here a tradition of treating landscape as 'matter of fact', a pre-existing object, is retained. For, as Mitchell attests, 'the "West" is an image of landscape so freighted with political meaning (not to mention more than 150 years of popular iconography) that the real places upon which those images have been built scarcely seem to matter' (1998: 12). I will return

in some detail to Mitchell's 'real places' later in this section. For the moment I will dwell on this 'encountering' of landscape as an unproblematized field of vision in order to draw out some of its implications for human and non-human living.

It was largely in terms of a negative response to the treatment of landscape as a 'timeless unity of form' (Cosgrove, 1984: 16), and through a positive reading of a wide body of writing in cultural studies, film theory and art criticism, that cultural geographers turned to 'interpret landscape not as a material consequence of interactions between a society and an environment, observable in the field by the more or less objective gaze of the geographer, but rather as a gaze which itself helps to make sense of a particular relationship between society and land' (Rose, 1993: 87). More than simply material relationships of society and land, landscaping set up particular modes of observation, worldliness and representation. Drawing on Law and Benschop's (1997) summary of the kinds of relationships that were in part constituted through the Renaissance humanism of fifteenth- and sixteenth-century Italy, these modes can be characterized in the following ways (direct quotes and paraphrasing from Law and Benschop, 1997: 160–1; see also Cosgrove, 1985):

1 The *observer*

   - is a *point* (constituted by the rules of perspective) at which matters are drawn together (a coherent point and a point of coherence)
   - is a point that is *not included in the world* it observes
   - is a point which is to some extent *in a relationship of control* with the world (depictions can be rearranged to re-present other worlds).

2 Meanwhile, the *world*

   - is *separate* from the observer
   - is a *volume containing objects* and is three-dimensional and Euclidean in character
   - exists prior to its depiction, *awaiting discovery*
   - *contains discrete objects* which pass through time with significant stability or differences, the latter of which are explicable in terms of determinable object interactions, collisions etc.
   - *has a need for narrative*, for stories that illuminate the character and displacement of objects in the world.

3   Finally, *representation*

  • is *illustrative*: the world and its narratives
    are already in existence, they simply
    require depiction.

Whilst there are subtle and not so subtle differences in landscape traditions (between, for example, southern and northern European practices:
see Alpers, 1989; Law and Benschop, 1997), it is
possible to suggest that this way of seeing was
intimately associated with two major supposi-
tions. First, the possibility existed for a neat,
centred 'subject'. Second, an equally neat, though
separate and possibly subordinate, solid 'object'
could exist. The implications for human/
non-human orders are legion. Cosgrove provides
a useful summary:

> Landscape distances us from the world in critical ways,
> defining a particular relationship with nature and those
> who appear in nature, and offers us the illusion of a
> world in which we may participate subjectively by enter-
> ing the picture frame along the perspectival axis. But
> this is an aesthetic entrance not an active engagement
> with a nature or space that has its own life. (1985: 55)

In other words, a settlement or division is performed through the act of framing landscape
(setting up, it could be added, the conditions necessary for an 'encounter' way of seeing). The
human subject, or a certain kind of human subject, is ideally distinguishable from a natural
object. Whilst, as I will take time to demonstrate
later on, this labour of division is always far from
a complete exercise, it nevertheless contributes
to the stabilization of certain relations of power
both between humans and between the human
and non-human worlds (for the former, see
Cosgrove, 1985; Daniels, 1989; Mitchell, 1996;
Rose, 1993; for the latter, see Fitzsimmons,
1988; Hinchliffe, 2000a; Latour, 1993; Whatmore,
1999). The constitution and enfranchising of
human *subjects* (and the political meaning of the
term is also relevant), and their estrangement
from human and non-human objects, provide a
setting for a recursive series of purification acts
(see Latour, 1993). To be a good subject (politically, aesthetically and scientifically) is to be as
distant as possible from the objects upon which
'he' gazed. The masculinization of observation
and the feminization (and racialization) of nature
(or the observed) contributed to a way of seeing,
or a modern epistemology, which dovetailed
neatly with a politics of representation (in terms
both of illustration and of suffrage).

To believe that this estrangement occurred, at
some point somewhere, is to accept a form of
modernism – with its sorry (masculine) tale of
inevitable fragmentation and romantic failure.
But narratives of loss should be beside the point.
What *is* important is the political work that the
practices of seeing can produce (sometimes
aided by the myths of a modern sensibility or
subject). So, for example, certain forms of the
scientific gaze are epistemological practices
which continue to labour aspects of this division
(see Haraway, 1989; Latour, 1993; Stafford,
1993).

A point, of course, of this way of seeing
approach to landscape production is to develop a
form of ideology critique. It is to denaturalize
this encounter, to demonstrate its exclusions and
its artifice. It is to highlight the political distributions that are performed through the labours of
division between subjects and objects, pure
humans and others, cultures and natures,
observers and observed, scientists and their
experiments (see Law and Benschop, 1997). One
particularly productive means of politicizing
these landscapes has been to focus upon and
historicize the practices of signification that have
contributed to their production. Commenting for
example upon the composition of landscape in
eighteenth- and nineteenth-century England, and
in particular its systematic placing of agricultural
labour to the background of picturesque rural
countryside, Williams suggests that landscape is
concerned with providing a relationship of
control between the owners and the workers of
land. He goes on to suggest that to landscape is
to distinguish between outside and inside – those
who can project and prospect (the outsiders) and
those who live in the scene (who are therefore
less likely to envision place and space in anything
like the same manner). For Williams, 'a working
country is hardly ever a landscape. The very idea
of landscape implies separation and observation'
(1973: 120). Williams is undoubtedly overstating
matters here, and I will shortly return to this problematic division between labouring and viewing,
but the point that landscaping is embroiled within
social relations of ownership, control, property
and a host of temporal and spatial relations, some
of which are neatly evoked by the multiple meanings of the word 'prospect', is well made
(Cosgrove, 1985; Hirsch, 1995).

In addition to denaturalizing landscape by
demonstrating its construction as one of a number
of ways of seeing, semiotic approaches really
come into their own when landscape meanings
are understood to be constituted through the
subject's reading of an arrangement of signs (and
the coincident re-enactment of those meanings
through the actions that they invite and condition). In this sense, landscape starts to be understood as a textual arrangement of signs (Barnes

and Duncan, 1992). And the power of the analytic of the textuality of landscape *potentially* resides in its foregrounding of the mutual construction of subjectivities and objectivities (for a clear treatment, see Curt, 1994). Cues are taken from a common reading of Foucault which enables a genealogy of forms to be reconstructed. Landscapes, like other matters, become the effects of a myriad of disciplines and delegations. Meanwhile, in case this sounds too much like a reworking of some form of structural determinism, subjects are portrayed as *involved* in the formulation of landscapes in ways that allow for a prospective and normative politics of resistance, play and subversion (Barnes and Duncan, 1992). Involvement is highlighted for a reason. The analytical import of textuality is first and foremost a means to disrupt a world of neat subjects and objects (without, it should be added, destroying the possibility of engaging with subjectivities). It is a means to move away from an analytical style which talks of encounters with landscape, or which speaks of humans and non-humans *interacting* with one another. Rather, textuality promotes a sense of *interpenetrating* subjectivities and objectivities. The collective subject Beryl Curt puts this way:

> The ways in which we 'experience' the world are wrapped up with our concerned engagement with 'the world'. The interpenetration is textuality.
>
> Textuality … is an analytic which serves to draw attention to the impossibility and futility of attempting to define something (some argument, life or whatever) as if it were fully self-present and self-sufficient – as if the world consisted of facts which, as the cliché has it, 'speak for themselves'. Textuality thus serves to trouble any arguments founded in the distinction between 'fact' and 'fictions', the 'discursive' and the 'real'. (1994: 36)

The textuality trick is easier said than done. Indeed, its utilization in cultural geography has on occasion managed to unsettle subject/object divisions, only to draw up a similarly firm distinction between 'texts' and their (intellectual) readers. Rose was quick to point out what she defines as an enduring, masculine desire for a solid looking object of analysis:

> The textual metaphor aims to stabilize disruptions and demonstrate learning and sensitivity: landscape textualized renders geographers' knowledge exhaustive. It performs as another example of aesthetic insecurity in geography. (1993: 101)

So, even if a large number of subject positions have been decentred through textuality – which broadly suggests that they no longer have the privilege of being the origin or source of meaning – paradoxically, there remained in cultural geography a platform on which to stand and view textual and intertextual landscapes (see also Burgess, 1990, for an early example of a geographer's objection to this form of elitism). So as Curt, again, insightfully comments (see also Hinchliffe, 1996, on this analytical ambivalence in technology-as-text metaphors):

> The interesting thing is that the words have changed (no longer is it subject/object, but reader/text) but the properties and powers attributed along the 'fault-line' of the dichotomy have remained the same. (1994: 42)

In short, there has been something of a tendency to reproduce an 'us and them' approach to landscape and textuality. This continuing labour of division (see Cooper, 1997) has been the subject of various critical interventions from feminist and Marxist strands of cultural geography. The former has drawn on psychoanalytical approaches and visual theory to emphasize the ambivalence between, and interpenetration of, observer and observed. So, for example, the 'pleasure and emotive force which landscapes may provide' (Nash, 1996: 149; see also Rose, 1993) disrupts any sense that a pre-formed subject (whether it is the country landowners of Williams' ideology critique, or the landscape geographer) can truly stand, distanced, from the scene. Landscapes are, then, emotional and passionate matters, made up of practices that are just as embodied for the observer as they are for Williams' romanticized workers. Rather than Williams' insiders and outsiders, we are all landscapers now (although the power to landscape and the powers of landscape remain uneven).

The feminist-inspired critique of the tendency to objectify the texts themselves starts to rematerialize the scene. And much of the remainder of this chapter draws on this project. But before I continue in this vein, I need to say something of the second productive critique of the landscape-as-text tradition derived from a Marxist-inspired engagement. If landscape geographers have been partially successful in decentring the subject (that point where meaning is reputedly gathered together), and feminist critiques have focused upon the tendency to recentre certain kinds of subjects, then the Marxist critiques have tended to express a fear of a decentring of the object. The landscape-as-text metaphor is rightly in some cases, and wrongly, I would argue, in others, suspected of dematerializing the world. The anxiety is generated by an aetherial space of textuality, 'a kind of pure cultura' (Curt, 1994: 25), which requires supplementing with some form of material production. As an example, Willems-Braun's (1997) study engages with the literary theory of postcolonialism to unpack the landscaping of the British Columbian forests.

The extratextual matter of this production of landscape is made clear in Willems-Braun's account of the nineteenth-century geologist George Dawson's travel texts (including surveys, diaries, photographs and so on). The latter are neatly regarded as engaging in a material framing of the world, one where nature and native land occupation are disaggregated (a common trope in the building of colonial and neo-colonial environmentalisms). But before this starts to sound as though Dawson was a mind-in-a-vat (see Latour, 1999b), or that 'his' texts were merely the outcome of a solely linguistic world of other texts, Willems-Braun rightly draws the attention to the distributed materiality of landscape production:

> Dawson's surveys and journals did not invent objects and landscapes in flights of fancy. These were material practices that engaged material worlds. Rather, in rendering the landscapes visible, the surveys constructed from what was *encountered* an ordered scene that could be read. Such practices ... were not simply textual, but highly material; they did not leave the land untouched. Instead they actively displaced and resituated landscapes within new orders of vision and visibility, and within regimes of power and knowledge that at once authorized particular activities and facilitated new forms of governmentality. (1997: 16, emphasis added)

Willems-Braun neatly argues for a material approach to cultural production, and advocates an understanding of the performance of texts in the material reproduction of landscapes (right up to the conservationists' approaches to British Columbia the following century). However, in its focus on *cultural* production, the materiality that the author evokes never seems to measure up to much more than a substrate, upon which meanings could be inscribed. If there is agency in these accounts then it fails to wander very far from Dawson's and his successors' admittedly material cultures. Nature tends to appear as a remainder, as something that is *encountered*, enrolled into human affairs, and is *given*, through the material practices of inscription and description, meaning. Nature and materiality figure in this account, and yet despite the will to render the scene as more than a human affair (and so rescue nature and the object from a pure *cultura*), a suspicion remains that they are not up to much.

A similar problem exists with a popular device used in cultural geography entitled the 'circuit of culture' (see Burgess, 1990; Johnson, 1986; Squire, 1994). The circuit, which focuses analytic attention on the production and consumption of landscape meanings, and sees particular cultural productions (like the landscapes and texts which Willems-Braun examines) as moments in a broad process, encourages researchers to engage with what Johnson (1986) calls 'acts' – matters underdetermined by existing textual inscriptions. The extratextual purity of acts, rather like the extratextual qualities of the materials in British Columbia, is, however, treated as something of a remainder that is left unexplored in this model of cultural production and consumption.

This remaindering of extratextual material and action is, seemingly, less of a problem for those Marxist-inspired analyses which fill the void with an account of the social production of landscape. Drawing on Marxist theories of labour value, landscape is produced through significatory practices as well as relations of labour which are 'embodied in any landscape' (Mitchell, 1998: 18). Mitchell, for example, seeks to combine theories of representation with theories of production in order to not only peel back layers of accreted meaning but also 'excavate the processes, including the processes of labour, that went into producing the actual form of the landscape' (1998: 21). This, it seems to me, is an eminently worthwhile project, particularly if the understanding of textuality continues to obscure materialities (although as I have hinted and as I will argue below, this need not be the case). And, perhaps more importantly, if it allows for the recovery of forms of agency that are normally obscured from landscaping practices. Indeed, Mitchell's (1996) writing on migrant workers and landscape production contributes to a series of political projects that relate to struggles over representation. However, Mitchell's project does tend to reproduce a subject/object distinction under another guise. In short, and despite some protestations to the contrary, Mitchell's project ends up, like the historians he criticizes, reproducing a strong division between human and non-human labour, and ultimately between cultures and natures. So, for example, when Mitchell approvingly cites Richard White's writing on the production of nature through embodied labour he tends to emphasize the human labours. In doing so, he misses White's (1996) potentially interesting sense of the creativity of human/non-human relations. What *is* recovered from the obfuscation of landscape is a strictly human sense of agency.

> The landscape, whether an English parkland, the view from an Italian villa, a California farm labour camp, the plains that constitute Chicago's hinterland or a Columbia River fish ladder, is a place structured *for* someone, *by* someone. (Mitchell, 1998: 22, emphasis in original)

There are two problems here. First, an assumption remains that landscapes can be read once and for all. Whilst the subject/object dichotomy re-emerges in admittedly politically useful ways – as I have accepted, such readings are strategically useful in terms of re-presenting a silenced majority – the ideology critique itself is expressed as a matter beyond worldly practices. As Whatmore has put it, 'such accounts share an inclination to exempt themselves from the representational moment, by variously claiming a privileged correspondence between concept and object, logic and process' (1999: 24). Second, in positing landscapes and power as functions of the intentionality of historically situated human subjects, there is a tendency to evacuate landscapes-as-lived. In other words, we are paradoxically left with an anaemic sense of landscape and agency. Mitchell, of course, should not be accused of writing bloodless histories and geographies, but in laying bare the *real* landscape, produced through various relations of labour, we are back to a pre-given order of things in which it is difficult to reimagine a place for different human/non-human orders. Despite their eye-opening quality, Mitchell's landscapes can be read as enclosed affairs, whose histories and geographies seem to follow set trajectories[2] (see Massey, 1999a, for a careful expansion of this argument with respect to dominant forms of progressive politics).

To be clear, this is not to deny the importance of power, fetishization, labour or even ideology in the production of landscape. Nor is it to empty the field of political commitment. Rather, it is to say that current formulations seem to rely upon a particular politics of representation and a particular form of inhabitation. In short, there is a tendency to treat the diversity and coexistence of non-human worlds as at best a matter of only passive interest and then in terms which strike out multiplicity with a universal natural body (in the form of transformations from first nature to second nature, or even to third nature: see Luke, 1996; and see Whatmore, 1999, for a critique). In the second half of this chapter I want to work towards another sense of landscape, one that can be more open to human and non-human difference and coexistence.

INHABITING LANDSCAPES GEOGRAPHICALLY

So how do we avoid the natural or cultural determinism that seems to follow on from some of the work that I have reviewed so far? How can non-human spaces be imagined and engaged without making them timeless and spaceless abstractions? How can we avoid centring landscape

meaning and value on certain humans and/or on humans alone? In this section I attempt to find some partial answers to such questions by pushing at what might be involved in inhabiting landscapes successfully. To be clear, the inhabitation that I want to push is not as cosy as it might at first sound. As I stated at the outset, there is no clear blueprint with which we can fall in line, no harmony to which we can adjust. Things are more dynamic than this adjustment model suggests (see Botkin, 1990; Zimmerer, 1994). For one thing, inhabiting human and non-human landscapes will produce changes to all parties (albeit to varying levels and to different degrees). And, as I stated at the outset, all landscape assemblages will remain somewhat out of joint. This is not a sense of inhabitation that can hope to cover all bases and produce a blanketed landscape which is reducible to one logic or schema. In this sense, landscape inhabitation will involve interrelations, but not necessarily interdependency.

In order to develop this sense of a connected though differentiated landscape, I want to explore the degree to which understanding landscaping as textual practice can reinvigorate a politics of inhabitation. The focus on textuality may seem counter-intuitive, especially given the tendency in recent cultural geography to talk of materialities, bodies, non-humans and so on as non- or extratextual matters. But, as I have argued, such a boundary drawing exercise is too quick and risks too much, especially if we are left more or less where we started, with an albeit deferred split between nature and culture (something that seems to me to be endemic in the tectonic and semiotic approaches that I have so far reviewed). The point of the argument here is to suggest that whilst there may be good reasons for being suspicious of textual models, it is nevertheless politically and intellectually important to avoid old pitfalls. So, before engaging with some of the main approaches that I have identified for developing a politics of landscape inhabitation, I want to make two points. First, rather than arguing for less text, textualities can actually be pursued for the work they do in producing an inhabitable and affective world. Second, there is a need to specify a little further what kinds of activities or practices are understood as textual.

Let me start with this reversal of the normal objection to cultural geography's treatment of landscapes and natures. We need more rather than less text. The normal objection, particularly from some forms of Marxist analyses and from environmentalists, is that we need less about texts and more about worlds. So, for example, in using the circuit of culture, analysts come across moments, or desire to find moments, when something or some process is outside the inscribing and de-scribing of text – moments that are 'underdetermined' by the current intertextual setup. I have hinted above that I am sympathetic to such an approach, and am similarly suspicious of anything that imagines the world is constructed, and interpreted, by humans, for humans and through linguistic constructions alone (what Whatmore rightly criticizes as the all too prevalent 'lexical cast of the cultural turn' 1999: 29). However, as Latour (2001) has noted, as soon as the phrase 'underdetermined' is used, there is a tendency to revert to a task of allocating between what humans say and what the rest of the world does. The aim therefore in the second half of this chapter is to refuse such an analytical process. So rather than looking for things that exist outside texts, the aim becomes one of gaining understanding of how texts (perhaps amongst other means) can enable what Latour calls a 'learning to be affected'. For Latour, the inscribing and describing activities of laboratory science and field science are not only productive of knowledgeable scientists. Learning to be affected is also a matter of engaging with a world that becomes *more highly differentiated as understanding proceeds*. This is not, it should be clear, a matter of simply becoming attuned to a pre-existing world (an explanation that simply reassembles the old binaries) but is a means through which humans and non-humans can add to the world. So, when Latour describes the textualization of smell through odour kits in the perfume industry, and the progressive refinement of testers' ability to differentiate fragrances, he notes how 'body parts are progressively acquired at the same time that world counter-parts are being registered in a new way' (2001: 2). Further, he argues that this is not simply a means by which testers find words to refer to the world. This would be the zero-sum game that many associate with the textual representation, or more accurately the linguistic capture, of the world. Rather, in learning to be affected – in articulating propositions – bodies, things and words all have the potential to become more than they were before the articulations began. So, for the sociologists of science,

> the pair human–nonhuman does not involve a tug-of-war between two opposite forces. On the contrary, the more activity there is from one, the more activity there is from the other. (Latour, 1999b: 147)

In other words, this kind of account moves away from texts as representatives and towards a sense of texts as habits, and as means to make connections. In doing this we unsettle the common

belief that human subjects are knowledgeable and (non-human) objects constitute simply what is known (or waiting to be known). The relationship is less one-sided. So, for example, when Hayles argues that a species extinction 'reduces the sum total knowledge about the world' (1995: 58), this is not because the living organisms that belong to that species are no longer available for study, but because 'it removes from the chorus of experience some of the voices articulating its [the world's] richness and variety' (1995: 58; see also Abram, 1996, for an attempt to convert this phenomenology of the senses to an environmental ethics, and my reservations of this project in the conclusion to the chapter).

This brings me to the second point. The model of text that I am starting to evoke here is perhaps a useful qualification to the one that Curt refers to (that of interpenetration), but it is certainly different to the one that cultural geography has, in the main, inherited from cultural and literary studies. It is a version of textuality that engages and enlivens the world rather than swamps it. The feminist theorist Elizabeth Grosz marks a significant distinction between a closed and overcoded textual model that she associates with a Derridean understanding of textuality, and one that is more characteristic, she argues, of Deleuze's open sense of textual activity. It is the latter, I will argue, that offers resources for inhabiting landscapes (see also Davies, 2000).

> Instead of a Derridean model of the text as textile, as interweaving – which produces a closed, striated space of intense overcodings, a fully semiotized model of textuality – a model that is gaining considerable force in architectural and urbanist discourses, texts could, more in keeping with Deleuze, be read, used, as modes of effectivity and action which, at their best, scatter thoughts and images into different linkages or new alignments without necessarily destroying their materiality. Ideally, they produce unexpected intensities, peculiar sites of indifference, new connections with other objects and thus generate affective and conceptual transformations that problematize, challenge, and move beyond existing intellectual and pragmatic frameworks. (Grosz, 1995: 126–7)

As I will hope to show, taking this latter sense of textuality along with Latour's affective bodies provides possibilities for extending some of Latour's interest in largely human schemata and world-making activities (albeit ones that rely on activities of non-humans), and takes us some way to developing transhuman geographies. As I will also suggest, it will be necessary to draw out what Latour's affected world and Grosz' linkages and alignments involve in order to situate this textual model in the landscape politics that I want to pursue here. For now, the idea that we need more rather than fewer texts (or articulations of propositions) in order to become affected, and that we should understand texts as actions that can but do not of necessity produce connections, together form the basis for thinking some more about landscapes, nature and inhabitation.

In the following subsections I use the suggestive framework provided by Grosz and Latour to introduce and then to qualify what might be involved in a material semiotic approach to inhabiting landscapes. The focus here will be on the limitations to an overly analytical approach to semiotics and so in the final subsection I draw out what an experimental or connective semiotics, more in line with Grosz' textual model, might involve.

Material semiotics

'Material semiotics' is employed to great effect in the writings of Latour, Haraway and others (see Akrich and Latour, 1992, for a useful introduction and Haraway, 1992, for an exposition). The term is used to emphasize that, far from being limited to the feared, pure cultura, meaning is just as much about material arrangements as it is about words on a page. For Whatmore, for example, material semiotics is a means to extend 'the register of semiotics beyond its traditional concern with signification as linguistic ordering, to all kinds of unspeakable "message bearers" and material processes, such as technical devices, instruments and graphics, and bodily capacities, habits and skills' (1999: 29). Material semiotics concentrates attention on the ways that stable meanings are built out of a wide range of actions and actants. The attraction to those in technoscience studies who have until recently been interested in the ways in which stable orders (like scientific truths and technological efficiencies) are produced is clear. Worlds are built through the more or less successful linking together of other worlds, and the longer and more robust the linkages, the more stable the construction (although see Munro, 1997, and Hinchliffe, 2000a, for criticisms of this equation of length and strength).

In terms of understanding landscapes, we start to open out a geography of *networked* relations. In these terms, landscapes are no longer simply human affairs (a reading based upon a fundamental division between subject and object). Material semiotics most significantly enables a recognition of human *and* non-human times and spaces and their roles in the co-constitution of worlds. This approach:

recognizes chains of translation of varying kinds and lengths which weave sound, vision, gesture and scent through all manner of bodies, elements, instruments and artefacts – so that the distinction between being present and being represented no longer exhausts, or makes sense of, the compass and possibility of social conduct. (Whatmore, 1999: 30)

Material semiotics, networks and weaving practices are all important to the politics of inhabitation. They start to enliven understandings of the importance of non-human and human acts in the making of worlds (and the spatialities that are implied in those activities). Likewise, they start to unsettle divisions between presence and absence and start to suggest a degree of openness to practice. Nevertheless, this openness is not always apparent in the growing body of work that calls itself actor network theory (ANT). Indeed, there is a danger (by no means inherent, but, given the way a good deal of actor network theory, in particular, has been operationalized – see the criticisms of Lee and Brown, 1994, and Law, 1999a – a real danger) that some of the more structural and totalizing elements of a semiotic approach can re-emerge in analysis (in Grosz' terms, this is in part the risk of a fully semiotized model of textuality).

Part of the problem may well be the route through which material semiotics has come to this area of geography. Haraway, Latour and Akrich all adopted the approach and terminology of A.J. Greimas (including his deployment of the term 'actant'). As Lenoir points out, Greimas' semiotics is 'an abstracting, ahistorical, structuralist semiotics aimed at looking for a logic of culture, proposing a structural explanation in terms of systems' (1994: 122), and even reducing textuality to deep biological structures. To be sure, Haraway's 'coyote grammar of the world' is very different to Greimas' ontology. But Lenoir worries that her adoption of numerous elements of his work, including 'actors, actants, narratives and the semiotic square' (1994: 132), requires a more stringent demonstration of how we can avoid his structural determinism.

Like Lenoir, I take it that the aim of engaging with semiotics is to avoid 'grids of actantial roles and thematic functions … [and] arid formalism' (1994: 136). Rather, it is to foreground the accidents and contingencies, the embodied and situated activities, as well as the consistencies and regularities, that make landscapes. To this end, the resident network topology of material semiotics is either being treated in more self-evidently open ways, emphasizing its active, practical (and therefore far from complete) usage in the verb 'to network' (see Whatmore, 1999), or being supplemented with other topologies. The point is to open analyses to those aspects of landscaping and other forms of ordering that are not so managerial and totalizing and which demonstrate awareness of the non-presences as well as the presences in landscapes. Examples include Mol and Law's (1994) use of fluid metaphors; Law's (1999b) interest in fire; Latour's (1999a) actant rhizomes; Haraway's (1994; 1997) game of cat's cradle; and Hetherington and Lee's (2000) blank spaces. All can be considered as attempts to abandon that tendency of actor network theory to be the final word (see Lee and Brown, 1994, for one of the first statements to this effect).

There are, in other words, various means to imagine ways of allowing a space for alterity in landscaping practices. Such spaces are necessary if we are to avoid some of the overinscription and overconfidence of, say, landscape semiotics or Marxist analyses. Meanwhile, even though some of this excitement has been generated through ANT's own, belated, preoccupations and troubles (compared say to feminist and poststructuralist engagements with alterity), Hetherington and Lee (2000) suggest that the totalizing tendency of social theory is more widespread. They suggest that the current 'relational turn' in geographical writing is in danger of inhabiting a similar political space to earlier versions of ANT. For example, relational theory tends to assume 'that all elements, regardless of their apparent ontological status, are open to being related one with another' (2000: 173). Hetherington and Lee argue that although relational theory manages to move away from human-centred versions of social theory, and thereby provides a basis for countering an 'ontology of division' (2000: 174) (between, say, human subjects and non-human objects), it does so only by constructing another, similarly constraining, ontology. For, in asserting that all elements may potentially be related, a commonality is supposed, 'in which all actants share a susceptibility to *force*, a susceptibility which provides the grounds on which they can become related to one another' (2000: 173–4). For Hetherington and Lee, then, a residual sameness remains in relational geographies. Despite its talk of openness, difference and the possibility for change (change which is not part of some predestined future: see Allen, 1999), the initial commonality, an ontology of force, makes 'it hard to see why there should be change at all' (2000: 174). In sum, we need to move away from 'a readiness to be ordered by virtue of shared human qualities [an ontology of division] or readiness to be related through human/nonhuman susceptibilities [an ontology of force]' (2000: 174). Instead,

the question of social order has changed from a question of shared properties or susceptibility to relation into a question of how relation may be forged at all. (2000: 175)

The answer to such a question lies, for Hetherington and Lee and for the philosopher Michel Serres, in a different semiotics where it is not only present elements which contribute to the building of landscapes of order. There is something other to the ensuing order which escapes characterization as a necessary element of that order, but which nevertheless does not necessarily exist outside the order (and therefore does not need to be brought 'in' through a conventional representational politics). These others, which are constitutionally indifferent to their placement in an order, and which can perform stabilization as well as change within an order (the authors use jokers in a game of cards, the figure zero in maths and angels as exemplars of this facility), are termed blank figures. Rather than representing the absence of presence in a landscape or order, blank figures do just the opposite. They are figures that are present absences. They are, the authors argue, absolutely vital to the process of ordering, but they are not easily dragged into an economy of representational signs. Meanwhile, their unearthliness is perhaps one way (although as I will suggest in the final subsection, not the only way) of rescuing landscape studies from a metaphysics of presence, and in particular, of earthly (land-locked) and territorial presences (a trait that Irigaray, 1997, associates with a peculiarly masculinist and romanticist approach to space, place and landscape; see also Thrift, 1999).

Now, despite their claims, Hetherington and Lee's notion bears a strong resemblance to the politics of difference that is at the heart of some versions of relational geography. It seems to me that, for example, Massey's power geometries do not boil down to an ontology of force. Indeed, without using the same language, Massey does insist on supplementing relationality with an openness to just the kind of surprise and uncertainty that intrigue Hetherington and Lee:

The relationality of space *together with its openness* means that space also always contains a degree of the unexpected, the unpredictable. As well as the loose ends then, space also always contains an element of 'chaos' (of the not already prescribed by the system). It is a 'chaos' which results from those happenstance juxtapositions, those accidental separations, the often paradoxical character of geographical configurations in which, precisely, a number of distinct trajectories interweave and, sometimes, interact. Space, in other words, is inherently 'disrupted'. (1999b: 37, emphasis added)

Figure 10.2 *Gunnar Theel, Nature's Laugh, New Jersey* (author's photo)

Together, Massey, Hetherington and Lee provide important reminders of the limits to representational politics. Landscape politics cannot simply be concerned with finding and drawing in the missing masses, for there will always be an unexpected component to the practical conduct of ordering or making space. The difference is that Massey draws us into this political realization without recourse to a set of what *might* be read as unworldly and politically indifferent figures. As I argue in the final subsection, this sense of practical contingency (even within the most successful of ordering regimes) helps to flesh out what might be involved in a politics of landscape inhabitation.

Experiments and connections

I have now sketched examples of attempts to maintain a sense of alterity in accounts of the ordering of worlds. For a politics of inhabitation and nature, this seems attractive. It is just that sense of surprise and strangeness, even, as Haraway has suggested, that 'independent sense of humour' (1991: 199), which is required for a positive inhabitation of landscape (see Figure 10.2). And yet, to a certain extent, these reimaginings

of textualities remain in many cases concerned with a particular style of inhabitation. They are concerned with building. Indeed, from tectonics (the study of how form is built), to semiotics (the study of how meaning is built) – and then to textualities of various kinds – the concern has ostensibly been to *account* for the construction of worlds. And even with the attempt to allow room for alterity, and thereby to avoid a crude representational politics, the building metaphor remains.

The problem is that holding onto a building metaphor runs the risk of returning to an admittedly more elaborate, knowing, academic gaze. Or to put this another way, whilst the figurations and elements of landscape may be changing, the form of knowing can seem to stay remarkably constant (see Hinchliffe, 2000b). Perhaps this is nothing more than a risk in Hetherington and Lee's work. Certainly they frequently remind readers that the project is not one of representing the formerly underrepresented. And there can be little doubt that these geographies start to open up possibilities for new connections and creativity. However, there seems to be a difference between building alterity into an account of the world, and accepting alterity as part of ongoing practice and changing our knowing practices to adopt what Deleuze characterized as a 'looser kind of sense' (Rajchman, 2000: 8). The most important lesson here is that, despite talk of indeterminacy and contingency, more work needs to be done if we are to stem the tendency to re-inscribe the division between intellectual and other forms of practice (what Napier has called the 'disconnected intellectual excitement', 1992: 65, that can be produced as we give life to mysterious objects).[3] One means of doing so is to take seriously a Deleuzian model of textuality which is, from the outset, oriented to active experimentation, and labours under no illusion of passive accounting.

Indeed, Deleuze's empiricism involves a rejection of any philosophical moves that require the mystical, the invisible or the absent as a means to critique conventional knowledge or thinking (Rajchman, 2000: 18). Following Whitehead, Deleuze sought the conditions under which something new is produced by 'putting one's trust not in some transcendence or *Urdoxa*, but rather in the world from which thinking derives and in which it becomes effective' (2000: 45). This is not to say that in dispensing with blankness, angels or other mystical figures, we are left with a straightforwardly knowable world (see Massey's point above). Indeed, the aim is not to know a world in which, in any case, 'virtual elements move too quickly for conscious inspection or close third person explanation' (Connolly, 1999: 24; see also Thrift, 2000b). What is crucial, and what angels and blanks can fail to underline, is the requirement to surrender some of the analytical baggage and experiment: to produce, in other words, 'a semiotics that would be diagrammatic or cartographic rather than symbolic or iconic, and diagnostic of other possibilities rather than predictive or explanatory' (Rajchman, 2000: 67).

In short, a shift in ways of knowing is being advocated, from a 'knowing what' to a 'know-how' (or, in Latour's terms, a 'learning to be affected'). As Thrift's (1996; 1999; 2000a) characterization of what he has termed a non-representational turn in intellectual labour has suggested, Deleuze, Latour and Grosz are not alone in urging for a shift away from attempts to match worlds and words. Likewise, there is a whole raft of work in feminist studies (see Grosz, 1995; Probyn, 1996), psychology (Bateson, 1973; Newman and Holzman, 1997; Shotter, 1993) and philosophy and neurobiology (Connolly, 1999; Varela, 1999) which urges us to do something other than provide accounts of the ways in which worlds are ordered or built. Nevertheless, and at the risk of closing too much down at this stage, let me mark out some of the

experimentalism that interested Deleuze from some of the other ways in which the building metaphor of social theory has been disclaimed.

Drawing on Heidegger's essay, 'Building, dwelling, thinking' (1971: 145–61), the anthropologist Ingold counterposes dwelling to building, as a means to highlight the notion that far from confronting the world (head on), humans live in amongst the world:

> the forms that people build, whether in the imagination or on the ground, arise within the current of their involved activity, in the specific relational contexts of their practical engagement with their surroundings … In short, people do not import their ideas, plans or mental representations into the world, since that very world, to borrow a phrase from Merleau-Ponty (1962: 24), is the homeland of their thoughts. Only because they already dwell therein can they think the thoughts they do. (1995: 76)

Ingold manages to unsettle any crude, cognitively based, distinction between human and non-human living forms by bringing to the fore the ongoing, practical engagements (and disengagements), contingencies and know-how that make living possible (see also Ingold, 2000). Humans, like many others, 'act to think' rather than think in order to act (see Thrift, 1999: 297). And yet, despite what is certainly an attractive means of reimagining landscape practices, there is a sense in which Ingold risks a rather 'earthly' romanticism by emphasizing the territorial qualities of 'dwelling' (not to mention 'homeland': see Thrift, 1999). In some ways, we are back to humans living and dwelling 'in' a landscape, which itself risks becoming ahistorical and, more importantly, ageographical. Indeed, it is the localism of these dwelt landscapes that remains problematic in this work. It is important, therefore, that dwelling does not become a means of returning to locality-based and 'presentist' senses of landscape and place. (See also Mitchell, 2001, on the dangers of mistaking landscapes as solely local achievements.)

Perhaps even more significantly, it is the combination of Heideggerian dwelling with Merleau-Ponty's notion of lifeworld that tends, despite claiming that Cartesian ontological priorities are being reversed (Ingold, 2000: 169), to reinstall human transcendence and so open up the old fault lines between humans and the rest. For Deleuze and Guattari, there is a piety in Merleau-Ponty's phenomenology which runs just this kind of risk (1994: 178). As Rajchman summarizes Deleuze's suspicions:

> Merleau-Ponty's notion of the flesh still harbours a strange piety, tied with a dream of an originary experience or *Urdoxa* … On the other hand the 'being of

sensation' that one extracts from common perceptions and personalized affects, or from the space of representation and the re-identification of objects, leads not to an intersubjective orientation in the world, but rather to a mad zone of indetermination and experimentation from which new connections may emerge. (2000: 8–9)

Another way of expressing this distinction is to highlight the quality of becoming in Deleuze's philosophy, as opposed to an orientation to being that is present in much phenomenological work (and which certainly comes to the fore in environmentalist attempts to articulate phenomenological work: see Abram, 1996).

In sum, the active textualities that are involved in making landscapes, in this Deleuzian sense, are experimental rather than analytical. Likewise, they are about ongoing and active engagement and connections/separations rather than cold and distant visions. And, furthermore, in being actively engaged, they are careful not to filter these engagements through phenomenological (pre)conditions: nothing, it is argued, need remain unchanged. This means that we need to learn to put our trust in the world which not only makes thought, but also makes thought effective. I want to finish by drawing out some of the implications of these arguments for landscape inhabitation.

CONCLUSION AND IMPLICATIONS

The first implication bears upon the ways in which human bodies and embodiments are thought and practised. There is a good deal here that is shared with the literature that speaks of embodied vision and which attempts to resensitize seeing by recasting it as bodily, sensuous experience. Seeing is never untouched by the sights and sites of vision. It is haptic and, in that sense, moving (see Taussig, 1993). But, it is important to add, embodying vision is not a simple matter of adding a ready-made human body to the eye or to the I of the subject. As Thrift has endeavoured to make clear, this is an embodiment which is certainly not fixed (and nor is it in any sense a reference to an essential body), but it is a process that incorporates a range of *specific* competences:

> This is, then, an embodiment which is folded into the world by virtue of the passions of the five senses and constant, concrete attunements to particular practices, which always involve highly attuned bodily stances as bodies move in relation to each other; ways of walking, standing, sitting, pushing, pulling, throwing, catching, each with its own cultural resources. (1999: 314)

The specifics (and the species) are important. It is right to say that being incorporated in a different body would be to live a different world (Hayles, 1995: 56). This is exactly what Latour's (2001) 'learning to be affected' attempts to evoke. As bodies (and presumably not just human bodies) engage with the world, so body parts and worldly counterparts are gained. It is important to clarify, therefore, that body specifics are far from being closed matters. Rather in the manner of Deleuze's suspicions surrounding phenomenology, there is a risk of returning to an ontology of division, based this time not on superior cognition or linguistic abilities but on embodied competence (see Callon and Law, 1995, for a review of the means through which speciesism is justified). Such a risk is, however, a problem only when the purpose of social science and of cultural geography remains exclusively analytical (continuing to ask, for example, what a body is, rather than working out what a body can do: see Probyn, 1996: 41). When embodiment is regarded as a practical and ongoing achievement, or even a political/ethical positioning, then we can return to *interaction* – but without a preordained notion of the boundaries that mark the interactants. This is, then, a different sense of interaction than the one that I attributed to landscape tectonics. This is a sense of natural and cultural difference without walls, a way of abandoning the foundational cartographies of autonomous political subjectivities without reducing the world to indifference (see also Whatmore, 1997). This is a possible opening for the deferral of natures which aren't universal, preordained, but which do maintain the capability to be different.

The second implication follows on from this argument. Non-human spaces are unlikely to be circumscribed by human actions (let alone thoughts). Nor do they exist 'out there', waiting to be re-presented in here. A more practical orientation would be to acknowledge that attempts to engage non-human spaces will always be marked by imperfect articulations, and will be matters out of joint. As Latour (2001) has skilfully demonstrated, non-human spaces can become entangled one moment only to develop, through their dynamic sociability, other kinds of spaces in the next. This has been particularly evident in modern industrial-agricultural food landscapes (see Whatmore, 1997) and in the risk landscapes marked by superconductive events (Clark, 1997). In everyday landscape practices, non-humans often object to the stories and roles that have been set for them (with disastrous results: see Hinchliffe, 2001, on the BSE crisis). The challenge for intellectual and political practice has been and will be to learn how to allow non-humans (along with those humans who are more used to being silent objects) to object more frequently in those settings that are not accustomed to other-than-human ingenuity. This, it should be stressed, is not a matter of representation, but is more akin to dialogical engagement (although without the sense that such engagement need necessarily lead to a consensus or agreement: see Mouffe, 2000). Achieving this sensibility requires the looser kind of sensing that was mentioned earlier, a building up of know-how and a learning to be affected.

The final issue is ethics. In landscape movements that have been largely informed by a politics of representation, the aim has often been to bring in the missing masses (the nature, the human labourers) or to reveal the artifice of social power. Environmentalists in particular have been keen in recent years to represent various absentees from landscape contests (including non-humans and the yet to be born). Once represented, the new political subjects can take part (even if remotely and through their spokespeople) in the deliberation over means and ends. Whilst it hasn't been the aim of this chapter to undermine such representational strategies, the argument has started to suggest that a politics of inhabitation may need to realize the limits to (and even what Deleuze called[4] the indignity of) speaking for and of others. This is in part because the complexities of living make such neat spatial encoding of self and other problematic (see Whatmore, 1997). It is also because, as Varela (1999) has made clear, representational politics tends to forget that most of our lived lives are characterized by skilled behaviours and practical ethical expertise rather than abstract ethical deliberations. A similar point is made by Abrams in his careful construction of environmentalist ethics, although a certain piety and sense of primordial nature is retained in this phenomenological excursion. Likewise, as the segmentations of the natural and the social become ever more difficult to sustain, so the idea of drawing up a 'natural contract' becomes more and more difficult to imagine (a problem that Serres' 1995 argument only partially resolves). Taking these points together starts to underline the Deleuzian injunction which is to 'go beyond our social identities and see society as experiment rather than contract' (Rajchman, 2000: 20).

The danger here is that a reasoning, abstract Cartesian subject can disappear only to be replaced by an equally abstract desiring, experimental,

individual (human) ethical agent (see Whatmore, 1997: 40). Such an imagination obscures 'the conditionality of dialogic engagement in terms of the mundane business of living' (1997: 40). It is these landscapes of the living that provide something like the distributed sense of agency, the sense of interrelatedness and partial dependencies, and the more experimental and diagnostic 'arts of connection' which break free from the entrenched cartographies of conventional, bounded, landscapes (territories, neighbourhoods, bodies, regions: see Whatmore, 1997). In this sense, inhabiting landscape requires an experimental geography, which works from landscapes as lived, and seeks to develop progressive forms of inhabitation through practical engagements. In terms of ethics, extension of concern derives from landscape practices and engagements – which, it should be stressed, proceed from the practical skills of landscaping and not from deliberation or the production of universal rules. Thus, 'the very relation of intellectuals to such "movements" or "processes of subjectivizations" must change, passing from a "representational" to an "experimental" role, freeing the "social imagination" from the representation of anything given, prior, original' (Rajchman, 2000: 101).

The trajectory I have taken through tectonics and semiotics to connections has enabled me, on the face of it, to remove a hyphen from the language of nature politics. In actual fact the hyphen is very probably irrelevant. What is important is the shift from a deferral to first nature (or for that matter to second nature), to a deferring and differing of natures. The latter takes us beyond a liberal democratic project of representation. To where it is less easy to communicate, although the ordinariness of living with natures suggests that there are more resources for inhabiting the landscapes of nature than we are perhaps prone to recognize. I take it that it is a task for cultural geography to engage with the everyday practices of animal, plant and geophysical natures, with all their geographical complexity, in order to recover what those resources are and how they might be instructive of other possibilities. Without, of course, seeking to have the final word.

NOTES

Thanks to Sarah Whatmore and my Open University colleagues for providing a number of challenging interventions in the writing of this.

1 By geographically and historically specific I don't mean to suggest that the spatiality and temporality of this way of seeing is easily located somewhere or periodized as some time (e.g. the modern period in

Europe). Indeed, geographies of seeing are rarely so bounded or neat: for an interesting cross- and transcultural approach to seeing landscapes, see Hirsch and O'Hanlon (1995).

2   Olwig's 'more substantive understanding of landscape' (1996: 631) is similarly an attractive and rich account of landscape semiotics which provides resources for avoiding a narrow focus on lexical issues, but he doesn't, it seems to me, escape the charges that I have made here. See also Olwig (1993).

3   Rose's warning, cited earlier, that the 'textual metaphor aims to stabilize disruptions and demonstrate learning and sensitivity' (1993: 101) could, it seems to me, just as easily be levelled at the more recent attempts to 'allow for' alterity.

4   Quoted from a conversation with Foucault. Cited in Rajchman (2000: 97).

REFERENCES

Abram, D. (1996) *The Spell of the Sensuous: Perception and Language in a More-Than-Human World*. New York: Pantheon.

Akrich, M. and Latour, B. (1992) 'A summary of a convenient vocabulary for the semiotics of human and nonhuman assemblies', in W. Bijker and J. Law (eds) *Shaping Technology/Building Society: Studies in Socio-Technical Change*. Cambridge, MA: MIT Press. pp. 259–64.

Allen, J. (with the collective) (1999) 'Afterword: open geographies', in J. Massey, J. Allen and P. Sarre (eds) *Human Geography Today*. Cambridge: Polity. pp. 323–8.

Alpers, S. (1989) *The Art of Describing: Dutch Art in the Seventeenth Century*. Harmondsworth: Penguin.

Barbour, M.G. (1996) 'Ecological fragmentation in the fifties', in W. Cronon (ed.) *Uncommon Ground: Rethinking the Human Place in Nature*. New York: Norton. pp. 233–55.

Barnes, T. and Duncan, J. (1992) 'Introduction: writing worlds', in *Writing Worlds: Discourse, Text and Metaphor in the Representation of Landscape*. London: Routledge.

Bateson, G. (1973) *Steps to an Ecology of Mind*. Herts: Paladin.

Berger, J. (1972) *Ways of Seeing*. Harmondsworth: Penguin.

Botkin, D. (1990) *Discordant Harmonies*. Oxford: Oxford University Press.

Burgess, J. (1990) 'The production and consumption of environmental meanings in the mass media: a research agenda for the 1990s', *Transactions of the Institute of British Geographers* 15: 139–61.

Callon, M. and Law, J. (1995) 'Agency and the hybrid collectif', *South Atlantic Quarterly* 94: 481–507.

Clark, N. (1997) 'Panic ecology: nature in the age of superconductivity', *Theory, Culture and Society* 14 (1): 77–96.

Connolly, W.E. (1999) 'Brain waves, transcendental fields and techniques of thought', *Radical Philosophy* 94: 19–28.

Cooper, R. (1997) 'The visibility of social systems', reprinted in K. Hetherington, and R. Munro (eds) *Ideas of Difference: Social Spaces and the Labour of Division*. Oxford: Blackwell. pp. 32–41.

Cosgrove, D. (1984) *Social Formation and Symbolic Landscape*. London: Croom Helm.

Cosgrove, D. (1985) 'Prospect, perspective and the evolution of the landscape idea', *Transactions of the Institute of British Geographers* 10: 45–62.

Cronon, W. (1991) *Nature's Metropolis: Chicago and the Great West*. New York: Norton.

Cronon, W. (1994) 'Cutting loose or running aground?', *Journal of Historical Geography* 20: 38–43.

Cronon, W. (ed.) (1996) *Uncommon Ground: Rethinking the Human Place in Nature*. New York: Norton.

Curt, B. (1994) *Textuality and Tectonics*. Milton Keynes: Open University Press.

Daniels, S. (1989) 'Marxism, culture and the duplicity of landscape', in R. Peet and N. Thrift (eds) *New Models in Geography*. London: Routledge. pp. 196–220.

Davies, B. (2000) *(In)scribing Body/Landscapes Relations*. Walnut Creek, CA: Altamira Press.

Deleuze, G. and Guattari, F. (1994) *What is Philosophy?* London: Verso.

Demeritt, D. (1994a) 'The nature of metaphors in cultural geography and environmental history', *Progress in Human Geography* 18 (2): 163–85.

Demeritt, D. (1994b) 'Ecology, objectivity and critique in the writings on nature and human societies', *Journal of Historical Geography* 20: 22–37.

Dobson, A. (1990) *Green Political Thought*. London: Unwin Hyman.

Duncan, J. (1995) 'Landscape geography', *Progress in Human Geography* 19 (3): 414–22.

Escobar, A. (1995) *Encountering Development: The Making and the Unmaking of the Third World*. Princeton: Princeton University Press.

Fairhead, J. and Leach, M. (1998) *Reframing Deforestation: Global Analysis and Local Realities: Studies in West Africa*. London: Routledge.

Fitzsimmons, M. (1988) 'The matter of nature', *Antipode* 21 (2): 106–20.

Goodwin, P. (1998) 'Hired hands or local voice: understandings and experience of local participation in conservation', *Transactions of the Institute of British Geographers* 23 (NS) (4): 481–501.

Grosz, E. (1995) *Space, Time, and Perversion*. London: Routledge.

Haraway, D. (1989) *Primate Visions*. London: Verso.

Haraway, D. (1991) *Simians, Cyborgs and Women: Reinventing Nature*. London: Routledge.

Haraway, D. (1992) 'The promises of monsters: a regenerative politics for inappropriate/d others', in L. Grossberg, C. Nelson and P. Treichler (eds) *Cultural Studies*. London: Routledge. pp. 295–337.

Haraway, D. (1994) 'A game of cat's cradle: science studies, feminist theory, cultural studies', *Configurations* 1: 59–71.

Haraway, D. (1997) *Modest_Witness@Second_Millennium: FemaleMan©_Meets_Oncomouse™*. London: Routledge.

Hawthorne, W.D. (1996) 'Holes and the sum of parts in the Ghanaian forest: regeneration, scale and sustainable use. Proceedings of the Royal Society of Edinburgh, 104B: pp. 75–176.

Hayles, N.K. (1995) 'Searching for common ground', in M. Soulé and G. Lease (eds) *Reinventing Nature? Responses to Postmodern Deconstruction*. Washington: Island.

Hecht, S. and Cockburn A. (1989) *The Fate of the Forest: Developers, Destroyers and Defenders of the Amazon*. London: Verso.

Heidegger, M. (1971) *Poetry, Language, Thought*. New York: Harper and Row.

Hetherington, K. and Lee, N. (2000) 'Social order and the blank figure', *Environment and Planning D: Society and Space* 18: 169–84.

Hinchliffe, S. (1996) 'Technology, power and space – the means and ends of geographies of technology', *Environment and Planning D: Society and Space* 14: 659–82.

Hinchliffe, S. (2000a) 'Entangled humans', in J. Sharp, P. Routledge C. Philo and P. Paddison (eds) *Entanglements of Power* London: Routledge pp. 219–37.

Hinchliffe, S. (2000b) 'Performance and experimental knowledge: outdoor management training and the end of epistemology', *Environment and Planning D: Society and Space* 18 (5): 575–95.

Hinchliffe, S. (2001) 'Indeterminacy in-decisions: science, politics and policy in the BSE crisis', *Transactions of the Institute of British Geographers,* 26 (NS) (2): 182–204.

Hirsch, E. (1995) 'Landscape: between place and space', in E. Hirsch, and M. O'Hanlon, (eds) *The Anthropology of Landscape: Perspectives on Place and Space*. Oxford: Oxford University Press. pp. 1–30.

Hirsch, E. and O'Hanlon, M. (eds) (1995) *The Anthropology of Landscape: Perspectives on Place and Space*. Oxford: Oxford University Press.

Ingold, T. (1995) 'Building, dwelling, living: how animals and humans make themselves at home in the world', in M. Strathern, (ed.) *Shifting contexts: Transformations in Anthropological Knowledge*. London: Routledge. pp. 57–80.

Ingold, T. (2000) *The Perception of the Environment: Essays in Livelihood, Dwelling and Skill*. London: Routledge.

Ingold, T. and Kurttila, T. (2000) 'Perceiving the environment in Finnish Lapland', *Body and Society* 6 (3–4): 183–96.

Irigaray, L. (1997) *Forgetting Air*. London: Athlone.

Johnson, R. (1986) 'The story so far: and further transformations', in D. Punter (ed.) *Introduction to Contemporary Cultural Studies*. London: Longmans. pp. 277–313.

Latour, B. (1993) *We Have Never Been Modern* London: Harvester Wheatsheaf.

Latour, B. (1999a) 'On recalling ANT', in J. Law and J. Hassard (eds) *Actor Network Theory and After*. Oxford: Blackwell. pp. 15–25.

Latour, B. (1999b) *Pandora's Hope: Essays on the Reality of Science Studies* Cambridge, MA: Harvard University Press.

Latour, B. (2001) 'Good and bad science: the Stengers–Desprest falsification principle', in M. Akrich and M. Berg (eds) *Bodies on Trial*. Durham, NC: Duke University Press.

Law, J. (1999a) 'After ANT: complexity, naming and topology', in J. Law and J. Hassard (eds) *Actor Network Theory and After*. Oxford: Blackwell. pp. 1–14.

Law, J. (1999b) 'Spatialities and technoscience'. Paper presented to the Open University Geography Discipline. Available from the author, Department of Sociology, University of Lancaster.

Law, J. and Benschop, R. (1997) 'Resisting pictures: representation, distribution and ontological politics', in K. Hetherington and R. Munro (eds) *Ideas of Difference: Social Spaces and the Labour of Division*. Oxford: Blackwell. pp. 158–82.

Lee, N. and Brown, S. (1994) 'Otherness and the actor network: the undiscovered continent', *American Behavioural Scientist* 37: 772–90.

Lenoir, T. (1994) 'Was the last turn the right turn? The semiotic turn and A.J. Greimas', *Configurations* 1: 119–36.

Luke, T. (1996) 'Liberal society and cyborg subjectivity: the politics of environments, bodies and nature', *Alternatives* 21: 1–30.

Massey, D. (1999a) 'Spaces of politics', in D. Massey J. Allen and P. Sarre (eds) *Human Geography Today*. Cambridge: Polity. pp. 279–94.

Massey, D. (1999b) 'Power-geometries and the politics of space–time' Hettner Lecture 2, Department of Geography, University of Heidelberg, Germany.

McGinnis, M.V. (1999) *Bioregionalism*. London: Routledge.

Merleau-Ponty, M. (1962) *Phenomenology of Perception*. London: Routledge and Kegan Paul.

Mitchell, D. (1996) *The Lie of the Land: Migrant Workers and the California Landscape*. Minneapolis: University of Minnesota Press.

Mitchell, D. (1998) 'Writing the Western: new Western history's encounter with landscape', *Ecumene* 5 (1): 7–29.

Mitchell, D. (2001) 'The lure of the local: landscape studies at the end of a troubled century', *Progress in Human Geography* 25 (2): 269–81.

Mol, A.M. and Law, J. (1994) 'Regions, networks and fluids: anaemia and social topology', *Social Studies of Science* 26: 641–71.

Moore, D.S. (1996) 'Marxism, culture and political ecology: environmental struggles in Zimbabwe's eastern highlands', in R. Peet and M. Watts (eds) *Liberation Ecologies: Environment, Development, Social Movements*. London: Routledge.

Mouffe, C. (2000) *The Democratic Paradox*. London: Verso.

Munro, R. (1997) 'Ideas of difference: stability, social spaces and the labour of division', in K. Hetherington and R. Munro (eds) *Ideas of Difference*. Oxford: Blackwell.

Napier, A.D. (1992) *Foreign Bodies: Performance, Art and Symbolic Anthropology*. Berkeley: University of California Press.

Nash, C. (1996) 'Reclaiming vision: looking at landscape and the body', *Gender, Place and Culture* 3 (2): 149–69.

Newman, F. and Holzman, L (1997) *The End of Knowing*. London: Routledge.

Olwig, K. (1993) 'Sexual cosmology: nation and landscape at the conceptual interstices of nature and culture; or what does landscape really mean?', in B. Bender, (ed.) *Landscape: Politics and Perspectives*. Oxford: Berg.

Olwig, K. (1996) 'Recovering the substantive nature of landscape', *Annals of the Association of American Geographers* 86 (4): 630–53.

Probyn, E. (1996) *Outside Belongings*. London: Routledge.

Rajchman, J. (2000) *The Deleuze Connections*. Cambridge, MA: MIT Press.

Rose, G. (1993) *Feminism and Geography: The Limits of Geographical Knowledge*. Cambridge: Polity.

Sauer, C.O. (1925) 'The morphology of landscape', reprinted in J. Leighly (ed.) *Land and Life: Selections from the Writings of Carl Ortwin Sauer*. Berkeley: University of California Press.

Sauer, C.O. (1966) *The Early Spanish Main*. Berkeley: University of California Press.

Serres, M. (1995) *The Natural Contract*. Ann Arbor: University of Michigan Press.

Shotter, J. (1993) *The Cultural Politics of Everyday Life*. Milton Keynes: Open University Press.

Squire, S. (1994) 'Accounting for cultural meanings: the interface between geography and tourism studies re-examined', *Progress in Human Geography* 18: 1–16.

Stafford, B.M. (1993) *Body Criticism: Imagining the 0*. Boston, MA: MIT Press.

Stott, P. (1998) 'Biogeography and ecology in crisis: the urgent need for a new metalanguage', *Journal of Biogeography* 25: 1–2.

Taussig, M. (1993) *Mimesis and Alterity: A Particular History of the Senses*. London: Routledge.

Thomas, W.L. (1956) *Man's Role in Changing the Face of the Earth*. Chicago: University of Chicago Press.

Thrift, N. (1996) *Spatial Formations*. London: Sage.

Thrift, N. (1999) 'Steps to an ecology of place', in D. Massey, J. Allen and P. Sarre (eds) *Human Geography Today*. Cambridge: Polity. pp. 295–322.

Thrift, N. (2000a) 'Afterwords', *Environment and Planning D: Society and Space* 18: 213–255.

Thrift, N. (2000b) 'Still life in nearly present time: the object of nature', *Body and Society* 6 (3–4): 34–57.

Varela, F. (1999) *Ethical Know-How* Stanford: Stanford University Press.

Whatmore, S. (1997) 'Dissecting the autonomous self: hybrid cartographies for a relational ethics', *Environment and Planning D: Society and Space* 15: 37–53.

Whatmore, S. (1999) 'Hybrid geographies: rethinking the human in human geography', in D. Massey, J. Allen and P. Sarre (eds) *Human Geography Today*. Cambridge: Polity. pp. 22–40.

White, R. (1996) 'Are you an environmentalist or do you work for a living? Work and nature', in W. Cronon (ed.) *Uncommon Ground: Rethinking the Human Place in Nature*. New York: Norton. pp. 171–85.

Willems-Braun, B. (1997) 'Buried epistemologies: the politics of nature in (post)colonial British Columbia', *Annals of the Association of American Geographers* 87 (1): 3–31.

Williams, R. (1973) *The Country and the City*. London: Chatto & Windus.

Wilson, A. (1992) *The Culture of Nature*, Cambridge, MA: Blackwell.

Worster, D. (1988) 'Doing environmental history', in D. Worster (ed.) *The Ends of the Earth: Perspectives on Modern Environmental History*. Cambridge: Cambridge University Press. pp. 289–307.

Worster, D. (1990) 'Seeing beyond culture', *Journal of American History* 76: 1142–7.

Zimmerer, K.S. (1994) 'Human geography and the "new ecology": the prospect and promise of integration', *Annals of the Association of American Geographers* 84 (1): 108–25.

Zimmerer, K.S. (2000) 'The reworking of conservation geographies: nonequilibrium landscapes and nature–society hybrids', *Annals of the Association of American Geographers* 90 (2): 356–69.

Landslides (Photo: Alex S. MacLean, reproduced with permission of the photographer)

Section 4

LANDSCAPE   Edited by David Matless

Introduction: The Properties of Landscape

David Matless

INTELLECTUAL PROPERTIES

Landscape is not the exclusive property of cultural geography. Like many *Handbook* topics, landscape has been a subject of understanding for many disciplines, and the sense of landscape as a territory with various proprietorial claims pertains in the academy as beyond. Landscape becomes a matter of political, economic and emotional value, and its capacity to move through different regimes of value lends it committed fascination, makes it an object of argument and care. The chapters in this section illustrate three variants of cultural geography's engagement with landscape, a topic which has helped define the subject over the past century. As cultural geography has sought to define landscape, so landscape has served to define cultural geography.

Landscape has consistently illustrated cultural geography's appetite for ideas of various intellectual affiliation, and in recent years the intellectual properties of landscape have become the object of critical enquiry for anthropology (Flint and Morphy, 2000; Hirsch and O'Hanlon, 1995), archaeology (Bender, 1993; Tilley, 1994), legal studies (Darian-Smith, 1999), art history (Andrews, 1999), cultural history (Schama, 1995), landscape history (Muir, 1999) and philosophy (Kemal and Gaskell, 1993), to name but a few academic

fields. An international conference in Edinburgh in March 2001 on 'Landscapes and Politics' illustrated such diversity well, cultural geographers moving in a minority among people from many disciplines and countries, landscape offering a common language. The work of organizations such as the Landscape Research Group in the UK indicates the capacity for landscape to bring together scholars, professionals and a wider band of 'practitioners' around a common ground, whether through publications, conferences, exhibitions or field events. Distinctions of scholarship and practice, of academics and others, are of course contentious here as elsewhere, suggesting on the one hand that scholarship is not itself an engaged practice, and on the other that non-academics might somehow lack ideas. The work of groups such as the LRG seeks in part to traduce that distinction, drawing together artists, architects, designers, geographers, planners, historians in an ostensibly non-hierarchical fashion. Landscape can be considered in terms of its capacity not only to bring together different regimes of value and move between disciplines, but to cross supposed epistemological hierarchies.

That said, the three chapters here are by cultural geographers, and it would be wrong to play down their distinct geographic perspective in favour of a blurred interdisciplinary terrain. If landscape crosses over boundaries, there is nevertheless a value in attending to the particular perspectives which proceed from differently situated academic knowledges. What then does cultural geography bring to the discussion? Landscape is one of the many transdisciplinary concerns of cultural geography, and here as elsewhere it is a matter less of claiming territory for a discipline than of showing how thinking geographically may alter the field, and of recognizing that the current disciplinary structure may dispose cultural geographers towards particular productive ways of thinking.

THE THREE CHAPTERS

The chapters here can be read as symptomatic and interrogative of particular perspectives within geography over recent decades. Don Mitchell extends his earlier Marxian analyses of the California landscape (Mitchell, 1996) to put forward a sense of landscape articulated through labour, in its production on the ground and as an analytic concept. Mitchell emphasizes 'the work of landscape' (2000: 91–119); the labour required to produce the Californian landscape, and the work done by certain visions of landscape to efface that labour from schemes of scenic value or promotional imagination. Mitchell's work here recalls John Berger and photographer Jean Mohr's (1967) study of an English country doctor and his patients, A Fortunate Man, a book beginning with a daytime photograph of a bucolic scene, over which Berger writes: 'Landscapes can be deceptive. Sometimes a landscape seems to be less a setting for the life of its inhabitants than a curtain behind which their struggles, achievements and accidents take place.' On the next page, over a darkened landscape with only the white paint of homes showing in the gloom, Berger continues: 'For those who, with the inhabitants, are behind the curtain, landmarks are no longer only geographic but also biographical and personal' (1967: 12–15; on Berger see Daniels, 1989). Mitchell's chapter in this volume works alongside his ongoing 'People's Geography Project' (www.peoplesgeography.org; cf. Samuel, 1981), which might be seen as an attempt to move the word 'geography' from the connotations

of Berger's first remark – distance, neutrality – to the next – connection, engagement, intimacy – while holding to the analytical framework offered by a geographical Marxism, with relations of power and interconnections of scale always to the fore.

Denis Cosgrove was a key figure in bringing such analyses into cultural geography, an issue reflected upon in his introduction to the 1998 edition of his *Social Formation and Symbolic Landscape* 1984. In his chapter for this volume Cosgrove focuses on the eye, and the close relationship in the western imagination between landscape, sight and geographical imagination and power. The analysis parallels Cosgrove's (2001) recent study of the earth in the western imagination, understanding how viewing the earth as a whole has figured the western sense of self, and informed cosmological, imperial, visionary geographical imaginations. Cosgrove brings out strongly the complexity of the visual, the many varieties of power it may bring, and its central place in landscape's epistemological field. Cosgrove here develops an argument strong in cultural geography over the last 20 years, namely the centrality of the visual to discourses of landscape and its implied relations of power, but in doing so acknowledges the critiques of earlier formulations of this analysis, in geography and beyond (Nash, 1996; Rose, 1993; and see Jay, 1993), drawing out the historical and political complexities of seeing.

Landscape in Sight is the title of a posthumous collection of writings by J.B. Jackson (1997), whose way of 'Looking at America' offers another visual culture of landscape, central to Tim Cresswell's chapter here. Jackson's emphasis on the ordinary, vernacular landscape has long been influential in cultural geography, especially in the USA (Groth, 1998; Groth and Bressi, 1997; Meinig, 1979), and provides a counterpoint to Cosgrove's discussion of power and the visual, and Mitchell's exploration of the landscapes of labour. Jackson provides a less evidently theoretical purchase on landscape; no obvious major social theory walks through his essays on garages, suburbs and highways. It is in part this sense of the vernacular and the ordinary, differently theorized though no less thoughtful, that draws Cresswell to Jackson's work. As for other figures such as English landscape historian W.G. Hoskins (Matless, 1993;

Meinig, 1979) and earlier 'aesthetic geographer' Vaughan Cornish (Matless, 1996), Jackson's work indicates that the engagement of cultural and historical geography (Williams, 1989) with landscape has a rich and complex social history in its own right, though engagements with that history have often been written to defend past writers against present trends (Muir, 1998). Detailed critical cultural historical geographies of cultural geography's engagement with landscape remain to be written; each chapter here provides a partial account within geographical literature through its particular focus, and elements of the story have been the subject of considerable dispute (on Sauer and contemporary cultural geography see Price and Lewis, 1993, and the subsequent responses by Cosgrove, Jackson and Duncan). Attention to earlier studies in geography and beyond can often show not only their difference to today, but the uncanny resonance of past and present concerns (Matless, 2001); we should certainly beware of having undue faith in our own historical novelty, and allow the past to give us pause. To briefly note one example, in a 1975 essay on 'Roads, office blocks and the new misery' the sociologist Fred Inglis critiqued certain forms of public and private development by asking: 'What, then, are the manifold relations between capital and landscape? Put the question in a more everyday fashion. More bluntly. What do you most notice these days about the English landscape as you travel it?' (1975: 172). The concerns of Mitchell and Cresswell, David Harvey and J.B. Jackson, appear adjacent in an argument for townscape from England in the 1970s. Histories of the landscape idea pertinent to cultural geography stretch well beyond Berkeley, indeed well beyond cultural geography.

Unlike Mitchell and Cosgrove, Cresswell writes here as one not known for publication on landscape. Cresswell's chapter extends his work on the social spaces of mobility and transgression (Cresswell, 1996), most recently developed in *The Tramp in America* (2001), to produce a sense of landscape emphasizing movement and practice. Cresswell uses earlier work on landscape and vision as a counterpoint to emphasize the practical aspects of seeing, moving away from oppositions between the visual and the practical, representation and embodiment, to stress instead the embodied and everyday 'doxic landscapes' which emerge

from Jackson's work on the one hand, and phenomenology on the other. What is striking here as in the other two chapters is the wide frame of reference informing cultural geographic study. Readers will also note recurrent themes, particular problematics, which crop up in landscape as elsewhere in cultural geography. Thus Cresswell ends with an issue one could also trace in Mitchell and Cosgrove's work, namely relationships of representation and experience, imagination and being (see also Crang, 1997). The implication would seem to be that oppositions between such terms become ever less helpful in understanding how landscape works; while we should always attend to how such dualities have come into play and had practical effects, they may be less useful in providing *a priori* structure for our analyses.

SHUTTLING THROUGH

If the three chapters exemplify and interrogate particular formulations of landscape within cultural geography, it is also possible to read across all three, against their grain, to draw out other questions of landscape. Two may be highlighted here, beginning with the issue of knowledge, which runs through each in different ways. The issues raised by Cosgrove surrounding the practice of representation, specifically the act of painting, may be seen as concerning the production of geographical knowledge, a matter taken up in another fashion by Crouch and Toogood (1999) in their work on the painter Peter Lanyon, seeking to understand his abstract landscapes in terms of the making of such knowledge through the processes of art. Cresswell's attention to phenomenological understanding is similarly echoed in recent work emphasizing knowledge of landscape through embodied experience, such as Foster's (1998) study of John Buchan and Wylie's (2002) study of Amundsen and Scott in the Antarctic, works which effectively attend to the historicity and spatiality of experience (Foucault, 1986). Mitchell's story of landscape and labour could be recast in terms of the power–knowledge of the various actors involved, thereby connecting to work linking landscape, science and governmentality

(Braun, 2000), and to social historical studies which stress the spatial knowledges and practices of workers within wider cultures of labour (Revill, 1994). Reading across the grain of these essays begins to reveal other connections of landscape.

If knowledge offers one running theme, time gives another. While all three chapters deploy a primarily spatial imagination, their concerns could be recast in terms of temporality and historicity, emphasizing the role of narratives of landscape history in informing present action, the matters of memory and history making up and triggered by visions of landscape, the emergent histories of doxic landscapes which enfold senses of past time into dreams and anxieties for the future. If in his recent *Progress in Human Geography* report on landscape Don Mitchell (2001) could look to the work of art historian Lucy Lippard (1997) to prompt thoughts on the 'lure of the local' and the relations of scale in landscape, Lippard's earlier work on contemporary art and the art of prehistory, *Overlay* (1983), equally prompts speculation on the theme of landscape as palimpsest, on issues of time, ritual and gender, and on the ability of aesthetic intervention to open up and regather the imprints of history. The issue is here not one of space against time, of the need to focus on one rather than the other, but of the ways in which landscape works through enfolding the spatial and temporal. One could argue that all studies of landscape entail historical geography, in all senses of that term.

Time is also present in contemporary geographical arguments over landscape in terms of the ways in which history may put the present into gear. In his essay on 'Recovering the substantive nature of landscape', Kenneth Olwig (1996) intervenes in current debates through a different genealogy of the term. Olwig begins by noting the confusion over landscape detected by Richard Hartshorne in *The Nature of Geography* (1939), whereby landscape's double meaning of land and its appearance implied a slippery concept best abandoned. Olwig responds not by embracing such doubleness but by offering a different history of landscape which asserts and cements its substance as a site of contested human habitation, 'a nexus of community, justice, nature, and environmental equity' (1996: 630–1). Rather

than the Renaissance Italian and Dutch origins of the term stressed by Cosgrove (1984), Olwig draws out a northern European concept expressive of custom, community and identity, and of independence from centralized power. Landscape for Olwig carries a substantive air of communal, independent thought and practice, a meaning he holds to have been usurped both historically and in contemporary cultural geography by a taste for and/or fascination with a classical ideal of landscape linked to new forms of landed and imperial power. Through a revised history of the word 'landscape', Olwig challenges current critical priorities, and reclaims Sauerian Berkeley School cultural geography as concerned with 'a substantive landscape in which issues of environment, economics, law, and culture are all important', a way of thinking in tune with the 'predominant American reality' of 'a vernacular landscape that tends to violate the visual aesthetics of perspective and harmony' (1996: 645).

Olwig's is an intriguing essay, a fine example of putting historical geography into play in contemporary academic context, and its historical nuances could be the subject of detailed discussion. A different history implies a different landscape, and a different way forward. Here though I would highlight and question Olwig's opening argument, which shares Hartshorne's suspicion of landscape's slipperiness in order to propose something more substantial. One could instead view the doubleness of the term as a virtue, as something which is both analytically productive and makes landscape so important a matter beyond the academy. If Olwig presents landscape's 'duplicitous meaning' (1996: 630) as a problem, others have seen it as an opportunity, as the defining feature of a powerful concept. Stephen Daniels suggests that it is the 'duplicity of landscape', as a cultural term carrying meanings of depth and surface, solid earth and superficial scenery, the ontological and the ideological, that gives it its analytical potential, 'not despite its difficulty as a comprehensive or reliable concept, but because of it' (1989: 197); 'We should beware of attempts to define landscape, to resolve its contradictions; rather we should abide in its duplicity' (1989: 218). Doubleness – representation and materiality, financial and emotional value – may make landscape hard to pin down, yet may also be at the heart of contests over it, and one thereby gains purchase on the topic by inhabiting its slippery nature. The power of landscape may reside in it being simultaneously a site of economic, social, political and aesthetic value, each embedded within and not preceding the other; landscape can be considered a term which productively migrates through regimes of value sometimes held apart (Matless, 1998). Such an argument resonates with W.J.T. Mitchell's suggestion that we approach landscape as a verb rather than a noun, 'a process by which social and subjective identities are formed', considering 'not just what landscape "is" or "means" but what it does, how it works as a cultural practice' (1994: 1). Indeed the question of what landscape 'is' or 'means' can always be subsumed in the question of how it works; as a vehicle of social and self identity, as a site for the claiming of a cultural authority, as a generator of profit, as a space for different kinds of living. Landscape entails movement through a complex philosophical and political field concerning control of and rights to land, definitions of pleasure and beauty, claims to authority over public and private space. Studies such as Lorimer's (2000) of the cultural politics of deerstalking in the Scottish Highlands may involve both the interrogation of historical claims to define the landscape as, for example, a site of scenic value, and the use of landscape as a way of operating through which one might understand such specific definitions within a wider field. Attention to the different claims to and definitions of landscape allows and demands a movement across physical and epistemological space, past and present.

Landscape thereby carries a relational hybridity, always already natural and cultural, deep and superficial, which makes for something inherently deconstructive. In Bruno Latour's terminology landscape might be regarded as a classic 'quasi-object', impossible to place on either side of a dualism of nature and culture, shuttling between fields of reference. Discussing environmental debate Latour asks, 'Can anyone imagine a study that would treat the ozone hole as simultaneously naturalized, sociologized and deconstructed?', and suggests that, 'In the eyes of our critics the ozone hole above our heads, the moral law in our hearts, the autonomous text, may each be

of interest, but only separately. That a delicate shuttle should have woven together the heavens, industry, texts, souls and moral law – this remains uncanny, unthinkable, unseemly' (1993: 5–6). Landscape, in cultural geography and elsewhere, might serve as such a delicate shuttle, weaving through matters often held apart. This section of the *Handbook* offers three versions of such a movement, demonstrating how landscape has been, is and will remain a rich field for enquiry.

REFERENCES

Andrews, M. (1999) *Landscape and Western Art*. Oxford: Oxford University Press.

Bender, B. (ed.) (1993) *Landscape: Politics and Perspectives*. Oxford: Berg.

Berger, J. and Mohr, J. (1967) *A Fortunate Man*. London: Allen Lane.

Braun, B. (2000) 'Producing vertical territory: geology and governmentality in late Victorian Canada', *Ecumene* 7: 7–46.

Cosgrove, D. (1984) *Social Formation and Symbolic Landscape*. London: Croom Helm.

Cosgrove, D. (1998) *Social Formation and Symbolic Landscape*. Madison: University of Wisconsin Press.

Cosgrove, D. (2001) *Apollo's Eye: A Cartographic Genealogy of the Earth in the Western Imagination*. Baltimore: Johns Hopkins University Press.

Crang, M. (1997) 'Picturing practices: research through the tourist gaze', *Progress in Human Geography* 21: 359–73.

Cresswell, T. (1996) *In Place/Out of Place: Geography, Ideology and Transgression*. Minneapolis: University of Minnesota Press.

Cresswell, T. (2001) *The Tramp in America*. London: Reaktion.

Crouch, D. and Toogood, M. (1999) 'Everyday abstraction: geographical knowledge in the art of Peter Lanyon', *Ecumene* 6: 72–89.

Daniels, S. (1989) 'Marxism, culture and the duplicity of landscape', in R. Peet and N. Thrift (eds) *New Models in Geography*, vol. 2. London: Unwin Hyman. pp. 196–220.

Darian-Smith, E. (1999) *Bridging Divides: The Channel Tunnel and English Legal Identity in the New Europe*. Berkeley: University of California Press.

Flint, K. and Morphy, H. (eds) (2000) *Culture, Landscape and the Environment*. Oxford: Oxford University Press.

Foster, J. (1998) 'John Buchan's "Hesperides": landscape rhetoric and the aesthetics of bodily experience on the South African Highveld, 1901–1903', *Ecumene* 5: 323–48.

Foucault, M. (1986) 'Preface to *The History of Sexuality*, Volume II', in P. Rabinow (ed.) *The Foucault Reader*. Harmondsworth: Penguin. pp. 333–9.

Groth, P. (ed.) (1998) 'J.B. Jackson and geography', *Geographical Review* 88 (4): i–iii, 465–527.

Groth, P. and Bressi, T. (eds) (1997) *Understanding Ordinary Landscapes*. New Haven: Yale University Press.

Hartshorne, R. (1939) *The Nature of Geography*. Lancaster: Association of American Geographers.

Hirsch, E. and O'Hanlon, M. (eds) (1995) *The Anthropology of Landscape*. Oxford: Clarendon.

Inglis, F. (1975) 'Roads, office blocks and the new misery', in P. Abbs (ed.) *The Black Rainbow: Essays on the Present Breakdown of Culture*. London: Heinemann. pp. 168–88.

Jackson, J.B. (1997) *Landscape in Sight: Looking at America*. New Haven: Yale University Press.

Jay, M. (1993) *Downcast Eyes: The Denigration of Vision in Twentieth-Century French Thought*. Berkeley: University of California Press.

Kemal, S. and Gaskell, I. (eds) (1993) *Landscape, Natural Beauty and the Arts*. Cambridge: Cambridge University Press.

Latour, B. (1993) *We Have Never Been Modern*. London: Harvester Wheatsheaf.

Lippard, L. (1983) *Overlay: Contemporary Art and the Art of Prehistory*. New York: New Press.

Lippard, L. (1997) *The Lure of the Local: Senses of Place in a Multicentred Society*. New York: New Press.

Lorimer, H. (2000) 'Guns, game and the grandee: the cultural politics of deerstalking in the Scottish Highlands', *Ecumene* 7: 403–31.

Matless, D. (1993) 'One man's England: W.G. Hoskins and the English culture of landscape', *Rural History* 4: 187–207.

Matless, D. (1996) 'Visual culture and geographical citizenship: England in the 1940s', *Journal of Historical Geography* 22: 424–39.

Matless, D. (1998) *Landscape and Englishness*. London: Reaktion.

Matless, D. (2001) 'Bodies made of grass made of earth made of bodies: organicism, diet and national health in mid-twentieth century England', *Journal of Historical Geography* 27: 355–76.

Meinig, D.W. (1979) 'Reading the landscape: an appreciation of W.G. Hoskins and J.B. Jackson', in D.W. Meinig (ed.) *The Interpretation of Ordinary Landscapes*. Oxford: Oxford University Press. pp. 195–244.

Mitchell, D. (1996) *The Lie of the Land*. Minneapolis: University of Minnesota Press.

Mitchell, D. (2000) *Cultural Geography*. Oxford: Blackwell.

Mitchell, D. (2001) 'The lure of the local: landscape studies at the end of a troubled century', *Progress in Human Geography* 25: 269–81.

Mitchell, W.J.T. (1994) 'Introduction', in W.J.T. Mitchell (ed.) *Landscape and Power*. Chicago: Chicago University Press. pp. 1–4.

Muir, R. (1998) 'Landscape: a wasted legacy', *Area* 30: 263–71.

Muir, R. (1999) *Approaches to Landscape*. London: Macmillan.

Nash, C. (1996) 'Reclaiming vision: looking at landscape and the body', *Gender, Place and Culture* 3: 149–69.

Olwig, K. (1996) 'Recovering the substantive nature of landscape', *Annals of the Association of American Geographers* 86: 630–653.

Price, M. and Lewis, M. (1993) 'The reinvention of cultural geography', *Annals of the Association of American Geographers* 83: 1–17.

Revill, G. (1994) 'Working the system: journeys through corporate culture in the railway age', *Environment and Planning D: Society and Space* 12: 705–25.

Rose, G. (1993) *Feminism and Geography*. Cambridge: Polity.

Samuel, R. (ed.) (1981) *People's History and Socialist Theory*. London: Routledge and Kegan Paul.

Schama, S. (1995) *Landscape and Memory*. London: HarperCollins.

Tilley, C. (1994) *A Phenomenology of Landscape*. Oxford: Berg.

Williams, M. (1989) 'Historical geography and the concept of landscape', *Journal of Historical Geography* 15: 92–104.

Wylie, J. (2002) 'Becoming icy: Scott and Amundsen's South Polar voyages, 1910–1913', *Cultural Geographies*.

11

Dead Labor and the Political Economy of Landscape – California Living, California Dying

Don Mitchell

> Dialectical thought arises because it is less and less possible to ignore the fact that civilization, in the very act of realizing some human potentials, also damagingly suppresses others. (Eagleton, 2000: 23)

CALIFORNIA LIVING

Sunday 12 May 1991 was one of those days that makes it obvious why everyone wants to live in California. The sun was shining brightly, mid-morning temperatures were in the low seventies. There was a light breeze that rustled the fresh green of the pear and eucalyptus trees in the back garden of my parents' house, where I was staying for a time. The vibrant flowers and lawns, the fragrant eucalyptus trees, and the remnant pear and walnut trees from the orchards that used to fill the valley between the golden-brown and oak-studded hills simply signaled the 'California dream' – or at least the suburban version of it. So out into the sun I went with coffee and the Sunday San Francisco *Examiner and Chronicle*, still then a rather urbane and witty paper with an excellent book review and good arts and music coverage.

Before I got to the reviews, I found buried in the first section of the paper a short article detailing the story of Francisco Bugarin. Bugarin was a 77-year-old strawberry farmer in the Salinas Valley, some 100 miles to the south of where I sat. A few days earlier, Bugarin had pleaded no contest to charges of 'operating an illegal labor camp and maintaining substandard camp buildings'

(San Francisco *Examiner and Chronicle* 12 May 1991). Some 40 temporary workers were housed in a variety of outbuildings on Bugarin's relatively small farm during the strawberry harvest a few months earlier. They lived in barns and sheds, sleeping among machinery, tools, and canisters of pesticides, the very implements they used to tend the fields during the day.

Such housing for strawberry pickers was hardly atypical, and in fact was almost luxurious compared to some that investigators uncovered in the region: in the nearby Monterey Valley during the 1980s, strawberry pickers working for José Ballín were found living in small caves that farmworkers themselves had dug out of the surrounding hillsides (Wells, 1996: 210–14). On another farm, the Monterey County Health Department in 1985 found:

> 50 to 60 farmworkers living in storage sheds, pick-ups, campers, makeshift cardboard and tin shacks, outhouses and truck bodies; no adequate or approved toilet facilities; no potable water from an approved water system; all food preparation areas were substandard; an accumulation of garbage, trash and refuse scattered throughout the complex … sleeping and living areas did not conform with the Uniform Housing Code, California Health and Safety Code and California Administrative Code; human waste was present inside and outside of the various living areas; pesticides, fertilizers and poison baits were stored within the living, sleeping, and cooking areas; open pesticide containers and spilled poison baits were within the living, cooking and sleeping areas; occupants ate their meals while sitting on and around pesticide containers; and hazardous materials, including pesticides were stored on the premises and not registered with the Monterey Country Health Department.[1]

Such conditions clearly are not the result of a lack of laws and regulations. At least since 1913 when the California Legislature passed the Labor Camp Sanitation Act and charged the State Board of Health together with the California Commission of Immigration and Housing with enforcing it, labor camp conditions have been the object of state oversight and regulation.

And yet, in the history and geography of farmworker housing in California, such conditions as those on the farm of Francisco Bugarin are more the norm than the exception (Mitchell, 1996). The living conditions on Bugarin's farm, while clearly illegal, were, by the standards of California agriculture in the early 1990s, actually fairly good. Workers at least had roofs over their heads and access to outhouse toilets. The conditions on Bugarin's farm only led him to jail because the barn in which many workers were sleeping caught fire and four workers were injured, two seriously. In light of the publicity of that fire, local authorities simply could no longer turn a blind eye to the illegal living conditions on Bugarin's farm. Bugarin, unlike the vast majority of his fellow small-farmers, and most of the large ones, was simply unlucky in that his rather typical housing conditions had attracted unwanted attention.

Nor, after Bugarin's plea in May 1991 did housing conditions for farmworkers improve across the state of California. In 1997, a reporter for the *New Yorker* found in San Diego hundreds of migrant workers living in cardboard shacks in brush-choked ravines, 'luckier' workers living in barely converted chicken sheds, and still countless others packing themselves, a dozen to a room, into overcrowded apartments in the cities and towns of the region. Many of the workers the reporter interviewed were Zapotecs who had been recruited during the 1980s to a San Diego flower farm from small villages in Oaxaca, and 'there they were enslaved: held in perpetual debt, frightened into submission by warnings about the Border Patrol, and forced to work sixteen hours a day' (Langerweische, 1998: 139). The flower farmer was eventually charged with slavery and plea-bargained a racketeering conviction. Like Bugarin, and in comparison to his colleagues and competitors, this farmer was simply unlucky. Citation for breaking labor and labor camp laws is exceedingly rare; arrest and conviction even more so. By contrast, debt bondage in California is not at all unusual, and is very rarely prosecuted.

Workers living in such conditions in the 1980s and 1990s toiled in an agricultural landscape that continued to grow increasingly seasonally intensive. Seasonal labor demand in California grew by about 20 per cent over the 1980s and 1990s (Bugarin and Lopez, 1998: 7). In the early 1990s, according to an estimate by the agricultural economist Philip Martin, some 800,000 to 900,000 farmworkers were needed on a seasonal basis to fill the equivalent of 300,000 to 350,000 year-round jobs (cited in Bugarin and Lopez, 1998: 9). During these same two decades, new farm labor contracting systems have been developed that have shifted the burden for compliance in housing, wages, work conditions, and immigration laws from many growers to transient labor contractors (Krissman, 1995; Thilmany and Martin, 1995). On California's 77,000 farms, there could be as many as 154,000 farm employers (since labor contractors are considered individual employers and are thus the 'site' of inspection), and in a typical year somewhere around only one-fifth of 1 per cent of farm workplaces are inspected (Bugarin and Lopez, 1998: 21).

Francisco Bugarin's plea-bargained conviction in 1991, then, was very much a rarity. But the worlds – the landscapes – that conviction represented, like the landscape of flowers, debt peonage and ravine shanties in San Diego, are not at all rare. They are all just as common as my parents' sunshine-filled, pear-tree-shaded backyard in the San Francisco Bay Area. But like my parents' suburban house, with its asphalt-shingle roof, neat painting, and New England colonial styling, and their garden, with its eucalyptus from Australia, its pampas grasses from Argentina, its turf grasses from Eurasia, and its pears from Europe, there is little purely 'natural' in the California agricultural landscape. Like the lawns and ornamental bushes in the subdivision in which I grew up, the California agricultural landscape is highly regulated, both through convention and by law. And, again, like that house in the suburban Bay Area, lovingly landscaped over 30 years, and by 1991 carefully tended by an army of 'landscape professionals' with their lawn mowers, leaf blowers, and gas-powered hedge clippers, the agricultural landscape is a place of work, of toil, of intensive, difficult labor. The California landscape, both in its suburban manifestations and in its agricultural districts, is a *social* more than a *natural* construction, a social more than a natural fact, a place that has been made and remade, that is constantly produced and reproduced, toiled in and toiled over.

And in California, the landscape – both suburban and agricultural – is a product of mobility: the physical mobility of farmworkers as they travel around the state looking for work (perhaps, even, eventually landing behind the handles of the lawn mower that kept my parents' yard so beautiful) no less than the upward mobility of professionals like my parents. It is tempting, therefore, to see the California landscape as fleeting, changing, ephemeral, as, along with Jean Baudrillard (1988), a mere simulacrum, in which *hyper*mobility constructs a 'space of flows' (Castells, 1996) in which 'all that is solid melts into air' (Marx and Engels, 1998: 38). In this view the landscape is never stable, never settled, never fixed.

Yet, as I read about Francisco Bugarin's no-contest plea on that Sunday morning, I knew that the landscapes within which he and his workers lived (and nearly died) were in many respects little different from the landscapes of pre-World War II agricultural California that I was then reconstructing from records in the archives at the Bancroft Library at the University of California. And so the question for me in May 1991, and the question that remains a decade later, is just how we can account for the *persistence* of the California landscape, for the persistence of such obvious injustice and deprivation, for the persistence of an agricultural landscape that is daily as violent as that experienced by Francisco Bugarin's farmworkers in that horrible midnight barn fire, and frequently even more so as workers have struggled, through organizing, strikes, and all manner of other forms of resistance, to change that landscape into something better, something more just. The answer to that question resides not just in a theory of the material production and reproduction of the landscape, which it will be the task of this chapter to develop, but also, more immediately, in the very landscape of suburban comfort that I was then living in when I read about Francisco Bugarin. Or, perhaps even more, the answer could be found, at least in rough outline, in the strawberries themselves that Bugarin's workers were picking that spring, the same strawberries, perhaps, that soon encouraged me to leave the sun of the back garden to go into the kitchen to wash and slice, so that they could dress up the waffles my parents and I would have for Sunday brunch.

DEAD LABOR

The strawberry, like any commodity, 'appears a trivial thing' which 'in reality, is a very queer thing, abounding in metaphysical subtleties and theological niceties' (Marx, 1987: 76) – not that I was thinking in *those* terms when I washed and sliced my strawberries on that Sunday morning in 1991. Rather, I was more concerned with the shape

and flavor of the fruit, its color and texture, none of which, it turns out, are at all 'trivial things'.[2] The strawberries I was working with may well have come from the Central Coast region that Bugarin farmed, since by early May, Central Coast berries dominate the market, not only in northern California, but across the country. Over the course of the post-World War II period, both the size and shape of strawberries, and the cultivation and labor practices that assured their place in my kitchen, have changed drastically. Berries have grown larger, more regular in shape, color, and texture, and less susceptible to spoilage during storage and transport. They have also grown less genetically capable of withstanding pests and disease as cleaner nursery practices and synthetic pesticides have been developed (Wells, 1996: 183). Or, more accurately, strawberry varietals have not so much *grown* as been *produced* to incorporate these characteristics. That is to say, the fruit I was preparing that morning – *its very shape and structure* – was the product of countless hours of labor (both physical and mental), countless years of experimentation, and countless shifts and transformations in the practices of planting, maintaining, and harvesting. While the strawberry may have *appeared* a trivial thing, there is nothing at all trivial about the millions of dollars of research conducted by the University of California, the years of industrial structuring and restructuring, and the untold hours that workers have spent, doubled over, planting and picking strawberries, all so that I could have big, juicy, uniform, cheap strawberries available to me nearly all year long.

The strawberry, as it is being washed and sliced, says nothing of the labor that makes it; it merely appears as just what it is, a complex bio-genetic entity – a berry. Nor does it say anything of the landscape within which it is produced. In California, this is a very complex landscape indeed. It is a landscape of a few large strawberry producers, hundreds of quite small ones, and a deeply concentrated marketing and shipping system. In strawberry growing, plants must be carefully tended, fields painstakingly prepared, and picking and packing closely supervised. Such a need for 'managerial involvement engenders diseconomies of scale', according to Wells (1996: 39). Thus strawberry farms tend to be relatively small. Small Central Coast strawberry farms are amazingly productive – and increasingly so. In California as a whole, per acre yield increased from 3.7 tons in 1946 to 24.2 tons in 1988. In that same period, California's share of the US fresh strawberry market increased from 6 to 74 per cent (Wells, 1996: 29).

Concurrently, fresh strawberry production has become spatially concentrated: California production is now almost completely (99 per cent) contained within coastal counties stretching from San Diego in the south to Santa Cruz in the north. Within those counties almost all strawberries are grown within three miles of the ocean, where temperatures are moderated year round by cool ocean currents (Wells, 1996: 31–2). The farms on which they are grown are as variegated as the coastal topography. Three main ethnic groups farm strawberries in the Central Coast area: Mexicans, Japanese, and 'Anglos'.[3] In the late 1980s, Mexicans were the most recent to move into the rank of farmers, were the most numerous, tended to have farms with the smallest acreage and the most marginal land, utilized the lowest amounts of technology, had the fewest improvements upon the land, picked the fewest crates per acre, and reaped the lowest value per crate of strawberries produced. Japanese, many of whose families began as sharecroppers in the inter-war period and moved back into strawberry production after their release from the World War II concentration camps, tended to have medium-sized farms and were the second most numerous. They tended to pay their workers more, and to provide more benefits, than did their Anglo and Mexican counterparts. While Anglo farmers were the fewest, they had the largest farms and were the most capital intensive. They received the highest return per crate of strawberries picked, and tended to occupy the best farming lands (Wells, 1996: 122–3). The Central Coast strawberry landscape was – and still is – thus a patchwork of farms that differ by size, capital intensity, and quality of production.

This patchwork of land and labor systems is held together by a network of purchasers, shippers, and marketers. In the late 1980s, seven firms (two grower-shippers, one cooperative, and four independent shippers) handled 60 per cent of the total fresh berry crop, and only two firms handled about 30 per cent. Despite deep concentration in marketing, the market remains relatively competitive, in part because demand for fresh berries, particularly in the traditional 'off-season' continues to grow. The growing and marketing of both processed and fresh strawberries are regulated and promoted by two 'advisory boards' established in the wake of market upheavals in the 1950s. These boards work to promote the competitive advantage of the California strawberry industry (including developing advertising campaigns, supporting research into new varietals and cultivation techniques, and regulating the dialectic of competition and cooperation necessary to the success of the industry).

If marketing organizations and the advisory boards hold together the network of strawberry producers, then this network, like the strawberry itself, is made possible only by physical labor. Tending and harvesting strawberries is painstaking, and painful, work. Strawberry growing is one of the most labor intensive of all California crops. As Wells notes, 'the care with which workers select, handle, and pack the fragile fruit is the greatest single determinant of market price' (Wells, 1996: 49). Harvest work is usually paid by piece-rate, but even so, pickers are expected to 'clean' the rows by picking and discarding misshapen, damaged, or spoiled fruit and dead leaves.

Workers call strawberries 'the fruit of the devil'. Strawberry picking and plant maintenance require that workers spend the day doubled over at the waist as they work their way down a row, often standing and stretching only when they reach the end or when they have filled a box. Back injuries are exceedingly common. So too are respiratory ailments from inhaling pesticides and dust (Wells, 1996: 169), severed fingers and hands, and progressively developed allergic reactions to strawberry juice, flowers, and leaves.

Strawberry picking is dangerous, and in that regard is not unlike agricultural work as a whole (which is the second most hazardous occupation in the country: Bugarin and Lopez, 1998: 25). Mortality in farming occupations is five and a half times the national average, and the average lifespan of farmworkers in America is only two-thirds that of other Americans (Bugarin and Lopez, 1998: 25; Myers and Hard, 1995). Acute and chronic injuries are common, and access to health care is spotty at best (Mobed et al., 1992). Danger does not only dwell in the fields. Rather violence, both chronic and episodic and often occurring quite distant from the point of production, is a fact of life, indeed an integral ingredient, of the agricultural economy (Mitchell, 2001). The vast majority of strawberry pickers, like agricultural workers as a whole in California, are foreign-born, and in the 1990s a majority may have been undocumented ('illegal').

During the 1990s, the militarization of the Mexico–US border has made the passage into the United States exceptionally dangerous. In the five years after Operation Gatekeeper – a program of fence-building and stepped-up border patrols in urban San Diego county – was instituted in 1994, deaths along the US–Mexico border in California have increased from around 20 per year to nearly 100 per year (Ellingwood, 1999; Gross, 1999; Nevins, 2001; Smith, 1999). Many of these deaths are from exposure or dehydration as undocumented immigrants seeking to cross the border have been pushed deeper into the mountains and desert. And even further from the field, economic restructuring in Mexico pushes migrants – mostly male – out of their home villages and north to the California fields, at the costs of seasonally and permanently disrupted families, disintegrating social networks, and greater class divisions (Rothenberg, 1998). As we have already seen, living conditions in the fields of California – and in the strawberry fields in particular – are, at best, deplorable. Tuberculosis is common, and poor housing conditions lead to outbreaks of communicable diseases (such as measles, which killed 33 children in two California agricultural counties in 1989–90) to which the larger population is now relatively immune (Bugarin and Lopez, 1998: 26). Lack of sanitary facilities both in the fields and in (formal and informal) labor camps leads to high rates of occupational skin diseases (Mobed et al., 1992: 370), and concerns about the safety of fruits and vegetables as they reach consumers.

So what then *were* those strawberries I was preparing? Like the knife I was using, the kitchen counter I was working at, and the newspaper I had left lying in the backyard, they were, simply, commodities. Or not so simply. As commodities they were the *embodiment* both of labor power and of the social relations that made the application of labor power to fruit growing in the fields of California possible in the first place. That labor power, however, and those social relations, are exceedingly complex. The *labor power* embodied in strawberries includes the labor of researchers at the University of California over the course of the twentieth century, as well as the immediate labor of the man or woman who picked the basket I purchased; it includes the labor of chemical workers who produced the pesticides sprayed on the strawberries (and, before them, the workers who constructed the chemical plant and its components) as well as the labor of the person who placed the berry plant in the ground the previous July or October; it includes, to some degree, the work of marketers, border patrol agents, buyers' agents, and (a very few) inspectors from the California Occupational Health and Safety Administration and the Department of Industrial Relations. The *social relations* embodied in strawberries include the international border that acts as a revolving door that both allows migrants in and pushes them back out again when their labor is no longer needed (Nevins, 2001; Robinson, 1999), the relative immiseration of strawberry workers (real wages have declined perhaps 40 per cent in the 1980s and 1990s), the adoption of sharecropping systems that shift risk from financiers and

landowners to impoverished small-time growers (like Francisco Bugarin), the lack of decent and affordable housing and sanitary facilities in and around the fields, the 'structural adjustment' of the Mexican economy, the successes and failures of farmworker unions in their struggles to organize the fields, and the uneven distribution of power that makes all this possible.

What is crucial, then, are the *conditions* under which labor power is applied, all so that I am able to eat cheap strawberries. My strawberries and Francisco Bugarin's impending jail term were intimately connected. The strawberries *internalized* or *reified* the social relations that made their cultivation and harvesting possible, turning them from a set of ongoing, struggled-over processes into a socially comprehensible (and very tasty) *thing*, a thing with a definite shape and structure, a morphology.

To put that another way, while produced by very much living labor, the commodity is itself *dead labor*, that is, labor ossified, concretized, materialized into a definite thing with a definite shape and a definable structure (which for strawberries would have something to do with the molecular structure of fructose and the cellular structure of the skin, flesh, and seeds). The question, therefore, is always one of *how labor is made dead*, and to what end. Marx argued that in the process of commodity production under capitalism 'dead labor' absorbs 'living labor' (1987: 293–4). The dead labor that is a commodity 'consumes' living labor 'as the ferment necessary for [its] own life-process, and the life-process of capital consists only in its movement as value constantly expanding'. In order for capital to constantly expand through commodity production the equation of surplus value extraction must be regulated so that, over the long haul, and under all manner of unpredictable and contingent historical exigencies, the return of unpaid labor to capital continues to expand (1987: 545–51). In this regard, when a worker in capitalism confronts the world (the world she or he has *made*) as a world defined by 'an immense accumulation of commodities' (1987: 43), she or he is confronting a world of *dead labor* in which her or his own exploitation is manifest – but also highly mystified. For in the world of commodities, 'a definite social relation between men [and women] … assumes, in their eyes, the fantastic form of a relation between things' (1987: 77): not relationships between living laborers, but relationships between labor now dead and, in fact, all but erased. The *materiality* of the commodity is a product of the 'mystical' transformation of living social relations into a tangible thing. Materiality in the world of commodities *is* 'dead labor'.

In California agricultural commodity production, however, this metaphor of dead labor needs to be understood in more than metaphorical terms. As I have already suggested, California agriculture is structured through violence, the violence done to bodies through incessant stoop-labor, pesticide exposure, and horrible living conditions, as well as the violence endemic at the border, in the home villages, in the campaigns to organize farmworkers (and resistance to these by growers), and in the cities and towns that farmworkers live in up and down the agricultural valleys of the state (see also Mitchell, 1996; 2001). When we want to understand how labor is 'made dead' in the fields of California, we need to focus not only on the immediate conditions and practices of production (as Miriam Wells does so well), not only on the technical transformations and innovations that mark the industry (as she also does so well), but also on the forms of violence (and other structuring relationships) that exist both at the point of production and along the whole of the networks through which farmworkers travel, daily and over the life course, to get to those fields (Mitchell, 2001). The strawberries I prepared and ate that morning in 1991 told me nothing about this violence, about those conditions that make labor dead (all too frequently literally). But the article about Francisco Bugarin hinted at it.

That article hinted at something else too. It hinted at the crucial role that the *landscape* plays in establishing the conditions under which labor is made dead in California. Farmworkers, those charged with the care and harvesting of my strawberries, spent their nights on the Bugarin farm, sleeping in barns and other outbuildings; on other farms they lived in little caves carved out of the hillside (Wells, 1996). Such conditions do not exist *despite* the awesome productivity and profitability of California agriculture, but are the very conditions of its existence. To understand this point we need to understand two others. First, the landscape, like the commodity (or rather, because it is a commodity), *is* dead labor. Second, the landscape, again like the commodity, mystifies the relationships that go into its making; it too is 'a very queer thing'. To make these points clear will take some explaining.

THE LANDSCAPE

'Landscape' entered the anglophonic geographical lexicon through Carl Sauer's explication and development of the German concept of *Landschaft*, first published in 1925.[4] For Sauer, following the German tradition, the 'naively given' object of study for geography was the landscape: the 'region' or 'area' and the 'peculiarly geographic association of facts' that defined that region or area (1963: 320). The geographer, for Sauer, was charged with understanding and explaining (not just describing) 'the phenomenology of landscape in order to grasp in all of its meaning and color the varied terrestrial scene' (1963: 320). This is still a vital project if, for example, we want to get 'behind' those strawberries and understand how they are *made*, for the 'terrestrial scene' is integral to contemporary strawberry production, as we have seen. But Sauer was little interested in contemporary political and economic issues. He was instead concerned to create a (historical) landscape geography that was fully objective, and that was shorn of the (naive) political orientation and subjectively induced fallacies of environmental determinism.[5] Yet, even so, Sauer's landscape geography remained normative in important ways. His overriding goal was to use the landscape as a heuristic tool to get at an understanding of the culture that made that landscape. Starting from what he called 'place facts' – the things extant in the landscape – Sauer argued that knowledge about *culture* could be grasped. Sauer's normative model was simple and elegant. He argued that in any region, a 'culture' went to work on the 'natural landscape' and transformed it into a 'cultural landscape'. That is, the cultural landscape – or what I am here calling simply 'the landscape' – was a stretch of humanely transformed nature, but nature transformed to serve a particular end: the needs and desires of the culture that made it. Working backwards from the fact of the cultural landscape, then, the geographer could see how nature was transformed and thus learn something about the culture that lived in and created the landscape: what that culture thought, what it wanted, how it lived. The landscape could be 'read' for clues about culture and cultural change.

How landscape is made

By extension the landscape can be understood to be a product of human labor, of people going to work on the land to make some *thing* out of it. This extension is, in fact, not made by Sauer, since he never sought to unpack 'culture', which in his argument – and much of the empirical work that supported it over the course of his long career – tended to operate as an assumed, or unexamined, 'thing' itself, a reified 'superorganic' entity with rules and logic of its own (Duncan, 1980). Behind that superorganic entity, as behind a commodity, is human labor – the intentional practices and social relationships that *make* it (Mitchell, 1995). To understand landscape, and to understand the 'culture' within which it exists, requires an examination of human practices – of forms of labor. Through labor the landscape is both made and made known. And in the process of working on and in landscape so too are people changed, whether that change is understood at the level of culture (as with Sauer, and as with the mystified world in which I can comfortably eat strawberries without knowing or really caring how I came by them) or at the level of the individual human body (as with the ruined backs of strawberry pickers). Labor then, is crucial, as Marx argues when he describes the fundamental activity that makes humans human – conscious labor:

> [Man] opposes himself to Nature as one of her own forces, setting in motion arms and legs, head and hands, the natural forces of his own body, in order to appropriate Nature's production in a form adapted to his own wants. By thus acting on the external world and changing it, he at the same time changes his own nature. (1987: 173)

And yet, what Marx so clearly recognized, but which Sauer (who anyway was most interested in archaic societies) never confronted, is that in the contemporary capitalist world the labor power that people bring to nature, like the things they produce from nature, is *alienated* – or at least *alienable* – from even those who possess it. 'An immeasurable interval of time,' Marx continues, 'separates the state of things in which a man brings his labour-power to market for sale as a commodity, from the state in which labour was still in its first instinctive stage' (1987: 173–4).[6] Or, to put that another way, *pace* Sauer, human *culture* never does anything: human 'culture' does not go to work on nature; people working under specific historical and geographical conditions do.

Those specific historical and geographical conditions – those conditions under which people oppose themselves to nature as one of its own forces – are themselves the product of past work, or past labor: they are the 'dead labor' that makes *specific* forms of living labor, specific labor practices, historically and geographically necessary. In these terms, landscape, as a focus and object of human labor, is a form of dead labor. Or as David Harvey has put it, the landscape should be 'regarded ... as a geographically ordered, complex, composite commodity' that, unlike a strawberry, is more or less permanently 'fixed' in place (1982: 233). The fixed landscape thus 'functions as a vast, humanly created resource system, comprising use values embedded in the physical landscape, which can be utilized for production, exchange and consumption'.

What landscape is for

The invocation of 'use values' is important. The 'use value' of the landscape is, as Harvey indicates, threefold. First, it is an instrument of production (as with the strawberry fields, packing sheds, and drip irrigation systems that comprise the visual scene of the Central Coast strawberry region). Second, the landscape serves as 'instruments of consumption' (Harvey, 1982: 229). Such instruments of consumption – which can be 'as diverse as cutlery and kitchen utensils, refrigerators, television sets and washing machines, houses, and the various means of collective consumption such as parks and walkways' – make it possible for labor to be reproduced (Mitchell, 1994). As Marx argued, such reproduction always possesses a definite 'historical and moral element' which often *appears* as a set of 'natural' and 'necessary wants' but which is, in fact, a product of past social struggle – dead labor (1987: 168). Finally, the landscape is important to exchange, both as fixed spaces through which capital, other commodities, and labor circulate (Henderson, 1999), and as an alienable 'product' (as property) which itself can be exchanged (Blomley, 1998). To put this last point another way, while the landscape as product and property can itself be exchanged, the material landscape also establishes the conditions under which circulation – a fundamental necessity of capitalism – can take place. The landscape both has, and is necessary for the production of, exchange value.

As Henderson (1999), drawing on and developing the influential 'Mann–Dickinson' thesis,[7] has so insightfully argued, processes of circulation are critical to capitalist production, and particularly to capitalist agricultural production. Drawing particularly on the second volume of *Capital*, Henderson shows that capitalist circulation must be understood as a series of 'barriers and interruptions to capital' itself (1999: 35). If we understand the most basic form of capitalist circulation to be the movement of money-capital into the production process and back out again in the form of commodities to be sold in the market for more money than the cost of their constituent parts and the labor applied in the process of making them – Marx's famous M-C-M – then it should be obvious that this process of circulation is anything but smooth. Capital is 'frozen in place', in the machinery used to produce commodities, in the land that machinery rests on, in the bodies of workers who apply labor to that machinery (and in the bodies of their families), in the houses and consumer goods that allow the laborer to reproduce her- or himself, and in the commodities themselves. Capital only begets more capital if it circulates (Harvey, 1982). But capital can only circulate if some portion of capital does not. At each stop on the path of circulation, at each moment capital is frozen, at each location where, in fact, labor is dead, risks accrue: commodity prices could tumble, new innovations by competitors could make production practices obsolete, labor could strike, finance capitalists could pull out and seek greater returns elsewhere, and in agriculture, pests could invade.

Crucially, as Mann-Dickinson posits, and as Henderson (1999: Chapter 2) explains so well, capitalist production is not *one* process of circulation, but many. Capital circulates in and out of workers' hands in the form of wages and commodities purchased. It circulates in and out of financial markets. It circulates in and out of land and machinery. Marx makes a distinction between 'work time' (the time that labor is actually applied to the production process), 'production time' (the time that capital is wrapped up in producing a commodity), and 'circulation time' (the time involved in getting commodities to market, selling them, and receiving payment in return) (Henderson, 1999: 35). The various circulation times – work, production, and circulation itself – organized in the production process (up to and including marketing) are highly uneven. They do not mesh easily, or often well.

This is especially the case in agriculture, where production time is highly discontinuous (crops, after all, take time to mature in the fields, and that lengthened time is a time of great risk) and therefore so too is labor time. Laborers' needs and desires, however, are not so discontinuous. The reproduction of agricultural labor power, therefore, is, in many if not all respects, qualitatively different from the reproduction of labor power in other more continuous industries. Since planting, weeding, and other non-harvest jobs often take considerably less labor than the harvest itself, agricultural labor is highly seasonal. Farmworkers are needed in huge numbers for often relatively short periods of time, but after that time, when little or no work is available, they become a burden, more a hindrance to capital circulation than its enabler. Solutions to this problem of uneven circulation times are many, but include the retention of family and peasant labor systems, the expansion of sharecropping (which shifts both risks and reproduction costs from financiers to poor, small-time farmers), and, particularly in California, *increased*

circulation – of laborers – by making agriculture both more seasonally intensive and more regionally diverse.

That is to say, the solution to uneven agricultural circulation times in the California landscape is, and long has been, in large part increased mobility, not so much on the part of capital (though that has been important), but crucially on the part of migratory workers, who have found migratory circuits both expanded and speeded up.[8] Immobilized workers become an unacceptable barrier to the circulation of capital. The solution is to set workers in motion, to mobilize them, to keep them moving (as we will soon see in more detail). One of the things the landscape is *for*, then, is the establishment of patterns of circulation, patterns of production and reproduction, or, most simply in California, patterns of crops and labor that are *profitable*. But that merely begs the question: just what *is* the landscape?

What landscape is

The landscape is a concretization or reification of the social relations that go into its making. It is the phenomenal form of the social processes and practices of production, consumption, and exchange, as complex as those may be. In theoretical terms, the landscape is a 'structured permanence', which Harvey, drawing on Whitehead (1925), defines as a relatively stable, solid 'thing … that we daily encounter in the world and without which physical and biological life would not and could not exist as we know it' (1996: 50). However, such 'permanences' may not be, indeed usually are not, naturally *necessary*. Rather, permanences, ranging from the institutional apparatus of the state to a child's red playground ball, are historically contingent, developed to solve some problem or fill some need (trumped up or not). The landscape is, in these terms, a 'complex moment in a system of social reproduction' (Mitchell, 1996: 35) – the social reproduction of capital and the social reproduction of people. 'Moment' is an important term because it indicates that any structured permanence remains permanent only to the degree that it is continually reproduced, and hence any moment can become a site of struggle. Indeed it *is*, by definition, a site of struggle, since it is an *internalization* (and concretization) of social relations (Harvey, 1996; Ollman, 1990). The question, then, is how these permanences, these complexes of social relations (like the landscape), are given form and sustained over time. How (by what social processes and struggles) does the 'varied terrestrial scene' come to be?

How is the labor that is embodied in it made dead? This is not an easy question to answer, because, like the commodities that comprise it, the landscape is fetishized and mystified, hiding the very processes that go into its making. Indeed, that is often precisely its point.

What landscape does

The best way to answer that question of how a landscape is sustained, therefore, is to change the question: if the landscape *is* a built, alienated, and fetishized *form* constructed through the labor of people under conditions established by the struggles of labor already dead, then what does the landscape *do*?[9] One of the things the landscape does is, as has already been indicated, to provide a stage upon which capital circulates (in and out of crops, for example), and upon and within which laborers (and their families) reproduce themselves. The landscape is the site for the production and reproduction of social life. Its homes, shops, roads, factories, and farms, its fields, forests, valleys, and ditch-banks, is *where* life is lived. But as Schein has suggested, the landscape functions 'as Bourdieu's (1977: 82, 79) *habitus*: history turned into nature through an amnesia of genesis' (1997: 663). Landscape naturalizes social relations and makes them seem inevitable. More accurately, landscape *reifies* social relations and creates not so much an 'amnesia of genesis' as a socially powerful *thing*, which in order to be transformed has to be taken apart, thrown into disarray, literally disintegrated as a landscape (Mitchell, 1996: Chapter 6). In this regard the landscape functions not only as a stage upon which life is lived, upon which the reproduction of capital and society occurs. It also functions as *reality*, as that which *is* (and hence, to some large degree, that which can be). That is to say, one function of the landscape is to display the normative order of the world.

Consider the case of José Ballín, whose workers lived in caves. 'As shocking as outsiders found these living conditions,' Miriam Wells reports,

neither the grower nor his workers found them particularly unusual … Not only did Ballín himself live in circumstances that many U.S.-born growers would have found unacceptable, but he saw his workers' shelter as evidence of his generosity. Ballín saw himself as his workers' *patrón*, and he offered many benefits in addition to wages. He let workers live on his land without charge, he made small loans to sustain some between paychecks, and – in response to phone calls from the border – he paid the *coyote* for several reliable returning workers and deducted it from their first paychecks. (1996: 212–13)

In addition, social workers and activists found it difficult to organize workers to press charges against Ballín, not only because conditions were not unusual, but also because of fear of deportation by many workers. In this regard, as Schein notes, the landscape must be understood 'as an articulated moment in networks that stretch across space'(1997: 663), networks that include the border and the border patrol, labor camps and hometowns, fields and caves (a point to which I will return). The landscape thus functions to make out of the extraordinary and contingent (caves for living in, debt peonage) that which is ordinary and necessary – and good (free housing, help in crossing the border).

The landscape, when uncontested, thus functions to establish the conditions under which surplus value is extracted. As I have previously argued (Mitchell, 1994: 13), the landscape defines what is 'natural' or 'rational' in a given place and in so doing materially affects the surplus value equation in a region. To the degree that the landscape is uncontested, to the degree that labor unrest can be stilled because of the sense that there simply is no alternative, then surplus value can be expanded. Landscape is thus a form of social regulation. The structured permanence that is the landscape both shapes and regulates social contest (helping to determine what is possible and what is not, displaying what is ordinary and expected, and what is not) at the same time as it is shaped through and regulated by social contest. This is the political economy of landscape.

What a landscape means

This point, however, leads to yet another question it is necessary to ask of the landscape. That question is the one of what a landscape *means*. What does it mean for a landscape to comprise fields and caves and barns used for sleeping, and what does it mean for that landscape to exist cheek-by-jowl with the sort of upper-middle-class suburban landscape like where I grew up? Here matters get quite complex. While Anglo-American geography borrowed its notion of landscape most directly from the German *Landschaft*, the history of the *idea* of landscape, as Cosgrove (1998) has detailed so brilliantly, is much deeper and more complicated than the geographical etymology lets on. With its roots in Renaissance Italy (among other places), 'landscape' denotes a particular 'way of seeing' (to use John Berger's felicitous phrase) wrapped up in a particular relationship to land understood as property. That is to say, while landscape as an areal association or assemblage of *things on the land* is a moment in processes of production and social reproduction, as an *ideology* it is a particular means of organizing and experiencing the *visual order* of those things on the land (Baker and Biger, 1992; Berger, 1971; Cosgrove, 1985; 1998; Daniels, 1993; Rose, 1993; Williams, 1973). Landscape is seen and experienced from outside – from beyond the limits of the landscape – as when one looks out and down upon the view from a viewpoint along a highway. In Cosgrove's words, 'in landscape we are offered an important element of personal *control* over the external world' (1998: 18). We can, to some degree, control the view, order it to emphasize what we find important or appealing. The 'insider's' view, thus, is not a landscape view: '"place" seems a more appropriate term' for such existential inside-ness (1998: 19).[10]

If there is an 'element of personal control', however, that element is itself deeply conditioned by the historical relationship of the landscape way of seeing to the commodification of land – to the transformation of land and place into *property*. Distance is achieved through alienation. To the degree that land can be set apart, alienated, and made exchangeable for other like and unlike pieces of land, then to that degree the distance can be achieved that allows land to be seen as landscape, as that which is 'out there' – as a separable *thing* which, in its specific morphology, can be viewed as a totality. In turn, this transformation of land into landscape, into something to be viewed, presupposes a specific division of labor in which the owner does not perform the labor that turns his land into a landscape; it presupposes a capitalist division of labor. The landscape is a specific way of seeing, a form of what Cosgrove (1998) calls a 'visual ideology', but it is so because the worked land that is landscape is a commodity. Like the strawberry, the look of the landscape reveals little, directly, about how it was made. Thus, *contra* Sauer, it is simply impossible to read 'culture' out of the landscape, at least not without first doing the hard work of determining just what sets of social relations constitute that 'culture' in the first place. The landscape mystifies.

This is the sense in which Raymond Williams (1973: 32–4) remarks that 'landscape' erases work, and the existence of workers. He does not mean that landscapes are unworked. Rather he means that the work that goes into making the landscape is both hidden from view and alienated from those who performed it (Daniels, 1989). Either that, or landscape serves to aestheticize work, to make it picturesque, which, of course, has the parallel effect of erasing not work itself, but the conditions under which work is done

(Williams, 1973: 56). Based on a technology of linear perspective that places the viewer in the 'divine' position of having the whole of the view structured and ordered for him (Cosgrove, 1985),[11] the landscape serves as a 'realistic' view that 'is in fact ideological …. Subjectivity is rendered the property of the artist and the viewer – those who control the landscape – not those belonging to it' (Cosgrove, 1998: 26). What is depicted and what is seen – what is understood to be meaningful – therefore, is determined from a point external to the landscape, and in the realm of (property-based) ideology: what Williams calls 'the explicit detached view of landscape' (1973: 56). As such, the *meaning* of the landscape, while constrained both by the 'place facts' that constitute the morphological landscape, and by the historically developed currents of ideology within which the landscape is viewed, is itself, therefore, a site of social contest. The meaning of the landscape is neither given nor ever stable; it is struggled over, its very 'naturalness' and 'realism' continually contested, its components continually reordered by interested social actors in their efforts to *use* the view to show the world new things, different aspects, better ways of seeing.

What do the strawberry fields of the Central Coast and my suburban hometown farther north therefore *mean*? That's an open question, the resolution of which is determined by and determines the way that social life – and the relations of production and reproduction – are played out over space and time. The meaning of landscape, in other words, is also a (contested) part of the political economy, and one of my goals in this chapter is to expose and contest, if only in a very small way, what the strawberry growing and suburban landscapes of California mean, to show how what look like pretty ordinary landscapes of agricultural production and suburban consumption are in fact complex, and inextricably linked, *places* defined by a geography of *injustice* that allowed me to live an easy California life only *because* others' labor – other dead labor – had made that life possible, even as the landscapes of production and consumption fetishize and mystify those relationships. All that is to say, the *meaning* of the landscape is a function of who has the power to *represent* that landscape.[12]

Where the landscape is

To make even that small claim, however, requires that I transform the meaning of landscape – or expand it in some particular ways. One of the standard definitions of landscape is a stretch of scenery seen from a single vantage point, as befitting an ideology wrapped up in the visual technology of perspective. The natural tendency, therefore, and one only encouraged by the long history of landscape as a specifically *visual* ideology, is to look for the facts of a landscape's production within the locally bounded landscape itself. That was the impetus behind Sauer's assumption that evidence of the 'culture' that made a landscape could be located within the 'place facts' of the extant landscape; it is also, arguably, the single defining characteristic that links together the whole of landscape studies in geography and related fields. The transformation and extension that needs to be made – and the very reason I have sought to show how a landscape both *is like* and *is* a commodity – is one of scale. Just as the social relations of production that make a commodity like the strawberry possible are to be found not only within the strawberry field itself, but also in the labor camps nearby, the boardrooms in the cities, the state capitol, the research laboratories of various University of California campuses, the produce markets and grocery stores across the country, the border (and hence Washington DC) and the villages, cities, and towns that workers come from in Mexico and Central America, so too must we look to these places to understand how the landscape we see in California has come to be, and, specifically, what it means for those 'belonging to it'. That is, to understand the landscape in all its material and ideological importance, we need to find better ways to link the strawberry fields to my parents' backyard. And we need to do so in a way that is sensitive to the complex material processes that operate not just locally, but across and within a wide variety of scales. We need to understand the landscape as constituted through processes that construct 'structured permanances' ranging from the bodies of workers, through the local 'place facts' and on to the regional, national and global economies and social relations within which it is embedded. And we need to show how landscapes in different places are linked together through the moving bodies of workers (and consumers), the moving flows of capital (and commodities), and the moving targets of regional, national, and global policy (and its violent enforcement). We need to show, that is, how the landscapes of Central Coast strawberry pickers and upper-middle-class suburbanites are not separate entities, but part of the same system, a system that may be highly differentiated and fully contradictory, but a system nonetheless. We need to understand that while the landscape is always physically *somewhere*, it is also socially constituted both there and *elsewhere*. We need to slice open the

landscape, just like I sliced open the strawberries, to see what it embodies, what it internalizes – and to locate the other places to which it is linked, to see *where* (else) labor is made dead so that it can continue to be.

CALIFORNIA DYING

When it is 'sliced open', the anatomy – the morphology – of the California landscape might best be described as being constituted by a series of 'points of passage' within a 'network of violence' (Mitchell, 2001). By violence, I do not mean only the metaphorical violence of 'dead labor' as Marx talked about it – the labor ossified in the commodity. I also mean real, bodily, physical violence: farmworkers' hands mangled by machinery; the gun and knife fights in the cities and labor camps that are often part of farmworkers' everyday lives; the remarkable violence visited upon migrants as they attempt to cross the border (rape, assault, and murder, as well as death by exposure in the deserts and mountains); and the violence implicit and explicit in economic dislocation, threatened starvation, and disrupted families and local ways of life in the source countries and villages of California farmworkers. Landscape – as a site or stage of production and reproduction – is knitted together by this network of violence. Landscape – as a 'way of seeing' that aestheticizes or erases the facts and relations of work – knits together this network of violence. That we are dealing with a *network* is made plain by circulations – of money, energy, workers – that make commodity production possible (Henderson, 1999).

Instead of seeing landscape as a localized 'thing' – the view from a single vantage point – landscape needs to be understood as a complex node or point of passage in a network (cf. Schein, 1997). From the perspective of migratory *workers* in California, these points of passage include the place of production, the labor camps and farm-country cities, the border, the various cities and villages of Mexico as workers trek north, and the home region or village. From the perspective of *capital*, the strawberry field is but one point through which capital circulates. Others are banks, farm implement factories, workers' bodies, and supermarkets. From the perspective of the agricultural production *system*, points of passage might be the soil and plants of the farm itself, the broader context of micro and regional climate, the regional marketing structures, the state research and taxing mechanisms and laws concerning labor and pesticide use, the national-scale laws and political economy, and global markets in which strawberries are increasingly sold. While all these different points of passage may appear external to the landscape, they are in fact what constitutes its *internal relations*, in the sense that Bertell Ollman (1990) develops that term in his study of Marxist dialectics (see also Harvey, 1996: Chapter 2).

Violence is a key to the landscape – and to the points of passage – for a simple reason. It is what made (and makes) my strawberries so cheap – and the back garden of my parents' house so inviting. More accurately, violence is what makes *labor* so cheap, and *thus* makes strawberries cheap. The age-old question wrapped up in the development of intensive seasonal and specialty agriculture in California is the one of labor: where it comes from, how to get more, how to get rid of it when it is no longer needed. California farms are capitalized – and have long been capitalized – on the assumption of a steady supply of cheap, often expendable labor (Fuller, 1939). At its inception in California, intensive farming was the child of violence (the violence that cheapens labor) – and investors know that (Daniel, 1981). Cheap labor – and more of it – is the mantra of California agriculture. But cheap labor is no natural commodity: it must itself be *made*, be conditioned and served up to growers when and where they need it, and, ideally, for no longer than they need it.

And so labor organizing is ruthlessly resisted and laborers are kept moving into and around the state by the threat of starvation or vigilante violence.[13] Newcomers, especially those from Mexico and points south, have to negotiate the increasingly militarized and dangerous border, a militarization and degree of danger that make it prohibitively expensive for workers to cross again if they are 'returned' to Mexico or Guatemala by the Immigration and Naturalization Service. The effect of *that* threat of continued violence is obvious: INS policies both make workers reluctant to report illegal work and living conditions to the state (and to seek medical care for injuries and illnesses sustained at work), and make workers do all they can to keep from getting caught, which has the combined effect of increasing worker powerlessness (Heyman, 1998). To produce the California landscape – the landscape that allows for strawberries to be grown so cheaply – requires that labor be made dead at a range of linked places stretching from the point of production across the globe.

But how, then, is that landscape of violence, that landscape constituted through a complex geography of violence, connected to my relative ease as I sit in the sun in the town I grew up in?

That town – Moraga is its name[14] – had once been the site of extensive pear and walnut orchards (pruned and picked, in their day, by armies of migratory workers). Owing to its proximity to San Francisco, Oakland, Berkeley, and the growing residential and commercial suburbs of Contra Costa County, Moraga was subdivided and developed as an upper-middle-class 'bedroom community' in the 1960s and 1970s.[15] The town exudes prosperous California suburbia. A limited mix of housing styles (colonial, hacienda, ranch) are set within maturing gardens, which in the early years were mostly tended by young families, but by the 1980s were more and more tended by teams of 'landscapers'. By the early 1990s, Moraga was happy to promote its lack of racial diversity (one publication of the time pointed with pride to the fact that it was more than 95 per cent non-Hispanic white) and the fact that, among incorporated towns and cities in California, it had the lowest crime rate. Even with a fairly severe recession in the California economy in the first half of the 1990s, houses were often selling for a half-million dollars or more.

For a kid like myself, Moraga was in many ways a great place to grow up (though by the time we became teenagers, its main virtue was being within an easy drive of Berkeley and San Francisco). We had the run of the yards and streets – and the surrounding hills, too, which were protected ranchlands that comprised a part of the local water district. The schools were excellent and the opportunities such schools opened up for us were vast. Moraga was the *promise* of California – or at least one version of it. It was a town of prosperous professionals and what used to be called petty bourgeoisie, whose idyll was the old idyll of comfortable living promised by the California dream. It was the suburban form of what Kevin Starr (1973), California's 'official historian', describes as an achievable Eden of middle-class respectability. Moraga was – and is – a world away from the fields and labor camps of Salinas and Monterey counties.

That is the genius of landscape. For, of course, Moraga is *not* a world away: it is only the flip side of Salinas and Monterey. Listen to how Kevin Starr describes the promise of the California landscape, as represented in the commodities it produced, in the first decades of the twentieth century:

> In the color of a plum or an apricot, in the luxuriance of a bowl of grapes set out in ritual display, in a bottle of wine, the soil and sunshine of California reached millions of Americans for whom the distant place would henceforth be envisioned as a sun-graced land resplendent with the goodness of the fruitful earth. (1985: 128)

The promise of California was that that 'fruitful earth' was easily within reach of the middle classes including all those 'easterners' hoping to make a new life out of the rural/suburban idyll at the edge of the continent. But such a dream is the deception of landscape.[16] For Starr, the development of 'fruit culture' out of the ruins of the worldwide nineteenth-century wheat boom

> nurtured the values of responsible land use, prudent capitalization, cooperation among growers in the matter of packing, shipping and marketing. Above all else, fruit culture encouraged rural civility in the care of homes, the founding of schools, churches and libraries, the nurturing of social and recreational amenities which stood in complete contrast to the Wild West attitude of wheat. (1985: 34)

This is not quite right. 'Fruit culture', as we have seen with the case of strawberries, did none of these things. Or, more accurately, the degree to which it did them was only by *separating* places like Moraga (where such things existed) and Salinas (where they did not, at least for strawberry pickers) in the realm of ideology, while at the same time closely *connecting* them in the realm of economic interdependence. The whole point of the landscape is to forestall the dialectical thinking necessary to uncover these connections, and to show, as Eagleton (2000: 23) puts it in the quotation with which I opened this chapter, that civilization (Starr's 'civility'), in allowing for human achievement and comfort, only does so by suppressing the dreams and aspirations, the very life chances, of countless others.

Or to turn that around, Moraga stands as *the proof that out of violence and repression comes a good* – the good of comfortable suburban living for so many. The difference between the patrician landowners Cosgrove studied, who oversaw the very invention of landscape so that they could live at their ease (encouraging rural civility), and my own family was only one of proximity, in both geographical and division of labor terms, to the point of production. Our ease has been spatially removed from the strawberry workers' repression, so now even the markers of it do not really need to be hidden; they can be obscured in plain sight as the suburban landscape tells us that our comfort never has to be interrupted by their bodily destruction. The landscape integrates by dividing, by separating, and by obscuring. And by distance: by placing oppression not just behind the well-tended copse of a manor house, but hundreds of miles away – or even across the globe. Yet Moraga is a landscape connected to the agricultural landscape in another sense. It is now actively cultivated and maintained by contingent workers, most of them Mexican or of

Mexican descent, and many now 'settled out' of the migratory stream that brought them into the US and the fields of California. It is possible that some of those who mowed and edged my parents' lawn – the lawn I set my chair on that morning in May 1991 – had been or would later become strawberry pickers. Or if not that, then harvesters and tenders of other fruits and vegetables, janitors in the nearby office parks and professional buildings that allowed Moraga to be such a successful bedroom community, or dishwashers in the newly trendy restaurants of nearby Lafayette and Walnut Creek. The Moraga landscape is *made possible* by many of the same people (or same types of people) as the strawberry landscape. Moraga is one of the points of passage and networks of violence through which immigrant workers pass as they knit together the California landscape.

But more than that, the Moraga landscape is made possible in another, perhaps less direct sense, by the networks of violence and points of passage that constitute the strawberry pickers' landscapes. Cheap fruit and vegetables, cheap staples too, in part make possible the hyperinflated house prices of the Bay Area economy. That is to say, the reproduction costs of families like mine (and thus the ability of profit to be made from *our* labor) are held down by cheap food costs. Our good California living is subsidized by the daily California dying (the violent repression of organizing, the dead labor integral to commodity production) that marks the California agricultural landscape.

CONCLUSION: DEAD LABOR AND THE POLITICAL ECONOMY OF LANDSCAPE

How, then, should 'the landscape' be conceptualized within geography? What is its ontological and epistemological status within our systems of knowing the world? Ontologically, landscape is best conceptualized first (though not exclusively) as *dead labor*. That is to say, the landscape is a social product, it is something made. And it is constantly remade. But to say the landscape is *constantly* remade is not to imply that it is always and fully in a state of flux. On the contrary, the landscape must also be understood as a 'structured permanence'. As Harvey develops Whitehead's philosophy of 'permanences', a permanence is not only the 'practically indestructible' object described earlier, but also, and crucially, a 'system of "extensive connection"' in which 'entities achieve relative stability in their bounding and their internal ordering of processes creating space, for a time' (1996: 261). The two key points here are, first, that a permanence is constructed out of *extensive connections*. That is, scale is important because processes operating at a variety of scales are what structure the permanence. But second, it is the permanence that *internalizes* these extensive connections, gives them shape and form, and turns them into a (relatively) stable thing, a thing that resides in the world and becomes an actor in ongoing social relations. These extensive connections include struggles over meaning as much as they include physical struggles over the conditions under which, for example, migratory farmworkers live. Part of what the landscape internalizes is what people make it mean. Intellectual labor as much as physical labor is made dead in and by the landscape. And crucially, *as* dead labor, as the *physical* embodiment of reified social relations, the landscape is a fetish. Or more accurately, the landscape necessarily, as part of its very being, fetishizes the labored over social relations that make it.

Epistemologically, then, the key to understanding (and therefore transforming) the landscape revolves around first understanding how – and especially where, by whom, and under what conditions – it is made. Second, the key question for landscape is not, therefore, so much why and how it is always changing (though this is crucial), but why and how it has *remained*. Just what is 'permanent' about it? What are the effects of that permanence? How is it maintained in the face of all those working to transform it? What is fetishized in the landscape and why? These are the questions we need always to ask of the landscape. But – and this is the genius of landscape as a way of seeing – they are the ones we hardly ever do ask.

A decade after that spring Sunday morning in California, I now live in upstate New York, and I am writing these words on a day, in its gentle warmth, not unlike that earlier one. Strawberries were selling this week at a nearby market for 99 cents a pint, a ridiculously cheap price. According to the crate at the store, the berries were picked on the Central Coast, and they are beautiful. My parents have since moved to the mountains of Virginia, and from the profit they made in selling their Moraga home (and a little perk in the tax code for the upper classes that allows capital gains taxes to be avoided on gifts to family members), my wife and I were able to purchase first a small condominium in Boulder, Colorado, and then a much larger old house in Syracuse, New York, where we now live. Instead of pear and eucalyptus, our yard is shaded by a big old

black walnut. The flowers are not California poppies, but the peonies so well suited to this wetter, colder climate. Down the road from our house, closer than the strawberry fields were to my parents' old place, is an incredibly productive, and incredibly beautiful, apple growing region. Every fall armies of migrant laborers come to the area to pick the ripe fruit. I know nothing about them. I know nothing about who they are, where they come from, how much they are paid, where they live both during and after the season. The apples we eat tell us nothing about these things, and the views from the roads and highways yield few clues either. The fruit pickers' lives might as well be a world away from my own. The landscape of my back garden, where the irises are in spectacular bloom, and a crew of landscapers have just built us a new, raised vegetable garden, is placid, quiet, bucolic. But so too was my parents' garden in Moraga.

NOTES

1 Action no. 31891, Municipal Court of California, County of Monterey, Salinas Judicial District, State of California, 16 December 1985, 7, quoted in Wells (1996: 211–12).
2 My account of strawberry production practices, labor relations, and the Salinas–Monterey landscape in this section draws heavily on Miriam Wells' impressive ethnography, *Strawberry Fields* (1996).
3 'Anglo' is the local term used for Americans of western and northern European origin.
4 That such is the case is a shame. Better that Anglo-American geography would have looked to the related old Dutch *Landskab* which denoted a form of regionally based social justice that stood against the hierarchical juridical forms of European feudalism (Olwig, 1996). *Landschaft*'s more restricted meaning – as a morphological area stripped of its normative implications – helped promote and institutionalize a remarkably depoliticized landscape geography, the 'old' cultural geography concerned with fenceposts and barn types that is so much the object of caricature by 'new' cultural geographers.
5 Sauer (1963) admitted, toward the end of his long essay, that aspects of the landscape were 'subjective', and that these aspects were important. But he also argued, in the very title of the section in which he makes this admission, that such subjectivity was 'beyond science'.
6 This passage, which goes on to note the differences between instinctive animal labor and conscious human labor, is usefully deconstructed in Harvey (2000).
7 The Mann–Dickinson thesis explains why seemingly 'archaic' labor practices, like peasant and family farming or sharecropping, remain important – and often expand – within capitalist agricultural production. The history of other industries would lead one to expect a

capitalist rationalization of labor (replete with detailed divisions of labor, new patterns of skilling and deskilling, etc.). In agriculture, however, older labor practices seem to be retained much longer, not so much as 'residuals' but as the very meat and marrow of the production process. Henderson (1999) develops a convincing theory of risk (to, for example, finance capital) to explain the persistence of these labor forms. See Mann (1990), Mann and Dickinson (1978), Wells (1996).

8   And often made more complex as workers move from agricultural work to suburban landscaping to restaurant work, or to whatever pickup work is available in the day-labor markets that have developed across the state.

9   As David Matless writes, 'the question of what a landscape "is" or "means" can always be subsumed under the question of how it works' (1998: 12).

10 Which is not to say that non-elites do not develop landscape sensibilities: of course they do.

11 I say 'him' because the landscape way of seeing is both historically and ideologically, in many respects, masculine (Rose, 1993; though see Nash, 1996).

12 The literature on the politics of landscape representation is now large. Two key works are Barnes and Duncan (1992) and Duncan and Ley (1993).

13 This fact of California agricultural production has been well documented (McWilliams, 1971; Daniel, 1981; Majka and Majka, 1982; Mitchell, 1996).

14 Named after an early Spanish colonial land-baron in California, Joaquin Moraga.

15 Moraga developed as a transitional community between the 'small-owner republic' of immediate post-war suburbia that Walker (1995) describes, and the monster-home suburbia of the 1990s.

16 Stephen Daniels (1989) has written of the ways that the landscape is 'duplicitous'. Such a duplicity does not imply that people are obviously 'dupes' of the landscape. Rather, it implies that part of the work – the socially intentional work – that landscape does is to hide consequences and connections from us, to fetishize.

REFERENCES

Baker, A. and Biger, G. (eds) (1992) *Ideology and Landscape in Historical Perspective*. Cambridge: Cambridge University Press.

Barnes, T. and Duncan, J. (eds) (1992) *Writing Worlds: Discourse, Text, and Metaphor in the Representation of Landscapes*. London: Routledge.

Baudrillard, J. (1988) *America*. London: Verso.

Berger, J. (1971) *Ways of Seeing*. London: Penguin.

Blomley, N. (1998) 'Landscapes of property', *Law and Society Review* 32: 567–612.

Bourdieu, P. (1977) *Outline of a Theory of Practice*. Cambridge: Cambridge University Press.

Bugarin, A. and Lopez, E. (1998) *Farmworkers in California*. Sacramento: California Research Bureau, California State Library, CRB-98-007.

Castells, M. (1996) *The Network Society*. Oxford: Blackwell.

Cosgove, D. (1985) 'Prospect, perspective, and the evolution of the landscape idea', *Transactions of the Institute of British Geographers* 10: 45–62.

Cosgrove, D. (1998) *Social Formation and Symbolic Landscape* (1984). Madison: University of Wisconsin Press.

Daniel, C. (1981) *Bitter Harvest: A History of California Farmworkers, 1870–1941*. Ithaca: Cornell University Press.

Daniels, S. (1989) 'Marxism, culture, and the duplicity of landscape', in R. Peet and N. Thrift (eds) *New Models in Geography*, vol 2. London: Unwin Hyman. pp. 196–220.

Daniels, S. (1993) *Fields of Vision: Landscape Imagery and National Identity in England and the United States*. Princeton: Princeton University Press.

Duncan, J. (1980) 'The superorganic in American cultural geography', *Annals of the Association of American Geographers* 70: 181–98.

Duncan, J. and Ley, D. (1993) *Place/Culture/Representation*. New York: Routledge.

Eagleton, T. (2000) *The Idea of Culture*. Oxford: Blackwell.

Ellingwood, K. (1999) 'Data on border arrests raise Gatekeeper debate', *Los Angeles Times* 1 October: A3.

Fuller, V. (1939) 'The supply of agricultural labor as a factor in the evolution of farm organization in California'. United States Senate, Subcommittee of the Committee on Education and Labor, Hearings on S. Res. 266, *Violations of Free Speech and the Rights of Labor*, Part 54, Exhibit 8762-A, pp. 19,777–19,898. Washington: Government Printing Office.

Gross, G. (1999) '5-year-old Gatekeeper is praised, denounced', *San Diego Union* 31 October: B1.

Harvey, D. (1982) *The Limits to Capital*. Chicago: University of Chicago Press.

Harvey, D. (1996) *Justice, Nature, and the Geography of Difference*. Oxford: Blackwell.

Harvey, D. (2000) *Spaces of Hope*. Berkeley: University of California Press.

Henderson, G. (1999) *California and the Fictions of Capital*. New York: Oxford University Press.

Heyman, J. (1998) 'State effects on labor exploitation: the INS and undocumented immigrants at the Mexico–United States border', *Critique of Anthropology* 18: 157–80.

Krissman, F. (1995) 'Farm labor contractors: the processors of new immigrant labor from Mexico for California agribusiness', *Agriculture and Human Values* 12: 18–46.

Langerweische, W. (1998) 'Invisible men', *New Yorker* 23 February and 2 March: 141.

Majka, L. and Majka, T. (1982) *Farmworkers, Agribusiness, and the State*. Philadelphia: Temple University Press.

Mann, S. (1990) *Agrarian Capitalism in Theory and Practice*. Chapel Hill: University of North Carolina Press.

Mann, S. and Dickinson, J. (1978) 'Obstacles to the development of capitalist agriculture', *Journal of Peasant Studies* 5: 466–81.

Marx, K. (1987) *Capital*, vol. 1. New York: International Publishers.

Marx, K. and Engels, F. (1998) *The Communist Manifesto*. London: Verso.

Matless, D. (1998) *Landscape and Englishness*. London: Reaktion.

McWilliams, C. (1971) *Factories in the Field* (1939). Santa Barbara: Peregrine Smith.

Mitchell, D. (1994) 'Landscape and surplus value: the making of the ordinary in Brentwood, California', *Environment and Planning D: Society and Space* 12: 7–30.

Mitchell, D. (1995) 'There's no such thing as culture: towards a reconceptualization of the idea of culture in geography', *Transactions of the Institute of British Geographers* 20: 102–16.

Mitchell, D. (1996) *The Lie of the Land: Migrant Workers and the California Landscape*. Minneapolis: University of Minnesota Press.

Mitchell, D. (2001) 'The devil's arm: points of passage, networks of violence, and the California agricultural landscape', *New Formations* 43: 44–60.

Mobed, K., Gold, E. and Schenker, M. (1992) 'Occupational health problems among migrant and seasonal farmworkers', *The Western Journal of Medicine* 157: 367–73.

Myers, J. and Hard, D. (1995) 'Work-related fatalities in the agricultural production and services sectors, 1980–1989', *American Journal of Industrial Medicine* 27: 51–63.

Nash, C. (1996) 'Reclaiming vision: looking at landscape and the body', *Gender, Place and Culture* 3: 149–69.

Nevins, J. (2001) *Operation Gatekeeper: The Rise of the 'Illegal Alien' and the Remaking of the U.S.–Mexico Boundary*. New York: Routledge.

Ollman, B. (1990) *Dialectical Investigations*. New York: Routledge.

Olwig, K. (1996) 'Recovering the substantive nature of landscape', *Annals of the Association of American Geographers* 86: 630–53.

Robinson, V. (1999) 'The social legitimation of state violence: stigmatization of migrant workers and the homeless in the United States'. Unpublished paper, Department of Geography, Rutgers University.

Rose, G. (1993) *Feminism and Geography: The Limits of Geographical Knowledge*. Minneapolis: University of Minnesota Press.

Rothenberg, D. (1998) *With These Hands: The Hidden World of Migrant Farmworkers Today*. New York: Harcourt Brace.

Sauer C. (1963) 'The morphology of landscape' (1925), in J. Leighly (ed.) *Land and Life: A Selection of the Writings of Carl Ortwin Sauer*. Berkeley: University of California Press. pp. 315–50.

Schein, R. (1997) 'The place of landscape: a conceptual framework for interpreting an American scene', *Annals of the Association of American Geographers* 87: 660–80.

Smith, C. (1999) 'Condemning migrant job seekers to death', *San Diego Union* 6 April: B7.

Starr, K. (1973) *Americans and the California Dream, 1850–1915*. New York: Oxford University Press.

Starr, K. (1985) *Inventing the Dream: California through the Progressive Era*. New York: Oxford University Press.

Thilmany, D. and Martin, P. (1995) 'Farm labor contractors play new roles in agriculture', *California Agriculture* 49 (5): 37–40.

Walker, R. (1995) 'Landscape and city life: four ecologies of residence in the San Francisco Bay Area', *Ecumene* 2: 33–64.

Wells, M. (1996) *Strawberry Fields: Politics, Class, and Work in California Agriculture*. Ithaca: Cornell University Press.

Whitehead, A. (1925) *Science and the Modern World*. New York: Macmillan.

Williams, R. (1973) *The Country and the City*. New York: Oxford University Press.

12

Landscape and the European Sense of Sight – Eyeing Nature

Denis Cosgrove

Landscape has a complex history as an organizing and analytical concept within cultural geography. Its usage has varied from reference to the tangible, measurable ensemble of material forms in a given geographical area, to the representation of those forms in various media such as paintings, texts, photographs or performances, to the desired, remembered and somatic spaces of the imagination and the senses. The complicated and contested evolution of landscape within twentieth-century geography and its complex relations with concepts such as place, region and area have been well documented, and will not be further rehearsed here (see Cosgrove, 1985; Daniels, 1989; Duncan and Duncan, 1988; Olwig, 1996). Here, I shall focus on a feature of landscape which cuts consistently across all its modern usages: landscape's connections with seeing and the sense of sight. Landscape, together with its cognates in other European languages, is by no means confined to visible topography. Indeed, the connections between the morphology of a territorially bounded region, and the identity of a community whose social reproduction is tied to usufruct rights and obligations over that area, lie at the root of the German *Landschaft* and its derivatives (Olwig, 1996). But there is a profound connection, forged over half a millennium, between the modern usage of landscape to denote a bounded geographical space and the exercise of sight or vision as a principal means of associating that space with human concerns. This usage undoubtedly relates to shifting modes of social appropriation and use of space, involving individual property rights and more atomistic constructions of self and identity (Cosgrove, 1998; Hirsch and

O'Hanlon, 1995). If geography is a discipline that examines relations between modes of human occupance and the natural and constructed spaces that humans appropriate and construct, then landscape serves to focus attention on the visual and visible aspects of those relations.

To connect the geographical landscape with the sense of sight is not to deny the significance both of other human senses and of rational cognition in shaping space, territory and meaning. Geographies of visually impaired people, for example, alert us to the significance of other senses, as well as fantasy, memory and desire, in shaping the relations between humans and the spaces of the material world (Porteous, 1990). Smell or sound can be much more powerful and immediate than sight in shaping emotional responses to a specific place. In the realms of dreams or memories, mood tends to dominate over somatic apprehension, and it may be difficult to recall or describe with precision the visual characteristics of the spaces encountered or experienced therein (Bell, 1997; Bishop, 1994; Park, 1994). Given the significance of imagined spaces and geographies for the ways that individual and collective worlds are actually shaped, recent theoretical dethroning of vision's primacy in western intellectual culture is not insignificant (de Certeau, 1988; Deutsche, 1991; Haraway, 1991; Rose, 1993; Women and Geography Study Group of the IBG, 1984). Western rationalism's tendency to align vision with knowledge and reason, captured in the common usage of the phrase 'I see' to imply both the physical act of vision and the cognitive act of reasoned understanding, has been attacked as a characteristic feature of modernity. In treating mind and body as distinct aspects of being, vision becomes the principal channel through which intellectual reason and the 'reason' or order of the sensible world can be mapped onto one another: the eye is rendered the window to a rational soul. These assumptions can be traced back to Aristotle and have been reinforced in scholastic, Cartesian and Enlightenment thought. Feminist and poststructuralist critics have challenged such dualistic thinking as masculinist, patriarchal and Eurocentric and have pointed both to the significance of non-visual forms of knowing and to the culturally encompassed nature of seeing itself (Howett, 1997; Merchant, 1990; Rose, 1997). This has forced a reassessment of the cultural primacy of vision, and with it a critical rethinking of the connections between landscape, geography and the sense of sight.

Neither space nor vision is a conceptually simple matter. Geographical spaces long remained framed and defined by the coordinates of Euclidean geometry, itself historically associated with studies of the physics of light (Kemp, 1990; Lefebvre, 1991). So long as geographical space remained absolute, conceptually rooted in the measurable materiality of a physical 'environment' external to the human body, the purest geographical landscapes were those outlined theoretically in spatial science. Such landscapes materialize collective, rational human action influenced by the frictional effects of distance, or describe empirically the ecological outcome of human occupation in delimited physical regions. 'Löschan landscapes' of hierarchical nodal points and polygonal territories exemplify the former; the palimpsest of cultural landscapes within the cadaster of the lower Mississippi Valley is a fine example of the latter (Bunge, 1966; Corner and MacIean, 1996). But geographical study today embraces various expressions of relative space, defined by culturally diverse coordinates of human experience and intention. Similarly, sight, vision and seeing – as such varied words imply – involve much more than a simple sense response: the passive, neutral imprint of images formed by light on the retina of the eye. Human sight is individually intentional and culturally conditioned. The lover sees only the loveliness of the loved; the city dweller from temperate latitudes is blind to the rich variety of snow surfaces that make up the polar landscapes inhabited by Inuit speakers (Sonnenfeld, 1994). Furthermore, sight in the modern world is increasingly prosthetic, directed and experienced through a vast array of mechanical aids to vision which radically extend the capacities of the unassisted eye: lenses, cameras, light projectors, screens and scopes. Central to my argument will be the coevolution in the modern west of spatial experience and conception and of the techniques and meanings of seeing. Cultural landscape may be regarded as one of the principal geographical expressions of this coevolution, whose critical examination is a current preoccupation within cultural geography.

To trace the connections between landscape and the western geographical imagination, I adopt here a broadly historical approach. Recognizing the west's long privileging of the sense of sight, I examine modes of vision (sight, gaze, insight, vision) and trace their connections to the different ways that space is apprehended, such as surface and depth, proximity and distance. I show how landscape images construct as well as reflect the geographical expression of individual and social identities. This reveals associations between landscape and such identifiers as gender, class, ethnicity and age. I consider too the territorial expression of social identities in landscape, exploring property, military, nationalist, imperial and colonial relations with land and its representations, for example in maps and paintings. Throughout, I emphasize that the evolution of landscape meanings in the west is as much a story of changing technologies of perception (cameras, lenses, film and screens) and modes of representation (perspective and colour theories) as of unmediated visual relations between the human viewer and material space.

VISION AND LANDSCAPE

Much of the revived interest among cultural geographers in recent decades has come from the simple but profound recognition that seeing is a culturally encompassed activity. We *learn* to see through the communicative agency of words and pictures, and such ways of seeing become 'natural' to us. But geographical dislocation or cultural change can disrupt the taken-for-grantedness of seeing, opening a space for more critical reflection on what is seen. The intellectual agenda of early-twentieth-century landscape study in cultural geography was set by concern for the urban-industrial erosion of what many took to be 'natural' relations between localized human communities and the physical environments in which they lived and worked. To many, those relationships seemed manifest in apparently immemorial patterns beyond city and railroad. A human 'ecology' was figured, for example in the triad 'place–work–folk' inherited by Patrick Geddes from Frederick Le Play (Matless, 1992; 1997). Whether in French Picardy or English

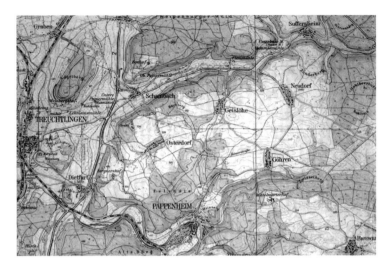

Figure 12.1   *Detail from 1:50,000 German topographic map showing settlement patterns*

Rutland, Virginia's Shenandoah Valley or Danish Jutland, the stability of relations that tied community unreflectively to a delimited territory seemed apparent in the visible forms of farms and villages, fields and fences, meadows and woods – in the 'morphology' of landscape whose usage remained close to *Landschaft*'s original Germanic usage (Cosgrove et al., 1996; Geipel, 1978; Olwig, 1984). Whether the label was landscape or *pays* or region, a principal concern of cultural geography became to describe and account for the ensemble of physical and human forms as they appeared in the field or on the topographic map, whose coloured and contoured image, scaled between 1:25,000 and 1:100,000, offered a synoptic impression of morphological permanence (Figure 12.1). Implicitly, the geographer's sovereign eye should disentangle the natural and cultural forces that had brought together land and life, offering a unique perspective to social science and history.

By the 1970s it was clear that across western Europe and North America, and increasingly the rest of the globe, the processes of modernization that had disconcerted earlier cultural geographers were continuous and unstoppable. In erasing the social processes that sustained 'cultural landscapes', modernity had destroyed their apparent naturalness. Even in France, for whose geographers the *tableau* of rural *pays* and peasant communities had been the visible expression of the nation's soul (Claval, 1995: 22–7), mass migration to the cities had left whole regions of emptied villages, abandoned farmhouses, enlarged fields and an ageing rural population. Within a generation these same landscapes

would be refilling with summer residents: the children and grandchildren of successful migrants to Paris and Lyon, English and German second-homers, romantic neoruralists and telecottagers (Matless, 1994). The arrival of such communities introduced new aesthetic and environmental pressures for preserving the visible elements of a redundant geographical order. Even in cities, the acceleration of 'creative destruction', whereby capital fixed in the form of an urban landscape of buildings and communications infrastructure is released for reinvestment and reconstruction, was removing any sense of ecological stability reflected in built form (Harvey, 1989: 4–124). These economic, demographic and cultural shifts were readily apparent across Europe and North America and increasingly elsewhere. The questions surrounding cultural landscapes in the late twentieth century reflected new ways of seeing as much as new ways of being, less to do with settled ecologies of land and life and more with environmental protection and aesthetics. The evolution of landscape study in cultural geography towards a critical examination of ways of seeing should be understood in this context.

From the late 1960s demands to identify and conserve 'landscape values' led geographers to examine how different individuals and social groups perceive the same rural or urban scene (Penning-Rowsell and Lowenthal, 1986). 'Amenity', to adopt a British term, was a well-honed political weapon within a politics of scenic preservation that elided visual with social order (Gruffudd, 1995; Matless, 1998; 1999). Landscape values denoted the aesthetic and

Figure 12.2   *Giorgione, 'Sleeping Venus', c. 1508* (Dresden Gemäldegaterie)

environmental concerns of a largely urbanized population for rural spaces known chiefly through recreational visits, sometimes coinciding and sometimes colliding with exclusionary property interests (Lorimer, 2000). 'Scientific' attempts to measure such values foundered on the realization that to speak of 'the same scene' itself presupposed an ability to produce a disinterested and objective visual image against which different perceptions of witnesses could be measured. Responses of sampled observers were measured to: viewing from a vantage-point, following an itinerary, examining photograph, map or film footage of a place or area. This generated useful information, for example about differences between residents and visitors, variously aged, gendered or ethnically ascribed people, and about apparently recurring recognition of significant physical features such as barriers, landmarks or pathways (Gold, 1980; Lynch, 1970). But surrogates for landscape were complex images, and measuring behavioural responses to 'real' landscapes presupposed the existence of shared assumptions about what made a 'landscape' in the first place, ignoring complex cultural and political associations between landscape and social relations. The briefest consideration of the difference in common usage of landscape between American and British English makes the point. For most Americans landscape is wild nature where the evidence of human presence is minimized, and preferably non-existent (Jackson, 1984: 9–56). In British English, landscape is very distinctly humanized, its garden-like qualities forming a significant criterion for aesthetic judgement (Daniels, 1993: 146–242). Such differences are the outcome of very different social relations with the land, expressed in property rights and

land ownership, class formation, and histories of settlement and resource exploitation (Cosgrove, 1998). But American and British landscapes do share a common status as seen objects, evaluated in large measure by conventions established in relation to pictorial images.

The term 'ways of seeing', coined by the art critic John Berger in 1969, neatly captured an idea long recognized among art historians that, in so far as it is meaningful, seeing is a learned ability (Berger, 1972). While there is no denying that sight is a physiological function, not present in all people and varying around 20/20 vision even among the sighted, the use of this function is learned. The child's attention is constantly directed in the process of establishing visual connections and giving names to specific groups of seen objects while ignoring others. These groups vary both culturally and according to more biologically related differences such as gender and age. The most elementary drawing lesson forces this recognition by reminding us of the limitations of learned sight, when one is asked, for example to trace the defining line of a seen object rather than drawing what 'you think you see'. Use of the sense of sight is shaped as much by images seen in the past, by individual experiences, memories and intentions, as by the physical forms and material spaces before our eyes. While it is obvious that much of learned seeing is personal, much too is social, governed by conventions about what may be seen, by whom, when and in what context, about the associations and meanings attributed to a given scene, and about its formal and compositional properties.

An example will clarify these cultural aspects of seeing. An early-sixteenth-century oil painting by the Venetian painter Giorgione depicts a nubile young woman lying naked in a 'landscape'

(Figure 12.2). This landscape is 'pastoral', composed of meadow grass and herbage beneath a tree, stretching beyond the foreground figure into a blue distance, with sheep and shepherd, and framed by blue mountain forms, visible against an evening sky. It draws upon a set of topographic associations traceable to ancient Greek and Roman poetry (Cafritz et al., 1988; Cosgrove, 1993: 222–251; Jenkins, 1998). The image has been copied and parodied by artists since its original painting. Giorgione is often regarded as a pioneer of the genre of secular landscape in western art. Beyond the obvious point that not all cultures would make the immediate connection between a surface pattern of variously pigmented oils on a canvas (or, in this reproduction, of ink dots on white paper) and a human female lying in an evening glade, my reproduction of Giorgione's image registers and activates whole sets of cultural responses in you, the viewer/reader. If my text made no reference to the image, if it were replaced by a colour photographic snapshot of an actual such scene, if the young woman were known to you, if the image was used to illustrate an erotic or sacred narrative, if you are a devout Muslim woman, or a 13-year-old American schoolboy – in every case the meaning of the image would be transformed according to altered conventions of seeing.

The Giorgione painting alerts us to the powerful connections between seeing and space. To make sense of the painting we must accept certain conventions for representing the external world on a flat surface. Among these are the three-dimensional depth of the space represented within the frame and its lateral extension beyond the frame, the perspective rules whereby smaller elements are assumed to be more distant, and aerial perspective conventions whereby more indistinct and blue-toned elements are assumed to be further away. I shall consider these representational conventions in more detail later. Beyond them we may consider the subject matter of this image. The presence of a nude human figure in an open field of grass upsets conventional connections between seeing and space. Powerful cultural norms confine human nakedness almost exclusively to private space, and define such privacy as removal from sight. What may be seen, by whom, and where, are among the most fundamental and contested cultural considerations in shaping social space. Geographies of what may be seen are generally more highly regulated that those of what may be heard, smelled, felt or tasted. The power that conventions of visibility exercise over location in the case of nudity is illustrated by how most of us might feel holding a business conversation over the telephone while naked. Barriers to vision and, conversely, vision's 'penetration' of space are significant determinants of material landscape. Such cultural conventions, activated by this image, have been subject to critical examination in recent cultural geography.

Language captures something of the rich cultural complexity of seeing. A glance is different from a stare, a sight is different from a vision. In considering the active use of the sense of sight most languages make a fundamental distinction between seeing and looking (in French *voir/regarder*, in Italian *videre/guardare*). The former suggests the passive and physical act of registering the external world by eye; the latter implies an intentional directing of the eyes towards an object of interest. In English, viewing implies a more sustained and disinterested use of the sense of sight; while witnessing suggests that the experience of seeing is being recorded with the intention of its verification or subsequent communication. Gazing entails a sustained act of seeing in which emotion is stirred in some way, while staring holds a similar meaning but conveys a sense of query or judgement on the part of the starer. Such complexity in the ordinary language of seeing suggests something of its cultural significance in our relations with the external world, both with physical objects and with other people. As the dual meaning of 'I see' indicates, connections between seeing and cognition are similarly complex. 'Insight' captures the human capacity to 'see' more than is immediately visible to the eye, the idea that humans may be capable of penetrating beyond physically registered surface to invisible meaning. 'Vision' is at once the physiological function and an imaginative capacity in which non-material phenomena are somehow witnessed.

Vision's connections with imagination suggest further cultural complexities to the sense of sight and acts of seeing. Imagination is the capacity to fashion images that have not previously existed in the material world of their maker. Imagination works with the raw materials of experience (it has none other available) to create and fashion new phenomena. Imagination is thus closely linked to human art and it finds expression in the realm of each of the senses: in heard music, tasted cuisine, bodily movements, smelled perfumes and graphic representations which appeal to the eye. The unique emotional power of visual images has always generated anxiety, prompting social control of their production and effects, from Plato's censure of painted images to religious iconoclasm, to secular concerns about pornography and violence on film. Social regulation points to a powerful connection between the sense of sight and the physical conduct of the body, between virtual and material worlds. We do not fully understand the nature of that connection, but it lies at the heart of a consistent historical attempt in western culture to bring the visual image and the material world into ever closer union.

PICTURING LANDSCAPE

Geographically, the idea of landscape is the most significant expression of the historical attempt to bring together visual image and material world; indeed it is in large measure an outcome of that process. The etymological roots of landscape lie in substantive connections between a human collective (denoted by the suffixes -schaft, -ship, -scape) and its common or usufruct rights over the natural resources of a bounded area (land) as recognized in customary law. But from its late-sixteenth-century appearance in English, such usage has always been subordinated to that of landscape as an area of land visible to the eye from a vantage-point (Cosgrove, 1998: 189–222; Helgerson, 1992; Turner, 1979). The vantage-point might be a high place, a hill or tower from which a 'prospect' might be enjoyed; it could be provided or enhanced by an instrument such as a looking glass or binoculars; it could be the medium of a drawing, painting, map or film (Charlesworth, 1999; Nuti, 1999). In every case, location serves to disengage the viewer physically from the witnessed geographical space. And, as 'vantage-point' denotes, landscape establishes a relationship of dominance and subordination between differently located viewer and object of vision (Appleton, 1996: 22–5). The vantage-point privileges the viewer of landscape in selecting, framing, composing what is seen; in other words, the viewer exercises an imaginative power in turning material space into landscape.

Implicit in this process is the idea of an aesthetic response to what is seen. 'Aesthetic' has two meanings: the neutral one of sense impression (still registered in its opposite, 'anaesthetic') and a more evaluative one of sensual pleasure and beauty. The sense impression is of a disembodied eye registering the formal and compositional qualities of a surface subtended to its gaze. The relationship between such sense impressions and the faculty of imagination was the subject of eighteenth- and nineteenth-century philosophical distinctions between sublime, beautiful and picturesque landscapes (Ballantyne, 1997; Barrell, 1980). Relationships between the landscape and the viewer are thus doubly distanciated, first by the physical distances between observation point and surface, and second by the separation of eye (body) and imagination (mind). This distanciation however also yields a power relationship privileging viewer over viewed. The authority offered to the viewer of landscape may be actual and material, as in the case of those English landowners of the eighteenth century who reorganized the fields, hedgerows, coppices and buildings on their estates to correspond to aesthetic conventions of landscape: 'pleasing prospects' (Daniels, 1999; Muir, 1999). More often today it is exercised over the choice of landscapes we experience and the judgements we make of them as tourists, hikers, photographers, moviegoers or visitors to galleries: 'enjoying landscape'.

Landscape thus denotes primarily geography as it is seen, imaged and imagined. This does not imply that landscape is intellectually superficial or insignificant as an object of geographical study and reflection, although vision obviously privileges surface and form over depth and process. But sight and action are intimately connected. A dramatic example is the designation, bounding and management of geographical areas as 'national parks'. This began in the United States at the turn of the twentieth century when certain visually dramatic, forested areas of the Rocky Mountains and western sierras caught the attention of dedicated naturalists. Many of these were initially drawn to these areas by their rendering in paintings, dioramas and photographs and were able to access them comfortably by means of newly built railroads (Morin, 1998; Novak, 1980). Today, such protected landscapes have been designated within almost every nation-state and the principal has been applied to the entire continent of Antarctica. Although concern for the continued existence of their flora and fauna has always been a powerful motivating force in the selection and designation of these areas, it is their visual appearance as landscape that has conventionally sustained their public appeal. Such areas have become sites of complex relations and sometimes contestation between scientific, social and aesthetic concerns surrounding issues of managing 'nature', accessing space and coding performance (Cosgrove, 1995; Grove, 1995; Neumann, 1995; 1998). Political implications are signalled in the designation of such zones as 'parks', a term whose entire history denotes the aesthetic appropriation of natural spaces: for hunting, recreation, pleasure. A majority of citizens may never have visited these landscapes, but know and treasure them through pictorial images. The American national park, for example, is known to many through the

Figure 12.3 *Ansel Adams, 'View of Yosemite'* (Ansel Adams Publishing Rights Trust)

purified surfaces of Ansel Adams' highly aesthetic photographs of western wilderness (Schama, 1995) (Figure 12.3). Connections between image and action and between the material and the imaginative in landscape have evolved historically in close connection with changing technologies of seeing and representing space.

TECHNOLOGIES OF VISION AND LANDSCAPE

Modern ideas and experience of landscape have evolved in close association not only with changes in the ownership and use of land but with technologies of seeing and representing space. In the economically progressive, urbanized regions of late medieval Europe – northern Italy, southern Germany, Flanders – the revival of population, trade and urban culture after the Black Death of 1350 saw the spread of new forms of rural property and production. Increasingly, urban capital and urban authority flowed between urban centres into their hinterlands, initiating processes of social and economic change that continue today. Urban investment progressively turned agriculture from a localized, largely self-sufficient and collective way of life into an industry through which capital mobilizes land and labour to yield profits. New modes of

exploiting nature and those who worked it demanded new ways of knowing and representing the natural world, locally and eventually across the globe. An example of this is the demand for accurate measure and recording of productive natural spaces for the purposes of establishing ownership and control within a land market. Thus we see the invention and use from the fifteenth century of surveying techniques including handbooks, instruments for measuring distance, angles, heights and areas, and the appearance of estate and cadastral maps (Figure 12.4). All these were in common use across Europe by the mid sixteenth century and would be refined and deployed over succeeding centuries to redraw the boundaries of enclosed estate lands, to drain and improve whole regions such as the Po Basin in Italy, the Netherlands, the Vendée in France or the English Fenlands, and to appropriate and colonize lands in overseas regions newly discovered to Europeans (Cosgrove, 1993: 123–35; Mariage, 1998).

From about 1500, in these regions' urban centres, for example in Nuremberg, Antwerp, Venice and Florence, merchants, scholars and craftsmen produced instruments, maps and pictures to regulate and celebrate the wealth, power and beauty of their native cities and regions. Jacopo de Barbari's bird's-eye view of Venice or Rosselli's map of Florence in which the artist includes himself seated on the hills at Fiesole (Figure 12.5), the coloured image of

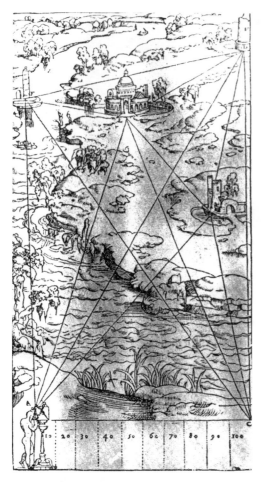

Figure 12.4 *Sixteenth-century engraving of 'The Surveyor'* (from E.G.R. Taylor, *Mathematical Practitioners*)

Nuremberg's towers and spires proudly set in open fields and royal forest, all date from this time (Nuti, 1999; Schulz, 1978; Söderström, 2000). They initiated a long tradition of celebratory urban landscape views and maps. The same merchants and patricians who thus celebrated their cities also commissioned 'chorographies' or detailed descriptions of their local regions, drawn and painted to offer an immediate visual impression of the lands in which their capital was invested. And they purchased 'cosmographies' to decorate their walls: small, jewel-like paintings, which offered panoramic scenes over the vast horizons beyond which their merchandise moved (Wood, 1993: 45–50). Painted images of city and country themselves offered opportunities for investing and displaying wealth and national pride: they required costly materials and great skill on the part of artists to produce (Alpers, 1984). The popularity of these painted scenes of nature, land

and urban space, given the new name of 'landscapes', spread rapidly in the sixteenth and seventeenth centuries, especially in Holland, England and Lombardy, the European regions of most rapid advance in capitalist forms of land tenure. From 1600 the invention and rapid development of lens technology in microscopes and telescopes opened new spaces to human vision and were enthusiastically embraced as aids to painting and mapping of space (Kemp, 1990). Technology served to enhance that identification between empirical observation, mathematical reasoning and knowledge which we call the scientific revolution. In the academic hierarchy of the fine arts, landscape long remained culturally inferior to oil paintings of sacred or historical events and portraiture. The taste for landscape was primarily bourgeois, and by the nineteenth century its practice, especially in water-colour, had become itself a mark of middle-class cultivation.

To make pictorial images of landscape requires compositional and drafting skills, including perspective, the ability to produce realistic appearances of three-dimensional space on a two-dimensional surface. Effective perspective learning demands a grasp of similar principals of geometry to those required for transforming physical nature: in architecture, water management, land survey, mapping, exploration and trade. To achieve realistic effects, artists have consistently embraced mechanical means such as the camera obscura, lenses and prisms, mirrors and polished surfaces, photographic equipment and plates, film and video. In landscape, the skills and techniques of the surveyor, the mapmaker, the planner and the artist overlap and have often been practised by the same individuals. This was especially true in the case of military mapping and art. The European states emerging from the processes of modernization initiated in the cities were large-scale territorial units whose survival and prosperity depended upon effective defence and effective administration of the realm (Heffernan, 1998: 170). Armies and navies undertook the first of these tasks, requiring detailed knowledge of topography and coastlines to perform their roles. Warfare has always been a principal stimulus to technological development, not only of the means of violence but of surveillance, strategic planning and battlefield operation. Military topographers – often recognized artists – trained officers in the skills of drawing, mapping and recognizing landscapes, while naval officers and ratings learned to sketch coastlines by eye and from memory (Martins, 1999).

The intimate connections between gathering and classifying spatial information, its accurate

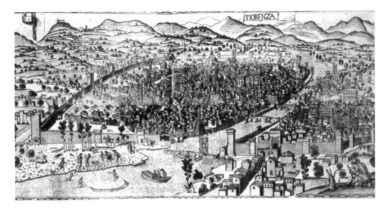

Figure 12.5 *Map with the Chain, woodcut map of Florence after Francesco Rossellini, c. 1500* (Staatliche Museen zu Berlin)

scaling and mapping, and its representation in visually realistic images have been continuously refined through the mechanization of vision. Technology has increased the significance of vision as the principal means of experiencing space. Among the most significant advances have been photography in the nineteenth century and powered flight in the early twentieth century. Photography's invention and development were intimately connected with making stage-sets, panoramas and moving dioramas – dramatic painted landscapes which were illuminated by spectacular chemical effects such as limelight. And the pictorial conventions of landscape painting were rapidly applied to its photography and later moving film. Powered flight further distanced the viewer from the land surface while offering to the observer the ability to view landscape at the scale and angle conventionally associated with the map. Invention of the airborne, automatic camera allowed World War I pilots to film extended strips of land, transforming local mapping and landscape appreciation (Cosgrove, 2001). It paralleled Hollywood's use of the movie camera to play landscapes of the American West as principal actors in a modernist epic of struggles over land and life. The aesthetic conventions of landscape have been continuously reinforced by developments in mechanical and prosthetic vision, which today dominates much of our waking lives through television, video, film and advertising images. Mechanically produced views and images of space as landscape have further evolved into satellite pictures, remote sensed images, interactive simulations and other advanced graphic technologies, offering strategists, planners and the private citizen a privileged eye across the globe's topography, scaled, located and manipulated virtually

at will. Visual reality (VR), used by military strategists, in CAD applications by architects and planners, in movies and personal entertainment packages, allows complex imaginary topographies and landscape morphologies to be constructed in the virtual space 'behind' the computer screen surface, and offers the illusion of entry and frictionless movement through these spaces (Hillis, 1994). The eye alone traverses VR landscapes, and while connected techniques allow other bodily senses to be stimulated, the physical action of the body's limbs is more or less eliminated in favour of a purely aesthetic experience.

MAKING THE PICTURED LANDSCAPE

The evolving relationship between vision, technology and landscape is not a morally or politically neutral affair. Privileging vision as the principal means of knowing the world devalues alternative modes of experience and cognition, and when the object of knowledge involves nature, as in the case of landscape, this privileging and devaluation extends broadly: geographically, socially and environmentally. Further, images do not merely represent a prior reality, they are powerful agents in shaping that reality (Cosgrove and Daniels, 1988; Mitchell, 1994). Thus, as the mechanization of vision has helped individuals look at actual scenes with eyes trained by pictorial images, so the patterns and forms of the external world have been altered to correspond to the conventions of pictorial landscape. I shall examine this process before turning to some of the consequences of landscape vision and action.

Among the most dramatic examples of how pictorial images of nature have influenced the ways that actual spaces are encountered is the early-eighteenth-century example of the Claude glass. Young aristocrats and the sons of newly rich merchants and financiers in Protestant Europe, sent to complete their education in the Classical *topoi* and urbane society of the south, found painted images of the Roman Campagna or of Mediterranean arcadia appealed to a taste shaped by Latin literature and imagined geographies of the exotic (Bermingham, 1986; Pugh, 1988; Said, 1993; Schama, 1995: 453–62). Landscape paintings by the artist Claude Lorraine became fashionable markers of wealth and status in polite English society, for example, commanding high prices for originals and spawning a host of imitations. Northern scenes too came to be framed, composed and illuminated by soft Mediterranean light. Painted images came to shape the vision of actual landscapes with the invention of a convex, circular instrument, the Claude glass, a highly polished copper surface through which actual views could be framed and tinted to resemble painted ones. The instrument's use required the viewer to face away from the scene, privileging the eye and distancing it from material nature as effectively as any movie screen or TV monitor.

While the fashion for the Claude glass was short-lived, its association with the transformation of physical spaces in conformity with pictorial tastes has been much more long-lasting. The desire to manipulate and reshape the natural world according to an image of perfection is widespread, and its fulfilment by individuals of wealth and power through the design of gardens is recorded historically in most civilizations (Hunt, 2000; Warnke, 1994). But there is a difference between the garden – conventionally a walled, fenced or otherwise bounded space whose sensual pleasures are as much to do with smell, sound, touch and taste as with vision – and designed *landscape*. 'Landscape gardening' involved removing the visual boundary between the spaces of recreation and production. The landscape garden is a pictorial illusion founded upon the capitalist property rights and investment strategies mentioned above. It commonly involved expunging pre-existing collective land rights and identity (denoted in the earliest meaning of 'landscape') in favour of 'picturesque' alterations to the land determined by aesthetic choices. Shaping actual landscapes according to pictorial images has been a foundation of landscape architecture. Humphry Repton's trademark approach to redesigning English landed estates involved coloured sketches with pasted flaps that could be turned to demonstrate the visual effects of his improvements (Daniels, 1999). Changing styles in landscape architecture and design have consistently paralleled those in the visual arts, and only in recent years has landscape architecture begun to examine critically the implications of its connections with seeing, paralleling in its attention to the ecological, social and political implications of its site selection and design the radical concerns of cultural geographers (Corner, 1999).

As the modern world witnesses the continuous substitution of usufruct by private property rights and regulation of land uses by state agencies, and as agriculture demands ever-smaller inputs of direct human labour, so the conscious manipulation of nature as landscape extends over wider geographical surfaces. Taste and fashion, formed in large measure by pictorial conventions, continue to be significant factors in shaping landscape, just as they are in shaping other consumption choices and in framing social identities. Because landscape is constituted from the everyday world in which we live, its naturalizing effects remain particularly powerful in obscuring the often unequal social relations which it expresses (Daniels, 1989). Yet the same social processes that remove people from direct dependence on productive land allow greater recognition of the inequalities and exclusions inscribed into landscape, and thus increased resistance to its regulatory effects. Landscape is increasingly recognized as a continuing process rather than a finished form. As process, nature is 'produced' and 'consumed' and its meanings 'naturalized' in landscape, especially through its association with vision as guarantor of truth.

LANDSCAPE AND SOCIAL PROCESS

Treating landscape as a process in which social relations and the natural world are mutually constituted in the formation of visible scenes, lived spaces and regulated territories democratizes and politicizes what would otherwise be a neutral and descriptive exploration of physical and cultural morphologies. It thus introduces into landscape study questions of identity formation, expression, performance and even conflict. These have been examined through the solidarities of class and ethnicity and through the ascription and experience of gender difference.

Landscape and class

It is currently unfashionable to use class as a significant social category in landscape study,

Figure 12.6   *Alex MacLean, 'California Agricultural Landscape'* (Rizzoli International Publications Inc.)

partly because of Marxism's tendency to reduce all culture to class consciousness (Cosgrove, 1998; Mitchell, 1996). But the history of western landscape that I have sketched commonly saw aesthetic design and appeals to 'nature' deployed to veil dramatic social inequality. In eighteenth-century England, for example, the association between expunging common rights over land and natural resources and physically expropriating communities on the one hand, and creating the pleasing vistas of landscape parks on the other was widely acknowledged at the time and has been intensively studied since (Barrell, 1980; Daniels and Seymour, 1990; Rosenthal et al., 1997). 'Model' villages such as Nuneham Courtney, Great Tew, Brocklesby or Elsinor were erected along turnpiked roads at the edge of landscaped parks to accommodate labourers relocated from homesteads inconveniently visible across his lordship's *landskip*, a process paralleled in the formal arrangement of great house and slave dwellings on plantations in the American South (Stewart, 1995). Village design often reflected the picturesque visual tastes of landlords more than the practical needs of tenants. The visible signs of social exclusion – in the form of fences, ornaments and plantings – were not infrequently the objects of attack and destruction by angry, dispossessed villagers. And the popularity among late-eighteenth-century English landowners of harmonious landscape scenes in which labour dissolves into the far distance has been interpreted to reflect the anxieties of landowners in the face of French Jacobinism.

Class and landscape moulded also the design of nineteenth-century urban and municipal parks and gardens. On both sides of the Atlantic, packing large numbers of poorly paid and inadequately housed factory workers into cities whose spatial structures were inadequate to service basic needs produced crises of sanitation, health and crime. These were apparent in mid-century typhoid and cholera epidemics, which directly threatened bourgeois lives and sensibilities. The middle-class response was to move to the suburban edges of the towns and to surround individual 'villas' with miniature landscapes based on pattern-book designs and using plants drawn from across a colonized globe, that were the progenitors of the modern suburban garden or yard (Preston, 1999). In public space, the layout of new cemeteries on the urban fringe and of municipal parks and gardens brought together picturesque design principals and the regulatory effects of 'rational recreation'. These enforced genteel strolling along serpentine paths and passive observation of form and colour in arboreta, flower beds and ornamental ironwork. Such essentially 'seen' landscapes were styled as contributions to both the physical health and the moral education of the industrial working class. But regulated landscape did not always fit the recreational demands of such groups. In Boston, New York and Chicago for example, baseball diamonds were favoured over flower beds among the immigrant industrial workers (Young, 1995).

Landscape's capacity to hide and soften visually the realities of exploitation and to 'naturalize' what is a socially produced spatial order continues today. California's agricultural landscape, long figured through utopian images of

Figure 12.7   *"…Searching for sea-halls; waves lap my wellington boots, carrying lost souls of brothers and sisters released over the ship side…" Pastoral, interludes* (Ingrid Pollard, Autograph, The Association of Black Photographers Ltd)

Edenic rurality featuring orange groves, strawberry fields, palms and roses set against distant blue mountains under a golden sky, obscures continuous and frequently brutal conflicts over land and water, and between landowners and migrant agricultural workers (Barron et al., 2000: 65–101; Mitchell, 1996). The latter, housed in trailer camps, watered with less care than the crops they cultivate, burned by the sunlight and chemicals that produce horticultural 'perfection', are as invisible in conventional images of California's garden landscape as they are from the state's freeways (Figure 12.6). Landscape's capacity to hide under a smooth, aesthetic surface the labour that produces and maintains it, is a direct outcome of its pictorial qualities and its identification with physical 'nature', placing the historical and contingent beyond critical reflection.

Landscape and ethnicity

Like class, differentiation of people by means of ascribed 'natural' or biological differences finds both expression and reinforcement in landscape. 'Race' is a mode of social differentiation based on visible differences between human bodies. An obvious and relatively innocuous example of its incorporation into landscape is the 'Chinatown' to be found in most metropolitan centres, marked by a standard repertoire of architectural and graphic symbols, often replacing much less innocent past forms of spatial marking and exclusion (Anderson, 1995; Lai, 1997). The concept of 'race', or more commonly today 'ethnicity', attributes significance to visible distinctions in skin colour, physiognomy and body form. A striking collection of photographs taken in 1980s England by the English artist Ingrid Pollard focused attention on the normalizing connections between landscape and ethnicity. As a black woman, Pollard intended her images to capture both a native attachment to English nature and a sense of being 'out of place', excluded from rural landscape. Implicitly and often explicitly, English culture places black people in cities, making them appear 'wrong' in the English landscape. Her photographs take their impact from challenging such visual expectations (Kinsman, 1995) (Figure 12.7).

Associations between landscape and ethnicity run much deeper than the visible presence of 'outsiders' within a landscape scene. Landscape conservation, design and appearance have drawn consistently on ecological theory and language to determine the appropriateness of landscape elements, drawing on ecology's authority as science to determine 'natural' occurrence and locational propriety. American nativism accounts in part for the influential 'Prairie School' of landscape architecture and its successors in California and Arizona. Like the landscape architect Willy Lange in Germany, adherents advocated using only 'native plants' in the public landscape of gardens and parks, justifying their selections on the grounds of natural ecological relations. In the light of our knowledge of continuous plant and animal evolution and migration, the concept of native species with exclusive rights to presence in landscape appears an entirely cultural product, whose roots lie in unexamined anxieties over identity and normalizing moral evaluations. A significant body of historical research in environmental history in recent years has demonstrated the close epistemological and practical connections between plant identification and classification, global ecological encounters, European imperialism, landscape transformation and modern environmentalism. The process bound together imaginative landscapes of Eden and Arcadia with the actual landscapes of tropical islands and botanical gardens (Grove, 1995).

Landscape and gender

The naturalizing power of landscape derives also from gendering nature. The historical connection between landscape and modernity and vision's epistemological privilege incorporate a shifting discourse of patriarchy. The long-lasting influence of Aristotle's theory of animation, which marked syllogistic reasoning as an adult male attribute, placed this denotation at the peak of a hierarchy of consciousness and life which reached down through females, children, barbarians, slaves, animals and plants to inanimate matter. Accordingly, culture was given 'male' and nature 'female' attributes. This patriarchal chain of being was reinforced in the early years of the seventeenth century, notably in Francis Bacon's formulation of empirical science as the means by which an active, masculine mind would subordinate a passive feminine matter. In genre classifications of the arts that endured into the twentieth century, images and representations of 'the great deeds of great men' – in tragic drama, epic poetry or history painting – commanded the highest respect, enacted and located in the public landscapes of urban power. Matters of the heart and private life, expressed in lyrical poetry or recorded in portraiture, were relegated to a domestic, gardened landscape populated by women and children, while fields and farms were the setting for peasant vulgarities expressed in unlettered, blank verse. Uncultivated wilderness remained the haunt of half-human satyrs, savages and monsters. Landscape images could be instantly coded according to this gendered social hierarchy.

Mapping gender alongside class onto space and nature opened the way to a more explicit exploitation as mathematical reason became bound to visual observation and the domination of nature by modern science. Giorgione's Venus, discussed above, is an early example of how the female body, presented as an object of erotic desire for the visual pleasure of an implicitly male subject (predictably both artist and patron were men), is painted in a 'state of nature'. This is the point of both the woman's nakedness and her otherwise absurd position in landscape. In the Classical narrative of such a 'discovery', the male viewer is punished for his voyeurism by being reduced to a state of nature. Thus Acteon was turned into a stag and killed by his own hounds for the mistake of witnessing Diana at her bath. In the modern formulation, however, the female body is fully aligned to nature and both are opened to an uncompromising and penetrating gaze as the passive property of men (Mulvey, 1989; Nash, 1996).

The strongly visual pleasure involved in representing and seeing nature as landscape has been criticized as an unreflective expression of patriarchal power expressed in a specifically heterosexual male eroticism. There is indeed a long history of associating the smooth topographic forms and serpentine lines 'of beauty' in picturesque landscape with the female body. The issue is more than merely representational: active exploitation of 'virgin' land by agriculture and large-scale extractive practices have long been legitimated by appeal to rational science in a language of conquest, control and subordination. Alternative connections with land to the rational knowledge and detachment associated with vision have been devalued in part through their gendered association with sentiment and soft-heartedness. The logic, language and imagery of twentieth-century landscape engineering, for example in the great dams of the western USA and former USSR, or Edward Teller's 'Project Ploughshare' proposal to use atmospheric nuclear explosions to dig canals and harbours, have distinct gender connections (Kirsch et al., 1998). While patriarchy adopts varied and indeed contradictory ways of figuring 'femininity' – irrational, wayward and wild as frequently as sentimental, soft and tame – the subordination of nature to the controlling power

Figure 12.8   *Caspar David Freidrich, 'The Cross in the Mountains'* (The National Gallery)

of male reason and ingenuity is a consistent trope. Feminist critics and artists have sought to rework modernist associations between landscape, the body and femininity, emphasizing the body as a social construction and its creative potential, and pointing out the possibilities of revisioning nature in terms of male bodies and alternative masculinities (Bunn, 1994; Norwood and Monk, 1987; Pratt, 1992).

LANDSCAPE: TERRITORY AND IDENTITY

If the social processes incorporated into landscape are normalized through its 'natural' qualities, the visible scene does more than simply reflect the imposition of prior cultural distinctions; it serves to regulate and order social relations. This disciplinary aspect of landscape has been intensively studied. It is most apparent in military or carceral landscapes, which involve the explicit threat or exercise of violence. J.B. Jackson (1984: 131–8) has written of the military landscapes he experienced as an intelligence officer in wartime France during 1944. Landscapes delicately limned by Vidalian geographers as medals struck in the likeness of their peasant cultivators were reduced to simplified sectors and zones, identified by the colour coding of flags and emblems. They came to be 'seen' by soldiers precisely in that way, and in many cases to be reduced by bombardment to little more than that (Clout, 1999; Gold and Revill, 1999). In

twentieth-century colonial wars, in South Africa, Malaya, Vietnam and Kenya for example, a consistent practice on the part of colonial forces was the removal of subject rural populations into the protective custody of 'defended' villages in order to prevent their infiltration by guerrilla troops and to clear space for conventional warfare (Sioh, 1998). This entailed removing villages, farmhouses and natural 'cover' in order to give advantage to vision and sight-based military technologies. 'Conventional' warfare actually refers to forms of military engagement developed on the agrarian landscapes of western Europe (whence the word 'campaign' is derived). Military landscapes might be regarded as the rawest expression of the characteristically modern landscape whose forms are determined by clearly marked linear spatial divisions, uniform vision and exclusionary practices. The most consistent expression of such territoriality is the nation-state, whose initial foundation was a concept of social collectivity expressed in an ecological concept of nationhood rooted in land. Its global spread is largely a result of European imperialism and colonization. As a geopolitical entity, the nation-state has drawn heavily upon both the naturalizing and the disciplinary powers of visible landscape.

Nature, nation, landscape

While the territorial state remains a primary basis of social identity for most of the world's peoples, contemporary processes such as decolonization, economic and cultural globalization, international labour migration and new communications technologies have reworked the bonds of allegiance between the state and many of its citizens. The process has permitted a critical examination of the traditional ways in which such allegiance was forged, especially within European nation-states, currently negotiating new relations between territory, citizenship and identity within the European Union (Agnew, 1998; Cosgrove et al., 1998). It is easier today to recognize the contingency of relations which had long seemed natural and permanent. Among the most powerful connections between nation and state is the material landscape. While nations are imagined communities, in which no citizen can ever be intimate with every fellow citizen, they are also imagined territories, since no citizen can ever know intimately the land of the whole state (Anderson, 1983; Hooson, 1994). Iconic images of nature and national landscape have thus played a powerful role in the shaping of modern nation-states as visible expressions of a claimed natural relationship between a people or nation and the territory or nature it occupied.

Perhaps the most dramatic example of this process is Germany, where academic geography played a key role within a discourse of territorial nationhood framed in terms of culture, landscape, community and home (*Heimat*). It was from German cultural geography that landscape emerged as a key concept, generating such techniques as visible 'landscape indicators' to distinguish cultural regions. In the formation of German national consciousness, ideas of both territorial boundary and physical landscape are deeply rooted. Humanists in fifteenth-century Nuremberg, Augsberg and Ulm, whose philological studies produced a German literary language, also promoted local mapping and landscape painting to celebrate belief in *Germania* (Schama, 1995: 75–120; Wood, 1993). Theirs was a distinct *Kultur*, unaffected by Roman interference – either imperial or papal – a bulwark of authentic Christian virtue in the face of Mediterranean decadence and Slavic barbarity. Germany's characteristic topography of hercynian rock, dense forest and open heaths became figured as a shaping force of Teutonic character. These elements were synthesized by early-nineteenth-century Romantic nationalists such as Caspar David Friedrich or the Grimm brothers into iconic landscapes of iron crosses set among crags, pine woods and dolmens (Figure 12.8). When Germany united as a nation-state in 1880 under the dominance of Prussia, *Berg und Wald* framed a muscular German self-image, so that the many grandiose public monuments erected to Count Bismarck deployed granite, oak leaves and iron crosses to commemorate the founder of the modern state (Lang, 1996; Michalski, 1998: 56–76).

Kulturlandschaft was a focus of geographical research in German universities in the late nineteenth century, reflecting anxieties within the new state over its territorial boundaries and cultural unity. Unlike France or Britain, the distribution of German speakers corresponded to no physically demarcated boundary. Geographers responded therefore to the cultural ideal of a profound connection between a German *Volk* and its soil, a social psychology captured in the concept of *Heimat* and visible in the unique settlement forms of the German village. Landscape indicators, such as house form, village morphology, field pattern and enclosure, defined the true German landscape as an ecological unity of nature and people (Sandner, 1994). This tradition of landscape and settlement geography yielded dire consequences in the 1940s with the replanning of captured eastern lands to resemble German *Kulturlandschäfte*. It continues to echo through German landscape conservation today. Less dramatic parallels to Germany's correlation of people and territory through landscape can be found in every European nation. They are evident in the iconic landscapes of downland, copses, hedgerows and village spires pictured on English topographic maps of the 1920s and in the tapestry of hill slopes dotted with fruit trees and vines gathered around the campanile of a walled city which the Italian Macchiolini painters produced in the years of Italian unification (Agnew, 1998; Graham, 1997). They are found too in the patchwork of tiny fields, whitewashed cottages and bare limestone washed by wild Atlantic breakers which Irish nationalists sought to preserve as the authentic landscapes of a Celtic, Catholic nation within the *Gaeltacht*. In every case pictorial images have served as vehicles conveying national pride and identity through specific, often geographically unrepresentative, landscapes. Even in the former USSR, despite its secular ideology and expressed belief in the human conquest of nature through socialism and communism, painters, film-makers, poets and novelists celebrated selected topographical features of Russian landscape as especially expressive of a collective Russian identity and purpose (Bassin et al., 2000).

Unsurprisingly, it has been material expressions of such iconic landscapes that have been subject to regulation as a means of preserving their visual appearance. Again, national parks offer obvious examples. Their very designation articulates a connection between a nation and a zone of characteristic nature. Originating national parks in the western United States date from the *fin-de-siècle* decades of intense nationalist self-definition through landscape. The cultural significance of wilderness as the characteristic American landscape, especially in the years following Frederick Jackson Turner's definition of the western frontier as the foundation of American democracy, has been closely studied (Cronon, 1996; Nash, 1982; Turner, 1894). American parks established a model, subsequently adopted in virtually every country in the world, of setting aside areas of the national territory on the basis of their landscape value and declaring them the nation's 'natural' heritage. Park scenery varies according to the natural and cultural landscape features deemed significant to the image of the nation. Thus Britain's are inhabited and farmed, but are almost exclusively located in upland regions characterized by extensive glaciated topography, moorland vegetation and sheep farming, a cultural framing specific to nationalist readings of landscape (Shoard, 1982). In Sri Lanka, by contrast, the principal national park at Yala in the south-east of the island is lowland 'dry jungle', a former colonial big game hunting region, now protected. But Yala's national significance derives also from its archaeological importance to the politically dominant Sinhala and their historical relations with the island's Tamil minority. It is a contested landscape, often closed to visitors because of the threat from guerrilla separatists who find refuge in its uninhabited spaces (Jazeel, 2000). The clearing, bounding and purification of space that produces and sustains such landscapes were a recurrent feature of twentieth-century European colonialism and are recognized by cultural geographers as expressions of the same processes that created aesthetic landscapes on newly reorganized private estates (Daniels and Seymour, 1990; Daniels et al., 1998).

The naturalness of national parks is culturally produced and maintained by strict land use management. Cultural markers are often absent; indeed, many of America's national parks had been forcefully cleared of occupants mere decades before they were designated. Among the boldest and also the most poignant inscriptions of nationhood into natural landscape is the monument carved into Mount Rushmore in South Dakota. Displaying the heads of four American presidents, it stands in a long tradition of monumental commemorative landscapes in which a distinctly white, male power is literally embodied. It is located in the region where the Plains Indians were finally defeated and confined to reservations. Controlling landscape is as much a symbolic as a material act, as the monuments raised on battlefields make visible. The capital cities of every nation-state are designed landscapes whose patterns of roads and open spaces, buildings and monuments invariably inscribe foundation myths, public memory, constitutional structures and heroic individuals into an iconography of nationhood (Atkinson and Cosgrove, 1998; Schama, 1995: 385–401; Warnke, 1994: 53–74). The iconography of such urban landscapes offers opportunities too for challenge, resistance and subversion of official meanings, as the fate of Lenin statues in nations of the former Soviet Union testifies (Bell, 1999; Michalski, 1998; Smith, 1999; Till, 1999).

Colonial landscape

W.T. Mitchell has called landscape 'the dreamwork of empire'. He is referring to the spatial and social perceptions, assumptions and practices that accompanied European expansion into non-European regions of the globe. Postcolonial study finds in landscape a valuable concept for examining the cultural aspects of colonialism. That colonization by definition involves appropriation and occupation of land permits a reworking of cultural geographers' long-standing interest in the overseas transfer, diffusion and simplification of European modes of occupance (Norton, 2000: 96–7). Colonization entailed a certain blindness to pre-existing cultural landscapes, evident for example in their figuring as 'new worlds', wilderness or discovered Edens. Prior occupants were connected to 'nature' by means of a limited number of landscape tropes derived from a stock of European stereotypes: golden-age innocents, wild savages, cannibals, nomadic hunters and gathers and pastoralists. In every case they were regarded as subjects of nature rather than its 'masters'. Their landscapes could not therefore be 'cultural'. A consistent feature in the European management of colonial space was the enforced sedentarization of native populations and the division of usufruct land into plotted and bounded properties. This was simultaneously a means of bodily control, of intensified economic exploitation and of accelerating what the colonizers took to be the cultural 'evolution' of native peoples (Noyes, 2000). The outcome was a new landscape whose visible order, signified by fenced property lines, geometrically distributed farms and villages, prompted comparison with picturesque European landscape (Doughty, 1982). Visible order in the landscape became to European eyes a justification of the colonizing mission. Evidence of previous native transformation of landscape and indeed of the native environmental and spatial knowledge necessary to initial European exploration, mapping and settlement was largely ignored or diminished in significance (Malcolm, 1998). Only recently have the complex aboriginal changes and adjustments to the landscape before and during the contact phase begun to be systematically studied by cultural geographers.

Imperial eyes gazed through European lenses, actually as well as metaphorically. The technologies of seeing which shaped rural Europe into landscape were applied to these 'other' spaces. Soldiers, sailors and scientific explorers as well as artists were trained in techniques of observing, surveying and sketching landscapes. Their representations were powerful elements in returning knowledge of exotic, 'other' places to imperial centres and they both framed and reinforced the imaginative geographies of empire. When examined critically these landscape images are revealed as hybrid creations, reflecting the meeting of conventions of seeing shaped at home and the need to record actually witnessed forms, phenomena and atmospheres for which those conventions were inadequate vehicles of expression (Martins, 1999). This realization is reinforced when we examine the processes of image creation rather than merely their finished surfaces. In landscape images as much as in the material shaping of colonial land scapes, geopolitical and economic imperatives often grated against moral principals, and the contradictions echo into the present. Thus, contemporary Malaysian insouciance over the disappearance of tropical rainforest, shocking to many westerners, can be understood in terms of the complex patterns of ownership, cultivation, management and conflict which shaped a plantation landscape and its bordering 'jungle' (Sioh, 1998).

THE LIMITATIONS OF LANDSCAPE VISION

The conventional emphasis on the visual and the visible in landscape forms and expressions is a logical outcome of its conceptual and historical evolution in the west. Many of the research techniques developed in the geographical study of landscape, from fieldwork to iconographic interpretation, maintain the focus on sight, vision and image. But the epistemological connection between observation, graphic representation and objectivity is now widely acknowledged to be deeply problematic. Vision's role in connecting knowledge and power has a complex and continuous history, and it continues to evolve. The penetrating capacities of sight expand radically as advanced optical technologies open new spaces of vision and shift the boundaries of public and private space. Thus critical landscape studies today emphasize the duality and provisional nature of sight: the returned gaze, the capacity of its subjects to manipulate, obscure, subvert or deform visual order. The democracy offered by modern image-making and image-manipulating technologies promotes historical re-evaluation too. Perhaps landscape was always a more open prison than we have recognized.

Today's extended prosthetics of vision parallel increased recognition of the human eye's embodiment. Vision is never entirely disconnected from the other sensual, cognitive and affective aspects of human conduct. While the permeable connections between external nature and the human body attract attention in the context of virtual worlds and cyborg spaces, they also have deep cultural and historical roots. Geomancy, for example, a key element in both landscape art and the design of gardens in China and Japan, and fundamental to the imaginative geography of the Korean peninsula, continues to shape the modernist landscape of Asian cities, even influencing modernist engineering projects such as the damming of great Chinese rivers (Jin, 2000). It is increasingly popular in the west as an aid to space design. *Feng-shui*, literally 'wind-water', refers to the animating elements of natural form, to the processes which shape landscape rather than its visible structures and patterns. According to this apprehension of land and life, the body of the earth and the bodies of those who inhabit the world should be brought into harmony if *Qi* is to flow smoothly through all things. Such concepts prompt us to move landscape beyond the confines of the visual towards more imaginative and encompassing embodiments that are at once sensual and cognitive.

All spatial activity is consciously or unconsciously performative. The sense of sight plays a critical but by no means exclusive role in bringing together the human and natural actors who perform landscape. While 'eyeing nature' has certainly been profoundly significant in shaping the cultural geography of the modern world and its study, revising the role and meaning of vision in landscape allows a more subtle and nuanced embrace of landscape's conceptual and empirical riches.

REFERENCES

Agnew, J. (1998) 'European landscape and identity', in B. Graham (ed.) *Modern Europe: Place, Culture and Identity*. London: Arnold. pp. 213–35.

Alpers, S. (1984) *The Art of Describing: Dutch Art in the Seventeenth Century*. Chicago: University of Chicago Press.

Anderson, B. (1983) *Imagined Communities: Reflections on the Origin and Spread of Nationalism*. London: Verso.

Anderson, K.J. (1995) *Vancouver's Chinatown: Racial Discourse in Canada, 1875–1980*. Montreal: McGill–Queens University Press.

Appleton, J. (1996) *The Experience of Landscape*. New York: Wiley.

Atkinson, D. and Cosgrove, D. (1998) 'Urban rhetoric and embodied identities: city, nation, and empire at the Vittorio Emanuele Monument in Rome, 1870–1945', *Annals of the Association of American Geographers* 88: 28–49.

Ballantyne, A. (1997) *Architecture, Landscape and Liberty: Richard Payne Knight and the Picturesque*. Cambridge: Cambridge University Press.

Barrell, J. (1980) *The Dark Side of the Landscape: The Rural Poor in English Painting, 1730–1840*. Cambridge: Cambridge University Press.

Barron, S., Bernstein, S. and Fort, I.S. (2000) *Made in California: Art, Image, and Identity, 1900–2000*. Los Angeles: County Museum of Art and University of California Press.

Bassin, M. et al. (2000) 'Landscape and identity in Russian and Soviet art', *Ecumene* 7: 249–336.

Bell, J. (1999) 'Redefining national identity in Uzbekistan: symbolic tensions in Tashkent's official public landscape', *Ecumene* 6 (2): 183–213(31).

Bell, M.M. (1997) 'The ghosts of place', *Theory and Society* 26: 813–36.

Berger, J. (1972) *Ways of Seeing*. London: BBC and Penguin.

Bermingham, A. (1986) *Landscape and Ideology: The English Rustic Tradition, 1740–1860*. Berkeley: University of California Press.

Bishop, P. (1994) 'Residence on earth: *anima mundi* and a sense of geographical "belonging"', *Ecumene* 1: 51–64.

Bunge, W. (1966) *Theoretical Geography*, Lund: Royal University, Dept. of Geography.

Bunn, D. (1994) '"Our wattled cot": mercantile and domestic space in Thomas Pingle's African landscapes', in W.J.T. Mitchell (ed.) *Landscape and Power*. London: University of Chicago Press. pp. 127–74.

Cafritz, R.C., Gowing, L. and Rosand, D. (1988) *Places of Delight: The Pastoral Landscape*. Washington Crown.

Charlesworth, M. (1999) 'Mapping, the body and desire: Christopher Packe's chorography of Kent', in D. Cosgrove (ed.) *Mappings*. London: Reaktion. pp. 109–24.

Claval, P. (1995) *La géographie culturelle*. Paris: Nathan Université.

Clout, H. (1999) 'Destruction and revival: the example of Calvados and Caen, 1940–1965', *Landscape Research* 24: 117–40.

Corner, J. (ed.) (1999) *Recovering Landscape: Essays in Contemporary Landscape Architecture*. New York: Princeton Architectural Press.

Corner, J. and MacLean, A.S. (1996) *Taking Measures across the American Landscape*. New Haven: Yale University Press.

Cosgrove, D. (1985) 'Prospect, perspective, and the evolution of the landscape idea', *Transactions of the Institute of British Geographers* 10: 45–62.

Cosgrove, D. (1993) *The Palladian Landscape: Geographical Change and its Cultural Representations in Sixteenth-Century Italy*. Leicester: Leicester University Press.

Cosgrove, D. (1995) 'Habitable earth: wilderness, empire, and race in America', in D. Rothenberg (ed.) *Wild Ideas*. Minneapolis: University of Minnesota Press. pp. 27–41.

Cosgrove, D. (1998) *Social Formation and Symbolic Landscape (1984)*. Madison: University of Wisconsin Press.

Cosgrove, D. (2001) *Apollo's Eye: A Cartographic Genealogy of the Earth in the Western Imagination*. Baltimore: Johns Hopkins University Press.

Cosgrove, D. and Daniels, S. (1988) 'Introduction: iconography and landscape', in D. Cosgrove and S. Daniels (eds) *The Iconography of Landscape*. New York: Cambridge University Press, pp. 1–10.

Cosgrove, D., Roscoe, B. and Rycroft, S. (1996) 'Landscape and identity at the Ladybower Reservoir and Rutland Water', *Transactions of the Institute of British Geographers*. 21: 534–51.

Cosgrove, D., Nash, C. and Doukellis, P. (1998) 'Cultural landscape', in T. Unwin (ed.) *European Geography*. New York: Longman. pp. 65–81.

Cronon, W. (1996) *Uncommon Ground: Rethinking the Human Place and Nature*. New York: Norton.

Daniels, S. (1989) 'Marxism, culture, and the duplicity of landscape', in R. Peet and N. Thrift (eds) *New Models in Geography*, vol. 2. London: Unwin Hyman. pp. 196–220.

Daniels, S. (1993) *Fields of Vision: Landscape Imagery and National Identity in England and the United States*. Princeton: Princeton University Press.

Daniels, S. (1999) *Humphry Repton: Landscape Gardening and the Geography of Georgian England*. New Haven: Yale University Press.

Daniels, S. and Seymour, S. (1990) 'Landscape design and the idea of improvement', in R.A. Dodgshon and R.A. Butlin (eds) *An Historical Geography of England and Wales*, 2nd edn. London: Academic Press. pp. 487–520.

Daniels, S., Seymour, S. and Watkins, C. (1998) 'Estate and empire: Sir George Cornewall's management of Moccas, Herefordshire and La Taste, Grenada, 1771–1819', *Journal of Historical Geography* 24 (3): 313–51.

de Certeau, M. (1988) *The Practice of Everyday Life*. Berkeley: University of California Press. pp. 91–130.

Deutsche, R. (1991) 'Boy's town', *Environment and Planning D: Society and Space* 9: 5–30.

Doughty, R. (1982) *Making Home in Texas*. Austin: University of Texas Press.

Duncan, J. and Duncan, N. (1988) '(Re)reading the landscape', *Environment and Planning D: Society and Space* 6: 117–26.

Geipel, R. (1978) 'The landscape indicators school in German geography', in D. Ley and M.S. Samuels (eds) *Humanistic Geography: Problems and Prospects*. London: Croom Helm.

Gold, J.R. (1980) *An Introduction to Behavioral Geography*. Oxford: Oxford University Press.

Gold, J.R. and Revill, G. (1999) 'Landscape of defence', *Landscape Research* 24: 229–320.

Graham, B. (1997) *In Search of Ireland: A Cultural Geography*. London: Routledge.

Grove, R.H. (1995) *Green Imperialism: Colonial Expansion, Tropical Island Edens, and the Origins of Environmentalism, 1600–1860*. Cambridge: Cambridge University Press.

Gruffudd, P. (1995) 'Propaganda for seemliness: Clough Williams-Ellis and Portmeirion, 1918–1950', *Ecumene* 2: 399–422.

Haraway, D. (1991) *Simians, Cyborgs and Women: The Reinvention of Nature*. London: Free Association.

Harvey, D. (1989) *The Condition of Postmodernity: An Inquiry into the Origins of Cultural Change*. Oxford: Blackwell.

Heffernan, M. (1998) *The Meaning of Europe: Geography and Geopolitics*. London: Arnold.

Helgerson, R. (1992) *Forms of Nationhood: The Elizabethan Writing of England*. Chicago: University of Chicago Press.

Hillis, K. (1994) 'The virtue of becoming a no-body', *Ecumene* 1: 177–96.

Hirsch, E. and O'Hanlon, M. (eds) (1995) *The Anthropology of Landscape: Perspectives on Place and Space*. New York: Oxford University Press. pp. 31–62.

Hooson, D. (ed.) (1994) *Geography and National Identity*. Oxford: Blackwell.

Howett, C.M. (1997) 'Where the one-eyed man is king: the tyranny of visual and formalist values in evaluating landscapes', in Paul Groth and T.W. Bressi (eds) *Understanding Ordinary Landscapes*. New Haven: Yale University Press. pp.99–110.

Hunt, J.D. (2000) *The Idea of the Garden*. Philadelphia: University of Pennsylvania Press.

Jackson, J.B. (1984) *Discovering the Vernacular Landscape*. New Haven: Yale University Press.

Jazeel, T. (2000) 'Exploring Srilankan identities through territory: Ruhuna National Park'. Unpublished paper, available from author.

Jenkins, R. (1998) *Virgil's Experience: Nature and History: Times, Names, and Places*. Oxford: Clarendon.

Jin, J. (2000) 'The influences of the idea of poongsoo on the traditional mapping of Korea'. Unpublished paper, available from author.

Kemp, M. (1990) *The Science of Art: Optical Themes in Western Art from Brunelleschi to Seurat*. New Haven: Yale University Press.

Kinsman, P. (1995) 'Landscape, race and national identity: the photography of Ingrid Pollard', *Area* 27: 300–10.

Kirsch, S., Millar, S., Mitchell, D., Frenkel, S. and Krygier, J.B. (1998) 'Nuclear engineering and geography', *Ecumene* 5 (3): 263–322.

Lai, D.C. (1997) 'The visual character of Chinatowns', in P. Groth and T.W. Bressi (eds) *Understanding Ordinary Landscapes*. New Haven: Yale University Press. pp. 81–4.

Lang, K. (1996) 'Monumental unease', in F.F. Hahn (ed.) *Imagining Modern German Culture, 1889–1910*. Washington: University Press of New England.

Lefebvre, H. (1991) *The Production of Space*. Oxford: Blackwell.

Lorimer, H. (2000) 'Guns, game and the grandee: the cultural politics of deerstalking in the Scottish Highlands', *Ecumene* 7: 403–31.

Lynch, K. (1970) *The Image of the City*. Cambridge: MIT Press.

Malcolm, G.L. (1998) *Cartographic Encounters: Perspectives on Native American Mapmaking and Map Use*. Chicago: University of Chicago Press.

Mariage, T. (1998) *The World of Andre Le Notre*. Philadelphia: University of Pennsylvania Press.

Martins, L.L. (1999) 'Mapping tropical waters: British views and visions of Rio de Janeiro', in D. Cosgrove (ed.) *Mappings*. London: Reaktion. pp. 148–68.

Matless, D. (1992) 'A modern stream: water, landscape, modernism, and geography', *Environment and Planning D: Society and Space* 10: 569–88.

Matless, D. (1994) 'Doing the English village, 1945–90: an essay in imaginative geography', in P. Cloke (ed.) *Writing the Rural: Five Cultural Geographies*. London: Chapman. pp. 7–88.

Matless, D. (1997) 'Moral geographies of English landscape', *Landscape Research* 22 (2): 141–56.

Matless, D. (1998) *Landscape and Englishness*. London: Reaktion.

Matless, D. (1999) 'The uses of cartographic literacy: mapping, survey and citizenship in twentieth-century Britain', in D. Cosgrove (ed.) *Mappings*. London: Reaktion. pp. 193–212.

Merchant, C. (1990) *The Death of Nature: Women, Ecology, and the Scientific Revolution*. New York: Harper and Row.

Michalski, S. (1998) *Public Monuments: Art in Political Bondage 1870–1997*. London: Reaktion.

Mitchell, D. (1996) *Lie of the Land: Migrant Workers and the California Landscape* Minneapolis: University of Minnesota Press.

Mitchell, W.J.T. (1994) 'Introduction' and 'Imperial landscape', in W.J.T. Mitchell (ed.) *Landscape and Power*. London: University of Chicago Press. pp. 1–34.

Morin, K.M. (1998) 'Trains through the plains: the Great Plains Landscape of Victorian women travellers', *Great Plains Quarterly* 18: 235–56.

Muir, R. (1999) *Approaches to Landscape*. London: Macmillan. pp. 149–81.

Mulvey, M. (1989) *Visual and Other Pleasures*. Basingstoke: Macmillan.

Nash, C. (1996) 'Reclaiming vision: looking at landscape and the body', *Gender, Place, Culture* 3: 149–69.

Nash, R. (1982) *Wilderness of American Mind*, 3rd edn. New Haven: Yale University Press.

Neumann, R.P. (1995) 'Ways of seeing Africa: colonial recasting of African society and landscape in Serengeti National Park', *Ecumene* 2: 149–70.

Neumann, R.P. (1998) *Imposing Wilderness: Struggles over Livelihood and Nature Preservation in Africa*. Berkeley: University of California Press.

Norton, W. (2000) *Cultural Geography: Themes, Concepts, Analyses*. New York: Oxford University Press. pp. 96–7.

Norwood, V. and Monk, J. (eds) (1987) *The Desert Is No Lady: Southwestern Landscapes in Women's Writing and Art*. New Haven: Yale University Press.

Novak, B. (1980) *Nature and Culture: American Landscape and Painting, 1825–1875*. New York: Oxford University Press.

Noyes, J. (2000) 'Nomadic fantasies: producing landscapes of mobility in German Southwest Africa', *Ecumene* 7: 47–66.

Nuti, L. (1999) 'Mapping places: chorography and vision in the Renaissance', in D. Cosgrove (ed.) *Mappings*. London: Reaktion. pp. 90–108.

Olwig, K. (1984) *Nature's Ideological Landscape: A Literary and Geographic Perspective on its Development and Preservation on Denmark's Jutland Heath*. London: Allen and Unwin.

Olwig, K.R. (1996) 'Recovering the substantive nature of landscape', *Annals of the Association of American Geographers* 86: 630–50.

Park, D.C. (1994) 'To the "infinite spaces of creation": the interior landscape of a schizophrenic artist', *Annals of the Association of American Geographers*. 84: 192–209.

Penning-Rowsell, E.C. and Lowenthal, D. (eds) (1986) *Landscape, Meanings and Values*. London: Allen and Unwin.

Porteous, J.D. (1990) *Landscape of the Mind: Worlds of Sense and Metaphor*. Toronto: University of Toronto Press.

Pratt, M.L. (1992) *Imperial Eyes: Travel Writing and Transculturation*. London: Routledge.

Preston, R. (1999) '"The scenery of the torrid zone": imagined travels and the culture of exotics in nineteenth-century British gardens', in F. Driver and D. Gilbert (eds) *Imperial Cities*. Manchester: Manchester University Press. pp. 194–214.

Pugh, S. (1988) *Garden, Nature, Language*. Manchester: Manchester University Press.

Rose, G. (1993) *Feminism and Geography*. Minneapolis: University of Minnesota Press.

Rose, G. (1997) 'Situating knowledge: positionality, reflexivities, and other tactics', *Progress in Human Geography* 21: 305–20.

Rosenthal, M., Payne, C. and Wilcox, S. (1997) *Prospects for the Nature: Recent Essays in British Landscape 1750–1880*. New Haven: Yale University Press.

Said, E. (1993) *Culture and Imperialism*, New York: Knopf.

Sandner, G. (1994) 'In search of identity: German nationalism and geography, 1871–1910', in D. Hooson (ed.) *Geography and National Identity*. Oxford: Blackwell. pp. 71–91.

Schama, S. (1995) *Landscape and Memory*. New York: Knopf. pp. 7–10.

Schulz, J. (1978) 'Jacopo de'Barbari's view of Venice: map-making, city views and moralized geography before the year 1500', *The Art Bulletin* 60: 425–74.

Shoard, M. (1982) 'The lure of the Moors', in J.R. Gold, and J. Burgess (eds) *Valued Environments*. Boston: Allen and Unwin.

Sioh, M. (1998) 'Authorizing the Malaysian rainforest: configuring space, contesting claims and conquering imaginaries', *Ecumene* 5: 144–66.

Smith, T. (1999) '"A grand work of noble conception": the Victoria Memorial and imperial London', in F. Driver and D. Gilbert (eds) *Imperial Cities*. Manchester: Manchester University Press. pp. 21–39.

Söderström, O. (2000) *Des Images pour agir: Le Visuel en urbanisme*. Lausanne: Editions Payot.

Sonnenfeld, J. (1994) 'Way keeping, way finding, way losing: disorientation in a complex environment', in K.E. Foot and P.J. Hugill (eds) *Re-reading Cultural Geography*. Austin: University of Texas Press. pp. 387–98.

Stewart, L. (1995). 'Louisiana subjects: space and the slave body', *Ecumene* 2: 227–46.

Till, K. (1999) 'Staging the past: landscape designs, cultural identity and *Erinnerungspolitik* at Berlin's *Neue Wache*', *Ecumene* 6: 251–83.

Turner, F.J. (1894) 'The significance of the frontier in American history', in F.J. Turner (ed.) *Frontier and Section: Essays by Frederick Jackson Turner*. El Paso: Texas Western College Press.

Turner, J. (1979) *The Politics of Landscape: Rural Scenery and Society in English Poetry, 1630–1660*. Oxford: Blackwell.

Warnke, M. (1994) *Political Landscape: The Art History of Nature*. London: Reaktion. pp. 39–52.

Women and Geography Study Group of the IBG (1984) *Geography and Gender: An Introduction to Feminist Geography*. London: Hutchinson.

Wood, C.S. (1993) *Albrecht Altdorfer and the Origins of Landscape*. London: Reaktion.

Young, T. (1995) 'Modern urban parks', *Geographical Review* 85: 535–55.

13

Landscape and the Obliteration of Practice

Tim Cresswell

Unlike the other contributors to this section, I do not use 'landscape' as a regular part of my academic vocabulary. Frankly I am not keen on the term. It is a concept too burdened with its own history – too fixated on origins. I have tended to focus on the concepts of place and space. I like place because of its everyday nature. Place, unlike space and landscape, permeates our everyday life and provides meaning in people's lives. Places are quite clearly 'lived'. At the other extreme space has an analytical quality about it. It rubs shoulders easily with social theory and allows conversations to occur with other social science disciplines. Landscape, on the other hand, does not have much space for temporality, for movement and flux and mundane practice. It is too much about the already accomplished and not enough about the processes of everyday life. It is too stuck in the humanities to make it amenable to the kind of critical theory that appeals to me. It seems altogether too quaint. I realize these are personal predilections. It is true that the concept can be made to grow and adapt, to colonize the dynamism of living geography, but I wonder what would be saved in the process. Why not, in other words, abandon the concept altogether and leave it as a dusty anachronism in the glossary of cultural geography? This chapter is about the limits of/to landscape.

Landscape is a very loosely used term. The relative voguishness of geography has ensured that it is no longer predominantly a concern of geography. Part of the problem is that landscape has resonance well beyond geography and even academia: it is part of our everyday vocabulary. There are landscapes of … just about anything.

In popular journalism it is not unusual to see reports on 'the changing leisure landscape' or 'the fantasy landscape'. This is sometimes fancifully changed to shortened forms such as 'mindscape' or 'pleasurescape'. Beyond geography in the academic world, landscape has become a well-worn metaphor. A quick look at a bibliographic database reveals a confusing array of uses for the term. Some are close to geography and have easily intelligible meanings such as 'mapping the criminal landscape', while others appear to use the term as a generic metaphor for a kind of intellectual survey. Examples include 'the legal landscape' and 'white flecks across the landscape of literary history', 'the political landscape of American schools', and 'Germany's changing religious landscape'. Still more take landscape out of the humanities and the social sciences and place it firmly in the lexicon of science, where theoretical physicists use something called 'landscape analysis' to discover the lattice structure of alpha beta transitions in protein folding and talk of 'free energy landscapes'. Urologists meanwhile are busy discussing the 'changing landscape of bladder augmentation'. These uses of the term 'landscape' actually tell us quite a lot about the term. 'The changing landscape of bladder augmentation' suggests an overview, a survey. It also suggests that the author of the paper is in a position to tell us all about it. He is authoritative. An idea such as 'free energy landscapes' points towards the importance of form – of topography. It also suggests we might be able to visualize energy. These are important factors in the notion of landscape we have inherited – an authoritative and surveying position for the viewer and a particular form for that which is being seen. We have become used to the importing of hard science into the social; perhaps there is room here for research into the migration of geography into theoretical physics and urology.

It is not the vagueness of the term that is my principal problem though. My problem with the idea of landscape can be illustrated through the term 'landscapes of practice'. The term is somewhat oxymoronic. Landscape, on the one hand, appears to encapsulate the notion of fixity – of a text already written – of the production of meaning and the creation of dominating power. Landscape is solid. Practice, on the other hand, is about fluidity, flow and repetition. It is about the negotiation between continuity and change. Practice has been seen by social and cultural theorists as an antidote to the representational – as an unexamined component of the everyday. So what could 'landscapes of practice' refer to? In many ways it resembles Raymond Williams' term 'structure of feeling'. Williams developed the notion of 'structure of feeling' in order to simultaneously bring the notion of structure and notion of feeling into question.

> If the social is always past, in the sense that it is always formed, we have indeed to find other terms for the undeniable experience of the present: not only the temporal present, the realisation of this and that instant, but the specificity of present being, the inalienably physical, within which we may indeed discern and acknowledge institutions, formations, positions, but not always as fixed products, defining products. And then if the social is the fixed and explicit – the known relationships, institutions, formations, positions – all that is present and moving, all that escapes or seems to escape from the fixed and the explicit and the known, is grasped and defined as the personal: this, here, now, alive, active, 'subjective'. (1977: 128)

The term 'structure of feeling' simultaneously brings the fixity of structure and the flow of feeling into account. In this formulation structure is not fixed and eternal but a part of a process. Feeling is not just 'personal' but moulded (not determined!) by the social. Structure and feeling do not inhabit two ends of a spectrum but are interconnected and productive of each other. These two terms ('landscapes of practice' and 'structures of feeling') are connected by more than their oxymoronic logic. Williams' language suggests further points of contact. The 'social' is notably 'fixed and explicit' while feeling is 'present and moving'. Williams relies upon an opposition of fixity and flow to make his point about

structure and feeling. Structure, the social, institutions and formations have been, in Williams' view, located in the past and seen as finished and coherent – static – while present experience, the lived, the here and now are seen as fleeting and subjective. His argument is that neither of these is the case. It is with this in mind that I explore the possibility that landscape and practice can say something to each other. Necessarily this involves a consideration of the role of vision in landscape studies and the tension between landscapes as seen (and scene) and landscapes as arenas of practice. This chapter, therefore, is structured around two terms and their relation to landscape – vision and practice. The chapter begins with a (brief) rehearsal of the use of the term 'landscape' in cultural geography, pointing out the elaborations, critiques and developments of the central role of vision in the definition of landscape. The tension between vision and practice is developed in the next section with reference to Raymond Williams, Michel de Certeau and J.B. Jackson. This leads into an extended discussion of theories of practice. The theme of landscape as *lived* is central to the final section which focuses on the work of recent archaeologists of landscape and the idea of 'doxic landscapes'.

WAYS OF SEEING LANDSCAPE

Landscape is clearly one of the central themes in contemporary cultural geography. The history of the term hardly needs elaborating given the vigorous reworking of the idea of landscape by what was once known as the 'new cultural geography' (now distinctly middle aged). Views of landscape within cultural geography can broadly be divided into three paradigmatic movements. In the early twentieth century, mostly under the influence of Carl Sauer and the Berkeley School, landscape was seen as a material artefact that was either natural or cultural. Cultural landscapes were made out of natural landscapes through the agency of a mysterious and overarching culture. Thus the landscape of an area was none other than the material expression of the (seemingly unified) group of people who lived in that region. In the 1970s humanistic geographers looked back on Sauer with some fondness (as well as W.G. Hoskin in the United Kingdom) and reformulated his view of landscape. Humanists such as Donald Meinig (1979a), Yi-Fu Tuan (1977) and Edward Relph (1976) argued for a view of landscape that took the human imagination into account. Rather than provide accounts of distinctive regional material landscapes, geographers began to see landscape as residing within the minds and eyes of beholders: landscape as a way of seeing. Finally, in the mid 1980s geographers such as Denis Cosgrove (1984; 1985; 1987), Stephen Daniels (1990; 1993) and Kenneth Olwig (1984), fired by the ascendancy of Marxist and radical accounts of the world, particularly those of Raymond Williams, argued that landscapes were material productions within which were coded particular ideologies. Landscapes were diagrams of power and influence that helped to reproduce the very power structures that produced them in the first place. In Daniels' (1990) terms, landscapes were duplicitous in that they could not be reduced to either their brute materiality or the more ideological notion of 'ways of seeing'.

Central to this potted history has been a critique of thinking of landscapes as purely material topographies. While there is no doubt that Sauer had a broader agenda than the straightforward mapping and description of different landscape morphologies, it is equally clear that the majority of work by his followers has been centred on just such projects. To Sauer landscape was geography's naively given phenomenon. We often forget that Sauer looked to phenomenology to construct his view of the discipline. 'Every field of knowledge is characterized by its declared preoccupation with a certain group of phenomena,' he declared, 'which it undertakes to identify and order according to their relations' (1965: 316). No other discipline could be said to have landscape as its 'object of consciousness', and it was (in Sauer's eyes) the unique and privileged role of geography to describe and understand (which amount to the same thing in much of Sauer's work) this phenomenon:

> The task of geography is conceived as the establishment of a critical system which embraces the phenomenology of landscape, in order to grasp in all its meaning and color the varied terrestrial scene. (1965: 319)

Importantly Sauer was keen to emphasize that landscape was not just a scene for the activities of people but something which included the works of people as a defining element. It was in this way that Sauer rejected the determinism of his predecessors, Semple and Huntingdon, and it is for this reason that Sauer's geography became known as 'cultural geography'. Unfortunately the activities of humans on the landscape became conflated with the action of an overarching and superorganic 'culture' on the land (Duncan, 1980).

Humanistic geographers inherited the concern with the issue of landscape but turned their attention away from the 'morphology' of landscape and towards the 'experience' of landscape. An important link here is J.B. Jackson (in many ways the hero of this story). Jackson followed the Berkeley line in his interest in the cultural landscape and his distrust of formal theory. But Jackson also began to explore the 'symbolic' aspects of landscape:

> [A] landscape is not a natural feature of the environment, but a synthetic space, a man-made system of spaces superimposed on the face of the land functioning and evolving not according to natural laws but to serve a community. A landscape is thus a space deliberately created to speed up or slow down nature. (1984: 8)

Jackson recognized the importance of art and emotion in the study of landscape as he attempted to teach his students to 'read' the landscape as a kind of text full of symbolic clues to the meaning that lies behind the bare morphology. In this sense his work influenced humanistic geographers, as is clear from the long dedication to him in the book *The Interpretation of Ordinary Landscapes* (Meinig, 1979b). The overwhelming theme of Meinig's volume is the way in which culture provides a lens to focus our idea of landscape. The attention throughout is on the way we see the landscape rather than what we see in the landscape. Vision becomes the central way of getting at landscape. Meinig's well-known paper 'The beholding eye' outlines 10 views of the same landscape which depend on the particular way in which individuals with different interests will interpret the same physical features.

> We may certainly agree that we will see many of the same elements – houses, roads, trees, hills – in terms of such denotations as number, form, diversity and color; but such facts only take on meaning through association, they must be fitted together according to some coherent body of ideas. Thus we confront the central problem: any landscape is composed not only of what lies before our eyes but what lies within our heads. (1979a: 34)

The same point is repeated by Yi-Fu Tuan in his essay 'Thought and landscape'. 'Landscape,' he argues, 'is not to be defined by itemizing its parts. The parts are subsidiary clues to an integrated image. Landscape is such an image, a construct of the mind and of feeling' (1979: 89). To many humanists landscape is an image. The focus is turned away from the form of the material landscape and towards the way we see landscape. We might even go further and say that some humanists believe that landscape exists only inside the heads of people as a perspective – a way of ordering the world.

The role of vision was not lost on cultural geographers such as Denis Cosgrove and Stephen Daniels who developed the theme of landscape still further. They agree that landscape is a 'way of seeing' but insist on the relationship between that way of seeing and the material conditions which overdetermine it. Cosgrove was one of the first geographers to suggest that geographers could benefit from the work of cultural studies and began to mix humanism with Marxism. He was worried that humanists had surrendered to 'idealism and subjectivism' in the 'intellectual examination of mind and matter' (1987: 87). The subjective reformulation of the external world (the landscape), Cosgrove tells us, is not the product of an autonomous mind but the product of culture which is itself an element of an over-determined superstructure. Landscapes as symbolic systems and as ways of seeing needed to be seen in relation to 'social formations'.

The canonical text on landscape in recent cultural geography is Denis Cosgrove's *Social Formation and Symbolic Landscape* (1984). Although the term 'landscape' had been used any number of times before by geographers, it had not been fully defined within a historical context. Cosgrove revealed the lineage of the term in a way that defined some crucial parts of it. He defines landscape as a 'way of seeing – a way in which some Europeans have represented to themselves and to others the world about them and relationships with it, and through which they have commented on social relations' (1984: 1). This first stab at a definition is followed by many others, all centred on a relationship between people and the world revolving around the act of seeing. Eight pages later, landscape is defined as 'the artistic and literary representation of the visible world, the scenery (literally that which is *seen*) which is viewed by the spectator' (1984: 9) which implied 'a particular sensibility, a way of experiencing and expressing feelings towards the external world, natural and man-made, an articulation of a human relationship with it. That sensibility was closely connected to a growing dependency on the faculty of sight as the medium through which truth was to be attained' (1984: 9). Cosgrove clearly agrees that the concept of landscape revolves around vision but, unlike humanists before him, he sought to place this 'way of seeing' in historical and social context. This contextualizing of vision is mirrored by Stephen Daniels in his outline of the 'duplicity' of landscape:

> [L]andscape may be seen as a 'dialectical image', an ambiguous synthesis whose redemptive and manipulative aspects cannot finally be disentangled, which can neither be completely reified as an authentic object in the world nor thoroughly dissolved as an ideological mirage. (1990: 206)

To Daniels the vagueness of landscape makes it an interesting term – one that holds together the material and the ideal, the solid and the superficial. It is still however primarily a visual order whether created through paintings or through country parks.

The other principal way of thinking about landscape in geography has been through the analogy of text. This is principally associated with the work of James Duncan. Duncan (1990) argues against the privileging of vision and suggests that we might think of landscapes as being constructed linguistically – as a signifying system that encourages some readings rather than others. As with a text, the producers write some meanings into a landscape but cannot finally control individual readings.

> The landscape, I would argue, is one of the central elements in a cultural system, for as an ordered assemblage of objects, a text, it acts as a signifying system through which a social system is communicated, reproduced, experienced, and explored. (1990: 17)

In Duncan's work landscape is not the detached object of a powerful gaze but a material symbolic system in which people live. Landscape is a representation of particular ideas produced by people on top of a social hierarchy. But as a text landscape is a space in which meaning and intention get pulled apart. The meaning is unstable because people act in it in ways that cannot be predicted. They read it differently. A key question for Duncan is how landscape acts as a communicative device to produce and reproduce a social and cultural order. Landscapes, he argues, can make subjects and objects appear as fixed and reified – can make what is cultural appear natural. They are the objectifier *par excellence*. They play this role through the use of textual devices such as metonymy, metaphor and allegory. Through these devices powerful people are able to tell morally charged stories about themselves, their relationships to others and their relations with the natural world.

> By becoming part of the everyday, the taken-for-granted, the objective, and the natural, the landscape

masks the artificial and ideological nature of its form and content. Its history as a social construction is unexamined. It is, therefore, as unwittingly read as it is unwittingly written. (1990: 19)

There is, of course, a problem with pitting text against vision. Texts, after all, have to be read. While it may be possible to read texts haptically, the most obvious way is by looking. Despite this critique Duncan's view of landscape is a more embodied and practised one than Cosgrove's. Cosgrove, adopting the classical definition, resists getting too involved. The landscape is seen from a distant, slightly elevated point. Landscape paintings are seen as part of the control of nature – the taming of chaos. The use of perspective arrests the temporal and makes a moment stand forever. The artist becomes the possessor and the people in the landscape – the world of practice is denied – are frozen out.

VISION AND THE OBLITERATION OF PRACTICE: THREE VIEWS FROM ABOVE

The 'gaze' at the centre of dominant conceptions of landscape has been critiqued for being masculinist. Another effect of this way of seeing is to remove the active subject (whether people in the world or the onlooker) from the world. This effect is explored in similar ways by Raymond Williams, Michel de Certeau and J.B. Jackson. In Raymond Williams' novel *Border Country* (1960) we see the hero Matthew Price returning to the place of his childhood in the Welsh borders after spending many years in England. What follows is an examination of the gap between the idea of the village as 'landscape' and the idea of the village as a lived and felt 'place'. As Matthew realizes he has become an outsider in his own village, he reflects on his change of perspective:

> He realized as he watched what had happened in going away. The valley as landscape had been taken, but its work forgotten. The visitor sees beauty, the inhabitant a place where he works and has his friends. Far away, closing his eyes, he had been seeing this valley, but as the visitor sees it, as the guide book sees it. (1960: 75)

Later in the novel Matthew gets back into the routine of the village: 'It was no longer a landscape or view, but a valley that people were using', yet from the top of a nearby hill 'The patch was not only a place, but people, yet from here it was as if no-one lived there.'

The contemplative gaze obliterates the world of the practical

The world of practice teases apart landscape in its orthodox form. The traditional image of landscape that haunts Matthew Price in *Border Country* is very much like the now poignant and not a little ironic view of New York from the top of the former World Trade Center described by Michel de Certeau in *The Practice of Everyday Life* (1984). From this viewpoint the city was comprehensible and masterable – it appeared laid out before the viewer as though the viewer could command it and take it in – just as the eminent urologist was able to provide a survey of all the latest developments in the field. The grid pattern of upper Manhattan was out there somewhere ordering the world nicely. The exclamation marks of other skyscrapers provided another dimension and sense of perspective to this landscape. When one is 'lifted out of the city's grasp,' de Certeau writes, 'One's body is no longer clasped by the streets that turn and return it according to an anonymous law; nor is it possessed, whether as player or played, by the rumble of so many differences and by the nervousness of New York traffic' (1984: 92). De Certeau does not mention landscape but he does reflect on all the ingredients that make up landscapes – fixity, vision, power.

> The totalizing eye imagined by the painters of earlier times lives on in our achievements. The same scopic drive haunts users of architectural productions by materializing today the utopia that yesterday was only painted. The 1370 foot high tower that serves as the prow for Manhattan continues to construct the fiction that creates readers, makes the complexity of the city readable, and immobilizes its opaque mobility in a transparent text. (1984: 92)

The view from the streets is slightly different. In the turmoil of Fifth Avenue or from the middle of Washington Square it is hard to keep hold of the word 'landscape'. Everything is always changing, in process, becoming. Skateboarders flip their tricks. Young men and women dance in front of the ice cream stand, the trees move in the breeze. The place is not finished, not obviously ordered and not easily framed. It is blurred at the edges. This is the space of practice. The view from on high is, as de Certeau puts it, based on the 'oblivion' and 'misunderstanding of practices' (1984: 93).

We do not have to read de Certeau to come across such observations. We need only to reopen texts we once read avidly in decades past. A third discussion of the obliviation of practice through vision is given by J.B. Jackson. Jackson remains one of the most astute observers of landscape and, indeed, practice:

> I was an early advocate of studying landscapes from the air. At one time I had a large collection of aerial views. But when flying became increasingly unpredictable I gave it up, and I recently drove from New Mexico to Illinois and Iowa in my pickup truck. It was a long trip with many monotonous hours, but I do not regret it. It broke the spell cast by the air-view of the grid system and reminded me that there is still much to be learned at ground level. What goes on within these beautifully abstract rectangles is also worth observing. (1997: 70)

Jackson was disturbed by the distancing involved in an essentially visual definition of landscape. This, to him, ignored the lived character of arrangements of space: 'we are *not* spectators; the human landscape is *not* a work of art. It is a temporary product of much sweat and hardship and earnest thought' (1997: 343). His critique of aestheticized looking permeates his annotations on landscape. In the 1960s he was an early observer of gentrification with its heritage areas, museums and coffee-table books.

> People, the magazines tell us, are once again beginning to appreciate the city. Books of glossy photographs of graffiti, ironwork details, and abstract landscapes are financially successful despite formidable prices. The Museum of Modern Art has its walking tours; restoration of old market districts is in full swing, with the end products economically viable, if often self-consciously precious ...
>
> Building watching, like bird watching, is a rewarding hobby. Both can refresh the person who practices them. Both can drive away apathy and ennui and reawaken a sense of vitality in the watcher by underscoring the beauty to be found in everyday life by one who will search for it. But building watching is a very small and isolated part of the urban experience. It bears the same relationship to living in a city as bird watching does to working a farm or ranch. (1997: 343–4)

Cosgrove, while (ironically) appreciating Jackson's acute powers of observation, is not happy with his definition of landscape because it does not recognize that 'landscape is indeed the view of the outsider, a term of order and control, whether that control is technical, political or intellectual'

(1984: 36). He argues that Jackson is not really talking about landscape because he is too inside it. It is true that J.B. Jackson had problems with the term 'landscape'. Although he is very closely associated with a journal of the same name and wrote paper after paper with landscape in the title, when it came down to it he found it too slippery and polysemous. He was too interested in practice and not satisfied with thinking of landscape in a restrictive sense, as a product of vision – as 'a portion of the earth's surface that can be comprehended at a glance' (1997: 306). Jackson, more than anyone, reveals the limits and tensions of the landscape idea. On the one hand he carefully described the origins and history of landscape in terms of vision, and on the other he enacted a critique of the term through an emphasis on everyday routines that produce and reproduce actual living landscapes. He asked those writing about landscape to be careful observers and simultaneously to undermine the priority given to the act of looking. Jackson's landscapes are ones that people inhabit and work in and they are landscapes that people produce through routine practice in an everyday sense.

J.B. Jackson and the practice of vision

Jackson's post-secondary education began in 1928 when he enrolled in the Experimental College of the University of Wisconsin in Madison where he took a broad liberal arts course covering art, literature and society in an interdisciplinary framework. After a year travelling in Europe he went on to Harvard and majored in history and literature. He was particularly inspired by Derwent Whittlesey's course 'Principles of geography'. He moved from there to the School of Architecture at the Massachusetts Institute of Technology in 1932 and withdrew after a year, going to an art school in Vienna where he learned drawing and draughting techniques. Again he left early and bought a motorcycle which he used to tour Europe for the next two years. On return to the United States he moved to New Mexico and became a cowboy. During the war he used his learning to become an intelligence officer – a role that involved the study and interpretations of maps. He saw active service in North Africa, Italy and France. While waiting for the Germans to surrender he occupied himself reading the works of French cultural geographers such as Vidal de la Blache whom he admired. After the war was over he returned to New Mexico and, following a riding accident, decided to match the European geography publications with an American version that was to become *Landscape*. The first issue was published in Spring 1951. In his early essays Jackson outlined the importance and pleasure of careful observation. 'I see things very clearly and I rely on what I see,' he explained, 'And I see things that other people don't see, and I call their attention to it' (Horowitz, 1997: xxiv). By 1956 Jackson was in close contact with the Geography Department at Berkeley, and Sauer (as well as Clarence Glacken, Lewis Mumford, Kevin Lynch and others) began to write for the journal, developing a multidisciplinary and maverick conception of the study of landscape. Jackson continued to criss-cross the United States on his motorbike and in his pickup truck and frequently wrote about the landscape from the perspective of a mobile observer. It is in these essays that the dualism of vision and practice breaks down.

If the equation that links landscape to vision has frequently erased practice, then J.B. Jackson's mobile view of landscape began to show how vision is a practice. J.B Jackson's way of looking is so much less reliant on that distanced gaze from above and so much more practised – more embodied. One of the best essays in this regard is 'The abstract world of the hot rodder'. He introduces us to the world of the pedestrian Sunday. Families would leave their homes on Sunday morning complete with all the accessories of really leisurely leisure. They would take the street car to some 'favorite patch' and walk into the distance 'strolling at a child's pace'. Jackson declares these Sunday excursions a lost treasure which inspired 'schools of painting and writing and music ... Untold minor scientific and artistic accomplishments came from the same abundant source; local botanical and geographical and historical descriptions, small books of nature verse, amateur sketches and compositions.' Despite this romantic notion of a weekend stroll he declares it mere nostalgia to think in those terms in the modern world because the nature of viewing – the way it is embodied – has changed.

> Much more has happened to us than the advent of the automobile; we have learned to see the world differently even on our holidays; we confront the familiar setting in a new manner. Broadly speaking, the former experience of nature was contemplative and static. It came while we strolled (at three miles an hour or less) through country paths with frequent halts for picking flowers, observing wildlife, and admiring the view. Repose and reflection in the midst of undisturbed natural beauty and a glimpse of something remote were what we chiefly prized. I do not wish to decry the worth of these pleasures; none were ever more fruitful in their time; but the layman's former relationship to nature ... was largely determined by a kind of classical perspective and by awe. A genuine sense of worship precluded any desecration, but it also precluded any desire for participation, and intuition that man also belonged. (1997: 202)

Jackson goes on to develop the idea of participation in an 'abstract world' brought on by technologies of speed such as the motorbike and increasingly popular outdoor pursuits such as skiing. Rather then standing back and looking, the hot rodders take part in the landscape. They experience it through embodied practice. A similar perspective is at the heart of his later essay 'The accessible landscape' (1988), in which Jackson describes the view through the window of his favourite pickup truck as he drives through America's vernacular heartland – not the static view from on high but a moving perspective more typical of the everyday practice of American life. Many years earlier in 'Other-directed houses' Jackson celebrated the highway strip as a place of pleasure and diversion for the motorist, pointing out, 16 years before Venturi et al.'s *Learning from Las Vegas*, that the buildings on the strip have to be designed for a kind of vision that travelled at 40 miles per hour – vision as a particular kind of historically specific practice.

Thinking Practice

'Practice' is a term which has begun to make a significant impact on cultural geography. Its origins are in critical theory, performance studies, feminism and post-Marxist social theory. On the whole it has been absent from landscape studies. It is already clear that to talk of 'practice' can mean quite different things to different people. I began this chapter with the observation that the notion of landscapes of practice was based on oxymoronic logic. Landscape, as we have seen, is predominantly thought of as a visual thing – as an image. It is also thought of as material and fixed – a kind of framing. Practice on the other hand is most frequently seen as radically unfixed and unfixable.

Theories of practice from Bourdieu to Judith Butler to Nigel Thrift are indebted to the work of Maurice Merleau-Ponty. He was particularly concerned with what he called the 'body-subject'. The work of Brentano and Husserl had led to the key idea that consciousness is always *consciousness of something*; in other words, there is a directedness to consciousness. The focus of intentionality, post-Husserl, had been the mind. Merleau-Ponty, in his conception of the 'body-subject', wanted to redirect intentionality to the body. Merleau-Ponty argues that 'being in the world' is a bodily thing (rather than mental), a pre-objective view. The fundamental relationship between subject and object exists at a bodily level that pre-exists congition. In Merleau-Ponty's words:

> The plunge into action is, from the subject's point of view, an original way of relating himself to the object, and is on the same footing as perception. (1962: 110–11)

Bodily practice, then, is the most basic form of intentionality: 'Consciousness is in the first place not a matter of "I think that" but of "I can"' (1962: 137). This phenomenological consideration of the body's movements makes a clear distinction between the world of embodied movement (motility) and objectivity and representation. Basic bodily movement is not 'thought about movement' (1962: 137) but embodied movement that take place in the world of the phenomenal. As examples, Merleau-Ponty points towards fairly complicated sets of bodily practices such as using a needle and thread, or typing on a typewriter. He notes how these operations are not thought about once they are learned. We do not calculate the positions of our hands in space and work out abstractly where they should go next. Rather, our hands act out of habit in a form of bodily consciousness.

> Consciousness is being towards the thing through the intermediary of the body. A movement is learned when the body has understood it, that is, when it has incorporated it into its 'world', and to move one's body is to aim at things through it; it is to allow oneself to respond to their call, which is made upon it independently of any representation. (1962: 138–9)

Independence from representation is the key point here. Merleau-Ponty wants to make sure that we understand that movement is not secondary to consciousness but a more primary form of consciousness.

Merleau-Ponty also considers the relationship between the movement of the body-subject and the world of space and time. His argument is that space and time are not mere backdrops to our movement but are 'inhabited' by movement – that there is a continuum between space/time and the body.

> [It] is clearly in action that the spatiality of our body is brought into being, and an analysis of one's own movement should enable us to arrive at a better understanding of it. By considering the body in movement, we can see better how it inhabits space (and, moreover, time) because movement is not limited to submitting passively to space and time, it actively assumes them, it takes them up in their basic significance which is obscured in the commonplaceness of established situations. (1962: 102)

The body inhabits space and time through motion. 'I am not in space and time,' writes Merleau-Ponty, 'nor do I concieve space and time; I belong to them, my body combines with them and includes them' (1962: 140).

Merleau-Ponty's phenomenology of bodily movement was taken up by David Seamon in the 1970s. Geographers such as Seamon were keen to arrive, through phenomenological enquiry, at the essence of geographical phenomena. Place was the central concept but, in Seamon's case, bodily mobility was a key component of the understanding of place. Like Merleau-Ponty, Seamon fixed on the 'everyday movement in space' – '*any spatial displacement of the body or bodily part initiated by the person himself*. Walking to the mailbox, driving home, going from house to garage, reaching for scissors in a drawer – all these behaviours are examples of movement' (1980: 148).

Through research groups Seamon came to the conclusion, in line with Merleau-Ponty, that most everyday mobile practices take the form of habit. People drive the same route to work and back every day without thinking about it. People who have moved house find themselves going to their old house and only realize it when they arrive at the door. People reach for scissors in the drawer while engaging in conversation. Such practices appear to be below the level of conscious scrutiny. The body-subject knows what it is doing: there is an

> inherent capacity of the body to direct behaviors of the person intelligently, and thus function as a special kind of subject which expresses itself in a preconscious way usually described by such words as 'automatic', 'habitual', 'involuntary', and 'mechanical'. (1980: 155)

Seamon invokes the metaphor of dance (ballet to be specific), in order to describe the sequence of preconscious actions used to complete a particular task such as washing the dishes. He calls such a sequence a body-ballet. When such movements are sustained through a considerable length of time he calls it a 'time–space routine'. This describes the habits of a person as they follow a routine path through the day – driving to work, leaving the kids at school, going to lunch, etc. Seamon also looks beyond the individual body movement to group behaviour. When many time–space routines are combined within a particular place, a 'Place-ballet' emerges which generates, in Seamon's view, a strong sense of place. The mobilities of bodies combine in space and time to produce an existential insideness – a feeling of belonging within the rhythm of life-in-place.

Where am I going with this? Thus far I have developed the idea of landscape as primarily visual and fixed and the idea of practice as about fluidity and embodiment. In other words I have provided a sketch of the contradictions implicit in the term 'landscapes of practice'. The word 'landscape' and the word 'practice' seem to be poles apart. When Raymond Williams developed the idea of a 'structure of feeling' he wanted to draw the two, apparently opposite, terms together – to make structure less fixed and certain and to make feeling less individual and irreducible. To end this chapter I want to do the same with landscape and practice. I want to make landscape seem less fixed, less reliant on the visual, less dependent on authoritative 'framing', and to make practice seem less free-floating and more connected to the forces that shape our lives. To do this I explore some ways forward arising out of recent literature on landscape: ideas of material culture developed by archaeologists and questions of 'moral geographies' developed within geography.

WAYS FORWARD: DOXIC LANDSCAPES

W.J.T. Mitchell, in *Landscape and Power*, makes an argument for landscape as a space in which we lose ourselves – a medium for life. He

> aims to absorb these approaches into a more comprehensive model that would ask not just what landscape 'is' or 'means' but what it *does*, how it works as a cultural practice. Landscape, we suggest, doesn't merely signify or symbolize power relations; it is an instrument of cultural power … Landscape as a cultural medium thus has a double role with respect to something like ideology; it naturalizes a cultural and social construction, representing an artificial world as if it were something given and inevitable, and it also makes that representation operational by interpellating its beholder

in some more or less determinate relation to its givenness as sight and site. (1994: 1–2)

We might think of such a landscape as a 'doxic landscape'. I take the term 'doxa' from Bourdieu's theorizations of practice. I have previously written about doxa in relation to place (Cresswell, 1996). Similar considerations can illuminate the idea of practised landscapes. Doxa refers to the realm of common sense – the taken-for-granted. Doxa is a product of practice. Because people act in certain preconscious ways, any given order tends to get re-established and reproduced owing to the 'naturalisation of its own arbitrariness' (Bourdieu, 1977: 164). In other words a world, or landscape, that is the product of a particular history is made to seem natural and thus becomes an important site for the reproduction of established ways of being. Such a landscape is very much a product and producer of practice. What might the concept of doxic landscapes mean in practice?

Some of the most suggestive work on landscape has come from recent archaeology. Like geography the discipline of archaeology has gone through a reaction against a positivist legacy. Part of this reaction is an increased interest in social theory and cultural geography. Leading exponents of theoretically informed archaeology include Mike Shanks, Chris Tilley and Barbara Bender. Predictably, perhaps, new ways of thinking about landscape have emerged from a critique of the politics of vision and an engagement with the politics of practice.

In Shanks and Tilley's work landscape falls under the more general rubric of 'material culture'. Their argument, crudely put, is that archaeologists studying material culture have simply looked at pottery shards and remnant landscapes with a view to mapping and describing – visualizing if you like. What they have not done is thought about their objects of study in terms of practice. They propose a new way of thinking about material culture that is informed by theories of practice. Material culture is theorized as

> an important element of Bourdieu's doxa, a representation of the given order of the world that constitutes an environment for living; as an effective force for social action; as ideologically informed due to its perceived simple functionality, concreteness and triviality which facilitate naturalisation and misrepresentation. (1987: 113)

Landscape, theorized as material culture in these terms, is not a product of vision in particular but a practised environment. Landscape cannot easily be 'decoded' as there is no code that it can be reduced to. Instead landscape is brought into being through social praxis as an objectified form.

> Material culture is an objectification of social being, a literal reification of that social being in the co-presences and absences embodied in the material form. (1992: 130)

Landscape seen this way is a practised landscape. Practices over time become embedded in the world and leave 'traces of varying degrees of solidity, opacity or permanence' (1992: 131). These traces are 'material culture'. Landscape as material culture is thus implicated in the process of social reproduction which involves the inter-connection between materiality, consciousness, action and thought.

Julian Thomas develops this in his analysis of the landscape surrounding Avebury. Thomas applies the kinds of ideas about material culture developed by Shanks and Tilley to landscape. Landscape in the western tradition, he argues, is a product of distance and position which con-struct 'a particular impression of the world, but are at the same time denied and the view is taken as universal, taking in everything' (1993: 22). The landscape thus becomes a passive object constructed through a process that is denied in the process of construction. Rather than thinking of archaeological landscapes as a distanced 'impression of the world', Thomas sets out to think through how views of landscape would change if they were instead thought of as bound together by the continuous flow of human con-duct. Utilizing a phenomenological approach it becomes possible to think of landscapes not as external objects but as sites of dwelling. Thomas reiterates a point made by Ted Relph that 'To think about the world or the entities within it as abstract things is to render them subject to obser-vation, to make them the object of casual curio-sity and distance oneself from them' (Thomas, 1993, citing Relph, 1985: 27). Rather than make landscape into an object marked by its existence at a distance, a phenomenological approach focuses on the experiential nature of the world around us. Heidegger's concept of 'dwelling' is the key here:

> Dwelling involves a lack of distance between people and things, a lack of casual curiosity, an engagement which is neither conceptual nor articulated, and which arises through *using* the world rather than through scrutiny. (1993: 28)

Thomas's agenda, in short, is to transform the archaeological landscape from a product of dis-tanced vision to an arena marked by its 'not-to-be-looked-at-ness' (1993: 28). Avebury, then, is understood quite differently when praxis is taken seriously. Thomas thinks through the processes of walking up to the ring of stones: his analysis is less about looking and more about doing, 'How was Avebury used?' not 'What did Avebury look like?'

More recently this idea has been developed on the interface between performance studies and archaeology by Michael Shanks and Mike Pearson (2000). Landscape is at the heart of their concerns. They build on the insights of Merleau-Ponty, focusing on the bodily experience of the land:

> We begin to walk. We feel the ground beneath our feet, the wind in our face. And as we do we leave traces. We are *involved* in the landscape ... We leave the prints of our body, the touch of flesh on metal and stone. We constantly wear things out, with our hands, our feet, our backs, our lips. And we leave the traces of singular actions: the unintentional, the random, the intimate, unplanned touch of history's passing: we break twigs, move pebbles, crush ants ... all the signs that trackers learn to read. (2000: 135)

Landscape becomes a palimpsest – a stratigraphy of practices and texts. In a performance piece called 'The first five miles' Pearson began a walk across Mynedd Bach near the village of Trefanter, south of Aberystwyth, wearing leather boots and gaiters, embroidered waistcoat, frock coat, top hat and lilac gloves and equipped with radio microphone, battery unit, earpiece, receiver and halogen lamp. He was accompanied by an assis-tant who carried a radio transmitter. As they walked they talked and the transmissions were sent to Aberystwyth and broadcast on Radio Ceredigion. These messages would interrupt a documentary about an attempt to enclose the peat bogs in the area in the 1820s by Augustus Brack-enbury and the various attempts to stop him made by local inhabitants in 'The War of the Little Englishman'. Some text was in English and some in Welsh and these were transmitted in stereo to the right and left respectively. As the walk progressed, accompanied by a fractured soundtrack, local memories intervened, telling stories of 'all those things which we might never regard as authentic history but which go to make up the *deep map* of the locale' (Shanks and Pearson, 2000: 144). The point of the perfor-mance was to illuminate a particular landscape – to create a work which:

has none of the dogmatism of the theatrical performance, or architectronics and that distanced aesthetic – framed up, laid out for our pictorial inspection and approval. So that the very *inauthenticity* of the performance allows room for manoeuvre, allows stances, of ownership, identity and interpretation, to be confirmed, challenged, confounded at the same time. (2000: 146)

The integration of landscape with practice inherent in the performances and writings of Shanks and Pearson is also a feature of some recent work on tourism. Tim Edensor's *Tourists at the Taj* (1998) considers the often-looked-at Taj Mahal as a landscape of practice. Four practices he focuses on are walking, gazing, photographing and remembering. Consider walking. Walking is a fairly basic feature of most people's lives – a routine and mundane practice. It is also a practice central to tourism. Most of us, I would guess, have spent time walking around sites while on holiday. Edensor suggests that in tourism 'Landscapes are criss-crossed and imprinted with the bodily presence of the visitor, and symbolic sites are negotiated via various paths' (1998: 105). The movements of tourists, he suggests, become sedimented in the landscape as a kind of 'place-ballet'. Indeed the places of most frequent passage become worn and grooved in the most material sense. The way these practices get to be sedimented is through repetition. Indeed, in the Taj, it seems, most walking is highly organized and choreographed.

At the Taj, obeying the instructions of guides and tour organisers, most package tourists follow prescribed paths, moving towards certain valorised spaces and features and not others. The performance of these disciplined collective choreographies constitutes a quite precise and predictable 'place-ballet'. Bodies are tutored and disciplined, kept together and directed by assumptions about what is deemed 'appropriate', by group norms, and principally by the orders of the guide. (1998: 107)

Edensor goes on to distinguish these kinds of regulated tourist practices from other ways of walking. Backpackers, apparently, are more likely to wander fairly randomly and unguided, while local visitors make a point of visiting the mosque as part of the itinerary. In each of these cases, though, his point is that the landscape is practised – it comes alive – in both regulated and unregulated ways.

It is notable the degree to which the work I have just summarized looks back to Merleau-Ponty for its inspiration. The habitual and embodied is used to critique a particular form of disembodied gaze. Edensor speaks of 'place-ballets' – a term which must have found its way from David Seamon into tourism studies by some circuitous route. Cultural geographers long ago enacted a trenchant critique of the idealism to phenomenology with its seeming inability to intuit power relations in the process of transcendental reduction. So while a rediscovery of forgotten strains of thought is always welcome, it is necessary to also pay attention to power and its relation to practice.

I think I am in agreement with Don Mitchell here. He has argued, as I do here, that landscape is very much a lived phenomenon. Cultural geography, in its turn to the world of representation, has looked away from material spaces and seen landscape as ideology, as representation. This critique has been framed as a critique of looking at landscapes as merely material forms. Landscape, in the process, has been turned into a rarified realm of art and gardens. By refocusing on the materiality of landscape it is possible for cultural geographers to interpret the role of landscape in the world of practice. As W.J.T. Mitchell has argued, 'an account of landscape has to trace the process by which landscape effaces its own readability and naturalizes itself' (1994: 2).

Recent work has connected the issue of landscape to that of 'moral geographies'. Very simply put, moral geographies concern the normative relationship between space and behaviour. What and who belong where and when? Such considerations necessarily bring together the landscape and body through practice. Cosgrove's landscape was not one to be in but one to be outside of. Landscapes were moral in so far as they could be read as a diagram of a particular ideology or social formation, whether that be feudalism, merchant capitalism or Buddhist kingship. David Matless' work on *Landscape and Englishness* (1998) is quite different. His landscapes are practised landscapes. He asks what practices are appropriate for an English landscape. He connects the way of seeing inherent in landscape to ways of dressing, of bodily comportment, of trespass and of mountain climbing. In this way the English landscape is connected to forms of practical belonging such as citizenship. The English landscape signified proper ways of being through the propaganda of a diverse range of interest groups such as ramblers, scouts, mystics and dancers. The practice of people in the English landscape was not divorced from questions of the visual. As Matless puts it:

Distinctions of citizenship and anti-citizenship turned on questions of appropriate conduct and aesthetic ability. Landscaped citizenship worked through a mutual constitution of the aesthetic and the social, the eye and the body. The aim of extending visual pleasure to the people was tempered by a desire to control potentially disruptive bodily effects. The education of the eye was to be accompanied by a self-control in the body. (1998: 62–3)

In Matless' moral geography of landscape, landscapes are inhabited, appropriately or otherwise, by people doing things. Like Jackson's landscapes these are not just seen at a distance but are used and lived in. Practice is not obliterated here. Rather landscape is the site of 'dialectical tensions of eyes and bodies, the visceral and the cerebral, pleasure and citizenship, ecstacy and organisation' (1998: 63). Matless and landscape archaeologists such as Shanks and Tilley have succeeded in putting flesh on the bones of J.B. Jackson's practised and everyday landscapes. Along with this return of the practising and mobile body has come a new interest in the everyday and unexceptional. The focus on vision tends to have led cultural geographers towards remarkable and, dare I say it, elite landscapes.

The challenge for cultural geographers of landscape is to produce geographies that are lived, embodied, practised; landscapes which are never finished or complete, not easily framed or read. These geographies should be as much about the everyday and unexceptional as they are about the grand and distinguished. Theories of practice can add theoretical weight to the wonderful insights of a geographer such as J.B. Jackson. What the idea of 'landscapes of practice' allows is an injection of temporality and movement into the static at the same time as practice is contextualized and given a frame.

REFERENCES

Bourdieu, P. (1977) *Outline of a Theory of Practice* Cambridge: Cambridge University Press.

Cosgrove, D. (1984) *Social Formation and Symbolic Landscape*. London: Croom Helm.

Cosgrove, D. (1985) 'Prospect, perspective and the evolution of the landscape idea', *Transactions of the Institute of British Geographers* 10: 45–62.

Cosgrove, D. (1987) 'Place, landscape and the dialectics of cultural geography', *Canadian Geographer* 22: 66–72.

Cresswell, T. (1996) *In Place/Out of Place: Geography, Ideology and Transgression*. Minneapolis: University of Minnesota Press.

Daniels, S. (1990) 'Marxism, culture and the duplicity of landscape', in N. Thrift and R. Peet (eds) *New Models in Geography*, Vol. 2. Boston: Allen and Unwin. pp. 177–220.

Daniels, S. (1993) *Fields of Vision: Landscape Imagery and National Identity in England and the United States*. Cambridge: Polity.

de Certeau, M. (1984) *The Practice of Everyday Life*. Berkeley: University of California Press.

Duncan, J. (1980) 'The superorganic in American cultural geography', *Annals of the Association of American Geographers* 70: 181–98.

Duncan, J.S. (1990) *The City as Text: The Politics of Landscape Interpretation in the Kandyan Kingdom*. Cambridge: Cambridge University Press.

Edensor, T. (1998) *Tourists at the Taj: Performance and Meaning at a Symbolic Site*. London: Routledge.

Horowitz, H. (1997) 'J.B. Jackson and the discovery of the American landscape', in J.B. Jackson *Landscape in Sight: Looking at America*. New Haven: Yale University Press. pp. ix–xxxiv.

Jackson, J.B. (1984) *Discovering the Vernacular Landscape*. New Haven: Yale University Press.

Jackson, J.B. (1997) *Landscape in Sight: Looking at America*. New Haven: Yale University Press.

Matless, D. (1998) *Landscape and Englishness*. London: Reaktion.

Meinig, D. (1979a) 'The beholding eye', in D. Meinig (ed.) *The Interpretation of Ordinary Landscapes*. Oxford: Oxford University Press. pp. 35–50.

Meinig, D. (ed.) (1979b) *The Interpretation of Ordinary Landscapes*. Oxford: Oxford University Press.

Merleau-Ponty, M. (1962) *The Phenomenology of Perception*. London: Routledge and Kegan Paul.

Mitchell, W.J.T. (ed.) (1994) *Landscape and Power*. Chicago: University of Chicago Press.

Olwig, K.R. (1984) *Nature's Ideological Landscape: A Literary and Geographic Perspective on its Development and Preservation on Denmark's Jutland Heath*. London: Allen and Unwin.

Relph, E. (1976) *Place and Placelessness*. London: Pion.

Relph, E. (1985) 'Geographical experiences and being in-the-world: the phenomenological origins of geography', in D. Seamon and R. Mugerauer (eds), *Dwelling places and Environment: Towards a Phenomenology of Person and World*. New York: Columbia University Press. pp. 24–36.

Sauer, C. (1965) 'The morphology of landscape', in J. Leighy (ed.) *Land and Life*. Berkeley: University of California Press.

Seamon, D. (1980) 'Body-subject, time–space routines, and place-ballets', in A. Buttimer and D. Seamon (eds) *The Human Experience of Space and Place*. London: Croom Helm. pp. 148–65.

Shanks, M. and Pearson, M. (2000) *Theatre/Archaeology*. London: Routledge.

Shanks, M. and Tilley, C. (1987) *Social Theory and Archaeology*. Oxford: Polity.

Shanks, M. and Tilley, C. (1992) *Re-constructing Archaeology: Theory and Practice*, 2nd edn. London: Routledge.

Thomas, J. (1993) 'The politics of vision and the archaeology of landscape', in B. Bender (ed.) *Landscape: Politics and Perspectives*. Oxford: Berg. pp. 19–48.

Tuan, Y.-F. (1977) *Space and Place: The Perspective of Experience*. Minneapolis: University of Minnesota Press.

Tuan, Y.-F. (1979) 'Thought and landscape', in D. Meinig (ed.) *The Interpretation of Ordinary Landscapes*. Oxford: Oxford University Press. pp. 89–102.

Venturi, R., Scott-Brown, D. and Izenour, S. (1972) *Learning from Las Vegas*. Cambridge, MA: MIT Press.

Williams, R. (1960) *Border Country*. London: Chatto and Windus.

Williams, R. (1977) *Marxism and Literature*. Oxford: Oxford University Press.

Harlem, 1934 by Edward Burra (© Tate, London 2002)

Section 5

PLACING SUBJECTIVITIES   Edited by Robyn Longhurst

Introduction: Subjectivities, Spaces and Places

Robyn Longhurst

Subjectivity grounds our understanding of who we are. It also grounds our claims to geographical knowledge. All geographical knowledge, whether it is spatial science, behavioural geography, Marxist geography, feminist geography or cultural geography, presupposes some theory of subjectivity. Different theories of subjectivity prompt different geographical knowledges. To put it crudely, 'doing' geography means 'doing' subjectivity (and vice versa). It is not surprising, therefore, that since the late 1960s and early 1970s geographers have shown a keen interest in the relationship between subjectivity and space.

Pile and Thrift explain: 'By the early 1970s, geographers had begun to experiment with different models of (what was then called) "man". These models of subjectivity were primarily drawn from the disciplines of cognitive psychology, political economy and philosophy' (1995: xii). For example, David Harvey in *Explanation in Geography* (1969) offered up behavioural geography as a corrective to the inert, lifeless mathematical models used in systems theory.

In the 1970s, humanist geographers also offered up new models of subjectivity. For example, Torsten Hägerstrand rejected a Cartesian separation between mind and body to explore subjects' 'lived' time-space geographies. Hägerstrand, and other humanist geographers, imagined subjects to be self-knowing, bounded and unique individuals. It was thought that subjectivity was contained within the body, which confined it and bound it as separate from what was outside, different and other (see Longhurst, 1995, and Rose, 1993, on the masculinist imperative inherent in this understanding of subjectivity).

In the early 1980s, Nigel Thrift explicitly addressed the issue of subject formation. Thrift (1983) did not fall on the side of either structure (that circumstances decide what people do) or agency (that people choose their own paths) but attempted to understand the subject in terms of both structure and agency. In the 1990s, the terms of debate on subjectivity were recast yet again. Geographers began to focus not on the structure/agency debate but on discourse and representation.

Cultural geographers, as well as others drawing on poststructuralist thought, began to examine some of the ways in which subjectivity and space can be understood as simultaneously real, imaginary and symbolic (Duncan, 1996; Keith and Pile, 1993; Pile, 1996; Pile and Thrift, 1995). One of the attractions of this approach for cultural, feminist, postcolonial and other critical geographers was that it offered a new way to begin to understand power relations. Poststructuralist theory led us to reject notions of a coherent subject, arguing that the many cultural, social, political and psychical processes that continually (re)construct the subject, make self-awareness and self-fashioning impossible (see Soper, 1986, on the differences between humanist and anti-humanist conceptions of the subject). Alcoff (1988) argues that it was Freud's initial questioning of the coherent and bounded subject that lay the ground for later rejections of it.

Currently, 'the subject' is a term often used to refer 'to the individual human being/agent, accenting both physical embodiment and the range of emotional-mental processes through which it thinks its place in the world' (McDowell and Sharp, 1999: 267). These days understanding the subject and subjectivity involves the negotiation of a whole series of interconnected terms such as the body, the self, identity and the person (Pile and Thrift, 1995). These terms 'are usually equivocal, often ambiguous, sometimes evasive and always contested' (1995: 6). Pile and Thrift, therefore, caution against attempts to provide precise definitions, preferring instead to provide 'a preliminary feel for the lie of the land' (1995: 6). There is little doubt that the debates about the subject and subjectivity are incredibly important. As Johnston et al. (2000: 802) point out, they are also vast and difficult to summarize. There are 'forests of literature on the subject' (Pile and Thrift, 1995: 1). In this introduction I can do little more than cut one narrow path.

Elsbeth Probyn begins her chapter by posing the question: 'what is subjectivity?' She makes the point that often subjectivity and identity get used interchangeably. Probyn, however, uses the term 'subjectivity' because it relates to the concept of the subject which, she argues, relates to the idea of ideology. She describes the French philosopher Louis Althusser's understanding of ideology as actively constituting or interpellating individuals as subjects. His understanding 'compels us to consider closely the material contexts which allow and delimit our individual and collective performance of selves'. Althusser's theory of the subject, explains Probyn, allows for this engagement with space and place, it allows us to examine 'the spatial imperative of subjectivity'. This 'spatial imperative of subjectivity' is a major theme in the chapters that follow.

Alastair Bonnett and Anoop Nayak, Michael Brown and Larry Knopp, and Liz Bondi and Joyce Davidson all attend to the intellectual trajectories that develop the relationship between aspects of subjectivity and space, seeing this as an important aspect of cultural geography. The argument is mounted that all forms of spatial thinking presuppose some theory of subjectivity and that, conversely, all forms of subjectivity presuppose some theory of space.

Brown and Knopp argue 'a variety of subjectivities are performed, resisted, disciplined and oppressed not simply in but through space'. Bonnett and Nayak make a similar point but link it to race and ethnicity. They claim: 'To speak the language of race and ethnicity is, very often, to talk geography.' It is impossible to use terms such as 'Asian', 'African', 'European' or 'western' without talking about territory, space and place.

Bondi and Davidson note: 'some important studies in cultural geography have argued that people and places are imagined, embodied and experienced in ways that are ... radically and inextricably intertwined with each other'. They cite Davidson (2001), Kirby (1996) and Nast and Pile (1998) as examples of work in which subjects and environments are treated as mutually constituted (also see Grosz, 1992). Bondi and Davidson explain: 'To be is to be somewhere, and our changing relations and interactions with this placing are integral to understandings of human geographies. Moreover, gender is inscribed deeply within these processes.'

All this is to suggest not that subjectivity can be mapped onto a fixed space as though it were a backdrop, but that the 'subject is mapped, and maps, into interminable dimensions of power which subsist at all points' (Pile and Thrift, 1995: 44). Subjects are 'entangled' (see Sharp et al., 2000) in multiple power relations that are simultaneously real, imagined and symbolic. These power relations may be differentially organized through varying relations of race, sexuality, gender and so on, but in all instances they are written on and through the subjectivities and spaces under discussion.

Given that subjectivity cannot be plucked from the spatial relations that constitute it, it is hardly surprising that a vast array of places and spaces, at a range of different scales, is represented in the chapters that follow. For example, Brown and Knopp, in articulating the complex relationship between sexuality, politics and place, refer to the home, the academy, gay bars, gyms, workplaces and chat rooms. Bondi and Davidson consider gendered environments such as the stereotypical building site, boxing ring and aeroplane cockpit, which they compare with the 'typing pool', the aerobic fitness centre and the aeroplane isle. This careful attention to the manifestation of power relations in local environments, however, does not necessarily preclude a careful eye being cast over broader territories. Bonnett and Nayak argue that 'white studies' needs to be global in its reach. While they realize this may, 'to some, have a slightly colonial ring to it', white identities are a global phenomenon that have global impacts. Whiteness continues to be constructed as a cultural and racial norm.

Not only is subjectivity always emplaced; it is also always embodied. This is the second theme that links the chapters in this section. Subjects have a weighty materiality. Probyn claims that there is a growing awareness amongst cultural theorists of a need to instil a sense of materiality into theorizations of subjectivity. She explains: 'this may be due to a collective turning away from some of the excesses of postmodernism and even post-structuralist thought which, it can be argued, diluted a clear sense of the constraints of context'.

Bonnett and Nayak display this awareness of a need to instil a sense of materiality into their theorizations. They argue that racial terms such as 'white', 'black' and 'people of colour' refer directly to the body and designate or 'fix' our flesh and blood in particular ways. Brown and Knopp reiterate that sexuality is not simply a social construct but a 'set of lived experiences'. In a discussion of the important role played by Bob McNee in putting sexuality on the geographical agenda in the mid 1980s, Brown and Knopp explain that his contributions represented 'an *embodied* insistence that oppression of gays and lesbians was real, systematic and fully present in the discipline of geography'.

Bondi and Davidson extend this point to argue that lived experiences and ideas are intimately interwoven. There is no clear distinction between 'doing' and 'theorizing' gender. In discussing the powerfulness, pervasiveness, and taken-for-grantedness of differentiating people by gender – by being 'male' or 'female' – Bondi and Davidson invite readers to reflect on their own surprise at discovering that the voice we heard as 'male' turns out to belong to a woman (or vice versa).

Probyn also moves seamlessly between 'doing' and 'theorizing'. She invites readers to consider Althusser's account of ideology alongside her own account of a 'happenstance' encounter with a young aboriginal woman in Redfern, Sydney. 'Doing' difference cannot be extracted from theorizing difference.

A third key theme evident in this section is the desire to challenge or resist (see Pile and Keith, 1997, on resistance and Thrift, 1997: 127, on the significance of the 'ion' in subjec*tion*) racist, heteronormative and/or masculinist regimes. Bondi and Davidson explain that their account arose from their 'own commitment to, and participation in, the development of feminist perspectives … in cultural geography'. This politics, argue Bondi and Davidson, may take the form of Enlightenment ways of thinking about human values, progress, inequalities, self-determination and so on, or it may take the form of post-Enlightenment critiques of metanarratives and universal knowledge: 'feminist perspectives need to deploy both Enlightenment and post-Enlightenment ways of thinking … and in so doing have the potential to foster productive paradoxes'.

There is much debate on what are the most effective ways to resist societal relations, processes and/or institutions that foster exploitation and domination (see Routledge, 1997: 68–72). Resistance is not simply the binary opposite of domination, nor is it something that can be plucked from spatiality. There are *geographies of resistance* (Pile and Keith, 1997). Like Bondi and Davidson, Brown and Knopp recognize that resistance is complex, contradictory and often paradoxical. They note, in relation to being simultaneously both inside and outside 'the closet', that it is possible to escape and resist one kind of oppression only to be engaged in new forms of oppression. 'The closet' itself can be read paradoxically as a space of lack but also as a space of 'creative, ingenious and transformative sexual, cultural and political resistances to heteronormativity'. Similarly, Bonnett and

Nayak point to contradictions and paradoxes around resistance. They cite the work of Kobayashi and Peake (2000) who argue that geographers need to embark upon a critical engagement with whiteness. The lives of dominantly white geographers, argue Kobayashi and Peake, 'are sites for the reproduction of racism, but they also hold the potential of being strategic sites of resistance' (2000: 399).

To sum up thus far, there are at least three key themes that run throughout the chapters in this section. The first and foremost is that subjectivities are performed not simply in but *through* space. Probyn concludes her chapter:

> We need to think of subjectivity as an unwieldy, continually contestable and affirmable basis for living in the world. Subjectivities are then simply a changing ensemble of openings and closings, points of contact and points which repel contact. In space, we orient ourselves and are oriented. That is the spatial imperative of subjectivities. (see p. 298)

The second theme to emerge is that subjectivity is grounded in the materiality of bodies, in everyday lived experiences, and in particular sites. Nast and Pile remind us 'we live our lives – through places, through the body' (1998: 1). They explain that there is a pressing need to examine the interconnections between bodies and places because the ways in which we live out these interconnections, these relationships, are political. This point leads to the third theme – that subjectivities and spaces cannot be separated from politics.

A commitment to rethinking questions of race, sex and gender with a view to opening up productive possibilities for change for the better is evident in all the chapters. The authors are keen to draw attention to the myriad spaces of political contestation and in particular, of resistance.

I now want to turn attention to the question: what exciting new work might lie just around the corner for cultural geographers interested in subjectivity and space? This question is difficult to answer. However, one possibility signalled by several authors in this section is that we might continue to see emerging research on the complex intersections between race, sexuality, gender, class and other facets of subjectivity.

One recent example of this type of work is Gibson-Graham, Resnick and Wolff's *Class and Its Others* (2000). The argument underlying this book is that individuals often participate in various class relations at different times: class subjectivities are multiple and unstable, interacting with other aspects of subjectivity in contingent and unpredictable ways. These subjectivities are constituted simultaneously at the interpersonal, community, national and global levels. We are increasingly seeing research that decentres one specific aspect of subjectivity as the primary category of analysis (see Longhurst, 2002). While it is difficult, and not always politically strategic, to examine simultaneously multiple axes of subjectivity, in some instances it may prove enlightening.

Readers will note that the chapters in this section (with the exception of Probyn's) are organized around three important aspects of subjectivity – 'race', sexuality and gender. This is not to suggest that these are the *only* important aspects of subjectivity. Issues such as body size and shape, disability, age and class are equally significant. 'Subjects are created in multiple positionings in material and discursive practices' (Walkerdine, 1995: 325). For example, Bondi and Davidson focus on gender but they recognize that 'gender is always bound up with other dimensions of human experience and subjectivity including those described by such terms as class, race, sexuality, age and so on'.

Brown and Knopp suggest that a 'queer geographical focus specifically on the interplay between sexualities and postcoloniality could provide some badly needed foundational knowledge for broader understandings of both colonialism and postcolonialism'. They cite Elder's (1998) work on the complicated intersections of race, class and sexuality in post-apartheid South Africa as an example of this kind of research. Drawing on a multidimensional matrix of differing axes of subjectivity might offer interesting possibilities. This matrix could include not just the conscious but also the unconscious. Bonnett and Nayak claim that Nast's (2000) work on black spatial containment, white 'flight' and urban renewal in Chicago offers a new route to understanding the socio-spatial processes of racialization. They claim that one of the most productive aspects of Nast's (2000) work is the explicit use of psychoanalysis (also see Pile, 1996).

Over the last five years we have seen what counts as the subject and subjectivity being extended (Pile and Thrift, 1995: 11). This trend may continue in the future. For example, interestingly, many of the chapters in this book (not just this section) can be read as contributions on the subject and subjectivity. I think this illustrates a growing recognition that the subject and subjectivity are at the heart of cultural geography. There have been numerous recent attempts to understand subjects as complex, multifaceted and interconnected with spaces. One such attempt is evidenced in geographers' recent interest in actor network theory (for reviews see Murdoch, 1997a; 1997b; also see Hetherington and Law, 1999).

One aspect of actor network theory is that it attempts to deconstruct the boundaries around which western knowledges are formulated, for example the boundaries between nature and culture (also see Haraway, 1991; Whatmore, 1999). In this way, the field of subjectivity can be seen to encompass 'the object world'. Questioning the boundaries between non-human and human, nature and culture, is leading to some very interesting research. For example, see Bankey's (2001) and Davidson's (2000; 2001) (cited in Bondi and Davidson's chapter) work on constructions of (irrational) agoraphobic women. The 'problematic' or 'phobic' experiences of these women bring into sharp focus the intertextuality of subjectivities and spaces. The authors focus quite literally on a blurring of boundaries between bodies and spaces. Susan Bordo (in Bordo et al., 1998: 80) discusses her experiences of agoraphobia, explaining that during a panic attack she became faint, but instead of putting her head between her knees 'as a normal person would do seeking to restore equilibrium through the trusted processes of one's own body' she responded like 'a drowning person', her only thought to find air. During panic attacks agoraphobics often feel as though their bodies collude with place to become one.

Prorok (2000) also attempts to understand the relationship between subjectivity and space through an examination of boundaries. She focuses on Espiritismo – 'a system of ritual healing indigenous to the Caribbean' (2000: 57). Prorok carried out fieldwork in an Espiritismo worshipping community in Manhattan, New York City for nearly three years. Her aim was to explore the boundaries between the human world and the spirit world, self and community, Catholic and non-Catholic, and male and female.

The kind of research cited above, and the contributions in this section, force us to consider questions surrounding boundaries – boundaries between minds and bodies, discourses and materiality, the conscious and the unconscious, different facets of subjectivity, and subjects and the object world. Increasingly, it seems that any consideration of subjectivity requires a consideration of boundaries (see Longhurst, 2001). I am suggesting not that boundaries always need to be removed or overcome, but that we need to think about what is at stake in the formulation and securing of particular boundaries for specific purposes in relation to subjectivity and space. We need to figure out which aspects of subjectivity and space matter when, and how coalitions can be constructed to create more emancipatory social relations.

Not only are questions about boundaries becoming increasingly important in understanding subjectivity, but so too are questions about the body and embodiment. Over the past few years the body and embodiment have come to occupy a prime position in cultural geographers' accounts of subjectivity. In 1995 Pile and Thrift noted:

> Nowadays, the subject and subjectivity are more likely to be conceived of as rooted in the spatial home of the body, and therefore situated, as composed of and by a 'federation' of different discourses/persona, united and orchestrated to a greater or lesser extent by narrative, and as registered through a whole series of senses. (1995: 11)

A quick glance at recent books indicates that there are many different perspectives on the body and embodiment (for a range of examples see Ahmed and Stacey, 2001; Bell et al., 2001; Butler and Parr, 1999; Duncan, 1996; Holliday and Hassard, 2001; Longhurst, 2001; Nast and Pile, 1998; Pile, 1996). One approach that has recently gained some popularity is 'non-representational theory' (a term coined by Thrift, 1996). Thrift and Dewsbury argue that non-representational theory 'emphasizes the flow of practice in everyday life as embodied, as caught up with and committed to the creations of affect, as contextual, and as inevitably technologised through language and objects' (2000: 415). Some of this work attempts to apply 'abstract' body theory to 'real' bodies (for example, Thrift, 1997, examines the bodily practices of dance and play).

This work has the potential for furthering understandings of body–subjectivity–space relations but I would caution cultural geographers against invoking bodies that appear to be mess and matter free. In talking about subjectivity we need to talk about the 'shape, depth, biology, insides, outsides, and boundaries of bodies placed in particular temporal and spatial contexts' (Longhurst, 2001: 2). The leaky, messy, awkward zones of the inside/outside of bodies and their resulting spatial relationships need to be examined. Focusing on a body that has no specified materiality (skin colour, body shape, genitalia, impairments, etc.) will not further feminist, socialist, anti-racist or disability activist agendas. Denying the weighty materiality of flesh and fluid will help preserve hegemonic bodily practices and politics.

To conclude, in this introduction I have not provided a point-by-point review of the chapters. Instead I have raised three key themes that I think are reiterated throughout the section. I haven't pointed to the ways in which the contributions differ. Such a reading of the chapters would have provided a different, perhaps better, introduction. In the final instance I leave you to tread your own path and to make your own (to use Probyn's phrase) 'connections and dis/connections' between the chapters that follow.

REFERENCES

Ahmed, S. and Stacey, J. (eds) (2001) *Thinking through the Skin*. London: Routledge.

Alcoff, L. (1988) 'Cultural feminism versus poststructuralism: the identity crisis in feminist theory', *Signs: Journal of Women in Culture and Society* 13: 405–36.

Bankey, R. (2001) 'La donna é mobile: constructing the irrational woman', *Gender, Place and Culture* 8 (1): 37–54.

Bell, D., Binnie, J., Holliday, R., Longhurst, R. and Peace, R. (2001) *Pleasure Zones: Bodies, Cities and Spaces*. New York: Syracuse University Press.

Bordo, S., Klein, B. and Silverman, M.K. (1998) 'Missing kitchens', in H.J. Nast and S. Pile (eds) *Places through the Body*. London: Routledge. pp. 72–92.

Butler, R. and Parr, H. (eds) (1999) *Mind and Body Spaces: Geographies of Illness, Impairment and Disability*. London: Routledge.

Davidson, J. (2000) '"… the world was getting smaller": women, agoraphobia and bodily boundaries', *Area* 32 (1): 31–40.

Davidson, J. (2001) 'Fear and trembling in the mall: women, agoraphobia and body boundaries', in I. Dyck, N. Lewis and S. McLafferty (eds) *Geographies of Women's Health*. London: Routledge.

Duncan, N. (ed.) (1996) *Bodyspace*. London: Routledge.

Elder, G. (1998) 'The South-African body politic', in H.J. Nast and S. Pile (eds) *Places through the Body*. London: Routledge. pp. 153–64.

Gibson-Graham, J.K., Resnick, S.A. and Wolff, W.D. (eds) (2000) *Class and Its Others*. Minneapolis: University of Minnesota Press.

Grosz, E. (1992) 'Bodies-cities', in B. Colomina (ed.) *Sexuality and Space*. New York: Princeton Architectural Press. pp. 241–53.

Haraway, D. (1991) *Simians, Cyborgs and Women: The Reinvention of Nature*. London: Free Association.

Harvey, D. (1969) *Explanation in Geography*. London: Arnold.

Hetherington, K. and Law, J. (eds) (1999) Special issue on actor-network theory and spatiality, *Environment and Planning D: Society and Space* 17.

Holliday, R. and Hassard, J. (2001) *Contested Bodies*. London: Routledge.

Johnston, R., Gregory, D., Pratt, G. and Watts, M. (eds) (2000) *The Dictionary of Human Geography*, 4th edn. Oxford: Blackwell.

Keith, M. and Pile, S. (eds) (1993) *Place and the Politics of Identity*. London: Routledge.

Kirby, K.M. (1996) 'Re: mapping subjectivity', in N. Duncan (ed.) *Bodyspace*. London: Routledge. pp. 45–55.

Kobayashi, A. and Peake, L. (2000) 'Racism out of place: thoughts on whiteness and antiracist geography in the new millennium', *Annals of the Association of American Geographers* 9 (2): 392–403.

Longhurst, R. (1995) 'The body and geography', *Gender, Place and Culture* 2 (1): 97–105.

Longhurst, R. (2001) *Bodies: Exploring Fluid Boundaries*. London: Routledge.

Longhurst, R. (2002) 'Geography and gender: a "critical" time?', *Progress in Human Geography* 26: 544–52.

McDowell, L. and Sharp, J.P. (eds) (1999) *A Feminist Glossary of Human Geography*. London: Arnold.

Murdoch, J. (1997a) 'Towards geography of heterogeneous association', *Progress in Human Geography* 21: 321–37.

Murdoch, J. (1997b) 'Inhuman/nonhuman/human: actor-network theory and the prospects for a non-dualistic and symmetrical perspective on nature and society', *Environment and Planning D: Society and Space* 15: 731–56.

Nast, H.J. (2000) 'Mapping the "unconscious": racism and the Oedipal family', *Annals of the Association of American Geographers* 90 (2): 215–55.

Nast, H.J. and Pile, S. (eds) (1998) *Places through the Body*. London: Routledge.

Pile, S. (1996) *The Body and the City: Psychoanalysis, Space and Subjectivity*. London: Routledge.

Pile, S. and Keith, M. (eds) (1997) *Geographies of Resistance*. London: Routledge.

Pile, S. and Thrift, N. (eds) (1995) *Mapping the Subject: Geographies of Cultural Transformation*. London: Routledge.

Prorok, C.V. (2000) 'Boundaries are made for crossing: the feminized spatiality of Puerto Rican Espiritismo in New York City', *Gender, Place and Culture* 7 (1): 57–79.

Rose, G. (1993) *Feminism and Geography: The Limits of Geographical Knowledge*. Minneapolis: University of Minnesota Press.

Routledge, P. (1997) 'A spatiality of resistances: theory and practice in Nepal's revolution of 1990', in S. Pile and M. Keith (eds) *Geographies of Resistance*. London: Routledge. pp. 68–86.

Sharp, J.P., Routledge, P., Philo, C. and Paddison, R. (eds) (2000) *Entanglements of Power: Geographies of Domination/Resistance*. London: Routledge.

Soper, K. (1986) *Humanism and Anti-Humanism: Problems in Modern European Thought*. London: Hutchinson.

Thrift, N. (1983) 'On the determination of social action in space and time', *Environment and Planning D: Society and Space* 1 (1): 23–57.

Thrift, N.J. (1996) *Spatial Formations*. London: Sage.

Thrift, N.J. (1997) 'The still point: resistance, expressive embodiment and dance', in S. Pile and M. Keith (eds) *Geographies of Resistance*. London: Routledge. pp. 124–51.

Thrift, N.J. and Dewsbury, J.-D. (2000) 'Dead geographies – and how to make them live', *Environment and Planning D: Society and Space* 18 (4): 411–32.

Walkerdine, V. (1995) 'Subject to change without notice', in S. Pile and N. Thrift (eds) *Mapping the Subject: Geographies of Cultural Transformation*. London: Routledge. pp. 309–31.

Whatmore, S.J. (1999) 'Hybrid geographies: rethinking the human in human geography', in D. Massey, J. Allen and P. Sarre (eds) *Human Geography Today*. Cambridge: Polity.

14

The Spatial Imperative of Subjectivity

Elspeth Probyn

In his *A Brief History of Time*, Stephen Hawking writes that 'when a body moves, it affects the curvature of space and time – and in turn the structure of space–time affects the way in which bodies move and forces act' (1988: 36). There's something so obvious about this learned comment. Obvious, that is, if we truly consider how we inhabit space. Can we ever conceive of ourselves outside the space we inhabit? Do we not at some level recognize the differing affects that play upon and within us as we move through different spaces? Profoundly, we experience our subjectivities, the ways in which we are positioned in regard to ourselves as subjects, in terms of both space and time. How can it be otherwise, given that our bodies and our sense of ourselves are in constant interaction with how and where we are placed?

In this chapter, I want to look more closely at how this happens. I will also argue that how we experience ourselves is deeply structured by historical processes that make us into subjects. I want to draw out theoretical models that help us to realize the complexity of the formation of subjectivity. What is it that drags upon us as we move through space? How are different spaces historically formulated as conducive to some subjectivities and not others?

To think about subjectivity in terms of space is evident yet relatively recent. Popular conceptions of our 'selves' commonly place them as somewhere deep within us. There is a long legacy in western thinking that places the core of ourselves as enclosed within. If subjectivity has been lodged away as a pristine entity untouched by the outward body, space too has a history of being conceptualized as bounded and contained. As Sue Best clearly argues, the pervasive metaphorization of space in terms of the feminine consolidates this 'persistent desire to

domesticate space, to bring it within a human horizon and, most importantly to "contain" it within this horizon' (1995: 183). Much of the research in cultural theory over the last decades has been directed at rethinking such conceptualizations. Thinking about subjectivity in terms of space of necessity reworks any conception that subjectivity is hidden away in private recesses. What we hold most dear, as an individual intimate possession, is in fact a very public affair. Thinking about how space interacts with subjectivity entails rethinking both terms, and their relation to each other.

Much of the most exciting work on subjectivity has been influenced by feminist perspectives. Feminists have raised crucial questions about the relations of power that permeate how subjectivities are constructed and experienced. Contrary to a long history in western thought that saw the body as troublesome and as an impediment to reason, feminists have argued that the body provides us with key knowledge about the working of our subjectivities. The body then becomes a site for the production of knowledge, feelings, emotions and history, all of which are central to subjectivity. As we'll see, the body cannot be thought of as a contained entity; it is in constant contact with others. This then provides the basis for considering subjectivity as a relational matter.

To start with a simple question: what is subjectivity? Often subjectivity and identity get used interchangeably. This is understandable because we think of ourselves as having an identity, or several. For the purposes of this chapter, however, I use the term subjectivity and rarely employ identity. This is because I want to outline the way subjectivity relates to the concept of the subject which, as we'll see, is also associated with the idea of ideology. While ideology is no longer as central as it was in cultural theory, it is nonetheless key to how the subject and subjectivity have been theorized.

In the early 1970s following the publication of an article by the French philosopher Louis Althusser, new ways of exploring ideology emerged within Anglo-American cultural theory. Althusser's theory was remarkable for the way in which it brought together a structural Marxist conception of society, and a Lacanian-influenced account of the structure of the psyche. Previously, ideology had been conceived of in the narrower sense of 'false consciousness'. This Marxist term referred to the ways that individuals were duped by ideology, and in vulgar terms, how we are brainwashed by capitalism. It is a simplification, yet we can say that ideology in this sense was seen as a force bearing down on passive individuals. Contrary to this tradition, Althusser advanced the question of how individuals are actively constituted as subjects through ideology. Crucially, Althusser laid the way for understanding ideology as sets of practices which engage us, and in which we are always engaged. And as I will describe in more detail, Althusser's theory of the subject is an influential cornerstone in thinking about the spatial nature of subjectivity.

While it has become commonplace to speak of identities and subjectivities as performative, Althusser's account of ideology compels us to consider closely the material contexts which allow and delimit our individual and collective performance of selves. Althusser may have fallen out of favour, but interestingly enough there is an increasing awareness in cultural theory of the need to instil some sense of the material within theorizations of the complex relations between individuals, selves, economic and structural forces, history and the present. This plays out in various directions but may be due to a collective turning away from some of the excesses of postmodernism and even poststructuralist thought which, it can be argued, diluted a clear sense of the constraints of context. It must be said that equally many of the theories grouped within these unwieldy categories have forcefully contributed to an expanded notion of what counts as material, or even as political. Nonetheless there is evidence of a return to something that might ground 'theory', that 'theory' must be put to some use. In the terms of a recent book edited by Butler, Guillory and Thomas, the question is 'what's left of theory?' Their introduction attests both to the need to make theory work in political contexts, and also to the unproductive divide that has operated in terms of who is seen as material and who is not:

If some of those who turn against theory in the name of politics do so by laying claim to referentiality and thematic criticism, then some of those who turn against politics in the name of theory do so by sacralizing the suspension of all reference to context. (2000: x)

While their concern is literary theory, a general rethinking of politics and theory can be seen across much of cultural theory. Equally it is interesting to note that the tide is turning against 'identity', and especially 'identity politics'. While much of this has been of a conservative ilk – that somehow 'identity' took us away from the proper study of disciplinary objects – it can be argued that there is a need to reground identity. For these reasons, a return to the notion of subjectivity offers us a way of thinking through anew some of the important questions of the past several decades.

In this sense Althusser is of note for several reasons. First and foremost, the combination of structural Marxism and an attention to the production of subjectivity allows his theory to be used across a wide spectrum of analyses. For instance, in the heyday of 'high theory' Althusser was used to analyse the operations of the filmic apparatus. Yet his insights also provide ways of understanding the interrelation between societal structures, their history and spatiality, and how they are experienced and incorporated.

At a very basic level, Althusser was interested in why society continues to run so smoothly even in the face of considerable inequities and inequalities amongst individuals. It's a question that in its simplicity continues to express some of the most pressing questions of our time. As we'll see, his model allowed for contradiction, a key notion of 1970s and early 1980s cultural theory that is still germane. For instance in terms of 'globalization' the contradictory ways in which we are placed and experience ourselves as subjects quite rightfully hold much current theoretical attention. And they are, of course, the subject of overt political action in the western world, and increasingly beyond. In both cases, there is a clear sense of the need to come to grips with the imbrication of the economic (which may be experienced, as Althusser argued, only in 'the final instance'), the cultural, ways of living and perceiving. In turn, and as I discuss later, this raises questions about how we are positioned in relation to each other: what are the relations of proximity that impinge on very differently experienced subjectivities?

Althusser's response to why the world keeps turning in the face of dissent, and experienced inequality, was to remark on the fact that while there are instances where people are kept in line through violence (what he called 'the repressive state apparatuses'), the main part of the work of getting us to accept our condition is through ideology. Althusser opens the term to what he called 'the ideological State apparatuses'. These include the family, education, religion and most of the legal procedures. Akin to Pierre Bourdieu's (1984) emphasis on class and education, Althusser's point was that we are informed at an early age by the work of these ideological apparatuses.

One of his central arguments is that we are 'interpellated' or 'hailed' by ideology. The classic example is the following scenario: you are walking on the street and a cop calls out 'hey, you'; seemingly we instinctively turn, 'what, me?' At that moment, says Althusser, we have gone from being an ordinary individual and have become a subject of and for the law. There are other scenarios that Althusser doesn't mention. For instance, if you are walking on the street and someone wolf-whistles, and you turn, more likely than not you are being interpellated as a ('pretty') woman. In other words, we may be walking along unconcerned whether we are male or female, black or white, straight or gay, when something happens that forces recognition of the fact that we are gendered, raced and sexed. To go back to the example of the policeman who hails you on the street: if you are, say, young, black and male the chances are that the interpellation of the law will strike more deeply than if you are white and middle class. In the case of the latter, you may not even recognize that you are being hailed. You may think that the apparatus of the law is there to serve you, not that you are a likely subject of its force.

One of Althusser's key points is that ideas about who is a good or a bad subject are always present in our society. Given the huge range of experiences it is surprising how limited are the choices in terms of good and bad subjects. These notions are not ephemeral but are stitched into us through our everyday practices. As Althusser states, 'an ideology always exists in an apparatus and its practices. This existence is always material' (1971: 155). In this way we can begin to understand that the ideas that a society has about what is feminine or masculine, what is 'normal', etc. do not just seep into our heads; these ideas are reproduced over and over again through the practices defined by different apparatuses, and then in our own practices. This is a more nuanced and much more pervasive view than was evident in the ways in which ideology was previously theorized. We are subjected to the practices of different ideological apparatuses, and we become subjects in terms of them. This leads to Althusser's argument that there is 'no ideology

except by and in an ideology; there is no ideology except by the subject and for the subject' (1971: 160). Further, ideology has the function (which defines it) of 'constituting' concrete individuals as subjects.

Althusser also forcefully raised the fact that we are all informed by ideology. Indeed, we all are *in* ideology, and to a certain extent the very fact of being within ideology is comforting. Althusser uses the example of religion as a perfectly hermetic system which gives its believers absolution. In giving yourself to God you are not only assured of a place in the afterlife; here on earth you will know your position. In Althusser's terms, you are a subject in as much as you are subjected to a higher subject, God. This higher subject guarantees your existence: 'Peace be with you.'

This structure of subject formation is also common outside religion. For example, in 12-step programmes modelled on AA, individuals give themselves over to a 'higher power' which then secures a subjectivity as 'a recovering alcoholic'. The 12-step system is a very simple ideological structure which allows us to see the process of subjection and subjectivity. The individual says to the group, 'Hi, my name is Fred, and I'm an alcoholic.' There is no last name because the system is not interested in other subjectivities you may bring to the group. The whole process is aimed at verbalizing, uttering and outing one subjectivity. This is secured by the promise that if you do articulate this subjectivity you can also give over to the higher subject all your other problems and worries. And that you will not drink. Ideological structures work on the concept of mutual recognition that by subjecting yourself to a higher subject, you exist.

To recap the points of this system of ideological recognition, we can state:

- Ideology interpellates individuals as subjects.
- Through practices they enact their subjection to the higher subject.
- This entails that there is a mutual recognition of subjects and higher subject, the subjects' recognition of each other, and finally the subject's recognition of him/herself.
- In turn, this provides the absolute guarantee that everything is so, and that on the condition that the subjects recognize what they are and behave accordingly, everything will be alright: 'Amen, so be it.'

The result of this process is that most individuals enact themselves as 'good' subjects. What emerges from this argument is that the category of the subject is absolutely central at the same time that it is ambiguous. We are free and accepting of our submission; we subject ourselves. In so doing we are allowed to forget the reality of being subjected to different ideological systems. In Althusser's terms, ideology represents 'not the system of the real relations which govern the existence of individuals, but the imaginary relation of those individuals to the real conditions in which they live' (1971: 165). Stuart Hall (1985) puts it more clearly when he argues that 'the problem is how to account for the fact that in the realm of ideas and meaning men can "experience" themselves in ways which do not fully correspond to their real situation'. In other words, what may seem intensely intimate and personal really is nothing more than a subject position we hold in relation to a larger system.

The important point is that the subjectivities that we build up in our practices of subjection are important to us as individuals. Another important consideration flowed from his theory. In Hall's (1985) famous statement, Althusser lets us live with difference. While I won't follow through on Hall's intricate argument, what I want to draw out is the fact that Althusser's theory can be used to think about how different spheres of subjectivity are enacted, more often than not by one individual. While it has become fashionable to talk about all subjectivity as fragmented, this can ignore the ways we are interpellated and inhabit sometimes quite conflicting subjectivities. For instance, if I am a gay male school teacher, my subjectivities will probably not be seamless. I will feel parts of my subjectivity in different contexts. Given the general homophobia of our culture, as a gay man I would be interpellated by the educational system as a 'bad' subject, even though I may be deeply invested in being a 'good teacher'. To return to the crucial point about the spatial configurations of subjectivities, we proceed from the basic idea that subjectivities are not abstract entities; they are always conducted *in situ*. They are also hard-won.

Teresa de Lauretis' (1988) argument about 'technologies of gender' is one of the most useful takes on Althusser's theory of ideology. Following Althusser, she also insists that ideology works fundamentally by means of its engagement of subjectivity. De Lauretis extends the scope of Althusser's argument by replacing 'ideology' with 'gender'. Her objective is to conceive of a new kind of subject, one that operates through gender, as in fact we all do in 'real life'. De Lauretis' proposal for thinking subjectivity centres on 'a subject constituted in gender though not by sexual difference alone, but rather across languages and cultural representations; a subject en-gendered in the experiencing of race and class, as well as sexual, relations; a subject, therefore, not unified but multiple, and not so much divided as contradicted' (1988: 1). Here again we hear the emphasis on subjectivity as the 'product and process' of practices. But contrary to Althusser's subject in ideology who cannot recognize himself as within ideology, de Lauretis argues that feminism, or the critical study of the ideology of gender, produces a subject who is aware of the workings of ideology. The subject, she argues, 'within feminism is one that is at the same time inside *and* outside the ideology of gender, and conscious of being so, conscious of that two-fold pull, of that division, of that doubled vision' (1988: 10).

De Lauretis argues that this is an uncomfortable position, but a necessary one. If we recall that one of the contentions of Althusser's description of ideology was that living within the system allowed for a sense that everything was alright, we can see that being inside and outside ideology would be disturbing. But precisely one of the important differences between Althusser and feminist arguments like de Lauretis' is the acknowledgement that everyday life throws up moments that intrude upon our senses of ourselves. Simply put, Althusser describes at a theoretical level how ideology as a system might work. De Lauretis, on the other hand, shuttles between 'life' and theory.

'Life' provides us with a critical entrance into theorizing subjectivity. De Lauretis uses a cinematic term, the space-off, to describe aspects of life that are outside the frame of dominant discourses. Consider the following quotation where she describes 'the space-off', 'the space not visible in the frame but inferable from what the frame makes visible' (1988: 26):

> [I]t is here [in the space-off] that the terms of a different construction of gender can be posed – terms that do have effect and take hold at the level of subjectivity and self-representation: in the micro-political practices of daily life and daily resistances that afford both agency and sources of power or empowering investments. (1988: 25)

For de Lauretis those micro-practices can be as diverse as feminist cultural representations or political practices that bring together 'the personal and the political'. To return to Althusser's description of being hailed or interpellated on the street by the police, we could also envision being interpellated by the sight of a woman on the street with a black eye, or a woman begging with her children. If we make the move to consider how these individual cases link to broader questions about domestic violence, or the poverty of large numbers of single mothers, we can understand what it means to be inside and outside the ideology of gender. At these moments, we may go from being 'just' an individual to recognizing ourselves as gendered subjects. To take another example, if you are not heterosexual (and maybe even if you are), the dominant representation of romance and family will at times irritate. At some level, the very fact of being at odds with culture is experienced like a visceral schism. In this case, chances are that your subjectivity will be keenly experienced as different from others. There is no doubt that this moment of misrecognition – when you do not feel hailed by dominant ideologies – can be painful. But it is also crucial to the production of another subjectivity, one that may be in the 'spaces-off' of mainstream culture.

De Lauretis provides us with a critical framework for thinking about subjectivities and space. She is very clear that when she speaks of the movement back and forth, she does '*not* mean a movement from one space to another beyond it, or outside' (1988: 25). In other words, she does not want us to think that there is ideology and there is 'reality', as if the latter were not inextricably caught with the former. Subjectivity is a process that is continually in play with 'reality' and 'ideology', dominant representations and our own self-representations. And as de Lauretis puts it, we all live with, and indeed within, 'the tension of contradiction, multiplicity, and heteronomy' (1988: 26).

Clearly then, subjectivity is not a given but rather a process and a production. It is also undeniable that the sites and spaces of its production are central. In other words, the space and place we inhabit produce us. It follows too that how we inhabit those spaces is an interactive affair. A jointly authored article published a few years ago argued that 'space is gendered and that space is sexed ... The reverse has also been shown: gender, sex and sexuality are all "spaced"' (Bell et al., 1994: 31–2). Their article presents a complex argument about sexual practices and space. In turn, the journal which published it (*Gender, Place and Culture*) asked several people to respond, including myself. I won't rehash my argument more than I already have, but I want to replay an example I used in order to extend the idea about subjectivity and space as interactive.

Consider this scene: your average type of pub somewhere (for some reason, a place in Kitislano, Vancouver comes to mind), the men are propped up on the bar, shoulder to shoulder, presenting a solid front of space gendered as masculine; they are men's men but certainly not gay. A single woman enters and she is checked over, chatted up or ignored. And if that space

feels stultifying, it is because she is walking into strata upon discursive strata that produce masculine space as the ground of differentiation and the grounds for their appropriation of women as Woman (which is to say, a man-made gender). (Probyn, 1994: 80)

What I wanted to raise here were the ways in which space presses against our bodies, and of necessity touches at our subjectivities. One of the important implications of thinking in terms of subjectivity rather than identity is that even in banal examples like this, the denseness, historicity and structural complexity become clear. There are of course lots of spaces that seem to be naturally masculine or feminine. For instance, the kitchen is held to be the woman's domain, and in our daily lives we may often experience this: from mothers cooking for families, to parties where the girls gather in the kitchen to talk. Historically, pubs have been designated as men's places. In western cultures until recently women were excluded either by law or by custom from entering the pub. In Quebec there are signs on the doors of brasseries that state: 'Women welcome.' This is because by law they now have to let women in. But it is a powerful reminder of how recent that change is. In Australia, women did not go to bars, and Aboriginal Australians were prohibited until recently. Indeed there are stories about how, during the Vietnam War, black American soldiers were allowed into bars and pubs, whilst Aboriginals were not allowed. The idea that the pub is a male-gendered space is not a myth but an actual historical construction.

So when I ask what happens when a woman goes into a bar, it is clear that she must confront at some level the fact that 'she does not belong here'. She will occupy that space quite differently from the men who are 'propped up on the bar'. She will be made to feel her gender subjectivity, whereas men may be able to forget that their subjectivities are also constructed through the interpellation of gender. This is a small example, but it may help us examine more closely how as individuals we inhabit space, and how space inhabits us.

In the example of the pub, I also wanted to bring out the ways that sexuality is highlighted in certain spaces. If the space of the pub is gendered as masculine, in my example it was also structured by heterosexuality. We can again ask the question of what happens when a woman goes into a bar, and complicate it by adding the fact that she is going to meet her girlfriend. In this scenario, not only will the women feel their gender, but they will also be made to feel their difference: that they are not heterosexual. This space reveals that parts of their subjectivities are caught up in a relation of being different, e.g. of not being 'like other women', of not being placed in a complementary opposition of man–woman, of being out of place. To an extent, they may make that space their own, but they are doing so across history and ideology. As Gill Valentine has clearly argued, 'As a result of this expression and representation of heterosexual relations in space, heterosexuals as a group are allowed to appropriate and take up space' (1993: 410).

It is hard to overemphasize the historical weight that the ideology of heterosexuality has had on defining space. However, we can also think about how straight men and women inhabit space that has been made queer. As David Bell has pointed out, geography and cultural theory were slow to get beyond the gender distinctions and 'recognise that different (and especially "non-conforming") men and women have different relations with space and that one element that conditions this is sexuality' (1991: 327). The queering of space has a complexity and a history as dense as those gendered masculine or feminine. For instance, many have credited the riots that took place at the Stonewall bar as the beginning of a modern gay movement. It may be an overly large claim, but Stonewall is interesting. The bar in New York's Greenwich Village was home to drag queen shows. In 1969 the police raided it, as they often did in regard to gay spaces. The riots that ensued were fuelled by courageous individuals who refused to hide their sexual preferences. Now all over the world there are bars named in Stonewall's honour. In terms of outing subjectivity, we can understand how brave those early protesters were. We also need to recognize the ways in which past practices become imbricated within present subjectivities. This example also gestures to the ways in which there may be common elements that are replayed in their contextual and temporal specificity. In this manner, they are continually rearticulated both as a ground of individual subjectivity and as a mode of linkage between and amongst individuals.

If gays and lesbians have been made to hide their sexual desires, a significant aspect of their subjectivities was constructed 'in the closet'. Eve Kosofsky Sedgwick has eloquently argued that the figure of the closet is central to western society's construction of knowledge and secrecy. From the end of the nineteenth century, we begin to see a new way of understanding and categorizing individuals:

What *was* new from the turn of the century was the world-mapping by which every given person, just as he or she was necessarily assignable to a male or a female gender, was now considered necessarily assignable as

well to a homo- or a hetero-sexuality, a binarized identity that was full of implications, however confusing, for even the ostensibly least sexual aspects of personal existence. (1990: 2)

The closet is an interesting spatial expression, although it allows for only two options: in or out. Moreover, as Sedgwick's argument indicates, as a figure it mandates that we be either homo- or heterosexual. In terms of Althusser's theory of interpellation, we can appreciate the sheer effort of continually deflecting our culture's ideology of heterosexuality. In de Lauretis' terms, to be queer is to construct yourself in the space-off of our society. As an ideological current, heterosexuality or heteronormativity pervades all aspects of life. It is also central to the apparatuses of the family, education, religion, the law, the media, to name but a few. Of course none of our subjectivities is constituted solely in regard to one factor. We are never only women or men, straight or queer. Our subjectivities are always situated at the nexus of gender and sexuality as well as class, ethnicity, social position, etc. And at each point we are faced with a complex set of binaries: are you a girl or a boy? Are you normal or queer? If you're not white, are you black?

Recently there have been interesting instances of straights inhabiting queer space. Of course this tends to be the case in urban and trendy places – the bars that line queer cores of cities and that sport signs with pink triangles declaring 'safe space here'. These relatively new queer spaces are therefore not the mainstream spaces of the family or the church. They are the result of manoeuvres and strategies. In some cases, parts of cities are claimed by queers and then become part of a generalized gentrification, quite often of inner city and previously working-class areas. They also attract commercial venues like cafés, delis, bars and clubs. In a gradual process, these areas become appealing to what might be called 'gay friendly' straights, normally a young and more adventurous set. In time however they become 'normalized', and they cease to be seen as only gay.

This for instance has happened in Sydney's downtown queer core of Darlinghurst which is now one of the premier sites for restaurants, bars and clubs. However during the Sydney Lesbian and Gay Mardi Gras the streets are closed for the huge parade of fabulous floats. At that moment, the 'gayness' of the space is highlighted, and hundreds of thousands of straights come to look. While there is a definite queer feel to the area, straight women now frequently habituate the gay bars of Oxford Street which runs the length of the queer strip. As Beverley Skeggs' (1999) research in Manchester's gay village shows, the attraction for straight women is that they do not have to endure the pickup routines of straight bars. Also they appreciate the style of the gay male clientele.

In Skeggs' research it became apparent that this straight invasion into queer space has repercussions on the queerness of identified queer space. In an interesting way, she argues that gender returns to trouble sexuality. This is especially so in regard to the relations between straight and gay women in queer space. Simply put, straight women may be attracted to gay men in terms of a non-threatening relationship that is still based in a gendered opposition of same–other. However, their relation to lesbians is quite different. To be blunt, lesbians are more threatening because they exist in a same–same yet different relationship to straight women. Heterosexual women may worry that they could be the object of desire for lesbians in ways that are impossible or at least less likely in their relationships with gay men. Conversely, Skeggs has also found that lesbians don't like the ways in which straight women appropriate space. There is an erasure of the fact of lesbians within queer space, as the space gets structured in terms of gay men to gay men, and straight women to gay men. This plays out in little but significant ways: for instance, Skeggs' lesbian informants complained that the toilets get filled with straight women doing their hair and makeup, and looking askance at the lesbians. Given the fact that lesbians have historically found it much more difficult than gay men to assert their sexuality outside of private spheres, this limits the free expression of sex in space. For instance, while public sex is accepted practice amongst gay men, what would the straight girls do if a couple of lesbians were having sex in the toilets?

This may seem like a trivial question but it does go to the heart of how space and subjectivity mutually interact. One of the defining divisions in our culture is that of private versus public space. In general, women have been only recently allowed to incorporate public space into their sense of self. Where one gets to do what with whom is therefore an important point. While it is often argued that the public penetrates more and more into the private, it is less common to hear how individuals' subjectivities are affected by the movement into the public or conversely into the private. In Kathy Ferguson's (1993) terms, this is why it is important to think about the mobility of subjects. She states that she has 'chosen the term *mobile* rather than *multiple* to avoid the implication of movement from one stable resting place' (1993: 158). In other words,

we need to conceptualize subjectivities in terms of not just the multiple positions we all hold, but how they get configured across space and places. In terms of the above discussion of sexuality and space, it's important not to conclude that there are hermetic spaces designated as queer and others as straight. There are places which act as nodes, or meeting points, but it's not as if we take off an identity as lesbian once we venture beyond them. As Geraldine Pratt argues, 'there is a deep suspicion about mapping cultures onto places, because multiple cultures and identities inevitably inhabit a single place (think of the multiple identities performed under the roof of a family home)' (1998: 27).

One of the important aspects of Pratt's work is the way she navigates between the excesses of seeing subjectivity as completely fragmented and errant, and a perspective that would place subjectivity as a side-effect of place. Pratt's research has focused on how migrant workers in North America inhabit their working spaces. In this sense, the workplace 'not only enable[s] but exact[s] the performance of particular gender, class, and racial identities' (1998: 28). In other research Pratt studied women employed in so-called non-skilled white-collar jobs. She argues that 'these women literally move through class locations during the day. At their jobs they are working class, at home they are middle class' (1998: 34). What close ethnographic work reveals is the fact that most individuals seek to anchor their senses of themselves. The women in Pratt's study obviously have an investment in both their jobs and their middle-class identities at home. Against much of the highly abstracted theoretical work on fragmentary, floating subjectivities, this returns us to the idea that we may be hailed by different ideological apparatuses, but we also seek some coherence even in the face of multiple interpellations. Speaking in terms of our increasingly multicultural and differentiated living conditions, Pratt states: 'It seems to me that efforts ... are not advanced by representations that conceive of cities as blurred, chaotic, borderless places.' I would add that our efforts to understand subjectivities also need to avoid celebrating subjectivity and identity as amorphous and as essentially boundless. Rather, as Pratt puts it, 'one must understand the multiple processes of boundary construction in order to disrupt them' (1998: 44).

At first sight this emphasis on boundaries seems to go against the prevalent direction in cultural geography that insists on the chaotic – on the fact that 'there is always an element of "chaos" in space' (Massey, 1999: 284). Doreen Massey, one of the more influential writers on space, defines this chaos as resulting from those happenstance juxtapositions, those accidental separations, the often paradoxical character of geographical configurations in which – precisely – a number of distinct trajectories interweave and, sometimes, intersect. Space, then, as well as having loose ends, is also inherently disrupted. (1999: 284)

The emphasis on the looseness and the potential for disruption is for Massey central in rethinking the politics of space, or rather the role of spatial thinking in renewing how we think about politics. The goal is to instil in our conception of politics recognition of 'the openness of the future, the interrelatedness of identities, and the nature of our relations with different others' (Massey, 1999: 292).

As in many accounts, the idea of relations, interrelations and proximities to others is key. As Gillian Rose writes, the question is how to imagine a space of relation (1999: 252). She asks: 'How are bodies positioned in relation to each other? What kinds of connection do these positions make possible, thinkable, visible, tangible? ... what kinds of space articulate what kinds of corporealized relation?' (1999: 252). While I agree with Massey that it is important to focus on the 'happenstance arrangement-in-relation-to-each-other' (1999: 282), we also need to bear in mind that we are produced in distinct ways because of how we are positioned, how we are interpellated.

To go back to an example I used in discussing Althusser's ideas, a young black man will be interpellated on the street in overdetermined ways. To extend that point, I may be walking along the street where I live and have a 'happenstance' encounter with a young Aboriginal woman. To contextualize this encounter, let me add that I live in an inner city area of Sydney that has the highest urban population of Aboriginal Australians. It also has one of the highest unemployment rates, the highest crime, the highest poverty levels, and is renowned for drug selling and use (mainly heroin). It is heavily policed by white cops, and because the university where I teach is down the road there is a constant parade of relatively affluent, mainly white students who seem to proceed *en masse* through the Aboriginal section, oblivious to the 'difference' that surrounds them. Their behaviour may be motivated by many reasons, including a protective bodily comportment in order to deter muggers. As I walk along in much the same manner, I come across the young Aboriginal woman who is crying. 'Are you OK?,' I ask. 'Ah sister, you wouldn't believe what happened.' She then details the death of an aunty and how she has come to Redfern to try to find her cousins. I try to offer consoling words, and wish her well as she continues on the street and I turn off to go home.

Now in terms of the questions raised above about the positions of bodies to bodies, what can we say of this brief encounter? Well in a fairly brutal manner, we'd have to say that the only way that such an encounter could occur would be in a happenstance way. We are relationally positioned as inhabiting different universes. In fact we could be seen as standing in binary opposition to each other: me white, she black; me affluent, she poor; me educated, she probably not; me the invader of her country, she the dispossessed. The list could go on and on. That our paths cross is also determined by the fact that Redfern is becoming gentrified, something that will not help her one jot and increases the pressures to remove all trace of the Aboriginal housing from this area. Further, our small encounter surely left her indifferent: she had more important things on her mind. For me, it registers because part of my evolving subjective processes involve the question of how to conduct myself as a white non-Australian within a geography of appalling racialized relations, and a history of violence. Bluntly put, I need her more than she needs me (Probyn, 2001).

This scenario captures for me some of the sheer difficulty of how to live in an interrelational framework. It also compels questions of how to think and conceptualize subjectivities in relation to others. In an early book, I posed the following question as one way to think connection:

In trying to speak within the tensions of 'who is she? and who am I?' I want to disrupt any certainty that we know the answers in advance, or that either a good or bad politics can be guaranteed by such a question. To engage our imaginations precisely opens us into a space where possibilities can be envisioned; a space where I may no longer recognise myself. (Probyn, 1993: 163)

Nice words, but I now think that this formulation relies too heavily on the optimism of openness. Quite simply, I'm not sure where that space would be in which I would no longer recognize myself. Nor do I see now why this is of necessity a 'good thing'. Even the guiding questions no longer satisfy. At the time, it allowed me a way to think through how we might put experience or, in the context of this chapter, our subjectivities to work. I envisioned a productive tension set up by the question of 'Who is she and who am I?' Equally I wanted to get away from a navel-gazing perspective on the question of subjectivity, which seemed to endlessly spiral around 'me'. In part, this may have been produced by an overly zealous insistence within some forms of feminism that white women should not attempt to speak for 'the other'. By now, hopefully, it is common sense that I cannot speak for an amorphous group, be it the other, or women, or whomsoever.

It also has to be said that I am no longer interested in the ins and outs of 'Who am I?' The broad brush depiction will do fine: white, female, relatively privileged, etc. I am, however, more than ever committed to thinking about how subjectivities can be thought of in terms of being both structured and porous, spatially determined, temporally heavy. This is why in this chapter I have returned to the basics of Althusser's structural theory of ideology. I suppose I could have equally deployed theories such as Bourdieu's who develops a notion of how social structures are incorporated. However there is something about the immediacy of Althusser's descriptions that attract me. They point to the multidimensional nature of how we produce ourselves, as well as how we live with difference.

Subjectivity is a question of sameness and difference, the near and the far. My preferred way of thinking about a wide range of issues is in terms of 'relations of proximity'. Dictionaries define 'proximity' as closeness: 'nearness in space, time, etc.' It is related to the Latin *proximus,* 'nearest'. Personally 'relations of proximity' bring to mind the near and the far, what cannot be rendered near, what is always produced as close. Furthermore, relations of proximity highlight the facts of connection or dis/connection. The term 'connection' has become widely used, and belongs in much the same frame as 'interrelation' or Massey's notion of 'arrangements-in-relation-to-each other'. Clearly her use of the hyphen emphasizes the connection between each term, and refers to possible connections amongst individuals. For me, this remains an important point even if, as I mentioned, I now want more ground upon which to base ideas of the types of connection that are possible. But logically, if we agree that we need to think about *possible* connections, then we must also address the conditions that will make them impossible, or at least difficult to enact.

In adjoining connection and dis/connection, I want to render central the facts that disable or render connection hard. These are the hard 'facts of life': conditions of inequality and non-commensurability due to economic power, class, social privilege, history, etc. They also return us to the ways in which we are interpellated differently: that we are hailed by different ideologies in different ways, and that the institutions that maintain relations of how we are hailed pose blocks to possible connections. In other words, subjectivities are differentially informed.

Emphasizing the absolute spatial nature of the processes of subjectivity should also remind us of where and how we are interpellated. Instead of plastering over those differences, we need to stop and address them. Sometimes that stopping will result in silence. And that slash between dis/connection should indicate a pause – a moment of non-recognition that may be expressed as simply as 'wow, you really are different from me'.

The point is not to stay caught in that moment of bewilderment or enchantment: that would only reinscribe difference as an exotic, fetishized or denied quality. In other words, this would be to replay the not-same as 'the other', which is to posit a relation of dubious connection. Nor is it to legitimate turning away, closing down in the face of non-connection. That would be to replay the history of how racialized, classed and other relations have tended to produce hermetic subjects. In Susan Willis' description, this would be a situation wherein 'To some extent, all [whites] are reified subjects, against whom it is impossible for blacks to mount passionate, self-affirming resistance or retaliation' (1989: 174). Conversely, it also renders it impossible for whites to have any connection to blacks except those of guilt, denial or retaliation. This is not the type of dis/connection I am thinking of, and cannot be because it is effectively no connection at all.

In terms of bringing together the different points of this chapter, in returning to Althusserian theory I have attempted to sketch out the ways in which space always informs, limits and produces subjectivity. Equally subjectivity connects with space, and it rearticulates certain historical definitions of space. In this sense, neither space nor subjectivity is free-floating: they are mutually interdependent and complexly structured entities. The interest in returning to the ideological underpinnings of the very notion of the subject is that it turns attention to the ways in which subjectivities are produced under very particular circumstances. This then can lead the way to rethinking the questions that press upon us: from the ways that globalization restructures every aspect of our lives, and interconnects us in visceral and symbolic ways with those 'far off', to the 'spaces-off' in which we perform new modes of subjectivity and rearticulate the limits of gender, sex, race and class.

We need to think of subjectivity as an unwieldy, continually contestable and affirmable basis for living in the world. Subjectivities are then simply a changing ensemble of openings and closings, points of contact and points which repel contact. In space, we orient ourselves and are oriented. That is the spatial imperative of subjectivities.

REFERENCES

Althusser, L. (1971) 'Ideology and ideological state apparatuses (notes towards an investigation)', in *Lenin and Philosophy.* New York: Monthly Review Press.

Bell, D. (1991) 'Insignificant others: lesbian and gay geographies', *Area* 23 (4), 323–9.

Bell, D., Binnie, J., Cream, J. and Valentine, G. (1994) 'All hyped up and nowhere to go', *Gender, Place and Culture,* 1: 31–47.

Best, S. (1995) 'Sexualising Space', in E. Grosz and E. Probyn (eds) *Sexy Bodies: The Strange Carnalities of Feminism.* London: Routledge.

Bourdieu, P. (1984) *Distinction: A Social Critique of the Judgement of Taste,* trans. Richard Nice. London: Routledge.

Butler, J., Guillory, J. and Thomas, K. (eds) (2000) 'Preface' in *What's Left of Theory? New Work on the Politics of Literary Theory.* London: Routledge.

De Lauretis, T. (1988) *Technologies of Gender: Essays on Theory, Film and Fiction.* Bloomington: Indiana University Press.

Ferguson, K. (1993) *The Man Question: Visions of Subjectivity in Feminist Theory.* Berkeley: University of California Press.

Hall, S. (1985) 'Signification, representation, ideology: Althusser and the post-structuralist debates', *Critical Studies in Mass Communication,* 10 (2): 91–114.

Massey, D. (1999) 'Spaces of politics', in D. Massey, John Allen and Philip Sarre (eds) *Human Geography Today.* Cambridge: Polity.

Pratt, G. (1998) 'Grids of difference: place and identity formation', in Ruth Fincher and Jane M. Jacobs (eds) *Cities of Difference.* New York: Guilford.

Probyn, E. (1993) *Sexing the Self: Gendered Positions in Cultural Studies.* London: Routledge.

Probyn, E. (1994) 'Lesbians in space: gender, sex and the structure of missing', *Gender, Place and Culture,* 2 (1): pp. 77–84.

Probyn, E. (2001) 'Eating Skin', in S. Ahmed and J. Stacey (eds) *Thinking through Skin.* London: Routledge.

Rose, G. (1999) 'Performing Space', in D. Massey, John Allen and Philip Sarre (eds) *Human Geography Today.* Cambridge: Polity.

Sedgwick, E.K. (1990) *Epistemology of the Closet.* Berkeley: University of California Press.

Skeggs, B. (1999) 'Matter out of place: visibility and sexuality in leisure spaces', *Leisure Studies* 18, pp. 213–32.

Valentine, G. (1993) '(Hetero)sexing space: lesbian perceptions and experiences of everyday spaces', *Environment and Planning D: Society and Space* 11: 395–412.

Willis, S. (1989) 'I shop therefore I am: is there a place for Afro-American culture in commodity culture?', in Cheryl A. Wall (ed.) *Changing Our Own Words: Essays on Criticism, Theory, and Writing by Black Women.* New Brunswick: Rutgers University Press.

15

Cultural Geographies of Racialization – The Territory of Race

Alastair Bonnett and Anoop Nayak

The categories we use to divide the world and its peoples are usually represented as uncontroversial and commonsensical. 'Asian', 'African', 'European', 'western' trip off the tongue as if they were the most natural of ideas. National, regional, religious and other ethnic classifications can seem just as obvious, to the extent that they are often attached to highly specific attributes and claims. Indeed, a lot of political and social conversation is made up of assertions such as 'What Russians really feel …' or 'Experience has taught the Africans that …'. Of course, communication relies on classification, which is in turn dependent upon generalization. It is when this process is racialized that it becomes problematic: in other words, when the kinds of terms mentioned above are employed to sustain the dangerous conceit that they refer to natural, 'presocial' and homogeneous entities with immutable attributes.

Perhaps surprisingly, it is only comparatively recently that social and cultural geographers have turned their attention to the construction of racial myths. It is surprising because a shared characteristic of the terms we have already mentioned is that they are territorial: to speak the language of race and ethnicity is, very often, to talk geography. Indeed, along with anthropology, geography is the most racialized of scholarly pursuits; a fact starkly evident from its institutional history (Livingstone, 1992; 1994). We take this to mean that *the critique of racialization* is one of the most pressing concerns for contemporary geographers. At least, it should be for those who wish to see the subject fulfil Kropotkin's (1996, first published 1885) hope that geography's inevitable involvement with questions of race and ethnicity can and should be based, not on racism, but on the ability to understand and challenge stereotype and prejudice.

It will be noticed that the most *explicitly* racial of racial terms, such as 'white', 'black' and 'people of colour', were not mentioned above. Such expressions are explicitly racial in the sense that they refer directly to the body; they are designed to connote those most seemingly natural things, our flesh and blood. In this way they 'fix' race, make it seem real. A paradox emerges: geography is based on race but the most racial of terms appear to lack territorial significance, to escape geography. Indeed, we would propose that the ability of such terms to appear as the *most* racial of racial categories (and hence to nudge other contenders into the amorphous terrain of 'ethnicity') is *related* to their capacity to make space seem irrelevant. This paradox should not tempt us into taking such categories 'on their own terms'. Rather it impels us to confront them as some of the most extreme examples of the way racial identities can be essentialized by being removed from history and geography.

Unfortunately, it is the most familiar and widely employed racial terms that have, traditionally, received the least critical attention. Whilst more 'exotic' identities have attracted geographers and anthropologists for many years, being white and/or European and/or western remain comparatively new objects of enquiry. Yet, we argue in this chapter, that it is only by understanding such terms – the ones against which all others are defined as exotic – that the wider system of racial demarcation and privilege can be brought into view. As this emphasis suggests, this chapter will not be following the conventional pathway, established since the 1960s, of identifying the geography of race with the discussion of the spatial patterns and cultural symbolisms of non-white residence within the west. Nor does it seek to ignore or denigrate such work. Rather we are concerned that such material be understood in relation to the construction of normative, often 'racially unmarked', identities. Without such a broadening of focus, 'racial and ethnic geography' is easily represented as a marginal subfield, that bit of geography that deals with 'the others'. Given that geography as a discipline is largely white and western, such an emphasis can easily slide into paternalism, an altruistic concern about 'problem communities' and eternally suffering 'victims'. By making it clear that white people are also the products of racialization, that their identities also have a history and a geography and, hence, are changeable, we can help collapse this kind of intellectual and political distance: transforming the critique of race and ethnicity from a 'subfield' into an essential and continuous theme running throughout a rigorous geographical education.

In the first part of this chapter we introduce a number of ways that the geography of territorial and social distinctions can be brought into focus and further explored, paying particular attention to the history of the terms 'Europe' and 'the west'. In the second part the focus shifts to the racialized nature of particular places and landscapes. As we shall see, the empirical work of geographers on the way certain streets, neighbourhoods and cities have been given racial meaning allows us to witness the relationship between race and territory in new ways and at a much more detailed level. This section concludes with the observation that, whereas traditionally such studies have concentrated almost exclusively on marginal and immigrant spaces, many are now turning towards more racially unmarked and normative places and landscapes.

This concern is developed in the third and fourth sections of this chapter. The former provides some theoretical context for new work within the area of cultural identity that has sought to move beyond essentialist notions of 'race' and black/white binaries. The fourth section looks at why and how the geography and history of whiteness have emerged as central concerns of scholars interested in the relationship between space and race.

THE CRITIQUE OF ETHNO-GEOGRAPHY

Questioning the geographical categories which we use to divide the world has a long history. However, the vast majority of this work has been designed not to problematize these divisions but, rather, to query the claims of certain groups to be admitted into the more privileged camps. For example, in *The Decline of the West* (1980, originally published 1918) Spengler was adamant that 'Europe' was no longer a useful expression, since it encouraged the view that the Russians existed within the same ethno-geographical sphere as western Europeans. 'It is thanks to this word "Europe" alone and the complex of ideas resulting from it,' he complained, 'that our historical consciousness has come to link Russia with the West in an utterly baseless unity' (1980: 233). Far from opening up racial ideas for critical inspection, this kind of distinction reinforces their authenticity. However, a far more careful, yet also radical, kind of approach can also be sought out.

Hay's little book *Europe: The Emergence of an Idea* (1957) provides one of the better known English-language examples. Hay's concern was not with protecting the concept of Europe from subversion, but with showing how a term now widely accepted as obvious and natural became hegemonic for political and social reasons. More specifically, Hay detailed how 'Europe' had very little purchase on the ancient and medieval imaginations and only developed as a self-definition for those within Europe in the modern period, gradually incorporating and eclipsing the older idea of Christendom. Although Hay does not detail the relationship, it is apparent from his studies that the idea of 'Europe' grew as part of a wider racializing project, a project often associated with the rise of racial science. From seeing themselves as simply English, French, Prussian and so on, Europeans came to map their existing religious affiliation (as Christians) and emerging colonialist identity on to a natural, biological entity, namely 'the white European race'. Hay's text preceded, by nearly 40 years, the slew of studies on the same theme that have been published over the last decade. Recent titles, such as *Inventing Europe* (Delanty, 1995), *The History of the Idea of Europe* (Wilson and Dussen, 1995) and *The Making of Europe* (Barlett, 1994), are testament to the new-found acceptance of this way of looking at ethno-geography. At an even larger scale, Lewis and Wigen have provided a general critique of *The Myth of Continents* (1997). As they observe, racial essentialism forms a core component of such divisions, with Europe – in spite of its physical status as a continent being the most clearly suspect – providing the standard against which other such regions are defined. 'While a few professionals may regard Europe as a mere peninsula of Asia (or Eurasia),' Lewis and Wigen note, 'most geographers – and almost all nongeographers – continue to treat it, not only as a fully-fledged continent, but as the archetypal continent' (1997: 36).

The boundaries and meaning of Europe were first asserted within colonial contexts or within those societies that saw themselves as peripheral to the ethnic centre of an emerging white, European identity. Exemplifying both forces, the eighteenth-century Russian geographer Vasilii Tatishchev's designation, in the 1730s, of the Urals and the Caucasus as, respectively, the eastern and southern termini of Europe is of interest, not merely because of its survival to the present day, but because it reflects the emergence of an almost paranoid desire amongst the Russian elite to establish Europe and Asia as highly meaningful and utterly separate entities. Amongst the Russian aristocracy of the eighteenth century, the notion was established that a discrete, identifiable culture of economic and technological advance and rationality existed to 'the west' and that, through 'westernization', the underdeveloped European qualities of Russia itself could be brought into view and any Asiatic trace could be erased or marginalized. Although, by 1917, this view was also common amongst Marxist revolutionaries, it was first established as part of elite Russians' attempts to cast themselves within a normative image of European culture and colonialism. As this suggests, in attempting to establish Russia as European, the westernizers were also fixing the majority of the country as peripheral, uncivilized and Asian. Emphasizing the colonial dynamic contained in this formula, Becker points out that what came to be seen as the proximity of Asia to Russia was understood,

not as a threat to Russia's European identity but rather as an opportunity to prove that identity. In bringing to her Oriental subjects the fruits of Western civilisation, Russia would be demonstrating her membership in the exclusive club of European nations. (1991: 50)

As these explanations imply, the uniting of the historical and the geographical imagination appears to be a necessary characteristic of serious study of 'taken for granted' ethno-geographical terms. In the late 1970s, *Orientalism* (1979) by Edward Said quickly emerged as the classic text of this kind of endeavour. Said's historical unpacking of the way the orient was invented within the west has proved an inspiration for geographers seeking to take their discipline in a more reflexive direction (Gregory, 2000). In one sense, though, Said provided a conventional departure point for explorations of identity, his focus being firmly on the way an exoticized and demonized other was excluded by an all-powerful western master narrative. By contrast, the recent

emergence of a literature on occidentalism (Carrier, 1995; Chen, 1995; Venn, 2000; see also Gogwilt, 1995; Lewis and Wigen, 1997) is indicative of a desire to situate and examine the west and westernization not as unstoppable, as an 'all-conquering' social and economic *fait accompli*, but as contingent and partial creations. Indeed the term 'occidentalism' was introduced into current debate by those who saw it as a counter-discourse of resistance again the west, a reverse objectification (see Hanafi, 1992; also Tonnesson, 1994).

The relationship between European hegemony and the demarcation of the world and its peoples has also provided a starting point for the exploration of other familiar categories. In *The Invention of Africa* (Mudimbe, 1988) and *Inventing Eastern Europe* (Wolff, 1994) it is shown how Africa and eastern Europe first emerged as west European designations, and how they played a crucial role in the designation of vast and diverse populations as having stereotypical, racial attributes. However, these works are also alert to the way this process has been taken up, adopted and adapted by different groups around the world. Africa has long since ceased to be a European idea, a fact seen particularly clearly within the rise of pan-Africanism and Afrocentrism. The intercultural nature of the contemporary use of ethno-geographical labels provides a corrective focus to the tendency to overemphasize European agency. Two studies that have been particularly influential in detailing the interaction of European and non-European classificatory systems are Pratt's *Imperial Eyes* (1992) (which concerns the 'transcultural' development of images and ideas of the Americas) and Thomas' *Entangled Objects* (1991) (a study of the way the symbolism of everyday artefacts from the Pacific was subject to interpretation and reinterpretation, back and forth, between the colonized and colonial powers). Yet, as Pratt stresses, an appreciation of transculturalism should not be used as a way of bypassing the fact of European dominance. The outcome and meaning of hybridity reflect the power relations of colonialism and neocolonialism. This also suggests that, however transcultural their present-day usage, ideas such as 'African', 'Asian', 'European' and so on are not easily extricated from the racial logic which first enabled their modern dissemination and acceptance.

REPRESENTATIONS OF RACE AND PLACE

For the past 40 years geographical discussions of race have tended to be focused on questions of settlement and residential segregation. More specifically, analysis of the spatial segregation of minority groups from majority populations has dominated debate (see, for example, Jones, 1978; Lee, 1977; Peach, 1975; Peach et al., 1981). Within North America and western Europe this focus has been narrowed further by the increasing association of the urban with 'multiracial' and immigrant populations. In such societies the geography of race has become equated with the study of the city. The 'spatialization of race', Cohen (1993) notes, has led to the inner city connoting blackness. Also drawing from the British context, Watt points out that this process 'reflects the racialised, and in certain cases, racist pathologisation of urban areas, as seen for example in the press reporting of the urban "riots" of the 1980s' (1998: 688; for discussion see Burgess, 1985).

Stanton argues that the spatialization of race is even more extreme within the USA: 'Conceptually the city is left to the poor and racially marginalised ... For the media, the barometer of the national consciousness, the American city is now the black city' (2000: 129). The darkening of the city's image is associated by Stanton with the devaluing of the city, its demotion to a 'hopeless' and irrational landscape. To illustrate this process Stanton offers the following anecdote:

Canal Street, the teeming main downtown New Orleans, was described as 'dead' by white residents when I arrived in the city. The energetic 'Third World' and African-American commercial presence there was not registered as a realm of the living. (2000: 129)

Since, in recent years, the partial 'revitalization' of many US urban cores has gone hand-in-hand with the 'return' of white people to the city (see Rubinowitz and Rosenbaum, 2000; though see Frey and Liaw, 1997), there is a slightly dated ring to Stanton's assessment. Moreover, it needs to be remembered that such narratives are far from universal and should never be accorded a paradigmatic status: the attempt to demote the city to 'urban jungle' has little currency outside certain countries within the west (most notably, the USA). Moreover, even in those countries where the 'urban jungle' became a ubiquitous discourse, white dominance remained an ever-present reality of city life.

Drawing on his studies of the connections between local and national 'racialized space' within Sweden, Pred emphasizes that race is concretized, made to appear real, through its fixing in space:

The social construction of race becomes one with the physical occupation of space. The racialized become the segregated, and racial meaning becomes inscribed upon space. The discursively Otherized become declared out of bounds, the physically Elsewhereized and Isolated. The categorically excluded become physically enclosed. The socially marginalized become out of reach, not easily socially knowable. The socially barred become locationally removed from opportunity-yielding social, economic, and political networks. The culturally distanced become the pushed out and areally stigmatized ... Another feat of ontological magic. The implacable taken-for-granteds and idea-logics of cultural racism are – abracadabra, hocus-pocus, simsalabim – concretized. (2000: 98–9)

The most interesting work in urban cultural geography has moved away from attempts to generalize about the implications of minority group settlement, and towards the deconstruction of the, often conflicting, representational strategies that surround particular racialized places and events. Significant geographical work on the kind of dynamics to which Pred alludes has also been provided by Jackson in his studies on the Notting Hill carnival (1988) and 'race' and crime in Toronto (1993; see also Jackson, 1987; Jackson and Penrose, 1993). Other important examples include Jacob's (1988; 1993; 1996) studies on racialized conflict over landscape use and symbolism in Australia; Smith's (1989a; 1989b) research on the racialization of residential space; Sibley's (1988; 1992; 1995) accounts of the geography of 'outsiders'; Robinson's (1996; 1998) analysis of the changing dynamics of race and space in South Africa; Anderson's (1993a) work on the 'Aboriginal space' of Redfern; and Keith's (1993) writings on the spatialized construction of 'race' and 'riots' in 1980s London. As Anderson stresses, the process of spatialized racialization needs to be understood as historically contingent. This point has also been made by Keith in his studies on the antagonistic 'racial' geographies of the police and 'black community' in London. Keith makes the valuable suggestion that the study of the generation of spatial meanings should be accompanied by the analysis of their 'closure' or contingent completion. However, Keith implies that the closure of these metaphoric, metonymic and syntagmatic associations is provisional, contingent and arbitrary. The meanings arrived at by different groups may reflect the unconscious, common-sense formation of prejudice or self-conscious, strategic forms of 'race-place' essentialism (for example, the ghetto as focus of Black Pride). Yet, in either case, they are geographically and historically mutable, liable to change, challenge and reformation. The suggestion that 'places are moments of arbitrary closure' (Keith, 1991a: 187) is supported by Keith in his book *Race, Riots and Policing: Lore and Disorder in a Multi-racist Society* (1993; see also Keith, 1987; 1988a;

1988b; 1991b; Keith and Pile, 1993; Back and Keith, 1999). This work focuses on the way that those areas of London 'associated with' the Afro-British community have been racialized in different ways by Afro-Britons and the police. Keith pays particular attention to the multiple symbolism of 'front lines' (that is, those streets, such as Railton Road in Brixton, that are seen to be at the front line in the conflict between the police and black people). These locations are analysed as being subject to metaphoric, metonymic and syntagmatic interpretation. We will elucidate Keith's argument by looking at each of these forms of meaning in turn.

Front lines, such as Railton Road, he explains, 'can be seen as metaphorically linked. They are not precise replications of each other, but in terms of the sign system involved they are almost mutually interchangeable' (1993: 165). Keith then proceeds to explain the difference between the metonymic and syntagmatic meaning of front lines. He suggests that 'black people' understand police action in front line areas from a historical vantage point; police action connotes (metonymically) a history of racism and police brutality. The police, on the other hand, possess a historically shallow perspective; the front line connotes (syntagmatically) day-to-day operational burdens. Thus,

[B]lack perceptions are constructed as a form of 'local knowledge' and are fundamentally metonymic in the reading of the social world; the police action is seen as part of a historical whole, invoking a 20 to 30 year history of Black experience in a particular 'place'. For the police, operational goals have priority and 'place' as a sign is read syntagmatically; the action is part of an expected sequence, an anticipated repertoire of behaviour that occurs wholly in the present ... It is this very structure of police practice in such areas which guarantees that the policing institution acts as a 'machine for the suppression of time' – history is lost. (1993: 166)

The symbolic processes at work within the racialization of place and space lend themselves to the vocabulary of semiotics. Yet the latter tradition is open to a variety of theoretical interpretations. In particular, it draws on and overlaps with the categories of repression and displacement made familiar by Freud and his followers. The potential of this latter approach has been indicated in a recent paper by Heidi Nast (2000). Nast offers a psychoanalytical analysis of the patterns of black spatial containment, white 'flight' and urban renewal apparent within Chicago. Thus, for example, she explains the 'sociospatial repression of black bodies, places and life' (2000: 232) by reference to the role of '[r]acist imaginary-symbolic renderings of black men as rapists' (2000: 231) within the white

psyche. More generally, her concern is with the way 'exteriorized landscapes and interiorized psyches have historically structured one another' (2000: 219). One of the most productive aspects of Nast's paper is that it situates itself within a *tradition* of psychoanalytically informed studies of spatial racialization. In particular, she discusses and draws on the attempts by White (1972) to provide a historical geographical translation and critique of Freud's non-historical and non-geographical account of the formation of the psyche. Central to White's analysis is the claim that the psyche was a product of the colonial construction and interiorization of the figure of the 'wild man'. In a passage cited by Nast (2000: 223), White notes that as 'wilderness was brought under control, the idea of the Wild Man was progressively despatialized. This despatialization was attended by a compensatory process of psychic interiorization.'

Such explicit use of psychoanalysis provides a useful challenge to the contemporary tendency to employ the language of repression and displacement without ever giving serious consideration to the psyche or, more specifically, the unconscious, as terrains in which the dilemmas of racialization are played out. Whether such a focus produces different conclusions on the material consequence and enactment of sociospatial racialization is, however, far less clear. Certainly, much recent work on what Anderson calls 'the interlocking semiotic and material processes' (1993a: 85) behind the racialization of urban space has managed to provide portraits of landscapes fraught with contradictory desires and processes without recourse to images of the archetypal psyche. Anderson's work is itself a good example. Anderson's (1987; 1988; 1991) earlier studies focus on constructions of Chinatown. More specifically, she provides a historical study of the different and dynamic urban constituencies which have helped shape the 'racial' boundaries and meanings associated with Vancouver's Chinatown. The 'space of knowledge called Chinatown', she explains, 'grew out of, and came to structure, a politically divisive system of racial discourse that justified domination over people of Chinese origin' (1988: 146). In some of her more recent studies Anderson (1993a; 1993b) has focused on the development of 'racial' meanings in the 'Aboriginal suburb' of Redfern in Sydney. 'Aboriginal Redfern,' Anderson comments, 'was constructed out of multiple and contradictory discourses and practices, the deconstruction of which clears the way for a non-essentialized theorisation of not only Aboriginal identity but also the place "Redfern"' (1993a: 87).

Anderson isolates two competing discourses that have racialized Redfern. The first emerges from 'Aboriginal rights' arguments that position the suburb at the heart of 'the Aboriginal community', a physical site that symbolically coalesces the multiplicity of indigenous voices into a 'pan-Aboriginal struggle against White Australia' (1993a: 86). Anderson concentrates her attention on how this 'cultural and political invention' has come into conflict with racist, white, constructs of Redfern and Aboriginality during the early 1970s. With the help of archive and interview material she shows how '[p]oliticians and officials … drew on an established (pejorative) set of images of Aboriginality, not out of any simple "prejudice", but in order to win the support of local White residents' (1993a: 86–7). This process, Anderson continues, helped 'construct a negatively racialized Redfern that has not eroded with time'. Thus Redfern became a central and disputed category in a socio-spatial conflict between white and black activists and sympathizers.

We believe that, however sophisticated its application, a focus on explicitly racialized places and peoples unintentionally runs the risk of normalizing the spatial dominance of the non-racialized majority. Certain critics have challenged this dominance from a different direction. For example, Shaw's (2000; 2001) recent ethnographic research in the Redfern district of Sydney takes whiteness as its principal focus:

> Away from the stark black/white racialised boundary near The Block [i.e. the Aboriginal identified area of Redfern], where the space of whiteness absorbs other ethnicities, whiteness appears to fade into ethnic neutrality. Away from the Aboriginal 'other', whiteness is not so visible. My observations of the spaces near The Block lead me to think about how whiteness strengthens and consolidates against the presence of The Block. (2001: 8)

A parallel development in those countries where the rural is connoted as white, such as Britain, is the new attention being given to non-urban environments. This focus brings into view the mutually reinforcing relationship between the racialized urban and the racially unmarked rural. Urry asserts that the '"racialisation" of the phenomenology of the urban works partly in England through the contrasting high valuation which is placed upon the English countryside which is taken to be predominately white' (1995: 27). It is important to note that this relationship does not necessarily turn on the presence or absence of non-white people. Indeed, it can be detected within visions of the degenerate and unnatural atmosphere of cities from the

nineteenth and early twentieth centuries. As we shall now see, the explicit interaction of discourses of class and race within the spatial-racial logic of this period makes it a particularly revealing illustration.

As a consequence of the large-scale migration of rural families to the city that characterized the late nineteenth century, the British (more especially the English) working class was often construed to be losing its national and racial rootedness. 'Traditional' rural folk were being lost to racial degeneracy. *In Rural England* (1902) Rider Haggard noted that this migratory flow 'can mean nothing less than the progressive deterioration of the race' (1976: 218). In *The Poor and the Land* the same author contrasted the 'puny pygmies growing from towns or town bred parents' with the 'blood and sinew of the race', the 'robust and intelligent' countryman (Haggard, 1905: xix). The racialized contrast between urban and rural relied, in part, on an association of the urban with immigrant labour. However, the perceived threat of the urban also drew upon an existing and specifically English discourse of national and racial romanticism which placed the essence of Englishness in 'the people[s] … natural breeding and growing grounds' (Lord Walsingham, quoted by Low, 1996: 19), the countryside. Indeed, the similarities between the stereotypes of the 'rosy cheeks' and 'healthy complexion' of the English peasant and the 'vigorous' nature of bourgeois whiteness are suggestive of Victorian middle-class writers' investment in rural nostalgia as a kind of origin myth of their own ascendance. This impression is strengthened by the fact that rural workers in the late nineteenth and early twentieth centuries tended not to be subject to horrified explorers but rather to reverential cultural retrieval. By 1911 folklore studies had been published on 29 of England's 40 counties. Colls' (1986; see also Howkins, 1986) observations of the development of 'folk study' in the period draw a direct contrast between the valued purity of the rural past and the racially degraded urban present.

Such historical context helps to further undermine the fallacy identified by Watt: 'One aspect of the importance of the inner-city discourse on race is that it tends to re-inforce the notion that racism only spatially occurs where black people live' (1998: 688). Watt's ethnographic study of the everyday geographies of white, Asian and black young people living in the 'commuter belt' area of the south-east of England illuminates the severely constricted spatial movement possible for racialized people within such a 'non-racialized' 'all-white' context. He found that:

The white middle class young people, who lived in upmarket commuter villages, were the least localist in orientation and had very little sense of belongingness in relation to where they lived ... Asians were the most localist in orientation amongst all of the young people we spoke to. Many of the young Asian men, in particular, felt a strong sense of loyalty to [one area] based upon a masculine-dominated street defensiveness: 'it's always filled with Asians here, it's like no one messes around with you, we just hang around with each other'. (1998: 692)

An irony born, at least in part, of middle-class mobility, that emerges from Watt's summary is that whilst racial exclusion demands a certain level of attachment to the local amongst minority residents of 'white space', many young white residents enjoy the 'privilege' of being (or at least professing themselves to be) unattached to the places where they live. The 'authentic England' may turn out to be populated by people who, far from being ideologically attached to local rural communities, are more akin to rootless middle-class nomads (see Murdoch and Marsden, 1994).

NEW THEORIES OF CULTURAL IDENTITY: BEYOND 'RACE'

As geographers have struggled to come to terms with the discipline's imperial legacy and to produce alternative, critical geographies of racialization, they have drawn inspiration from new developments in the fields of cultural studies, postcolonial literature and the broader sociology of race and ethnic studies. Peter Jackson, whose *Maps of Meaning* (1995) forms a key part of this interdisciplinary bridge-building exercise, declared his intention to reinvigorate the stasis of cultural geography. This could be achieved, he believed, by combining 'some of the most important ideas from cultural studies with some recent developments in human geography, seeking alternative approaches to the geographical study of culture from the traditional obsession with landscape' (1995: 3).

To illustrate the changing politics of cultural identity Stuart Hall, referring to Paul Gilroy's book *There Ain't No Black in the Union Jack* (1987), considered that until recently he 'didn't care, whether there was any black in the Union Jack. Now not only do we care, we must' (1993: 258). That blackness could no longer be seen as 'alien', other and essentially 'unBritish', is indicative of a new 'ethnic assertiveness' apparent in recent generations of minority youth.

However, by rejecting the concept of race as anything other than a social artefact, anti-essentialist scholars have also brought into question the very basis of black identity. For if race is little more than a social concept discursively mapped upon the bodies of black people through the process of racialization, how useful is it to politically organize around this collective identity in the first place?

Increasingly, anti-racist writers and activists have begun to question the value of colour-based alliances. Strategically useful they may have been, but the extent to which a racial dualism can adequately articulate the historical and geographical complexity of cultural identity appears limited. These issues point to a move away from western binary relations of racism (which are currently black/white, though historically have tended to centre on a geographical axis – east/west, orient/occident – or a religious affiliation, such as Christian/Muslim, civilized/heathen). The new politics of cultural identity now suggests composite forms of discrimination and lends itself to the consideration of internal gradations within 'blackness' or 'whiteness'. Hall has suggested that the implosion of 'black' could in turn lead to the production of 'new ethnicities':

> What is at stake here is the recognition of the extraordinary diversity of subject positions, social experiences and cultural identities which compose the category 'black'; that is, the recognition that 'black' is essentially a politically and culturally constructed category, which cannot be grounded in a set of fixed transcultural or transcendental racial categories and which therefore has no guarantees in Nature. What this brings into play is the recognition of the immense diversity of differentiation of the historical and cultural experiences of black subjects. This inevitably entails a weakening or fading of the notion ... of 'race'. (1993: 254)

According to Hall, this new politics of difference may lead to 'the end of the innocent notion of the essential black subject' (1993: 254) and engender a greater appreciation of plurality (see Mercer, 1994).

The manner in which multiple identities are clumsily collapsed and conflated is an issue that has not been lost on those working in the field of public policy. Thus Carrington et al., in their research on the recruitment and retention of ethnic minority new teachers in England and Wales, found that the available ethnic categories 'are perceived to be ambiguous, anachronistic and discrepant with commonly held subjective definitions' (2001: 44) of interviewees. Tariq Modood has also emphasized problems with imported forms of ethnic classification. His critique of black identity centres upon the manner in which the term 'black' has functioned to exclude different ethnic minority groups from the UK policy debate, as his example of South Asian subjects clearly reveals (see Modood, 1988; 1994). This work demonstrates how particular minority ethnic groups may not necessarily identify with the dominant discourses of either racism or anti-racism, but instead prefer to mobilize around religious or cultural forms of identity. Moreover, the implosion of a black/white racial dualism has also had the effect of laying open to inspection the category of whiteness. The social geographer Peach only exaggerated slightly when he recently declared, 'If the attempt to impose a single "black" identity on diverse ethnicities was a bad fault of the past, in the 1990s, homogenising "whiteness" is now considered to be worse' (2000: 621).

Postcolonial writings on cultural syncretism and hybridity may also be claimed to have provided a 'third space' (Bhabha, 1990) that transgresses the black/white racial dualism. For Homi Bhabha, cultural hybridity encourages a radical proliferation that 'gives rise to something different, something new and unrecognisable, a new area of negotiation and representation' (1990: 211). This new 'third space' emerges from the colonial encounter and the more recent historical processes of globalization, migration and settlement. Kevin Robins notes: 'Globalisation, as it dissolves the barriers of distance, makes the encounter of colonial centre and colonised periphery immediate and intense' (1991: 25). The supposedly new hybrid ethnicities that are flourishing in many urban areas (Back, 1996) are the consequence of this postcolonial, globalizing dynamic. The third space, or what Mary Louise Pratt has called the 'contact zone', may thus be cast as a space where 'the spatial and temporal co-presence of subjects previously separated by geographic and historical junctures, and whose trajectories now intersect' (1992: 7) can come together. In doing so it gives rise to new, hitherto unimagined cultural identities that are hybrid translations of their former antecedents.

However, critics have suggested that the focus on cultural identity that characterizes these assessments of contemporary hybridity has been at the expense of political and social concerns. For those left unconvinced by the rhetoric of postmodern possibilities, racism remains a far more salient site of concern than new ethnicities (Cohen, 1999). This cautious approach to the claims associated with Bhabha and other postcolonial and new ethnicities scholars indicates an unwillingness to allow discourses of cultural hybridity to subsume and marginalize the way racial demarcation and division are reproduced.

Globalization and postmodernism have not led to the implosion of identifications with nations, regions or localities. Moreover, the 'immediate and intense' nature of the urban ethnic mix is, surely, as old as cities themselves: the third space is a tradition not a novelty. This, in turn, suggests that, although the established binaries of western racial discourse may have become suspect, the collapsing of old distinctions should not be confused with the dawn of a new post-racist era.

A final concern with the new ethnicities approach remains a geographical tendency for researchers to focus upon urban areas in developed countries. This has led some geographers to write of 'the hegemonic status of the inner-city discourse in relation to race and space' (Watt, 1998: 688), and others to assert that 'These geographies are offered as colourful empirical demonstrations of the cultural cosmopolitanism that is the contemporary Western moment' (McGuinness, 2000: 225–6). Notwithstanding these criticisms, research on cultural identities and new ethnicities has offered some useful insights into subjectivity and race. However, to counteract some of the geographical limitations of this work we will now turn our attention to recent, global studies on whiteness and white places undertaken by contemporary social and cultural geographers.

HISTORIES AND GEOGRAPHIES OF WHITENESS

The most influential group of writers in the field of what we may tentatively call 'white studies' has emerged from a Marxist tradition of historical scholarship in the US. These new labour historians have applied their knowledge of the history of labour movements to understand how whiteness has become, in the words of Theodore Allen, 'the overriding jet-stream that has governed the flow of American history' (1994: 22). As David Roediger has demonstrated in *The Wages of Whiteness* (1992) and *Towards the Abolition of Whiteness* (1994), social class and ethnic privilege are mutually enforcing components in the historical making of the American working class. For Roediger whiteness fulfilled its role as a divisive form of cultural capital that served to separate established immigrants from 'new' immigrants, Irish peoples, southerners, slaves and those yet to be fully assimilated into the sacred pantheon of 'white' American citizenship. For Roediger, the US has watched unfold 'The sad drama of immigrants embracing whiteness while facing the threat of being victimised as nonwhite' (1992: 180).

Scholars in geography are also now beginning to critically investigate whiteness, though mainly from a representational, feminist or deconstructivist position. Yet it remains a historical irony that such contemporary investigations often fail to recognize that, within geography at least, whiteness is not a new topic of research (for example, Trewartha, 1926; Woodruff, 1905). The imperial geographies of the late nineteenth and early twentieth centuries constantly strove to map out and legitimize white supremacy. Clearly, it would be wrong to imagine that white dominance is necessarily subverted by simply openly talking about, or 'outing', whiteness. Whiteness was very much out (and proud) for a long time in geography. Trewartha reported on the unsuitability of white Europeans to muscular labour in the 'wet tropics', concluding that 'the brown man is superior to the white in his economy of sweating' (1926: 472). By 1931 Dane Kennedy felt able to comment upon the 'perils of the midday sun' in the colonial interior, leaving him to ponder why 'natural laws drove the white race to control, but prevented them from populating the tropics' (1990: 123).

However, whilst whiteness was once treated as a natural and stable identity, it is now increasingly viewed as something that has a history and a geography, an impermanent social formation that can be changed and challenged (Bonnett, 1993; 1997; 2000a; 2000b). An important starting point remains Ruth Frankenberg's qualitative analysis of 30 white Californian women in the US, *White Women, Race Matters: The Social Construction of Whiteness* (1994). A chapter entitled 'Growing up White: the Social Geography of "Race"' is an attempt to show how whiteness is enacted in small American towns at a neighbourhood scale. Thus, Frankenberg examines 'the interlocking effects of geographical origin, generation, ethnicity, political orientation, gender and present-day geographical location' (1994: 18) on the lives of her white respondents. Elsewhere, Frankenberg has depicted her work to be a 'Racial social geography' that involves 'the racial and ethnic mapping of a landscape in physical terms, and enables also a beginning sense of the conceptual mapping of self and other with respect to race operating in white women's lives' (1993: 54). The intertwining of geography with whiteness is also apparent in the introduction – entitled 'Local Whiteness, Localising Whiteness' – to Frankenberg's edited volume *Displacing Whiteness: Essays in Social and Cultural Criticism* (1997).

Schech and Haggis (1998) have discussed postcolonial forms of whiteness in a time and place where the circulating currents of globalization simultaneously give way to a monocultural defensiveness, exemplified by Pauline Hanson's One Nation Party in Australia. Using Hanson's landslide victory in the working-class suburb of Brisbane in order to assess the popularity of the One Nation Party and its appeal to whiteness, their study explores how Australian identity has been challenged from without by the 'powerhouse' economies in the Asian–Pacific rim, and also from within by the claims to Aboriginal land rights from indigenous peoples. The One Nation Party has been seen to be successful in articulating a sense of white unease through claiming that whites are now the marginalized victims of society. In this context the authors conclude that, 'Perhaps postwhiteness is a necessary requisite for postcoloniality' (1998: 627).

Another study that combines a national and local exploration of whiteness is Kobayashi and Peake's (2000) examination of press reportage of the shootings that occurred in Columbine High School in Littleton, Colorado, USA. Littleton is portrayed as a 'safe' neighbourhood, far removed (at least in the white imagination) from the vividly racialized inner-city 'hood' and ghetto. According to the authors the racialized representations of these areas show that 'Place does matter both because social processes such as whiteness are bounded, and because the complex feelings of both racism and antiracism are highly evocative of particular landscapes' (2000: 396). Kobayashi and Peake go on to remark on the need for geographers to embark upon a critical engagement with whiteness. They note how 'the lives of dominantly white geographers, are sites for the reproduction of racism, but they also hold the potential of being strategic sites of resistance' (2000: 399). Indeed, in a rare disclosure of white subjectivity the feminist geographer Gillian Rose has considered how whiteness has empowered her work in the discipline: 'I may feel marginalised in geography as a woman,' she opines, 'but my whiteness has enabled my critique of geographical discourses by allowing me to get close enough to them to have a good look' (1993: 15).

Peter Jackson's study of shopping patterns has led him to declare that 'constructions of whiteness should be traced at a variety of scales from the nation to the neighbourhood' (1998: 100). A fine example is Paul Watt's (1998) previously mentioned interview-based study with 70 young people (15–21 years) of 'white', Asian and Afro-Caribbean background in the south-east of England (see also Dwyer and Jones III, 2000). Watt reveals how it is not only white people but 'white places' that have escaped the geographical gaze. This perspective is also proffered by McGuinness (2000) as he seeks to persuade geographers to recast their gaze from cosmopolitan locales such as Kilburn High Road to 'Middle England' and other less marked zones of ethnic enquiry. Others have used the method of ethnography to examine whiteness and the geography of racist violence in English suburbs such as 'Kempton Dene' in the west midlands (Back and Nayak, 1999). This work demonstrates how a skinhead gang were able to transform their neighbourhood into a 'white space'. At a macro level this was partially achieved through economic transformations in the region, 'white flight' from the city and the city council housing policies aimed at dispersing ethnic minorities throughout the conurbation. However, the constitution of the Kempton Dene estate as 'white' was further secured at a local micro-political level through intimidation, harassment and the perpetuation of racist graffiti. The geographical analysis of suburban whiteness is also a feature of some of Frances Widdance Twine's recent work. In one study she examines the lives of 16 female university students of African American descent who were 'raised white' on the suburban outskirts of America. For these 'brown skinned white girls', as Twine describes her respondents, the suburban privileges of whiteness are seen to act as a type of 'comfort zone' (1996: 215) which enables their ethnic inclusion into mainstream practices at material, cultural and even psychic levels. 'Currently, some contemporary social geographers are deploying an analysis of whiteness to re-think how race, class and gender are mutually constituting categories, articulated through and against one another in complex, and at times contradictory ways' (Haylett, 2001; Nayak, 2002).

There is much to learn from such small-scale, empirically grounded studies about the theoretical complexity of ethnicity in people's everyday lives. However, it is a concern that geographers have combined this focus with a parochial geographical horizon, rarely lifting their sights above the familiar terrain of Britain, North America and Australia.[1] The narrow geographical limits of 'white studies' act to reinforce the tendency towards insularity that has characterized much recent work within ethnic and racial studies in the west. The notion that this area of enquiry should have global ambitions may, to some, have a peculiar colonial ring to it. After all, from the 1950s onwards, ethnic and racial studies have travelled in precisely the opposite direction, rejecting imperialist anthropology for the dissection of racism in western nations. Whilst we are in agreement with the political motivations behind this project, it has had the effect of making debates on race in the west highly confined

and claustrophobic. Moreover, the coherence of abstracting particular national narratives from the world economy – of studying 'French racism' or 'British anti-racism' – is increasingly questionable. White identities are, if nothing else, global phenomena, with global impacts. Indeed, the nature and implications of their local manifestations only come into view when they are understood as global. This approach has led Bonnett (2000a; 2000b) to develop a historical and geographical analysis of whiteness. A key concern of this work is to show how the history of how whiteness became racialized is also the history of how groups previously identified as white (such as the Chinese) began to call themselves something else and how Europeans began to believe that they were the world's only true whites. Bonnett approaches this issue with reference to material largely from the Middle East, China and Japan, societies where white identification amongst elite groups existed well into the nineteenth century. Bonnett (2000a; 2000b) also traces more recent associations between Europeans, white skin and neoliberal, consumer-based cultural globalization. Drawing on popular culture from Latin America and Japan, he interrogates the way the 'fun, free and flexible' lifestyles of neoliberalism have been connoted as white and western. This process has ensured the reproduction and reformation of the role of the white as a key symbol of success, modernity and wealth. Thus despite the almost universal abandonment of explicit doctrines of white supremacy, and the adoption of anti-racist rhetoric as the lexicon of legitimacy by institutions the world over, whiteness continues to be reified as a racial and cultural norm. The patterns and paths of resistance to this process are diverse, yet if any one attribute of the white racial norm stands out from the last century it is its capacity for adaptation and survival.

CONCLUSIONS

We have argued in this chapter that it is only by understanding such normative terms as 'white' and 'western' – the ones against which others are defined as exotic – that wider systems of racial privilege can be brought into view. By making it clear that categories such as whiteness are also the products of racialization, that they too have a history and a geography and, hence, are changeable, we can help transform the critique of race and ethnicity from a 'subfield' into an essential theme running throughout a rigorous geographical education. This transformation provides a challenge to the familiar colonial role of 'racial studies' (as supplier of anthropological information on colonial subjects) as well as to its more contemporary incarnation (as the social analysis of the predicament of racialized minorities in the west). This also implies that the tradition of emphasizing the racialization of localities as the principal research focus of 'racial geography' needs to be widened to enable a fuller appreciation of the intersections of race and geography.

Geographers cannot avoid issues of racial and ethnic division: they are part and parcel of the discipline's history and its most basic vocabularies. However, perhaps because of this, geographers are well placed to provide a more insightful and considered judgement than social critics for whom ideas like 'European', 'western', 'African' and so on are less clearly visible as contingent and mutable creations.

NOTE

1   But see Lambert's (2001) recent geography of 'racial reinscription' (p. 335) in colonial Barbados.

REFERENCES

Allen, T.W. (1994) *The Invention of the White Race.* London: Verso.

Anderson, K. (1987) 'Chinatown as an idea: the power of place and institutional practice in the making of a racial category', *Annals, Association of American Geographer* 77: 580–98.

Anderson, K. (1988) 'Cultural hegemony and the race-definition process in Chinatown, Vancouver: 1880–1980', *Environment and Planning D: Society and Space* 6: 127–49.

Anderson, K. (1991) *Vancouver's Chinatown: Racial Discourse in Canada* 1875–1980. Montreal: McGill–Queens University Press.

Anderson, K. (1993a) 'Constructing geographies: "race", place and the making of Sydney's Aboriginal Redfern', in P. Jackson and J. Penrose (eds) *Constructions of Race, Place and Nation.* London: UCL Press.

Anderson, K. (1993b) 'Place narratives and the origins of the Aboriginal settlement in inner Sydney, 1972–1973', *Journal of Historical Geography* 9 (3): 314–35.

Back, L. (1996) *New Ethnicities and Urban Culture: Racisms and Multiculture in Young Lives.* London: UCL Press.

Back, L. and Keith. M. (1999) '"Rights and wrongs": youth, community and narratives of racial violence', in P. Cohen (ed.) *New Ethnicities, Old Racisms.* London: Zed.

Back, L. and Nayak, A. (1999) 'Signs of the times? Violence, graffiti and racism in the English suburbs', in

T. Allen and J. Eade (eds) *Divided Europeans: Understanding Ethnicities in Conflict.* The Hague: Kluwer.

Barlett, R. (1994) *The Making of Europe: Conquest, Colonization and Cultural Change 950–1350.* London: Penguin.

Becker, S. (1991) 'Russia between East and West: the intelligentsia, Russian national identity, and the Asian borderlands', *Central Asian Review* 10 (4): 47–64.

Bhabha, H.K. (1990) 'The third space: interview with Homi Bhabha', in J. Rutherford (ed.) *Identity: Community, Culture, Difference.* London: Lawrence and Wishart.

Bonnett, A. (1993) 'Forever 'white'? Challenges and alternatives to a 'racial' monolith', *New Community* 20 (1): 173–80.

Bonnett, A. (1997) 'Geography, "race" and Whiteness: invisible traditions and current challenges', *Area* 29 (3): 193–9.

Bonnett, A. (2000a) *White Identities: Historical and International Perspectives.* Harlow: Pearson.

Bonnett, A. (2000b) *Anti-racism.* London: Routledge.

Burgess, J.A. (1985) 'News from nowhere: the press, the riots and the myth of the inner city', in J. Burgess and J.R. Gold (eds) *Geography, the Media and Popular Culture.* Kent: Croom Helm.

Carrier, J. (ed.) 1995 *Occidentalism* Oxford: Oxford University Press.

Carrington, B., Bonnett, A., Demaine, I., Hall, I., Nayak, A., Short, G., Skelton, C., Smith and Tomlin, R. (2001) *Ethnicity and the Professional Socialisation of Teachers.* London: Teacher Training Agency.

Chen, X. (1995) *Occidentalism: A Theory of Counter-Discourse in Post-Mao China.* New York: Oxford University Press.

Cohen, P. (1993) *Home Rules: Some Reflections on Racism and Nationalism in Everyday Life.* London: University of East London.

Cohen, P. (1999) 'Through a glass darkly: intellectuals on race', in P. Cohen (ed.) *New Ethnicities, Old Racisms.* London: Zed.

Colls, R. (1986) 'Englishness and the political culture', in R. Colls and P. Dodd (eds) *Englishness: Politics and Culture 1880–1920.* London: Croom Helm.

Delanty, G. (1995) *Inventing Europe: Idea, Identity, Reality.* Basingstoke: Macmillan.

Dwyer, O. and Jones III, J. (2000) 'White socio-spatial epistemology', *Social and Cultural Geography* 1 (2): 209–22.

Frankenberg, R. (1993) 'Growing up white: feminism, racism and the social geography of childhood', *Feminist Review* 45: 51–84.

Frankenberg, R. (1994) *White Women, Race Matters: The Social Construction Of Whiteness.* Minnesota, Minneapolis University Press.

Frankenberg, R. (ed.) (1997) *Displacing Whiteness.* Durham: Duke University Press.

Frey, W. and Liaw, K. (1997) 'Immigrant concentration and domestic migrant dispersal: is movement to non-metropolitan areas "white flight"?', *Professional Geographer* 50 (2): 215–32.

Gilroy, P. (1987) *There Ain't No Black in the Union Jack.* London: Routledge.

Gogwilt, C. (1995) *The Invention of the West: Joseph Conrad and the Double-Mapping of Europe and Empire.* Stanford: Stanford University Press.

Gregory, D. (2000) 'Edward Said's imaginative geographies', in M. Crang and N. Thrift (eds) *Thinking Space.* London: Routledge.

Haggard, R. (1905) *The Poor and the Land.* London: Longmans, Green.

Haggard, R. (1976) 'Town versus country', in P. Keating (ed.) *Into Unknown England, 1866–1913: Selections from the Social Explorers.* Manchester: Manchester University Press.

Hall, S. (1993) 'New ethnicities', in J. Donald and A. Rattansi (eds) *'Race', Culture and Difference.* London: Sage.

Hanafi, H. (1992) *Muquaddima fi ilm al-Istighrab.* Cairo.

Hay, D. (1957) *Europe: The Emergence of an Idea.* Edinburgh: Edinburgh University Press.

Haylett, C. (2001) 'Illegitimate subjects?: abject whites, neoliberal modernisation, and middle-class multiculturalism', *Environment and Planning D: Society and Space* 19: 351–70.

Howkins, A. (1986) 'The discovery of rural England', in R. Colls and P. Dodd (eds) *Englishness: Politics and Culture 1880–1920.* London: Croom Helm.

Jackson, P. (ed.) (1987) *Race and Racism: Essays in Social Geography.* London: Allen and Unwin.

Jackson, P. (1988) 'Street life: the politics of carnival', *Environment and Planning D: Society and Space* 6: 231–7.

Jackson, P. (1993) 'Policing difference: "race" and crime in metropolitan Toronto', in P. Jackson and J. Penrose (eds) *Constructions of Race, Place and Nation.* London: UCL Press.

Jackson, P. (1995) *Maps of Meaning: An Introduction to Cultural Geography.* London: Routledge.

Jackson, P. (1998) 'Constructions of "whiteness" in the geographical imagination', *Area* 30 (2): 99–106.

Jackson, P. and Penrose, J. (eds) (1993) *Constructions of Race, Place and Nation.* London: UCL Press.

Jacobs, J. (1988) 'Politics and the cultural landscape: the case of Aboriginal land rights', *Australian Geographical Studies* 26: 249–63.

Jacobs, J. (1993) '"Shake 'im this country": the mapping of the Aboriginal sacred in Australia – the case of Coronation Hill', in P. Jackson and J. Penrose (eds) *Constructions of Race, Place and Nation.* London: UCL Press.

Jacobs, J. (1996) *Edge of Empire: Postcolonialism and the City.* London: Routledge.

Jones, P. (1978) 'The distribution and diffusion of the coloured population in England and Wales 1961–1971', *Transactions of the Institute of British Geographers* 3 (4): 515–32.

Keith, M. (1987) '"Something happened": the problems of explaining the 1980 and 1981 riots in British cities', in P. Jackson (ed.) *Race and Racism: Essays in Social Geography.* London: Allen and Unwin.

Keith, M. (1988a) 'Racial conflict and the "no-go areas" of London', in J. Eyles and D. Smith (eds) *Qualitative Methods in Human Geography.* Cambridge: Polity.

Keith, M. (1988b) 'Riots as "social problem" in British cities', in D. Herbert and D. Smith (eds) *Social Problems in the City.* Oxford: Oxford University Press.

Keith, M. (1991a) 'Knowing your place: the imagined geographies of racial subordination', in C. Philo (comp.) *New Words, New Worlds: Reconceputualising Social and Cultural Geography.* Lampeter: Social and Cultural Geography Study Group of the Institute of British Geographers. pp. 178–92.

Keith, M. (1991b) 'Policing a perplexed society?: No-go areas and the mystification of police–Black conflict', in E. Cashmore and E. McClaughlin (eds) *Out of Order? Policing Black People.* London: Routledge.

Keith, M. (1993) *Race, Riots and Policing: Lore and Disorder in a Multi-racist Society.* London: UCL Press.

Keith, M. and Pile, S. (eds) (1993) *Place and the Politics of Identity.* London: Routledge.

Kennedy, D. (1990) 'The perils of the midday sun: climatic anxieties in the colonial tropics', in J.M. MacKenzie (ed.) *Imperialism and the Natural World.* Manchester: Manchester University Press.

Kobayashi, A. and Peake, L. (2000) 'Racism out of place: thoughts on whiteness and an antiracist geography in the new millennium', *Annals of the Association of American Geographers* 9 (2): 392–403.

Kropotkin, P. (1996) 'What geography ought to be' (1885), in J. Agnew D. Livingstone and A. Rogers (eds) *Human Geography: An Essential Anthology.* Oxford: Blackwell.

Lambert, D. (2001) 'Liminal figures: poor whites, freedmen, and racial reinscription in colonial Barbados', *Environment and Planning D: Society and Space* 19: 335–50.

Lee, T. (1977) *Race and Residence: The Concentration and Dispersal of Immigrants in London.* Oxford: Clarendon.

Lewis, M. and Wigen, K. (1997) *The Myth of Continents: A Critique of Metageography.* Berkeley: University of California Press.

Livingstone, D. (1992) *The Geographical Tradition.* Oxford: Blackwell.

Livingstone, D. (1994) 'Climate's moral economy: science, race and place in post-Darwinian British and American geography', in A. Godlewska and N. Smith (eds) *Geography and Empire.* Oxford: Blackwell.

Low, G. (1996) *White Skins, Black Masks: Representation and Colonialism.* London: Routledge.

McGuiness, M. (2000) 'Geography matters? Whiteness and contemporary geography', *Area* 32 (2): 225–30.

Mercer, K. (1994) *Welcome to the Jungle: New Positions in Black Cultural Studies.* London: Routledge.

Modood, T. (1988) '"Black", racial equality and Asian identity', *New Community* 14 (3): 397–404.

Modood, T. (1994) 'Political blackness and Asian identity', *Sociology* 28 (4): 859–76.

Mudimbe, V. (1988) *The Invention of Africa: Gnosis, Philosophy, and the Order of Knowledge.* Bloomington: Indiana University Press.

Murdoch, J. and Marsden, T. (1994) *Reconstituting Rurality.* London: UCL Press.

Nast, H. (2000) 'Mapping the "unconcious": racism and the oedipal family', *Annals of the Association of American Geographers* 90 (2): 215–55.

Nayak, A. (2002) 'Last of the 'Real Geordies'?: white masculinities and the subcultural response to deindustrialisation', *Environment and Planning D: Society and Space* 6.

Peach, C. (1975) 'Introduction: the spatial analysis of ethnicity and class', in C. Peach (ed.) *Urban Social Social Segregation*. London: Longman.

Peach, C. (2000) 'Discovering white ethnicity and parachuted plurality', *Progress in Human Geography* 24 (4): 620–26.

Peach, C., Robinson, V. and Smith, S. (eds) (1981) *Ethnic Segregation in Cities*. London: Croom Helm.

Pratt, M.L. (1992) *Imperial Eyes: Travel Writing and Transculturation*. London: Routledge.

Pred, A. (2000) *Even in Sweden*. Berkeley: University of California Press.

Robins, K. (1991) 'Tradition and translation: national culture in the global context', in J. Corner and S. Harvey (eds) *Enterprise and Heritage: Crosscurrents of National Culture*. London: Routledge.

Robinson, J. (1996) *The Power of Apartheid: State, Power and Space in South African Cities*. Oxford: Butterworth-Heinemann.

Robinson, J. (1998) 'Spaces of democracy: remapping the apartheid city', *Environment and Planning D: Society and Space* 16: 533–48.

Roediger, D. (1992) *The Wages of Whiteness*. London: Verso.

Roediger, D. (1994) *Towards the Abolition of Whiteness*. London: Verso.

Rose, G. (1993) *Feminism and Geography*. Oxford: Polity.

Rubinowitz, L. and Rosenbaum, J. (2000) *Crossing the Class and Color Lines: From Public Housing to White Suburbia*. Chicago: University of Chicago Press.

Said, E. (1979) *Orientalism*. New York: Vintage.

Schech, S. and Haggis, J. (1998) 'Postcolonialism, identity and location: being white Australian in Asia', *Environment and Planning D: Society and Space* 16: 615–29.

Shaw, W. (2000) 'Ways of whiteness: Harlemising Sydney's Aboriginal Redfern', *Australian Geographical Studies* 38 (3): 291–305.

Shaw, W. (2001) 'Way of whiteness: negotiating settlement Agendas in (post)colonial inner Sydney'. Unpublished PhD thesis, University of Melbourne.

Sibley, D. (1988) 'Survey 13: purification of space', *Environment and Planning D: Society and Space* 6: 409–21.

Sibley, D. (1992) 'Outsiders in society and space', in K. Anderson and F. Gale (eds) *Inventing Places: Studies in Cultural Geography*. Melbourne: Longman Cheshire.

Sibley, D. (1995) *Geographies of Exclusion: Society and Difference in the West*. London: Routledge.

Smith, S. (1989a) *The Politics of 'Race' and Residence*. Cambridge: Polity.

Smith, S. (1989b) 'Race and racism', *Urban Geography* (10): 593–606.

Spengler, O. (1980) *The Decline of the West* (1918). New York: Knopf.

Stanton, M. (2000) 'The rack and the web: the other city', in L. Lokko (ed.) *White Paper, Black Marks: Architecture, Race, Culture*. London: Athlone.

Thomas, N. (1991) *Entangled Objects: Exchange, Material Culture, and Colonialism in the Pacific*. Cambridge, MA: Harvard University Press.

Tonnesson, S. (1994) 'Orientalism, occidentalism and knowing about others', *Nordic Newsletter of Asian Studies* 2: 1–8.

Trewartha, G. (1926) 'Recent thoughts on the problem of White acclimatisation in the wet tropics', *Geographical Review* 16: 467–78.

Twine, F.W. (1996) 'Brown skinned white girls: class, culture and the construction of white identity in suburban communities', *Gender, Place and Culture* 3 (2): 205–24.

Urry, J. (1995) *Consuming Places*. London: Routledge.

Venn, C. (2000) *Occidentalism: Modernity and Subjectivity*. London: Sage.

Watt, P. (1998) 'Going out of town: youth, 'race', and place in the South East of England', *Environment and Planning D: Society and Space*, 16: 687–703.

White, H. (1972) 'The forms of wilderness: archaeology of an idea', in E. Dudley and M. Novak, (eds) *The Wild Man Within: An Image in Western Thought from the Renaissance to Romanticism*. London: Macmillan.

Wilson, K. and Dussen, J. (eds) (1995) *The History of the Idea of Europe*. London: Routledge.

Wolff, L. (1994) *Inventing Eastern Europe: The Map of Civilization on the Mind of the Enlightenment*. Stanford: Stanford University Press.

Woodruff, C. (1905) *The Effects of Tropical Light on White Men*. New York: Rebman.

16

Queer Cultural Geographies – We're Here! We're Queer! We're Over There, Too!

Michael Brown and Larry Knopp

During the last quarter-century, disciplinary orthodoxies in geography have been subject to devastating criticisms, leaving geography with less of a center than ever. Sexuality studies and queer theory have been an especially potent force amidst projects that challenge the exclusions of geographers' views and explanations of the world and its places. Because sexuality is an enormously diverse and elusive set of lived experiences, as well as a social construct that defies, in immediately apparent ways, efforts to impose order on it, its consideration has led to new and specifically 'queer' ways of thinking about difference (and related concepts, such as identity, space and power). Neither 'objective' behavioral nor 'subjective' psychological approaches, nor structuralist theories, can do sexuality justice; nor can either strictly cultural or biological perspectives. Furthermore sexuality is intimately linked both to profoundly social exercises of power and to highly individuated experiences of desire, which themselves are interlinked and variable across time and space (Foucault, 1980).

Serious engagements with sexuality, then, necessitate a careful reconsideration of some fundamental ontological, epistemological and methodological issues. These include the relationship between nature, society and human agency;[1] the nature of identity;[2] problems of naming and counting;[3] of drawing inferences and conclusions;[4] of the roles of qualitative and quantitative methods in social science (how can we understand the social consequences of sexualities without understanding them as lived experiences?); objectivity and subjectivity (can sexualities ever be understood as strictly objective or subjective phenomena?); and more.

On at least some of these issues consensus emerged. Sexual identities are now most fruitfully seen as culturally and ideologically constructed subjectivities and significations that serve and resist dominant forms of power. Power, meanwhile, is seen as working through discourses and representations as much as through more conventional material practices (such as coercion backed by violence), such that even academic work itself becomes highly (and self-consciously) politicized. And space has been discovered by academics in a wide variety of disciplines as a concept (if not always a very well-theorized one) that helps them to understand and communicate the processes whereby various forms of difference *and* power are ontologically constructed, reproduced and resisted.

Yet queer theory, the strand of theorizing that has emerged from sexuality studies, ironically tends to question and problematize notions of consensus, stability or privileged argument. In this way, its aim to rethink social life from the standpoint of sexual dissidents is intertwined with a postmodernism that is deeply suspicious of metanarratives or Archimedean perspectives. This built-in contradiction means queer geography is often difficult to characterize and subject to internal debate. Nevertheless, as contributors to this volume our charge is to attempt just that. Accordingly we showcase here some of the tensions, contradictions and milestones of this emerging field within geography and signal their promise and pitfalls. Our chapter proceeds in three steps. We first offer a very brief temporal overview of work in the area. We then shift to a more in-depth consideration of the particular spaces and subjectivities that geographers have considered in this work. These range from the closet to cyberspace and from the individual to the globe. Finally, we imagine several new queer geographies that have yet to be written. We want to stress at the outset that what follows is by no means an exhaustive or definitive survey of the field. Our hope is that this chapter helps others to at least begin to find their bearings and consider ways their own interests might jive with queer geographies – because we are absolutely certain that they can and must!

QUEER GEOGRAPHY AND ITS PRECURSORS

While the publication of *Mapping Desire* (Bell and Valentine, 1995) is often heralded as the beginning of sexuality and space studies, even its own editors remind us that there is a longer (and courageous) legacy to appreciate. For example, a number of disparate activities began to take shape in the 1970s that put sexual minority identities and communities on the discipline's map (at least for those who were willing to see). The simple act of arranging meetings of gay and lesbian geographers at Association of American Geographers' meetings precipitated extraordinarily nasty public (and published) denouncements from established and secure figures in the discipline (Carter, 1977). Meanwhile, a small number of researchers, mostly in urban, cultural and economic geography and often linked to the nascent gay and lesbian rights movement, began drawing attention to gays and lesbians in other ways. Barbara Weightman (1980), for example, sought to bring into the open the significance of gay bars as social spaces. Christopher Winters (1979) and Bill Ketteringham (1979; 1983) noted the important role played by gays and lesbians in inner-city commercial and residential 'revitalization'. And Bob McNee (1984; 1985) examined the role of *oppression* in creating distinct but marginalized gay and lesbian spaces and networks in cities, at the same time as he valiantly challenged the discipline's own homophobia and sexism both in and out of print. In this respect McNee was far ahead of his time. His contributions represented an *embodied* insistence that oppression of gays and lesbians was real, systematic and fully present in the discipline of geography – this at a time when critical reflection on academic practice (except perhaps in the realm of research and teaching 'ethics') was virtually unheard of! In his written work on the subject McNee focused on where, how and why gays and lesbians are and are not able to express our *embodied difference* from heterosexuals,

particularly our same-sex desire and practice but also other forms of gender non-conformity (for example, drag) and affiliations (for example, with prostitution) that tend to make middle-class professionals (like geographers) 'squeamish'. McNee backed these assertions up with what were at the time very controversial efforts to make space at professional conferences for gays and lesbians to *be* homosexual (not just to network). He attended at least one session in drag himself, led an informal field trip to the gay/lesbian and red-light entertainment district of Denver during an Association of American Geographers' conference there, and helped to organize the first informal gay and lesbian caucus of the Association of American Geographers by posting invitations for gay and lesbian geographers to meet.

But with the possible exception of McNee, none of this 'first wave' of geographers dealing with sexuality consciously challenged the positivist epistemology that underlay most human geography at the time. Still, it quickly became clear that this was inevitable. Similar (and certainly not coincidental) developments were taking place in feminist geographic quarters where issues of gender-based power relations were being pushed, as well as in more traditional radical circles where political economists (mostly, but not exclusively, Marxists) were pressing issues of class. More recently, anti-racists have experienced essentially the same challenge in the context of 'race'. It seems dominant paradigms were ill-suited to all of these issues.[5]

For sexuality studies, the result has been a series of efforts to deploy the theoretical tools of feminism and political economy (and anti-racism) as well as engagements with postmodern, poststructuralist and queer theories. In an early effort involving one of us (Lauria and Knopp, 1985), Larry used the combination of Marxian-inspired theories of organizations and urban land use, feminist approaches to gender and sexuality, and some early lesbian/gay social theory to understand the role of gay communities in urban redevelopment. Tim Davis (1991; 1995) similarly employed Marxian-inspired theories of social movements, augmented by feminist theory, to understand gay and lesbian political activism. British geographers Peter Jackson (1989), Gill Valentine (1993a), Gillian Rose (1993), David Bell (1995) and Jon Binnie (1993), meanwhile, brought questions of representation, desire and performance to the fore in otherwise similar discussions. Eventually the challenge represented by these latter contributions (which itself was simultaneously a major contributor to, and product of, the cultural turn in human geography as a whole) led to a proliferation of geographical sexuality studies that were theoretically anti-structuralist, anti-modernist and very self-consciously 'queer' (for example, Brown, 1997; 2000; Callard, 1996; Elder, 1999; Knopp, 1999; 2000; Nast, 1998). Now questions of representation (and the politics of representation), performativity (à la Butler, 1990; 1993), citizenship/belonging, culture generally, and cultural politics in sexuality and space studies have all but supplanted the more traditional subject matter of social area analysis, social movements and urban development. However, this diverse array of topics and approaches has not evolved nearly as smoothly as the foregoing suggests, and it is in this context that the contradictions we mentioned earlier have begun to manifest themselves.

SPACES, SCALES AND THEIR DISCONTENTS

The languages (and paradoxes) of poststructuralism and queer have emerged as particularly robust in the contexts of the so-called 'new' cultural geography. Perhaps the archetypal construct emerging from this field is the closet, and it is a concept with which we and several others have worked extensively. As a spatial metaphor the closet conveys the sense of denial, erasure and concealment that is at the heart of sexual oppression. The humanities-based Diana Fuss (1991) and Eve Sedgwick (1990) have stressed the need to deconstruct the inside/out dualism implicit in the metaphor. What they are referencing is the liminality and non-Euclidean nature of the lived closet experience. They demonstrate that one *can* in fact be simultaneously inside and outside the closet (for example, in different contexts or even ontologically, in the sense that the closet itself entails an epistemology of 'knowing by not knowing': Sedgwick, 1990). It is also evidenced by the more material reality that in escaping one kind of oppression (that which ensues from having no words or language with which to name one's desire) one must engage with new forms of oppression associated with a naming of desire that is always partial and to some degree marginalizing in its capacity to represent actual lived experience. This point resonates with the broader theme of this section of this book that subjectivities are never static but rather always in the process of *becoming*.

So it is surprising that the closet has received so little explicit (and critical) attention from geographers (but see Davis, 1991; Knopp, 1994; and most significantly Brown, 2000). Recently

Michael conceptualizes the gay man's closet as a spatial practice of power–knowledge, and not just a metaphor (Brown, 2000). He examines how the closet is constructed both materially and discursively at four spatial scales: the body, the city, the nation and the globe. At the level of the body, he looks at how the closet is often a space for the performativity of sexuality. At the level of the city, he explores the ways in which the closet allows the commodification of sexual desire through the construction of commercial spaces of sexual consumption. National-scale closeting, meanwhile, is examined through a study of the effects of the categorizations and quests for 'validity' and 'reliability' that necessarily, as features of positivist epistemology, undergird national censuses. And at the global scale, he shows that the closet is 'not so much a lack, but a productive if occluded space' (2000: 22). That is, it is not always a disempowered, abject artifact, but can also be the setting for creative, ingenious and transformative sexual, cultural and political resistances to heteronormativity. In another piece of our work (Knopp, 1994), Larry similarly argues that closeting is a patently contradictory collective process of privatization and alienation as well as of resistance and empowerment. It involves the socialization of certain kinds of experience as 'private' and their marking as 'alien'. It both protects queer people and makes us vulnerable, through the development of subtle codes and cues that can be deployed strategically by queers as well, potentially, as homophobes (including queer homophobes). And because it relies on queers ourselves to police our own desires, closets are often constructed as threatening to a heterosexualized dominant culture, which then justifies heterosexist campaigns of violence, harassment and intimidation as a defensive measure.

Queer subjectivities, however, do not exist in disembodied forms or only in closets. Performance artist Kate Bornstein (1998) emphasizes, by example as well as in writing, the importance of location to successful gender-bending bodily performances. Among geographers, Bell et al. (1994) very provocatively make this point by considering the capacity of bodies that defy visual and behavioral expectations to disrupt the shared meanings of public space. In particular, they demonstrate how hypermasculine gay men and hyperfeminine lesbians can ironically subvert the hegemony of the heterosexual presumption in everyday environments. Certain other geographers, however, caution that the appropriation and parodying of masculinity and femininity as constructed in/by/for heterosexuality – at least in the ways described by Bell et al. – do little or nothing, in a practical political sense, to undermine either heterosexual hegemony or the 'tyranny' of heterosexually–constructed gender (Kirby, 1995). Larry has further argued that the deconstruction of sexualities alone is inadequate (and indeed dangerous) as a political strategy, since it leaves unanswered crucial questions of value and ethics, such as the implications for race- and gender-based power relations when certain highly sexualized (and potentially sexist) and racialized (and potentially racist) practices are engaged in by sexual minorities (Knopp, 1995).

Lynda Johnston (1996), meanwhile, queers the spaces of the gym and the female body by demonstrating how disruptive to heterosexual norms female body-builders' physical presence can be. The importance of physical interventions in bodily processes such as menstruation and birth control have also been analyzed and interpreted in terms of their significance to the construction of genders and sexualities (Cream, 1995). And in the context of the central London banking industry workplace, Linda McDowell (1994; 1995) explores the role that gendered and sexualized spaces can play in disciplining bodily comportment (for example, styles of dress, physical build, hairstyle, makeup, stance, manner of self-presentation). These works are synecdochical of a spate of recent literature that focuses on bodies as a frame for examining the relationship between subjectivity and space (Duncan, 1996; Nast and Pile, 1998; Pile, 1996).

Given the coding of sex and sexuality as 'private' matters, it is not surprising that some geographers have begun also to queer the space of the 'home'. Whether it is seen as a space of capital accumulation, social reproduction, caregiving, double oppression or just a haven in a heartless world, the home is increasingly viewed also as a site of heteronormative structure by these scholars. Several examples come to mind. Valentine (1998) describes vividly how the 'private' space of her home was usurped and disciplined by the homophobic predations of an anonymous harasser. In the process she demonstrates convincingly how the disciplining power of heteronormativity works through home space, and in collaboration with a wide range of cultural assumptions. Valentine (1993a) and Mackenzie and Rose (1983) make the brilliant but simple point that residential architecture and design in the late twentieth century have presumed the norm of a heterosexual 'nuclear' family. In terms of resistance to this, the capital gains on domestic property through gentrification have been shown by a number of geographers (including Larry: see Knopp, 1990b; Peake, 1993) to be due

in part to a complex interplay between market forces and resistance to oppression on the part of gay and lesbian owners. This includes an acknowledgment of the very problematic role of some gay men and lesbians in what can at times be quite predatory forms of gentrification. Michael, meanwhile, explored how incredibly complicated the home geographies associated with the lives of people living with HIV/AIDS can be (Brown, 1997), while Jay (1997) stresses that the politics of domesticity among lesbian and gay parents are not as similar to those of heterosexual parents as it might at first appear.

Spaces more traditionally constructed as 'public' have similarly been subject to the critical eye of queer theory and queer studies, including the sometimes highly charged spaces of teaching and research. In 1997 and again in 1999, for example, the *Journal of Geography in Higher Education* featured symposia dealing with the practice and politics of teaching and researching sexualities. Both of these were strongly influenced by queer theory. Special sessions at conferences have also focused on the politics of teaching and researching sexualities, again from primarily queer perspectives. Topics covered have ranged from how queer theory and experiences can be used to shed light on subjects such as borders, boundaries, nation-states and capitalism, through the ways in which racism, sexism and heterosexism can be deconstructed in the classroom, to thorny issues of curriculum, institutional and pedagogical priorities, and the evaluation of educational 'output' (including faculty performance).

Ironically, consensuses (of sorts) have emerged around at least three points. First, all human relations – including those in the academy – are sexualized and, in most contemporary cultural contexts, characterized by the processes associated with homophobia and heterosexism (most significantly, closeting). This means that even the most 'private' and mundane of spaces are sites in which dominant relations of power are reproduced and (potentially) resisted. Second, the embodied experiences of real human beings are almost always queer in at least one dimension or another. They often involve, for example, iterative performances that reveal the constructedness of taken-for-granted everyday experiences and artifacts (for example, borders, nation-states, economies). This being the case, a queer methodology that reveals the power relations and mechanisms behind these constructions is potentially very effective. Third, sexual politics (and the politics of sexuality) are particularly apposite topics in colleges and universities. Attention to these can be touchy, but is important both politically and for its ability (potentially) to provoke learning (or 'unlearning',[6] as the case may be).

Perhaps the oldest and most developed body of literature dealing with sexuality and space (and one in which both of us have been deeply involved) is that which addresses sexuality in the context of 'the urban'. Both within and beyond the discipline of geography, scholars and activists have written quite widely about processes of gay community development, territoriality, neighborhood change, gentrification, social movements, urban politics and the cultural politics of urban space (as these pertain to sexuality). Most but not all of this work focuses on gay, lesbian and other sexual minority identities and communities. Because it has a nearly quarter-century history, this literature, as a whole, is less dominated by queer and other forms of post-structuralist theory than some of the more contemporary literatures discussed above. Some of it is in fact quite descriptive and empiricist: for example, Ketteringham's (1979; 1983) work on the role of gay business enterprises in revitalizing the Broadway corridor of Long Beach, CA, and Winters' (1979) consideration of the role of gay people in the social identities of evolving neighborhoods. Other efforts address gay and lesbian territoriality from the perspectives of symbolic interaction, oppression/resistance and anarchism (Levine, 1979; McNee, 1984; 1985; Murray, 1979; Weightman, 1980). And much of the work produced in the 1980s and early 1990s proceeded from a Marxian or neo-Marxian urban political economy perspective, with focuses on gay gentrification, the roles of gay and lesbian interest groups and social movements in urban politics, and the emergence of gay and lesbian residential and commercial spaces in cities (Castells, 1983; Castells and Murphy, 1982; Davis, 1995; Knopp, 1990a; 1990b). But beginning in the early 1990s, poststructuralist and queer theories – responding in part to the rise of queer politics and AIDS activism, and in part to the demonstrated inadequacies of more structuralist paradigms – began to inform work on sexuality and space even while the empirical focus remained heavily on urban areas and experiences (for example, Bech, 1993; Bell, 1994; Bell and Valentine, 1995; Binnie, 1993; Brown, 1994; 1999; Ingram et al., 1997; Knopp, 1998; Seebohm, 1994; Valentine, 1993b; but see Kramer, 1995). Attention began to be paid to the erotic significance of urban public space, conflicts over how space is constructed, coded and used through sexualized performances (for example, in parades), and spatial strategies for both enforcing and resisting heterosexism and other forms of

sexual control. 'Sexuality', meanwhile, finally began to be construed much more broadly, to include various constructions of heterosexuality as well as minority sexualities (Nast, 1998). Non-geographers have also been insightful in exploring the intersection of sexuality and the city, as evinced by the work of Bill Leap (forthcoming). Indeed, his innovative work on how sexualities intersect with race, class and gender through fragments of metropolitan Washington, DC has the potential to queer classic regional geography!

Even more recently a body of work has begun to appear that focuses on non-metropolitan and rural queer sexualities (Phillips et al., 2000). Because it is so new, this work is, not surprisingly, heavily influenced by queer theory. Interestingly, it seems focused primarily on getting non-metropolitan and rural queer issues 'on the map' in the discipline of geography, in much the same way that the urban work that preceded it seemed interested in getting sexual minority issues addressed. What awaits future researchers is the very exciting project of exploring the links between sexuality issues in different places and at different scales (see below).

At a somewhat broader scale, a small but growing body of explicitly geographical work now addresses the mutual constitutions of sexualities and discourses of nationhood, citizenship and the state.[7] Bell (1995) and Binnie (1997), for example, consider the significance of laws regarding minority sexual practices and their enforcement through borders and immigration policies for notions of European citizenship and individual European national identities (for example, British, Dutch). Michael, meanwhile (Brown, 1997; 2000), has yoked modes of citizenship and Foucauldian notions of governmentality by exploring challenges to state-centered discourses of citizenship implicit in much AIDS activism and the closeting effect of the positivist epistemology underlying the British and US national censuses. And Nast (1998; 1999) has explored the ways in which discourses of nationhood are not only always sexualized, but classed, gendered and racialized as well (again, in a mutually constitutive way).

One area that is as yet still underdeveloped within geography is global-scale studies of sexuality and space (but see Kidron and Segal's 1984 *The New State of the World Atlas* and Seager's 1997 *The State of Women in the World Atlas*). Non-geographers, such as Robert Aldrich (1993), Neil Miller (1992), Lars Ebenstein (1993) and Dennis Altman (1997) have looked at issues such as the globalization of western 'gay' culture, the allure of certain world regions to generations of gay and lesbian travelers, and experiencing a queer unity in diversity through traversing the globe as 'gay'. But clearly there is much more that could be done from a queer perspective at this scale (we offer a few modest suggestions below).

Another space, whose scale (and, perhaps, subjectivity) is arguably high and low, all and none at the same time is that of cyberspace. Winckapaw (1999) has considered the important role of the internet in shaping and empowering new lesbian subjectivities and resistances, as has Bornstein (1998). This area could be fleshed out much more by considerations of such phenomena as sex cruising in chat rooms, the forging of networks and communities in and between rural areas by online queers, transnational human rights and other activism, and the role of cyberspace in reshaping gender, sexuality and sex play.

QUEER GEOGRAPHIES AND BEYOND

Despite the promise evidenced by all of this engagement with queer theory and a wide variety of spaces and places, there are limits and dangers to note. These include the political and personal ennui of constantly queering each other's work, the construction of what amount to new orthodoxies despite a self-conscious political opposition to orthodoxy, or, alternatively, an unwillingness to acknowledge *any* political commitments at all, which can lead to power vacuums.[8]

In addition, work on sexuality and space in geography remains largely peripheral even within cultural geography. This is because the discipline's traditional corpus has been largely untouched by a queer sensibility. Such subfields as cultural ecology, cultural diffusion, geographies of language, religion, travel and tourism, recreation, and even landscape interpretation (among others) still largely ignore the issue of sexuality (much less incorporate elements of a queer epistemological critique) in their approaches (but see Jackson, 1989; Mitchell, 2000).

In working to address these twin (and somewhat contradictory) problems, we would suggest beginning modestly, by opening geography's own closet door. There is a sociology and a politics to the production of geographic knowledge that involves personal relationships, desire and power. Anyone who has participated in academic culture for any period of time knows this. As in all endeavors, who desires whom (or what), who acts on this, who doesn't, and how these relationships and desires are negotiated make a huge difference, not just to the way in

which academic work proceeds but to what gets produced and legitimized (or delegitimized) as well. For example, Smith (1987) brings to light very sensitively the link between homophobia and Cold War geopolitics in the power struggle between Isaiah Bowman and Derwent Whittlesey at Harvard in the 1940s that ultimately led to the closing of that and most other Ivy League departments of geography. Still, this piece of the puzzle clearly needs to be explored more deeply, so as to more clearly specify the ways in which sexualized threats to nationhood and 'national security' are constructed discursively *and* materially. Similarly, while Elder et al. (forthcoming) discuss in somewhat more detail the heterosexed history of Geography as a discipline, their interpretation tends towards the psychoanalytical. A more materialist analysis would complement and strengthen their argument immensely. And Gillian Rose (1993), in her deep critique of the discipline's masculinism, also links this to its heterosexism – though it has been argued that her own perspective could also benefit from a more careful queering (Binnie, 1997).

Of course, engaging in this kind of disciplinary soul-searching and dirty laundry airing is more easily said than done, and as we have said can lead to a kind of ennui and dispiritedness. A host of ethical and political dilemmas immediately present themselves once this history starts being researched, which may then be applicable to understanding both the benefits and the limits of queering geography's content as well as its history as a discipline. The most obvious is the issue of 'outing' – when, to whom and how to reveal matters that may be at once profoundly personal (and potentially hurtful) *and* crucial to understanding the history of the discipline. Related to this is the issue of separating salacious gossip from legitimate (and important) biographical and historical detail. And of course even the asking of certain questions can be harmful to any community – including a community of scholars – whose dynamics might be done more harm than good. The point here is that queering, as an activist scholarly enterprise, is by no means unproblematic. Of course, similar ethical and political dilemmas arise whenever a project of critically examining power relations unfolds, but in the context of western cultures, where so much social power is exercised through sexually charged anxieties and hostilities, we would suggest that particular care must be taken.

In terms of more topical agendas, we would ask all cultural geographers to recognize the centrality of sexuality to all aspects of culture. Sexuality is an always present aspect of the human experience. As such it is always implicit, if not explicit, in cultural constructs. We think it crucial, therefore, that cultural geographers be open to exploring this dimension of whatever their topical concerns may be. In the context of spatial interaction, for example, very little if any attention has been paid to the sexualized aspects of communication, transportation, trade and colonial/postcolonial (and other) relations. Yet it does not require too much 'thinking beyond the boxes' to conclude that sexuality may be very important indeed in understanding these phenomena. For example, at the nexus of cultural and economic geography, interest in spaces of culture industries and the gendered and racialized dimensions of labor markets are a vibrant zone of research. Yet sexuality remains largely ignored here. The idea that there may be occupational segregation by sexuality in industries such as travel and tourism is almost taken for granted in both the contemporary travel industry and many gay/lesbian subcultures. Not only is there strong anecdotal evidence suggesting that customer services in many industries (not just travel and tourism) are often staffed by women and gay men,[9] it is clear from the rise of sex tourism in places such as Thailand that the segmenting of markets by sexuality is also growing in importance. What does this rather explicit sexualizing (and, potentially, queering) of the travel and tourism industries mean for our understandings of what travel and tourism are all about? While travel and tourism have long been recognized as vehicles for acting on desire for 'the other', this has rarely been explored in any kind of detail (though see del Casino and Hanna, 2000). A queer perspective in particular, with its refusal to 'fix' sexualities or sexual identities, would be extraordinarily useful in tackling this project. At the same time, researchers in this or similar areas should be mindful that while a relentless queering and deconstruction of identities, places, etc. may be insightful, it can also be very occluding and dangerous, particularly in the context of cross-cultural research.

Cultural turns are also bending the often-quantitative work done in population geography, especially around issues of migration. Here again, sexuality has the potential to invigorate scholarship. Mark Ellis (1996), for instance, looked at the migration patterns of HIV-positive people around access to healthcare services. Michael noted how coming out is often spatialized as migration (Brown, 2000). The potential for exploring queer migrations and perhaps even diasporas, is fascinating.

Similarly, despite the increasingly important roles played by telecommunications media in producing and feeding sexual desire[10] there is precious little research by geographers on this issue (Winckapaw, 1999, notwithstanding). Such an endeavor would likely reveal much in the way of sexual conflicts and contradictions (witness struggles over control of sexual content on the internet). And given recent events in Europe regarding the EU's and European Parliament's controversial (and contradictory) attempts to influence member states' laws regarding sexual minorities, and the broader contexts of economic unions, 'free trade' and globalization within which these have taken place, it is equally surprising that the whole issue of trade has not been queered. In addition to the impact of internal state policies regarding sexuality on trade, geographers could look at the ways in which divisions of labor in trade are sexualized (for example, in ports, on docks, in commodity markets and exchanges, etc.), discourses and practices of sexuality in trading institutions (for example, GATT, OPEC), and sex itself as a commodity that is traded. Once the sexual logics and practices underlying these phenomena are made clearer, the project of understanding these logics will be that much easier.

Sexualized power relations are also clearly important in both the history of colonialism and contemporary post colonial economic and political relations (witness the global traffic in women as sex workers and the rise, again, of sex tourism in places like Thailand). A few geographers (for example, Domosh, 1991; Gregory, 1995; Nast, 1998; Phillips, 1999; Rothenberg, 1994) have recognized this and begun to explore it. But by and large this project has been the domain of postcolonial writers and theorists in the humanities (especially feminist postcolonialists). They tend to explore it in the context of either direct colonial practices of domination or diasporic subjectivities and other postcolonial cultural products. A queer geographical focus specifically on the interplay between sexualities and postcoloniality could provide some badly needed foundational knowledge for broader understandings of both colonialism and postcolonialism (so long as it does not become an end in itself). Glen Elder's (1998; forthcoming) work on the complicated intersections of race, class and sexuality in post-apartheid South Africa provides an important path in this direction.

Despite the relative scarcity of such work, studies of diasporic postcolonial subjectivities constitute one of the very few areas of study that brings *any* kind of contemporary critical sensibility to the issue of spatial interaction. Most approaches to spatial interaction still employ largely positivist epistemologies and fairly narrow and mechanistic notions of culture (as, for example, a *thing* rather than a *process*). Not surprisingly, then, the way in which 'culture' is seen as evolving spatially in this kind of work is via distinct 'carriers' and along (usually hierarchical) networks. And while virtually no work explicitly addresses the diffusion of, or encounters between, sexual cultures in space, this theme is implicit (as we have argued above) in a great deal of the critical sexuality and space literature. We would suggest, then, that in addition to new focuses on communication, transportation, trade and colonialism/postcolonialism, a queered approach to the issue of spatial interaction might focus as well on the connections between queer cultures and subjectivities in different places, on the interactions between and among dominant and subordinate sexualities at different scales, and on ways in which processes of sexualization are transmuted as they evolve spatially. For example, one might look at the roles of migration and communication in shaping queer cultures and forms of resistance in different places; at how the policing of sexualities at supranational, national and subnational scales affect each other; and at how the meanings of 'straight', 'gay', etc. have changed as they have crossed space and encountered other ways of conceptualizing and accounting for human (sexual) experience.

Having said all this, it is important that we interject at this point some additional cautionary comments regarding the epistemology of desire and sexuality with which we are working. Clearly, our notions of these human experiences are mediated by our own social locations as middle-class white academics in the 'west'. While we obviously would not argue that this disqualifies us (or others like us) from looking at desire and sexuality beyond our own cultural frames of reference, we do think it important that people like us engage in careful self-criticism in the process, so as to avoid, to the extent possible, participating in the hegemonic globalization of our own notions. Even more importantly, we think that a corrective will be needed in the form of non-western geographers producing their own queer geographies (while we in the west listen).

In parallel fashion, it is important to note that our own notions of the closet (Brown, 2000; Knopp, 1994) are not the only ones worth examining from a queer perspective. Valentine's (2000) new edited collection on lesbian geographies provides the beginnings of a badly needed corrective to the bias towards studies of gay male issues in queer geography. Indeed, Nast (forthcoming) challenges us to consider the ways in which patriarchies have historically worked in and through gay male culture. But other sexual and sexualized minorities (and, indeed, others)

also have closets that urgently require unpacking. We are thinking here not only of bisexuals, sex workers, fetishists of various kinds, and others who might identify (or be identified, culturally) by their sexual 'difference', but individuals and groups for whom 'non-sexual' pleasures or practices (for example, food, music, dance, ethnicity, language, religion) or even more culturally ascribed characteristics (for example, race, nationality) might produce unique but culturally powerful ways of knowing by not knowing (for example, closets). What of the 'quadroons' and other mixed-race women of antebellum Louisiana who at times 'passed' as white? The Jews and half-Jews who survived the holocaust by posing as non-Jews? The political dissidents who survive purges by concealing their histories of activism? Surely queer insights into the workings of closets generally will have some purchase in these contexts as well.

As we have already argued (and contrary to the assertions of some gay cultural critics – for example, Savage, 1999; 2000), once closet doors are opened, the other side can still be a very dangerous place. In some parts of the world, being a sexual dissident can still be a capital offense. Valentine's (1998) personal exposé illustrates what the fear of the same kind of violence can do to people even in allegedly safer societies. Related to this is the growing iconographic significance of sites of violence (tragic and triumphant) to queer people. The Stonewall bar in New York City was recently designated as a National Historic Landmark, and a national gay and lesbian magazine noted that people from all over are now making pilgrimages to the remote fence in Wyoming where gay college student Matthew Shepard was brutally beaten and murdered (*The Advocate*, 2000). These examples highlight the need for geographers to investigate in further detail the dangerous spaces of abuse, harm and bashing that threaten the queer body quite materially.

At the same time, we think it very important that queer geographies not be equated solely with geographies of queerness (or non-heterosexuality). Queerness is as much an intellectual and epistemological perspective as a set of subjectivities organized around sexuality. Accordingly, we wholeheartedly endorse calls instigated by Nast (1998) and others for queer geographies to begin taking *heterosexuality* and its myriad expressions (and contradictions) seriously. As a privileged set of identities and practices, heterosexuality has been taken for granted even while being constructed monolithically as nothing less than the basic value underlying 'the family' and, by extension, 'civilization'. But of course heterosexuality is in fact an enormously diverse set of practices and relations that can even include (seemingly contradictorily) same-sex activity (for example, 'male rape' and other forms of sexual engagement by heterosexually identified men with other men), not to mention eroticized interactions of dizzying varieties. While it would be impossible to provide an exhaustive agenda for geographers interested in heterosexualities here, a starting point might be consideration of the processes producing spaces, places and environments that are strongly coded as heterosexual. For example, while suburbanization has been analyzed in terms of capitalism, racism and patriarchy, it is clearly also a product of heterosexism (see Knopp, 1990a, for a brief acknowledgement of this). Those interested in capitalist land markets, housing policies and racist/patriarchal social relations cannot ignore the heterosexual dimensions to these phenomena. Spaces and places designed specifically to facilitate heterosexual sex, courtship and marriage might also be examined from a queer perspective. What of 'straight' porn shops and cinemas where heterosexually identified men, consuming heterosexually oriented pornography, have sex (ironically, at times, with each other)? Or debutante balls, prom dances, 'lovers' lanes', 'singles bars' and weddings where men and women learn the rituals of dating, courting and having sex? Alternatively, one might consider many of the issues we raised above (communication, colonialism/postcolonialism, trade, transportation closets) from a distinctly heterosexual (but still queer) perspective.

Given the rise in geography of perspectives like political ecology and nature–society studies, we think it important that queer perspectives on nature and environment be on cultural geographers' agendas as well. Queer epistemological critiques generally, as well as the consideration of queer sexualities, could surely inform approaches to nature–society debates, just as feminist critiques and perspectives on gender have done in defining fields like *feminist political ecology* (Rocheleau et al., 1996). The key would be to look at the links between cultural/political systems as a whole (but from a perspective that foregrounds especially issues of sexuality and associated ways of knowing), social constructions of 'nature', and biological systems. One might ask, for example, how heterosexism is related to masculinist and patriarchal approaches to environmental management, and how it is inscribed in landscapes and physical systems that have been transformed by 'man'. More generally, how might the closeting of queer desire in western societies discipline men's *and* women's

constructions and uses of 'nature'? In the context of 'non-western' cultures, we might ask why it is that so many people who in the west might be considered 'queer' are ascribed special status as spiritual naturists (for example, the *berdache* of some native North American cultures). These questions (and many others like them) would seem to be particularly rich avenues for inquiry by new generations of broadly trained cultural geographers interested in sexuality and queer theory.

Finally, we would call for a continuation and deepening of self-reflexivity in queer geographies of the future. Incumbent in any postmodern scholarship, queer perspectives cannot just challenge and critique others; they must also self-critique as well. After all, queer theory insists that no position is completely innocent or unproblematic. Kim England's (1994) now-classic essay on her 'failed' research attempt as a straight woman investigating lesbian networks in Toronto shows the power and truths such auto-critiques can generate. Furthermore, they lead to a mutually supportive and constructive engagement with each other's work that draws on the affinities of queer theory and feminist politics by challenging the masculinist academic politics of trenchant critique and backbiting. Cultivating such an academic culture seems especially important in a small but geometrically growing field like queer geography.

CONCLUSION

As this volume's perspective is to see cultural geography as less a cogent scholarly object and more a critical way of approaching a wide variety of spaces and places, this chapter has focused on how topics deriving from the consideration of sexuality shape that view. We have offered a brief history of how issues of and debates surrounding sexuality have affected human geography generally, from simple mapping and decloseting linked to nascent gay-rights movements to a more critical and anti-positivist scholarship informed by Marxist and feminist perspectives. The postmodern turn in the discipline has augmented and broadened queer geography, linking scholarship on a wide variety of issues with more thoroughly critical and deconstructive questions of representation, epistemology and ontology. These developments have, admittedly, precipitated their own tensions and debates (for example, around the (im)possibility of a clear and committed politics within a thoroughly deconstructivist project of queering: see our discussion of Bell et al., 1994, above), but overall we see them as broadening and enriching activist scholarship.

We have also tried to showcase the ways in which, as a consequence of these latest developments, a variety of spaces have already begun to be queered, as well as the limits of this queering project. From the closet to the body, to the city, to the nation and to the globe, new queer cultural geographies show us that a variety of subjectivities are performed, resisted, disciplined and oppressed not simply in but *through* space. However awkward and choppy our scale-fixing may be (admittedly, queers inhabit bodies and the globe simultaneously!), our intention has been to demonstrate the truly impressive myriad locations on the globe, debates within the academy, and political issues in 'the real world' in which queer issues, experiences *and* geographies are implicated.

As a perpetual deconstructive process, the project of queering (cultural) geography will never be finished (nor should it be). In our view this openness makes it one of the most exciting and intellectually promising areas of enquiry in the entire discipline. Consequently, we closed the chapter with some of our own suggestions for queer cultural geographies we'd like to see in the future, along with some cautionary comments about potential dangers. The list is by no means complete or commanding, but rather is illustrative of where we have been, where we are and where we *might* go. No doubt even more fascinating geographies, whose 'queerness' exceeds far beyond our current imaginations, have yet to be written by current students of cultural geography!

NOTES

1 Questions like: are sexualities learned behaviors, expressions of desire, genetic predispositions, some combination of these, or something else?

2 Questions like: are sexual identities ascribed, claimed or socially constructed?

3 Questions like: whom do we count as 'gay', 'straight' or 'bi', anyway; how do we know these subjectivities; and how do we even know which categories to use?

4 Questions like: how do we sample; which case studies are apposite; when are hermeneutics more appropriate? See Brown (1995) and England (1994) for examples.

5 Especially in light of the perceived failures of 'scientific' liberalism – e.g. Keynesian economics – and the rise of neoconservative ideologies and social policies.

6 We refer here to calls for the 'unlearning' of oppressive behavior patterns and thought processes that are so ingrained as to be 'second nature' (as in John Singleton's 1995 film on race, gender and sexual politics on a university campus titled *Higher Learning*).

7 Geographers actually lag in this regard; George Mosse (1988), a historian, and Andrew Parker (1992), a literary critic, began exploring this issue much earlier.

8 These can be exploited by quite sinister forces with no regard whatsoever for ethics, values or social justice (Knopp, 1995).

9 Perhaps in part because of a long history in western societies of women and gay men being coded as bearers of culture and protectors of aesthetic values (Betsky, 1997; Ingram, 1993).

10 The range is impressive: from erotica, pornography and prostitution to online dating services, chat rooms, cell phones.

REFERENCES

Aldrich, R. (1993) *The Seduction of the Mediterranean*. London: Routledge.

Altman, D. (1997) 'Global gaze/global gays', *GLQ: A Journal of Lesbian and Gay Studies* 3: 417–36.

Bech, H. (1993) 'Citysex: representing lust in public'. Paper presented at Geographies of Desire Conference, Netherlands' Universities Institute for Coordination of Research in Social Sciences, Amsterdam.

Bell, D. (1994) 'Bisexuality: a place on the margins', in S. Whittle (ed.) *The Margins of the City*. Aldershot: Ashgate. pp. 129–42.

Bell, D. (1995) 'Pleasure and danger: the paradoxical spaces of sexual citizenship', *Political Geography* 14: 139–53.

Bell, D. and Valentine, G. (eds) (1995) *Mapping Desire*. London: Routledge.

Bell, D., Binnie, J., Cream, J. and Valentine, G. (1994) 'All hyped up and no place to go', *Gender, Place and Culture* 1: 31–47.

Betsky, A. (1997) *Queer Space: Architecture and Same-Sex Desire*. New York: Morrow.

Binnie, J. (1993) 'Invisible cities/hidden geographies: sexuality and the city'. Paper presented at Social Policy and the City Conference, Liverpool.

Binnie, J. (1997) 'Invisible Europeans: sexual citizenship in the new Europe', *Environment and Planning A* 29: 237–48.

Bornstein, K. (1998) *My Gender Workbook*. New York: Routledge.

Brown, M. (1994) 'The work of city politics: citizenship through employment in the local response to AIDS', *Environment and Planning A* 26: 873–94.

Brown, M. (1995) 'Ironies of distance: an ongoing critique of the geographies of AIDS', *Society and Space* 13: 159–83.

Brown, M. (1997) *RePlacing Citizenship: AIDS Activism and Radical Democracy*. New York: Guilford.

Brown, M. (1999) 'Reconceptualizing public and private in urban regime theory: governance in AIDS politics', *International Journal of Urban and Regional Research* 23: 70–87.

Brown, M. (2000) *Closet Space: Geographies of Metaphor from the Body to the Globe*. London: Routledge.

Butler, J. (1990) *Gender Trouble*. London: Routledge.

Butler, J. (1993) *Bodies that Matter*. London: Routledge.

Callard, F. (1996) 'The body in theory'. Paper presented at the Annual Meeting of the Association of American Geographers.

Carter, G. (1977) 'A geographical society should be a geographical society', *The Professional Geographer* 29: 101–2.

Castells, M. (1983) *The City and the Grassroots*. Berkeley: University of California Press.

Castells, M. and Murphy, K. (1982) 'Cultural identity and urban structure: the spatial organization of San Francisco's gay community', in N. Fainstein and S. Fainstein (eds) *Urban Policy under Capitalism*. Beverly Hills: Sage.

Cream, J. (1995) 'Women on trial: a private pillory', in S. Pile and N. Thrift (eds) *Mapping the Subject*. London: Routledge. pp. 158–69.

Davis, T. (1991) '"Success" and the gay community: reconceptualizations of space and urban social movements'. Paper presented at the First Annual National Graduate Student Conference on Lesbian and Gay Studies, Milwaukee.

Davis, T. (1995) 'The diversity of queer politics and the redefinition of sexual identity and community in urban spaces', in D. Bell and G. Valentine (eds) *Mapping Desire*. London: Routledge. pp. 284–303.

Del Casino Jr, V. and Hanna, S. (2000) 'Representations and identities in tourism map spaces', *Progress in Human Geography* 24 (1): 23–46.

Domosh, Mona (1991) 'Toward a feminist historiography of geography', *Transactions of the Institute of British Geographers* 16: 95–104.

Duncan, N. (1996) 'Renegotiating gender and sexuality in public and private spaces', in N. Duncan (ed.) *BodySpace*. London: Routledge. pp. 127–45.

Ebenstein, H. (1993) *Volleyball with the Cuna Indians*. New York: Penguin.

Elder, G. (1998) 'The South-African body politic', in S. Pile and H. Nast (eds) *Places through the Body*. London: Routledge. pp. 153–64.

Elder, G. (1999) '"Queerying" boundaries in the geography classroom', *Journal of Geography in Higher Education* 23: 86–93.

Elder, G. (forthcoming) *Malevolent Geographies: Sex, Space, and the Apartheid Legacy*. Columbus: Ohio University Press.

Elder, G., Knopp, L. and Nast, H. (forthcoming) 'Sexuality in geography', in C. Wilmot and G. Gaile (eds) *Geography in America at the Dawn of the 21st Century*. Oxford: Oxford University Press.

Ellis, M. (1996) 'The post-diagnosis mobility of people with AIDS', *Environment and Planning A* 28: 999–1017.

England, K. (1994) 'Getting personal: reflexivity, positionality, and feminist research', *The Professional Geographer* 46: 80–9.

Foucault, M. (1980) *The History of Sexuality. Vol. 1: An Introduction*. Harmondsworth: Penguin.

Fuss, D. (1991) *Inside/Out: Lesbian Theories, Gay Theories*. New York: Routledge.

Gregory, D. (1995) 'Between the book and the lamp', *Transactions of the Institute of British Geographers* 20: 29–57.

Ingram, G. (1993) 'Queers in space: towards a theory of landscape, gender and sexual orientation'. Paper presented at Queer Sites Conference, University of Toronto.

Ingram, G., Bouthillette, A. and Retter, Y. (1997) *Queers in Space: Communities/Public Places/Sites of Resistance*. Seattle: Bay.

Jackson, P. (1989) *Maps of Meaning*. Boston: Unwin Hyman.

Jay, E. (1997) 'Domestic dykes: the politics of IN-difference', in G. Ingram, A. Bouthillette and Y. Retter (eds) *Queers in Space: Communities, Public Places, Sites of Resistance*. Seattle: Bay.

Johnston, L. (1996) 'Flexing femininity: female bodybuilders refiguring the "Body"', *Gender, Place and Culture* 3: 327–40.

Ketteringham, W. (1979) 'Gay public space and the urban landscape'. Paper presented at the Annual Meeting of the Association of American Geographers.

Ketteringham, W. (1983) 'The Broadway Corridor: gay businesses as agents of revitalization in Long Beach, CA'. Paper presented at the Annual Meeting of the Association of American Geographers, Denver.

Kidron, M. and Segal, R. (1984) *The New State of the World Atlas*. New York: Penguin.

Kirby, A. (1995) 'Straight talk on the Pomo-Homo question', *Gender, Place and Culture* 2: 89–95.

Knopp, L. (1990a) 'Some theoretical implications of gay involvement in an urban land market', *Political Geography Quarterly* 9: 337–52.

Knopp, L. (1990b) 'Exploiting the rent-gap: the theoretical significance of using illegal appraisal schemes to encourage gentrification in New Orleans', *Urban Geography* 11: 48–64.

Knopp, L. (1994) 'Social justice, *sexuality* and the city', *Urban Geography* 15: 644–60.

Knopp, L. (1995) 'If you're going to get all hyped up, you'd better go *somewhere*!', *Gender, Place and Culture* 2: 85–8.

Knopp, L. (1998) 'Sexuality and urban space: gay male identities, communities and cultures in the U.S., U.K. and Australia', in R. Fincher and J. Jacobs (eds) *Cities of Difference*. New York: Guilford. 149–76.

Knopp, L. (1999) 'Introduction' and 'Out in academia: the queer politics of one geographer's sexualization', *Journal of Geography in Higher Education* 23: 77–9; 116–23.

Knopp, L. (2000) 'A queer journey to queer geography', in P. Moss (ed.) *Engaging Autobiography: Geographers Writing Lives*. Syracuse: Syracuse University Press.

Kramer, J. (1995) 'Bachelor farmers and spinsters: gay and lesbian identities and communities in rural North Dakota', in D. Bell and G. Valentine (eds) *Mapping Desire*. London: Routledge. pp. 200–13.

Lauria, M. and Knopp, L. (1985) 'Toward an analysis of the role of gay communities in the urban renaissance', *Urban Geography* 6: 152–69.

Leap, B. (forthcoming) *Gay City*. Minneapolis: University of Minnesota Press.

Levine, M. (1979) 'Gay ghetto', *Journal of Homosexuality* 4: 363–77.

Mackenzie, S. and Rose, D. (1983) 'Industrial change, the domestic economy and home life', in T. Anderson and S. Duncan (eds) *Redundant Spaces in Cities and Regions?* London: Academic. pp. 155–200.

McDowell, L. (1994) 'Performing work: bodily representations in merchant banks', *Society and Space* 12: 727–50.

McDowell, L. (1995) 'Body work: heterosexual gender performances in city workplaces', in D. Bell and G. Valentine (eds) *Mapping Desire*. London: Routledge. pp. 75–98.

McNee, B. (1984) 'If you are squeamish…', *East Lakes Geographer* 19: 16–27.

McNee, B. (1985) 'It takes one to know one', *Transition* 14: 2–15.

Miller, N. (1992) *Out in the World: Gay and Lesbian Life from Buenos Aires to Bangkok*. New York: Vintage.

Mitchell, D. (2000) *Cultural Geography: A Critical Introduction*. Oxford: Blackwell.

Mosse, G. (1988) *Nationalism and Sexuality: Middle-Class Morality and Sexual Norms in Modern Europe*. Madison: University of Wisconsin Press.

Murray, S. (1979) 'The institutional elaboration of a quasi-ethnic community', *International Review of Modern Sociology* 9: 165–77.

Nast, H. (1998) 'Unsexy geographies', *Gender, Place and Culture* 5: 191–206.

Nast, H. (1999) '"Sex", "Race", and multiculturalism: critical consumption and the politics of course evaluations', *Journal of Geography in Higher Education* 23: 102–15.

Nast, H. (forthcoming) 'Queer patriarchy', *Antipode*.

Nast, H. and Pile, S. (eds) (1998) *Places through the Body*. London: Routledge.

Parker, A. (ed.) (1992) *Nationalisms and Sexualities*. London: Routledge.

Peake, L. (1993) '"Race" and Sexuality: challenging the patriarchal structuring of urban social space', *Society and Space* 11: 415–32.

Phillips, R. (1999) 'Writing travel and mapping sexuality', in J. Duncan and D. Gregory (eds) *Writes of Passage: Reading Travel Writing*. London: Routledge. pp. 70–91.

Phillips, R., West, D. and Shuttleton, D. (eds) (2000) *De-Centering Sexualities: Politics and Representations Beyond the Metropolis*. London: Routledge.

Pile, S. (1996) *The Body and the City: Psychoanalysis, Space and Subjectivity*. London: Routledge.

Rocheleau, D., Thomas-Slayter, B. and Wangari, E. (1996) *Feminist Political Ecology: Global Issues and Local Experiences*. London: Routledge.

Rose, G. (1993) *Feminism and Geography*. Minneapolis: University of Minnesota Press.

Rothenberg, T. (1994) 'Voyeurs of imperialism: *The National Geographic Magazine* before World War II', in A. Godlewska and N. Smith (eds) *Geography and Imperialism*. Oxford: Blackwell. pp. 155–72.

Savage, D. (1999) 'Too chicken to come out?', *Out* 63: 34.

Savage, D. (2000) 'Everything', *The Stranger* 9: 39.

Seager, J. (1997) *The State of Women in the World Atlas*. New York: Penguin.

Sedgwick, E. (1990) *Epistemology of the Closet*. Berkeley: University of California Press.

Seebohm, K. (1994) 'The nature and meaning of the Sydney Mardi Gras in a landscape of inscribed social relations', in R. Aldrich (ed.) *Gay Perspectives II: More Essays in Australian Gay Culture*. Sydney: University Printing Service. pp. 193–222.

Smith, N. (1987) 'Academic war over the field of geography', *Annals of the Association of American Geographers* 77: 155–72.

Valentine, G. (1993a) '(Hetero)sexing space: lesbian perceptions and experiences of everyday spaces', *Society and Space* 11: 395–413.

Valentine, G. (1993b) 'Negotiating and managing multiple sexual identities: lesbian time–space strategies', *Transactions of the Institute of British Geographers* 18: 237–48.

Valentine, G. (1998) 'Sticks and stones may break my bones: a personal geography of harassment', *Antipode* 30: 305–32.

Valentine, G. (ed.) (2000) *From Nowhere to Everywhere: Lesbian Geographies*. Binghamton: Haworth.

Weightman, B. (1980) 'Gay bars as private places', *Landscape* 24: 9–17.

Winkcapaw, C. (1999) 'The virtual spaces of lesbian and bisexual women's electronic mailing lists', *Journal of Lesbian Studies* 4: 45–59.

Winters, C. (1979) 'The social identity of evolving neighborhoods', *Landscape* 23: 8–14.

17

Troubling the Place of Gender

Liz Bondi and Joyce Davidson

LOCATING GENDER

This chapter examines gender as an aspect of subjectivity that is both taken for granted and extraordinarily elusive. We illustrate the power of gender categories both experientially and theoretically, arguing that some kind of gender binary is very deeply woven into the fabric of cultural life. At the same time we illustrate how gender is always bound up with other dimensions of human experience and subjectivity including those described by such terms as class, race, sexuality, age and so on. That gender is inseparable from these contributes to its elusiveness: the meaning(s) of gender cannot be isolated from the specificities of class, race and so on. Moreover, although the categories 'men' and 'women' seem readily distinguishable from one another, on closer inspection it turns out to be impossible to locate the source of this distinction unambiguously. Gender therefore poses us with puzzles.[1] It is a profoundly influential concept through which our lives are marked and lived, but if we scratch the surface of its meanings perplexing questions and confusions are revealed.

The combination of familiarity and elusiveness that characterizes gender can be understood in terms of the influence of claims to knowledge that purport to be universal in scope, which are characteristic of dominant traditions of western, anglophone thought. Such claims have been subject to extensive criticism by feminists among others, who have pointed to biases implicit in statements about people 'in general' that treat as universal the experiences of some (Gilligan, 1982), and to the very particular kind of human subject (typically white, male, western, affluent, heterosexual, able-bodied, adult) invoked in the making of such claims (Lloyd, 1984; Nicholson, 1990).

Criticisms do not, however, demolish patterns of thought overnight. Indeed critiques of universal knowledge claims necessarily draw on the very traditions they criticize. Nowhere is this clearer than in feminist discussions of gender: feminists cannot avoid using gender categories but at the same time repeatedly question their validity and meaning (Weir, 1997).

One of the most influential responses to this dilemma is to 'situate' and pluralize claims to knowledge, including those advanced by others and by ourselves (for a classic statement see Haraway, 1988). Situating or locating gender (and any other potentially or purportedly universal concept) requires both a cultural and a geographical imagination: it requires that we attend to particular meanings in particular contexts, and it problematizes conceptualizations of space as well as gender. This chapter aims to illustrate and to extend the application of such an imagination, and it does so by drawing out some of the ideas about human subjects and the spatialities of subjectivities invoked in cultural geography's engagements with gender. We begin this task by elaborating and situating our claim about the pervasive and problematic qualities of gender. We then explore in greater depth four themes that emerge from this account, and which illustrate the range of, and some key directions in, current research about gender in and around the specialism of cultural geography. Our account is necessarily a positioned one and arises specifically from our own commitment to, and participation in, the development of feminist perspectives in human geography in general and in cultural geography in particular (Bondi et al., 2002). As we will illustrate, 'feminism' is itself multifaceted and contested, but we assume throughout that concepts and practices of gender are bound up with the exercise of power.

The differentiation of human beings by gender is so powerful and pervasive that it is very difficult to imagine (or write about) a person without drawing upon one of the two labels 'male' and 'female' either implicitly or explicitly. In western societies (and in many non-western societies) most of us allocate people to one category or the other so routinely and so habitually that we become aware of our practice only when confronted by a body or a voice or a name that we cannot categorize with confidence. Think, for example, of the curiosity aroused by chance encounters – maybe walking on a city street or through a mall – with individuals whose appearance prompts us to question whether they are male or female. Or think of the surprise at discovering that the voice heard as 'male' turns out to belong to a woman (or vice versa). Note also how particular contexts are called into play by our injunction to imagine a particular person, indicating how our ascription of gender categories is necessarily situated culturally and geographically. Reflecting on gender in these ways also renders untenable any clear distinction between 'doing' and 'theorizing': gender is equally about our ideas, practices and reflections about ourselves and others in a way that weaves these aspects together (Ahmed, 1998). This understanding of theory as practice is very influential within feminist politics and feminist geography (Katz, 1994; 1996; Kobayashi, 1994; McDowell, 1992), and is increasingly influential in discussions about reflexivity and activism in cultural geography (Blomley, 1994; Maxey, 1999).

The taken-for-grantedness of gender categories has often been just as powerful within research and scholarship as elsewhere. But this has not gone unchallenged. Feminist geographers have worked hard to denaturalize gender, with important consequences throughout human geography including the field of cultural geography. Denaturalizing gender has entailed demonstrating implicit and naturalizing assumptions within existing work, thereby opening up new questions for investigation. Since the mid 1970s, feminist geographers have pointed to two consequences of the widespread failure to problematize gender. First, there has been a strong tendency to neglect the perspectives and concerns of certain groups, either by omission or by stereotyping, especially those different from, other than and disadvantaged relative to the groups from whom the majority of geographers are drawn. Gender is one aspect of this: most geographers, and particularly 'powerful' geographers, are men, and women have been marginalized in terms of the substance and practice of geographical research (Hayford, 1973;

McDowell, 1979; McDowell and Peake, 1990; Monk and Hanson, 1982; Tivers, 1978). Secondly, as well as reinforcing the disadvantage and devaluation underpinning this neglect, failure to problematize gender has undermined geographical thought quite generally, generating 'androcentric' or 'masculinist' claims to knowledge that purport universality (Rose, 1993; Women and Geography Study Group of the IBG, 1984).

These critiques have paved the way for a substantial and wide-ranging body of work which focuses in a variety of ways on the gendering of human lives and human geographies. Much of this work has been informed, and often inspired, by political and moral commitments, for example to challenge gender and other inequalities, or to contribute to emancipatory or liberatory goals. Such commitments are suggestive of 'Enlightenment' ways of thinking about issues of equality, justice, self-determination and so on, in the sense of invoking universal ideas about human life and human values. This prompts interpretations of feminism as a modern political movement dependent on metanarratives of progress. But against this are the poststructuralist and post-Enlightenment perspectives with which critiques of universal knowledge claims are more commonly associated, which have been deeply influential in cultural geography and which we have also suggested are influential in understandings of gender. We would argue that feminist perspectives need to deploy both Enlightenment and post-Enlightenment ways of thinking (Bondi, 1990; Bondi and Domosh, 1992; Deutsche, 1991; Massey, 1991; McDowell, 1991; Morris, 1988; Soper, 1990) and in so doing have the potential to foster productive paradoxes (Rose, 1993).

In the remainder of this chapter we examine in more detail the notions of subjectivity invoked in analyses of gender within and around the field of cultural geography. There already exist several pertinent reviews, which offer particular accounts of conceptualizations of gender. For example, McDowell (1993a; 1993b; also see McDowell, 1999; McDowell and Sharp, 1997) traces the development of feminist perspectives in geography, emphasizing the growing influence of poststructuralist ideas and associating this with the so-called cultural turn. Implicit in McDowell's account is a view that feminist geography has progressed from important but necessarily less sophisticated understandings towards more powerful and nuanced engagements with gender. The Women and Geography Study Group (1997) stress the existence of diverse uses of the concept of gender among feminist geographers, and acknowledge some of the political tensions around its theorization. We aim to explore important aspects of such debates by organizing our remarks in terms of four themes or strands of feminist politics, concerned with gender equality, women's autonomy, multiple differences and the deconstruction of categories, which between them straddle some of the tensions between Enlightenment and post-Enlightenment ways of thinking (Warner, 2000). We provide an overview of the knowledges generated within the domain of cultural geography by each of these themes, drawing out the conceptualizations of the (gendered) human subjects invoked. We also attend to the concepts of space brought into play by different ideas about gender, thereby linking the themes of equality, autonomy, difference and deconstruction associated with feminist politics to the problematics of cultural geography.

RE-PLACING GENDER

We have been using the term 'gender' in a way that is peculiar to the English language and which cannot readily be translated into other languages, even those with words deriving from the same linguistic root, such as Spanish and French (Haraway, 1991; Moi, 1999; Widerberg, 1999). Moreover, it is a usage traceable to particular contexts and purposes, and redolent of particular ways of thinking about the politics and subjects of feminism. To elaborate, in English the term 'gender' has become deeply and problematically bound up with the idea of a distinction between sex and gender. In feminist writings, this distinction is widely attributed to research by Robert Stoller (1968) concerned with psychological aspects of transsexualism. In his work as a psychiatrist he gathered information about individuals seeking surgery to change sex, who typically reported experiencing a profound mismatch between their biological categorization as male (or more rarely as female) and their own sense of themselves. Theorizing this mismatch, he argued that our understandings of ourselves as male or as female arise culturally through processes independent of the biology or morphology of our bodies. While these different dimensions of definition often correspond, there is nothing necessary or causal about this. He suggested that the cultural form of differentiation and identification be termed 'gender', while the biological difference be termed 'sex'. Throughout his work on transsexualism he insisted on the psychological primacy of (cultural) gender over (biological) sex (Stoller, 1985).

Second-wave feminist scholarship began to emerge at much the same time as Stoller's original research, and the notion that differences between women and men were attributable primarily to cultural processes rather than to biological givens was argued widely and powerfully, often drawing on Simone de Beauvoir's claim that 'One is not born but rather becomes a woman' (1997: 295). A common theme in feminist arguments that sought to denaturalize assumptions about women and men concerned the absence of any binary distinctions in biological attributes associated with 'sex': feminists drew on evidence about chromosomal, hormonal and other patterns to argue that the mutually exclusive categories of 'male' and 'female' are imposed on, rather than given by, nature (Cream, 1995; Greer, 1971). In parallel with this, feminists pointed to various sources of cross-cultural and historical evidence that revealed wide variations in the allocation of tasks between women and men, and in the attributes imputed to women and men, evidence that endorsed the argument that differentiation between male and female has much more to do with culture than biology (Moore, 1988; 1994). Early anglophone feminist writers did not use the term 'gender' (Firestone, 1972; Mitchell, 1971), until Ann Oakley (1972) took up Stoller's notion of a distinction between (biological) sex and (cultural) gender. Thereafter this understanding of the term 'gender' swiftly took hold, so that by the time the Women and Geography Study Group of the IBG published their landmark text *Geography and Gender* this usage was offered unproblematically:

> We use the term 'gender' to refer to *socially created* distinctions between masculinity and femininity, while the term 'sex' is used to refer to biological distinctions between men and women. (1984: 21, emphasis in original)

This line of argument has been of profound importance in the development of western, anglophone feminism(s). If the major differences between women and men are cultural rather than biological in origin, then the ways in which women are disadvantaged relative to men are not given in nature but are cultural in origin, maintained through the exercise of power, and can be modified through social and political means. In this way the sex–gender distinction has been crucial for feminists working for the emancipation of women.

The demands for equality associated with this application of the sex–gender distinction can be traced historically to the eighteenth-century Enlightenment thinking of Mary Wollstonecraft and the tradition of liberal humanism. Within feminist versions of this tradition, what women and men share by virtue of their common humanity is of much greater importance than the ways in which they differ, and constitutes the basis for a politics of equality, including, for example, equal citizenship rights. Liberal humanism advances an understanding of the adult human subject as a coherent, integrated, bounded, autonomous being capable of self-direction, self-control and so on: liberal feminists from Wollstonecraft onwards have argued that these characteristics apply as much to women as to men but that women have been prevented from realizing their full potential by outmoded assumptions expressed in limitations on their opportunities for education, for economic independence and so on.

Arguments for equality between men and women have, of course, been enormously influential within cultural geography. Moreover, while the particular understanding of the human subject associated with liberal humanism has been subject to criticism, the notion of equality has not been relinquished. On the contrary, the politics of equality remain of the utmost importance in the cultural world with which geographers engage, and within the practice of cultural geography.

Two closely related features of the vision of subjectivity invoked by the theme of equality have been particularly productive for cultural geography, namely the question of where gender is understood to reside, and the way the relationship between subject and environment is conceptualized. According to the liberal feminist tradition, gender contributes nothing fundamental to human subjectivity. Instead it operates as a kind of adornment, but one which is prone to be regarded as more significant than it is (which is why gender inequalities persist in many arenas and settings, and why feminists continue to argue for equality). Gender is thus understood as something external to the core of the human subject, as something imposed on but not residing within the essential nature of human being. At the same time as 'externalizing' gender, this understanding of subjectivity invokes a radical separation between the subject and the environment in which s/he exists: the human subject occupies a position of self-determining, rational agent independent of, and external to, that environment, including both its social and its physical facets.

These two elements of the liberal humanist model of human subjectivity have together invited us to think about gender as belonging at least as much to environments – the spaces and places in which we live our lives – as to people. Put another way it has helped us to think

about gender as inscribed on 'natural' and built environments, as well as, and as a way of, marking and adorning bodies. Cultural geography has thus sought to reveal the variously gendered constitution of a diverse array of environments, including for example in workplaces (Hanson and Pratt, 1995; Massey, 1995; McDowell, 1995; 1997), homes (Domosh, 1998; Dowling and Pratt, 1993; Gregson and Lowe, 1995; Marston, 2000; Roberts, 1991), residential neighbourhoods (Bondi, 1998a; Hubbard, 1998; 2000), retail spaces (Blomley, 1996; Domosh, 1996; 1998; Gregson and Crewe, 1998), leisure and sports environments (Johnston, 1995; 1996; McEwan et al., 2002; Morin et al., 2001) and cyberspace (Wakeford, 1999). Each of these places has become gendered in accordance with the kind of activities that take place there, and with the ways in which these activities have traditionally been conceptualized and 'marked' as either masculine or feminine.

Several of the studies cited argue that gender and sexuality are inextricably intertwined within the various spaces on which they focus, and refuse to be bound by the separation between sex and gender associated with the emergence of feminist perspectives in geography. This conceptualization of gender initially inhibited explicit engagement with questions of sexuality, as well as with embodiment, and has subsequently been extensively critiqued within cultural geography (Bell and Valentine, 1995; Binnie and Valentine, 1999; Bondi, 1998b; Domosh, 1999; Longhurst, 1995; 1997; Nast, 1998). We would argue, however, that the conceptual separation of subject and environment invoked by the sex–gender distinction has survived such critiques and has been highly productive in fostering research about the geographical and cultural inscription of gendered subjectivities understood in a wide variety of ways.

To elucidate this notion of the inscription of gender on environment, we draw on some everyday and supposedly 'common-sense' examples. Think, for instance, of the stereotypical building site, boxing ring and aeroplane cockpit, and the starkly contrasting gender associations of secretarial spaces (the traditional 'typing pool'), the aerobic fitness centre and the aeroplane aisle (Grimshaw, 1999; Hochschild, 1983; Pringle, 1989). To be sure, such correlations are neither rigid nor static, but the gendering of such places is clearly illustrated by the way the gender identities of those who transgress the conventions are viewed. Female boxers and brickies are likely to attract comment as unwomanly or 'mannish', while male flight attendants and secretaries are often characterized as effeminate or effete. Precisely because these various places bear such powerful gender inscriptions, men and women put their masculine or feminine cultural credentials in question through their mere presence in those environments (Cresswell, 1996). The extent to which it is considered unacceptable to inhabit spaces culturally coded in terms of the other('s) gender is illustrated clearly by the negative connotations of the gendered adjectives applied to such suspect subjects. Even when, or where, gender conventions weaken, 'transgressions' remain noteworthy, illustrated in terms like 'male nurse' (distinguished from 'nurse') or 'female pilot' (distinguished from 'pilot'): the gender of terms without adjectival specification can, apparently, be assumed.

Feminist theorists have sought to unearth the roots of such placings of gender, in order to ascertain on what grounds the gendered associations are made. Attempts have been made to identify patterns that reveal why one environment, and the associated activities that colour its conceptualization, is coded masculine, and another feminine, and often the patterns (whether 'found' or 'imposed') are of a dualistic nature. Many and various dualisms have been identified and linked to cultural and geographical imaginings of gender. For example, it has been argued that environments are gendered according to whether they are conceptualized as 'public' or 'private', with linkages to binaries of 'culture' and 'nature', 'mind' and 'body', 'reason' and 'emotion', 'active' and 'passive', which are all structured hierarchically, the first terms of these pairings not only inscribed as masculine but privileged as well.

The literature of cultural geography provides abundant empirical evidence of the operation of such dichotomies. But the gender patterning of space almost invariably turns out to be complicated, disrupting as well as reproducing gendered binaries. Natural environments, for example, are often feminized and viewed as passive, rendering them available for conquest and control by radically separate masculine subjects (Kolodny, 1975; Rose, 1993; Seager, 1993; Women and Geography Study Group, 1997; Woodward, 1998). Mountaineering is thus traditionally coded as 'masculine', and the places associated with mountaineering – its peopled environments – are similarly masculinized even if the land itself is coded as feminine. The urge to climb and to conquer mountains can thus be understood in terms of a particular, heterosexual version of masculinity (Rose, 1993). However, mountaineering also cuts across associations between masculinity and 'rationality', largely driven, as mountaineers acknowledge, by passions rather than logic, in addition to which

the muscular strength and fitness required entails a fixation with the bodily which in other contexts might be seen as 'unmanly' (Longhurst, 2000; Morin et al., 2001). What this (stereotypical) example illustrates is that the gendering of spaces and places is rarely straightforward. Dualisms, and the value of their constitutive parts, shift depending on context, and they do not themselves 'explain' or provide reliable analytical tools for understanding the projection of a gender binary onto 'natural' and built environments. Indeed, what emerges most clearly from the way the relationships between dualisms operate is that they serve to protect gender inequalities by privileging all things male (Bondi, 1992; 1998a).

Another example reinforces the point: the work of cooking, cleaning and caring in general tends to be linked with women, especially when unpaid and carried out in domestic spaces. Where such activity takes place beyond the domestic space of the home, it has the potential to become managed, professionalized and masculinized. The more men become involved in a particular place, the more respected that place becomes; and conversely, the more valued the place becomes, the more likely it is to become a 'male preserve'. Think of the chef's kitchen, as opposed to the housewife's; the consultant surgeon's common room, as opposed to that of the nurse. Any place where 'men's business' is conducted is likely to be esteemed as a space of action and achievement, command and control (Spain, 1992). In contrast, the more servile and superficial business of 'women's space' is likely to be trivialized, denigrated and domesticated, that is, excluded from the peculiarly defined 'public' sphere of influence and importance.

Kevin Hannam and Pamela Shurmer-Smith recognize the pervasiveness and tenacity of such gendered geographical divides, and suggest that (predominantly male) geographers have been complicit in strengthening this tautologous link between women and the private sphere, through the practice of

> classifying the domestic realm as the place where women and children are permitted and the public realm as the place where they are wholly or partially excluded. [Thus] in much of Africa we find the domestic realm extending to the cultivated fields where women work, but not necessarily to the front of people's houses where men talk. (1994: 109; compare Monk and Hanson, 1982)

Bondi and Domosh (1998) further elaborate the argument by showing how distinctions between public and private, characteristic of modern urban societies, can be understood in terms of a coalition of gender and class interests, in the service of which spaces are defined and redefined.

This analysis illustrates how accounts of subjectivity associated with the theme of equality enable us to think of gender as residing in environments external to human subjects. However, we have also shown how attempts to trace the sources of gender inequalities in such environments founder, and return instead to the power relations through which (gendered) geographies are produced. Later in this chapter we will argue for approaches to subjectivity that do not view subjects and environments as separable in this way. However, at this point we note how examination of the gendering of spaces and places illuminates important aspects of the persistence of gender inequalities and in so doing provides resources with which to challenge deep-seated or 'sedimented' assumptions about the meaning and placing of gender categories.

While assessments of the impact of feminist politics in particular contexts vary (Mitchell, 1984; Rowbotham, 1989; Scott, 1999), the optimistic view that distinguishing between cultural gender and biological sex would have a straightforward liberatory effect (via the notion of equality) has largely faded (Evans, 1994). In the late twentieth and early twenty-first centuries few feminists would claim that the cultural is intrinsically easier to change than the biological. Indeed in some ways such a claim had always been deeply ironic given that the sex–gender distinction had come from a psychiatrist who argued that it was easier (and more emancipatory) to intervene surgically than to change the gender identifications of his patients (Shapiro, 1991). Another strand of feminist politics has emphasized the need for separate spaces for women, within which the resistances of gender are acknowledged. We turn to this argument next, elaborating its impact on cultural geography.

RECONFIGURING GENDER

Closely related to the theme of equality is that of autonomy. Conceptually, autonomy underpins liberal feminist arguments for equality because the capacity for self-determination or self-rule is seen as intrinsic to human nature, regardless of gender. Moreover, Simone de Beauvoir's inspiring analysis drew strongly on the argument that women are socially constructed as (men's) 'other' and consequently denied autonomy. Thus, women's autonomy may be viewed as a prerequisite for gender equality. But as we will show, the politics of autonomy has not been defined solely in terms of equality with men.

Whereas the logic of equality points towards opportunities for, and treatment of, men and women 'regardless' of gender, autonomy has required the development of women-focused practices, and the production or 'performance' of differently sexed space (Rose, 1999). Since its emergence in the mid 1970s, feminist geography has been associated with efforts to practise geography differently as well as to counteract and correct omissions and biases in geographical knowledges. In the academic discipline of geography such practices are epitomized by the emergence of groups and networks committed to the development of feminist perspectives, such as the Women and Geography Study Group (a group within the Royal Geographical Society and Institute of British Geographers), the Geographic Perspectives on Women Specialty Group (a group within the Association of American Geographers) and Geogfem (an electronic discussion list for feminist geography). None of these exclude men, not least because the principles of equality operated by professional organizations prohibit the use of gender as a criterion for membership. However, they successfully create spaces in which women are generally more 'vocal' than men and which are often temporarily occupied only by women (Delph-Januirek, 1999; McDowell, 1990; Nairn, 1997).

Attempts to create new kinds of spaces of and for knowledge production have, however, often proven problematic because of deep-seated tensions between academic conventions and feminist commitments (Bondi, 2002; Domosh, 2000; Hanson, 2000; Nast and Pulido, 2000; Penrose et al., 1992; Seager, 2000). Such tensions sometimes revolve around the cooperative, non-hierarchical ways of working espoused by feminists which sit uneasily within the highly individualistic systems of reward and recognition characteristic of university cultures. Such cultures typically proclaim the high degree of autonomy available to academics, but, as feminists have shown, the form of autonomy espoused is implicitly and specifically gendered masculine (Aisenberg and Harrington, 1988; Grosz, 1990; Hekman, 1992; Lloyd, 1984; Morley and Walsh, 1996). Feminists have therefore argued and agitated for the scope to redefine autonomy.

As well as underpinning the efforts of feminist geographers to influence spaces and practices within the academy, the theme of autonomy has generated a body of work in cultural geography concerned with parallel efforts in other settings. Whereas a concern with equality has generated research about the inscriptions of spaces as both masculine and feminine, a concern with autonomy has prompted a specific focus on women's spaces, thereby 'correcting' a profound substantive and conceptual neglect. A small number of studies focus on women-only living and working spaces (Taylor, 1998; Valentine, 1997). More numerous are analyses of women's efforts to 'own', embody or reclaim spaces, in the context of the dominance of normative, heterosexual versions of masculinity (Bondi, 1990; Duncan, 1996; Koskela, 1997; Valentine, 1996). These studies often emphasize the importance of countering representations of women as victims at the same time as acknowledging the profound difficulty of challenging hegemonic versions of gender. What emerges is a commitment to simultaneously 'undo' and 're'do' gender.

Ali Grant (1998), for example, writes about the politically motivated exclusion of men from spaces of feminist activism such as 'Take Back The Night' marches (against violence against women), the development of refuges for women fleeing domestic violence, and the work of rape crisis centres. Feminist activists have long argued that the creation and maintenance of women-only 'safe spaces' is a crucial component of feminist projects. However, this tradition's 'sin of disloyalty to men' has provoked anger and accusations of 'man-hating' and lesbianism, the latter being 'an old but still effective way of maintaining a strict gender hierarchy' (1998: 50). Such accusations can be particularly damaging when they originate from or take hold within locally influential organizations who, through funding and other means, effectively regulate, restrict, 'domesticate' and devalue the potential achievements of feminist activists, resulting in stark choices between closure and deradicalization. The latter route has frequently entailed the progressive institutionalization of what began as feminist projects and spaces, in the course of which women-only principles have been relinquished in order to retain funding. Grant argues that this is in fact the purpose of the hostility against feminist activism because women-only spaces challenge hegemonic understandings of appropriate gender relations:

> Women who contest and transgress gender materially and symbolically expose the manufactured nature of the category 'women' … [T]ransgressive females have tended to be marginalised, as spaces of resistance have been institutionalised. Removing these transgressive bodies and ideas … reduces the possibilities in place for the continuation of the development of radical counter-hegemonic identities. (1998: 53)

Grant's analysis illustrates how, from the perspective of hegemonic versions of gender relations, 'doing' gender without men is deemed wholly inappropriate and unacceptable. But the creation of non-institutional spaces, *different* environments where difference is allowed to emerge – different ideas, attitudes and behaviours – is essential for feminist activists in order to *undo* the dominant system of gender (compare Shugar, 1995). For this reason, it is important who is and who is not involved in those spaces. The 'women-only' spaces that activists opposing violence against women aim to create cannot be understood within our usual ways of 'thinking gender', and therefore challenge established patterns of thought as well as widespread patterns of behaviour.

Quoting from her interviews with feminist anti-violence activists, Grant (1998) demonstrates just how enlightening many had found spending time with other women in spaces of resistance, and how essential this experience was to their developing sense of their different (often lesbian and in Grant's terms 'UnWomanly') subjectivities. She argues persuasively that such critical collective spaces were and are crucial in processes of 'consciousness raising' and politicization, through which the category 'woman' may be redefined beyond its current restrictively gendered and heterosexist basis and bias. This argument is, therefore, not about equal access to spheres of activity and subject positions traditionally dominated by men, but about spaces through which to realize women's autonomy and difference from men. It is about women occupying and traversing spaces boldly and confidently (Koskela, 1997); it is about women dressing and behaving on their own terms rather then in relation to men (Skelton, 1998); it is about performing and producing spaces differently.

Challenges to the power of regulatory 'fictions of gender' (Butler, 1990) have taken various forms, with attempts to construct and *imagine* sexed and spatial difference being a prominent feature in feminist utopian literature. Such projects present images of how gender relations could be, and thereby stretch imaginations and aspirations beyond the 'common-place' of here-and-now towards the 'no-place' of utopia, where the subjectivities of women are freed from patriarchal constraints. Often, this entails a radical separatist agenda, where women live on their own and on their own terms. Feminist utopian novels such as Marge Piercy's *Woman on the Edge of Time*, Joanna Russ' *Whileaway* and Monique Wittig's *Guérillères* proffer more or less single-sex self-defining cultures where men figure minimally or not at all (also see Munt, 1998; Valentine, 1997). By imagining a cohesive and autonomous social identity for women, such writings highlight the gap between our experiential realities and our political ideals, providing a sharp and stimulating contrast between existing and aspirational social and symbolic orders. Lucy Sargisson writes that such texts describe cultures of women which

> contain a material or cultural memory of oppression along gender lines ... [which] can be connected to the common search for subjectivity and identity as women, and to the renunciation of historico-cultural traditions that have constructed Woman as an artifice to complement (make perfect) Man. (1996: 206–7)

The non-fictional practice of *l'écriture feminine*, most closely associated with the French feminist philosophy of, for example, Luce Irigaray and Hélène Cixous, can also be seen as a 'utopian' attempt to strengthen and simultaneously recreate autonomous feminine subjectivities. *L'écriture feminine* attempts to think sexual difference anew, and suggests that 'imagining how things could be different is part of the process of transforming the present in the direction of a different future' (Whitford, 1991: 19). Such writing thus seeks to bring about change in the present, rather than mapping out the future in advance. It is therefore a form of dynamic or fluid utopianism that works to reach beyond women's potential for autonomous relations with men to create a new ethics of sexual *difference*. As Kevin Hannam and Pamela Shurmer-Smith note,

> *l'écriture feminine* is not to everyone's taste and it can certainly be very hard work to read, but, since it resides at the edge of experiment in generating new ways of thinking the world and abounds in spatial metaphors, it really cannot be ignored by cultural geographers. (1994: 119; also see Davidson and Smith, 1999)

Indeed, Gillian Rose (1996; 1999) has in recent years engaged with Irigaray in an imaginative and performative attempt to rethink the very nature of space.

The theme of autonomy has provoked a great deal of controversy within and beyond feminist geography. For some, the women-only or women-centred practices and spaces associated with a feminist politics of autonomy presume an essential sameness among all women (and implicitly among all men). According to this interpretation, in contrast to the notion of gender as a superficial adornment associated with the theme of equality, gender is understood to constitute a fundamental and dualistic aspect of human subjectivity. Such essentialism has attracted intense and sometimes vitriolic criticism (Fuss, 1989; Spelman, 1988). However, it is an interpretation which depends upon and reproduces the binary thinking it criticizes: it constructs an opposition between gender as either superficial or fundamental. As our account illustrates,

engagements with women's autonomy carry more potent challenges to conceptualizations of subjectivity. By questioning and exploring the limits of the assumed universality of normative fictions of rational, unitary and self-directing subjects, the possibility of difference comes into view. The theme of autonomy emphasizes gender as a source of difference and in so doing focuses on women, women's experiences and women's spaces. But, as Minnie Bruce Pratt (1984) and Caroline Ramazanoglu (1989) amongst others have argued, gender is one amongst a multitude of fluid and power-laden differences that shape subjectivities and spaces, which have been profoundly influential in cultural geography's engagements with gender.

FRACTURING GENDER

The processes of categorization we described in the first section of this chapter are not unique to gender. We make equally routine and habitual assumptions about people's age, about their ethnicity and/or 'race', about their socio-economic position, about the shape, capacities and requirements of their bodies, about their sexuality and so on. In some ways the existence of these taken-for-granted assumptions is essential to our ability to interact with one another, at least in contexts where daily routines bring 'strangers' into close proximity with one another. We cannot approach each and every interaction as if we know nothing of what might be similar or different between ourselves and the lives of others. Rather, we need to be able to 'place' the people we come across in our daily lives within a framework of meaning, which we do by drawing upon culturally specific repertoires of interpretation. Thus, we use local knowledges that enable us to interpret visual, aural and other signs (pertaining to contexts as much as to people) in terms of distinctions of gender, age and so on. We also need to be able to 'place' or 'situate' ourselves in relation to others, which we do in part by 'identifying' ourselves as belonging within particular categories and outside others, from which we 'disidentify' (Skeggs, 1997). These processes of identification and disidentification draw on whatever ideas and resources are available to us to think about ourselves.

The categories we so frequently apply unthinkingly or unproblematically relate to our own and other people's lives in complex ways. Contrary to much everyday usage, they do not express uncontentious, easily definable or constant attributes of the lives to which they refer.

For example, there are many different ways of being 'women' or 'men'. Indeed as soon as adjectives are added we imagine 'women' and 'men' in many different ways: consider for example how you imagine 'young women' or 'old men' or 'a white woman' or 'a homeless man', and so on. In other words, gender categories are analytical abstractions. Moreover, many of us resist particular labels or the assumptions attaching to them at least some of the time, but at other times we use them of ourselves without question. For example, whether as students, research workers or lecturers, those of us involved in higher education are likely sometimes and in some places to experience our gender as irrelevant but at other times and in other places as important to what we do, and we are likely sometimes to accept and sometimes to question assumptions others make about us under the heading 'gender'. This does not depend upon an explicit awareness of issues of gender: it is evident in unspoken responses of the kind 'but I'm not like that' or 'I'm not sure if that applies to me'. In such responses, although gender may be an aspect of what is disputed, it may be impossible (and irrelevant) to discern whether what is rejected pertains to gender or to another kind of category, illustrating that the notion of gender as a discrete facet of human experience is a way of thinking that relates problematically to the experiences to which it refers. Indeed, as many feminist scholars have argued persuasively, categories of gender, race, class, age and so on cannot be 'disentangled' or 'separated' from one another, so we cannot isolate something called 'gender' from other aspects of lived experience (Battersby, 1998; Grosz, 1994; Pratt, 1984; Ramazanoglu, 1989; Spelman, 1988; West and Fenstermaker, 1995; Young, 1990).

Differences between women have been an important theme and focus of attention within feminist politics and feminist thought. We would argue that, despite all its limitations, the sex–gender distinction has encouraged an association between gender and difference. To elaborate, whereas 'sex' is typically defined in terms of two mutually exclusive categories, male and female, feminists deployed 'gender' in order to foster the possibility of multiple versions of femininities and masculinities. In this sense 'gender' highlights and opposes stereotypical representations of, and assumptions about, women and men that abound, whether in academic disciplines (in relation to geography, see Monk and Hanson, 1982; Rose, 1993; Tivers, 1978) or in wider cultural and political contexts (Coward, 1984; Williamson, 1980), and which underpin a wide variety of exclusionary practices.

One way of challenging such representations and assumptions has been to demonstrate the existence of multiple femininities (and masculinities), which entails bringing into question general claims about the nature of 'women' (and 'men'). But overgeneralized claims about 'women' (and 'men') are not the sole preserve of 'mainstream' academia, 'mainstream' politics or 'mainstream' culture: in advancing the case for women's liberation, feminists have sometimes, often unwittingly, reproduced the same exclusionary practices for which they have criticized others. In particular, white, middle-class, heterosexual women have sometimes written or spoken as if representing women 'in general' but in so doing have eclipsed and effectively silenced women unlike themselves. Critiques advanced by black women, by working-class women, by lesbians, by Third World women, by disabled women and so on, have all, in a wide variety of ways, ensured that differences among women have become central to the politics of feminism (Bhavnani, 2001; Carby, 1982; Chouinard, 1999; Chouinard and Grant, 1995; hooks, 1989; Mohanty, 1988; Mohanty et al., 1991; Wittig, 1992).

Such critiques have been associated with struggles around cultural identities and representations, which Nancy Fraser (1995; 1997) describes as a 'politics of recognition', and which she argues are analytically distinct from a 'politics of redistribution', that is, struggles around material inequalities. Fraser's distinction has generated a good deal of debate, much of which revolves around the value of making such analytical distinctions (Fraser, 2000; Phillips, 1997; Young, 1997). However, neither Fraser nor her critics disagree that in practice political struggles often address issues of poverty *and* of representations, as several studies in cultural geography have illustrated. Geraldine Pratt and Susan Hanson, for example, examine how geography constructs interconnecting differences of gender, race and class which have significant material consequences:

> Gender is constituted differently in different places, in part, because residents in those places differ in class or racial or other social variables; that is, places are sites where particular sets of social relations are experienced and compressed. But geography is implicated in a deeper way. Class itself is a process that is mediated by place and space ... The experiences of class and gender are structured by local resources, including locally available forms of paid employment, as well as local cultures. These resources develop synergistically in relation to the social characteristics of existing residents. (1994: 11–12)

Drawing on evidence from neighbourhoods in Worcester, Massachusetts, they tease out several dimensions of the spatiality of daily lives, emphasizing how the 'imagined geographies' of employers and workers combine with workplace cultures and social networks to reinforce intersecting differences of gender, class and ethnicity (also see Hanson and Pratt, 1995). Illustrating the intense geographical constraints experienced by many of those living in poor neighbourhoods (who include first-generation immigrants who have travelled large distances before arriving in Worcester), they argue 'that there is a stickiness to identity that is grounded in the fact that many women's lives are lived locally' (Pratt and Hanson, 1994: 25), which often leads to a strong sense of differences between women, and which masks interdependencies and commonalities.

As many studies have shown, mobility, especially commuting times and distances, varies markedly between different social groups (for a review concerned with gender in relation to other factors see Law, 1999). Put another way, the scale at which daily lives are lived varies very significantly; indeed the idea of 'scale' is itself a social construct that generates and sustains interlocking differences – gender, class, race and so on (Marston, 2000; Ruddick, 1996; N. Smith, 1993).

Space and spatialities clearly fracture (and are fractured by) gender and other categories and therefore contribute to a proliferation of versions of feminine and masculine subjectivity. In so far as the politics of identity grounds political interests in particular experiences, this proliferation has resulted in a splintering of feminist politics and a recognition of multiple feminisms. Taken to an extreme, this implies a conceptualization of subjectivities as corresponding directly to positions people occupy within the frameworks producing difference, including those of gender, class, race and place. But people are not bound to pre-given subject positions in this way: experiences of gender, class, race, place and so on are themselves social constructs, and influence rather than determine people's political affiliations and actions. Moreover the notion of correspondence between a particular form of subjectivity and a particular structural position 'buys into' the notion of individuals as centred, bounded, autonomous entities.

We would argue that while the imposition of socially constructed and therefore 'fictional' conceptual labels (such as race, gender and so on) can be used for politically dubious purposes, and so must be questioned and never treated as 'natural' (Irigaray, 1993), they have a long history of usage and cannot be simply abandoned by emancipatory political projects. Moreover,

such separations and categorizations can be used strategically, for example to highlight those parts of the webs of our identities (Griffiths, 1995) around which changes most urgently need to be made (in relation, for example, to experience of racism, heterosexism, ageism and so on). But, there are still potential problems linked with feminist usage and understanding of the concept 'woman': being a woman can mean different things at different times in different places to different people, and no one definition of womanhood would be appropriate across the broad board of women's experience. Why then would we wish to argue for and present a 'definition' at all?

While it is clear that women in no sense constitute a unitary group, a constructive and usable understanding of the concept of womanhood is still required for the purposes of feminist theory and practice. (After all, for whom are feminists concerned, if not 'women' as a group?) Wittgenstein's anti-essentialist philosophical approach, and particularly his notion of 'family resemblance', is potentially useful here. Following Davidson and Smith's (1999) feminist interpretation of this notion, we would contend that concepts such as 'woman' need not be understood and employed representationally, as referring to certain essential features shared by all members of the group 'women'. Rather, much as Wittgenstein argues that we identify members of a family in terms of, for example, resemblances in eye colour, build or patterns of speech, and not through a particular feature or features common to all, concepts such as 'games' or, according to Davidson and Smith, 'women' can be understood productively in terms of a 'complicated network of similarities overlapping and crisscrossing' (1994: 74). That is to say, when we use the term 'woman' we need not use it in a way that specifies and depends upon any particular shared experiences or features for its sense. In fact, what we understand and what we communicate by the term will alter according to the context within which it is used, and this shifting, contextualized sense of how a concept operates in practice is extremely, multiply, useful for feminist theorists and cultural geographers.

There are powerful reasons why we should resist the temptation to produce a precise definition that attempts to 'pin down' exactly what being a woman entails. Such exact conceptual boundaries are likely to be restrictive, exclusive and counter-productive politically. In so far as we do however require a working definition of 'woman', we would argue in favour of 'strategic imprecision', that we have no need to settle or 'tie down' its meaning once and for all. For, as Wittgenstein makes plain, it is not 'always an advantage to replace an indistinct picture with a sharp one ... Isn't the indistinct one often exactly what we need?' (1988: no. 71; cited in Davidson and Smith, 1999: 77). Wittgenstein's reference to a degree of usefulness that might change signals his recognition of and emphasis on the importance of context, and provides valuable insights into the shifting and 'unnatural' nature of gendered being: it helps to clarify the ways in which 'even' our biology is not set in stone but conceptualized, mediated and fractured by language and culture.

Such anti-essentialist and strategically imprecise definitions of 'woman' pervade feminist work within and beyond cultural geography. To take one example, in Ann Phoenix's (1995) study 'Mothers Under Twenty' the respondents' various circumstances and social positioning(s) presented the researchers with conceptual and practical challenges and dilemmas, and Phoenix's account illustrates some of the benefits and difficulties of working with any concept or definition of women which inevitably directs attention towards a specific 'group' identified, or rather 'constructed', for the purposes of the study. Despite the obvious similarities or we might say, recognizable 'family resemblances' between these women in terms of age and 'maternal status', Phoenix (1995) makes plain that individual respondents differed markedly in their most immediate experiences and concerns: even within this group of women who shared apparently significant aspects of their lives 'in common', their categorizations as women and as young mothers turn out to be rather particular aspects that coexisted with a multitude of highly salient differences between them. While the hardships and prejudices they faced in their daily lives may be tied up with their gender and/or youth and/or parental responsibility, such experiences are not immediately attributable in this way, or to any other identifiable or isolatable features of their identities.

Phoenix (1995: 59) writes, for example, about the complex and varied experience of white mothers of mixed-race children in undeniably racist circumstances; of respondents living in obvious poverty, evidenced by 'the emptiness of food cupboards' or 'lack of milk for tea' or simply the 'wintry cold' of their flat. While each of the respondents' lives is uniquely complex, for certain purposes it is useful to draw attention to experiential and material similarities that can be found among and between a number of women. This is not to essentialize the experience of this group, to delimit their subjectivity through specific terms and 'conditions' (such as motherhood). It is rather to draw a temporary, purposeful and strategic conceptual boundary around them, to pick out, highlight and explore 'family resemblances' in order to better understand the needs and experiences of this 'group'. Drawing attention to particular points of contact does not, however, mean that we are oblivious to other resemblances and distinctions. And, despite clear commonalities beyond gender (relative youth and motherhood), any definition of 'woman' capable of including and speaking to the experience of all of even this small group of respondents would have to be 'imprecise'. Above all, we could not assume that gender was a matter of utmost priority in these women's lives, and it should be clear that, for alternative political purposes, we might wish to draw attention to and prioritize aspects of their lives as women over and above their gender.

The extent to which issues of subjectivity can be brought to the fore, or fade to the background, depends on complex and inconsistent contextual factors and we may wish to elide certain differences and emphasize others for a particular reason. Cultural geographers have therefore found they need to approach the categories 'women' and 'men' flexibly, and that while gender may be a primary concern of the researcher, in order to engage with respondents' own priorities it may need to be viewed *through* the more predominant prism of, for example, race or class. The definition of the gendered subject that is employed therefore needs to be imprecise enough to allow diversity at some times but a form of specificity at others, for example, when we engage with issues of particular relevance to young black single mothers, or grandmothers who are disabled and working class. Conceptualized in terms of family resemblance, the categories 'women' and 'men' allow for such shifts to take place. Moreover, they remain faithful to the ways in which such concepts tend to operate in the contexts of everyday life.

In a variety of ways the theme of difference has proven to be both contentious and productive. It has had a powerful and a profound influence on analyses of gender that deploy geographical and cultural imaginations, through which a complex fracturing of gendered subjectivities has been illuminated. As our discussion has shown, attending to difference has fostered an understanding of subjectivities and spaces as mutually constituted. Moreover the strategic imprecision for which we have argued suggests an inherent ambiguity and a necessary fuzziness about gender and other categories. A wide range of spatial metaphors has been used to explore these ideas, and the theme of difference can be thought of as invoking complex topologies and multidimensional spaces (Pratt, 1992; Price-Chalita, 1994). While this suggests a much more complicated relationship between subjectivities and environments than emerged from the theme of equality, it still implies that, however complexly interwoven, the two are analytically separable. In the next section we outline ways of thinking about gender that further problematize this boundary.

DE-LIMITING GENDER

In the second section we argued that the sex–gender distinction enabled feminists to insist that the categories 'women' and 'men' are produced within a cultural as opposed to a biological realm; in the third section we showed how women's autonomy has the potential to challenge normative views of subjectivity; and in the fourth section we argued that the concept of gender has served as a flexible container for difference. In this section we discuss ways in which the theme of deconstruction serves to unsettle the dualistic underpinnings of such conceptualizations.

As we have stressed, the understandings of gender associated with a politics of equality stem from the adoption of a distinction between sex and gender. This distinction sets up a series of binary oppositions, in which gender is associated with the cultural and with multiplicity, while sex is associated with biology and with homogeneity. This dualistic framework has become deeply problematical for feminists (Edwards, 1989). Distinguishing between sex and gender opens up the thorny question about the relationship between them and this is transposed onto the relationship between biology (or 'nature') and culture (including 'nurture'). Within the framework of binary oppositions, there are two broad possibilities: either gender can be understood as the cultural elaboration or embellishment of what is given in nature, or gender radically transcends the matter of nature (Connell, 1985). Both these possibilities suffer serious consequences. If gender is produced by the cultural elaboration of an underlying biology then we are drawn back repeatedly to arguments about precisely where sex ends and gender begins, with biology (and

therefore sex) understood as existing independently of, and prior to, culture. If gender floats free of the matter of nature, this implies that gender is an attribute of the mind while sex is an attribute of the body. But why, if the mind transcends the body, should the mind be divided by gender? In practice, the notion of a mind operating untrammelled by bodily concerns has been forcefully critiqued by feminists as a fiction, which covertly positions attributes of maleness as human norms, and through which gender inequalities are enforced and reinforced (Battersby, 1998; Butler, 1990; Grosz, 1994; Hekman, 1992; Jay, 1981; Lloyd, 1984; 1989; Moore, 1994; D. Smith, 1993).

Analytically, therefore, gender categories are highly unstable or contingent (Butler, 1991; Felski, 1997; Harding, 1986). Lacking any stable content, the categories 'women' and 'men' acquire meaning only through their use in particular contexts: they are 'produced' through our living of them. In this sense the categories are always 'fictions' upon which we draw routinely (see Butler, 1990; Riley, 1988).

This understanding situates gender in everyday lives and everyday contexts: we live our lives as 'women' or as 'men' through our routine practices played out in the places we inhabit. Put another way, the power of particular fictions of gender resides in our enactment of particular ways of being 'women' or 'men' in particular places, and dominant understandings of gender may be unsettled or brought into question if our routine practices diverge from these fictions in significant ways. Within this framework gender is inseparable from its contexts. It is, moreover, impossible to imagine 'men' and 'women' except as embodied – the examples in the first section of encountering individuals whose appearances or voices leave us uncertain about whether they are 'male' or 'female' assume and make central the bodies of those with whom we interact – and feminist geographers have therefore explored potentially disruptive embodied performances of gender in a variety of settings including corporate workplaces (McDowell, 1995; McDowell and Court, 1994) and gyms (Johnston, 1996) amongst many others.

Robyn Longhurst (2000), for example, demonstrates the extent to which pregnant women are expected to behave in public space in a manner 'becoming' to their 'condition', and correspondingly, how disruptive challenges to such expectations can be. Her respondents describe how their behaviour is judged against exacting standards of 'appropriateness', and how their actions are restricted and censured accordingly. 'Heavily' pregnant women in particular are seriously discouraged from entering certain environments (such as bars), taking part in certain activities (such as rugby), or 'displaying themselves', that is entering 'public' space, in clothing considered 'revealing'. Women who breach such normative and normalizing standards of expectant feminine decency can find themselves subject to moral outrage and indignation, and this is illustrated pointedly by Longhurst's (2000) account of a New Zealand 'pregnant bikini competition'. Letters to the editor of a newspaper that reported the event used evocative terms such as 'spectacle', 'abhorrent' and 'shame' to express their response to photographs of these 'pregnant women with attitude' (2000: 463). One respondent was moved to 'draw frocks' on them, presumably to reinstate their misplaced modesty. What this episode reveals is that atypical and unexpected acts can test and potentially stretch the limits of 'public decency', and the modes of 'doing' gender that it is un/able to contain. The 'scantily clad' and pregnant presence of women in a city centre is a contestatory and controversial act capable of subverting accepted ideas of how a particular manifestation of gender – that of a woman during pregnancy – 'ought' to be performed.

This example shows how both regulatory norms and performances of gender are context specific. It also implies a separability between gendered subjectivities and contexts: the places of public space are construed as 'external' to the gendered subjects who perform and negotiate the norms through which spaces, bodies and subjectivities acquire and articulate gendered meanings. But some important studies in cultural geography have argued that people and places are imagined, embodied and experienced in ways that are more radically and inextricably intertwined with each other (Davidson, 2000a; 2000b; Kirby, 1996; Nast and Pile, 1998). That is to say, subjects and environments exist, not as definitively bounded entities in proximity with each other, but as dynamically interconnected and mutually constitutive.

This intertwining of subject and environment is not 'normally' brought into conscious awareness, but 'problematic' or 'phobic' experiences of space, including for example agoraphobia and vertigo, serve to highlight what is usually taken for granted (Bordo et al., 1998; Kirby, 1996). The experiences drawn upon in such studies are themselves highly gendered. Of agoraphobia Esther da Costa Meyer writes '[i]ts connection to women is beyond dispute', and she argues that it can be understood to 'allegorize the sexual division of labour [hence the notion of agoraphobia as the 'housewife's disease'] and the inscription of social as well as sexual difference in urban space … Agoraphobia represents a virtual parody of twentieth-century constructions of femininity' (1996: 141, 149).

In this context more detailed analyses of experiences of agoraphobia have explored how (gendered) bodies and spaces are experienced in panic attacks, which precipitate and help to maintain agoraphobic geographies (Bankey, 2001; Bekker, 1996; Davidson, 2001). Panic can be understood to express a radical and horrifying disintegration of boundaries around selves which ordinarily provide subjects with a sense of being safely delimited and demarcated from surrounding spaces. When such boundedness is thrown into question by problematizing phenomena such as panic or vertigo, the fragility and imagined nature of taken-for-granted boundaries become excruciatingly evident (Davidson, 2001). Such sensory realization can be deeply disturbing to selves for whom the regulatory fiction of existing as bounded, autonomous isolates, in proximity with but separable from their environs, is deeply felt.

Experience of such boundary crises typically prompts sufferers to seek out environments likely to engage soothingly, rather than disturbingly, with their sense of themselves. For many who suffer agoraphobic panic, this entails avoidance of populous and otherwise sensorially stimulating places such as restaurants, shopping malls and public transport, where the volume and intensity of interactions with sentient and non-sentient others is often experienced as profoundly corrosive of already fragile boundaries presumed to contain the subject (Davidson, 2001). Ironically, given the popular misconception that agoraphobia involves a fear of open spaces, safe spaces for the agoraphobic subject are often found in quiet countryside or parkland, or at night, when the potential for others to impinge in this way is reduced. More often (and stereotypically), however, sufferers will feel safest in the more predictable spaces of their homes, whose walls can serve to reinforce their fragile and weakened boundaries. In a sense, the very bricks and mortar are imbued with and have power to imbue feelings of safety and security for the subject within.

Agoraphobic panic can be understood to constitute a kind of boundary dispute, a phenomenal discordance that occurs in and through the relations between (gendered) people and places. It is thus well placed to reveal something of the nature of this 'betweenness', given that, in so far as the source and sense of panic are locatable at all, they arise neither internally nor externally in relation to the subject, but rather from the boundary between. As a boundary crisis, panic undermines the distinction between subjects and environments, underscores their inescapable interrelation and reveals the 'normal' sense of separateness from our surrounds to be a partly fictitious, imagined distinction. Our sense of this discrimination – between what is self and what is not – is maintained by habit and confidence that the world will behave according to our projections and expectations, that is, by a kind of 'ontological security'. Once this has been disrupted, our 'imaginings' inevitably alter.

This research on agoraphobic geographies and subjectivities contributes to the deconstruction of conceptual separations between subjects and environments by showing that there is no possibility of subject or experience without situation. Whether we are aware of this or not, each is continually co-constitutive of the other. To *be* is to be some*where*, and our changing relations and interactions with this placing are integral to understandings of human geographies. Moreover, gender is inscribed deeply within these processes. So, even as dualistic separations between people and places are brought into question, binaries persist, in this instance producing family resemblances in the mutual imbrication of subjectivity and space.

CONCLUSION: ON THE PARADOXES OF GENDER

In this chapter we have explored the cultural and geographical imaginings of gender in relation to four themes which have been influential in feminisms within and beyond cultural geography. We have emphasized throughout that binary gender categories are both deeply problematic and extremely pervasive. On the one hand, to think about gender productively, we need to escape from the confining straitjacket of binary oppositions, whether of sex and gender, or body and mind, or biology and culture, or subject and environment, and so on. But on the other hand, it is important to acknowledge that the concepts we deploy, such as gender, are often, if not always, linked to such binaries. In other words,

dualistic thinking necessarily clings to the concept of gender. Consequently we are not arguing that cultural geographers need to move 'beyond' binary thinking, or 'beyond' the themes of equality, autonomy, difference and deconstruction. We wish instead to problematize the very idea of going 'beyond', and in so doing to encourage further discussion of the political and spatial commitments implicit in the various conceptualizations of gender deployed in cultural geography.

Notions of progress are very influential in academic research and scholarship. Reviews of existing bodies of literature often conclude by identifying gaps or research 'frontiers', or by arguing for new research agendas. Although such arguments are not necessarily presented as normative claims about a field of research, they nevertheless encourage a view of knowledge as developing in a discernible direction. Several characteristics follow from this, including the tendency for specialist subfields to emerge successively, and a view of the scholar or researcher as capable of (re)viewing a subfield as a whole. We would argue that any review, such as this one, is paradoxical: it is necessarily situated, and yet to perform the function of a review it also claims the capacity to delimit and thematize a body of knowledge. Tensions or paradoxes of this kind are pervasive in academic scholarship. We would argue that fostering and working with such paradoxes is a highly productive way of doing cultural geography.

In conclusion we therefore argue for the acknowledgement and strategic 'acceptance' of paradoxes on a number of different levels. We wish to encourage cultural geographers to reflect on the interplay, tensions and contradictions between them, to recognize that we need to work with paradoxes, and that their inherently problematic nature can be used to positive advantage, enabling forms of academic practice that simultaneously perform and subvert established ways of thinking and writing about gender. 'Doing' and theorizing gender in such ways will necessarily entail 'doing' and theorizing space and place as well. For example, as we have shown, conceptualizing subjects and environments as separate has been enormously fruitful for understanding cultural geographies of gender, but so too are understandings of subject and environment as inseparable. Exploring (rather than resolving) the contradiction between separability and inseparability is likely to generate creative ways of thinking about spaces and subjectivities.

NOTE

1 The first person plural is an ambiguous form of address. We refuse the self-alienation of the third person when we are numbered among those to whom we refer. But 'we', 'us' and 'our' provoke questions about who is included and excluded. We do not claim to speak for humanity in general; we do not presume to speak for our readers. Rather we invite you to identify yourself in relation to our voice, whether with, against or elsewhere. Through this text, but beyond our awareness, we enter into the production of our own and other subject positions (Bondi, 1997; Card, 1991).

REFERENCES

Ahmed, S. (1998) *Differences that Matter: Feminist Theory and Postmodernism*. Cambridge: Cambridge University Press.

Aisenberg, N. and Harrington, M. (1988) *Women of Academe: Outsiders in the Sacred Grove*. Amherst: University of Massachusetts Press.

Bankey, R. (2001) 'La donna è mobile: constructing the irrational woman', *Gender, Place and Culture* 8 (1): 37–54.

Battersby, C. (1998) *The Phenomenal Woman: Feminist Metaphysics and the Patterns of Identity*. Cambridge: Polity.

Bekker, M.H.J. (1996) 'Agoraphobia and gender: a review', *Clinical Psychology Review* 16 (2): 129–46.

Bell, D. and Valentine, G. (eds) (1995) *Mapping Desire*. London: Routledge.

Bhavnani, K.-K. (ed.) (2001) *Feminism and Race*. Oxford: Oxford University Press.

Binnie, J. and Valentine, G. (1999) 'Geographies of sexuality: a review of progress', *Progress in Human Geography* 23: 175–87.

Blomley, N. (1994) 'Activism and the academy', *Environment and Planning D: Society and Space* 12: 383–5.

Blomley, N. (1996) 'I'd like to dress her all over: masculinity, power and retail space', in N. Wrigley and M. Lowe (eds) *Retailing, Consumption and Capital*. Harlow: Longman. pp. 239–56.

Bondi, L. (1990) 'Feminism, postmodernism and geography: space for women?', *Antipode* 22: 156–67.

Bondi, L. (1992) 'Gender symbols and urban landscapes', *Progress in Human Geography* 16: 157–70.

Bondi, L. (1997) 'In whose words? On gender identities and writing practices', *Transactions of the Institute of British Geographers* 22: 245–58.

Bondi, L. (1998a) 'Gender, class and urban space', *Urban Geography* 19: 160–85.

Bondi, L. (1998b) 'Sexing the city', in R. Fincher and J.M. Jacobs (eds) *Cities of Difference*. New York: Guilford. pp. 177–200.

Bondi, L. (2002) 'Gender, place and culture': paradoxical spaces?', in P. Moss (ed.) *Feminist Geography in Practice: Research and Methods*. Oxford: Blackwell. pp. 80–6.

Bondi, L. and Domosh, M. (1992) 'Other figures in other places: on feminism, postmodernism and geography', *Environment and Planning D: Society and Space* 10: 199–213.

Bondi, L. and Domosh, M. (1998) 'On the contours of public space: a tale of three women', *Antipode* 30: 270–89.

Bondi, L., Avis, H., Bankey, R., Bingley, A., Davidson, J., Duffy, R., Einagel, V.I., Green, A.-M., Johnston, L., Lilley, S., Listerborn, C., McEwan, S., Marshy, M., O'Connor, N., Rose, R. and Vivat, B. (2002) *Subjectivities, Knowledges and Feminist Geographies*. London: Rowman and Littlefield.

Bordo, S., Klein, B. and Silverman, M.K. (1998) 'Missing kitchens', in H.J. Nast and S. Pile (eds) *Places through the Body*. London: Routledge. pp. 72–92.

Butler, J. (1990) *Gender Trouble*. London: Routledge.

Butler, J. (1991) 'Contingent foundations: feminism and the question of "postmodernism"', *Praxis International* 11: 150–65.

Carby, H. (1982) 'White woman listen! Black feminism and the boundaries of sisterhood', in Centre for Contemporary Cultural Studies (ed.) *The Empire Strikes Back: Race and Racism in 70s Britain*. London: Hutchinson. pp. 212–35.

Card, C. (1991) 'The feistiness of feminism', in C. Card (ed.) *Feminist Ethics*. Kansas: University Press of Kansas. pp. 3–31.

Chouinard, V. (1999) 'Life at the margins: disabled women's explorations of ableist spaces', in E.K. Teather (ed.) *Embodied Geographies: Spaces, Bodies and Rites of Passage*. London: pp. 142–56.

Chouinard, V. and Grant, A. (1995) 'On not being anywhere near the "project": revolutionary ways of putting ourselves in the picture', *Antipode* 27: 137–66.

Connell, R.W. (1985) *Gender and Power*. Cambridge: Polity.

Coward, R. (1984) *Female Desire: Women's Sexuality Today*. London: Paladin.

Cream, J. (1995) 'Re-solving riddles: the sexed body', in D. Bell and G. Valentine (eds) *Mapping Desire*. London: Routledge. pp. 31–40.

Cresswell, T. (1996) *In Place/Out of Place*. Minneapolis: Unversity of Minnesota Press.

da Costa Meyer, E. (1996) 'La donna è mobile: agoraphobia, women, and urban space', in D. Agrest, P. Conway and L. Kanes Weisman (eds) *The Sex of Architecture*. New York: Abrams. pp. 141–56.

Davidson, J. (2000a) ' "... the world was getting smaller": women, agoraphobia and bodily boundaries', *Area* 32 (1): 31–40.

Davidson, J. (2000b) 'A phenomenology of fear: Merleau-Ponty and agoraphobic life-worlds', *Sociology of Health and Illness* 22 (5): 640–60.

Davidson, J. (2001) 'Fear and trembling in the mall: women, agoraphobia and body boundaries', in I. Dyck, N. Davis Lewis and S. McLafferty (eds) *Geographies of Women's Health*. London: Routledge. pp. 213–30.

Davidson, J. and Smith, M. (1999) 'Wittgenstein and Irigaray: gender and philosophy in a language (game) of difference', *Hypatia* 14 (2): 72–96.

de Beauvoir, S. (1997) *The Second Sex* (1949), trans. by H.M. Prashley. London: Vintage.

Delph-Januirek, T. (1999) 'Sounding gender(ed): vocal performances in English university teaching spaces', *Gender, Place and Culture* 6 (2): 137–53.

Deutsche, R. (1991) 'Boys' town', *Environment and Planning D: Society and Space* 9: 5–30.

Domosh, M. (1996) 'The feminized retail landscape: gender ideology and consumer culture in nineteenth century New York City', in N. Wrigley and M. Lowe (eds) *Retailing, Consumption and Capital*. Harlow: Longman. pp. 257–70.

Domosh, M. (1998) 'Geography and gender: home, again?', *Progress in Human Geography* 22: 276–82.

Domosh, M. (1999) 'Sexing feminist geography', *Progress in Human Geography* 23 (3): 429–36.

Domosh, M. (2000) 'Unintended transgressions and other reflections on the job search process', *Professional Geographer* 52: 703–8.

Dowling, R. and Pratt, G. (1993) 'Home truths: recent feminist constructions', *Urban Geography* 14: 464–75.

Duncan, N. (1996) 'Renegotiating gender and sexuality in public and private spaces', in N. Duncan (ed.) *BodySpace*. London: Routledge. pp. 127–45.

Edwards, A. (1989) 'The sex/gender distinction: has it outlived its usefulness?', *Australian Feminist Studies* 10: 1–12.

Evans, J. (1994) *Feminist Theory Today*. London: Sage.

Felski, R. (1997) 'The doxa of difference', *Signs: Journal of Women in Culture and Society* 23: 1–21.

Firestone, S. (1972) *The Dialectic of Sex*. London: Paladin.

Fraser, N. (1995) 'From redistribution to recognition? Dilemmas of justice in a "post-socialist" age', *New Left Review* 212: 68–93.

Fraser, N. (1997) *Justice Interruptus: Critical Reflections on the "Post-Socialist" Condition*. London: Routledge.

Fraser, N. (2000) 'From rethinking recognition', *New Left Review* 3: 107–20.

Fuss, D. (1989) *Essentially Speaking*. London: Routledge.

Gilligan, C. (1982) *In a Different Voice: Psychological Theory and Women's Development*. Cambridge, MA: Harvard University Press.

Grant, A. (1998) 'Unwomanly acts: struggling over sites of resistance', in R. Ainley (ed.) *New Frontiers of Spaces, Bodies and Gender*. London: Routledge. pp. 50–61.

Greer, G. (1971) *The Female Eunuch*. London: Paladin.

Gregson, N. and Crewe, L. (1998) 'Dusting down *Second Hand Rose*: gendered identities and the world of second-hand goods in the space of the car boot sale', *Gender, Place and Culture*. 5: 77–100.

Gregson, N. and Lowe, M. (1995) ' "Home"-making: on the spatiality of daily social reproduction in contemporary middle-class Britain', *Transactions of the Institute of British Geographers* 20: 224–35.

Griffiths, M. (1995) *Feminism and the Self: The Web of Identity*. London: Routledge.

Grimshaw, J. (1999) 'Working out with Merleau-Ponty', in J. Arthurs and J. Grimshaw (eds) *Women's Bodies: Discipline and Transgression*. London: Cassell. pp. 91–118.

Grosz, E. (1990) 'The body of signification', in J. Fletcher and A. Benjamin (eds), *Abjection, Melancholia and Love: The Work of Julia Kristeva*. London and NewYork: Routledge. pp. 80–104.

Grosz, E. (1994) *Volatile Bodies*. Bloomington: Indiana University Press.

Hannam, K. and Shurmer-Smith, P. (1994) *Worlds of Desire, Realms of Power: A Cultural Geography*. London: Arnold.

Hanson, S. (2000) 'Networking', *Professional Geographer* 52: 751–8.

Hanson, S. and Pratt, G. (1995) *Gender, Work and Space*. London: Routledge.

Haraway, D. (1988) 'Situated knowledges: the science question in feminism and the privilege of partial perspective', *Feminist Studies* 14: 575–99.

Haraway, D. (1991) *Simians, Cyborgs and Women: the Reinvention of Nature*. London: Free Association.

Harding, S. (1986) 'The instability of the analytical categories of feminist theory', *Signs: Journal of Women in Culture and Society*. 11: 645–64.

Hayford, A. (1973) 'The geography of women: an historical introduction', *Antipode* 5: 26–33.

Hekman, S.J. (1992) *Gender and Knowledge: Elements of a Postmodern Feminism*. Cambridge: Polity.

Hochschild, A.R. (1983) *The Managed Heart*. Berkeley: University of California Press.

hooks, b. (1989) *Talking Back*. Boston: South End Press.

Hubbard, P. (1998) 'Sexuality, immorality and the city: red-light districts and the marginalisation of street prostitutes', *Gender, Place and Culture* 5 (1): 55–76.

Hubbard, P. (2000) 'Desire/disgust: mapping the moral contours of heterosexuality', *Progress in Human Geography* 24: 191–217.

Irigaray, L. (1993) *Ethics of Sexual Difference*. London: Athlone.

Jay, N. (1981) 'Gender and dichotomy', *Feminist Studies* 7: 38–56.

Johnston, L. (1995) 'Reading the sexed bodies and spaces of gyms', in H.J. Nast and S. Pile (eds) *Places through the Body*. London: Routledge. pp. 244–62.

Johnston, L. (1996) 'Flexing femininity: female body builders refiguring the body', *Gender, Place and Culture* 3: 327–40.

Katz, C. (1994) 'Playing the field: questions of fieldwork in geography', *The Professional Geographer* 46: 67–72.

Katz, C. (1996) 'Towards minor theory', *Environment and Planning D: Society and Space* 14: 487–99.

Kirby, K. (1996) *Indifferent Boundaries*. New York: Guilford.

Kobayashi, A. (1994) 'Coloring the field: gender, "race", and the politics of fieldwork', *The Professional Geographer* 46: 73–80.

Kolodny, A. (1975) *The Lay of the Land*. Chapel Hill: University of North Carolina Press.

Koskela, H. (1997) 'Bold walk and breakings: women's spatial confidence versus fear of violence', *Gender, Place and Culture* 4 (3): 301–19.

Law, R. (1999) 'Beyond "women and transport": towards new geographies of gender and daily mobility', *Progress in Human Geography* 23 (4): 567–88.

Lloyd, G. (1984) *The Man of Reason*. London: Methuen.

Lloyd, G. (1989) 'Woman as other: sex, gender and subjectivity', *Australian Feminist Studies* 10: 13–22.

Longhurst, R. (1995) 'The body and geography', *Gender, Place and Culture* 2: 97–106.

Longhurst, R. (1997) '(Dis)embodied geographies', *Progress in Human Geography* 21: 486–501.

Longhurst, R. (2000) '"Corporogeographies" of pregnancy: "bikini babes"', *Environment and Planning D: Society and Space* 18: 453–72.

Marston, S. (2000) 'The social construction of scale', *Progress in Human Geography* 24 (2): 219–42.

Massey, D. (1991) 'Flexible sexism', *Environment and Planning D: Society and Space* 9: 31–57.

Massey, D. (1995) 'Masculinity, dualisms and high technology', *Transactions of the Institute of British Geographers* 20: 487–99.

Maxey, I. (1999) 'Beyond boundaries? Activism, academia, reflexivity and research', *Area* 31 (3): 199–208.

McDowell, L. (1979) 'Women in British geography', *Area* 11: 151–4.

McDowell, L. (1990) 'Sex and power in academia', *Area* 22: 323–32.

McDowell, L. (1991) 'The baby and the bath water: diversity, deconstruction and feminist theory in geography', Geoforum 22: 123–33.

McDowell, L. (1992) 'Doing gender: feminism, feminists and research methods in human geography', *Transactions of the Institute of British Geographers* 17: 399–416.

McDowell, L. (1993a) 'Space, place and gender relations: Part I. Feminist empiricism and the geography of social relations', *Progress in Human Geography* 17: 157–79.

McDowell, L. (1993b) 'Space, place and gender relations: Part II. Identity, difference, feminist geometries and geographies', *Progress in Human Geography* 17: 305–18.

McDowell, L. (1995) 'Body work: heterosexual gender performances in city workplaces', in D. Bell and G. Valentine (eds) *Mapping Desire*. London: Routledge. pp. 75–95.

McDowell, L. (1997) *Capital Culture*. Oxford: Blackwell.

McDowell, L. (1999) *Gender, Identity and Place*. Cambridge: Polity.

McDowell, L. and Court, G. (1994) 'Performing work: bodily representations in merchant banks', *Environment and Planning D: Society and Space* 12: 727–50.

McDowell, L. and Peake, L. (1990) 'Women in British Geography revisited: or the same old story', *Journal of Geography in Higher Education* 14: 19–30.

McDowell, L. and Sharp, J.P. (eds) (1997) *Space, Gender, Knowledge*. London: Arnold.

McEwan, S. et al. (2002) 'Crossing boundaries: gendered spaces and bodies in golf', in L. Bondi et al. (eds) *Subjectivities, Knowledges and Feminist Geographies*. Lanham: Rowman and Littlefield. pp. 90–105.

Mitchell, J. (1971) *Woman's Estate*. Harmondsworth: Penguin.

Mitchell, J. (1984) *Women: The Longest Revolution*. London: Virago.

Mohanty, C. (1988) 'Under Western eyes: feminist scholarship and colonial discourses', *Feminist Review* 30: 61–88.

Mohanty, C.T., Russo, A. and Torres, L. (eds) (1991) *Third World Feminism and the Politics of Feminism*. Bloomington: Indiana University Press.

Moi, T. (1999) *What is a Woman? And Other Essays*. Oxford: Oxford University Press.

Monk, J. and Hanson, S. (1982) 'On not excluding half of the human in human geography', *The Professional Geographer* 34: 11–23.

Moore, H. (1988) *Feminism and Anthropology*. Cambridge: Polity.

Moore, H. (1994) *A Passion For Difference*. Cambridge: Polity.

Morin, K. with Longhurst, R. and Johnston, L. (2001) '(Troubling) spaces of mountains and men: New Zealand's Mount Cook and Hermitage', *Social and Cultural Geography*. 2: 117–39.

Morley, L. and Walsh, V. (1996) *Breaking Boundaries*. London: Taylor and Francis.

Morris, M. (1988) *The Pirate's Fiancée*. London: Verso.

Munt, S. (1998) 'Sisters in exile: the lesbian nation', in Rosa Ainley (ed.) *New Frontiers of Space, Bodies and Gender*. London: Routledge. pp. 3–19.

Nairn, K. (1997) 'Hearing from quiet students: the politics of silence and voice in geography classrooms', in J.P. Jones III, H.J. Nast and S.M. Roberts (eds) *Thresholds in Feminist Geography*. Lanham: Rowman and Littlefield. pp. 93–115.

Nast, H.J. (1998) 'Unsexy geographies', *Gender, Place and Culture* 5 (2): 191–206.

Nast, H.J. and Pile, S. (eds) (1998) *Places through the Body*. London: Routledge.

Nast, H.J. and Pulido, L. (2000) 'Resisting corporate multiculturalism: mapping faculty initiatives and institutional-student harassment in the classroom', *Professional Geographer* 52: 722–37.

Nicholson, L. (ed.) (1990) *Feminism/Postmodernism*. London: Routledge.

Oakley, A. (1972) *Sex, Gender and Society*. London: Maurice Temple Smith.

Penrose, J., Bondi, L., McDowell, L., Kofman, E., Rose, G. and Whatmore, S. (1992) 'Feminists and feminism in the academy', *Antipode* 24 (3): 218–37.

Phillips, A. (1997) 'From inequality to difference: a severe case of displacement', *New Left Review* 224: 143–53.

Phoenix, A. (1995) 'Practising feminist research: the intersection of gender and "Race" in the research process', in M. Maynard and J. Purvis (eds) *Researching Women's Lives From a Feminist Perspective*. London: Taylor and Francis. pp. 49–71.

Pratt, G. (1992) 'Spatial metaphors and speaking positions', *Environment and Planning D: Society and Space* 10: 241–4.

Pratt, G. and Hanson, S. (1994) 'Geography and the construction of difference', *Gender, Place and Culture* 1 (1): 5–30.

Pratt, M.B. (1984) 'Identity, skin blood heart', in *Yours in Struggle: Three Feminist Perspectives on Anti-Semitism and Racism*. Brooklyn: Long Haul. pp. 11–63.

Price-Chalita, P. (1994) 'Spatial metaphor and the politics of empowerment: mapping a place for feminism and postmodernism in geography?', *Antipode* 26: 236–54.

Pringle, R. (1989) *Sectretaries Talk*. London: Verso.

Ramazanoglu, C. (1989) *Feminism and the Contradictions of Oppression*. London: Routledge.

Riley, D. (1988) *Am I That Name?* London: Macmillan.

Roberts, M. (1991) *Living in a Man-Made World*. London: Routledge.

Rose, G. (1993) *Feminism and Geography*. Cambridge: Polity.

Rose, G. (1996) 'As if the mirror had bled: masculine dwelling, masculinist theory and feminist masquerade', in N. Duncan (ed.) *Body Space: Destabilizing Geographies of Gender and Sexuality*. London: Routledge. pp. 56–74.

Rose, G. (1999) 'Performing the body', in D. Massey, J. Allen and P. Sarre (eds) *Human Geography Today*. Cambridge: Polity. pp. 247–59.

Rowbotham, S. (1989) *The Past Is Before Us*. London: Pandora.

Ruddick, S. (1996) 'Constructing difference in public spaces: race, class, and gender as interlocking systems', *Urban Geography* 17 (2): 132–51.

Sargisson, L. (1996) *Contemporary Feminist Utopianism*. London: Routledge.

Scott, J.W. (1999) 'Some reflections on tender and politics', in M. Marx, J. Lorber and M.B. Hess (eds), *Revising Gender*. London: Sage. pp. 70–98.

Seager, J. (1993) *Earth Follies: Feminism, Politics and the Environment*. London: Earthscan.

Seager, J. (2000) '"And a charming wife": gender, marriage and manhood in the job search process', *Professional Geographer* 52: 709–21.

Shapiro, J. (1991) 'Transexualism: reflections on the persistence of gender and the mutability of sex', in J. Epstein and K. Straub (eds) *Body Guards*. London: Routledge. pp. 248–79.

Shugar, D.R. (1995) *Separatism and Women's Community*. Lincoln: University of Nebraska Press.

Skeggs, B. (1997) *Formations of Class and Gender*. London: Sage.

Skelton, T. (1998) 'Ghetto girls/urban music: Jamaican ragga music and female performance', in R. Ainley (ed.) *New Frontiers of Space, Bodies and Gender*. London: Routledge. pp. 142–54.

Smith, D. (1993) *Texts, Facts and Femininity: Exploring the Relations of Ruling*. London: Routledge.

Smith, N. (1993) 'Homeless/global: scaling places', in J. Bird, B. Curtis, T. Putnam, G. Robertson and L. Tickner (eds) *Mapping the Futures*. London: Routledge. pp. 87–119.

Soper, K. (1990) 'Feminism, humanism and postmodernism', *Radical Philosophy* 55: 11–17.

Spain, D. (1992) *Gendered Spaces*. Chapel Hill: University of North Carolina Press.

Spelman, E. (1988) *Inessential Woman: Problems of Exclusion in Feminist Thought*. Boston: Beacon.

Stoller, R. (1968) *Sex and Gender*. London: Hogarth.

Stoller, R. (1985) *Presentations of Gender*. New Haven: Yale University Press.

Taylor, A. (1998) 'Lesbian space: more than one imagined territory', in R. Ainley (ed.) *New Frontiers of Space, Bodies and Gender*. London: Routledge. pp. 129–41.

Tivers, J. (1978) 'How the other half lives: the geographical study of women', *Area* 10: 302–6.

Valentine, G. (1996) '(Re)negotiating the "heterosexual street": lesbian productions of space', in N. Duncan (ed.) *BodySpace: Destabilizing Geographies of Gender and Sexuality*. London: Routledge. pp. 146–55.

Valentine, G. (1997) 'Making space: separatism and difference', in J.P. Jones III, H.J. Nast and S.M. Roberts (eds) *Thresholds in Feminist Geography*. Lanham: Rowman and Littlefield. pp. 65–76.

Wakeford, N. (1999) 'Gender and the landscapes of computing in an internet café', in M. Crang, P. Crang and J. May (eds) *Virtual Geographies: Bodies, Space and Relations*. London: Routledge. pp. 178–201.

Warner, S. (2000) 'Feminist theory, the women's liberation movement and therapy for women: changing our concerns', *Changes* 18 (4): 232–43.

Weir, A. (1997) *Sacrificial Logics*. London: Routledge.

West, C. and Fenstermaker, S. (1995) 'Doing difference', *Gender and Society* 9: 8–37.

Whitford, M. (1991) *Luce Irigaray: Philosophy in the Feminine*. London: Routledge.

Widerberg, K. (1999) 'Translating gender', *NORA: Nordic Journal of Women's Studies* 6: 133–8.

Williamson, J. (1980) *Consuming Passions: The Dynamics of Popular Culture*. London: Marion Boyars.

Wittgenstein, L. (1988) *Philosophical Investigations*. Oxford: Blackwell.

Wittig, M. (1992) *The Straight Mind and Other Essays*. Hemel Hempstead: Harvester Wheatsheaf.

Women and Geography Study Group of the IBG (1984) *Geography and Gender*. London: Hutchinson.

Women and Geography Study Group (1997) *Feminist Geographies: Explorations in Diversity and Difference*. Harlow: Addison Wesley Longman.

Woodward, R. (1998) '"It's a man's life!": soldiers, masculinity and the countryside', *Gender, Place and Culture* 5 (3): 277–300.

Young, I.M. (1990) *Throwing Like a Girl and Other Essays in Feminist Philosophy and Social Thought*. Bloomington: Indiana University Press.

Young, I.M. (1997) 'Unruly categories: a critique of Nancy Fraser's dual systems theory', *New Left Review* 222: 147–60.

Section 6

AFTER EMPIRE   Edited by Jane M. Jacobs

Introduction: After Empire?

Jane M. Jacobs

In my former hometown of Melbourne, Australia, the audaciously modern lines of the recently opened Melbourne Museum (2000) sit stridently alongside the proud Victorian architecture of the Royal Exhibition Building (1880). Separated by over a century, these two buildings obviously reflect the different architectural sensibilities that adhere to their times. But their architectural difference belies a set of commonalities that tell us much about the idea of empire, and particularly as it expressed itself through the familiar cases of European imperial expansion from the seventeenth through to the early twentieth century. The exhibition and the museum are emblematic institutions of the age of empire (Driver and Gilbert, 1999). Exhibitions were forward looking performatives that celebrated the futures promised by imperial expansion. Museums, on the other hand, collected and displayed all that needed to be salvaged in the wake of civilization's spread. Nowadays Melbourne's Royal Exhibition Building houses a motley range of consumer display events, while its new museum is a key tourist attraction. Such are the fortunes of the institutions of empire.[1]

Being postcolonial, the Melbourne Museum can no longer comfortably display the objects of 'others' as an imperial museum might have. Instead it self-consciously curates stories and things animated by the experiences and postcolonial agendas of those 'others'. The visitor to this museum, keen to see something of 'traditional indigenous culture', may well be drawn to a display that, at first glance, seems to offer a modest collection of spears. The display does not, however, tell the viewer anything about the 'origins' of those spears in an anthropological sense. These spears are part of a self-conscious display that tells the story of the museum's own imperial collecting practices: who collected, how they collected, and the transnational circuits of trade and exchange that existed within the museum (as opposed to the indigenous) world. The display reflects the ambivalence created when an imperial institution, like the museum, finds itself in postcolonial times. The story it tells of collecting derives from the kind of analytical perspective delivered into contemporary scholarship primarily by way of Said's critique of cultures of imperialism. The objects at the centre of the exhibit (anthropologists, their collecting equipment and their traded artifacts) occupy the self-reflexive analytic space created by anthropology questioning its ethnographic authority.[2]

Much that needs to be said about the museum as an imperial institution is said through this exhibit. Yet what of the spears that first draw the visitor into the visual orbit of this story? For all of its critical cleverness, this display tells us little of those spears as indigenous possessions. Of course, armed with postcolonial theory we might argue that there is no such thing as a discrete 'Aboriginal object' anymore, and so no need (or way) to place such objects neatly back into an indigenous frame. We might, for example, approve of the way in which this exhibit so usefully displays the unruly afterlife of ethnographic things taken up into global circuits of collecting and consumption (Appadurai, 1996; Thomas, 1994). Yet, for those invested in a postcolonial politics which tries to give 'voice' back to the colonized, the incorporation of these objects into the story of the born-again museum may feel an only too familiar 'silencing'. And if the message of this exhibit is not ambiguous enough, then we need only begin to ask, as Clive Barnett (1997: 139–40) usefully does, what 'silence' might exactly mean in such a context. Is the 'silence' that is registered around the indigenous life of these objects the result of the noise an imperial institution makes when it is being self-reflexive? Or is it a strategic silence resulting from an indigenous choice to stop satisfying non-indigenous romances about traditional Aboriginal life? Are these objects mute, or are they animated by a new language of resistance? Is this a postcolonial museum, or is it an imperial institution whose powers have been enhanced by certain postcolonial practices?

A decade ago Cindi Katz (1992) opened her essay on postcolonialism, new anthropology and geography with a reading of the politics of a museum/art display. In a similar move, I wish to use this example of a postcolonial museum display to bring into sharp relief a range of issues associated with the after-effects of the postcolonial turn upon the discipline of geography – and specifically cultural geography. In the first instance, we might ask how imperial institutions (like the museum) and knowledge frames (like geography) decolonize themselves. What do the kinds of critical historical frameworks embodied by this postcolonial museum display reveal about imperial pasts and what do they continue to conceal? How necessary is something called 'postcolonial theory' to such a process? Who is in charge of such rearrangements of knowledge and

power, and specifically, whose political, ethical and moral objectives are being served by these critical approaches? Is it possible to break structures of 'othering' that so characterize the cultural logic of imperialisms of all kinds? And is this something those 'others', whose postcolonial futures may now depend on claiming 'cultural' difference, actually want?

POSTCOLONIALISM AND GEOGRAPHY

For a discipline like geography, whose past has been so closely linked to empires of old, how it might move towards geographies that have alternative assemblages of power and space remains vexed. The chapters to follow chart the varied ways in which geographers, and others whose interests intersect with the explicitly spatial concerns of the discipline, have attempted such a task. They also indicate some of the difficulties of this kind of process for a disciplinary field (and an academy) that is overdetermined by its practical and institutional histories, its language hegemonies, and its cultures of knowing and reportage. In what follows readers will find generous intellectual pathways through this scholarship. They will also find useful measures of enthusiasm and skepticism about the intellectual field of postcolonial theory that has become so central to the project of decolonizing the discipline.

It will be clear from the chapters to follow that the term 'postcolonialism' refers to much more than formal (political) processes of decolonization. As Brenda Yeoh's contribution shows, postcolonial states remain complexly entangled in their colonial pasts, just as they may be negotiating forms of neo-imperialism, or launching themselves into new transnational possibilities. The persistence of various kinds of colonialism led Anne McClintock (1992: 87) to issue the salutary warning that the term 'postcolonial' was 'prematurely celebratory'. Imperialism is not something that belongs back then, nor is it confined to specific sovereign configurations; nor, for that matter, is it something that necessarily even produces a geopolitical unit called 'empire'.[3] James Sidaway has noted that current geographies exhibit 'an array of postcolonialisms' (2000: 592).

Working primarily in a political geographical register, he goes on to provide a dense compendium of the different types of contemporary internal colonialisms and transnational imperialisms and multiple postcolonialisms. It is both useful and necessary for this kind of diversity to be brought into view. This is especially so when one thinks of the uneven way postcolonial theory has been taken up by anglophonic geography – often in the historical mode, and often in relation to British imperialism. In the context of such diversity it is necessary to think of postcolonialism as a diffusely expressed and ongoing set of effects that come into being alongside of, and in relation to, imperialisms of various kinds (be they new or old, territorially defined or diffuse). That there is the need to offer such clarification around the term 'postcolonialism' is, of course, a symptom of the term's 'elasticity' and, subsequently, its limits as an analytic tool (Moore-Gilbert, 1997: 11), a set of issues comprehensively accounted for by Anthony King's chapter in this section.

CULTURES OF IMPERIALISM

David Scott (1999: 11) usefully defines the 'problem space' of postcolonial studies to be the epistemological assumptions upon which anti-colonial struggles of various kinds depend. This is why, he argues (1999: 18), 'culture' comes so centrally into the critical frame of postcolonial theory. Within the discipline of geography the interest in the 'cultures of imperialism' was largely taken up in response to the analytical template provided by Edward Said (1978) in his study of orientalism. Said's concept of 'culture' is quite precisely defined in the opening pages of Culture and Imperialism, where he uses it to refer to 'all those practices, like the arts of description, communication, and representation, that have relative autonomy from the economic, social, and political realms and that often exist in aesthetic forms' (1993: xii–xiii). He goes on to specify that this includes popular 'lore' as well as specialized knowledges about distant parts of the world, a definition of direct relevance to traditional understandings of geography. There is much we might quibble about in

relation to Said's definition of culture, not least its supposed 'autonomy' from politics and economy, or its emphasis on what, by implication, is 'high' culture. We might also note the way in which Said avoided obviously anthropological definitions of culture, presumably because he saw that disciplinary field to be complicit in the fabrication of orientalism. It was a version of Said's concept of culture that influenced a specific strand of geographical scholarship concerned with rethinking imperialism, and specifically the European imperialisms that drew to a close in the middle years of the twentieth century.

Said's historical (and Foucauldian) approach fell on the fertile ground of the sub-disciplinary area of historical geography (for a range of examples, see Graham and Nash, 2000). Dan Clayton's chapter to follow provides a careful account of this influence and how it was consolidated by Said's spatial sensibilities. Enlivened by this critical framework, historical geographers have investigated the imaginaries, logics and practices of imperialism, diagnosing their racisms, charting their parts in the enframing of difference, and revealing their role in the deployment of imperial power. Most significantly, this scholarship tackled the 'historical amnesia' (Gandhi, 1998: 7) of the discipline of geography. It did this by investigating the European geographical knowledges that both gave sense to, and made sense of, imperial expansions: the speculative geographies of explorers, the making of maps, the scientific theories of climate and race, and the pragmatic spatialities of colonial governance and settlement.

What emerges from these historical approaches is a peculiar sense of the ways in which heterogeneous and often contradictory cultural logics cohered to produce quite emphatic and seemingly singular empire effects.[4] As Driver noted, to suppress such 'divergences' is to do 'a violence ... to the history of the colonial encounter' (1992: 34). Consequently, geographical research on the history of imperialism has assiduously set about the task of producing accounts that attest to its complexities and contradictions. The landmark collection by Anne Godlewska and Neil Smith (1994) showcased the diverse range of themes taken up in this scholarship. Nowhere has the will to diversify the understanding of empire been more evident than in recent feminist historical geographies.[5] These geographies have brought into view the women whose lives were entangled with the masculinist project of the making of empires (see, as examples only, Blunt, 1994; Blunt and Rose, 1994; McEwan, 1996; Morin, 1998). The travel diary and personal journal are the primary archive of this scholarship because women of this time were largely excluded from official and scientific forms of knowledge production. The 'map of empire' such feminist geographies produce shows just how internally contradictory were the cultures of imperialism. On the one hand, women were accessories to the masculinist project of empire building, often drawing on vectors of racial difference in order to assume a position of superiority denied them within their own patriarchal social settings. On the other hand, the very positioning of women as peripheral to the privileged spheres of knowledge and action associated with empire building often placed them into relations with the colonized that unsettled those lines of difference and distinction (for example, Lester, 2002; Morin and Berg, 2001).

The overall legacy of these historical geographies for the project of decolonizing the discipline is somewhat unclear. One of the key collective contributions of the now extensive body of critical historical geographies of empire has been to point to the complexities of colonialism as a lived event: the ambiguities of colonial authority, the porosity of the boundaries between colonizer and colonized, and the diversity of views about difference and superiority. This kind of historical revisionism can have decolonizing effects, but, much as in the case of the self-reflexive museum display with which I opened this editorial introduction, it is essential to know what the political register of that effect might be. Is such scholarship offering a radical decolonization of the discipline or simply a therapeutic moment that makes contemporary geography feel better about itself? Does such critical revision decentre authorization or refine the discipline's pre-existing ways of knowing? The chapters by Dan Clayton and Anthony King to follow, both tackle these challenging questions.

In thinking through such questions one might note the introversion of such geographies and how they invariably turn our attention back to European knowledge fields, and to those figures who translated these into ways of colonizing. They recentre Europe in the name of decentring that place and that idea. In doing so, they enact a disciplinary specific version of the epistemic violence so famously diagnosed by Spivak in her essay 'Can the subaltern speak?' (1988). Spivak's question brings into our sights one of the key limits of many critical historical geographies of imperialism: their failure to adequately account for what was happening on the other side of the frontier and thereby bridge the intersubjective space that is 'contact'. Dan Clayton's chapter to follow reflects, in part, on his own valiant attempts to hear the 'native voice' through the incomplete and partial perspectives of the imperial archive that frames it.[6]

Despite any shortcomings we might detect in the now extensive scholarship on the historical geography of imperialism, one question remains viable: would it be possible for modern geography to effectively decolonize its practices without this kind of critical revisionist scholarship? Such reformulations are a part of a necessary space clearing gesture. As Smith and Godlewska argued, 'only through a full acknowledgement of [geography's] past can we begin to understand the role of geographers in the maintenance of a certain privileged order of things' (1994: 8). More optimistically, I would add, it is only through such looking back that the discipline might look forward to imagine alternative (postcolonial) geographies. As Clayton's chapter to follow argues, whatever shape these alternative geographies might take, they cannot be wholly outside the histories (and historicist structures and representational modes) of the geographies that preceded them. This does not mean that geographical knowledges cannot be renewed 'from and for the margins' in a project that Dipesh Chakrabarty (2000: 16) describes as 'provincializing'. This is the task to which geography must now attend.[7]

POSTCOLONIAL CULTURAL GEOGRAPHIES?

Many geographers have, for some time, operated in an ethical and political framework with anti-colonial objectives. In this sense the discipline of geography has been provincializing itself for some time and through theoretical perspectives other than that offered up by the framework of postcolonial theory. Most obviously we might think of radical (i.e. post-developmentalist) 'development geographies', as represented in part in Section 7 in this volume, edited by Jennifer Robinson. We might also think of the work that has proceeded through the framework of geographies of race and racism.[8] Within both development geography and geographies of racism, certain critical threads have raised relevant suspicions around the very term 'culture' now apparently so central to emerging postcolonial geographies. These are suspicions we need to remain mindful of even if the culture concepts at the heart of current postcolonial geographies are quite distinct from those that were active at earlier times. For example, James Blaut's (1992; 1993) sustained critique of Eurocentrism has included an account of the way in which 'biological racism' was supplanted by 'cultural racism'. This racism, he argued, prolonged the life of European assumptions of superiority by replacing scientific theories of environmentally determined racial character with claims about the so-called 'uniqueness' of European 'mentality and culture' (1992: 294). Furthermore, he argues that 'cultural racism' served as the foundation upon which twentieth-century modernization theory was built, a theory which itself legitimated an entirely new period of western intervention in places deemed to be less developed. Clearly, geographers need to carefully historicize how they use and account for a concept like 'culture' and nowhere more so than in the project of making the discipline aware of its imperial sympathies.

Similar reservations might be expressed in relation to the culture concept upon which Sauerean cultural geography relied. This seemingly arcane question has recently been reactivated in the wake of claims that a revised 'Sauerean tradition' might provide a template for a decolonized cultural geography. Sauer is available for this recuperative project in part because his recognition of precolonial ways of life and landscape provide an important counterpoint to colonialist assumptions about colonized lands being 'pristine' or 'empty'

(Sluyter, 1997: 701). Some of this work has attempted to mount a 'postcolonial' reinterpretation of contemporary landscapes on the basis that precolonial landscapes quite literally shaped, and continue to inhabit, the colonial landscapes that followed (Sluyter, 1999; 2001). Certainly, Sauer's own intellectual project was set against modernism and the various forms of destruction (of cultures and natures) he felt it wrought. And, as Solot (1986) has demonstrated, Sauer's culture areas chorology was emphatically against the kinds of evolutionist frameworks of human development that legitimated Europe's task of delivering 'civilization' to other 'cultures'. In keeping with the revisions of anthropological thinking about culture set in train by Franz Boas, Sauer's concept of a 'culture area' signified a point at which the discipline of geography departed from the 'scientific' theories of climate and race that were so much a feature of the geographies of empire. This said, it is not always clear that Sauer's culture areas approach was entirely outside some of the developmentalist assumptions of evolutionism. Although his interest in cultural change was not determined by any 'general law of progress', he still conceded that there were 'progressive cultures and others that show almost no sign of change' (Solot, 1986: 515). In short, in Sauer's work the boundary between what we might think of as a self-evidently imperialist, meta-evolutionist perspective (which saw the present of other cultures as Europe's past) and a seemingly post-imperialist, synchronic relativism (which imply coevalness and particularity) are not always clearly marked.

Before Sauerean approaches are embraced in decolonizing revisions of cultural geography, another kind of historical space clearing gesture is needed. This entails asking what kinds of social and political effects 'culture area' geographies had on the relations between those people whose 'presence' was being defined by 'culture' (and specifically cultural artifacts) and those geographers who had the privilege to do such defining. Such a project is in keeping with Barnett's suggestion that the 'cultural turn' in the discipline needs to account for the 'formations of the cultural in different institutional situations and ... distinctive forms of social regulation' (1998: 622–3). For example, while the 'culture area' concept helped refute evolutionary accounts of different ways of life (i.e. it functioned *against* imperialist knowledge frames), it was this very idea that was a tool in the classification of ethnographic collections by museums (i.e. it functioned *as* an imperialist knowledge frame). Put simply, knowledge fields that understood other 'cultures' through a relativist (as opposed to evolutionist) model were not automatically located outside colonial logics. As Patrick Wolfe (1999: 55) suggests in relation to anthropology, cultural relativism had specific geopolitical value for the consolidation of power in late colonial settler societies. This issue is hinted at by Bruce Braun (Willems-Braun, 1997: 705) when he argues that it remains unclear what kind of relationship Sauer's nostalgic and romantic reconstruction of pre-contact cultures had to the contemporary politics of land in Latin America.

Of course, it is important to acknowledge that colonized peoples involved in contemporary postcolonial struggles have their own nostalgias to which the frame of 'culture' is often very central. 'Culture' and 'cultural difference' have gained a new legitimacy in the governance of many states that imagine themselves to be postcolonial and/or multicultural. Under such conditions, the ability of marginalized groups to garner some form of recognition (and material benefit) depends entirely upon performing their needs and aspirations by way of 'cultural difference'. Sauerean-style cultural geographies may well offer useful 'tools for a politics of decolonization' in such contexts (1997: 705).[9] For example, a geography that is attentive to the embedded accretions and ongoing (and adjusting) associations between land and life can provide much needed proofs of evidence for adjudications of the prior rights of indigenous peoples (see, as example, Baker, 1999). In such contexts, the essentialist concepts of culture that remain attached to such geographies can be put to strategic use. Whether such geographies are radical or reactionary in their effects – and here we might compare the land rights context I have just mentioned with certain kinds of nationalisms and exclusivist regionalisms – remains an open question, and not always one under the control of those scholars who produce these geographies.

Brenda Yeoh's thoughts on postcolonial memory work in the chapter to follow adds

to our questioning of this notion of 'strategic essentialism' (see also Jackson and Penrose, 1993; Jacobs, 1996; Penrose, 1995). She shows how in places like Singapore the present is not simply defined by and against a colonial past. Contemporary identity in Singapore is constituted out of a complex mix of migration and settlement that both preceded European colonization and postdated decolonization. What stands for 'Singaporean tradition' in such a hybridized space is necessarily complex. Consequently, any effort to reclaim the past for the present nation has its own internal politics in which various subsets of the colonized jostle to have their traditions recognized and sanctioned by the state. Yeoh's chapter is also very suggestive of the need to think beyond cynical accounts of 'heritage' and 'nostalgia' that emerged in cultural geography (and elsewhere) in the 1980s. In settings like Singapore the dislocations that arise from colonialism and contemporary transnationalisms mean that nostalgia is not just a yearning for something lost. The making of heritage and the re-enactment of traditions can also be what Debbora Battaglia refers to as a practiced 'vehicle of [self] knowledge' (1995: 77), a means by which subjects occupying multiple sites in culture and history continue to know themselves. Such identity questions have become central to cultural geographical investigations into postcolonial circumstances. A good example is provided by Richa Nagar's (1997) research with South Asian communities in Tanzania, which shows that simplistic identity models that are confined to the structure of 'colonized' and 'colonizer' are no longer sufficient. Yeoh's chapter to follow extends this scholarship. In the first instance, it usefully locates the concept of 'culture' that underscores the scholarship on identity, within an academic frame associated with the analytic tools furnished by cultural studies. In the second instance, it shows, by way of the case of Singapore, how questions of identity are structured through complex intersections of varied axes of difference.

Arjun Appadurai (2000) argues that the simplistic reproduction of a binarized logic is unhelpful in thinking about culture in contemporary times. He is especially suspicious of theories of globalization which, he argues, presuppose a form of 'trait geography' in order to account for the people and places whose lives are depicted as being overrun by global forces. As an alternative, Appadurai proposes 'process geography' that will 'name and analyze ... mobile civil forms' and chart 'various kinds of action, interaction and motion' (2000: 7). Yeoh's chapter shows that nations that are cross-hatched by transnational flows of people can use that movement as something against which the security (and boundedness) of the nation is imagined, sometimes joyously through the frame of multiculturalism, and sometimes anxiously through the frame of xenophobia. For Yeoh such diasporic and hybrid spaces are genuinely agonistic, throwing up perplexing new forms of racism and new levels of self-reflexivity.[10] As Kay Anderson (2000) suggests, the challenge facing contemporary geography is to give analytical contours to these kinds of transnational processes and the social spaces they produce. In searching for the tools by which to conduct such a geography it is crucial that our scholarship does not simply take for granted the concept of culture being activated. As this editorial has tried to show, the way we conceptualize culture is not given and, in fact, the question of how it is conceptualized is central to the postcolonial potentials of our scholarship.

Let me finish by offering a few thoughts on the politics of doing cultural geographies 'after empire'. It would limit the prospect for a radical shift in the politics of the discipline's knowledge field if revision were confined to the matter of what topics were researched. Decolonizing the discipline relates also to modes of research and representation. I am not simply suggesting that an attention to 'discourse' be replaced with more material geographies – that the imperial archive be forsaken for postcolonial ethnography (see Hansen and Steppuat, 2001). Ethnographies of postcolonial formations can suffer from their own omissions and elisions. Nowhere is this more evident than in thinking through the identity categories and processes that come to be associated with something called 'resistance'. For example, David Slater (1998) has argued in relation to indigenous struggles against internal colonialisms in Central and South America that the idea of the 'indigenous' is often deployed in a way that masks the heterogeneity within such a category. This

is a version of what Sherry Ortner would describe as 'dissolving the subject ... into a set of "subject effects"' (1995: 183). Ortner offers a most relevant consideration of how to research and represent resistance. She argues that there has been a romantic impulse to 'sanitize the internal politics of the dominated' and labels this a form of 'ethnographic refusal' (1995: 179). She concludes that glossing over the prior and ongoing politics among subalterns leads to an impoverished understanding of resistance itself. One might add to this observation another specification that returns us to the problematic of 'culture' itself. When resistance is understood as operating in the logic of 'culture', what precisely does that mean? As Barnett (1998) so provocatively asks: what concept of 'culture' is being activated in geographies that claim to chart the 'cultural politics' of resistance?

When thinking about how cultural geography might conduct itself postcolonially, we need to be mindful of ourselves as 'situated actors engaged in the political work of representation and the production of knowledge' (Katz, 1992: 496). In this regard a thoroughly decolonized geographical praxis may have as much to do with our forms of representation as it does the postcolonialness of our theories. Crang (1992), for example, suggested that geographers construct more 'polyphonic' geographies as a way of reconfiguring academic authority in relation to 'research subjects' (see, as an example, Huggins et al., 1995). Katz's own reflections on her field experience in Sudan reveal that neither theory nor writing strategy alone can structure a fully decolonized research process. As Katz notes, she was, after all, 'the one defining the terrain of the questions and looking for illumination in the practices of others' (1992: 502). There are some signs for optimism, and here I think of the action research of Richard Howitt (1998; Howitt and Jackson, 1998) in which the research questions are determined collaboratively with the indigenous communities with which he works. But there are also good reasons to remain sceptical about the capacity of the discipline to fully move on from the frameworks of knowledge and power bequeathed by the idea of empire. For example, Tony King's chapter to follow provides a sobering reminder that decolonizing a disciplinary field requires reshaping the structure of the academy itself: who is employed, who gets to speak, in what languages one speaks (both literal and theoretical) and, of course, who cares to listen. As my account of the postcolonial museum display I opened with sought to demonstrate, even when we comport ourselves differently around the object of 'culture', we cannot be sure if the effects of the geographies we produce will be as postcolonial as we might wish.

NOTES

I would like to thank Stephen Cairns, Dan Clayton, Tony King, Brenda Yeoh and the editors for their feedback on earlier versions of this editorial.

1   Word restraints prohibit me from citing extensively relevant examples of every style of geography that I touch upon in this introduction. Further detail is usually provided by the chapters to follow. In the main I reference one indicative case that can provide a trail scholars can follow.
2   For a specific geographical discussion of this see the special issue of *Environment and Planning D: Society and Space*, 1992.
3   Most radically and controversially we might think of Michael Hardt and Antonio Negri's (2000) suggestion that the logics of globalization have a scope and intensity (an unboundedness) that have produced empire effects in excess of anything that the process of imperialism might have delivered.
4   A useful discussion of this is in the introduction to Bell et al. (1995).
5   For a comprehensive analysis of the relationship between postcolonial theory and feminism see Cheryl McEwan's chapter in this collection (Chapter 21).
6   Speaking about contemporary South Africa, Crush (1994: 346–7) further opens out the problematic of how 'subaltern' voices might be brought into historical and geographical narratives. See also Crush (1986; 1992), and Simon (1998).
7   Challenging the 'centrisms' of the discipline has been a persistent concern in recent years. See, as examples, Blaut (1993), Katz (1995), McEwan (1998), Mitchell (1998) and Sparke (1994).
8   An exemplary body of self-consciously decolonizing geographical work is the considerable scholarship that has emerged in response to the racist spatial logic of settlement in Southern Africa (e.g. Crush et al., 1982).
9   Just as, say, GIS and other mapping techniques can offer useful technologies of translation and legitimation for postcolonial land claims.
10 My own work in this area has taken a psychoanalytic approach to analyzing these aspects of colonialism's aftermath (see also Gelder and Jacobs, 1998; Gooder and Jacobs, 2000; Jacobs, 1997).

REFERENCES

Anderson, K. (2000) 'Thinking "postnationally": dialogue across multicultural, indigenous and settler spaces', *Annals of the Association of American Geographers* 90 (2): 381–91.

Appadurai, A. (1996) *Modernity at Large: Cultural Dimensions of Globalization*. Minneapolis: University of Minnesota Press.

Appadurai, A. (2000) 'Grassroots globalization and the research imagination', *Public Culture* 12 (1): 1–20.

Baker, R. (1999) *Land Is Life: From Bush to Town – The Story of the Yanyuwa People*. Sydney: Allen and Unwin.

Barnett, C. (1997) '"Sing along with the common people": politics, postcolonialism and other figures', *Environment and Planning D: Society and Space* 15: 137–54.

Barnett, C. (1998) 'Cultural twists and turns', *Environment and Planning D: Society and Space* 16: 631–4.

Battaglia, D. (1995) 'On practical nostalgia: self-prospecting among urban Trobrianders', in D. Battaglia (ed.) *Rhetorics of Self-Making*. Berkeley: University of California Press. pp. 77–96.

Bell, M., Butlin, R. and Heffernan, M. (eds) (1995) *Geography and Imperialism, 1820–1940*. Manchester: University of Manchester Press.

Blaut, J.M. (1992) 'The theory of cultural racism', *Antipode* 24 (4): 289–99.

Blaut, J.M. (1993) *The Colonizer's Model of the World: Geographical Diffusionism and Eurocentric History*. New York: Guilford.

Blunt, A. (1994) *Travel, Gender and Imperialism: Mary Kingsley and West Africa*. New York: Guilford.

Blunt, A. and Rose, G. (eds) (1994) *Writing Women and Space: Colonial and Postcolonial Geographies*. New York: Guilford.

Chakrabarty, D. (2000) *Provincializing Europe: Postcolonial Thought and Historical Difference*. Princeton: Princeton University Press.

Crang, P. (1992) 'The politics of polyphony: reconfigurations in geographical authority', *Environment and Planning D: Society and Space* 10: 527–49.

Crush, J. (1986) 'Towards a people's historical geography for South Africa', *Journal of Historical Geography* 12: 2–3.

Crush, J. (1992) 'Beyond the frontier: the new South African historical geography', in C. Rogerson and J. McCarthy (eds) *Geography in a Changing South Africa*. Cape Town: Oxford University Press.

Crush, J. (1994) 'Post-colonialism, de-colonization, and geography', in A. Godlewska and N. Smith (eds) *Geography and Empire*. Oxford: Blackwell. pp. 333–50.

Crush, J., Reitsma, H. and Rogerson, C. (1982) 'Decolonising the human geography of Southern Africa', *Progress in Human Geography* 7: 203–31.

Driver, F. (1992) 'Geography's empire: histories of geographical knowledge', *Environment and Planning D: Society and Space* 10: 23–40.

Driver, F. and Gilbert, D. (eds) (1999) *Imperial Cities: Landscape, Display and Identity*. Manchester: Manchester University Press.

Gandhi, L. (1998) *Postcolonial Theory: A Critical Introduction*. St Leonards: Allen and Unwin.

Gelder, K. and Jacobs, J.M. (1998) *Uncanny Australia: Sacredness and Identity in a Postcolonial Nation*. Melbourne: University of Melbourne Press.

Godlewska, A. and Smith, N. (eds) (1994) *Geography and Empire*. Oxford: Blackwell.

Gooder, H. and Jacobs, J.M. (2000) '"On the borders of the unsayable": the apology in postcolonizing Australia', *Interventions: International Journal of Postcolonial Studies* (August): 230–48.

Graham, B. and Nash, C. (eds) (2000) *Modern Historical Geographies*. Harlow: Pearson.

Hansen, T.B. and Stepputat, F. (eds) (2001) *States of Imagination: Ethnographic Explorations of the Postcolonial State*. Durham, NC: Duke University Press.

Hardt, M. and Negri, A. (2000) *Empire*. Cambridge, MA: Harvard University Press.

Howitt, R. (1998) 'Recognition, respect and reconciliation: steps towards decolonization?', *Australian Aboriginal Studies* 1: 28–34.

Howitt, R. and Jackson, S. (1998) 'Some things do change: indigenous rights, geographers and geography in Australia', *Australian Geographer* 29 (2): 155–73.

Huggins, J., Huggins, R. and Jacobs, J.M. (1995) 'Kooramindanjie: place and the postcolonial', *History Workshop Journal* 39: 165–81.

Jackson, P. and Penrose, J. (1993) 'Placing "race" and nation', in P. Jackson and J. Penrose (eds) *Constructions of Race, Place and Nation*. London: UCL Press. pp. 1–26.

Jacobs, J.M. (1996) *Edge of Empire: Postcolonialism and the City*. London: Routledge.

Jacobs, J.M. (1997) 'Resisting reconciliation: the secret geographies of (post)colonial Australia', in S. Pile and M. Keith (eds) *Geographies of Resistance*. London: Routledge. pp. 203–18.

Katz, C. (1992) 'All the world is staged: intellectuals and the projects of ethnography', *Environment and Planning D: Society and Space* 10 (5): 495–510.

Katz, C. (1995) 'Major/minor: theory, nature, and politics', *Annals of the Association of American Geographers* 85: 164–8.

Lester, A. (2002) 'Obtaining the "due observance of justice": the geographies of colonial humanitarianism', *Environment and Planning D: Society and Space*, 20 (3).

McClintock, A. (1992) 'The Angel of Progress: pitfalls of the term "post-colonialism"', *Social Text* 10 (2/3): 84–98.

McEwan, C. (1996) 'Paradise or pandemonium? West African landscapes in the travel accounts of Victorian women', *Journal of Historical Geography* 22 (1): 68–83.

McEwan, C. (1998) 'Cutting power lines within the palace? Countering paternity and eurocentrism in the "geographical tradition"', *Transactions of the Institute of British Geographers* 23 (NS): 371–84.

Mitchell, K. (1998) 'Lingua franca', *Environment and Planning D: Society and Space* 16: 505–8.

Moore-Gilbert, B. (1997) *Postcolonial Theory: Contexts, Practices, Politics*. London: Verso.

Morin, K.M. (1998) 'British women travelers and constructions of racial difference across the nineteenth-century American West', *Transactions of the Institute of British Geographers* 23 (NS): 311–30.

Morin, K.M. and Berg, L.D. (2001) 'Gendering resistance: British colonial narratives of wartime New Zealand', *The Journal of Historical Geography* 27 (2): 196–222.

Nagar, R. (1997) 'Communal places and the politics of multiple identities: the case of Tanzanian Asians', *Ecumene* 4 (1): 3–26.

Ortner, S.B. (1995) 'Resistance and the problem of ethnographic refusal', *Society for the Comparative Study of Society and History* 37: 173–93.

Penrose, J.M. (1995) 'Essential constructions? The "cultural bases" of nationalist movements', *Nations and Nationalism* 1 (3): 391–417.

Said, E. (1978) *Orientalism*. London: Routledge and Kegan Paul.

Said, E. (1993) *Culture and Imperialism*. London: Chatto and Windus.

Scott, D. (1999) *Refashioning Futures: Criticism after Postcoloniality*. Princeton: Princeton University Press.

Sidaway, J.D. (2000) 'Postcolonial geographies: an exploratory essay', *Progress in Human Geography* 24 (4): 591–612.

Simon, D. (1998) 'Rethinking (post)modernism, postcolonialism and posttraditionalism: South–North perspectives', *Environment and Planning D: Society and Space* 16: 219–45.

Slater, D. (1998) 'Post-colonial questions for global times', *Review of International Political Economy* 5 (4): 647–78.

Sluyter, A. (1997) 'On "buried epistemologies: the politics of nature in (post)colonial British Columbia": on excavating buried epistemologies', *Annals of the Association of British Geographers* 87 (4): 700–2.

Sluyter, A. (1999) 'The making of the myth in post-colonial development: material and conceptual landscape transformation in sixteenth-century Veracruz', *Annals of the Association of American Geographers* 89 (3): 377–401.

Sluyter, A. (2001) 'Colonialism and landscape in the Americas: material/conceptual transformations and continuing consequences', *Annals of the Association of American Geographers* 91 (2): 410–28.

Smith, N. and Godlewska, A. (1994) 'Introduction: Critical histories of geography', in N. Smith and A. Godlewska (eds) *Geography and Empire*. Oxford: Blackwell. pp. 1–8.

Solot, M. (1986) 'Carl Sauer and cultural evolution', *Annals of the Association of American Geographers* 76 (4): 508–20.

Sparke, M. (1994) 'White mythologies and anemic geographies', *Environment and Planning D: Society and Space* 12: 105–23.

Spivak, G.C. (1988) 'Can the subaltern speak?', in C. Nelson and L. Grossberg (eds) *Marxism and the Interpretation of Culture*. Basingstoke: Macmillan. pp. 271–313.

Thomas, N. (1994) *Colonialism's Culture: Anthropology, Travel and Government*. Melbourne: Melbourne University Press.

Willems-Braun, B. (1997) '"Reply" on cultural politics, Sauer, and the politics of citation', *Annals of the Association of American Geographers* 87 (4): 703–8.

Wolfe, P. (1999) *Settler Colonialism and the Transformation of Anthropology: The Politics and Poetics of an Ethnographic Event*. London: Cassell.

18

Critical Imperial and Colonial Geographies

Daniel Clayton

This chapter considers geographers' current fascination with the imperial/colonial past and traces the impact of postcolonialism on their interpretive sensibilities.[1] Over the last 10 years there has been an explosion of interest in the links between geography and empire. Geographers have become interested in the imperial genealogy of their discipline, the spatiality of colonialism and empire, and how we might revisit imperial and colonial geographies from postcolonial perspectives.

Some have sought to evaluate the ways in which geography has worked as an imperial discipline and discourse, and have used their findings to pluralize and politicize understanding of what David Livingstone (1992) has called 'the geographical tradition'. Critical attention has been drawn to the imperial roles played by diverse producers and arbiters of geographical knowledge (explorers, naturalists, cartographers, surveyors, photographers, geographical societies, professional geographers and so on), and to geography's Eurocentric moorings (see Driver, 1995). Others have posed more general questions about the geographies of colonialism and empire. There are fast-growing literatures on how imperialism was shaped by spatial formations of knowledge and power, how empire was invested with geographical meaning through diverse cultural media (e.g. travel narratives, museums and school curricula) and how imperialism was tied to the fabrication of insidious locational imaginaries such as the orient, 'darkest Africa' and the tropics (see Driver and Yeoh, 2000; Duncan and Gregory, 1999; Gregory, 2000b). Still others have dealt more explicitly with colonial geographies and the ongoing extension of colonial power around the world. There is a range of work on the production and representation of colonial space, on how colonial spaces were built around the axes of class, race, gender and religion, and on how different natural environments and indigenous peoples (or natures and cultures) impacted on colonial projects and encounters (see Butzer, 1992; Kenny, 1999). Some of this literature also expresses a strong postcolonial concern with how geographers might support current anti-colonial struggles and processes of decolonization (see Howitt, 2001).

Geographical research on imperial/colonial issues has become very popular in Anglo-American geography, but is less prominent in non-English-speaking countries (though see Bruneau and Dory, 1994; Claval, 1998; Lejeune, 1993; Siliberto, 1998). And while geographers do not necessarily assume that all empires have been or are western, their work on geography and empire deals almost exclusively with the history and consequences of modern western colonization. There is also a heavy emphasis on the nineteenth century, and the British empire and its successor states.

The term 'critical imperial and colonial geographies' is meant to capture geographers' diverse interests. It covers their attempts to: (1) show that the discipline of geography, and a broader set of geographical discourses and practices, played a critical – or vital – role in empire; (2) criticize these vital geographies and move the discipline beyond their binds and conventions; (3) treat the links between geography and empire as symptomatic of the relations of power that inhere in the production of geographical knowledges; and (4) give geography a niche in wider postcolonial debates about colonialism and western dominance.

The chapter is divided into three sections and a number of subsections that sketch (what I see as) the key themes in this burgeoning area of geographical inquiry, and point to some of the ways in which geographers' critical endeavours can be called postcolonial. The first section places the geographical literature in an encompassing intellectual setting and sketches geographers' mixed reaction to the advent of postcolonialism in the western academy. The second section outlines the diverse ways in which we can construe the links between geography and empire; and the third section raises some questions about geographers' critical aims.

GEOGRAPHY AND POSTCOLONIALISM

Geographers have a long-standing critical interest in imperialism and colonialism, but the post-1980s literature considered here is characterized by a number of new trends. Much of it has emerged in critical dialogue with postcolonialism, which has become a trendy (if troublesome) buzzword for a range of critical practices that grapple with what it means to work 'after', 'beyond' and 'in the knowledge of' colonialism (see Gregory, 2000a). Much of it displays an anti-essentialist concern with the social construction of knowledge and identity, and the machinations of knowledge and power. And much of it treats geography as an eclectic, shifting and contested body of concepts, knowledges and practices rather than as an autonomous discursive field or tightly defined discipline. The bulk of the chapter surveys these changing ideas about geography. But we cannot fully understand how and why geographers are turning to the imperial/colonial past unless we first place their work in a wider postcolonial intellectual context. It is important to so situate geographers' work for numerous reasons, but let me make two sets of observations that are pertinent to the discussion that follows.

The power of postcolonialism

First, it has become commonplace to observe that the postcolonial world has placed new demands upon western theory and scholarship. There are demands to listen to the other, to appreciate claims to difference, to incorporate minor histories into mainstream history, and to come to terms with the cultural politics of academic knowledge. Western academics have become more attuned to the Eurocentric assumptions embedded in their disciplinary visions, more sensitive to issues of otherness and cultural diversity, and more alert to the idea that the

universals enshrined in European (and especially post-Enlightenment) thought are at once indispensable and inadequate tools of critique. It 'is now unacceptable to write geography in such a way that the West is always at the centre of its imperial Geography,' Trevor Barnes and Derek Gregory (1997: 14) declare in a recent geography textbook, and scholars from other disciplines are spouting similar messages. 'For scholars and teachers of my generation who were educated in what was an essentially Eurocentric mode,' the influential Palestinian-American literary critic Edward Said has written, 'the landscape and topography of literary study have ... been altered dramatically and irreversibly ... [S]cholars of the new generation are much more attuned to the non-European, genderized, decolonized, and decentred energies and currents of our time' (2001: 65).

Among other things, this new – postcolonial – generation has pressed home the idea that the configuration of Europe as the self-contained fount of modernity and sovereign subject/centre of world history is a powerful fiction that obscures the reciprocal constitution of Europe and its others. Postcolonial critics return to the past to reveal that identities, cultures, nations and histories have long been hybrid and intertwined, and never self-sufficient or mutually exclusive, with a select group of cultures being innately superior over others. In this sense, postcolonialism works as a critical perspective on the west which shows that 'colonisation was never simply external to the societies of the imperial metropolis ... [but] was always deeply inscribed within them' (Hall, 1996: 246). Europe 'was constructed from outside in as much as inside out' through processes of 'transculturation', Mary Louise Pratt remarks, beginning with the metropole's 'obsessive need to re-present its peripheries and other continually to itself' (1992: 4–7).

But postcolonialism does not simply amount to a 'writing back' to the west, or to a politics of recognition, that debunks Eurocentric knowledge and the denial of cultural and cognitive equality that lay at the heart of the west's spirit of domination. Postcolonial criticism is also driven by the recognition that the freedom to take control of the means of self-representation that independence presented to colonized peoples did not create some instantaneous freedom from the burdens of colonial history. Leela Gandhi describes postcoloniality as 'a condition troubled by the consequences of a self-willed historical amnesia' – by a desire to forget the past and the west – and suggests that 'the theoretical value of postcolonialism inheres, in part, in its ability to elaborate forgotten memories of this condition' (1998: 3–17). Crucially, 'postcoloniality must be made to concede its own part or complicity in the terrors – and errors – of its own past.' We should not turn a blind eye to the seductions of modernity and colonial power. We need to recover the lines of mutual desire between self and other that crossed the colonial world as well as those of coercion and mutual antagonism, and we need to evaluate the ongoing influence of European habits and categories of thought. In this sense, Gandhi suggests, postcolonialism can be seen as an 'ameliorative' and 'therapeutic' project that *necessarily* returns to the colonial past in order to help postcolonial subjects deal with 'the gaps and fissures in their condition.' It grapples with the spectre of belatedness and incompleteness that haunts decolonization and anti-colonial struggles: the spectre of only arriving on the scene of autonomy after the west, and of struggling to be modern yet different. Postcolonial energies are focused on the extent to which colonialism had a binary (dichotomous, exclusionary and systematic) or ambivalent (differentiated, fretful and contradictory) cast. The critical elucidation of colonialism as a conceptual totality with some transhistorical traits and grossly unequal effects is tempered by work that emphasizes colonialism's diversity, hybridity and susceptibility to deformation (see Loomba, 1998).

Lastly in this subsection, postcolonialism is centrally concerned with the connections between culture and power. Economic and political factors and explanations for colonialism and empire are not ignored so much as integrated into new cultural interpretive frameworks that explore the creation and circulation of meaning, and the binaries of self/other, centre/periphery, modernity/tradition, coloniser/colonized and so on, that shaped (some would say overdetermined) metropolitan-colonial relationships. 'Colonialism was made possible, and then sustained and strengthened, as much by cultural technologies of rule as it was by the more obvious and brutal modes of conquest,' Nicholas Dirks (1996: xi) suggests (also see Thomas, 1993). Culture and power are often connected via the concept of discourse, and a short digression into the work of Edward Said is appropriate at this juncture, because it is seen as pivotal to the development of postcolonialism as an academic project that recovers the political significance of culture.

As Robert Young usefully notes, it was Said's elaboration of 'the idea of Orientalism as a *discourse* in a general sense that allowed the creation of a general conceptual paradigm through which the cultural forms of colonial and imperial ideologies could be analysed,' and colonialism could be seen 'as an ideological production across different kinds of texts produced historically from a wide range of different institutions, disciplines and geographical areas' (2001: 343). Said argued that we cannot fully understand how imperialism and colonialism work unless we examine the discursive means by which the west arrogated to itself the power to grant (and deny) cultural respect to others and authorize what counts as truth (and what does not). His pathbreaking book *Orientalism* (1978) shows how western power was exercised through a particular kind of language (discourse – a term that Said borrowed from Foucault) that was replete with cultural attitudes of superiority and dominance. Said exposed the west's propensity to demean and dominate the other, and emphasized the binary (essentialist, exclusionary and self-consolidating) cast of colonial discourse. He examined how the orient constructed and manipulated in orientalist discourse served as Europe's 'surrogate and even underground self' – as a 'distorting mirror' in which Europe defined itself and celebrated its superiority (1978: 27; Washbrook, 1999: 598).

Orientalism 'opened the floodgates of postcolonial criticism,' Gyan Prakash (1995: 201) recounts, by challenging taken-for-granted oppositions between western knowledge and western power, scholarly detachment and worldly motives, and representation and reality. Said's treatment of colonialism as a discourse that produces, fixes and encodes knowledge in diverse forms and locations inspired a new generation of scholars to re-examine the knowledges and identities authorized by colonialism, and to explore how western power hinged on discursive strategies of cultural projection, incorporation, debasement and erasure. In short, Said drew out the discursive (or epistemic) violence of colonization.

The pitfalls of postcolonialism

But second, and as some of this implies, there is no consensus about the appropriate aims and scope of postcolonial studies. Postcolonialism has become an intellectual battleground for competing philosophies, and one that pitches the politics of totality, solidarity and universalism against those of location, belonging and relativism. 'Eclectic,' 'fragmented' and 'agonistic' are the words that perhaps best describe the field of postcolonial studies, and postcolonial work in geography is far from cohesive.

Geographers are embracing and developing postcolonial perspectives with a mixture of excitement and caution. On the one hand, what they often simply term 'the postcolonial critique' is bolstering the so-called 'cultural turn' in human geography, reaffirming the importance of historical perspectives within the discipline, and bringing many new objects of study into critical play (see Graham and Nash, 2000). Said has a talismanic status in all of this, and not just because of his ideas about discourse, which now have a standard place in geography. His work has been doubly important to geographers because of his insistence that imperialism and colonialism should be conceptualized geographically – as constellations of power that are intrinsically concerned with land, territory, displacement and dispossession (see Gregory, 1995). Much of Said's work is based on the idea of 'imaginative geography' – 'the invention and construction of a geographical space called the Orient, for instance, with scant attention paid to the actuality of the geography of its inhabitants' (Said, 2000: 181; see also Said, 1993). This elastic critical motif now frames myriad geographical studies, and geographers use it to spatialize (if not always carefully historicize) the idea of colonial discourse 'in a general sense'. But Said is not the only postcolonial thinker with geographical interests. Indeed, he has nurtured a spatial turn in postcolonial studies, and geographers have drawn on the work of scholars such as Paul Carter and Timothy Mitchell who are keenly interested in the spatiality of colonialism and empire (see Gregory, 1994: 15–208).

On the other hand, there are complaints about the type of work that postcolonialism is encouraging within and beyond geography. Much postcolonial work, it is suggested, is thin on detail, hung up on questions of discourse and the agency of the colonizer, marred by textualism and wanton generalization, too tightly based on the colonial experience of particular parts of the world (particularly India), and imbued with forms of intellectual tourism that keep us within the imperial trajectory of the west. The most common complaints are that postcolonial critics lose sight of the diversity and materiality of colonialism and empire, and only partially realize their commitment to the postcolonial subject because they fixate on the projection of a western will to power and tend to reduce colonialism to matters of discourse. Geographical writing on colonialism and empire generally retains a much stronger concern with bodily experiences and material practices, the physicality of movement and interaction, and the creation of networks of power than much postcolonial work that emanates (especially) from literary and cultural studies. It is partly for these reasons that geographers sometimes represent literary/cultural

postcolonialism as an alien body of ideas that needs to be recontextualized wherever it is taken. And it is important to note that geographers are not simply drawing on postcolonial theory. They have also engaged the new historiography of western science (e.g. the work of Bruno Latour and Stephen Shapin), feminist philosophies, French poststructuralist theory, and scholarship in the fields of imperial history and cultural anthropology.

The critical imperial and colonial geographies that we will now explore in more depth have emerged in the midst of these developments and debates. Postcolonialism can be described as a powerful interdisciplinary mood in the social sciences and humanities that is refocusing attention on the imperial/colonial past, and critically revising understanding of the place of the west in the world. Yet different disciplines have been implicated in empire in different ways and do not have identical postcolonial concerns.

GEOGRAPHY, COLONIALISM AND EMPIRE

Geographical research on colonialism and empire takes diverse forms, but it is possible to identify two main orientations in the literature. We can distinguish between research that concentrates on what Felix Driver (1992) has dubbed 'geography's empire' and that which explores what Derek Gregory (2001a) calls 'colonizing geographies'. The former body of work has a more or less exclusively metropolitan-disciplinary focus, whereas the latter is more concerned with the historical-geographical diversity of colonialism and empire.

Geography's empire

In the early 1990s, geographers started to question self-contained and in-house narratives of the history of geography, prise open the western biases enshrined in geographical thought, and explore geography's historical imbroglio with empire. When Driver's paper 'Geography's empire' appeared in 1992, there was hardly any critical reflection on the discipline's historical complicity in empire. 'Some might regard … [this] as a sign of the strong hold that the colonial frame of mind has upon the subject,' he wrote. 'It is as if the writings of our predecessors were so saturated with colonial and imperial themes that to problematise their role is to challenge the status of the modern discipline. Yet this is perhaps the very thing that needs to be done if geographers are to exploit present intellectual and political opportunities' (1992: 26). Driver was referring to the opportunities presented (mainly) by Said's work, and over the last 10 years there has been an explosion of interest in how imperialism and Eurocentrism both activated and were activated by geographical imaginations and practices.

In a formative collection of essays, Anne Godlewska and Neil Smith noted that while 'geography has always pursued a wide range of intellectual agendas simultaneously and ... not all of these can be traced directly to the concerns of empire,' it is clear that 'the very formation and institutionalization of the discipline was intricately bound up with imperialism' (1994: 1–8). For others, too, geographers 'were the essential midwives of European imperialism. They provided both the practical information necessary for overseas conquest and colonization and the intellectual justification for expansion through their increasingly elaborate "theoretical" writings on geo-politics and the impact of climatic and environmental factors on the evolution of different races' (Bell et al., 1995: 6).

Work on geography's empire challenges the self-confident and assertive narratives of exploration, conquest, settlement and rule – with error giving way to truth, science conquering myth, modernity supplanting tradition, and civilization being imposed on savagery – that pervade the annals of geography and imperial historiography. Geographical approaches to the world that were once viewed as enlightened and disinterested are now seen as powerful constructions that induced and sustained imperial relationships, and stories of triumphal and uncontested western progress are now told as halting (and sometimes haunting) tales of human struggle. Geographers have shown how many of their discipline's founding and distinctive knowledges and practices – its narratives of exploration and travel, maps and resources inventories, and systems of spatial comparison, classification and planning – worked as tools of material and intellectual dispossession. They have been especially concerned with the images of empty and undeveloped space awaiting the transformative hand of the west that became central to the view that geography is about finding a certain type of order in the world. It is now argued that this order – a Eurocentric and Cartesian order – was made rather than given (or there all along and waiting to be found), and often made in ambivalent ways.

In line with many postcolonial critiques, 'empire' is conceived as a distorting mirror within which the discipline of geography came to define and champion itself. Work on geography's empire works as a critique of the west that is telescoped through a particular set of disciplinary lenses. A great deal of attention has been paid to the spaces of knowledge (e.g. the field and the study) and sites of study in which knowledges were produced, and the physical and institutional effort it took to draw order out of chaos (to travel, collect, map, represent, possess and survive). Geographers have re-examined the activities of individuals, ranging from well-known figures in the history of geography such as Alexander von Humboldt and Halford Mackinder, to lesser figures such as Eric Dutton (a geographer and colonial administrator in Africa) and James Rennell (who surveyed India for the British) who, it is argued, should be included in a critical historiography of geography and empire (see Ryan, 1997; Myers, 1998).

Such case studies feed into wider discussions of geo-imperial discourses, and empirical vignettes are connected to an encompassing body of theory. This range of work reveals how tensions emerged between different modes of knowledge production, how the creation of true and trustworthy (universal and reliable) geographical knowledge depended on the adjudication of boundaries between credible and incredible knowledge, and how geography was constructed from the outside in, through the amassing of data about foreign lands and the creation of geographical categories separating 'us' from 'them' (see Heffernan, 2001; Withers, 2000). For example, in an avowedly postcolonial reading of the 'Africanist discourse' of the London-based Royal Geographical Society (RGS) between 1831 and 1871, Clive Barnett tries to show that

> The actual conditions of cross-cultural contact upon which the production of nineteenth-century geographical knowledge depended are retrospectively rewritten [for metropolitan audiences] to present ['racially unmarked'] European subjects as the singular sources of meaning. ... Without the use of local guides and interpreters, the exploits of men represented as untiringly persevering, independent and self-denying seekers of the truth [and nothing but] would have been impossible. But this routine practical dependence on local knowledges and information is not accorded any epistemological value. Local knowledge is refashioned as a hindrance, as a barrier to the arrival of the truth ... Indigenous geographical meanings and knowledges are admitted into this discourse on the condition of being stripped of any validity independent of European definitions of scientific knowledge ... The knowledge of non-European subjects is represented ... as the confusion and noise against which European science takes shape and secures its authority. (1998: 244–5)

This passage trades on the epistemic violence of geography's empire, and Barnett specifies the importance of science and reason as duplicitous vectors of inscription (also see Anderson, 1998).

There is a strong focus in the literature on 'official' geographical ventures that were sanctioned by the state, learned societies and geography's professional research culture. But geographers such as Teresa Ploszajska (1996) and Avril Maddrell (1998) have drawn our attention to the role that geographical education (school textbooks and field trips) in Britain played in shaping imperial assumptions among the young. And in recent years there has been a spate of work on the forms of travel, leisure and consumption that nurtured imperial attitudes among the middle and lower classes (e.g. Phillips, 1997).

Felix Driver's *Geography Militant* (2000) is perhaps the most accomplished account to date of British geography's nineteenth-century imperial heritage. Driver charts the formation of a Victorian 'culture of exploration' that centred on Africa, involved the mobilization of a variety of material and imaginative resources (equipment, guides, patronage, publicity, authority, scholarship, myths and so on), and hinged on the creation of new spaces of knowledge. He stresses the importance of examining both the production and the consumption of geographical knowledges, reading both official and popular texts, and thinking about the site-specific negotiation of meaning and power. The image of 'darkest Africa' that was presented to the British public was put together by a gentlemanly network of scholars, politicians and philanthropists who made the RGS an authoritative site for the promotion of exploration and dissemination of geographical knowledge. Yet this imaginative geography of Africa was also moulded in public spaces of knowledge such as the museum, exhibition hall and advertising billboard, and in popular accounts of exploration (such as those of Henry Morton Stanley) that were deemed to be sensational by the geographical authorities.

Driver effectively pluralizes understanding of the geographical tradition, and politicizes work on exploration and empire by showing how many of the discursive tropes that entered into the geographical construction of 'darkest Africa' (scientific rhetoric, and a thirst for adventure and the exotic) are being recycled in a variety of contemporary cultural forms. 'The notion of geographical knowledge as the preserve of a modern university-based profession (the "discipline" of geography) is clearly anachronistic for the nineteenth century,' he argues, and 'inappropriate for the twentieth' (2000: 7, 202, 216). Britain's growing and changing imperial presence in the world over the course of the nineteenth century became central to geography's public image, and Britain's imperial past is still at large in our geographical imaginations.

Work on geography's empire debunks what Gillian Rose (1995) has dubbed the 'specular spatiality' of the geographical tradition. We are encouraged to challenge the way that geography has worked as a disciplinary space 'into which some are gathered and from which others are exiled', and which imposes order (historically in the guise of reason and science, and in the name of civilization and progress) on an uncharted and unruly world. Few geographers would dissent from the view that geography has always been a practical science, and many now insist that we look at a greater range of geographical knowledges and practices than the ones that geography's historiographers have deemed central to the definition and development of the discipline. Geographers are pointing to other voices and ways of knowing that geography's empire has appropriated or placed out of bounds, and thinking about what a more inclusive history of geography might look like. And 'geography' is treated as both a practical (embodied, investigative, instrumental) pursuit and a discursive (conceptual, textual, institutional, pedagogic) enterprise. Critical geographers look for signs of ambivalence, contradiction and the assertion of power in the geographical archive, and are showing that empire was deeply inscribed within the discipline of geography.

Colonizing geographies

Yet the fashioning of 'critical imperial and colonial geographies' cannot simply be whittled down to a revamped disciplinary history. Derek Gregory has urged us to break out of the parameters of a 'contextual' historiography, which recovers the historical contingency and mutability of geographical ideas and practices, and devise a 'spatial analytics' that 'discloses the implication of spatiality in the production of power and knowledge' (1998: 11). He uses the term 'colonizing geographies' to signpost the myriad ways in which geography and colonialism work into one another, and the myriad critical positions from which such workings might be approached. I will touch on four such approaches with some examples from the literature: by analytical focus (e.g. the production and representation of space), by politico-intellectual position (e.g. feminism), by substantive theme (e.g. cartography) and by location – though there are other ways of characterizing the literature.

First, geographers' interest in geography's empire is tied to broader analytical concerns – with, for example, representation, abstraction, visualization and embodiment – that inevitably reposition how the term 'geography' is viewed both historically and epistemologically. In a string of books and articles, Gregory has explored 'colonizing productions of nature and space' and what he calls 'topographicalization' (the spatial modalities through which imperial/colonial encounters, practices and representations worked). For a start, he argues, the modern discipline of geography was not simply 'a constitutively European science' propelled by reason, as some have claimed, but also a 'profoundly Eurocentric science' (cf. Livingstone and Withers, 1999). Moreover, Eurocentrism is imbued with 'a system of geo-graphs [or modes of earth-writing] that order its representations' – geo-graphs that absolutize space, objectify the world, normalize the subject, and abstract nature and culture in imperialist terms (Gregory, 1998: 3–40, 60–7). He has also explored the formative spaces – or 'topo-logies/graphies' – that are carried *within* texts and modalities of travel: spaces with no centre ('rhizomatic space'), complex spaces with a centre ('labyrinthine space') and ordered, linear spaces ('striated space') that shaped how cultural meanings were made and remade through travel (Gregory, 2000b).

Gregory tries to tease out the conceptual orders that imbued the empirical work undertaken by geography's imperial/colonial agents. To borrow Foucault's terminology, he starts to provide an archaeology of the geographies of imperialism and colonialism that underwrites the genealogies produced by students of geography's empire. He does so with an eclectic body of theory, and shows how imperial access and colonial control revolved around the creation of material and discursive vantage points (or 'spaces of constructed visibility'). He has focused on western travellers in Egypt, but has recently become interested – as have many others – in the colonial production of nature (see Gregory, 2001b).

Others have contributed to debates about the spatiality of colonial discourse by focusing on representations of nature and space. Consider David Livingstone's thesis that geographical texts and contexts are 'reciprocally constituted in the midst of the messy contingencies of history' (1991: 414), and Richard Phillips' argument that the nineteenth-century explorer Richard Burton treated 'geography' as a starting point for explorations of sexuality 'in which all is fluid and boundaries are set up only to be crossed' (1999: 73). Livingstone discusses how western

scientists (including geographers) devised 'moral geographies' of racial superiority that revolved around scientific observations and truth claims about the links between climate, virtue and social development. Climate, particularly, 'became an exploitable hermeneutic resource to make sense of cultural difference and to project moral categories onto global space', with the temperate world being exalted over the tropical world (Livingstone, 2000b: 93). Scientists 'helped [to] produce in the minds of the Victorian public an imagined region – the tropics – which was, at once, a place of parasites and pathology, a space inviting colonial occupation and management, a laboratory for natural selection and racial struggle and a site of moral jeopardy and trial' (Livingstone, 1999: 109). Phillips, by contrast, shows how, in Burton's mind, 'travel, translation, geography and sexology were intimately related', and how this explorer produced dynamic, ambivalent and travelling 'sexual geographies' that both exposed and subverted the dominant heterosexism of the imperial centre – England. Both Livingstone and Phillips read 'geography' as an imperial *discourse* – as a way of delimiting and encoding knowledge, producing a colonial other, and inciting power and desire – but their projects of historical retrieval and critical revision invest discourse with very different textual and contextual meanings.

Second, there are specialist literatures on particular geo-imperial/colonial knowledge practices. The map has a special place in geographers' critical deliberations and postcolonial studies. It now seems obvious that cartography played a crucial role in the imperialists' self-legitimizing construction of space as universal, measurable and divisible. And as Graham Burnett notes, the history of cartography has provided 'an exemplary arena for exploring how the representational production of empire … [created] a stage for dramatic imperial gestures' (2000: 6). Scholars have followed the critical lead of the late Brian Harley, who started to explore the connections between maps, knowledge and power in the 1980s, and there is now an enormous literature on the intertwined histories of empire and cartography (see Jacob, 1992). Much of this literature emphasizes the power of maps. Matthew Edney, for example, explores how the Great Trigonometrical Survey of India (started in 1817) helped the British 'to reduce India's immense diversity to a rational and ultimately controllable structure' (1997: 14–35). Surveying and mapmaking were central to the creation of 'a conceptual image that consciously set the Europeans apart from the Indians they ruled … [and] a cartographic image of the

[Indian] empire as a single territorial and political entity'. And Burnett (2000: 38–52) has looked at the power of cartographic *metalepsis* – how explorers, surveyors and mappers both invoked and remapped the authoritative (and often mythic) texts of their predecessors in order to advance territorial claims, bound colonial space, and secure their own reputations. Yet these and other scholars are also concerned with how we might interrupt and subvert the spatial certainty of the map. We might recover moments of ambivalence in the cartographic record, probe the local knowledges that western travellers used and erased, and delve into the fraught physical and cross-cultural circumstances in which cartographic knowledges were made (e.g. Bravo, 1999; Clayton, 2000b). There is also a plethora of work on alternative – aboriginal and post-colonial – mappings that are based on different cultural and epistemological premises than the 'abstract projective, co-ordinate geometries' of western cartography (Lewis and Woodward, 1997: 537).

Third 'colonizing geographies' are amenable to critical examination from a variety of politico-intellectual positions. Feminism is one such position, and one that is leaving a deep mark on geographers' engagement with identity and difference. Alison Blunt and Gillian Rose (1994: 14) have stressed the need to question 'imperial cartography', entertain more fluid conceptions of subject formation, and explore 'paradoxical spaces' that resist the 'transparent' and 'homogeneous' space mapped by masculine forms of knowing. They treat 'imperial cartography' as a concept metaphor for all of those strategies (including cartography proper, of course) that inscribe gender differences as spatial differences by constructing some spaces as essentially feminine and others as definitively masculine.

Some of this feminist-geographical literature has an acutely disciplinary orientation, but much of it examines the broader gendered spatial boundaries that were authored and authorized by colonialism, and their articulation with constructions of race and class. Sara Mills (1999) and Judith Kenny (1995) have discussed the ways in which British and Indian men and women negotiated the 'confined spaces' of colonial India (its hill stations and cantonments), and wonder about the adequacy of confinement (or exclusion and transgression) as ways of organizing understandings of femininity and masculinity in colonial contexts. Alison Blunt (1994; 2000) and Cheryl McEwan (2000) have explored how the subject and viewing positions of white women travellers changed as they moved between 'home' and 'away' and presented themselves as women,

scientists, explorers, writers and agents of empire before 'civilized' and 'savage' audiences. And James Duncan (2000) has started to unpack the construction of colonial masculinity in natural environments that were radically different than the ones from which colonizers hailed. He shows how, in the coffee plantations of the Kandyan Highlands of Ceylon, moral discourses on tropical climate and nature were connected to a second discourse of 'moral masculinity' that was inflected by the narrative conventions of the masculine adventure novel. Planters represented the tropical highlands as a physical and psychical adversary that tested their manliness and moral fibre, and against which their stories of heroic (and sometimes ignominious) struggle, and fear of losing their masculinity, gained textual momentum.

Fourth, geographers are concerned with how we might devise an alternative postcolonial geography that is not just concerned with disciplinary issues but is more widely interested in the nature of colonialism and decolonization in different parts of the world. Work in this vein emphasizes the contextually located nature of colonialism and explores what Jane Jacobs (1996) has called 'the politics of the "edge"' (the subversive influence of the margin on centrist practices of spatial demarcation, and the discordant postcolonial politics of countries such as Australia). Much of this literature treats colonialism as an intersubjective (if unequal) relationship between colonizing and colonized groups, and stresses the need to distinguish between Eurocentric and nation-centred imperial visions, and the different logics of power enshrined in settler and dependent colonial formations. Much of it is also concerned with the ways that global forces and local exigencies were articulated in the making of particular historical geographies, and how specific postcolonial predicaments frame the questions that are asked about the colonial past and the postcolonial theories that are used (see Crush, 1994).

Regionally focused studies of colonialism's geographies eschew any essentialized vision of either western power or native agency, and many of them show that colonial discourses were skewed and subverted by their material positioning in the colonies. Alan Lester (1998) has shown how British discourses on southern Africa were shaped by the competing visions of colonial officials, humanitarians and settlers, and the different kinds of spaces and conflicting geographies they created – missions, stations, farms, government spaces of control, and social spaces of segregation. And one of the main themes of my work on native–western contact in the Pacific north-west at the end of the eighteenth century is that western explorers and traders and the metropolitan politicians and pundits who turned native space into imperial territory engaged native people in markedly different ways (Clayton, 2000a).

Historical work on colonialism's geographies contains some rich, empirically developed insights into colonial hybridity, ambivalence and anxiety (three key postcolonial tropes). Geographers argue that these dynamics stemmed as much from the messy pragmatics of cultural contact as they did from colonialism's inherent contradictions (such as the colonizer's need to 'civilize' its others yet keep them in everlasting otherness). This scattered body of geographical work on different colonial regions also fills out the postcolonial idea – expressed forcefully by Prakash (1996) – that imperial projects became somewhat hinged as they left metropolitan space and came into contact with alien peoples and environments. Geographers are showing that imperial expansion kick-started diverse and often unpredictable interactions between 'Europe', the natural environment and indigenous peoples, and argue that the geographer's traditional interest in the associations between land and life should lie at the heart of geographical studies of colonialism (see Harris, 2002; Sluyter, 2001).

As these remarks imply, work on colonialism's geographies raises difficult questions about whether it is possible to generalize about colonialism and empire in geographical terms (and whether geographers should want to do so), and about the relative weight we should accord to the power of western representations of people and territory relative to more material and embodied forms of imperial assertion and colonial dispossession that involved native peoples. I have argued that we should think about how particular places and local circuits of contact were articulated with the global imaginings, networks and flows of empire, and develop multiscaled geographies of these local-global connections that account for the interplay between 'the material' and 'the discursive'.

This, in outline, is how questions of colonialism and empire have appeared in geography over the last 10 years. On the one hand, we have an academic centre bent on decentring, and on the other a range of geographical scholarship that is concerned with colonialism's multifarious geographies. These two orientations are not poles apart. Geographical studies of the imperial/colonial past have multiple critical trajectories, and there is cross-fertilization between the subthemes I have identified. There have also been some impressive attempts – I think, especially, of the

work of Jane Jacobs (1996), Jonathan Crush (1994) and Alan Lester (2001) – to bring metropolitan-disciplinary and colonial-'edge' agendas into a unitary analytical field and explore the tensions between them. But if my attempt here, largely for the sake of convenience, to draw a distinction between work on geography's empire and colonizing geographies has any merit, then it is because it is important to think about how geographers position themselves in relation to colonialism and empire. If, as Gregory insists, 'we are the creatures and creators of situated knowledges' (1998: 57), then we must think about the production and politics of positionality. Postcolonial theory teaches us to historicize our work with reference to its site(s) of production, think about our geographies of intellectual labour, and remain alert to the implications of our acts of interpretation.

It is this question of positionality that I will now take up a little more directly, beginning with some questions about geographers' critical aims and moving on to their vexed engagement with other voices.

GEOGRAPHERS' POSTCOLONIAL PREDICAMENTS

Decentring the west and decolonizing geography

What should we make of the industriousness with which geographers are dredging up onerous representations of foreign peoples and places? Are they contesting images and discourses that are best left in the past, or ones that need to be unpacked and challenged because they influence the present? Is work on geography's empire meant to constitute some sort of enlightenment for the discipline? Are geographers documenting a deleterious disciplinary past in order to demonstrate that 'we' now do things differently? Are they trying to mark the discipline's contemporaneity by making geographical modes of analysis that come from other periods exist in the same time (in our research and teaching) but remain in different moral, political and epistemological worlds? Geographers have raised these and other crucial questions about their critical endeavours, but only in fits and starts, and their answers to them are far from uniform and in some ways paradoxical.

David Livingstone (1998: 15), for instance, reports that he has been accused of reinforcing racism by re-presenting images of race in his writing. He says that he finds this a puzzling charge, because he sees his work as an attempt to elucidate 'the intellectual sources of racism', but this spat over the politics of postcolonial representation opens up a number of important issues. Jane Jacobs rehearses the widely held concern that 'despite its postcolonial leanings', revisionist work on the imperial/colonial past 're-inscribes the authority of the events, networks and people that it seeks to decentre and revise' (2001: 730). It does so in a number of ways, but Nicholas Thomas argues that work on colonial discourses and representations is especially problematical because it frequently privileges the power of inscription over material practices and portrays western concepts and visions as 'impervious to active marking or reformulation by the "Other"' (1993: 3, 105).

This leads us to a related concern: that much postcolonial work within (and beyond) geography that seeks to identify with the subject positions of the colonized remains stuck in a Eurocentric mould. Mary Louise Pratt has criticized recent work on travel writing for its fixation with European experience. The experience of travel, she complains, is 'examined from within the self-privileging imaginary that framed the travels and travel books in the first place'(2001: 280). European sensibilities remain of intrinsic interest, and while ideas of cultural negotiation are explored in methodological terms, they are rarely pursued in great substantive depth. Scholars working in this area are teaching us a great deal about how Europeans 'staged' foreign places for inspection and imperial consumption, how identities were renegotiated as Europeans travelled from here to there and remade distinctions between 'home' and 'away', and about the stresses involved. But this literature tells us far less about the non-European peoples and places that supposedly infiltrated and splintered a sovereign European self and imperial subject. We learn a great deal about how Europeans envisioned the other and the faraway, but much less about how the staging of place worked in negotiation with other places and peoples themselves.

This fixation – if it is that – is not misguided in itself. It only becomes a bone of critical contention when scholars working in this way claim that they are also bringing the perspectives and impress of the colonized more clearly into view. Postcolonial critics have responded to – or at least defused – these kinds of charges by claiming that their 'critical apparatus does not enjoy a panoptic distance from colonial history but exists as an aftermath, as an after – after being worked over by colonialism. Criticism formed as an aftermath acknowledges that it inhabits the structures of Western domination that it seeks to

undo' (Prakash, 1994: 1475). Eurocentric habits and categories of thought are very much part of this aftermath, Prakash argues, and we need to question 'the comfortable make-believe' that there exists a critical position outside the historical configurations of colonialism from which a postcolonial future (or decolonized discipline) will emerge. We will not find a true or authentic 'native' perspective that is uncontaminated by the experience of colonization, or a timeless or unitary European worldview that can be deconstructed. Prakash (1996) insists that we critique colonialism *in media res* – from inside a story that has not ended – and Bill Ashcroft notes that 'the most intransigent problem to face postcolonial states today is (still) the challenge of reconstructing inherited institutions and practices in a way that adheres to the demands of local knowledges, makes use of the benefits of local practices, and maintains an integrity of self-representation' (2000: 23).

Gregory (1998) also makes this sort of point, suggesting that we can easily fall into the assumption that the geographical knowledges and practices we are placing under the critical spotlight belong to the past. Historical work on the geographies of colonialism and empire, however piously or unwittingly Eurocentric it may be, has been effective in revealing that demeaning and domineering representations of the other are still alive in western cultural and geographical imaginations, and that 'the fatal attractions of colonial nostalgia [for "timeless scenes" of, and windows on to, "ancient" worlds like Egypt] are inscribed in contemporary forms of travel' (Gregory, 2001c: 113). Similar claims have been made about the political economy of colonial nostalgia (the imbrications of wealth, stability and empire). And among other (appalling) things, the recent tragic events in America highlight the enduring power of imaginative geographies, imbued with imperial symbolism, that glide over and crash into the complexities of cultural difference.

Such ideas sharpen the political edge of 'critical' work on empire that is written from the metropolitan-theoretical heartlands of the discipline, and geographers like Gregory rightly see their work as a critique of the present. But questions remain about geographers' critical aims. Clive Barnett, for instance, thinks that 'the value of history in the relativized historiography of geography remains largely unproblematized' (1995: 414). And he suggests that work on the imperial/colonial past has become popular not because it necessarily has a bearing on geography's present but because it is 'a convenient arena in which we get to practise with different sorts of difficult theory'. One wonders, too, about the extent to which geographical work on the imperial/colonial past helps postcolonial subjects to come to terms with 'the gaps and fissures in their condition'. Studies of geography's empire surely decentre geographical thought, may satiate a poststructuralist thirst for multiplicity and dispersal, and may even be ameliorative for geography and therapeutic for geographers. But in what ways are they postcolonial?

Significantly, geographers' attempts to impute a critical distance between 'then' and 'now' work by a different historicist light than the one that guides postcolonial work in countries such as India. In India especially, Partha Chatterjee argues, where people are daily reminded of their subjection, 'it is precisely the present from which we feel we must escape,' and 'our desire to be independent and creative is transposed on to our past' (1997: 281). In European post-Enlightenment thought, by contrast, the present is conceived of as the site of one's escape from the past. 'This makes the very modality of our [Indian] coping with modernity radically different from the historically evolved modes of Western modernity.' As he implies, postcolonial work that is rooted in the experience of subjection is likely to be different from that which stems from a sense of guilt, or historic injustice, or what have you. We should not think of the former type of postcolonial work as more truly postcolonial than the latter. Rather, my point is that we cannot talk about imperial/colonial history without thinking about the locations (academic, intellectual, cultural, geographical) from which we historicize our investment in the past and anthropologize our investment in the other.

Other voices and native geographies

Let us now turn to a particular knot of questions within this critical fabric – questions of otherness. There has been a flurry of work by geographers on processes of othering, but little of it delves very deeply into the critic's relationship with the other (see Staum, 2000). Colonial discourse analysis in geography, like that in other disciplines, often works at a great distance from its objects of discourse – its others. It is difficult to get at 'the native' side of the story from thoroughly lopsided archives that do not render knowledge about 'them' on 'their' terms. But geographers exacerbate such problems by focusing exclusively on the white/western historical record. Geographical studies that conceptualize colonial encounters as negotiated, situated, intersubjective, contested or

anxiety-ridden often work much better in theory than in practice.

For example, in an essay on 'British women travellers and constructions of racial difference across the nineteenth-century American West', Karen Morin (1998) links the travellers' meetings with and representations of Native Americans to gendered colonial discourses, and tries to 'decentre' such discourses by thinking through 'the social relations inherent in the multiple contact zones [in this case, mainly railroad stations] within which the encounters took place'. Like many other similar studies, the analysis of texts and representations is conceptually sophisticated but the idea that colonial discourses *respond* to the stresses and strains of the contact zone – Thomas' point – gets abstracted away because the other is viewed solely through the filters of a white/western record.

The aim here is not to single out Morin's essay for criticism, but to suggest that it points to a widespread interpretive problem in the geographical literature (and postcolonialism more generally): that otherness is dealt with through the determining pressure of western discourses. The colonial world is deconstructed according to the word of the west. Some geographers confront this problem by recoiling from the analysis of native agency (by not trying to speak for the other) and sticking to the task of showing how dominant knowledges were put together. But such pared-down lines of enquiry can come at a price. They can romanticize the other, or make erroneous assumptions about how natives responded to newcomers. Without any 'native' testimony to go on, they can make imperialism and colonialism look too austere (and thus exaggerate the power of the west) or too anxiety-ridden (and thus overinflate the agency of the critic who chooses to see this trait in the colonial record, or chooses to equate knowledge and power). Much work that passes as postcolonial within geography hinges on the trials and tribulations of colonizers in other spaces rather than on the intersubjective contours of the contact zone in question.

Brenda Yeoh (2000) points out that work on the historical geography *of* colonialism overshadows the difficult but crucial task of uncovering 'the historical geographies of the colonised world'. As difficult and time-consuming as it may be (not what academics under pressure to publish quickly want to hear), she argues, it is vital that geographers complement their deconstructive work on (and in) 'the centre' with research on (and at) the margins of empire and the agencies of the colonized. Yeoh does this in her work on colonial Singapore, and there are pockets of historical-geographical research that deal with native agendas and try to listen to the other. I have used archaeological and ethnographic records as well as historical sources to explore how the native – Nuu-chah-nulth – groups of Vancouver Island incorporated British and American fur traders into their own conflicts, strategies of colonization and systems of the world (Clayton, 2000a). In this contact zone (and I suspect others), native peoples felt anything but possessed or inferior to westerners during the early years of contact. At the same time, the story I tell of native tribal competition, conflict and territorial change hardly squares with images of the ecological Indian living in harmony with nature and his/her neighbours, and in traditional territories from time immemorial, that have played an important role in white-liberal sympathy for native causes (a certain romanticism) and the defence of native land claims in the law courts (a certain strategic essentialism).

My experience raises more general concerns. Critical human geographers may find other voices and start to redress the biases and erasures of colonialist historiography, but how far can they go with them, especially if they hold the view that all voices, identities and narratives are constructions (fictions) of sorts? Do you apply one set of deconstructionist techniques to the white record, and some other set to the native record? And if you apply the same set to both records, are you not likely to diminish your ability to decolonize history? There are no simple answers to such questions of theoretical probity. Critical geographers run the risk of subordinating other voices to the secular codes and conventions of western academic discourse (to rational criteria over the use of evidence that underpin the social sciences and humanities). Barnett notes that many attempts to restore hitherto excluded voices to our accounts still conform to a western model of representation and criticism that 'construct[s] texts as having "voices" hidden within them which await rearticulation through the medium of the critic' (1997: 145). This model 'inscribe[s] colonial textuality within a quite conventional economy of sense which ascribes to voice and speech the values of expressivity, self-presence, and consciousness, and understands the absence of such signs as "silence", as an intolerable absence of voice, and therefore as a mark of disempowerment'. And postcolonial debates about historical discourse have taught us that academic history runs a fine line between rewriting history from 'other' perspectives and longing for lost objects – for a radically heterogeneous world and/or coeval colonial ethnography (see Chakrabarty, 2000).

Finally, as much of this discussion shows, 'critical imperial and colonial geographies' are intellectual constructs that are implicated in the construction of the objects that they apprehend. Work that finds its critical feet by 'displacing', 'interrupting' and 'subverting' demeaning and domineering knowledges is based on retrospective understandings of the relations between knowledge, power and geography. Such work *works*, in part, by the gravity of the imperial/colonial geographies it creates and conjures with – by its ability to jog us out of complacency, and expose and criticize previously unseen and taken-for- granted ideas (see Jacobs, 2000). Power and dominance are rendered as the partly real and partly imagined templates on which geographers stencil their critical commitment to postcolonialism. One of the basic problems with postcolonial attempts to augment understandings of difference (as diversity, multiplicity, otherness) is that they can also homogenize understandings of sameness – or translate a history of the other into the history of the same. Geographers write of *colonizing* geographies, *normalizing* discourses, *insidious* imaginaries in order to take them apart, and to some degree depend on such standardized images of what colonialism was all about to make their critiques work. They depend on powerful words that enable them to make powerful critical gestures, and we need to think about what is lost and gained as they pin colonialism up as a totality (e.g. as a system of epistemic violence) or inherently contradictory and compromised project of power.

In an essay on Said's ideas about the role of the intellectual, Bruce Robbins (1998) notes that models of an inclusionary or democratic oppositional criticism that challenges power in its many guises depend on processes of 'intellectual rarefaction'. Critical intellectual authority with regard to matters of exclusion and marginality, and dominance and hegemony, stems from 'the presumed rarity or scarcity of those willing to confront non-intellectual authority'. It is the *restrictiveness* of this group that gives it its ethico-political legitimacy, Robbins argues, and 'an ethical scarcity defined by opposition will be indistinguishable from a social scarcity that is a potential source of profit and prestige'. The intellectual who faithfully inverts the authority of power is dependent on and prized by that power. In other words, critical human geographers, like postcolonial intellectuals, have a certain investment in cultivating questions of difference and power, and cultivating their own scarcity, if you will, as well as challenging the legacies of colonialism.

CONCLUSION

There can be no simple summary to a chapter like this, which ranges over a wide critical terrain. But I will end with three points – two of them derived from the literature and a final point of my own. First, Barnes and Gregory note that work on the geographies of colonialism and empire has a central stake in 'the worlding of human geography' (1997: 14–23). This range of work is challenging the view that geography is a field of study that is capable of producing an impartial and independent body of knowledge. The intertwined histories of geography and empire that are currently being explored underscore the notion that geographers are socialized into a discipline and discourse 'whose assumptions, concepts and ways of working are always and everywhere earthed in material grids of power'. Geography is not simply a way of finding order in the world; it is also about the creation and command of order.

Second, Livingstone suggests that work on 'geography's historical geographies' is 'relativising our definition of geography', 'pluralising our conception of geography', and prompting us to 'particularise our own practice of geography' (2000a: 7). We acknowledge that 'what geography is cannot be uncovered in isolation from the conditions of its making' in different times and places, and we 'now admit, even celebrate, the impossibility of laying aside our own particularity in cognitive and practical projects'. We are caught up in 'the retaliation of the situated', and postcolonialism is one of its key manifestations. Indeed, there are few signs that the range of work reviewed here will become denigrated by a sort of geographical prose of counterinsurgency that restores order and objectivity to geographers' research and writing. Explicitly *historical*-geographical work on postcolonial matters will no doubt continue to grow and change, in part as the wider field of postcolonial studies changes, but it is not likely to be displaced from debates about the historicity of human geography or its cultural politics.

Yet third, there are clearly problems with this literature. In my view, and as Hall (1996: 249) notes about postcolonialism more generally, geographers' descent into discourse and focus on geography's empire can easily become an alibi for deconstructive work that falls into the trap of assuming that the theoretical critique of essentialism necessarily entails its political displacement, and in a sense bypasses the postcolonial world beyond Europe altogether. The critique of colonialism can become a seductive but sanitized western intellectual pastime that may serve the professional needs of oppositional academics – who, as David Scott (1999) has observed, are suffering from the loss of familiar and stable political objects – but that barely connects with the practical predicaments of formerly colonized peoples and places. This is not to say that work on 'real-world' postcolonial problems is better than work that deconstructs empire from its metropolitan-disciplinary pavilions. But I do want to end by calling for more dialogue between geographers working within the different orientations identified above, and geographers working in different parts of the world.

NOTES

I would like to thank Charles Withers and his postgraduate group at Edinburgh for their detailed comments on an earlier and much longer draft of this chapter, Jane Jacobs for her patience and comradely editorial advice and Joe Doherty for commenting on a penultimate draft.

1 Unfortunately, there is little space here to discuss the ways in which questions of geography and space are dealt with by postcolonial scholars in other disciplines. Suffice it to say that geographical concepts and metaphors have become coveted critical commodities.

REFERENCES

Anderson, K. (1998) 'Science and the savage: the Linnean Society of New South Wales, 1874–1900', *Ecumene* 5 (2): 125–43.
Ashcroft, B. (2000) '"Legitimate" post-colonial knowledge', *Mots Pluriels* 14. 13–27.
Barnes, T. and Gregory, D. (eds) (1997) *Reading Human Geography: The Poetics and Politics of Inquiry.* London: Arnold.
Barnett, C. (1995) 'Awakening the dead: who needs the history of geography?', *Transactions of the Institute of British Geographers* 20 (NS): 417–19.
Barnett, C. (1997) '"Sing along with the common people": politics, postcolonialism, and other figures', *Environment and Planning D: Society and Space* 15: 137–54.
Barnett, C. (1998) 'Empire and worldly geography: the Africanist discourse of the Royal Geographical Society, 1831–73', *Transactions of the Institute of British Geographers* 23 (NS) (2): 239–52.
Bell, M., Butlin, R. and Heffernan, M. (eds) (1995) *Geography and Imperialism, 1820–1940.* Manchester: Manchester University Press.
Blunt, A. (1994) *Travel, Gender and Imperial: Mary Kingsley and West Africa.* London: Guilford.
Blunt, A. (2000) 'Spatial stories under siege: British women writing from Lucknow in 1857', *Gender, Place and Culture* 7 (3): 229–46.

Blunt, A. and Rose, G. (eds). (1994) *Writing Women and Space: Colonial and Postcolonial Geographies.* London: Guilford.

Bravo, M.T. (1999) 'Ethnographic navigation and the geographical gift', in David N. Livingstone and Charles W.J. Withers (eds) *Geography and Enlightenment.* Chicago: University of Chicago Press. pp. 199–235.

Bruneau, M. and Dory, D. (eds) (1994) *Géographies des colonisations XV–XX siècles.* Paris: Hatmann.

Burnett, D.G. (2000) *Masters of All They Surveyed: Exploration, Geography, and a British El Dorado.* Chicago: University of Chicago Press.

Butzer, C. (ed.) (1992) Special issue on the Columbus quincentenary, *Annals of the Association of American Geographers* 83 (2).

Chakrabarty, D. (2000) *Provincializing Europe: Postcolonial Thought and Historical Difference.* Princeton: Princeton University Press.

Chatterjee, P. (1997) *A Possible India.* Delhi: Oxford University Press.

Claval, P. (1998) *Histoire de la géographie française de 1870 à nos jours.* Paris: Nathan.

Clayton, D. (2000a) *Islands of Truth: The Imperial Fashioning of Vancouver Island.* Vancouver: UBC Press.

Clayton, D. (2000b) 'On the colonial genealogy of George Vancouver's chart of the northwest coast of North America', *Ecumene* 7 (4): 371–401.

Crush, J. (1994) 'Post-colonialism, de-colonization, and geography', in A. Godlewska and N. Smith (eds) *Geography and Empire.* Oxford: Blackwell. pp. 333–50.

Dirks, N. (1996) 'Foreword' to Bernard Cohn *Colonialism and Its Forms of Knowledge.* Princeton: Princeton University Press.

Driver, F. (1992) 'Geography's empire: histories of geographical knowledge', *Environment and Planning D: Society and Space* 10: 23–40.

Driver, F. (ed.) (1995) 'Geographical traditions: rethinking the history of geography', *Transactions of the Institute of British Geographers* 20 (NS): 403–4.

Driver, F. (2000) *Geography Militant: Cultures of Exploration and Empire.* Oxford: Blackwell.

Driver, F. and Yeoh, B. (eds). (2000) Special issue on Tropicality, *Singapore Journal of Tropical Geography* 21.

Duncan, J. (2000) 'The struggle to be temperate: climate and "moral masculinity" in mid-nineteenth-century Ceylon', *Singapore Journal of Tropical Geography* 21: 34–47.

Duncan, J. and Gregory, D. (eds) (1999) *Writes of Passage: Reading Travel Writing.* London: Routledge.

Edney, M. (1997) *Mapping an Empire: The Geographical Construction of British India, 1765–1843.* Chicago: University of Chicago Press.

Gandhi, L. (1998) *Postcolonial Theory: A Critical Introduction.* Edinburgh: Edinburgh University Press.

Godlewska, A. and Smith, N. (eds) (1994) *Geography and Empire.* Oxford: Blackwell.

Graham, B. and Nash, C. (eds) (2000) *Modern Historical Geographies.* London: Prentice Hall.

Gregory, D. (1994) *Geographical Imaginations.* Oxford: Blackwell.

Gregory, D. (1995) 'Imaginative geographies', *Progress in Human Geography* 19 (4): 447–85.

Gregory, D. (1998) 'Explorations in critical human geography'. Hettner Lectures 1, University of Heidelberg.

Gregory, D. (2000a) 'Postcolonialism', in R.J. Johnston, D. Gregory, G. Pratt and M. Watts (eds) *Dictionary of Human Geography*, 4th edn. Oxford: Blackwell. pp. 612–15.

Gregory, D. (2000b) 'Cultures of travel and spatial formations of knowledge', *Erdkunde* 54: 297–309.

Gregory, D. (2001a) *The Colonial Present.* Oxford: Blackwell.

Gregory, D. (2001b) '(Post)colonialism and the production of nature', in Bruce Braun and Noel Castree (eds) *Social Nature.* Oxford: Blackwell.

Gregory, D. (2001c) 'Colonial nostalgia and cultures of travel: spaces of constructed visibility in Egypt', in Nezar AlSayyad (ed.) *Consuming Tradition, Manufacturing Heritage: Global Norms and Urban Forms in the Age of Tourism.* London: Routledge.

Hall, S. (1996) 'When was "the post-colonial"? Thinking at the limit', in I. Chambers and L. Curtis (eds), *The Post-colonial question: Common Skies, Divided Horizons.* London: Routledge. pp. 242–60.

Harris, C. (2002) *Making Native Space: Colonialism, Resistance and Reserves in British Columbia.* Vancouver: UBC Press.

Heffernan, M. (2001) '"A dream as frail as those of ancient Time": the in-credible geographies of Timbuctoo', *Environment and Planning D: Society and Space* 19: 203–25.

Howitt, R. (2001) 'Frontiers, borders, edges: liminal challenges to the hegemony of exclusion', *Australian Geographical Studies* 39 (2): 233–45.

Jacob, C. (1992) *L'Empire des cartes: Approche théorique de la cartographie à travers l'histoire.* Paris: Albin Michel.

Jacobs, J.M. (1996) *Edge of Empire: Postcolonialism and the City.* London: Routledge.

Jacobs, J.M. (2000) 'Difference and its other', *Transactions of the Institute of British Geographers* 25 (NS) 4: 403–8.

Jacobs, J.M. (2001) 'Touching pasts', *Antipode* 33 (4): 730–4.

Kelly, J. (1999) 'Maori maps', *Cartographica* 36 (2): 1–30.

Kenny, J. (1995) 'Climate, race and imperial authority: the symbolic landscape of the British hill station in India', *Annals of the Association of American Geographers* 85: 694–714.

Kenny, J. (ed.) (1999) 'Colonial geographies: accommodation and resistance', thematic issue of *Historical Geography* 27.

Lejeune, D. (1993) *Les sociétés de géographie en France et l'expansion coloniale au XIXième siècle.* Paris: Albin Michel.

Lester, A. (1998) 'Reformulating identities: British settlers in early nineteenth-century South Africa', *Transactions of the Institute of British Geographers* 23 (NS) (4): 515–31.

Lester, A. (2001) *Imperial networks: Creating Identities in Nineteenth-century South Africa and Britain.* London: Routledge.

Lewis, M. and Woodward, D. (eds) (1997) *The History of Cartography*, vol. 2, book 3, *Cartography in the Traditional African, Arctic, Australian, and Pacific Societies*. Chicago: University of Chicago Press.

Livingstone, D.N. (1991) 'The moral discourse of climate: historical considerations on race, place and virtue', *Journal of Historical Geography* 17: 413–34.

Livingstone, D.N. (1992) *The Geographical Tradition: Episodes in the History of a Contested Enterprise*. Oxford: Blackwell.

Livingstone, D.N. (1998) 'Reproduction, representation and authenticity: a rereading', *Transactions of the Institute of British Geographers* 23 (NS): 13–20.

Livingstone, D.N. (1999) 'Tropical climate and moral hygiene: the anatomy of a Victorian debate', *British Journal of the History of Science* 32: 93–110.

Livingstone, D.N. (2000a) 'Putting geography in its place', *Australian Geographical Studies* 38 (1): 1–9.

Livingstone, D.N. (2000b) 'Tropical hermeneutics: fragments for a historical narrative', *Singapore Journal of Tropical Geography* 21 (1): 92–8.

Livingstone, D.N. and Withers, C.W.J. (eds) (1999) *Geography and Enlightenment*. Chicago: University of Chicago Press.

Loomba, A. (1998) *Colonialism/Postcolonialism*. London: Routledge.

Maddrell, A.M.C. (1998) 'Discourses of race and gender and the comparative method in geography school texts, 1830–1918', *Environment and Planning D: Society and Space* 16: 81–103.

McEwan, C. (2000) *Gender, Geography and Empire: Victorian Women Travellers in West Africa*. London: Ashgate.

Mills, S. (1999) 'Gender and colonial space', *Gender, Place and Culture* 3 (2): 125–47.

Morin, K.M. (1998) 'British women travellers and constructions of racial difference across the nineteenth-century American West', *Transactions of the Institute of British Geographers* 23 (NS): 311–30.

Myers, G.A. (1998) 'Intellectual of empire: Eric Dutton and hegemony in British Africa', *Annals of the Association of American Geographers* 88 (1): 1–27.

Phillips, R. (1997) *Mapping Men and Empire: A Geography of Adventure*. London: Routledge.

Phillips, R. (1999) 'Writing travel and mapping sexuality: Richard Burton's sotadic zone', in J. Duncan and D. Gregory (eds) *Writes of Passage: Reading Travel Writing*. London: Routledge. pp. 70–91.

Ploszajska, T. (1996) 'Constructing the subject: geographical models in English schools, 1870–1944', *Journal of Historical Geography* 22 (4): 388–98.

Prakash, G. (1994) 'Subaltern studies as postcolonial criticism', *American Historical Review* 1475–90.

Prakash, G. (1995) '*Orientalism now*', *History and Theory* 34 (3): 199–212.

Prakash, G. (1996) 'Who's afraid of postcoloniality?', *Social Text* 49: 187–203.

Pratt, M.L. (1992) *Imperial Eyes: Travel Writing and Transculturation*. London: Routledge.

Pratt, M.L. (2001) 'Review of Duncan and Gregory (eds) *Writes of Passage'*, *Journal of Historical Geography* 27 (2): 279–81.

Robbins, B. (1998) 'Secularism, elitism, progress and other transgressions', in Keith Pearson, B. Parry and J. Squires (eds) *Cultural Readings of Imperialism: Edward Said and the Gravity of History*. London: Lawrence & Wishart.

Rose, G. (1995) 'Tradition and paternity: same difference?', *Transactions of the Institute of British Geographers* 20 (NS): 414–16.

Ryan, J. (1994) 'Visualizing imperial geography: Halford Mackinder and the Colonial Office visual instruction committee, 1902–11', *Ecumene* 1 (2): 157–76.

Ryan, J. (1997) *Picturing Empire: Photography and the Visualisation of the British Empire*. Chicago: University of Chicago Press.

Said, E.W. (1978) *Orientalism*. New York: Random House.

Said, E.W. (1993) *Culture and Imperialism*. New York: Knopf.

Said, E.W. (2000) 'Invention, memory, and place', *Critical Inquiry* 26: 175–92.

Said, E.W. (2001) 'Globalizing literary study', *Publications of the Modern Language Association of America* 116 (1): 64–8.

Scott, D. (1999) *Refashioning Futures: Criticism after Postcoloniality*. Princeton: Princeton University Press.

Siliberto, M.C. (1998) 'Il mundus novus di Amerigo Vespucci: fra discipline geographie, storiche e filologiche', *Rivista Geographica Italiana* CV (2/3).

Sluyter, A. (2001) 'Colonialism and landscape in the Americas: material/conceptual transformations and continuing consequences', *Annals of the Association of American Geographers* 91 (2): 410–28.

Staum, M.S. (2000) 'The Paris geographical society constructs the other, 1821–1850', *Journal of Historical Geography* 26 (2): 222–38.

Thomas, N. (1993) *Colonialism's Culture*. Princeton: Princeton University Press.

Washbrook, D. (1999) 'Orients and occidents: colonial discourse theory and the historiography of the British Empire', in R.W. Winks (eds) *The Oxford History of the British Empire, vol. V Historiography*. Oxford: Oxford University Press. pp. 596–611.

Withers, C.W.J. (2000) 'Voyages et crédibilité: vers une géographie de la confiance', *Géographie et Cultures* 33: 3–17.

Yeoh, B.S.A. (2000) 'Historical geographies of the colonised world', in B. Graham and C. Nash (eds) *Modern Historical Geographies*. London: Longman, pp. 146–66.

Young, R. (2001) *Postcolonialism: An Historical Introduction*. Oxford : Blackwell.

19

Postcolonial Geographies of Place and Migration

Brenda S.A. Yeoh

For those of us writing from the margins (of one form or another), few would not have been attracted by the term 'postcolonial' and wondered about the promise and possibilities it may contain. This is particularly the case in the face of statements such as Hall's that the world today is 'incontrovertibly *post-colonial*', as

> colonisation so refigured the terrain that, ever since, the very idea of a world of separate identities, of isolated or separable and self-sufficient cultures and economies, has been obliged to yield to a variety of paradigms designed to capture these different but related forms of relationship, interconnection and discontinuity. (1996: 257, 252–3)

Hall goes on to claim that the 'postcolonial' touchstone offers 'an alternative narrative, highlighting key conjunctures to those embedded in the classical narrative of Modernity … [a] re-narrativisation [that] displaces the "story" of capitalist modernity from its European centring to its dispersed global "peripheries"'(1996: 249). Much indeed has been claimed for the power of the postcolonial critique in cutting right through to the very terms by which knowledge is constructed and the world mapped. As Anthony King argues,

> Compared to other representations of the contemporary global human condition, postcolonial studies may be said to restore history and colonialism into presentist theories of globalization, and contest representations of the contemporary world in terms of Eurocentric notions of postmodernism. (1999: 99)

Such is what Jacobs calls the 'fantastic optimism of the "post" in postcolonialism' (1996: 24). Yet its value and impact as a critical and emancipatory discourse within geography and beyond cannot be taken for granted. Critics have already alerted us to the dangers. Drawing on his work

on nationalism and postcolonial identity in Sri Lanka, Perera (1998: 6) warns that just as the application of the category 'precolonial' to societies prior to their incorporation into European political and economic systems tends to fix the 'colonial' as the main point of reference, so adding the prefix 'post-' may also impose 'the continuity of foreign histories' and 'subordinate indigenous histories'. Reflecting on the techniques of subjugation and violence applied by 'pro-Indonesia militias' in the recent annexation of East Timor, Kusno concludes that 'Behind the postcolonial [present] can lurk the spectre of a future more sinister than the colonial past itself' (2000: xii). Critics have thus argued that not only is postcolonial discourse out of touch with post-colonial realities, it may itself serve to mask and at the same time perpetuate the presence of a Eurocentric pall over current efforts at (re-)-constituting the world in discursive and material terms. As Sidaway points out, 'any postcolonial geography must realize within itself its own impossibility, given that geography is inescapably marked (both philosophically and institutionally) by its location and development as a western-colonial science' (2000: 593).[1]

Would it then be possible to steer between the seduction of optimistic claims as to postcoloniality's 'possibilities' and the disabling gridlock of critiques as to its 'impossibilities'? I shall argue that starting places leading to possible paths may be found by taking on board the view that the 'postcolonial' is not a totalizing or monolithic discourse representing one half of any simple west/non-west bifurcation of the world, but in fact a highly mobile, contestatory and still developing arena where opportunities for insight may be gained at multiple sites. Its redemptive features as a means of resisting colonialisms of all forms and its manipulative aspects as a vehicle for colonialism to reproduce itself cannot be totally disentangled, but I argue that the way forward is not to accept the paralysis of such an impasse but to take advantage of the 'shape-shifting instability of the concept' (Hau, 2000: 78) and to strategically and critically mine this variegated field[2] for insights and impulses. It is by encouraging multiple points of entry into the discourse and the presence and participation of a wider range of subjects, scholars and activists that one may hope to chisel at the edges of this epistemological empire and carry the ground away from the current western-centric loci (both philosophical and institutional, as Sidaway observes) of its imagining. This is no easy task, and perhaps the most difficult questions revolve around how we may move beyond 'iconoclastic talk about "domination" of alien models and theories' to the construction of alternative frameworks and metatheories which reflect 'indigenous' worldviews and experiences (Atal, 1981: 195). Shamsul points out that 'to have an academic discourse beyond "orientalism" and "occidentalism" is rather a tall order as long as we cannot break away from and become totally independent of colonial knowledge' (1998: 2). Not only did the colonial project invade and conquer territorial space, it has systematically colonized indigenous epistemological spaces, reconstituting and replacing these using a wide corpus of colonial knowledge, policies and frameworks. With decolonization, ex-colonies have regained (sometimes partial) political territory, but seldom the epistemological space. Yet, surely the more constructive response here is not to reject western discourse as tainted and hence disabling, but to use 'its very own tools of critical theory ... not only to dismantle colonialism's signifying system but also to articulate the silences of the native by liberating the suppressed in discourse' (Zawiah Yahya, quoted in Alatas, 1995: 131).

I would also argue that these endeavours are more likely to achieve transformative effect if postcolonialism is interrogated as part of a serious and sustained engagement with 'material practices, actual spaces and real politics' (Barnett, 1997: 137). As King notes, 'postcolonialism (or postcoloniality) exists as a concept for representing particular conditions in the contemporary world, especially (though not only) in regard to issues of identity, meaning, and consciousness and, not least, the material forms and spaces in which they are embodied' (1999: 101). Yet, many discussions of 'the making of postcolonial subjects as hybrid, contradictory, and ambivalent' tend to be 'noticeably unmediated by the material properties of space' (Kusno, 2000: 211). If the

main limits of postcolonial theories lie in their mistaken 'attempt to transcend in rhetoric what has not been transcended in substance' (Ryan, 1994: 82), then an important starting place in overcoming some of these limitations would be to dissect postcoloniality as threaded through real spaces, built forms and the material substance of everyday biospheres in the postcolonial world. By overlaying and etching the complex contours of the postcolonial debate onto a specific space with both material and imagined dimensions, geographers in particular are well positioned to grasp the substance along with the critique and avoid the navel-gazing tendencies of certain forms of postcolonial studies, which seem reluctant to go much further beyond theorizing 'the meaning of the hyphen in post-coloniality'.[3]

My purpose then is to begin exploring the postcolonial terrain on foot, so to speak, not attempting to cover a lot of ground as expected in an extensive survey, but recognizing my own inescapable location in material space. I begin with postcolonial Singapore before ranging further afield where other paths may be discerned. In these explorations, I will first turn to the way postcolonial nations search for 'groundings' to locate, define and solidify a sense of identity before moving on to consider the way these nations cope with 'unmoorings' – fluidities and mobilities which not only transgress the borders of the nation, but further trouble the inviolability of its body.[4]

POSTCOLONIAL MEMORY AND THE LOCALIZING OF IDENTITY

> Singapore, in many ways, is the product of forgettings. Singapore occurred, and continues to sustain itself, as a result of recurrent acts of forgettings. Forgetting is the condition of Singapore. (Devan, 1999: 22)

The engagement with memory and identity in postcolonial nations such as Singapore is fraught terrain, woven around the politics of inclusion and exclusion, of 'remembering' and 'forgetting' where both acts are not just accidental or ignorant acts but more often than not 'structural necessities' (1999: 22). For its people, knowing what to remember (and what to forget) in order to arrive at a sense of self-identity is a complicated business involving being able to trace a line of sight through multiple prisms refracting the nation's history in different directions, for Singaporeans 'inherit [an] Asian identity through Westernization (which for some is almost identical as modernization) via colonialism' (Koh, 1999: 46). And because the past contains radical breaks and unresolved contradictions compressed within a relatively short space and time, it is prone to simplifications by those such as the agencies of the state which are 'tempted to confer upon it an ideal history, a proper genealogy' (Devan, 1999: 33) for the sake of building the nation and producing the 'ideal of the postcolonial citizen' (Srivastava, 1996: 406).

In this endeavour, the power of the landscape as 'a vast repository out of which symbols of … ideology can be fashioned' (Duncan, 1985: 182) may be harnessed. Amidst the enormous pressures of forging an independent nation out of the raging political and socio-economic fires of the 1960s, Singapore's postcolonial political leaders did not forget to draw on the power of landscape spectacle when confronted with 'a complex, multiracial community with little sense of common history, with a group purpose which is yet to be properly articulated … in the process of rapid transition towards a destiny which we do not know yet' (Goh Keng Swee, then Minister of the Interior and Defence, quoted in Chew, 1991: 363). In 1966, Singapore's first National Day Parade (enacted every year since) was staged at the Padang, an expanse of green flanked by British municipal and religious institutions and which served as both cricket and ceremonial ground (a quintessentially British combination) in the colonial days. As the sea of green vanished beneath the feet of thousands of parade participants arranged in serried ranks and wielding military and musical instruments, flags and other paraphernalia, synchronized displays of parade motifs asserting the joys of living and working together as 'one people, one nation and one Singapore' appropriated what was once the locus of British colonial power and civic pride and reinscribed it with equally ostentatious meanings congruent with nationhood (Kong and Yeoh, 1997; Rajah, 1999). The architectural spectacularity of the colonial past and the animated spectacularity of the momentous present were drawn together and fused in a collective act of remembrance and amnesia, which also then served as a vehicle to envision what *should* lie beyond.

Such occasions where a simultaneous remembering and forgetting of the colonial past are capitalized upon to prescribe a new beginning and a utopian future are replicated in a variegated number of ways in the postcolonial struggle for identity. Beyond the cultural politics of creating landscapes of spectacle, struggles as to how to deal with 'not-so-hidden histories and not-so-absent geographies of imperialism' also continue to be played out in everyday postcolonial landscapes. In Singapore, this is clearly seen, for example, in the strategies to rewrite the colonial toponymic text and inscribe nationhood: presented on independence with an official network of street and place names rooted in the colonial imagination – commemorating British royalty, governors, heroes and dignitaries, honouring European city fathers and public servants, recalling linkages with Britain and the British empire, and racializing places by separating the colonized into distinct segregated districts by race – the new architects of the Singapore landscape soon got to work experimenting with new significations better tailored to project the new order. 'Old colonial nuances, British snob names of towns and royalty' were deliberately avoided, and a slew of 'rewritings' transformed the landscape: first, a Malayanizing to signal Singapore's allegiance to the Malay as opposed to the colonial world in the 1960s, followed closely by the introduction of a multiracial logic in street naming in accordance with the foundational racial arithmetic of the new nation, and moving on in the 1970s and 1980s to the use of 'mathematical naming', 'pinyinization' (a Mandarin system of romanizing Chinese characters seen to be superior to the haphazard translations from Chinese dialect bequeathed by the British) and 'bilingualism' to give the toponymically Anglicized city an 'Asian feel' (Yeoh, 1996a).

In the symbolic (re)production of the landscape, the postcolonial strategy here is not so much to erase the colonial imprint but to recolonize with a different script, a script which destabilizes the logic of colonial imaginings by offering its own accents in counterpoint to what was there before. Clearly, postcolonial strivings for a new identity do not completely banish the colonial past but involve the selective retrieval and appropriation of indigenous and colonial cultures to produce appropriate forms to represent the postcolonial present. As Kusno observes, postcolonial identity is 'ironic', 'contradictory' and anxious about 'inauthenticity', constituted by both a 'relatively unproblematic identification with the colonizer's culture, *and* a rejection of the colonizer's culture'(1998: 550).

Architectural design provides us with further everyday material forms to examine the 'relationships between the memorialisation of the past and the spatialisation of public memory' (Johnson, 1995: 63; see also Johnson 1996; 1999) in the postcolonial context of nation-building. For example, in his comparative analysis of the design of parliamentary complexes in postcolonial states in Asia and the Middle East, Vale

(1992; see also Kironde, 1993; Perera, 1998) singles out the design of the capitol complex as emblematic of the state's desire to use spectacular and monumental means to deliver and prescribe national identity. It is unfortunate that in the complex postcolonial enterprise of cultivating national identity, the balance between 'cultural self-determination and international modernity' in the design of these monumental forms was not always particularly imaginative, sometimes reduced to a question of how to be 'Western without depending on the West' (Vale, 1992: 53, quoting Edward Shils), or, even worse, resulting in what one Arab intellectual dismisses as 'slums of the West'.[5] To take a different example, in the context of the architectural design of university campuses at Bandung and Jakarta, Kusno (1998: 565) argues that the object of architectural desire in postcolonial new order Indonesia is fundamentally split between a denial and displacement of 'colonial origins' on the one hand, and a recitation of the coloniality of 'Indonesian architecture' on the other (see also Oduwaye, 1998, on the development of university campuses in postcolonial Nigeria). These two strains are perpetually contradictory and yet indissolubly intertwined, giving rise to attempts to 'anaesthetize the pain of this contradiction … by a continuous attempt to recover and imagine [the new order's] own "tradition"' (Kusno, 1998: 572).

The contested 'heritagizing' of specific elements of the landscape inherited from the colonial past is particularly salient in illuminating the spatialized cultural politics at work in postcolonial nations. Coming back to the case of Singapore, part of the postcolonial exercise in forgetting involved making deep and thorough excisions in the landscape to remove all that is thought to be obsolete or retrogressive, and to make way for embedding 'new' memories appropriate to the state's construction of the national self. If the first two decades of the nation's development were dictated by systematic amnesia and the erasure of the past through major state-driven programmes of urban renewal and redevelopment, the next two saw a more concerted attempt to recover memory loss and in so doing fashion an appropriate genealogy which would constitute the nation's legitimacy and which is clearly marked, signposted and concretized in the landscape. 'Remembering' emerged at a specific time and place in the nation's development, both as an inevitable condition of the cycle of progress and loss and as a deliberate strategy of forging the nation's future. Chua (1995) argues that 'nostalgia' and a harking back to the past – a past portrayed as a 'foreign country' where 'they do things differently'

(Hartley, quoted in Lowenthal, 1985: xvi) – during the 1980s and 1990s were rooted in the wider critique of and resistance to the relentless drive towards economic development, the frenetic pace of life, high stress levels, the corruption of new-found materialism and the consequent 'industrialization of everyday life'. The nostalgia of the nation is hence a postcolonial critique of postcolonial success – the emergence of the nation from the jaws of colonialism to have miraculously 'arrived' in an economic and material sense, only to find a place bristling with efficiency and productivity but bereft of a certain depth of memory and history. The work of salvaging and heritagizing remnant landscapes – resurrecting once-obsolete shophouse districts in Chinatown, Kampong Glam and Little India as 'heritage districts' and 'ethnic quarters' (a particularly colonial construct) or repackaging the civic and cultural district into heritage trails offering the best of colonial Singapore (Chang, 1997; 2000; Chang and Yeoh, 1999; Huang, et al., 1995; Yeoh and Huang 1996; Yeoh and Kong, 1994; 1996) – is symptomatic of the state's response to such a critique. The response itself is however highly problematic, for not only does it 'forget' the earlier attempts at excising the colonial past to make space for creating a modern Singapore with a breakaway trajectory leading to a different future, it also appears unaware of the contradictions between the two postcolonial impulses of straining to 'forget' and needing to 'remember'. Symptomatic of these tensions is the fact that the Urban Redevelopment Authority in Singapore, in charge of renewal and redevelopment of the city's physical fabric under a 'demolish and rebuild' philosophy, is also the national conservation authority overseeing the preservation and protection of buildings signifying the 'history and memory of the place' (www.ura.gov.sg).

The difficulties of postcolonial remembrance and amnesia are further compounded by the fact that what constitutes 'history' in multiethnic postcolonial nations is a major minefield. This is because 'the [postcolonial] text speaks with a multitude of languages' (Cleary, 1997: 28), mixing colonial idioms with the postcolonial in indissoluble ways, making it difficult to sieve out what belongs to the pure, non-colonized 'self', and troubling attempts to either break from, or draw on, the colonial past as 'other'. This is also because drawn into the postcolonial crucible are a multitude of different interest groups and alliances alongside the postcolonial state and commercial ventures, each staking a different claim on the nation's heritage, and a right over what it should not 'forget to remember', as well as what it should 'remember to forget' (Devan, 1999: 22). Hindsight is hence perpetually unstable, shifting with each perspective. What is valorized and mapped as 'heritage' in official and popular imaginative geographies becomes locked into questions such as who controls (and benefits from) the whole process of transforming 'history' into tangible presences (and hence also absences) on the landscape and for what purposes (such as group identity formation, nationalism or tourism) (Bonnemaison, 1997; Chang, 1997; Jones and Bromley, 1996; Jones and Varley, 1999; Kwok et al., 1999; Parenteau et al., 1995; Shaw and Jones, 1997).

While the postcolonial state in many instances (as in Singapore: see Kong and Yeoh, 1994) has exercised heavy-handed control over the definition of what constitutes the nation's memory and the actual work of heritagizing, other agencies, including marginal groups, have also played a part in some of these struggles over place. Even in Singapore, in responding to state-envisioned heritage landscapes, there are clearly alternative readings and resistances within the body of the postcolonial nation against such hegemonic intentions, although little expressed in confrontational style. Some have clearly found state-propelled conservation and preservation efforts superficial, with little penetrating beneath the veneer of commercialization to creatively connect with the past. As such, so-called heritage landscapes designed by state agencies have been dismissed by some as 'a piece of kitsch ... some kind of feeble confection' (architect Tay Kheng Soon, *The Straits Times,* 18 February 2000) and by others as somewhat bland and disengaged from the development of a sense of national identity.

Elsewhere in the postcolonial world, the politics of what constitutes heritage continue to unfold as nations search to define their identities. In the state-designated 'historic city' of Melaka, for example, Portuguese Eurasians resist being excised from official accounts by their 'chameleon-like abilities' in repositioning their 'tradition' and 'heritage' within Malaysian debates about postcolonial national identity (Sarkissian, 1997). In the same city, Cartier (1993; 1997: 555) examines how place-based constructions of cultural identity and representations of state nationalism are drawn into the politics of space surrounding Bukit China, a monumental traditional Chinese burial ground, and details the protracted struggles over its transformation into a 'nationscape, a site-specific distillation of half a millennium of Malaysian history'.

POSTCOLONIAL MIGRATIONS AND THE 'MIGRANCY OF IDENTITY'

> [T]he postcolonial [highlights] the complexities of diasporic identification which interrupt any 'return' to ethnically closed and 'centred' original histories ... in the global and transcultural context ... It made the 'colonies' themselves, and even more, large tracts of the 'post-colonial' world, always-already 'diasporic' in relation to what might be thought of as their cultures of origin. The notion that only the multi-cultural cities of the First World are 'diaspora-ised' is a fantasy which can only be sustained by those who have never lived in the hybridised spaces of a Third World, so-called 'colonial', city. (Hall, 1996: 250)[6]

That postcolonial nations are 'always-already diasporic' and constitute 'hybridised spaces' comes as no surprise to those of us living in once-colonized cities such as Singapore. As Harper writes of the polyglot city once constituted by streams of immigrants from China, India, the Malay archipelago and other far-flung places and dominated by a small European imperial diaspora:

> Singapore is a child of diaspora. Its history embodies many of the tensions of blood and belonging that the concept evokes. Singapore testifies to the difficulties of creating a modern nation-state on a model inherited from Europe in a region where history mocks the nation-state's claims to cultural and linguistic exclusiveness. The post-colonial experience of Singapore has been dominated by the attempts of the state – an artifact of British rule – to surmount these constraints and to create a national community bounded by a common culture and a sense of place, and bonded by individual allegiance. (1997: 261)

As colonialism reached far and deep into once-localized societies, it generated a multitude of mobilities across borders, coalescing into what M.L. Pratt calls 'contact zones' *par excellence* which invoke the 'spatial and temporal copresence of subjects previously separated by geographic and historical disjunctures, and whose trajectories now intersect' (1992: 7). Diasporas of all hues – imperial diasporas, labour diasporas, trade diasporas, cultural diasporas among them (see Cohen, 1997, for a typology) – quickened in response to the demands of empire, criss-crossed, interlocked and produced hybridized spaces arranged in kaleidoscopic disarray. As earlier discussed, one of the primary tasks of postcolonial nation-building is to transform a motley crew of diasporic orphans, whose emotional homelands diverge from their physical locations as well as from each other, into a settled people who belonged to a single home nation in every way.

Even as postcolonial nation-building attempts to territorialize and naturalize diasporic encounters produced by colonialism and coax stable social formations out of them, the forces of globalization have thrown up further mobilities of people across borders, detaching them from the home nation and inserting them elsewhere, this time facilitated by the space–time compression of an even more interconnected globe wrought by modern transportation and communications technology. Terms such as 'diaspora' and 'transnationalism' have stirred the imagination of commentators describing the transience and ambivalence of movements across spaces today, as illustrated in John Lie's description:

> The idea of diaspora – as an unending sojourn across different lands – better captures the emerging reality of transnational networks and communities than the language of immigration and assimilation … It is no longer assumed that emigrants make a sharp break from their homelands. Rather, premigration networks, cultures, and capital remain salient. The sojourn itself is neither unidirectional nor final. Multiple, circular, and return migrations, rather than a singular great journey from one sedentary space to another, occur across transnational spaces. People's movements, in other words, follow multifarious trajectories and sustain diverse networks. (1995: 304)

As many authors have recently argued (see, for example, Anthias, 1998; Brah, 1996; Clifford, 1997; Gilroy, 1993; Hall, 1995; articles in Huang, et al., 2000), the concepts of diaspora and diasporic identities provide for a less essentialized and more historically and analytically informed framework to understand not only the large and complex range of transmigrant movements that is taking place today, but also the ability to challenge our existing conceptions of culture, place and identity as closed, fixed and unchanging.

The kaleidoscope of diasporic spaces produced under the mobilities associated with colonialism, and further subjected to the disciplining gaze of nationalism with its concerns over territoriality and the inviolability of the social body within the nation's borders, is hence once more shaken up by globalizing forces. Not only must nations and nationalisms be problematized in the context of colonial and postcolonial experiences, as Winichakul (1994) argues in the case of Asia in general and Thailand in particular, migrations and diasporas which disrupt the nation's 'geo-body' must also be understood in a similarly multifaceted context. Even as 'colonialism's

geographies' are already highly complex – 'overlain with other cartographies of indigenous exchange, dependency, accommodation, appropriation, and resistance', as Anderson (2000: 384) notes – each turn of the kaleidoscope continues to fracture previous patterns and introduce new instabilities. This leads Cohen to conclude that 'globalization and diasporization are separate phenomena with no necessary causal connection … [but] they do "go together" extraordinarily well' (1997; 175), albeit in ways which obfuscate the ironic twists in history. As Kang notes,

> It is an irony of history that Japan, the former colonial power and aggressor in Asia, has been reborn as an 'ethnically homogeneous' [sic] nation-state, while the victims of colonialism and fascism have been sundered apart and separated. However, with globalization and the gradual breakup of the Cold War, the history and memories of the colonized peoples who have been sundered apart [and scattered around Asia] have emerged in the form of history with a small 'h'. (2001: 137)

Identity formation and its postcolonial 'migrancy' have to be understood not as captive, or traced in some unilinear fashion, to colonialism, but instead constituted by a skein of tangled threads, of which the ambitions of colonial empires and the diasporas of their imperial subjects form one significant strand. Indeed if 'roots' always precede and are inextricably intertwined with 'routes', as Clifford (1997: 3) observes, then it has to be added that these 'roots' are intricately rhizomatic ones. Returning to the case of Japan, despite the fact that the Japanese empire was 'once burdened with the complexities and inequalities of an ethnically and culturally mixed population' and complicit in the creation of 'diasporic existence' among people of Korean descent, for example, post-war national history has attempted to reduce these complexities to 'the history of a single ethnic identity' played out within the geography of the four 'home islands', thereby forgetting the other peoples of empire (Kang, 2001: 141). It is against this history of selective remembering and forgetting, and as part of the unfolding tensions between 'empire'/'nation' on the one hand and 'diasporas' on the other, that contemporary transnational migrations are taking place, troubling the complexly fractured but concealed history of the nation.

This is central to Vera Mackie's (2002) analysis of the 'spaces of difference' formed as a result of the insertion of immigrant others into the fabric of Japanese society, once thought to be homogeneous. She argues that 'the relationships between immigrants and their relatively privileged hosts in Japan have been shaped by a history of imperialism and colonialism and the features of the contemporary political economy of East Asia'. The enforced military prostitution of the Second World War and the colonial period (which resurfaced as the 'comfort women' issue in the 1990s) foreshadowed a time in the 1970s when the Japanese nation-state could still assume that embodied encounters with 'difference' in the form of South East Asian women could be safely displaced offshore (as played out in sex tourism and other sexualized practices of 'gazing' on the rest of Asia). In more recent decades, however, the state has had to confront the presence of these 'others' within its own boundaries. Filipino women who enter Japan through labour or marriage migration, for example, are often marked by sexualized images (Mackie, 1998; Suzuki, 2000), a construction in which Japanese immigration policy is complicit in producing, for immigrant female workers are barred from being employed as domestic workers and are limited to entering the country under the legal category of 'entertainer', which is often a mask for the provision of sexualized activities from singing and dancing, waitressing and hostessing, to prostitution. It is interesting to note that the sexualized body – the most 'irreducible locus for the determination of all values, meanings, and significations' and 'the measure of all things' (Harvey, 2000: 97–8; see also Law, 2000) – continues to bear the marks of coloniality even when geopolitical forms of colonization have been dismantled. Diasporic spaces 'of the other' are hence (re-)emerging from within the social body of the nation; these can no longer be externalized. As Iwabuchi (1998) puts it, the '[once-colonized] subject is already within Japan and not just "out there" '. The politics which inhabit these spaces are double edged: while it would appear that the 'notion of discrete territory of the nation' and 'the transgressive fact of migration' are counterpoints to each other, van der Veer notes that 'self' and 'other', 'transgressors' and 'the established', are also 'structurally interdependent' and that 'nationalism (which has its basis in the control of space or territory) needs this story of migration, the diaspora of others to establish the rootedness of the nation' (1995: 2, 6). What is also important here is that encounters between 'nation' and 'diaspora' are understood in and against a postcolonial context 'created through the histories which connect people in different nations' (Mackie, 2002).

These connections between colonial and postcolonial encounters are often multifarious and ramifying. As Lisa Law (2002) notes in her discussion of transnational activism among female migrant worker advocacy groups in Hong Kong and the Philippines and the emergence of a 'post-national, diasporic public sphere' – transnational labour migration today propelling millions to 'transgress national borders in search of greener pastures' – is more provisional, and 'less decipherable in terms of clear colonial or imperial histories'. The colonial imprint is however present if not always distinct. As Keiko Yamanaka (2000) shows, while the presence of a small Nepalese transnational community within the borders of Japan creating a 'space of difference' may be more immediately explained by the relative prosperity of East Asian economies and chronic labour shortages in Japan's manufacturing and construction industries, it also has its roots in a distinctive 'culture of emigration' and 'remittance economy' forged out of the longstanding British colonial tradition of designating 'martial races' to serve as Gurkha soldiers in the British and Indian armies. 'Global warriors' who used to service the needs of one historic empire have reinvented themselves as 'global workers' responding to the rising labour demands in another domain – Japan and other 'tiger' economies in Asia, including the former British colony of Hong Kong (2000: 70). While there is no need to argue that these 'warriors' and 'workers' are umbilically tied to the same or singular logic, it is useful to note that colonial and postcolonial 'migrancies' are indissolubly if complexly intermeshed, sometimes with unexpected outcomes. Others such as Michael Samers (1997) trace a much more clearly delineated line connecting the 'production of diaspora' to colonialism and neocolonialism; the emergence of what Samers calls an 'automobile diaspora' centred around a Renault factory in France and comprising Algerian migrant workers is explained in terms of the erosion of pre-capitalist modes of production in Algeria by French colonialism and the subsequent expansion of the French economy in the post-war period.

As in Japan, coming to terms with 'foreigners in our midst' has recently become a major preoccupation in Singapore (one in every four persons within Singapore's borders is a non-citizen), even as the city-state aspires to join the global league as a 'cosmopolitan city' and a crucial 'brains service node' for business and information industries in the new 'knowledge-based economy'. A globalizing city does not only entail the presence of multinational corporate headquarters, transnational elites of the professional and managerial class (referred to in Singapore as 'foreign talent'), and hi-tech, cultural and tourism industries, but has to be sustained by an underbelly of low-skilled, low-status 'foreign workers'[7] who minister to the needs of the privileged in residential, commercial and industrial settings (Yeoh, et al., 2000).

As the presence of foreigners – whether female live-in domestic workers (mainly Filipino, Indonesian and Sri Lankan) inserted into the sanctity of the family and privatized home space, or male construction and manual workers (mainly Bangladeshi, Indian, Thai and PRC nationals) gathering to form 'weekend enclaves' in public arenas – becomes increasingly felt (Yeoh and Huang, 1998), it is interesting that while Singapore had found it relatively easy to forget (or at least to selectively remember) its colonial roots in many ways, public discourse[8] attempts to grapple with new diasporas which have washed ashore recently by harking back to the colonial past. For example, against claims that foreign workers pollute the physical and social landscape and should be tightly controlled, some counsel a degree of empathy, for if Singaporeans should 'look into the mirror of their ancestral past', they would remember their immigrant 'forefathers [who] made their way to the south seas from China and India to seek salvation' (*The Sunday Times, 27 July 1997*).

A recent furore over the banning of foreign maids from dining in social clubs such as the Singapore Cricket Club (once the quintessentially British bastion of white privilege) and swimming in condominium pools sparked off charges that 'some Singaporeans are behaving like their former British colonial masters' ('Are Singaporeans behaving like the white Raj?', asks a columnist, *The Straits Times*, 2 August 2000). It has also been said that Singaporeans' attitudes towards foreigners have been expressed in three ways (*The Straits Times*, 21 September 1997): 'looking up to them' (the colonial mentality that the white expatriate is always right), 'looking down on them' (colonial notions of superiority and inferiority in the allocation of '3D' – dirty, dangerous and difficult – jobs to foreign manual workers), and 'fear of them' (fear of their physical presence and their 'taking our jobs, our children's places in schools, and marrying our daughters' which mirrors the colonial obsession with 'sanitation' 'moral hygiene' and 'racial purity').

In dealing with migrant others, there is some sense in which the relationship between 'nation' and 'migration' continues to be interpreted (and critiqued) within colonial frames of reference drawn from Singapore's history as the product of overlapping diasporas. This is because, as Aguilar (1996: 6) observed in a different context, in scripting national history (what is popularly known as 'the Singapore story'), migrants perform a 'reflexive and refractory function' which operates partly through the discourse of 'race/ ethnicity' (for other examples of postcolonial migrations and identity negotiations, see Nagar, 1997; and the essays in Fincher and Jacobs, 1998). While such reflections signal a growing awareness, on the part of some, of the need to get past the 'colonial' in the conduct of contemporary social life, the extent to which the nation can succeed in crafting a truly multicultural, cosmopolitan social fabric depends on whether it is able not only to reflect on deep-seated colonial hierarchies and mentalities within society but also to confront and defeat them. This entails not only recognizing the marks of coloniality – its binary categories and cartographies – in the present, but also possessing the will to undo the conceptual infrastructure behind such thinking and to imagine the nature and quality of social and political encounters between people differently.

This is no easy task. As Jacobs reminds us, recognizing the traces of empire and the presence of postcolonial politics may be 'a mark of being beyond colonialism', but could well also signify 'the persistent "neo-colonial" relations within the "new" world order'(1996: 25). In the metropolitan core cities of now dismantled empires such as London, postcolonial migrations following in the wake of the empire's ebb coalesce into diasporic communities which are 'thrust together with anxiously nostalgic ones', giving rise to 'a politics of racism, domination and displacement which is enacted, not on distant shores, but within the very borders of the nation-home' (1996: 24; see also Western, 1993, on the symbolism of place for Barbadian Londoners; Smith, 1996, on the politics of Starbucks coffee as the 'empire filters back'; Rex, 1997, on migrations from postcolonial societies to Britain; Driver and Gilbert, 1998, 1999, on landscape politics in London and other post-imperial cities; Samers, 1998, on immigrants and ethnic minorities as 'postcolonial subjects' in the context of the European Union).

The marks of colonial ideology continue to 'underscore the definitions of "self" and "other" that lay at the heart of spatially diverse and contradictory understandings of nation, whiteness, power, subjection, Commonwealth' as well as shape the 'imagined geographies' and 'identity politics' of postcolonial diasporas (Keith and Pile, 1993: 17). It is hence not sufficient to recognize these marks for what they are but to translate awareness into tactics and practice. As Abbas argues, 'postcoloniality begins … when subjects find themselves thinking and acting in a certain way … finding ways of operating under a set of difficult conditions that threatens to appropriate us as subjects, an appropriation that can work just as well by way of acceptance as it can by rejection' (1997: 10).

Moving back to the 'edge of empire' (Jacobs, 1996) and in rethinking the in-betweenness of Australia as a place which 'belongs to neither its Anglo-centred past nor to an assuredly post-colonial or Asian future', Anderson (2000: 381, 383; see also Schech and Haggis, 1998, on post-colonial understandings of the 'white self' in Australia) argues for 'historiciz[ing] the nation-state within the global relations of European modernity and colonialism, recognizing that the very concept of the "nation-state" was itself an export of Europe'. This opens the way to denatu-ralizing claims on the part of Anglo-derived white settlers to ownership of a 'national' or 'core' culture *vis-à-vis* other 'minority' groups conventionally categorized as 'migrants' and 'indigenous' people. Anderson goes on to con-tend that the politics of majority–minority status positionings are not confined within national bor-ders but are more usefully mapped onto a broader transborder terrain to take into account 'diaspora relationships' (2000: 386). It should be added that as much as the nation-state, along with its mater-ial borders and metaphorical boundedness, has its genesis within European colonialism, trans-national flows which criss-cross the world today are also rooted in, and inflected by, the same con-ditions (these flows presuppose the presence of national borders to be crossed in the first place).

The 'postnational' imaginary – 'imagining and feeling geopolitical connections across and beyond national borders' – that Anderson (2000: 385) advocates therefore cannot pre-clude, and perhaps must be heralded by, a sense of the 'postcolonial'. It is by historicizing our understandings of 'nation' and 'diaspora', as well as the space of their encounter, within the power relations spun by colonialism that we problematize and come to grips with the simul-taneous logics of nation-building and transna-tional flows in a globalizing world. In short, we should recognize that the localizing of identities (that nations strive after) and the migrancy of identities (that transgress the nation) – as well as the ways in which they collide, collude or contradict – are both part of the same post-colonial conundrum.

NOTES

This chapter would never have seen the light of day without the encouragement and patience of Jane Jacobs and Steve Pile.

1 Some of these 'inescapable markings' are obvious if taken for granted, including the 'growing hegemony of English as the language of geography' (Short, et al., 2001: 1) as well as 'academic dependency' of a practical nature in the postcolonial world perpetuated by 'the relative abundance of Euroamerican funding for research and training, the high levels of prestige attached to publishing in British and American schol-arly journals, the greater value attached to a Western university education' (Alatas, 2001: 26).

2 That there are 'multiple postcolonial conditions' has already been noted, for example by Sidaway (2000) who maps a range of 'postcolonialisms' in counter-point to a number of colonialisms, quasi-colonialisms, neocolonialisms, internal colonialisms, breakaway settler colonialisms, etc.

3 This does not imply that the question of the hyphen is a trivial one. Mishra and Hodge (1991: 399, 407) take the compound word 'post-colonial' to refer to 'some-thing which is "post" or after colonial' and go on to further distinguish many forms of postcolonialism when the hyphen is dropped, including a key distinc-tion between 'oppositional postcolonial', forged in its most overt form in post-independent colonies at the historical phase of 'post-colonialism', and 'complicit postcolonial', which is 'an always present "underside" within colonization itself'.

4 I owe the heuristic distinction between 'groundings' and 'unmoorings' to Jane Jacobs.

5 When asked to consider Singapore, Hong Kong and Seoul as successfully liberalized economies and soci-eties in East Asia, an elderly Arab intellectual was reported to have replied: 'Look at them. They have sim-ply aped the West. Their cities are cheap copies of Houston and Dallas. That may be all right for fishing villages, but we are heirs to one of the great civilisations of the world. We cannot become slums of the West' (Fareed Zakaria, editor of *Newsweek International*, reproduced in *The Sunday Times*, 21 October 2001).

6 The 'migrancy of identity' is Rapport and Dawson's term, which they use to signal that socio-cultural places are not 'coherent' or 'localized' 'universes of meaning' and that questions of identity must be 'treated in rela-tion to, even as inextricably tied to, fluidity or move-ment across time and space' (1998: 4–5).

7 In Singapore, the term 'foreign workers' is usually applied to unskilled or semi-skilled migrant workers while the term 'foreign talent' is reserved for the highly skilled.

8 It should be noted that civil society in Singapore is still largely constructed on the basis of a fixed centred polity, not on the reality of a fluid, fractured landscape which Singapore as a global city typifies, or what Appadurai (1990) calls the 'ethnoscape', the 'land-scape of persons who constitute the shifting world in which we live: tourists, immigrants, refugees, exiles, guestworkers and other moving groups'. In particular, by virtue of being women, domestics and non-citizens, foreign domestic workers are usually excluded from public discourse and debates on the shape of civil space in Singapore (Yeoh and Huang, 1999). Mackie

(2002) makes a similar observation in the context of Japan that while male foreign workers in the construction and manufacturing industries are often discussed in terms of labour policy, women immigrant workers are discussed in terms of morality and policing, and remain beyond the pale of discourse on citizenship. It is hence noteworthy that the limited discourse in Singapore which has developed in the public arena on foreign domestic workers tends to be where colonial metaphors surface most.

REFERENCES

Abbas, A. (1997) *Hong Kong: Culture and the Politics of Disappearance.* Minneapolis: University of Minnesota Press.

Aguilar, F.V. Jr (1996) 'Filipinos as transnational migrants: guest editor's preface', *Philippine Sociological Review* 44: 4–11.

Alatas, S.F. (1995) 'The theme of "relevance" in Third World human sciences', *Singapore Journal of Tropical Geography* 16 (2): 123–40.

Alatas, S.F. (2001) 'Introduction: alternative discourses in Southeast Asia', in S.F. Alatas (ed.) *Reflections on Alternative Discourses from Southeast Asia.* Singapore: Centre for Advanced Studies and Pagesetters Services. pp. 13–31.

Anderson, K. (2000) 'Thinking "postnationally": dialogue across multicultural, indigenous, and settler spaces', *Annals of the Association of American Geographers* 90 (2): 381–91.

Anthias, F. (1998) 'Evaluating "diaspora": Beyond ethnicity?', *Sociology* 32 (3): 557–80.

Appadurai A. (1990) Disjuncture and difference in the global cultural economy', in M. Featherstone (ed.) *Global Culture: Nationalism, Globalization and Modernity.* London: Sage. pp. 295–310.

Atal, Y. (1981) 'The call for indigenization', *International Social Science Journal* 33 (1): 189–97.

Barnett, C. (1997) '"Sing along with the common people": politics, postcolonialism and other figures', *Environment and Planning D: Society and Space* 15: 137–54.

Bonnemaison, S. (1997) 'Encounter with the past: design work for a postcolonial commemoration', *Antipode* 29 (4): 345–55.

Brah, A. (1996) *Cartographies of Diaspora: Contesting Identities.* London: Routledge.

Cartier, C.L. (1993) 'Creating historic open space in Melaka', *Geographical Review* 83: 359–73.

Cartier, C.L. (1997) 'The dead, place/space, and social activism: constructing the nationscape in historic Melaka', *Environment and Planning D: Society and Space* 15: 555–86.

Chang, T.C. (1997) 'Heritage as a tourism commodity: traversing the tourist–local divide', *Singapore Journal of Tropical Geography* 18: 46–68.

Chang, T.C. (2000) 'Singapore's little India: a tourists attraction as a contested landscape', *Urban Studies* 37 (2): 343–66.

Chang, T.C. and Yeoh, B.S.A. (1999) '"New Asia – Singapore": communicating local cultures through global tourism', *Geoforum* 30: 101–15.

Chew, E.C.T. (1991) 'The Singapore national identity: its historical evolution and emergence', in E.C.T. Chew and E. Lee (eds) *A History of Singapore.* Singapore: Oxford University Press. pp. 357–68.

Chua, B.H. (1995) 'That imagined space: nostalgia for *kampungs*', in B.S.A. Yeoh and L. Kong (eds) *Portraits of Places: History, Community and Identity in Singapore.* Singapore: Times Editions. pp. 222–41.

Cleary, M.C. (1997) 'Colonial and post-colonial urbanism in north-west Borneo', in R. Jones and B.J. Shaw (eds) *Contested Urban Heritage.* Aldershot: Ashgate. p. 28.

Clifford, J. (1997) *Routes: Travel and Translation in the Late Twentieth Century.* Cambridge, MA: Harvard University Press.

Cohen, R. (1997) *Global Diasporas: An Introduction.* London: UCL Press.

Devan, J. (1999) 'My country, my people', in K.W. Kwok, C.G. Kwa, L. Kong and B. Yeoh (eds) *Our Place in Time: Exploring Heritage and Memory in Singapore.* Singapore: Singapore Heritage Society. pp. 21–33.

Driver, F. and Gilbert, D. (1998) 'Heart of empire? Landscape, space and performance in imperial London', *Environment and Planning D: Society and Space* 16: 11–28.

Driver, F. and Gilbert, D. (1999) (eds) *Imperial Cities: Landscape, Display and Identity.* Manchester: Manchester University Press.

Duncan, J.S. (1985) 'Individual action and political power: a structuration perspective', in R.J. Johnston (ed.) *The Future of Geography.* London: Methuen. pp. 174–89.

Fincher, R. and Jacobs, J. (1998) (eds) *Cities of Difference,* New York: Guilford.

Gilroy, P. (1993) *The Black Atlantic: Modernity and Double Consciousness.* Cambridge, MA: Harvard University Press.

Hall, S. (1995) 'New cultures for old', in D. Massey and P. Jess (eds) *A Place in the World? Places, Cultures and Globalization.* Oxford: Oxford University Press. pp. 175–213.

Hall, S. (1996) 'What was "the post-colonial"? Thinking at the limit', in I. Chambers and L. Curti (eds) *The Post-Colonial Question: Common Skies, Divided Horizons.* London: Routledge. pp. 242–60.

Harper, T.N. (1997) 'Globalism and the pursuit of authenticity: the making of a diasporic public sphere in Singapore', *Sojourn* 12 (2): 261–92.

Harvey, D. (2000) *Spaces of Hope.* Edinburgh: Edinburgh University Press.

Hau, C. (2000) 'Colonialism, communism, and nation-state formation: the haunting of Asia and Asians', in K.-W. Kwok, I. Arumugam, K. Chia and C.K. Lai (eds) *'WeAsians': Between Past and Future.* Singapore: Singapore Heritage Society. pp. 78–97.

Huang, S., Teo, P. and Heng, H.M. (1995) 'Conserving the civic and cultural district: state policies and public opinion', in B.S.A. Yeoh and L. Kong (eds) *Portraits of Places: History, Community and Identity in Singapore.* Singapore: Times Editions. pp. 24–45.

Huang, S., Teo, P. and Yeoh, B.S.A. (2000) 'Special issue of Women's Studies International Forum on "Diasporic subjects and identity negotiations: women in and from Asia"', *Women's Studies International Forum*. 23 (4): 391–8.

Iwabuchi, K. (1998) 'Pure impurity: Japan's genius for hybridism', *Communal/Plural: Journal of Transnational and Crosscultural Studies*. 6 (1): 71–85.

Jacobs, J.M. (1996) *Edge of Empire: Postcolonialism and the City*. London: Routledge.

Johnson, N.C. (1995) 'Cast in stone: monuments, geography, and nationalism', *Environment and Planning D: Society and Space* 13: 51–65.

Johnson, N.C. (1996) 'Where geography and history meet: heritage tourism and the big house in Ireland', *Annals of the Association of American Geographers*. 86 (3): 551–66.

Johnson, N.C. (1999) 'Framing the past: time, space and the politics of heritage tourism in Ireland', *Political Geography* 18 (2): 187–207.

Jones, G.A. and Bromley, R.D.F. (1996) 'The relationship between urban conservation programmes and property renovation: evidence from Quito, Ecuador', *Cities* 13 (6): 373–85.

Jones, G.A. and Varley, A. (1999) 'The reconquest of the historic centre: urban conservation and gentrification in Puebla, Mexico', *Environment and Planning A* 31 (9): 1547–66.

Kang, S. (2001) 'Post-colonialism and diasporic space in Japan', *Inter-Asia Cultural Studies* 2 (1): 137–44.

Keith, M. and Pile, S. (1993) 'Introduction part 1: the politics of place …', in M. Keith and S. Pile (eds) *Place and the Politics of Identity*. London: Routledge. pp. 1–21.

King, A.D. (1999) '(Post)colonial geographies: material and symbolic', *Historical Geography* 27: 99–118.

Kironde, J.M.L. (1993) 'Will Dodoma ever be the new capital of Tanzania?', *Geoforum* 24 (4): 435–53.

Koh, T.A. (1999) 'Commentary: our place in time – re-creating ourselves and our selves', in K.W. Kwok, C.G. Kwa, L. Kong and B. Yeoh (eds) *Our Place in Time: Exploring Heritage and Memory in Singapore*. Singapore: Singapore Heritage Society. pp. 38–46.

Kong, L. and Yeoh, B.S.A. (1994) 'Urban conservation in Singapore: a survey of state policies and popular attitudes', *Urban Studies* 31 (2): 247–65.

Kong, L. and Yeoh, B.S.A. (1997) 'The construction of national identity through the production of spectacle: an analysis of National Day Parades', *Political Geography* 16: 213–39.

Kusno, A. (1998) 'Beyond the postcolonial: architecture and political cultures in Indonesia', *Public Culture* 10: 549–75.

Kusno, A. (2000) *Behind the Postcolonial: Architecture, Urban Space and Political Cultures in Indonesia*. London: Routledge.

Kwok, K.W., Kwa, C.G., Kong, L. and Yeoh, B. (eds) (1999) *Our Place in Time: Exploring Heritage and Memory in Singapore*. Singapore: Singapore Heritage Society.

Law, L. (2000) *Sex Work in Southeast Asia: The Place of Desire in a Time of AIDS*. London: Routledge.

Law, L. (2002) 'Sites of transnational activism: Filipino non-government organisations in Hong Kong', in B.S.A. Yeoh, P. Teo and S. Huang (eds) *Gender Politics in the Asia–Pacific Region*. London: Routledge. pp. 205–22.

Lie, J. (1995) 'From international migration to transnational diaspora', *Contemporary Sociology* 24 (4): 303–6.

Lowenthal, D. (1985) *The Past Is a Foreign Country*. Cambridge: Cambridge University Press.

Mackie, Vera (1998) 'Japayuki Cinderella Girl: containing the immigrant other', *Japanese Studies* 18: 45–63.

Mackie, V. (2002) '"Asia" in everyday life: dealing with difference in contemporary Japan', in B.S.A. Yeoh, P. Teo and S. Huang (eds) *Gender Politics in the Asia–Pacific Region*. London: Routledge. pp. 181–204.

Mishra, V. and Hodge, B. (1991) 'What is postcolonialism?', *Textual Practice*, 5: 399–413.

Nagar, R. (1997) 'The making of Hindu communal organizations, places, and identities in postcolonial Dar es Salaam', *Environment and Planning D: Society and Space* 15: 707–30.

Oduwaye, A.O. (1998) 'Urban landscape planning experience in Nigeria', *Landscape and Urban Planning* 43: 133–42.

Parenteau, R., Charbonneu, F., Toan, P.K., Dang, N.B., Hung, T., Nguyen, H.M., Hang, V.T., Hung, H.N., Binh, Q.A.T. and Hanh, N.H. (1995) 'Impact of restoration in Hanoi's French colonial quarter', *Cities* 12: 163–73.

Perera, Nihal (1998) *Society and Space: Colonialism, Nationalism, and Postcolonial Identity in Sri Lanka*. Boulder: Westview.

Pratt, Mary Louise (1992) *Imperial Eyes: Travel Writing and Transculturation*. London: Routledge.

Rajah, A. (1999) 'Making and managing tradition in Singapore: the National Day Parade', in K.W. Kwok, C.G. Kwa, L. Kong and B. Yeoh (eds) *Our Place in Time: Exploring Heritage and Memory in Singapore*. Singapore: Singapore Heritage Society. pp. 66–75.

Rapport, N. and Dawson, A. (1998) 'Introduction', in N. Rapport and A. Dawson (eds) *Migrants of Identity: Perceptions of Home in a World of Movement*. New York: Berg. pp. 3–18.

Rex, J. (1997) 'The problematic of multinational and multicultural societies', *Ethnic and Racial Studies* 20 (3): 455–73.

Ryan, M. (1994) *War and Peace in Ireland: Britain and the IRA in the New World Order*. London: Pluto.

Samers, M. (1997) 'The production of diaspora: Algerian emigration from colonialism to neo-colonialism', *Antipode* 29 (1): 32–64.

Samers, M. (1998) 'Immigration, "ethnic minorities", and "social exclusion" in the European Union: a critical perspective', *Geoforum* 29 (2): 123–44.

Sarkissian, Margaret (1997) 'Cultural chameleons: Portuguese Eurasian strategies for survival in postcolonial Malaysia', *Journal of Southeast Asian Studies* 28 (2): 249–62.

Schech, S. and Haggis, J. (1998) 'Postcolonialism, identity, and location: Being white Australia in Asia',

Environment and Planning D: Society and Space 16: 615–29.

Shamsul, A.B. (1998) 'Arguments and discourses in Malaysian studies: In search of alternatives'. Paper presented at the International Workshop on Alternative Discourses in the Social Sciences and Humanities: Beyond Orientalism and Occidentalism, National University of Singapore, 30 May to 1 June 1998.

Shaw, B.J. and Jones, R. (eds) (1997) *Contested Urban Heritage: Voices from the Periphery*. Aldershot: Ashgate.

Short, J.R., Boniche, A., Kim, Y. and Li, P.L. (2001) 'Cultural globalization, global English, and geography journals', *Professional Geographer* 53 (1): 1–11.

Sidaway, J.D. (2000) 'Postcolonial geographies: An exploratory essay', *Progress in Human Geography* 24 (4): 591–612.

Smith, M.D. (1996) 'The empire filters back: consumption, production and the politics of Starbucks coffee', *Urban Geography* 17 (6): 502–24.

Srivastava, S. (1996) 'Modernity and post-coloniality: the metropolis as metaphor', *Economic and Political Weekly* 17: 403–12.

Suzuki, N. (2000) 'Between two shores: transnational projects and Filipina wives in/from Japan', *Women's Studies International Forum* 23 (4): 431–44.

Vale, L.J. (1992) *Architecture, Power, and National Identity*. New Haven: Yale University Press.

Van der Veer, P. (1995) 'Introduction: the diasporic imagination', in P. van der Veer (ed.) *Nation and Migration: The Politics of Space in the South Asian Diaspora*. Philadelphia: University of Pennsylvania Press. pp. 1–16.

Western, J. (1993) 'Ambivalent attachments to place in London: twelve Barbadian families', *Environment and Planning D: Society and Space* 11: 147–70.

Winichakul, T. (1994) *Siam Mapped: A History of the Geo-Body of the Nation*. Honolulu: University of Hawaii Press.

Yamanaka, K. (2000) 'Nepalese labour migration to Japan: from global warriors to global workers', *Ethnic and Racial Studies* 23 (1): 62–93.

Yeoh, B.S.A. (1996a) 'Street-naming and nation-building: toponymic inscriptions of nationhood in Singapore', *Area* 28: 298–307.

Yeoh, B.S.A. (1996b) *Contesting Space: Power Relations and the Urban Built Environment in Colonial Singapore*. Kuala Lumpur: Oxford University Press.

Yeoh B.S.A. and Huang, S. (1996) 'The conservation–redevelopment dilemma in Singapore: the case of Kampong Glam Historic District', *Cities* 13: 411–22.

Yeoh, B.S.A and Huang, S. (1998) 'Negotiating public space: strategies and styles of migrant female domestic workers in Singapore', *Urban Studies* 35 (3): 583–602.

Yeoh, B.S.A and Huang, S. (1999) 'Spaces at the margin: migrant domestic workers and the development of civil society in Singapore', *Environment and Planning* A 31: 1149–67.

Yeoh, B.S.A. and Kong L. (1994) 'Reading landscape meanings: state constructions and lived experiences in Singapore's Chinatown', *Habitat International*, 18 (4): 17–35.

Yeoh, B.S.A. and Kong, L. (1996) 'The notion of place in the construction of history, nostalgia and heritage in Singapore', *Singapore Journal of Tropical Geography* 17 (1): 52–65.

Yeoh, B.S.A., Huang, S. and Willis, K. (2000) 'Global cities, transnational flows and gender dimensions: the view from Singapore', *Tijdschrift Voor Economische En Social Geografie (Journal of Economic and Social Geography)* 2 (2): 147–58.

20

Cultures and Spaces of Postcolonial Knowledges

Anthony D. King

Thinking through the idea of 'after empire' one is automatically drawn to the extensive body of thought that, in the last two decades of the twentieth century, came into being under the rubric of postcolonial theory. In the obvious sense, 'post colonial' simply means 'after the colonial' and, until the early 1980s, was used 'to describe a condition referring to peoples, states and societies that have been through a process of formal decolonization' (Sidaway, 2000: 594). In the flurry of literature since that time, the scope of the term has widened. In 1990, Robert Young suggested that the analysis of colonial discourse 'itself forms the point of questioning Western knowledge's categories and assumptions' (1990: 11); it demanded, in Mongia's words, 'a rethinking of the very terms by which knowledge has been constructed' (1995: 2).

These claims are reflected across the disciplines, including geography. Thus, Gregory views postcolonialism as 'a critical politico-intellectual formation that is centrally concerned with the impact of colonialism and its contestation on the cultures of both colonizing and colonized peoples in the past, and the reproduction and transformation of colonial relations, representations and practices in the present' (2000: 612). As 'there have been many colonialisms', the 'post' brings into consciousness especially those of the sixteenth to the twentieth centuries 'in such a way that our understanding of the present is transformed' (2000: 612). Postcolonial approaches, according to Sidaway, in some accounts 'aim to transcend the cultural and broader ideological legacies and presences of imperialism' [sic] (2000: 594). Does this hint at an intellectual (or social) revolution from below?

In these conceptualizations, both Gregory and Sidaway implicitly connect a longer tradition of critical writings on colonialism with work more recently established as 'postcolonial studies' (Loomba, 1998: xv). Given the proliferation of valuable accounts of postcolonial studies at the turn of the millennium, by both geographers and others (Ashcroft, 2001; Gregory, 2000; Loomba, 1998; Robinson, 1999; Sidaway, 2000; Yeoh, 2001; Young, 2001) my first aim will be to address certain key issues which, because of the failure to combine insights from different disciplines, have been neglected. Moreover, recognizing the need for postcolonial theory to engage with 'material practices, actual spaces and real politics' (Yeoh, 2001: 457; see also Barnett, 1997; Loomba, 1998: 94) rather than simply analyses of representation and discourse, I also take up Driver's suggestion to explore the 'relatively unexamined ... role of space in a whole variety of modern aesthetic, cultural and political discourses beyond a narrow definition of social theory' (1992: 25), though particularly in relation to architecture, urban form and built and spatial environments, not least as these are critically impacted by such socially constructed notions as 'the tropics' as a climatic and natural environment. To do this adequately, attention must be paid to postcolonial studies and paradigm(s), undertaken prior to the publication of Edward Said's *Orientalism* (1978), frequently taken as the starting point of the current phase of postcolonial studies (for example, Ashcroft, 2001; Ashcroft et al., 1998; Loomba, 1998; Williams and Chrisman, 1994)[1].

The chapter also addresses perhaps the most neglected question in the literature, namely:

why, when, where and by whom has this new knowledge paradigm been developed? (See also Slater, 1998: 648.) This suggests not only a particular spatialized history but also one inflected by issues of social and political power in the academy. Following some initial definitions, the chapter therefore begins by addressing questions in the geography (as well as history and sociology) of knowledge. What are the spatial, political and social *conditions* accounting for both the production and the consumption of postcolonial studies? How have meanings and values that construct the world we know as postcolonial (including the intellectual practices, cultures and social networks that constitute 'postcolonial studies', if not actual postcolonial conditions) become established and institutionalized? Other than new electronic communication technologies, what do these social and intellectual movements have in common with, for example, the global indigeneity movements where worldwide connections have been made between what were once highly localized struggles of indigenous peoples?

TERMINOLOGY AND CONCEPTS

'Postcolonialism' is frequently used in the literature in a relatively loose way to refer to phenomena in both the postcolony and the postmetropolis. In this chapter, I conform with Loomba's (1998: 6) definition and adopt a more precise spatial conception that highlights the distribution of power. Imperialism (or neo-imperialism) refers to a phenomenon that originates in the metropolis; what happens in the colonies, as a result of imperial domination and control, is colonialism, or neocolonialism.

Similar variability exists in the literature concerning the term's periodization. Ashcroft et al. state that 'post-colonialism deals with the effects of colonization [*sic*] on cultures and societies and has generally had a clearly chronological meaning designating the post-independence period' (1998: 186), the meaning adopted here. However, 'post-colonialism as it has been employed in recent accounts has been primarily concerned to examine the processes and effects of, and reactions to, European colonialism from the sixteenth century up to and including the neo-colonialism of the present day' (1998: 188), that is, from the moment colonization began (see also Ashcroft, 2001). The distinction between postcolonial and postimperial can also be ambivalent, according to the positionality and location of the author. Describing London as a (technically) postimperial city (King, 1990) foregrounds its

earlier imperial role without necessarily invoking imperial contexts. For postcolonial migrants from Jamaica who live in London, however, it may be seen as postcolonial. As Australian Jane Jacobs points out, 'in settler dominions like Australia, it is the colonist who is imperialist' (1996: 37), and historical references to the British colonial empire (Sabine, 1943) are common.

As colonialism impacts the metropolitan society and culture as much as it does the colonial (if not in the same ways) it is clear that, while distinguishable, the phenomena (and their analysis) are inseparable. No discussion of the facts of race, space and place in the historically imperial, postimperial or neo-imperial city, such as Paris or Brussels, is complete without reference to the colonial, postcolonial or neocolonial city of Algiers or Kinshasa (Driver and Gilbert, 1999; Jackson and Jacobs, 1996; King, 1990; Yeoh, 2001). Theories of the postcolonial, as also of globalization or the world system, envisage the singular nation-state as an inappropriate unit of analysis.

While the bulk of the literature restricts the scope of the paradigm to the impact of European imperialisms between the periods stated, Sidaway's (2000: 596) valuable essay reconsiders 'different and diverse demarcations' of the term, exploring the possibility of extending the scope to the 'multiplicity of postcolonial conditions', and investing a wider meaning in the oft-stated comment that 'postcolonialism doesn't describe a single condition' (Dirks, 1992; Loomba, 1998; McClintock, 1992). As much of Europe has been subject to imperial rule, he considers the use of the analytical paradigm for Roman, Hapsburg, Ottoman, English, French and other empires in Europe, as also for the Soviet successor states and ex-Yugoslav republics, and develops a suggestive categorization of multiple *post*colonial conditions. These include colonialisms, quasi-colonialisms and neocolonialisms; second, internal colonialisms (see also McClintock, 1992); and third, breakaway settler colonialisms. Whether addressing issues of identity, the ethnic and racial composition of societies, political and spatial organization, architectural culture, language or other cultural phenomena, these categories have obvious analytical utility.

CRITIQUES OF THE 'POSTCOLONIAL'

The most trenchant critique of the 'postcolonial' is probably that of McClintock, who argued that the term 're-orients the globe once more around a single, binary opposition: colonial/post-colonial ... [such that] Colonialism returns at the moment of its disappearance' (1992: 85–6). For Boyce Davis (1994), among others, it is too premature, too totalizing, and 'recentering resistant discourses of women'. Geographer Jenny Robinson has suggested that postcolonial studies 'has replaced the "other" and "other" places as the object of study of Western academics, simply replaying the same relations of domination which characterized colonialism' (1999: 210). From a more conventionally Marxist perspective, James Blaut (1993) has argued that notions of the postcolonial misrepresent the realities of neocolonialism which structure the contemporary world. Thus, in Dirlik's terms, 'The global condition implied by postcoloniality appears at best as a projection onto the world of postcolonial subjectivity and epistemology – a discursive constitution of the world' (1994: 336). It is an analytical category that began, in his (only partially facetious) view, 'When Third World intellectuals ... arrived in First World academe' (1994: 329). If postcolonial studies has become institutionalized in the anglophonic western academy, especially as colonial discourse analysis, we need to know what, where and who are its political, social and cultural referents. And as implied in Dirlik's comments, we also need to know what the material and political implications, and real effects, of postcolonial studies actually are.

In this chapter I shall argue for postcolonial studies and criticism not only as relevant for understanding the contemporary world but, more especially, for the relevance of material, spatial, architectural and urban geographical studies for the development and strengthening of recent 'literary-based' postcolonial theory and criticism from which, with some few exceptions (Ashcroft, 2001; Said, 1993), it is conspicuously absent (for example, Loomba, 1998; Young, 2001). Moreover, the neglect of the cultural realities and cultural politics of postcolonialism and colonialism in conceptualizations of 'globalization' or the 'world system', and in theories of 'development' over the last four decades, has been one of the most profound omissions in the public perception as well as the academic study and understanding of the modern world. The most visible aspects of this omission are the inequities of economic, cultural and political power, both material and symbolic, manifest in the territories and especially the cities of the 'postimperial' as well as 'postcolonial' world (as land and racial disputes in Zimbabwe, or race riots in the north of England, demonstrated in the early twenty-first century).

UNDERSTANDING POSTCOLONIAL WORLDS: QUESTIONS OF KNOWLEDGE AND POWER

At the root of all postcolonial theory and criticism are questions of knowledge and power. These are essentially concerned with issues of agency, representation and, especially, the representation of culture(s) under asymmetrical political and social conditions. And as Dirlik (1994: 334), Chakrabarty (1992) and others have stated, it takes 'the critique of Eurocentrism as its central task'.

In thinking about the representation and understanding of issues in the contemporary world, culture becomes important in two ways. First, theoretical and historical representations of imperialism, colonialism and the contemporary global condition are culturally and historically constructed. In stating this we recognize that 'culture', used in the anthropological sense, is itself a particular and powerful construct that accompanied imperial expansion, a lens or container through which other peoples were known and delivered back to the west. As Clifford points out, the 'anthropological definition of culture ... emerged as a liberal alternative to racist classifications of human diversity ... a sensitive means for understanding different and dispersed "whole ways of life" in a high colonial context of unprecedented global interconnection' (1988: 234). The concept has nevertheless remained. Second, in their interpretations of social, political and spatial relations, different representations give greater or lesser attention to specific cultural phenomena such as language, religion, aesthetics or other symbolic and representational practices. In this context, the concept of the postcolonial has become of particular significance in regard to issues of identity, meaning and agency and, not least, in regard to the material forms and spaces in which they are embedded (Yeoh, 1996; 2001). It is these which act as vehicles for the exercise of subaltern agency.

In the following sections where I explore explanations for the earlier relative invisibility of postcolonial studies and then, from the 1980s, their subsequent explosion into view, I trace two traditions of 'postcolonial' scholarship, located broadly in the social sciences and in the humanities, but separated both in historical time and by geographical and disciplinary space, whether in regard to production or consumption. The need to combine insights from both approaches is implicit in the account.

POSTCOLONIAL WORLDS:
THE SOCIAL SCIENCES

Critical and theoretically informed studies of modern European colonialism as a political, economic, socio-spatial and cultural process, were produced in the final decades of, or just after, the formal ending of colonial rule by metropolitan sociologists, anthropologists, geographers and others, whether French (Balandier, 1966), Dutch (Furnivall, 1948; Wertheim 1964), American (Cohn, 1987, from the 1950s; Turner, 1971), British (Langlands 1969; McGee 1967; Smith, 1965) or others. Based on fieldwork in the colonial or postcolonial territory, their audience was presumed to be in that territory or the metropolis. Critical writings of Fanon (1967; 1968), Mannoni (1956) and Memmi (1965) remain the outstanding texts of the colonized, addressing, among other issues, the psychological traumas experienced by the colonially oppressed.

Significant here, however, are those studies which focused, not on representations of the colonized, but on the institutions and cultures of the colonizers themselves, particularly their forms of knowledge or particular *practices* which may, or may not, have had written texts to accompany them but which are, as Rabinow (1989) illustrates, all part of the system of discursive power: for example, medicine, cartography, agriculture, urban planning or architectural design. Subsequent to Said's use of Foucault's concept of discourse in *Orientalism* (1978), this has been termed 'colonial discourse' and refers to cultural forms and practices developed in the very specific context of the colonial situation with its particular distribution of power.[2] Ashcroft et al., write, 'Colonial discourse is greatly implicated in ideas of the centrality of Europe, and thus in assumptions that have become characteristic of modernity: assumptions about history, language, literature and "technology" … It is the system of knowledge and beliefs about the world within which acts of colonization take place' (1998: 41–2).

A prime example of such knowledge, and one having momentous influence on decisions about the location of settlements, the appropriation of space, socio-spatial forms, practices of urban development and architectural culture was the so-called 'miasmic' theory which perceived the origins of disease as determined by emanations from the ground, particularly in the so-called 'tropics'. Here, 'the tropics' are to be understood as perhaps the most foundational concept in the 'imagined geographies' of imperialism, a colonial construct of climate and nature. In this sense, making colonies was also about remaking nature,

not just in terms of assessing and exploiting the economic value of colonized lands as resource value, but also in terms of other (scientific) values and risk (disease). Encountering 'the tropics' was about fear and the risk and threat of disease and madness, signs of not being in control. From this developed the myths and paranoias of the tropics, such as the notion of the 'white man's grave' or, in eighteenth-century India, 'two monsoons' as the expectation of life. Combined at different times with various levels of racism and policies of colonial social control, this notion of 'the tropics' was to be a dominant factor influencing colonial social, spatial and political practices as well as relations.

The European construction of the concept of 'the tropics' and the 'tropical' has received attention from both geographers and others (Arnold, 2000; Driver and Yeoh, 2000; Livingstone, 1991; 1999; 2000). As Driver and Yeoh suggest, there is a need 'to raise questions about the multiple practices through which the tropics were known, by no means all of them articulated in discursive terms' (2000: 3). 'Identification of the northern temperate regions as the normal and the tropics as altogether other – climatically, geographically, and morally other – became an enduring imaginative geography which continues to shape the production and consumption of knowledge' (Arnold, 2000: 7). In examining the history of ideas of tropical nature in general and tropical geography in particular, Arnold emphasizes 'how important scientific ideas and academic authority were to the construction of the northern idea of the tropics' (2000: 3, 7). The discourse of tropicality, according to Arnold (2000: 7), is sufficiently important to parallel the idea of orientalism as a cultural and political construction of the west, as discussed by Edward Said.

The material implications of these tropical discourses are best registered in relation to practices of medicine, housing, architecture, planning, dress and the body (that is, the specific construction of cultural knowledge and practice oriented to the habits of behaviour of Europeans in 'the tropics'). In an age prior to the understanding of bacteriological theories of infection, or Ross' 1890s connection of malaria to the anopheles mosquito, miasmic theories dominated every colonial settlement and daily living decision. As polluted air was taken as the cause of disease, and 'native settlements' and 'habits' the origin of this, colonial settlements in India (and elsewhere in 'the tropics') were invariably located windward of, in front of, at a distance from, and preferably at higher altitudes than, the former. Throughout South and South East Asia, as well as West Africa, massive local resources were expended on the construction of hill stations (Freeman, 1999; Frenkel and Western, 1988; Kennedy, 1998; Kenny, 1995; 1997; King, 1976; Spencer and Thomas, 1948), road and rail access to them, at cooler, higher altitudes, distant from 'native towns' on the plains, primarily for the rest and recuperation of colonial military and administrative personnel, and later, the comfort of colonial wives.

In cantonments, the extent of spatial appropriation, design and orientation of colonial barrack rooms, the allotted space for each occupant, and air-circulating design features were determined by reference to the temperature, flow and volume of air (King, 1976). Special headgear and protection for the white, European body were the subject of extensive scientific research and Royal Commissions (King, 1976; Renbourne, 1961). 'The tropics,' as Arnold states, 'were created as much as discovered' (2000: 10), not least with the help of the ideological and material apparatus disseminated by imperialism, including the two most globally recognized tropical symbols, the solar topee and the tropical bungalow. As forms of protection for the body, they shared an uncanny similarity of structure: the high, 'double-layered' crown and the double-layered roof; the broad rim of the hat and the broad roof of the verandah, shading the head or body; the ventilation holes in the crown and the high windows permitting cross-ventilating currents of fresh air; the raised crown not touching the head and, using stilts or pillotis, the raised floor of the bungalow, high above the 'dangerous emanations' from the ground. These all resulted from the same 'scientific' medical principles: protecting bodies from the sun's heat and encouraging the circulation of 'fresh air' (King, 1984; Renbourne, 1961).

Combined over decades with changing racial attitudes, cultural (and capitalist) notions of property, and planning principles based on colonialist rationalities, miasmic theories of disease structured the shape of colonial cities, and European sectors in them, in ways recognizable long after the end of colonial rule. The less densely developed, cleaner and greener, windward sector of the colonial city was to become the site for post-independence suburbs of the elite and multinational tourist hotels.

Notions of hybridity and mimicry which characterize colonial discourse (Bhabha, 1994) also have their referents in spatial and building form. They also provide opportunities for thinking about hybridities in a cultural geographical sense, not least in relation to hybrid cartographies (Jacobs, 1993) and different types of hybrid architectural forms (Jacobs et al., 2000). The concept of the bungalow, 'the one perfect house for all tropical countries' (King, 1984: 200),

the outcome, over time, of 'interactions with indigenous peoples and places' (Driver and Yeoh, 2000), with the help of tropical discourses, moves into different regions of the globe.

The transformation of these tropical knowledge paradigms into the 'sanitary syndrome' of the twentieth century and subsequently into the 'scientific' basis of much contemporary urban planning worldwide is their bequest. That the tropics 'continues to shape the production and consumption of knowledge' (Arnold, 2000: 10) is evident from South East Asia to Brazil, where the concept of 'tropical architecture' or 'tropical modernism' is rematerialized through the lenses of 'regionalism', national culture and international capital (Kusno, 2000; Stepan, 2000; Yeang, 1987). In this sense, the new 'tropical urbanisms' mark out newly developed regional or national identities while simultaneously permitting the urban elite to stay within the transnational space of modernism.

On the general topic of knowledge and power, in the context of colonialism, some of the most pioneering work was undertaken by anthropologist Bernard S. Cohn. In the words of Dirks,

> Long before the powerful theoretical proposals of Michel Foucault made knowledge a term that seemed irrevocably linked to power, and before Edward Said so provocatively opened up discussion of the relations between power and knowledge in colonial discourses and Orientalist scholarship, Bernard Cohn had begun to apply an anthropological perspective to the history of colonialism and its forms of knowledge in a series of essays written between the mid 1950s and early 1980s. (1996: ix)[3]

If Cohn stands out as a pioneer in addressing questions of representation in colonial societies, there are other scholars whose ethnographic, descriptive, critical and theoretical explorations of colonial communities, colonial space and colonial cities between the 1950s and early 1970s laid essential foundations for any subsequent research in the field.[4] This raises important issues of whether a new knowledge paradigm only comes into existence when, like 'postcolonial theory', it is designated with a distinctive label. I suggest below the different social and spatial conditions under which the phenomenon of 'postcoloniality' became (belatedly) recognized and widely studied in the western academy from the early 1980s.

To summarize: in what might be called (compared with the 1980s) the 'pre-global' 1960s and early 1970s, both the audience for, and interest in, such culturally oriented postcolonial studies in the metropolitan academy were restricted to specific disciplinary cells, whether in anthropology, development studies, or among those scholars with first-hand experience of one-time colonies. From the late 1960s, radical movements in the west spurred a surge of interest in Marxism and, not least in urban and regional studies, the advent of a neo-Marxist urban political economy. Cultural issues, as understood in current postcolonial discourses, referring especially to issues of identity, cultural autonomy and subjectivity, were not part of these discourses. In David Slater's retrospective view, 'the failure to theorize subjectivity and identity' (1995: 71) was the major weakness of such Marxist accounts. We might say that, in the 1960s and early to mid 1970s, compared with the situation in the 1990s, the social sciences were much less interested in culture, and the humanities (apart from an apolitical, 'commonwealth literature' discourse: Moore-Gilbert, 1997: 2) much less interested, if at all, in colonialism. One can search the geographical (and other) literature in vain at this time looking for research on issues of culture and identity in 'postcolonial' or 'developing countries'. The surge of interest in these questions which surfaced in the early to mid 1980s appears overwhelmingly in the anglophone west rather than the postcolonial societies themselves (see also Young, 2001: 62). As I discuss below, it is the outcome of selective social and spatial processes of globalization.

POSTCOLONIAL WORLDS: THE HUMANITIES

What might be termed the 'second' or 'humanities' phase of postcolonial criticism seems, from the frequency of publication, to have taken off in the early 1980s,[5] its practitioners being both metropolitan and 'Third World' intellectuals (Dirlik, 1994). The earlier (social science) generation of 'colonialism' scholars were generally male, European or North American, were occasionally financed through 'aid' arrangements and 'displaced' from the metropole to the colony or postcolony, and had a research focus on the politics, culture, society and space of the colonial society. In contrast, scholars associated with the second phase (specifically termed 'postcolonial studies') have generally been from the humanities, and are as likely to be female as male, to be black (or non-white) as white, to have their origins in the one-time colonies (either exploitative, for example, India, the Caribbean, or settler, for example, Australia, New Zealand, South Africa) as in the metropolis, and to have been 'displaced' from the colony or postcolony to the metropolis

or vice versa. To various degrees, their critique is informed by cultural or literary theory, particularly feminism and poststructuralism, and the objects of analysis are primarily texts – literary (histories, travel writing, letters, diaries, manuals, etc.) as well as graphic, photographic or cartographic. As suggested above, the direct or indirect starting point of this humanities phase of postcolonialism has, in many if not all cases, been Said's *Orientalism* (1978). While much, if not all, of the postcolonial geographical work dealing with 'empire' is in this mode of analysing texts and representations, it is worth noting that, in terms of location, much of it is by 'postimperial' scholars rather than the one-time 'colonized indigenes' or 'settlers' suggested above. In geography, as well as other disciplines, it is also evident that the work from postimperial as well as postcolonial (both previously colonized as well as settler) scholars has very different priorities, politics and agendas. Here, it is worth mentioning Western's work on Cape Town (1997, first published in 1981) which, exceptionally for that time, deals with issues of identity and the politics of exclusion.

In recognizing that postcolonial theory and criticism comprise 'a variety of practices many of them pre-dating the period when the term "postcolonial" began to gain currency' (Dirlik, 1994; see also Moore-Gilbert, 1997; Williams and Chrisman, 1994), both the latter accounts begin their histories in the early years of the twentieth century with reference to the anti-racist and/or anti-colonial writings of W.E.B. Dubois, Sol Platje and, subsequently, the Harlem Renaissance, C.L.R. James, Frantz Fanon and others. Moore-Gilbert continues: 'While postcolonial criticism has apparently a long and complex history outside Europe and America, it arrived only belatedly in the Western academy and British university departments more particularly' (1997: 2). Here, Moore-Gilbert documents the development of the field in English studies, emerging from conferences held in the UK in the early 1980s (for example, 'Literature and Imperialism' in 1983, 'Europe and its Others' in 1985), paying attention to the earlier, anglocentric and apolitical paradigm of 'commonwealth literature' whose sponsors had held their first conference in 1964.

Where Moore-Gilbert traces the chronology of the postcolonial paradigm in literary studies in the UK, less attention is given to the geography, sociology and demography of its development. No mention is made, for example, of the virtual absence, among British university populations in the 1970s, whether humanities faculty or students, of non-white, British-born subjects. Moore-Gilbert is also silent in regard to his own origins (born in Tanzania, son of a game warden in colonial Tanganika, which became Tanzania in 1961) or his geographical location – like Paul Gilroy, the author of perhaps the earliest proto-postcolonial monograph in British cultural studies, *There Ain't No Black in the Union Jack* (1987),[6] also subsequently based in the University of London's Goldsmiths' College in south-east London, a major site of multiethnic communities, especially from the Caribbean and Africa, in the postimperial metropolis. In 1991, approximately one-fifth of London's 6.5 million inhabitants were classified as ethnic minorities, the large majority either born in or descended from parents born in the one-time colonial empire. Nearly half of Britain's ethnic minority population lived in London in 1991 (Starkey, 1994: 24-5).[7]

While Dirlik (1994) is partly correct in maintaining that the postcolonial paradigm had resulted from a postcolonial, English- (and native-language-) speaking diaspora from east to west, i.e. 'when Third World intellectuals have arrived in First World academe', it clearly also needed audiences, situated in appropriate locations. New forms of knowledge need receptive subjects and, where teaching relies on monographs, textbooks and journals to publish academic research, all three began to appear in the (postimperial?) global city of London, the major publishing centre of the anglophone world in the late 1980s.[8] With minority populations predominantly formed, from the early 1960s, by immigration from the one-time colonial empire (Ireland, India, Pakistan, Bangladesh, the Caribbean, East and West Africa, Hong Kong), the UK can more obviously be represented as 'postcolonial'. Conversely, the United States, with a much greater diversity in its ethnic minority origins (though with obvious colonial relationships with Puerto Rico and its indigenous Native American peoples), has been represented under this label in relation to its (internal) history of slavery and recent immigration. The widespread adoption of the paradigm in the United States academy from the late 1980s also rests on an increasing acknowledgement that most of America's 'ethnic minorities' – Hispanics, Chicano/as, Asian Americans and others – are in the United States as the result of colonial wars (in the nineteenth century with Spain and Mexico, and in the twentieth in Korea, Vietnam, El Salvador and Grenada, amongst others). The adoption of the postcolonial paradigm is also associated with the civil rights movements from the 1960s and the resultant ideological construction of a multicultural society (Geyer, 1993).

The growth of minority faculty numbers in the American academy, and an increasing audience of minority students in US higher education in the later 1980s and early 1990s, primarily in public institutions, also provided an educational market for postcolonial studies (Finkelstein et al., 1998).[9] The paradigm was also boosted by the increased internationalization of graduate education, not least students from decolonized states. Under the category of the 'postcolonial', such students have acquired a framework with which to recognize their historic colonial connections with the USA and with postcolonial confrères elsewhere, but also an identity within 'the west' to contest a western, white (and as I discuss below, frequently male) cultural hegemony as well as the facts of social and academic racism. In this sense, postcolonialism has enabled an alliance between 'Third World' students and those of African American, Latina/o, Asian American or Native American US minorities. These material factors affecting the cultural politics of education also need to be seen alongside intellectual developments, including the emergence of 'an explicit theoretical sensibility (which was largely foreign to mainstream history); and secondly, an attempt to recover the political significance of culture and the epistemic violence of colonialism' (Gregory, 2000: 613), if the rapid growth of postcolonial theory and criticism is to be understood.

It is evident from these accounts that different postcolonial critiques, irrespective of the geographical or disciplinary location in which they emerged, have shared a common objective in deconstructing epistemologies and reformulating the objectives and methodologies of knowledge production. Because they operate under different social, historical and geographical conditions, however, their referents, as well as their politics and agendas, are understandably not the same. This is apparent from the critical work that has been undertaken on interrogating and deconstructing 'imperial geographies'.

Geography, understood literally as 'a writing of the (surface of) the earth' (*OED*), is inherently a universalizing science, though developed within specific political and historical conditions. Can there ever be, therefore, a 'postcolonial geography' that allows people and places to represent themselves in their own terms? Such a question raises issues not only about the histories of geographical knowledge (Driver, 1992) but also about its geographies – an issue which, in academic and popular imaginations, is more frequently grasped through their varied cartographies (Harley, 1987; 1988) and the different worldviews they represent. In examining 'the interplay between colonial power and modern geography during the "age of empire"', Driver (1992: 27) cites Hudson's (1977) pioneering account tracing the close correlation between the birth of modern geography and the emergence of a new phase of capitalist imperialism in the 1870s. Geography developed, in Hudson's view, 'to serve the interests of imperialism ... including territorial acquisition, economic exploitation, militarism and the practice of race and class domination'. This radical and empirical geographical tradition is also manifest in *The Geography of Empire* (1972) where Buchanan suggests that the material impoverishment wrought by European and US imperialism 'may well prove to be of less significance than the undermining of indigenous cultures' through 'the pillage of brains', the 'infiltration of Third World education systems' and similar cultural practices (1972: 19).

However, as Driver is at pains to point out, '[G]eography during the "age of empire" was more than simply a tool of capitalism, if only because imperialism was never merely about economic exploitation. There are significant aspects of the culture of imperialism ... which deserve much more attention from the historians of modern geography than they have yet received' (1992: 27).

Many of these issues are addressed in the essays in Godlewska and Smith's collection *Geography and Empire* (1994), the major merits of which are not only its wide geographical scope, addressing the origins and intellectual and material implications of imperial geographies in the main European imperial societies, as well as Japan, the USA and elsewhere, but also the fact that it draws attention to a critical postcolonial geographical tradition, largely independent (with the exception of the introduction and the chapter by Crush) of work developed in the light of the postcolonial theory of the late 1980s. The editors nonetheless fully recognize the contribution of Said's important works, *Orientalism* (1978) and *Culture and Imperialism* (1993). In highlighting Said's 'vivid geographical sensibility' they draw attention to his discussions of 'imagined political geographies' and particularly how literary scholars have 'failed to remark the geographical notation, the theoretical mapping and charting of territory that underlies Western fiction, historical writing and philosophical discourses of the time' (Said, 1993, cited by Godlewska and Smith, 1994). Yet in recognizing the 'brilliant vista' of Said's work they also draw attention to its ambivalence 'towards geographies more physical than imagined, a reluctance to transgress the boundaries of discourse and to feel the tangible

historical, political and cultural geographies he evokes' (1994: 6–7). This last is an important point and one to which I return below. However, irrespective of the objects, methods, scope and, not least, the possibilities of a reformed and critical geographical knowledge in a western-oriented world of scholarship where English has, for better or worse, become the commonly accepted international language, there exists the barrier (or facility?) of what I shall call the 'post-coloniality of English'.

THE POSTCOLONIALITY OF ENGLISH

The extent to which contemporary (anglophonic) human geography is still (unwittingly) a post-colonial project is manifest in a paper by two Madrid-based Spanish geographers, researching 'the extent to which international journals of human geography are really international'. Language is probably the most basic constitutive feature of culture, occasionally used as a metaphor for culture itself. In the postcolonial literature, not least in geography, surprisingly little attention has been devoted to its most fundamental and constitutive feature, namely the anglophonic nature of the discourse itself.

On the assumption that English is the common language of communication of international journals, the study examines the national origins (by institutional affiliation) of papers published, the composition of editorial boards according to national locations, and related indices, of 19 prominent English-language geography journals. The authors show, among other results, that over 86 per cent of articles come from either the one-time metropolitan country (UK) or countries of what they refer to as the 'Anglo-Saxon world' (largely, the USA, and also Canada, Australia and New Zealand); similarly, the editorial boards are made up of scholars, over 80 per cent of whom are from these countries. Either curiously or expectedly, in stating their conclusions, first, that such journals 'are not, in fact, very international' and also that 'human geography is still fragmented into national or linguistic communities' (Gutierrez and Lopez-Nieva, 2001: 67), the authors do not refer to these findings as an outcome of politically, geographically and culturally influenced colonial and 'postcolonial' histories, nor do they make any use of that term in their analysis. Yet combining this information with what is common knowledge about the extensive circulation of anglophone geographers (as well, of course, as many other disciplinary academics) between appointments in Australia, Canada, the UK, the USA, Singapore, the Caribbean and elsewhere, the stereotypically weak language skills of (especially British, American and Australian) academics, and the relative lack of familiarity with developments in, for example, German and Russian, let alone Chinese or Arabic, scholarship, would seem to confirm the persistently postcolonial nature of anglophone scholarship, from which the structural impediments of language (and the lack of it) offer little escape. This naturally leads to the question of academic diasporas.

Unlike other major modern migrations, such as that of European Jewish intellectuals at the turn of the twentieth century, and subsequently in the face of Nazi persecution in the 1930s, the migration and transmigration of postcolonial anglophonic intellectuals within, and across, the space of the anglophonic world have constructed what until recently may be characterized as disaggregated histories. Anglophone postcolonial intellectuals in New York are linked with those in India, the Middle East, Canada, South Africa, Australia, South East Asia, the Caribbean and elsewhere. This creation of different postcolonial identities – a result of a 'technical' postimperialism – from nation states with widely differing standards of economic prosperity, even though (some) intellectuals within poorer states may be educationally and socially privileged, has constructed a cultural space linked across, and by means of, a particular anglophonic language world. It is this anglophonic postcolonial subjectivity which, as the opening quotes from Young and Mongia suggest, is behind the suggestion to transform the nature of established, metropolitan-based systems of knowledge.

An examination of at least some postcolonial texts can reveal a combination of citations drawn from authors originally from a vast array of postcolonial as well as other countries (see, for example, the extensive bibliographies in Williams and Chrisman, 1994, or Young, 2001). This (apparently) theoretically coherent assembly of diasporically related scholars, increasingly cutting across the disciplinary differences of the humanities and social sciences suggested above, would have been rare, at least on such a scale, as little as four or five decades ago. On the other hand, it would be necessary to recognize not only the dominance of English as the medium of this literature and the absence of other major world languages (Mandarin, Hindi, Spanish, Arabic), but also the metropolitan location where much of this knowledge is being produced, where it is being consumed, and by whom. What this indicates is how specifically postcolonial and

anglophonic, or perhaps 'post-' or 'neo-imperial', and very definitely not 'global', these particular discourses are.

POSTCOLONIAL GEOGRAPHIES: MATERIAL AND SYMBOLIC

Reference has already been made to the need 'to feel the tangible, historical, political and cultural geographies that Edward Said "evokes" rather than addresses' (Godlewska and Smith, 1994, above) and suggestions have been made elsewhere as to why the current 'humanities' phase of postcolonial studies has, until the mid to later 1990s, largely ignored the realm of colonial urban, spatial and built environment studies.[10] In ignoring the physical, spatial, architectural, urban and landscape realities in which many of these various colonial discourses developed, these accounts erase (by ignoring) the essential *material* conditions, and mental referents, without which other cultural practices and forms of representation (in addition to architecture, planning and urban design) – writing, mapping, ethnography, film, photography, painting – would have been impossible. In an analogy with Said's comment (cited above) that literary scholars 'have failed to remark the geographical notation, the theoretical mapping and charting of territory that underlies Western fiction, historical writing and philosophical discourses', we can also say that literary (and many other) scholars have failed to remark the physical, spatial, symbolic, visual and material environments in which everyday life actually occurs. Moreover, such spatial and built form arrangements are not simply signifiers of power and control, they also materially affect the life chances of those who live within them. As Brenda Yeoh has argued, they were also real spaces in which the authority of colonial power and control was effectively contested (Yeoh, 1996). It is to the materialities, built forms and physical spaces of the city and how these affect and help to produce and reproduce social relations, identities, memories and subjectivities that I turn.

The real spaces of the city (whether seen primarily as representative of social relations of power or as the lived everyday space of socially, racially and ethnically differentiated populations) are essentially *dynamic*. As they change over time, they represent transformations (or reaffirmations) in the social distribution of power, in access to resources, not least in postcolonial times. One of the most frequently cited texts in postcolonial studies is Frantz Fanon's

The Wretched of the Earth (1968: 37–40), particularly, his representation of the 'colonial world' as being epitomized by the racially and spatially segregated colonial city: 'The colonial world is a world divided into compartments ... of native quarters and European quarters, of schools for natives and schools for Europeans, etc.'

This description would *seem to* provide both a clear benchmark from which to examine the extent to which such one-time divided colonial cities which fitted this description have been transformed in the decades since independence, but also the opportunity for an assessment of such transformations by postcolonial (and other) subjects (King, 1976: 282; Yeoh, 2001). However, as Yeoh, among others, points out, the situation is far more complex. In addressing criticism of the postcolonial paradigm, Moore-Gilbert writes, 'The objection of many (neo-Marxist) critics to postcolonialism rests on the assumption that material forms of oppression are the only ones worth bothering about and that postcolonialism's characteristic focus on the cultural or textual levels of the West's relationship with the "Rest" is a symptom of its ignorance or inadequate attention to the "real" dynamics of that relationship' (1997: 64).

The opposition which Moore-Gilbert suggests here between the 'material' and the cultural or textual levels of oppression is a false dichotomy and one resulting from the unfortunate hiatus between what I have characterized as the earlier, 'social science' scholarship on colonialism and this second 'humanities' phase. Most obviously it is the absence, in the latter, of attention to the phenomenon/a of material *space* which allows Moore-Gilbert not to recognize, or perhaps not to acknowledge, that material forms of oppression are also cultural and textual as well as vice versa. Here, it is not the metaphorical 'space' which 'historians, anthropologists and literary theorists of colonial discourse' frequently evoke, but real, material space that needs attention: 'many of their works barely touch on the real, physical consequence of colonialism's spatial tactics, or indigenous responses to them' (Myers, 1999: 27; see also Cairns, 1998).

Myers refers to 'the vast and growing literature by geographers engaged in the critical analysis of interrelationships between colonialism, geography and 'discourse' broadly inclusive of travel writings, map-making and academic representations of formerly colonized places' (1999: 50). An archaeology of geographical knowledges (Driver, 1992) would map out not only the critical turning points in the development of geographical representations, but how such geographies have been mobilized to counter

material oppression (for example, Crush, 1994; Jacobs, 1993; 1996; Jackson and Jacobs, 1996). In the remainder of this section I draw selectively on research in these various fields in order to illustrate some of the different ways in which studies on the materiality of space (including issues of land, territoriality, architectural cultures, the built environment, planning and urban design) have illuminated issues of identity, subjectivity and resistance in a variety of contexts.

A broad distinction might be made here between two types of research and writing. The most common, 'critical colonial histories' use colonial discourse analysis to interrogate historical archives in relation to spatial environments, the objective being to raise political and social awareness around issues of representation and equity for the postcolonial present, and including issues of gender and colonial space (Mills, 1991). Included here is the now extensive body of scholarship addressed to the analysis and deconstruction of colonial travel writing, much of it, from a critical feminist perspective (Blunt, 1994; Blunt and Rose, 1994; Duncan and Gregory, 1999; Mills, 1991; 1994a and b; Pratt, 1992). The second type addresses contemporary situations in *supposedly* postcolonial states, focusing on continuing inequities and injustices based on race, class and other markers of difference as part of an ongoing cultural politics of space. Illustrative of this second approach are case studies explored in *Edge of Empire: Postcolonialism and the City* in which Jane M. Jacobs traces ways in which 'the cultural politics of colonialism and postcolonialism continue to be articulated in the present' (1996: 10). Documenting City of London redevelopments in the 1980s, she demonstrates how the continuity of discourses over sites seen as 'the economic centre of Empire as well as the spiritual and otherworldly sense of Empire' were mobilized to influence subsequent decisions about urban design. 'Place,' she writes, 'plays an important role in which memories of empire remain active' (1996: 40). Addressing the settlement of Bangladeshis in East London's Spitalfields, and highlighting the continuity between colonial and postcolonial situations, Jacobs writes, 'Many of the new labour arrangements of global cities like London (produced by postcolonial migrations) quite literally re-work people already categorized as available for exploitation under colonial economies' (1996: 71). In her native Australia, Jacobs (1998) speaks to a contemporary politics of space, demonstrating, for example, the inadequacy of the 'paradigmatic imperial technical instrument of "mapping"' to establish the authentic Aboriginal sacred, and charting, as an outcome of the 'map's' ineffectivity, a shift in the Australian national discourse from the secular to the sacred. Other studies trace the outcome of policies of urban aestheticization which gesture towards – yet stop short of – a full recognition of Aboriginal difference in the city (Jacobs, 1998).

What is crucially important to recognize here, however, is the degree of public acknowledgement in Australia of a multicultural and postcolonial *consciousness* and a cultural politics willing to act on it. The postcolonial paradigm here is not (as elsewhere) confined to cells within the academy but operates on the streets. What this reinforces is a point already made, namely that how a postcolonial critique operates, and what it does as a result, depends on local contingencies, that is, how it shapes people's thoughts and conditions as specifically postcolonial.

Essays in *Postcolonial Space(s)*, by an international group of architects and urbanists, bring yet another angle to an understanding of the postcolonial. In the words of the editors, 'Postcolonial space is a space of intervention into those architectural constructions that parade under a universalist guise and either exclude or repress different spatialities of often disadvantaged ethnicities, communities or peoples' (Nalbantoglu and Wong, 1997: 7). While representative chapters (for example Nalbantoglu on the carved dwelling in Turkey or Cairns on the traditional Javanese house) successfully contest the boundaries of what 'architecture' is usually taken to be, the commitment of contributors to exploring innovative poststructuralist, psychoanalytic, feminist and other approaches, characteristic of postcolonial theory, needs also to be read alongside accounts that acknowledge the persistent and powerful material factors of uneven development worldwide (Smith, 1984) and the corporate *knowledge* influences of global capitalism.

Whether postcolonial criticism can have real political effects can be seen by comparing some of the Australian cases cited above with the account of Chatterjee and Kenny (1999) who argue that, despite five decades of independence, attempts to bridge the vast spatial, social, economic and infrastructural inequities, as well as religious, cultural and lifestyle differences, between old and new Delhi, the legacy of hegemonic colonial planning, and to create a single capital symbolizing the unity and identity of the nation, have yet to be resolved. In suggesting reasons for this, the authors point to the *ambiguities* of the postcolonial, the fact that 'the replacement of previous hierarchies of space, power and knowledge has not been complete'; 'Muslim, Hindu and western socio-cultural norms co-exist, albeit uneasily, in Delhi's built environment' (1999: 93). Multiple identities produce a multiplicity of spatialities. The issue of dealing with space and identity in postcolonial (as well as other) cities deserves a monograph in itself, not least because the *recognition* of postcoloniality apparently agitates outside commentators more than indigenous governments and professionals. Again, this demonstrates how different local circumstances of materiality produce different conditions for the possibility of thinking and being postcolonial.

Yeoh's (2001) overview of geographers' research on postcolonial cities in the 1990s is organized around four themes: identity, encounters, heritage and an interrogation of the relevance of the postcolonial paradigm itself. A cluster of studies has addressed the different ways in which postcolonial states have endeavoured to engage with, but also distance themselves from, their colonial pasts, hopefully cultivating new national citizen subjectivities in the process: new capitals (Gilbert, 1985; Holston, 1989; Perera, 1998) or capital complexes (Vale, 1992), toponymic reinscription (Yeoh, 1996), spectacular towers (King, 1996) and transformative modernisms (Holston, 1989; Kusno, 2000), among others. In Kusno's telling phrase, the spatial reformulations result from what 'the colonial and postcolonial have done to each other' but under very different political regimes. Yet while, in Yeoh's words, 'architecture and space can … be interrogated for [their] embodiment of colonial constructions and categories in order to reveal the postcolonial condition' (2001: 459), attempting to read architectural meanings without the discourses that accompany them is a notoriously ambiguous project. As colonialism and imperialism are two sides of the same coin, also important here are the essays in Driver and Gilbert's *Imperial Cities* (1999). These address the architectural and urban spaces constructed by imperialism in Europe's imperial metropoles – Paris, London, Rome and Vienna, among others – demonstrating in the process the urbanistic competition between these capitals which European imperialisms generated.

Under 'encounters', Yeoh also reviews some of the literature on the distinctive social, ethnic and racial characteristics of postcolonial cities, and the opportunities they offer for newly emergent practices and social identities (echoing themes in the classic essay of Redfield and Singer, 1954), yet also noting the resilience of older colonial representations. Such 'imagined pluralism … is drawn upon in positive ways to position the postcolonial nation as a cosmopolitan society' (2001: 460).

Issues of heritage in the postcolonial city are buffeted between the cultural politics of different regimes and cyclical shifts of the world economy; what were once symbols of colonial control or authority get demolished or refurbished according to the priorities that tourism policy commands. The contested 'heritage' of various elements in the postcolonial city and their investment with symbolic capital is an important theme of this literature. But one-time residential areas (and houses) of the colonial elite can also become, in postcolonial times, a contested terrain for different social and ethnic populations, both materially and in battles over nomenclature. Focusing on the postcolonial as conventionally understood, Yeoh (2001) also cites studies that show how, in the Asia–Pacific region, old colonial 'core centres' of the global economy are being spatially transformed by new shifts in the spatial economy, not least in regard to the economic power of Japan (see also Perera, 1998).

These issues of identity, encounters and heritage which Yeoh highlights here, with their emphasis on indigenous agency (and well represented in literary form by Khatabi, 1993, writing on the North African medina), focus attention once more on the different meanings of 'postcolonial'. The questions which Yeoh addresses relate to the possibility and methods (however contested) of 'hearing or recovering the experience of the colonized' (Sidaway, 2000: 594) during or after colonization. Not all postcolonial places or nation-states permit this possibility. As Myers writes in relation to postcolonial urban planning in East Africa, 'Although draped in the banners of socialism rather than imperialism, the post-colonial state inherited from the colonial regime an obsession with spatial order as an ideological tool and means of civic control' (1995: 1357). The result is that projects are 'activated in the interests of powerful leaders more than in the interests of the people for whom (they) ostensibly were planned'.

What emerges from these studies, therefore, is the critical importance of the presence or absence of *local*, indigenous perceptions of 'postcoloniality', what is implied by it, and who has the power of political control. The problem of postcolonial studies, according to Kusno (2000), has been a tendency to undertake a critique of colonial discourse around a broad, often undifferentiated critique of 'the west', without acknowledging that colonialisms come in many forms. Recognizing historical differences between various colonial states, Kusno argues that, until the demise of Suharto's 'new order' in 1998, Indonesia in fact continued as a colonial regime in all but name. Departing from a key theme in much postcolonial criticism, he argues

that colonialism in Indonesia did not bring about a displacement of indigenous culture. As Indonesians were encouraged by Dutch orientalist discourse to remain 'Indonesian', they never thought of themselves as part of a colonial legacy.

Where many previous studies of colonial urbanism have examined ways in which colonial power attempted to use spatial discourses to exert control over the indigenous population (for example, Metcalf, 1989; Mitchell, 1988; Rabinow, 1989; Wright, 1991), more recent research has documented the role of indigenous agency in either resisting, accommodating or problematizing the nature of that control (Hosagrahar, 2000; Raychaudhuri, 2001; Yeoh, 1996). The major exception is Yeoh's (1996) innovative study of Singapore, demonstrating the extent to which indigenous inhabitants retained power over their own space. Also absent in most postcolonial urban and architectural studies are attempts to understand the connection between the built environment and the construction of the subject. In this context, Kusno explores how different spatial discourses and practices under Sukarno helped shape the national imagination and subjectivity; and under Suharto, how the making of two social and spatial categories (the underclass kampung and the middle-class 'real estate' suburb) helped to create two divided national subjectivities in the capital city.

POSTCOLONIAL PROSPECT

As an interpretive paradigm, the use of postcolonialism is surely destined to spread. The evidence lies not only in the ever-increasing literature, or the new and imaginative ways the concept is being used (Sidaway, 2000), but also because the political, cultural, social, demographic and also religious conditions (King, 2000a and b) in which it has been established will certainly grow. Briefly stated, its achievement has been to put colonial and postcolonial realities at the centre of understandings of the contemporary global condition. As a way of giving voice to and recognizing the agency and identities of subaltern others, it is increasingly being taken up across the disciplines, irrespective of geographical location and language (Vacher, 1997; Young, 2001: 62; see also Clayton, Chapter 18 in this volume). Its value is in complementing (and also contesting) other representations of the contemporary world – cultural or political economic theories of globalization, postmodernism, dependency or post-dependency theories and others (Featherstone et al., 1995; Simon, 1998; Yeoh, 2001). Each of these can be of value but in specific and limited contexts. The dangers are in attempting to use the concept in a totalizing fashion or, indeed, attempting to explain *everything* from a postcolonial framework or perspective.

As for the world of cities, we conclude by asking two questions. How persistent are post-colonial (imperial or neoimperial) forces in impacting urban space, not only in postcolonial and postimperial cities but in other cities elsewhere? And are these conceptual categories the most useful, valid or appropriate to describe contemporary urban transformations? Some scholars have suggested that the 'postcolonial city' is an 'unusual and transitory experience' (Yeoh, 2001: 462). In this case, Yeoh asks, can the 'postcolonial' endure as a meaningful category? (See also Simon, 1998.) Less frequently asked is whether a new form of diasporic colonial city (i.e. a social and spatial formation ascribable to imperialism) is being re-established in the one-time metropole (Philo and Kearns, 1993) or in urban settlements in other parts of the one-time colonial empire, not least the USA, generated not only from European imperialisms but rather by US imperialism itself (King, 2000b).

NOTES

Many thanks to Jane Jacobs for her thoughtful editing and innumerable perceptive suggestions. Abidin Kusno and Steve Pile also gave many helpful comments on earlier drafts.

1　'Postcolonialism' and 'postcolonial studies' are used here to cover different theoretical and methodological orientations including colonial discourse analysis, postcolonial theory and postcolonial criticism, and the use of the hyphenated 'post-colonial' (see Ashcroft et al., 1998; Moore-Gilbert, 1997; Williams and Chrisman, 1994). The use of the hyphen is not consistent in the literature.

2　What I have referred to, drawing on the work of sociologist John Useem, as those of the colonial third culture (King, 1976: 58–66).

3　For other comments on Cohn's work and influence on the power/knowledge issue, see the preface in Rabinow (1989). Ranajit Guha, cofounder of the Subaltern Studies group, in his introduction to an earlier collection of Cohn's essays, cites a passage from one of them: 'Anthropological "others" are part of the colonial world. In the historical situation of colonialism, both white rulers and indigenous peoples were constantly involved in representing to each other what they were doing. Whites everywhere came into other people's worlds with models and logics, means

of representation, forms of knowledge and action, with which they adapted to the construction of new environments, peoples, by new "others". By the same token those "others" had to restructure their worlds to encompass the fact of white domination and their own powerlessness.' Guha goes on to comment: 'It follows, therefore, that according to this approach, the interpenetration of knowledge and power constitutes the very fabric of colonialism' (1987: xx). This is a key to the main point of this chapter.

4　I list some 40 of these in the 20 years following 1954: see King (1976: 22).

5　Homi Bhabha (1983); an earlier reference reflecting Foucauldian influences is Peter Hulme (1981). In the humanities, Edward Said's utilization of Foucault's notion of discourse in his *Orientalism* (1978) is used by a number of authors to suggest this title as the foundational text in the analysis of 'colonial discourse'. Williams and Chrisman note, for example, that 'it is perhaps no exaggeration to say that Edward Said's *Orientalism*, published in 1978, single-handedly 'inaugurates a new area of academic inquiry, colonial discourse, also referred to as colonial discourse theory or colonial discourse analysis' (1994: 5). The writings of Gayatri Spivak (e.g. 1985; 1988; 1990) are central to the canon.

6　The primary aim of Gilroy's book is to address the representation of the black presence in Britain, and to provide 'a corrective to the more ethnocentric dimensions of [cultural studies]' (1987: 11–12). In a penultimate chapter on 'Diaspora, utopia and the critique of capitalism', Gilroy states, 'it is impossible to theorize black culture in Britain without developing a new perspective on British culture *as a whole*. This must be able to see behind contemporary manifestations into the cultural struggles which characterized the imperial and colonial period' (1987: 156). Gilroy's second major work, *The Black Atlantic*, setting out the space of cultural studies (especially black music) delineated by his title, was published in 1993. Confirming his thesis, Gilroy subsequently crossed the black Atlantic to a position in the American academy.

7　Starkey (1994: 24–5) gives figures of 20–27 per cent black African population, and of 10–12 black Caribbean, for the boroughs of Lewisham and Southwark, the immediate environs of Goldsmiths' College. Data on the percentage of non-white faculty in individual British universities has only become available in the 1990s. In 1998, 4.8 per cent of female academics were black (1.2 per cent) or Asian (3.6 per cent), and 6.1 per cent of male academics (1.1 per cent black, 5.0 per cent Asian) (Young, 1998: 17).

8　An early text, Bill Ashcroft et al. *The Empire Writes Back: Theory and Practice in Postcolonial Literature* (1989), was followed a few years later by two readers (Williams and Chrisman, 1994; Ashcroft et al., 1995), a minor marker in the institutionalization of new academic paradigms. Some conception of the extensive publication through the 1980s and 1990s is indicated in these texts, as well as the bibliography here.

9 Finkelstein et al. (1998) argue that the changing
   composition of new, full-time faculty, hired between
   1988 and 1992, containing more women, minorities
   and foreign-born, is leading to a significant change in
   academia – even though many of the minority faculty
   are 'ghettoized' by discipline. The authors show that
   racial/minority representation in the American profes-
   soriate rose from about 6–7 per cent (1970s) to about
   10 per cent (1987). In the cohort of new entrants
   (1988–92) they represent 17 per cent (Asian/Pacific
   islander 2.2; Black/not Hispanic 5.7; Hispanic 3.1;
   American Indian/Alaskan native, 0.5). The figures for
   the first three of these categories in the humanities are
   6.0, 4.3, 4.5; in the social sciences, 4.3, 6.8, 3.4 (1998:
   29, 125). Said points out that 'few European or
   American universities devoted curricular attention to
   African literature in the early 1960s' (1993: 288).
10 'The early exponents of postcolonial criticism
   focused on a critique of literary and historical writings and ... were located in the humanities of the
   Western academy. Subsequently, the objects of the
   deconstructive postcolonial critique expanded to
   include film, video, television, photography, painting,
   all examples of cultural praxis that are portable,
   mobile, and circulating in the West ... Why did it not
   address, in any significant way, the impact of imperialism on the design and spatial disciplines of architecture, planning and urban issues more generally,
   whether in the colony or, indeed, in the metropole? ...
   not only because they are different disciplines (and difficult to handle) but because the cultural products on
   which imperial discourses are inscribed – the spaces of
   cities, landscapes, buildings – unlike literary texts,
   films, and photography, are, for these postcolonial critics and their Western audiences, not only absent and
   distant, they are also not mobile. Critics have to take
   their own postcolonial subjectives halfway around the
   world to experience them' (King, 1995: 544).

REFERENCES

Arnold, D. (2000) '"Illusory riches": representations of
   the tropical world, 1840–1950', *Singapore Journal of
   Tropical Geography* 21 (1): 6–18.
Ashcroft, B. (2001) *Post-Colonial Transformation*.
   London: Routledge.
Ashcroft, B., Griffiths, G. and Tiffin, H. (1989) *The
   Empire Writes Back: Theory and Practice in Post-
   Colonial Literatures*. London: Routledge.
Ashcroft, B., Griffiths, G. and Tiffin, H. (1995) *The Post-
   Colonial Studies Reader*. London: Routledge.
Ashcroft, B., Griffiths, G. and Tiffin, H. (1998) *Key
   Concepts in Post-Colonial Studies*. London: Routledge.
Balandier, G. (1966) 'The colonial situation: a theoretical
   approach', in I. Wallerstein (ed.) *Social Change: The
   Colonial Situation* (1951). New York: Wiley.
Barnett, C. (1997) 'Sing along with the common people:
   politics, postcolonialism and other figures', *Environ-
   ment and Planning D: Society and Space* 15: 137–54.

Bhabha, H. (1983) 'The Other question: the stereotype and
   colonial discourse', *Screen* 24: 18–36.
Bhabha, H. (1994) *The Location of Culture*. London:
   Routledge.
Blaut, J.M. (1993) *The Colonizer's View of the World:
   Geographical Diffusions and Eurocentric History*.
   New York: Guilford.
Blunt, A. (1994) *Travel, Gender and Imperialism: Mary
   Kingsley and West Africa*. London: Guilford.
Blunt, A. and Rose, G. (1994) *Writing Women and Space:
   Colonial and Postcolonial Geographies*. London:
   Guilford.
Boyce Davis, C. (1994) *Black Women, Writing and
   Identity: Migrations of the Subject*. London: Routledge.
Buchanan, K. (1972) *The Geography of Empire*.
   Nottingham: Russell.
Cairns, S. (1998) 'Postcolonial architectonics',
   Postcolonial Studies 1 (2): 211–35.
Chakrabarty, D. (1992) 'Postcoloniality and the artifice of
   history: who speaks for "Indian" pasts?', *Represen-
   tations* 32 (Winter): 1–27.
Chatterjee, S. and Kenny, J. (1999) 'Creating a new capi-
   tal: colonial discourse and the decolonization of Delhi',
   Historical Geography 27: 73–98.
Clifford, J. (1988) *The Predicament of Culture*.
   Cambridge, MA: Harvard University Press.
Cohn, B.S. (1987) *An Anthropologist among the Historians
   and Other Essays*. Delhi: Oxford University Press.
Crush, J. (1994) 'Post-colonialism, de-colonization
   and geography', in A. Godlewska and N. Smith
   (eds) *Geography and Empire*. Oxford: Blackwell:
   330–50.
Dirks, N. (1992) 'Introduction', in N. Dirks (ed.) *Colonial-
   ism and Culture*. Ann Arbor: University of Michigan.
Dirks, N. (1996) 'Foreword', in B.S. Cohn (ed.)
   Colonialism and Its Forms of Knowledge. Princeton:
   Princeton University Press. pp. 1–4.
Dirlik, A. (1994) 'The postcolonial aura: Third World criti-
   cism in the age of global capitalism', *Critical Inquiry*
   20: 328–56.
Driver, F. (1992) 'Geography's empire: histories of
   geographical knowledge', *Environment and Planning
   D: Society and Space* 10: 23–40.
Driver, F. and Gilbert, D. (eds) (1999) *Imperial Cities:
   Landscape, Display and Identity*. Manchester: Manchester
   University Press.
Driver, F. and Yeoh, B.S.A. (2000) 'Constructing the trop-
   ics: an introduction', *Singapore Journal of Tropical
   Geography* 21 (1): 1–5.
Duncan, J.S. and Gregory, D. (1999) *Writes of Passage:
   Reading Travel Writing*. London: Routledge.
Fanon, F. (1967) *Black Skins, White Masks*. New York:
   Grove.
Fanon, F. (1968) *The Wretched of the Earth*. New York:
   Grove.
Featherstone, M., Lash, S. and Robertson, R. (eds) (1995)
   Global Modernities. London: Sage.
Finkelstein, M.J., Seal, R. and Schuster, J.H. (1998) *The
   New Academic Generation*. Baltimore: Johns Hopkins
   University Press.

Freeman, D.B. (1999) 'Hill stations or horticulture? Conflicting imperial visions of Cameroon Highlands, Malaysia', *Journal of Historical Geography* 25 (1): 17–35.

Frenkel, S. and Western, J. (1988) 'British tropical colony: Sierra Leone', *Annals of the American Association of Geographers* 78 (2): 211–28.

Furnivall, J.S. (1948) *Colonial Policy and Practice*. London: Cambridge University Press.

Geyer, M. (1993) 'Multiculturalism and the politics of general education', *Critical Inquiry* 19 (3): 499–533.

Gilbert, A. (1985) 'Moving the capital of Argentina', *Cities* August: 234–42.

Gilroy, P. (1987) *There Ain't No Black in the Union Jack: The Cultural Politics of Race and Nation*. London: Hutchinson.

Gilroy, Paul (1993) *The Black Atlantic: Modernity and Double Consciousness*. Cambridge, MA: Harvard University Press.

Godlewska, A. and Smith, N. (eds) (1994) *Geography and Empire*. Oxford: Blackwell.

Gregory, D. (2000) 'Post-colonialism', in R. Johnston, D. Gregory and D.M. Smith (eds) *The Dictionary of Human Geography*, 4th edn. Oxford: Blackwell.

Guha, R. (1987) 'Introduction', to B.S. Cohn *An Anthropologist among the Historians and Other Essays*. Delhi: Oxford University Press. pp. i–xxiv.

Gutierrez, J. and Lopez-Nieva, P. (2001) 'Are international journals of human geography really international?', *Progress in Human Geography* 25 (1): 53–70.

Harley, J.B. (1987) 'The map and the development of the history of cartography', in J.B. Harley and D. Woodward (eds) *The History of Cartography*, vol. 1. Chicago: University of Chicago Press. pp.1–43.

Harley, J.B. (1988) 'Maps, knowledge and power', in D. Cosgrove and S. Daniels (eds) *The Iconography of Landscape*. Cambridge: Cambridge University Press. pp. 277–312.

Holston, J. (1989) *The Modernist City*. Chicago: University of Chicago Press.

Hosagrahar, J. (2000) 'Mansions to margins: modernity and the domestic landscapes of historic Delhi, 1874–1910', *Journal of the Society of Architectural Historians* 60 (1): 24–45.

Hudson, B. (1977) 'The new geography and the new imperialism, 1870–1918', *Antipode* 9 (1): 12–19.

Hulme, P. (1981) 'Hurricanes in the Caribee: the constitution of the discourse of English colonialism', in F. Barker et al. (eds) *1641: Literature and Power in the Seventeenth Century*. Proceedings of the Essex Conference on the Sociology of Literature, Colchester, July 1980.

Jackson, P. and Jacobs, J.M. (1996) ' Postcolonialism and the politics of race', *Environment and Planning D: Society and Space* 14: 1–3.

Jacobs, J.M. (1993) '"Shake 'em this country". The mapping of the Aboriginal sacred in Australia: the case of Coronation Hill', in P. Jackson and J. Penrose (eds) *Constructions of Race, Place and Nation*. Minneapolis: University of Minnesota Press pp. 100–20.

Jacobs, J. (1996) *Edge of Empire: Postcolonialism and the City*. London: Routledge.

Jacobs, J. (1998) 'Staging difference: aestheticisation and the politics of difference in contemporary cities' in R. Fincher and J.M. Jacobs (eds) *Cities of Difference*. London: Guilford. pp. 252–78.

Jacobs, J., Dovey, K., and Lochert, M. (2000) 'Authorising aboriginality in Australia', in Lesley N.N. Lokko (ed.) *White Papers, Black Masks: Architecture, Race, Culture*. London: Athlone Press. pp. 218–35.

Kennedy, D. (1998) *The Magic Mountain: Hill Stations and the British Raj*. Berkeley: University of California Press.

Kenny, J. (1995) 'Climate, race, and imperial authority: the symbolic landscape of the British hill station in India', *Annals of the Association of American Geographers* 85 (4): 694–714.

Kenny, J. (1997) 'Claiming the high ground: theories of imperial authority and the British hill stations in India', *Political Geography* 16 (8): 655–73.

Khatabi, A. (1993) 'A colonial labyrinth', *Yale French Studies* 83 (2): 5–11.

King, A.D. (1976) *Colonial Urban Development: Culture, Social Power and Environment*. London: Routledge and Kegan Paul.

King, A.D. (1984) *The Bungalow: The Production of a Global Culture*. London: Routledge and Kegan Paul (2nd edn, Oxford University Press, 1995).

King, A.D. (1990) *Global Cities: Postimperialism and the Internationalization of London*. London: Routledge.

King, A.D. (1995) 'Writing colonial space: a review essay', *Comparative Studies in Society and History* 37: 541–54.

King, A.D. (1996) 'Worlds in the city: Manhattan transfer and the ascendance of spectacular space', *Planning Perspectives* 11: 97–114.

King, A.D. (2000a) 'Postcolonialism, representation and the city', in S. Watson and G. Bridge (eds) *Blackwell Companion to the City*. Oxford: Blackwell. pp. 261–9.

King, A.D. (2000b) 'Reworlding the city', *Planning History: Bulletin of the International Planning History Society* 22 (3): 5–16.

Kusno, A. (2000) *Behind the Postcolonial: Architecture, Urban Space and Political Cultures in Indonesia*. London: Routledge.

Langlands, B.W. (1969) 'Perspectives on urban planning for Uganda', in M. Safier and B.W. Langlands (eds) *Perspectives on Urban Planning for Uganda*. Department of Geography, Makere University College, Uganda.

Livingstone, D.N. (1991) 'The moral discourse of climate: historical considerations on race, place and virtue', *Journal of Historical Geography* 17 (4): 413–34.

Livingstone, D.N. (1999) 'The spaces of knowledge: contributions towards an historical geography of science', *Environment and Planning D: Society and Space* 13: 5–34.

Livingstone, D. (2000) 'Tropical hermeneutics: fragments for a historical narrative. An afterword', *Singapore Journal of Tropical Geography* 21 (1): 92–8.

Loomba, A. (1998) *Colonialism/Postcolonialism*. London: Routledge.

Mannoni, O. (1956) *Prospero and Caliban: The Psychology of Colonialism*. London: Methuen.

McClintock, A. (1992) 'The angel of progress; pitfalls of the term "post-colonialism"', *Social Text* 31/32: 84–98.

McGee, T.G (1967) *The Southeast Asian City*. London: Bell.

Memmi, A. (1965) *The Colonizer and the Colonized*. New York: Orion.

Metcalf, T.R. (1989) *An Imperial Vision: Indian Architecture and Britain's Raj*. Berkeley: University of California Press.

Mills, S. (1991) *Discourses of Difference: An Analysis of Women's Travel Writing and Colonialism*. London: Routledge.

Mills, S. (1994a) 'Gender, knowledge and empire', in A. Blunt and G. Rose, (eds) *Writing Women and Space: Colonial and Postcolonial Geographies*. London: Guilford. pp. 29–50.

Mills, S. (1994b) 'Gender and colonial space', *Gender, Place, Culture* 3: 125–47.

Mitchell, T. (1988) *Colonizing Egypt*. Cambridge: Cambridge University Press.

Mongia, P. (1995) *Contemporary Postcolonial Theory: A Reader*. London: Arnold.

Moore-Gilbert, B. (1997) *Postcolonial Theory: Contexts, Practices, Politics*. London: Verso.

Myers, G.A. (1995) 'A stupendous hammer: colonial and post-colonial reconstructions of Zanzibar's other side', *Urban Studies* 32 (8): 1345–59.

Myers, G.A. (1999) 'Colonial discourse and Africa's colonized middle: Ajit Singh's architecture', *Historical Geography* 27: 27–55.

Nalbantoglu, G.B. and Wong C.T. (eds) (1997) *Post-colonial Space(s)*. Princeton: Princeton University Press.

Perera, N. (1998) *Society and Space: Colonialism, Nationalism and Postcolonial Identity in Sri Lanka*. Boulder: Westview.

Philo, C. and Kearns, G. (1993) 'Culture, history, capital: a critical introduction to the selling of places', in C. Philo and G. Kearns (eds) *Selling Places: The City as Cultural Capital, Past and Present*. Oxford: Pergamon.

Pratt, M.L. (1992) *Imperial Eyes: Travel Writing and Transculturation*. London: Routledge.

Rabinow, P. (1989) *French Modern: Norms and Forms of the Social Environment*. Boston: MIT Press.

Raychaudhuri, S. (2001) 'Colonialism, indigenous elites and the transformation of cities in the non-western world: Ahmedabad (western India), 1890–1947', *Modern Asian Studies* 35 (3): 677–726.

Redfield, R.S. and Singer, M. (1954) 'The cultural role of cities', *Economic Development and Cultural Change* 3: 53–73.

Renbourne, E.T. (1961) *Life and Death of the Solar Topee: Protection of the Head from the Sun*. London: War Office, Directorate of Physiological and Biological Research, Report 117.

Robinson, J. (1999) 'Postcoloniality/postcolonialism', in L. McDowell and J.P. Sharp (eds) *A Feminist Glossary of Human Geography*. London: Arnold. pp. 208–10.

Sabine, N. (1943) *The British Colonial Empire*. London: Collins.

Said, E. (1978) *Orientalism*. London: Penguin.

Said, E. (1993) *Culture and Imperialism*. London: Vintage.

Sidaway, J. (2000) 'Postcolonial geographies: an exploratory essay', *Progress in Human Geography* 24 (4): 591–612.

Simon, D. (1998) 'Rethinking (post)modernism, postcolonialism, and posttraditionalism: South–North perspectives', *Environment and Planning D: Society and Space* 16: 219–45.

Slater, D. (1995) 'Trajectories of development theory: capitalism, socialism and beyond', in R.J. Johnston, Peter J. Taylor and Michael Watts (eds) *Geographies of Global Change*. Oxford: Blackwell. pp. 63–76.

Slater, D. (1998) 'Post-colonial questions for global times', *Review of International Political Economy* 5 (4): 647–78.

Smith, M.G. (1965) *The Plural Society in the British West Indies*. Oxford: Oxford University Press.

Smith, N. (1984) *Uneven Development: Nature, Capitalism and the Production of Space*. Oxford: Blackwell.

Spencer, J.E. and Thomas, W.L. (1948) 'The hill stations and summer resorts of the Orient', *Geographical Review* 39 (4): 637–51.

Spivak, G.C. (1985) 'Three women's texts and a critique of imperialism', *Critical Inquiry* 18 (4): 756–69.

Spivak, G.C. (1988) *In Other Worlds*. London: Routledge.

Spivak, G.C. (1990) *The Post-Colonial Critic: Interviews, Strategies, Dialogues*. London: Routledge.

Starkey, M. (1994) *London's Ethnic Minorities: One City, Many Communities. An Analysis of the 1991 Census Results*. London: London Research Centre.

Stepan, N.L. (2000) 'Tropical modernism: designing the tropical landscape', *Singapore Journal of Tropical Geography* 21 (1): 76–91.

Turner, V. (ed.) (1971) *Colonialism in Africa*. London: Cambridge University Press.

Vacher, H. (1997) *Projection coloniale et ville rationa-lisée. le Rôle de l'espace colonial dans la constitution de l'urbanisme en France 1900–1931*. Aalborg: Aalborg University Press.

Vale, L.J. (1992) *Architecture, Power and National Identity*. New Haven: Yale University Press.

Wertheim, W.F. (1964) *Indonesian Society in Transition*. The Hague: van Hoeve.

Western, J. (1985) 'Undoing the colonial city?', *Geographical Review* 73 (3): 335–57.

Western J. (1997) *Outcast Cape Town*, 2nd edn. Berkeley: University of California Press.

Williams, P. and Chrisman, L. (1994) *Colonial Discourse and Postcolonial Theory*. New York: Columbia University Press.

Wright, G. (1991) *The Politics of Design in French Colonial Urbanism*. Chicago: University of Chicago Press.

Yeang, K. (1987) 'Regionalist design intentions', in K. Yeang *Tropical Urban Regionalism: Building in a Southeast Asian City*. Singapore: Mimar. pp. 12–33.

Yeoh, B. (1996) *Contesting Space: Power Relations and the Urban Built Environment in Colonial Singapore*. Oxford: Oxford University Press.

Yeoh, B. (2001) 'Postcolonial cities', *Progress in Human Geography* 24 (3): 456–68.

Young, L. (1998) 'The colour of ivory towers', *The Times Higher* 5 June: 17.

Young, R. (1990) *White Mythologies: Writing History and the West*. London: Routledge.

Young, R.J.C. (2001) *Postcolonialism: An Introduction*. London: Blackwell.

Section 7

BEYOND THE WEST   Edited by Jennifer Robinson

Introduction: Cultural Geographies
Beyond the West

Jennifer Robinson

In 1999, a new episode in the popular British spy series, James Bond, was released: *The World Is Not Enough*. Since 1962, when Bond found himself in Jamaica tracking down the archvillain Dr No, he had been sent to save the world in a repertoire of countries and cities whose geographical reach puts the British empire in the shade – from New Orleans to Rio, Haiti to New York, Istanbul to Moscow, Los Angeles, Berlin, Casablanca, Calcutta ... but, in 1999, the action opened in London. Bond crashed out of the very openly secret headquarters of MI6 (see Pile, 2001) on the Thames in pursuit of the villain who had just blown a hole in its side, and careered down the usually placid river. In Q's latest design model speedboat, which handily turned into a car, the chase moved into the City of London and the back streets of the East End and former Docklands. James Bond's city-wrecking tactics had finally come home – just in time to christen the vast Millennium Dome by sliding down its roof. It was the beginning of the film, so he never caught up with the baddie. But the Bond genre had caught up with the times. The adventures and character of James Bond have shifted over time to cope (sort of) with détente, the end of the Cold War, new kinds of enemies, and tetchy cooperation with effective and tough female agents (and villains).

The Bond genre draws the world into a particular geography in its space of action, glamour and prejudice of various kinds. The inclusion of London as a site to be trashed in the course of saving the planet marked, for me, a symbolic moment in which the action of British interests drew closer in, turned back on itself as the ordering of the world shifts once again. As we seek to address long-standing forms of cultural imperialism embedded in the very concepts which are deployed in the service of academic research and writing, we could do worse than to keep in mind the ageing yet perennially popular figure of Bond. Not simply as a reminder that reinvention is possible – new lead actors, new sexual politics, new geopolitical orders – but, and equally instructive, always with the same plot: to prevent someone (else?) taking over the planet! So, a cautionary tale for postcolonializing intellectual enterprises, which, it has been noted, might so easily reinscribe relations of power and domination into their revisionist efforts. But the Bond genre also works as a vivid metaphor of how geography is embedded in the ways in which as scholars we come to know about and engage with the world.

Geographers have, like Bond, travelled the planet, but to produce knowledge about it rather than to save it. Initially this was about extracting information about other places to draw them into the expanding database (and economic and political empires) of the west. Over time, and by the late twentieth century, geography has settled into the western academy, but with some profound divisions in its production of knowledge. For example, as Jackson and Jacobs noted in 1996, geography conferences saw almost no overlap between audiences in sessions on racism and those on postcolonialism. Sessions on racism confronted contemporary political issues, while those on postcolonial geographies were (and still are) embedded in analyses of the past. The irony here is that one of the key imperatives of a broader postcolonial critique in cultural studies and social theory has been to help scholars 'attend to the complex ways that the past inheres in the present' (1996: 3). It is in this spirit that this section turns from the more usual round of geographic engagements with the historical phenomenon of the colonial or postcolonial moment, and suggests ways in which the postcolonial gaze might be cast back on the production of (western) cultural geography itself. And here, other divisions, in conferences and in the literature, are also significant: the great distance, for example, which has arisen between many geographers who work on and in different regions of the world; and the circulations of a hegemonic, unmarked and apparently unlocated realm of geographic theory, which is in fact profoundly tagged by its production in the dominant Anglo-American 'heartland' of graduate schools, research funds and publication outlets (Yeung, 2001).

There is a geography, then, to how cultural geographic knowledge is produced, the sites of production of its theory, the routes it tracks as it travels and is transported across the globe, the places it never reaches, and the vast zones of the world which never inform its imagination. Like James Bond's adventures, though, the western geographical imagination follows certain restricted pathways around the world, enabled and inscribed by the geopolitical moment. Its journeys also track a form of imperialism and are shaped by deeply entrenched global divisions and inequalities. But it certainly does not always have its own way (unlike Bond, who usually lives to see off another baddie). So while this section sets out

some of the ways in which cultural geography's knowledges have been and still are embedded in western hegemony, our ambition is equally to demonstrate how the empire has already been assailed from a range of different places and perspectives. The empire has indeed written back – and like the villain in The World Is Not Enough, have taken their grievances to the heart of the empire. Geography's knowledges are profoundly shaped by colonial pasts and geopolitical presents. But they have also already been shaped, as we will demonstrate, by the demands and challenges of people in poor countries (women, rulers, activists, NGOs, popular movements); in the west (black and working-class women, diasporic intellectuals, internationalist activists); and in the academy (the field of area studies, scholars working outside the western academy, an emerging postcolonial critique). So while there is a long way to go before western geography redresses its inheritance of neo-imperialist practice, in some fields of the discipline, we argue, there have already been important changes, which could perhaps serve as an inspiration for those working in other fields.

We have chosen to look in this section at feminist cultural geography; culture and development; and cultures of democracy. In each of these fields, we track the routes whereby knowledge and academic and political practice have been dislocated from their hegemonic western centres. The authors show how the roots of imperial knowledge are often found in other places, yet denied (as with democracy); how challenges to the limited scope of theory and politics have fundamentally changed international intellectual and political agendas (in relation to feminism); and how the 'objects' of powerful forms of knowledge and institutional capacities have variously embraced, criticized and even rejected them (in the case of development studies). In all three cases such challenges have certainly changed both theory and practice. The three authors, and the three fields they have chosen to engage with, each come to the politics of postcolonializing cultural geography from somewhat different positions. But all of them track ways in which thinking spatially (or geographically) about the circuits and tracks of knowledge brings into view not only the persisting power relations in these academic and institutional fields, but also the sources of potential alternatives and opposition to hegemonic forms of geographic knowledge and practice.

In this introduction I will draw out how an emergent postcolonial form of cultural geographic knowledge is being produced in the realm of theory and politics across the three examples covered by the chapters in this section. And here, I suggest, geography has an important potential contribution to make to efforts to postcolonialize western academic practices. To a broader cultural studies, geography can offer a nuanced way to think about the spatiality of how knowledge is produced, circulated and transformed. Importantly for this book, then, the intersection between cultural studies and geography is a particularly fertile zone in which to explore the potential to move beyond western hegemony in the production of knowledge. For the geography of culture, as Gupta and Ferguson (1999) so clearly point out, has to be imagined in quite different ways from the mosaic of nationally bounded and discrete units which predominated in cultural anthropology until recently. And, following Clifford (1997), the object of study of anthropology can no longer (if it ever was) appropriately be thought of as the bounded unit of the village or community. Rather, we need to attend to what he calls the 'discrepant cosmopolitanisms' of people everywhere, whose routes and connections beyond their place of abode have been and perhaps increasingly are as important as any 'local' culture or social network. In an era where diasporic links and transnational relations of all kinds are increasingly definitive of cultural dynamics, the idea that academic knowledge, or the political practices which so often flow from such knowledge, can remain bounded in unreflective and hegemonic national entities, or only attach themselves to and follow the tracks of dominant forms of transnationalism, is quite problematic.

The ambition of this section is to demonstrate some of the ways in which prominent forms of western cultural geography have been challenged to acknowledge their locatedness; and to engage with alternative forms of transnational connectedness. In so doing, the close association between geographical knowledge and hegemonic geopolitical formations has, at least to some extent, been disrupted. As all three authors acknowledge, these power relations are not so easily displaced – but their chapters highlight the importance of ongoing efforts to 'decolonize the disciplinary gaze' (Jacobs, introduction to Section 6 in this volume). They do not suggest that dominant forms of knowledge are easily overturned, or that alternatives are inevitably going to succeed (see also Sidaway, 2000). Through the three chapters, each of the authors treads a close path between acknowledging the capacities of powerful institutions and ideas, and exploring the already existing alternatives which might displace, if not dislodge, forms of patriarchy, capitalism and western hegemony.

A significant contribution of these chapters, then, is to trace some of the 'discrepant' intellectual cosmopolitanisms which have shaped important areas of cultural geographical knowledge, in the fields of feminism, democratic politics and development.

For development studies, the 'object' of its enquiry has always suggested a broad geographic scope for the production of its knowledge. But this field has historically been profoundly divided between a dominant base in western academies and agencies, and the subordinate fields of practice and application. Michael Watts shows that while appreciating the continued significance of these divisions and power relations (as his conclusion suggests), there are many different trajectories along which the field of development studies has travelled. Modernizing elites in poor countries have embraced in their own ways the ambition of progress and industrialization; and intellectuals in 'developing' countries have critically reflected on the practice of development and its impacts, to find it substantially wanting. Underdevelopment theory and post-developmentalism are just two examples of trends in development which have been led by scholars arguably 'beyond' the west, inspired by and challenged to respond to widespread opposition to the consequences of development around the world.

But like the transnational form taken by the anti-globalization movement's opposition to the latest rounds of development interventions (neoliberalism, genetic modification, environmental 'protection'), the tracks of academic knowledge are more complex than simply

'Third World' opposition to 'western' hegemony. That is because many of the scholars whose work has been crucial in this field have been on the move, tracking paths into the west, or from western bases to political and intellectual centres in poorer countries. The grounds for a postcolonial critique of development are possibly immanent in 'local' alternatives to western hegemony (Escobar, 2001). But they are also apparent within the circuits of hegemonic knowledge fields, amongst those working within the World Bank, as Watts explores. Transformations in knowledge can and do emerge amongst those speaking in the voice of modernization to challenge it, or engaging with western theory to dislocate it (Gilroy, 1994). New ways of thinking about and intervening in the world are potentially found in all of these 'places'. Moreover, as Watts shows in some detail, development ideas and practices are a result of the complex dynamics of intersecting knowledges and practices in specific places and institutions. This undermines any possibility of holding to a politics which pits a hegemonic development ideology against a populist resisting alterity. Tracking these complexities, or enumerating already existing alternatives to hegemonic development practices, is however, in his view, in danger of underplaying the actually existing and very powerful forms of capitalist exploitation which continue to frame economic growth and livelihoods across the globe.

Geographies of feminist cultural politics are instructive here, because they raise important questions about cultural incommensurability, and a politics of difference in the production of new forms of knowledge. Cheryl McEwan traces some of the ways in which western feminism has been challenged by the political and intellectual voices of black and working-class women in the west, and by 'Third World' feminism. Both of these initiatives have challenged dominant understandings of what feminism involves, and also insisted on exposing white women's complicity in forms of neocolonial power relations within the field of feminist scholarship (classically, Mohanty, 1989).

Western feminism has been changed, both politically and in terms of theoretical endeavours, as a result. It is impossible to assume, for example, that accounts of 'home' or public/private divides can be based solely on white western women's experiences (see, for example, Rose, 1993). But Cheryl McEwan is suggesting that the geographic tracks of feminist knowledge and practice need to be complicated even further. The field of feminist politics, she suggests, is pluralized, and the existence of different forms of feminism in different contexts perhaps marks the limits of engagement. Western feminism, then, rather than origin or exemplar, is just one amongst many different kinds of feminism, albeit with overlapping and intersecting lineages. But to truly decentre feminism (and other kinds of knowledge and politics too?), following Chakrabarty (2000), she proposes a 'provincialization' of the west. And here the caution of many postcolonial writers – that the postcolonial critique might simply reinscribe western dominance – is also relevant. To insist that the west should necessarily engage with, learn from, write about, other places in the course of decolonizing itself, is also to open the door to a new phase of neo-imperialism in which the west remains at the centre. Moreover, in relativist vein, pluralizing feminisms suggests an equivalence which the power relations of institutions, economics and academic publishing belies. To counter this, McEwan proposes acknowledging and supporting contestation: acknowledging the fragile associations across different interests and positions within the field of international feminisms, for example, and encouraging political links and intellectual comparisons which bypass the west. Once again, a geographical imagination of routes, the connections and tracks across and between different positions and places which make those places what they are, could cut a path through the perennially intractable incommensurability and difference/universality distinctions which plague this field, and others.

As all three authors note, though, simply reimagining the discursive field – of feminism, development or democracy – is not going to make the deeply embedded power relations of the present, and the inheritances of the past, disappear. Perhaps most pressingly the circulation of certain ideas of democracy, deeply marked by their association with US exceptionalism, reminds us of this. At large in the world, the idea of democracy and the sovereign rights of people (as opposed to rulers) has supported numerous geopolitical insurgencies on the part of powerful nations, in the poorest countries around the world. David Slater sketches out some of the ways in which theories of democracy have been tied historically to ideas and practices of exclusion, racism and slavery. 'Renarrativizing' (following Watts) democracy's histories is one path to postcolonializing this central concern of cultural and political geography. But being alert to the contemporary contestations and transformations in the practices of democracy around the world is another, which ties closely to McEwan's suggestion that it is political contestation that is most likely to provoke change. Slater concludes his chapter with an assessment of contemporary political sources of a 'beyond' to Euro-Americanism in the field of democracy. Alternative indigenous practices of democracy, new kinds of accommodations between multiple traditions of democracy within changing national contexts, or emergent oppositional forces which draw on a range of influences in constituting their forms of democracy, all speak of already existing alternatives.

But while the west's form of democracy has travelled the globe through various eras of geopolitical dominance, these alternatives, although certainly circulated internationally (e.g. the practices of the Zapatistas, or South African experiments in constitutional democracy), have no privileged circuits through which to extend or impose their agendas around the world. Slater concludes by suggesting that 'one of the key problems we face in the West is to find ways of expanding our geographies of reference and learning so we do not reproduce the arrogance and ignorance of self-contained visions of superiority'.

Writing on the relations between postcolonialism and South African geography, Jonathan Crush asks: does 'the decolonization of the discipline require a rupture with the knowledge industry of the western heartlands of geographical enterprise, or is there room for a productive, postcolonial interface?' (1993: 62). His review shows that South African geography has followed a course which ties it to intellectual trends in Anglo-American geography (especially a radical Marxism), but also to, for example, Indian historiography, local trade unions and political movement intellectuals. In addition, the politics of anti-apartheid conflict (and later post-apartheid governing) have profoundly shaped the direction of geography there. In many ways, South African geographers have carved a path between being a 'sink for Euro-American thinking' (1993: 63) and remaining connected to its intellectual dynamics. Part of this is because in the course of contributing to local politics, and responding to the demands of a social history inspired by subaltern studies and a Marxism committed to working with and for 'the people', esoteric international publications are not that useful. But, as Yeung (2001) points out in relation to South East Asia, the institutional demands for 'international' publications place scholars around the world in a position of having to engage with western scholarship. The practices of western-based refereeing and editorial review need interrogation here if, on a practical level, western cultural geography is to engage with and learn from scholars in other parts of the world. But the question which Crush poses on behalf of South African geography continues to be relevant. Why should geographers, writing in and about different contexts, whose intellectual worlds are shaped by a diversity of historical and contemporary influences, of which western scholarship is just one, engage with the west?

The growing interest within western cultural geography in responding to challenges to decolonize their imagination – acknowledging past influences on what is both a hybrid and a provincial Euro-Americanism, and attempting to learn from and engage with alternative and related traditions – might not always be reciprocated. Except that the geopolitics of economic and cultural dominance affect academia too, and scholars outside the west are disciplined in various ways to continue to seek the rewards attached to engagement. The entwining of dominant forms of knowledge with oppositions, outsides and alternatives, in the fields of feminism, development and democratic theory, are significant exemplars of the challenges and opportunities facing the mainstream of western geography, as well as other geographies. Not only can scholars seek to decentre and provincialize Anglo-American geography, but – and to the extent that geographers working in and on other regions choose to engage with this intellectual tradition – there is a real opportunity to enrich and diversify the field of western cultural geography. In the words of Chakrabarty, this might ensure that the 'world may once again be imagined as radically heterogeneous' (2000: 46).

As Sidaway notes, 'at their best and most radical, postcolonial geographies will not only be alert to the continued fact of imperialism, but also thoroughly uncontainable in terms of disturbing and disrupting established assumptions, frames and methods' (2000: 607). The task, he suggests, is to find a path between a necessary continuing engagement with fields of knowledge which for historical and geopolitical reasons have been dominated by the west (Chakrabarty, 2000), and a search for 'forms and directions that will at the very least relocate (and perhaps sometimes radically dislocate) familiar and often taken-for-granted geographical narratives' (2000: 607).

Such an engagement, though, cannot be on the terms set by western geography. Speaking back to the self-appointed centre of the discipline will involve challenging the familiar tracks of publication and distribution through the wealthiest countries in the world, reinscribing a range of ways of writing, sources of inspiration, criteria of excellence and, most significantly, broadening the grounds of theoretical reflection. Moreover, while unlearning privilege as loss (Spivak, 1990), western geographers cannot expect reciprocation from their counterparts elsewhere. But, as the chapters which follow make clear, the conversations have already begun. And perhaps most importantly, we might return here to consider our hero Bond's family motto, 'The World Is Not Enough'. The point is not to replace one megalomaniac vision of global dominance with another: in this case, to gain the world as a resource for western geographical scholarship is both not enough, and most definitely not what is being proposed here! Rather, it is to transform the ethics, politics and geographies of scholarly engagement.

REFERENCES

Chakrabarty, D. (2000) *Provincializing Europe*. Princeton: Princeton University Press.

Clifford, J. (1997) *Routes: Travel and Translation in the late Twentieth Century*. Cambridge, MA: Harvard University Press.

Crush, J. (1993). 'The discomforts of distance: postcolonialism and South African geography', *South African Geographical Journal* 75 (2): 60–8.

Escobar, A. (2001) 'Culture sits in places: reflections on globalism and subaltern strategies of localization', *Political Geography* 20: 139–74.

Gilroy, P. (1994) *The Black Atlantic: Modernity and Double Consciousness*. Cambridge: Harvard University Press.

Gupta, A. and Ferguson, J. (1999) *Culture, Power, Place: Explorations in Cultural Anthropology*. London: Duke University Press.

Jackson, P. and Jacobs, J. (1996) 'Editorial: postcolonialism and the politics of race', *Society and Space* 14 (1): 1–3.

Mohanty, C. (1989). 'Under western eyes: feminist scholarship and colonial discourses', in C.T. Mohanty (ed.) *Third World Women and the Politics of Feminism*. Bloomingdale: Indiana University Press. pp. 51–80.

Pile, S. (2001) 'The un(known) city … or, an urban geography of what lies buried below the surface', in I. Borden, J. Kerr, A. Pivaro and J. Rendell (eds) *The Unknown City: Contesting Architecture and Social Space*. Cambridge, MA: MIT Press. pp. 262–79.

Rose, G. (1993) *Feminism and Geography*. Cambridge: Polity.

Sidaway, J. (2000) 'Postcolonial geographies: an exploratory essay', *Progress in Human Geography* 24 (4): 591–612.

Spivak, G. (1990) *The Post-Colonial Critic: Interviews, Strategies, Dialogues*. New York: Routledge.

Yeung, H. (2001) 'Editorial: redressing the geographical bias in social science knowledge', *Environment and Planning A* 33 (1): 1–9.

21

The West and Other Feminisms

Cheryl McEwan

Most recent commentators on cultural geography recognize the significant impact of feminist theory (Baldwin et al., 1999; Crang, 1998; Jackson, 1989; Mitchell, 2000). This is reflected both in the subject of inquiry and in the theoretical underpinnings of cultural geography. Concerns with the cultural politics of gender, identity, sex and sexuality at a variety of different scales (from body to nation) and in a variety of different spaces (homes, communities, cities and imaginative spaces) point toward a reinvigorated and politically relevant cultural geography. Against this backdrop, this chapter explores the implications of recent debates and developments within international feminism for cultural geography. At an abstract level, assumptions by western feminists about what their political project entails have been called into question by a range of criticisms concerned with dislocating western centrism. Encounters with different feminisms and different gender relations have raised issues about what exactly it means to be feminist and have ensured that a western-centric political vision is no longer desirable. Criticism of white western feminism from commentators from both north and south[1] has been significant in breaking down western centrism and has major implications for what feminist cultural geographies might look like, both in the west and in 'other' places.

In what follows, I review the major debates between 'western' and 'other' feminisms, tracing the emergence of a powerful body of criticism from the south and, more recently, from post-communist countries. Of particular significance are the moves towards cultural explanations within feminism, how these have often been inspired by and related to the challenges posed by anti-western-centric approaches, and the

debates arising from these developments. Critiques of western feminism have exposed its previously unacknowledged ethnocentrism by focusing on discourse and representation, exploring the ways in which culturally located western feminisms perpetuate entrenched power relations and the 'othering' of non-western women and non-western modes of thought. Having first outlined the major points of these critiques, in the subsequent sections of the chapter I explore the impacts of the debates inspired by them on feminist theory, on the possibilities of formulating an international feminist agenda, and finally on feminist cultural geographies. I trace the beginnings of attempts to dislocate the western centrism of cultural geography, explore some of the differences this has made to rethinking specifically feminist geographical issues, and also reflect on how cultural geographies might contribute to these feminist debates. I make some suggestions for new areas of research arising from, through and beyond the dilemmas raised by these debates, which challenge some of the assumptions at the heart of western-based feminist cultural geographies. Finally, I map out the most pressing issues that future feminisms have to contend with, and how these might inform cultural geographies.

THE TURN TOWARDS CULTURE: FEMINISMS AND RESPONSES FROM BEYOND THE WEST

Western feminisms are largely engaged in critiquing the Enlightenment and its offshoots in modernity. These theoretical contestations have been the standard battleground for much of today's social, political and social science rhetoric and a site of substantial theoretical exchange and deliberation. Until the 1980s, there was a tendency to assume a commonality in the forms of women's oppression and activism worldwide (for example, Morgan, 1984). Many western feminists assumed that their political project was universal, and that women globally faced the same universal forms of oppression. However, divisions among women based on nationality, race, class, religion, region, language and sexual orientation have proved more divisive within and across nations than western theorists acknowledged or anticipated. Indeed, it was the assumption of sameness, which many assumed reflected an ethnocentric and middle-class bias, that incurred the resentment of many 'Third World' women. This surfaced at the UN conferences on women in Mexico City (1975) and Copenhagen (1980), where deep divisions were generated between women from the north and south. Heated debates at these conferences highlighted the profound differences amongst women across the global divides of north and south as well as within south and north along class and political lines. The theoretical fallout from these debates is an emphasis on difference as opposed to universalism. Political economy approaches have been condemned and largely rejected for stereotyping 'Third World' women as passive victims of global exploitation, and there has been a general turn towards cultural explanations of gender within feminism. In addition, critiques of western feminism are not confined to the west and its former colonies. There is now a significant body of criticism emanating from women in post-communist countries of eastern Europe and the former Soviet Union (see Drakuliç, 1993; Einhorn, 1993; Funk and Mueller, 1993). A number of core issues underpin these critiques and the broader shift towards cultural explanations, the consequences of which have been efforts to develop new ways of feminist theorizing and practice, which in turn inform feminist approaches within geography.

Destabilizing dominant discourses

Critics have stressed the need to destabilize the dominant discourses of imperial Europe, such as history, philosophy, linguistics and feminism. These discourses are unconsciously ethnocentric, rooted in European cultures and reflective of a dominant western worldview. Alternative approaches problematize the very ways in which the world is known, challenging the unacknowledged and unexamined assumptions at the heart of European and American disciplines that are profoundly insensitive to the meanings, values and practices of other cultures. Since the 1980s, black feminists, in particular, have explored the ways in which feminism is historically located in the dominant discourses of the west, a product of western cultural politics and therefore reflecting western understandings of sexual politics and gender relations. Indeed, in many cultures (particularly in the south) feminism is associated with cultural imperialism. In a germinal essay in 1984, for example, Valerie Amos and Pratibha Parmar trace the historical relationship between western feminism and imperial ideologies, institutions and practices. They argue that like gender, the category of feminism emerged from the historical context of modern European colonialism and anti-colonial struggles; histories of feminism must therefore engage with its imperialist origins.

Western feminism's unbecoming past was first exposed in the 1970s. Critics of the racism inherent in US women's suffrage movements, such as Angela Davis (1982) and Ellen Dubois (1978), inspired an outpouring of critical work by black feminists (Anzaldua and Moraga, 1981; Lorde, 1984; Rich, 1986). Black feminist theory and politics began to have a significant influence on feminism. In the British context, careful literary and historical work has made it impossible to refute the claim that white British women's historical experience, in all its complexity and variation, was bound up culturally, economically and politically with imperial concerns and interests.[2] As Burton argues, however, the original intention of Amos and Parmar's essay was 'not to clear the way for a more politically accountable historiography of Euro-American women's movements, but rather to make space for histories of black women, women of colour, and anti-colonialist and nationalist women' (1999: 218). She contends that,

> Before the 1980s, it was possible for even some of the most accomplished feminist historians in the West to express surprise that there had been women's movements and feminist cultures outside Europe and North America before the 1960s, even as they failed to realise the neocolonialist effect this kind of ignorance was having on the production of postcolonial counter-histories. (1999: 218)

Chandra Mohanty (1991) offered an enormously influential analysis of the insufficiency of western epistemological frameworks for recovering, let alone understanding, the cultural and historical meanings of women's experiences and structural locations outside the west. Mohanty's criticism of the invisibility of black and Third World women in histories of feminism precipitated an outpouring of publications. These focused especially on Indian and Egyptian women's movements (Badran, 1995; Baron, 1994; Jayawardena, 1995; Southard, 1995), but also on countries and cultures having less self-evident (or less well-known) relationships to European empires, such as Iran (Afary, 1996; Kandiyoti, 1991; Shahidian, 1995). The outcome of this feminist and anti-imperialist scholarship has been an attempt to reorient western feminisms, such that they are perceived no longer as exclusive and dominant but as part of a plurality of feminisms, each with a specific history and set of political objectives. As discussed, contrary to the widespread belief that the inspiration, origins and relevance of feminism are bourgeois or western, related to a particular ideology, strategy or approach, it is now recognized that feminism does not simply originate in the west. There are many incidents of precolonial women's movements around the world and various forms of feminism have existed and continue to exist across cultures.

Black feminist critiques have also offered more profound examinations of the racism and ethnocentrism at the heart of western feminisms. As bell hooks argues,

> All too frequently in the women's movement it was assumed one could be free of sexist thinking by simply adopting the appropriate feminist rhetoric; it was further assumed that identifying oneself as oppressed freed one from being an oppressor. To a grave extent such thinking prevented white feminists from understanding and overcoming their own sexist-racist attitudes toward black women. They could pay lip-service to the idea of sisterhood and solidarity between women but at the same time dismiss black women. (1984: 8–9)

The relationship between western and 'other' feminisms has often been adversarial. This is partly because of the failure of white women to recognize the power that structures relationships with black women that is also a legacy of imperialism, and partly because the concepts central to feminist theory in the west become problematic when applied to black women. One example is the explanation given for the inequalities in gender relations. Many black and 'Third World' feminists object to western feminists who see men as the primary source of oppression. Assumptions at the heart of white western feminisms do not reflect the experience of black women (Carby, 1983; Nain, 1991). This is because for black women there is no single source of oppression; gender oppression is inextricably bound up with 'race' and class. Furthermore, in many cultures black women often feel solidarity with black men and do not advocate separatism; they struggle with black men against racism, and against black men over sexism. This debate has generated theories that attempt to explain the interrelationship of multiple forms of black women's oppression, such as race, class, imperialism and gender, without arguing that all oppression derives ultimately from men's oppression of women.

Similar criticisms have been levelled at understandings of the public–private dichotomy. A large part of western feminist (and feminist geographical) literature is dedicated to critiquing the separation of public and private spheres, arguing that it devalues women's contribution to society, and that it has been used to confine women and inhibit their input. A major problem with this kind of criticism is that it ignores the contentions of some feminists in other parts of the world that the private realm does indeed exist separately from the public one, but that both domains are needed and political. For example, instead of motherhood being a private occupation forced on some women, which limits their political inputs or contributions, it is actually reconstructed as a chosen political occupation with important social and economic repercussions. The activities of some Islamist feminists and the Argentinian Mothers of the Disappeared are examples of where women have sought an empowering 'private' function, challenging western feminist assumptions about the home, family and motherhood being sites of oppression. These assumptions have also begun to be challenged within geography (in particular, Gillian Rose's 1993 discussion of home and family, and the work of Sarah Radcliffe and Sallie Westwood, 1996, on Ecuador). However, there is a tendency in much feminist geography to implicitly rely upon western-centric notions of public–private spheres. More broadly, many poor women of the south resent the bourgeois preoccupations of western feminisms. Economic exploitation and political oppression, as well as provision of basic needs such as clean water and children's education, are seen as more pertinent than issues of sexual politics and gender oppression which often motivate middle-class feminism in the north (Schech and Haggis, 2000: 88).

Similar tensions exist between women in eastern and western Europe. As Funk (1993: 319) argues, these tensions cause tremendous bitterness and suspicion on both sides, and are especially marked in reunified Germany. They are generated by the real structural power and economic imbalances between eastern and western European women, but also by power imbalances at the level of discourse, where western notions are hegemonic within feminism, risking the suppression and distortion of post-communist women's concerns. Eastern European women resent the imposition of western cultural and economic values, which extend even down to the level of fashions and the way different women dress. There are negative stereotypes on both sides, and differences in culture, socialization and personality. As Funk (1993: 320) argues, like their southern counterparts, women in former state socialist countries appear to be more oriented than western feminists toward children and family. They have different attitudes toward the individual and the collective and to authority, and are more sceptical of the benefits of paid work. They often have different attitudes toward men and toward collective action. A moralistic rejection by some western feminists of post-communist cultural differences risks a failure to recognize that the family was often a refuge from state control, in much the same way that the family in other cultures has been a refuge from slavery and imperialism. Differences in tradition, culture, personality, beliefs and desires, therefore, demand the interrogation and destabilization of dominant western feminist discourses. There is a need to 'provincialize' (Chakrabarty, 1992) western feminisms rather than see them as a paradigmatic form of feminism *per se*.

The politics of speaking and writing

Critics have also challenged the experiences of speaking and writing by which dominant discourses come into being, focusing on the problematic relationship between women in the north and south (and especially between white women and those in the former colonies). For example, a term such as 'the Third World' homogenizes peoples and countries and carries other associations – economic backwardness, the failure to develop economic and political order, and connotations of a binary contest between 'us' and 'them', 'self' and 'other' (Darby, 1997: 2–3). These practices of naming are not innocent. Rather they are part of the process of 'worlding' (Spivak, 1990), or setting apart certain parts of the world from others, with roots located historically in imperialism. Edward Said (1978) has shown how knowledge is a form of power, and by implication violence; it gives authority to the possessor of knowledge. Knowledge has been, and to large extent still is, controlled and produced in the west. As we have seen, feminism is not innocent of this. The power to name, represent and theorize is still located here, a fact

that 'other' feminisms seek to disrupt by challenging perceived western arrogance and ethnocentrism, and incorporating the voices of marginalized peoples.

Women in the south have been particularly concerned with contesting the power to name, including the use of terms such as 'primitive', 'native', 'traditional' and 'Third World women'. Their complaint is that western feminism has the power to speak for women elsewhere. This has not changed since colonial times; black women are still denied a voice and the authority to represent themselves. As Mohanty (1991) argues, black and southern women are constructed as 'other', located outside white, middle-class norms. Diversities among women (in terms of class, ethnicity, culture, religion, sexuality and so on) are erased by monolithic and singular epithets such as 'Third World women'. T. Minh-Ha Trinh (1989) describes the exclusionary tactics of western feminism that make the concerns of 'Third World women' 'special' because they are not 'normal', because they are other, and because they are not written by white women. She writes:

> Have you read the grievances some of our sisters express on being among the few women chosen for a 'Special Third World Women's Issue' or on being the only Third World woman at readings, workshops and meetings? It is as if everywhere we go, we become someone's private zoo. (1989: 82)

It is still the case that white western women are empowered (economically and socially) to make women in other cultures the object of their investigations, when the reverse is often neither possible nor feasible. For example, Sinith Sittirak describes her experiences as a Thai woman studying in Canada:

> Officially, there are no regulations to prevent me from exploring Canadian or any other ethnic groups. However, like many other 'international students' who received scholarships from development projects, it implicitly seemed that we 'should' focus on our own issues in our homes. That is the way it is. At that moment, I did not question as to why a Thai student had to focus on Thai issues, while Canadian students had much more academic privilege and freedom to study and speak about any women's issues in any continent from around the world. (1998: 119)

The consequence of these criticisms is that the presumed 'authority' and 'duty' of western academics to represent the whole world is increasingly being questioned both from within and from without its ideological systems (Duncan and Sharp, 1993). Western feminists have increasingly begun to recognize that international feminism is constituted by a multiplicity of voices, including those of women in the south. The challenge for feminist scholarship is to transcend the colonizing boundaries of modernist discourse, which demands the recognition of difference and the multiplicity of axes and identities that shape women's lives. Greater emphasis is now placed on the 'positionality' of the researcher in relationships of power. As Duncan and Sharp (1993) argue:

> It is much more than a question of being culturally sensitive or 'politically correct'... it requires a continual and radical undermining of the ground upon which one has chosen to stand, including, at times, the questioning of one's own political stance.

Black feminists and women in the south are fighting for spaces in which to articulate their own demands and shape their own political agendas. Furthermore, marginalized women are resisting their representation by elite women from within their own cultures, many of whom are now located within the western academy. As one scholar comments,

> Frankly, I'm very tired of having other women interpret for us, other women sympathise with us. I'm interested in articulating our own directions, our own aspirations, our own past, in our own words. (Skonaganleh: R'a, in Sittirak, 1998: 135)

Taking into account the criticisms that black feminists in particular have articulated regarding the exclusionary tactics of white feminism, constant reflection on the creation and production of knowledge remains important. As bell hooks argues,

> if we do not interrogate our motives, the direction of our work, continually, we risk furthering a discourse on difference and otherness that not only marginalizes people of color but actively eliminates the need of our presence. (1990: 132)

Gayatri Chakravorty Spivak (1990) takes this argument a step further by arguing that western feminists need not only to acknowledge the situatedness of their knowledges (i.e. their cultural specificity and, therefore, their partiality), but also to 'unlearn' their privilege as loss. This involves recognizing that privileges (race, class, nationality, gender, and so on) may have prevented the attainment of certain knowledges, not simply information not yet received, but the knowledge rendered incomprehensible by reason of specific social and cultural positions. In order to unlearn these privileges western feminists need to work hard at gaining some knowledge of others who occupy those spaces most closed to their view. It also means recognizing the importance of attempting to speak to those others in such a way that they might be able to answer back.

Critiques of western spatial metaphors and temporalities

Related to these issues of power and knowledge is an explicit critique of the spatial metaphors and temporality employed in western discourses. Whereas previous designations of the 'Third World' signalled both spatial and temporal distance – 'out there' and 'back there' – attempts to dislocate western centricity insist that the 'other' world is 'in here' (Chambers, 1996: 209) and that the modalities and aesthetics of the south have partially constituted western languages and cultures. This attempt to rewrite the hegemonic accounting of time (history) and the spatial distribution of knowledge (power) that constructs the 'Third World' has certainly been significant within feminism.

As discussed above, the ways in which western women represent their southern counterparts, and the power relationships inherent in this, have increasingly been brought under scrutiny. As the 'Third World' is frozen in time, space and history, so this is particularly the case with 'Third World women' (Mohanty, 1991). Carby (1983) writes:

> Feminist theory in Britain is almost wholly Eurocentric and, when it is not ignoring the experience of black women at 'home', it is trundling 'Third World women' onto the stage only to perform as victims of 'barbarous', 'primitive' practices in 'barbarous', 'primitive' societies.

Western feminists often universalize their own particular perspectives as normative, and essentialize women in the south as tradition-bound victims of timeless, patriarchal cultures (Mohanty, 1991: 71). In so doing, western feminist scholarship reproduces the colonial discourses of mainstream, 'male-stream' scholarship. What Mohanty (1991: 72) calls the 'colonialist move' arises from the bringing together of a binary model of gender, which sees 'women' as an *a priori* category of oppressed, with an 'ethnocentric universality', which takes western locations and perspectives as the norm. The effect is to create a stereotype – 'Third World woman' – that ignores the diversity of women's lives in the south across boundaries of class, ethnicity and so on, and reproduces 'Third World difference'. As suggested, this is a form of othering, a reprivileging of western values, knowledge and power (hooks, 1984; Ong, 1988;

Spivak, 1990; Trinh Minh-Ha, 1989). Mohanty argues that western feminism is too quick to portray women in the south as 'victims', to perceive all women as oppressed and as the subjects of power.

More recently, Uma Narayan (1997; 1998) has shown how feminist writings about women in the south not only misunderstand and disguise the constructed nature of 'tradition', but also fall into the trap of cultural essentialism. In trying to account for difference between women,

> seemingly universal essentialist generalisations about 'all women' are replaced by culture-specific essentialist generalisations that depend on totalising categories such as 'Western culture', 'Non-western cultures', 'Western women', 'Third World women', and so forth. The resulting portraits of 'Western women', 'Third World Women', 'African women', 'Indian women', 'Muslim women', 'Post-Communist women', or the like, as well as the picture of the 'cultures' that are attributed to these various groups of women, often remain fundamentally essentialist. They depict as homogeneous groups of heterogeneous people whose values, interests, ways of life, and moral and political commitments are internally plural and divergent. (1998: 87–8)

The consequence is 'an ongoing practice of "blaming culture" for problems in "non-western" contexts and communities' (1997: 51).[3]

Recovering the voices and agency of the marginalized

In response to what Spivak (1985) refers to as the 'epistemic and actual violence' of western discourses, critics have demanded an attempt to recover the lost historical and contemporary voices of the marginalized, the oppressed and the dominated, through a radical reconstruction of history and knowledge production (Guha, 1982). Reflecting on the Eurocentrism of histories of imperialism, Spivak (1985: 338) argues that the agency and voices of colonized peoples were deliberately erased by the colonizers, giving the impression of a unidirectional cultural imperialism dominated by western powers. The concern with revealing the epistemic and actual violence of writing out the historical agency of colonized peoples seems to have a great deal of resonance with the contemporary concerns of international feminism.

Black women and feminists around the world are contesting the authority of western women to represent their lives, and are fighting for spaces in which their voices can be heard and their stories told. For example, Ifi Amadiume (1997) challenges western anthropologists and other social scientists to recognize their own complicity in producing a version of Africa that is more a reflection of their own class-based patriarchal thought. Her work uncovers a hidden matriarchal history of Africa that continues to empower women in political struggle. This is part of a larger struggle by Africans to construct alternative, anti-racist and anti-imperialist epistemologies of self-representation and self-generated ideals. Furthermore, there are now well-established debates concerning representation, essentialism and difference, which have made researching and writing about gender relations and 'women', especially outside one's own cultural milieu, an incredibly complex topic (this is discussed in greater detail below).

In the face of this sustained criticism, many western feminists are now acutely sensitive to the intersections of power with academic knowledge and their privilege in relation to 'other' women, and are developing more ethical ways of researching and writing. However, as Spivak (1990) argues, there is still a need for greater sensitivity to the relationship between power, authority, positionality and knowledge. The implications of western feminists writing about women outside their own cultural milieu must be considered in the context of the global hegemony of western scholarship: in other words, western domination of the production, publication, distribution and consumption of information and ideas.

THE CULTURAL TURN: ELITIST AND APOLITICAL?

The shift towards cultural explanations and concerns with discourse and representation outlined above have been ridiculed by many activists (primarily in the south and post-communist contexts, but also from within western feminism) as elitist and removed from reality. The problem is often posed as a schism between theory and practice, or the gap between western feminist theorizing and the practical needs of women globally. Theoretical preoccupations are not easily translated into direct politics, and are accused of shifting the focus away from the material problems of women's lives. Many critics argue that organizing and obtaining women's human rights cannot be removed from ensuring a better life for men and women in societies characterized by poverty and a lack of freedom and democratic norms; the turn towards culture has been charged with ignoring these issues. Concerns with representation, text and imagery are perceived as too far removed from the exigencies of the daily lives of millions of impoverished people. One response has been a rejection by some critics of the turn towards culture, and an objection to the emphasis on difference and discourse away from material conditions. These objections are based on the notion that 'poverty is real'. Cecile Jackson (1997: 147) is one such scathing critic of 'postist' feminist understandings of poverty and gender, 'where culture, ideas and symbols are discursively interesting and constitutive of power, whilst materiality is of questionable status, and at least suspect', and where poverty becomes 'largely a state of mind' rather than a matter of material struggle for survival. She argues that real women and the challenges facing them get lost in the morass of text, image and representation. The turn towards culture, therefore, is in danger of 'chucking the baby out with the bathwater' (Udayagiri, 1995: 164): in dismissing the universalist assumptions of political economy, the material problems of the daily existences of many women are also erased.

An alternative response is a move by some critics towards cultural relativism. Proponents of this approach suggest that the solution to imperialism and universalism is through respect of difference in a plurality of identity politics. This respect for cultural difference, however, offers little assistance in terms of dealing with some of the complex issues confronting international feminist movements. By refusing to theorize cultural dominance, relativists implicitly evaluate all cultural positions as equal. This gives them no basis for making moral judgements about social justice in terms of feminist aspirations to deal with gender inequality and patriarchal power. It also ignores differences (class, regional, religious, ethnic) between women within specific cultural locations. The problems with cultural relativism crystallize around issues such as female circumcision. As Susanne Schech and Jane Haggis argue,

> How do cultural relativists distinguish between those African women who argue female circumcision is an important culturally specific part of being a woman in those cultures where it is practised, and those African women who argue that it is a dangerous and damaging patriarchal practice? (2000: 110)

In addition, the 'culture' that is preserved through such respect is often a patriarchal one that preserves male privilege at women's expense. As Anne Marie Goetz (1991: 146) argues, at the UN International Conferences, 'official' feminisms (often allied to and representing national governments and their political agendas) have used arguments about 'cultural respect' to block more radical, 'unofficial' feminisms that pose a greater threat to the status quo.

The underlying problem is that relativist arguments share a view of cultures and identities as bounded, coherent and autonomous. Such notions have been rejected by geographers (see Crang, 1998; Jess and Massey, 1995; McEwan, 2000b) and by cultural and feminist theorists (Butler, 1990; Hall, 1995) alike, not least because this replicates notions of culture informing conservative fundamentalisms in a variety of contexts. Moreover, replacing universal sameness with cultural difference does not disrupt colonial power relations between women that persist into the present; cultural difference can be used to deny any possibility of 'different' (for example, formerly colonized) women 'becoming the same' (i.e. achieving equality with women in the north). Women in the south are always marked by difference, since cultural difference is also racialized (Frankenberg and Mani, 1993; Narayan, 1998). Feminist scholarship has warned against this simple plurality of feminisms organized around some absolute conception of national and/or cultural difference. As Rey Chow (1990) argues, 'it is when the West's "other women" are prescribed their "own" national and ethnic identity in this way that they are most excluded from having a claim to the reality of their existence'. The real challenge for contemporary feminism lies in finding an alternative to false universalisms that subsume difference under hegemonic western understandings, and to relativism that would abandon any universalist claim in favour of reified and absolute conceptions of difference.

Black and 'Third World' feminisms have made important contributions in theorizing both power and knowledge and the significance of discourse, which generates very real interventions with very real effects (Rajan, 1993; Rose, 1987). They demand that we are able to see, responsibly and respectfully, from another's point of view. However, they could perhaps engage more with material issues of power, inequality and poverty, and resist focusing on text, imagery and representation alone. Strategies must be found for an active feminism that can make a difference. This involves combining the material with the symbolic and encourages the building of coalitions across differences. It demands, firstly, a material analysis 'to point to the consequences and inter-relations of different sites of oppression: class, race, nation and sexuality' (Goetz, 1991: 151) and, secondly, a recognition of the partial and situated quality of knowledge claims (Haraway, 1991). Therefore, western feminisms have to be seen as simply partial and local knowledges, constrained by their boundaries and the limited nature of viewpoint. Strategies to dislocate western-centric approaches perceive all knowledge as contestable, in contrast to the hands-off 'respect' of cultural relativists. Certain issues will unite women cross-culturally (for example, sexist oppression); other struggles, such as those for racial justice or national liberation, might mean confrontation between women. Stereotypes and generalizations need to be problematized. For example, as Narayan argues,

> there is no need to portray female genital mutilation as an 'African cultural practice' or dowry murders and dowry related harassment as a 'problem of Indian women' in ways that eclipse the fact that not all 'African women' or 'Indian women' confront these problems, or confront them in identical ways, or in ways that efface local contestations of these problems. (1998: 104)

The challenge is to produce something constructive out of disagreement, and to combine material concerns and emphasis on local knowledges with postcolonial and poststructuralist dismantling of knowledge claims. Ferguson (1998: 95) theorizes this as a new 'ethico-politics'. She suggests that the problem that western feminists need to confront is that they are located in the very global power relations that they might aspire to change; hence there is a 'danger of colluding with knowledge production that valorises status quo economic, gender, racial and cultural inequalities'. There is a need for self-reflexivity, recognition of the negative aspects of one's social identity and devaluation of one's moral superiority to build 'bridge identities' across difference. This allows other knowledges to talk back, and creates a 'solidarity between women that must be struggled for rather than automatically received' (1998: 109). This does not mean generalizations cannot be made, but it puts the emphasis back on how they are made. As Schech and Haggis (2000: 113) argue, these postcolonial feminist approaches are not simply about deconstructing western feminisms. Rather they provide a more comprehensive project of remoulding a conceptual framework 'capable of embracing a global politics of social justice in ways which avoid the "colonizing move"'.

THEORY INTO PRACTICE AT THE INTERNATIONAL SCALE?

There are now many instances of women around the world forging bonds of solidarity across difference at a variety of different scales. However, global power relations still pose problems for international feminism at the global scale. The UN conferences on women are a clear example where the theoretical developments outlined above often fail to translate into feminist practice. As discussed, the Mexico City (1975) and Copenhagen (1980) UN conferences were significant in highlighting differences between women. Better communications were established at the 1985 Nairobi conference once the myth of global sisterhood was abandoned and the profound differences in women's lives and the meanings of feminism across cultures were acknowledged (Basu, 1995: 3). Recognition of these differences is a product of insights gained from changes in the global order, changes in the forms of women's activism and the complicated, often conflictual, interchanges between local and global feminisms.

Diversity is now at the core of international feminism. As Karam (2000: 176) argues, this is not just about going beyond a singular identity for feminism, but about levels of identification and different feminist convictions and ideas among different generations of women with the bodies of feminist knowledge that have emerged recently. The need to identify different feminisms is now an acceptable theoretical premise; the diversity of feminist strategies means that there are also different priorities. However, as Amadiume (2000: 10) argues, familiar problems resurfaced at the fourth UN Conference on women, in Beijing, 1995. She argues that the 'Platform for Action' document produced at this conference was a unique achievement in pressing policy-makers to take action on women's issues and forcing governments seriously to address these issues. They include poverty, global economics, women's human rights, armed conflict, violence against women, political and economic participation, power-sharing, institutional mechanisms, media, access to healthcare and education, environment and protection of girl children. However, Amadiume argues that this is a 'laundry-list approach to women's issues', encouraging

> European women to return from Beijing with an illusion of a truly global process and a harmonious global sisterhood, with all women saying the same thing in spite of diversity. I even heard a few bourgeois women saying that women's differences have finally been resolved and that women are now the same everywhere. (2000: 12–13)

She suggests that well-meaning as these global concerns are, they should continue to be assessed in the context of western economic, political and cultural imperialism.

The 'Platform for Action' was criticized by indigenous women's groups, including Native American women who contest dominant notions of women's community roles, economic development, traditional lifestyles and family values. It was also criticized by coalitions, such as the International Network of Women of Colour and various 'Third World' women's groups, as being Eurocentric and privileged, especially on issues such as abortion, sexuality, marriage, motherhood and reproductive rights. Islamic groups criticized it for trying to impose western European notions of modernity on Islamic countries, ignoring the role of religion and the importance of moral and spiritual values in all aspects of life in these cultures. Less privileged women were more concerned with issues such as the negative effects of structural adjustment, global economics, basic needs like the right to land, citizenship, clean water, food and shelter, education and primary healthcare. Amadiume argues that elite women within poorer countries, who attend these conferences and help set the agenda, have become divorced from the real issues affecting less privileged women within their own countries and are therefore complicit in maintaining global power relations. The shift from grassroots-articulated focus to professional leadership imposed from above means that goals and issues have become repetitive in a fixed global language, and are controlled by paid UN and other donor advisers, consultants and workers (2000: 14–15).

On occasion, feminist theoretical concerns have been translated into practice, and sometimes they are not solely the product of dominant western discourses. This has certainly been evident in the gender, geography and development literature. The way in which the binary model of 'practical' and 'strategic' gender interests became central to gender and development approaches is but one example. This evolved from the strong grassroots women's movements developed in Brazil in the 1980s around initiatives aimed at structural change at specific points (Alvarez, 1990; Alvarez and Escobar, 1992; Jelin, 1990). These notions were then adopted (some might say appropriated) by western theorists and seemed to predetermine what feminist concerns were during the early 1990s (Moser, 1993). Therefore, despite the persistent problems in attempting to produce a global agenda for international feminisms, some commentators are more positive than Amadiume. They argue that it is through creating politics of difference, and politics of inclusion, that feminist networks have managed to create the bridges and multilevel connections from local organizations to international networks. As Braidotti (1995: 188) argues,

this is at the core of emancipatory globalization. In addition, while the turn to cultural politics within international feminism might seem to reinforce western-centric feminisms, a sense of the possibilities for a different kind of alliance among feminisms is also signalled by critiques from within western discourses. One example is Judith Butler's (1990) critique of representation in feminist politics. Butler critiques the representation of an entity called 'woman' and the idea that feminist theory often assumes that the female subject already exists rather than being produced in and through the cultural politics of identities, and produced differently in different places, always with the possibility of contestation. These radical notions of identity suggest that there is a need for a stronger sense of what a feminist politics might look like in different places.

Assuming that feminism is a cultural construct that does not accept unquestioned transference of thoughts and answers from one area of the world to another, the key question for Lavrin (1999: 175) is: is it possible to save its 'international' character without losing the wealth generated by its internal diversity? Race and class remain the biggest sources of division within national and international feminisms, and there are major divisions between academics and women without formal education in most cultures. Lavrin argues:

> The articulation of the personal, the regional, and the national into a universal formula understood by the largest number of women remains the most elusive objective of the feminist search for an international consensus. Yet there is hope. While in the past the difficulty of global communication hindered the search for mutual recognition, today we have much better tools to engage in the process of understanding the differences among the multiple manifestations of women's activities and the place that 'feminism' occupies in their agenda. (1999: 186–7)

The international feminist movement has taken up new information and communications tools to support global networking. The Beijing conference was a catalyst for women's electronic activism, particularly in setting an agenda for a global communications network for women (Harcourt, 1999). There are obvious problems that women in poorer countries might be further marginalized by these new technologies, since they do not share the same access. However, by linking women's practical experience with an increasing need to influence national and global policy, the new communications provide a means through which women across the world can improve and enhance their attempts to bring about a more gender-equitable global culture. In the words of Donna Haraway, these new approaches to gender identities and new technologies of communication between women might allow feminists to:

> negotiate the very fine line between appropriation of another's (never innocent) experience and the delicate construction of the just-barely-possible affinities, the just-barely-possible connections that might actually make a difference in local and global histories ... These are difficult issues, and 'we' fail frequently ... But 'our' writing is also full of hope that we will learn how to structure affinities instead of identities. (1991: 113)

DISLOCATING WESTERN CENTRISM AND CULTURAL GEOGRAPHY

Intersections between the efforts to dislocate western-centric knowledges and feminisms have important implications for cultural geographies. As discussed previously, these debates have made a difference to how certain specifically feminist geographical issues might be thought. The ways in which public and private spheres are conceptualized is one example. Related to this, debates emanating from cultural contexts outside the west might also bring into question the spaces in which politics are articulated, beyond the formal public sphere into homes, communities and neighbourhoods. In addition, encounters with traditions, ideas and criticisms beyond the west have shifted fundamentally the work of many cultural geographers. The embracing of global perspectives is, in part, responsible (McDowell, 1994: 147).

There is now a wealth of contemporary research exploring issues such as cross-cultural processes of global change and development,[4] cultural politics of postcolonial spaces in former imperial metropoles,[5] cultural geographies of commodity chains that connect peoples, regions and countries in the north and south,[6] national identities and nationalisms in specific cultural contexts,[7] and imaginative geographies of colonial and postcolonial landscapes and cultures.[8] It is perhaps at the margins of cultural geography, where there are intersections with feminist approaches in historical, development, political and economic geographies, that the lessons of anti-ethnocentrism have been most observed. Debates about positionality and representation, for example, are now well established in feminist geographies (see Madge, 1993; McDowell, 1992; Radcliffe, 1994; Robinson, 1994; Rose,

1997). Many feminist geographers are acutely aware of the intersections of power with academic knowledge, and this is often articulated through recognition of their own privilege in relation to the people they study. This privilege means greater access to resources, the power to produce knowledges, the luxury of professional status (Kobayashi, 1994), and the power of interpretation of the voices and opinion of others (Gilbert, 1994). There is now greater sensitivity within geography to recovering the agency and voices of others, both within archives (Barnett, 1998; McEwan, 1998) and through interactions with the researched (Pratt, 2000; Robinson, 1994; Townsend, 1995).

Despite these developments, the author of a recent textbook on cultural geography argues that 'cultural geography (especially new cultural geography) has been overweeningly Eurocentric' (Mitchell, 2000: xvii). To be sure, the core concerns of feminist cultural geographies (for example, re-evaluations of masculinities and femininities, landscapes, sexuality, public and private spheres, gender roles) are often perceived as side issues by feminists and women activists in societies outside the west. As discussed previously, the kind of feminism alluded to by all manner of feminist theorists, which has also set the agenda for feminist cultural geographies, is seen as largely removed from the exigencies of daily life of women living in poverty (Karam, 2000: 176).[9] Mitchell is perhaps more progressive than many commentators in recognizing the situatedness and partiality of *his* cultural geography and the problems of implicit ethnocentrism. However, his prescriptions on 'what a feminist cultural geography is, or could be' are also partial, and the question needs to be asked: what relevance would these feminist cultural geographies have outside the west? Mitchell's feminist cultural geography is informed by 'direct political imperatives', including 'the construction and reproduction of gender, as it is encoded in the spaces of cities and the space of the body', the 'evolution of domestic architecture, the development of "pink collar ghettos"', and 'debates about whether and how much mothers should work outside the home' (2000: 229).

Many of these issues are rooted deeply within western feminism and western cultural politics, with their specific understandings of gender politics, spatiality and the goals of feminist struggles. They do not necessarily translate in other contexts, particularly within Asia, Africa and Latin America, where the majority of the world's women happen to be located. The contemporary 'culture wars' that Mitchell discusses are also relevant to struggles between different feminisms in different parts of the world. As discussed, this has long been recognized within feminist and anti-colonial debates, but perhaps less so within cultural geography. The failure to engage in these debates, which are essentially about power and knowledge, (re)produces the ethnocentric cultural geographies that Mitchell castigates,[10] but then does little to address.

A politically relevant feminist cultural geography needs to engage with cultural politics, but these are not merely confined to western contexts. Instead, there should be a focus on cultural politics across cultures, a global as well as a local perspective on these politics, and recognition of the existence of diverse cultural geographies. It is significant also that the theoretical and empirical shifts outlined in this chapter have had some impact within cultural geography and some feminist geographers are producing precisely the kind of research that disrupts problematic ethnocentricities. In particular, the work on translocal and transnational geographies and the experiences of diasporic groups of women is significant. Gerry Pratt (1999), for example, uses poststructuralist theories of the subject and discourse analysis to explore the experiences of Filipinas in Vancouver. She examines how discursive constructions of 'Filipina' influence identities and impose limitations on occupational options, resulting in highly skilled and educated women becoming domestic workers on migration. Similarly, Alison Blunt (forthcoming) explores the geographies of home and identity of Anglo-Indian women. Blunt considers the spatial politics of home and identities on domestic, national and transnational scales, extending feminist and postcolonial theories through an exploration of the intertwined and contested geographies of 'domicile and diaspora' and their embodiment by Anglo-Indian women in India, Australia and Britain. Her approach is a productive engagement with feminist and postcolonial theories and a significant contribution to the exploration of the 'messiness of actual race politics' (Jackson and Jacobs, 1996: 3) and their material geographies.

Clearly, the lessons learned from the engagement between western and 'other' feminisms are beginning to inform contemporary cultural geographies and open up exciting avenues for research. From a historical perspective, there are possibilities for revealing alternative cultural histories that challenge the selective memory of parochial and univocal history and recognize the imbrications of these alternative histories in the global social formations fashioned by imperialism and colonialism. Feminist cultural geographies are also beginning to contribute to understandings of common global problems, cultural politics and the lessons drawn from international feminism. They are beginning to draw, for example, on the numerous instances of contemporary mobilizations of women and feminists in various women's and new social movements around common global problems (see, for example, Clark and Laurie, 2000; Laurie, 1997; Nagar, 2000). These are less interested in relations between nations and have the capacity to cross borders in their analyses and demands, whether these borders are those of gender, race, class or culture. Examples include the common bonds being forged by women workers in the global economy, ecological and environmental degradation, broad alliances against various forms of religious chauvinisms and fundamentalisms, and international campaigns around women's rights as human rights. Cultural geographies have the potential to explore examples of 'globalization from below', where there is a linking together of both diverse histories and the potential for cross-cultural alliances. In the realm of global cultural politics, there is also potential to explore alternative south–south linkages, not only because of their increasing significance within these contexts, but also to disrupt the hegemonic position of the west as frame of reference (John, 1999: 202).

By drawing on feminist insights and attempts to dislocate western centrism, cultural geographers also have the potential to contribute to new understandings of international feminism. As Alexander and Mohanty argue, the problem with the term 'international' in international feminism is that:

> To a large extent, underlying the conception … is a notion of universal patriarchy operating in a transhistorical way to subordinate all women... 'International', moreover, has come to be collapsed into the culture and values of capitalism. (1997: xix)

As recent research suggests (Pratt, 1999), feminist cultural geographies can contribute to new ways of thinking about women in similar contexts across the world, in *different* geographical spaces, rather than as *all* women across the world. Feminist research in global contexts involves shifting the unit of analysis from local, regional and national culture to relations and processes across cultures, and cultural geography is ideally placed for this kind of analysis. Grounding analyses in particular, local feminist praxis is necessary, but there is also a need to understand the local in relation to larger, cross-national processes. Feminist cultural geographers are beginning to respond to Alexander and Mohanty's call for a 'comparative, relational feminist praxis that is transnational in its response to and engagement with global processes'. This involves acknowledging, and working through, the productive tension between the 'centrifugal force of discrepant feminist histories and the promising potential of political organising across cultural boundaries' (Sinha et al., 1999: 1). It also requires working with women at grassroots level in different cultural contexts, breaking down hierarchies of knowledge/power that privilege the expert/outsider, undermining western universalisms and providing a basis for a new understanding of global diversity (Marchand and Parpart, 1995: 19). Here there are clear parallels with approaches in anthropology and models of local hybrid cultures, which challenge the orthodoxies of western thinking by bringing local knowledge to the fore in ways that dismantle the *a priori* categories of feminist theory.

The tensions and dilemmas in new ways of conceiving a cross-cultural feminist politics should both inform, and be informed by, feminist cultural geographies. Criticism from black women and feminists in the south has had a considerable impact on gender studies within geography, and a politically relevant cultural geography would gain a great deal from engaging further with these critiques. A more global perspective would help to countervail Eurocentrism. It would facilitate engagement by feminist cultural geographers in arguments about 'why women are important, and why gender is an indispensable concept in the analysis of political-cultural movements, of transition, and of social change' (Moghadam, 1994: 17). It would also allow cultural geographers to contribute to the critical exploration of relationships between cultural power and global economic power. Moreover, cultural geographical themes might potentially contribute to these feminist debates. How space and spatiality are conceptualized is particularly significant, for if localities are produced through links elsewhere, then something of the divide between difference and the international can be framed differently. Alison Blunt's (forthcoming) work on the spatial politics of home and identity at a range of spatial scales is one example of a productive engagement with these ideas. This chapter has explored how feminism has practically built a form of knowledge and cultural politics that perhaps transcends the locatedness that is evident in much of contemporary cultural geography. The increasing influence of these ideas within cultural geography suggests that it is ideally placed to engage with more radical notions of identity and developing a stronger sense of what a feminist politics might look like in different places.

NOTES

1 I use these terms to distinguish between the former European colonial powers and their former colonies, and between Anglo-European (often referred to as 'western') countries and non-Anglo-European countries. I consider north–south preferable to other epithets (First World, Third World; developed, developing), while recognizing that such binaries can hide the real causes of oppression and exploitation and are not strictly geographically accurate (Australia and New Zealand, for example, are predominantly white and western but located in the southern hemisphere).

2 See, for example, Ferguson (1992), Lewis (1996), Melman (1992), Midgley (1992), Ware (1994). This more critical understanding of feminist historiography has also informed the work of cultural geographers on gender and imperialism (Blunt, 1994; McEwan, 2000a; Morin, 1998).

3 See also El Saadawi (1997), Schech and Haggis (2000: 103–4).

4 See, for example, McEwan (2000b), Schech and Haggis (2000), Skelton and Allen (1999).

5 See, for example, the work of Kay Anderson (1989; 1990; 1991) on Vancouver's Chinatown, and Jane Jacobs (1996) on postcolonialism in British and Australian cities; also see Anderson and Jacobs (1997), Driver and Gilbert (1999).

6 See Barnaby (2000), Cook and Crang (1996), Cook and Harrison (1998).

7 For example, the exploration of gender and national identity in Ecuador by Radcliffe and Westwood (1996).

8 See, for example, Driver (1992), Gregory (1995), Nash (1999).

9 Similarly, cultural geography has been charged with becoming too theoretical and removed from the materialities of everyday life, of being too culturalist and neglecting political economy and sociological explanations. Aspects of life such as poverty are neglected by much of cultural geography and it therefore lacks political influence and, indeed, relevance (Thrift, 2000: 1–2).

10 Key texts in cultural geography are predominantly Anglocentric (for example, Crang, 1998; Cosgrove and Daniels, 1988; Daniels, 1993; Jackson, 1989).

REFERENCES

Afary, J. (1996) *The Iranian Constitutional Revolution, 1906–1911: Grassroots Democracy, Social Democracy, and the Origins of Feminism*, New York: Columbia University Press.

Alexander, M.J. and Mohanty, C.T. (1997) 'Introduction: genealogies, legacies, movements', in M.J. Alexander and C.T. Mohanty (eds) *Feminist Genealogies, Colonial Legacies, Democratic Futures*. London: Routledge. pp. xiii–xlii.

Alvarez, S. (1990) *Engendering Democracy in Brazil: Women's Movements in Transitional Politics*. Princeton: Princeton University Press.

Alvarez, S. and Escobar, A. (1992) 'Conclusion: theoretical and political horizons of change in contemporary Latin American social movement', in A. Escobar and S. Alvarez (eds) *The Making of Social Movements in Latin America: Identity, Strategy and Democracy*. Boulder: Westview.

Amadiume, I. (1997) *Reinventing Africa: Matriarchy, Religion and Culture*. London: Zed.

Amadiume, I. (2000) *Daughters of the Goddess, Daughters of Imperialism: African Women, Culture, Power and Democracy*. London: Zed.

Amos, V. and Parmar, P. (1984) 'Challenging imperial feminisms', *Feminist Review*. 17: 3–19.

Anderson, K. (1989) 'Cultural hegemony and the race-definition process in Chinatown, Vancouver: 1880–1980' *Environment and Planning D: Society and Space* 6 (2): 127–51.

Anderson, K. (1990) ' "Chinatown re-oriented": a critical analysis of recent development schemes in a Melbourne and Sydney enclave', *Australian Geographical Studies* 28 (2): 137–54.

Anderson, K. (1991) *Vancouver's Chinatown: Racial Discourse in Canada, 1875–1980*. Montreal: McGill–Queens University Press.

Anderson, K. and Jacobs, J. (1997) 'From urban Aborigines to Aboriginality and the city: one path through the history of Australian cultural geography', *Australian Geographical Studies* 35 (1): 12–22.

Anzaldua, G. and Moraga, C. (eds) (1981) *This Bridge Called My Back: Writings by Radical Women of Colour*. New York: Kitchen Table Women of Colour Press.

Badran, M. (1995) *Feminists, Islam and Nation: Gender and the Making of Modern Egypt*. Princeton: Princeton University Press.

Baldwin, E., Longhurst, B., McCracken, S., Ogborn, M. and Smith, G. (1999) *Introducing Cultural Studies*. London: Prentice Hall.

Barnaby, W. (2000) 'When a meal out changes lives', interview with Ian Cook *Times Higher Education Supplement* 5 May: 32.

Barnett, C. (1998) 'Impure and worldly geography: the Africanist discourse of the Royal Geographical Society, 1831–73', *Transactions of the Institute of British Geographers* 23 (2): 239–52.

Baron, B. (1994) *The Women's Awakening in Egypt: Culture, Society and the Press*. New Haven: Yale University Press.

Basu, A. (1995) 'Introduction', in A. Basu (ed.) *The Challenge of Global Feminisms: Women's Movements in Global Perspective*. Boulder: Westview.

Blunt, A. (1994) *Travel, Gender and Imperialism: Mary Kingsley and West Africa*. New York: Guilford.

Blunt, A. (forthcoming) *Domicile and Diaspora: Anglo-Indian Women and the Spatial Politics of Home*. Oxford: Blackwell.

Braidotti, R. (1995) 'Afterword: forward looking strategies', in R. Buikema and A. Smelik (eds) *Women's Studies and Culture: A Feminist Introduction*. London: Zed.

Burton, A. (1999) 'Some trajectories of "Feminism" and "Imperialism"', in M. Sinha D. Guy and A. Woollacott (eds) *Feminisms and Internationalism*. Oxford: Blackwell. 214–24.

Butler, J. (1990) *Gender Trouble: Feminism and the Subversion of Identity*. London: Routledge.

Carby, H. (1983) 'White women listen! Black feminism and the boundaries of sisterhood', in Centre for Cultural Studies *The Empire Strikes Back*. London: Hutchinson.

Chakrabarty, D. (1992) 'Provincialising Europe: postcoloniality and the critique of history', *Cultural Studies* 6 (3): 337.

Chambers, I. (1996) 'Waiting on the end of the world?', in D. Morley and K.-H. Chen (eds) *Stuart Hall: Critical Dialogues in Cultural Studies*. London: Routledge. pp. 201–11.

Chow, R. (1990) *Women and Chinese Modernity: The Politics of Reading between East and West*. Oxford: University of Minnesota Press.

Clark, F.C. and Laurie, N. (2000) 'Gender, age and exclusion: a challenge to community groups in Peru', *Gender and Development* 8 (2): 80–8.

Cook, I. and Crang, P. (1996) 'The world on a plate: culinary culture, displacement and geographical knowledges', *Journal of Material Culture* 1 (2): 131–53.

Cook, I. and Harrison, M. (1998) 'Getting with the fetish? Manufacturing Jamaican hot pepper sauces for the UK market'. Paper presented at the RGS/IBG Economic Geography Research Group conference 'Geographies of Commodities', Manchester University.

Cosgrove, D. and Daniels, S. (eds) (1988) *The Iconography of Landscape*. Cambridge: Cambridge University Press.

Crang, M. (1998) *Cultural Geography*. London: Routledge.

Daniels, S. (1993) *Fields of Vision*. Cambridge: Polity.

Darby, P. (ed.) (1997) *At the Edge of International Relations: Postcolonialism, Gender and Dependency*. London: Pinter.

Davis, A. (1982) *Women, Race and Class*. New York: Vintage.

Drakuliç, S. (1993) *How We Survived Communism and Even Laughed*. London: Vintage.

Driver, F. (1992) 'Geography's empire: histories of geographical knowledge', *Environment and Planning D: Society and Space*. 10: 23–40.

Driver, F. and Gilbert, D. (eds) (1999) *Imperial Cities: Landscape, Display and Identity*. Manchester: Manchester University Press.

Dubois, E.C. (1978) *Feminism and Suffrage: The Emergence of an Independent Women's Movement in America, 1848–1869*. Ithaca: Cornell University Press.

Duncan, N. and Sharp, J.P. (1993) 'Confronting representation(s)', *Environment and Planning D: Society and Space* 11: 473–86.

Einhorn, B. (1993) *Cinderella Goes to Market: Citizenship, Gender and Women's Movements in East Central Europe*. London: Verso.

El Saadawi, N. (1997) *The Nawal El Saadawi Reader*. London: Zed.

Ferguson, A. (1998) 'Resisting the veil of privilege: building bridge identities as an ethico-politics of global feminisms', special issue on Border Crossings:

Multicultural and Postcolonial Feminist Challenges to Philosophy, Part 2, *Hypatia* 13(3): 95–114.

Ferguson, M. (1992) *Subject to Others: British Women Writers and Colonial Slavery, 1670–1834*. London: Routledge.

Frankenberg, R. and Mani, L. (1993) 'Crosscurrents, crosstalk: "race", postcoloniality and the politics of location', *Cultural Studies* 7 (2): 292–310.

Funk, N. (1993) 'Feminism East and West', in N. Funk and M. Mueller (eds) *Gender Politics and Post-Communism: Reflections from Eastern Europe and the Former Soviet Union*. London: Routledge. pp. 318–30.

Funk, N. and Mueller, M. (eds) (1993) *Gender Politics and Post-Communism: Reflections from Eastern Europe and the Former Soviet Union*. London: Routledge.

Gilbert, M. (1994) 'The politics of location: doing feminist research "at home"', *Professional Geographer* 46: 90–6.

Goetz, A.M. (1991) 'Feminism and the claim to know: contradictions in feminist approaches to women in development', in R. Grant and K. Newland (eds) *Gender and International Relations*. Bloomington: Indiana University Press. pp. 133–57.

Gregory, D. (1995) 'Imaginative geographies', *Progress in Human Geography* 9: 447–85.

Guha, R. (ed.) (1982) *Subaltern Studies*. New Delhi: Oxford University Press.

Hall, S. (1995) 'New cultures for old', in D. Massey and P. Jess (eds) *A Place in the World? Places, Cultures and Globalization*. Oxford: Oxford University Press. pp. 175–213.

Haraway, D. (1991) *Simians, Cyborgs and Women*. London: Free Association.

Harcourt, W. (ed.) (1999) *Women@Internet: Creating New Cultures in Cyberspace*. London: Zed.

hooks, b. (1984) *Feminist Theory from Margin to Centre*. Boston: South End.

Jackson, C. (1997) 'Post-poverty, gender and development?', *Institute of Development Studies (IDS) Bulletin* 28 (3): 145–53.

Jackson, P. (1989) *Maps of Meaning*. London: Unwin Hyman.

Jackson, P. and Jacobs, J.M. (1996) 'Postcolonialism and the politics of race', *Environment and Planning D: Society and Space*. 14, 1–3.

Jacobs, J.M. (1996) *Edge of Empire. Postcolonialism and the City*. London: Routledge.

Jayawardena, K. (1995) *The White Women's Other Burden: Western Women and South Asia during British Rule*. London: Routledge.

Jelin, E. (ed.) (1990) *Women and Social Change in Latin America*. London: Zed.

Jess, P. and Massey, D. (1995) 'The contestation of place', in D. Massey and P. Jess (eds) *A Place in the World*. Oxford: Oxford University Press.

John, M. (1999) 'Feminisms and internationalisms: a response from India', in M. Sinha D. Guy and A. Woollacott (eds) *Feminisms and Internationalism*. Oxford: Blackwell. pp. 195–204.

Kandiyoti, D. (1991) *Women, Islam and the State*. Philadelphia: Temple University Press.

Karam, A. (2000) 'Feminist futures', in J.N. Pieterse (ed.) *Global Futures*. London: Zed. pp. 175–86.

Kobayashi, A. (1994) 'Coloring the field: gender, "race" and the politics of fieldwork', *Professional Geographer* 46: 73–80.

Laurie, N. (1997) 'Negotiating femininity: women and representation in emergency employment in Peru', *Gender, Place and Culture* 4 (2): 235–52.

Lavrin, A. (1999) 'International feminisms: Latin American alternatives', in M. Sinha D. Guy and A. Woollacott (eds) *Feminisms and Internationalism*. Oxford: Blackwell. pp. 175–90.

Lewis, R. (1996) *Gendering Orientalism: Race, Femininity and Representation*. London: Routledge.

Lorde, A. (1984) *Sister/Outsider: Essays and Speeches*, Boston, MA: Crossing.

Madge, C. (1993) 'Boundary disputes: comments on Sidaway (1992)', *Area* 25 (3): 294–9.

Marchand, M. and Parpart, J. (eds) (1995) *Feminism/Postmodernism/Development*. London: Routledge.

McDowell, L. (1992) 'Doing gender: feminism, feminists and research methods in human geography', *Transactions of the Institute of British Geographers* 17 (4): 399–416.

McDowell, L. (1994) 'The transformation of cultural geography', in D. Gregory, R. Martin and G. Smith (eds) *Human Geography: Society, Space and Social Science*. London: Macmillan.

McEwan, C. (1998) 'Cutting power lines within the palace? Countering paternity and eurocentrism in the "geographical tradition"', *Transactions of the Institute of British Geographers* 23 (3): 371–85.

McEwan, C. (2000a) *Gender, Geography and Empire: Victorian Women Travellers in West Africa*. Leicester: Ashgate.

McEwan, C. (2000b) 'Geography, culture and global change', in M.J. Bradshaw, P. Daniels, M. Bradshaw, D. Shaw, and J. Sidaway (eds). *Human Geography: Issues for the 21st Century*. London: Longman.

Melman, B. (1992) *Women's Orients: Englishwomen and the Middle East, 1718–1918*. London: Macmillan.

Midgley, C. (1992) *Women against Slavery: The British Campaigns, 1780–1870*. London: Routledge.

Mitchell, D. (2000) *Cultural Geography*. Oxford: Blackwell.

Moghadam, V. (1994) 'Introduction: women and identity politics in theoretical and comparative perspective', in V. Moghadam (ed.) *Identity Politics and Women: Cultural Reassertions and Feminisms in International Perspective*. Boulder: Westview. pp. 3–26.

Mohanty, C.T. (1991) 'Under Western eyes: feminist scholarship and colonial discourses', in C.T. Mohanty, Russo and L. Torres et al. (eds) *Third World Women and the Politics of Feminism*. Bloomingdale: Indiana University Press. pp. 51–80.

Morgan, R. (1984) *Sisterhood Is Global: The International Women's Movement Anthology*. New York: Anchor.

Morin, K. (1998) 'British women travellers and constructions of racial difference across the nineteenth-century American West', *Transactions of the Institute of British Geographers* 23 (3): 311–30.

Moser, C. (1993) *Gender Planning and Development: Theory, Practice and Training*. London: Routledge.

Nagar, R. (2000) 'Mujhe Jawab Do! (Answer me!): women's grassroots activism and social spaces in Chitrakoot (India)', *Gender, Place and Culture* 7 (4): 341–62.

Nain, G.T. (1991) 'Black women, sexism and racism: black or anti-racist feminism?', *Feminist Review* 37: 1–22.

Narayan, U. (1997) *Dislocating Cultures: Identities, Traditions and Third World Feminism*. New York: Routledge.

Narayan, U. (1998) 'Essence of culture and a sense of history: a feminist critique of cultural essentialism', special issue on Border Crossings: Multicultural and Feminist Challenges to Philosophy, Part 1, *Hypatia* 13 (2): 86–107.

Nash, C. (1999) 'Landscape', in P. Cloke, P. Crang and M. Goodwin (eds) *Introducing Human Geographies*. London: Arnold. pp. 217–25.

Ong, A. (1988) 'Colonialism and modernity: feminist representations of women in non-western societies' *Inscriptions* 3/4: 79–104.

Pratt, G. (1999) 'From registered nurse to registered nanny: discursive geographies of Filipina domestic workers in Vancouver, B.C.', *Economic Geography* 75 (3): 215–36.

Pratt, G. (2000) 'Research performance', *Environment and Planning D* 18 (5): 639–53.

Radcliffe, S. (1994) '(Representing) post-colonial women: authority, difference and feminisms', *Area* 26 (2): 25–32.

Radcliffe, S. and Westwood, S. (1996) *Remaking the Nation*. London: Routledge.

Rajan, R.S. (1993) *Real and Imagined Women: Gender, Culture and Postcolonialism*. London: Routledge.

Rich, A. (1986) *Blood, Bread and Poetry: Selected Prose, 1979–1985*. New York: Norton.

Robinson, J. (1994) 'White women researching/representing 'others': from anti-apartheid to postcolonialism?', in A. Blunt and G. Rose (eds) *Writing Women and Space*. New York: Guilford. pp. 197–226.

Rose, G. (1993) *Feminism and Geography*. Cambridge: Polity.

Rose, G. (1997) 'Situating knowledges: positionality, reflexivities and other tactics', *Progress in Human Geography* 21 (3): 305–20.

Rose, J. (1987) 'The state of the subject (II): the institution of feminism', *Critical Inquiry* 29 (4): 9–15.

Said, Edward (1978) *Orientalism*. London: Peregrine.

Schech, S. and Haggis, J. (2000) *Culture and Development: A Critical Introduction*. Oxford: Blackwell.

Shahidian, H. (1995) 'Islam, politics, and the problems of writing women's history in Iran', *Journal of Women's History*. 7: 113–44.

Sinha, M., Guy, D. and Woollacott, A. (1999) 'Introduction: why feminisms and internationalism?', in M. Sinha, D. Guy and A. Woollacott (eds) *Feminisms and Internationalism*. Oxford: Blackwell. pp. 1–13.

Sittirak, S. (1998) *The Daughters of Development*. London: Zed.

Skelton, T. and Allen, T. (eds) (1999) *Culture and Global Change*. London: Routledge.

Southard, B. (1995) *The Women's Movement and Colonial Politics in Bengal, 1921–1936*. New Delhi: Manohar.

Spivak, G.C. (1985) 'Subaltern studies: deconstructing historiography', in R. Guha (ed.) *Subaltern Studies IV*. New Delhi: Oxford University Press. pp. 330–63.

Spivak, G.C. (1990) *The Post-Colonial Critic: Interviews, Strategies, Dialogues*, ed. S. Harasym. New York: Routledge.

Thrift, N. (2000) 'Dead or alive?' in I. Cook, D. Crouch, S. Naylor and J. Ryan (eds) *Cultural Turns/Geographical Turns*. London: Prentice Hall. pp. 1–6.

Townsend, J.G. (1995) *Women's Voices from the Rainforest*. London: Routledge.

Trinh, T. Minh-Ha (1989) 'Difference: "a special Third World women issue"', in *Woman, Native, Other: Writing Post-Coloniality and Feminism*. Bloomington: Indiana University Press.

Udayagiri, M. (1995) 'Challenging modernisation: gender and development, postmodern feminism and activism' in M. Marchand and J. Parpart (eds) *Feminism/Postmodernism/Development*. London: Routledge. pp. 159–77.

Ware, V. (1994) *Beyond the Pale: White Women, Racism and History*. London: Verso.

22

Beyond Euro-Americanism – Democracy and Post-colonialism

David Slater

BY WAY OF AN OPENING

Although critical human geography has witnessed a growing interest in the conceptual and thematic concerns emanating from the post-colonial turn (see, for example, Gregory, 1995; Sidaway, 2000), and although more analytical attention has been recently given to questions of the spatiality of democracy (Massey, 1995; Robinson, 1998; Slater, 1998), so far little work has been developed around the spatiality or geopolitics of democracy and democratization from a post-colonial perspective. For this chapter my use of the term 'post-colonial perspective' refers to three interconnected objectives: (1) to provide a critical approach to certain influential western visions of democracy; (2) to emphasize the mutually constitutive, although unequal, role played by colonizer and colonized or globalizer and globalized within the domain of west/non-west encounters, including here an awareness of the crucial links between 'inside' and 'outside'; and (3) within this same domain, to briefly touch on certain facets of the politics of conceptual knowledge, for example, how global are our geographies of reference, how ethnocentric are western theorists of democracy, and what can we learn from other non-western theorists? I shall return to these points at the end of the chapter.

The continuing critique of westocentrism or Euro-Americanism has been nurtured by post-colonial perspectives, but not infrequently these critical appraisals have tended to bypass the ways political theory in general and democratic theory in particular are constructed and deployed, and similarly political geography has tended not to be a predominant site for the application of post-colonial theory. In this chapter I plan to examine two interconnected issues. First, I want to discuss what might be meant by Euro-Americanism and how such a prevailing vision has become rooted in western social science; and second, I intend to illustrate the argument developed in this first section by briefly examining certain problems with the western theorization of democracy and democratization. These two sections will then lead me to conclude by drawing out one or two implications for the way the geographies of the cultural and the political might be reframed.

TOWARDS A SPECIFICATION OF EURO-AMERICANISM

To begin with I shall not differentiate Eurocentrism from Euro-Americanism, but rather marshall my points from the literature in general, a literature which overall tends to employ the former term much more frequently than the latter, and which often fails to specify the possible differences between the two, giving rise to the possible implication that the differences if discernible are not particularly significant. Before considering the distinction between these two terms, and in particular in the context of west/non-west relations, let us begin by specifying three constituent elements of Euro-Americanism which can give us a working definition of the concept.

First, Euro-Americanist theoretical imaginations or historical interpretations emphasize the posited leading civilizational role of the west in modern times by reference to some *special* and *primary* feature of its own socio-economic, political and cultural life. This positing of the special and the primary has been reflected in a variety of ways. From Max Weber western thought has inherited the notion that the occident is the 'distinctive seat of economic rationalism' (1978: 480) or that outside Europe neither scientific nor artistic nor political nor economic development entered upon that 'path of rationalization, which is peculiar to the Occident' (1992: 25). From the annals of critical Marxist theory we read in Gramsci's *Prison Notebooks* that European culture is the *only* historically and concretely universal culture (Gramsci, 1971: 416, emphasis added). More recently, and in a not dissimilar fashion, Žižek (1998), from a post-Marxist position, argues that Europe is the key home of democracy and democratic theory. From other sources, which focus on materiality, emphasis falls on the posited historical superiority of western science and technology (for a critique see Adas, 1990), and in the ethico-political realm, the west has been constructed as the cradle of human rights, progress, enlightened thought, reason and philosophical reflection. Specifically, in the domain of philosophy, western culture has been customarily portrayed as being the only culture capable of self-critique and evaluation (see, for example, Castoriadis, 1998: 94) or the only culture to claim an 'experimental success' through bringing people into some degree of comity and of 'increasing human happiness, which looks more promising than any other way which has been proposed so far' (Rorty, 1999: 273).

Second, the special or primary feature or essential set of features uniquely possessed by the west are regarded within Euro-Americanism as essentially *internal* to European and American development. This set of features or overall model is enframed in a way that assumes the existence of an independent logic and dynamism of Euro-American development. There is certainly no sense of such development being a copy or the result of a process of cross-cultural encounters. Moreover, and rather critically, there tends to be an ethos of superiority – neatly captured in Foucault's notion of the 'historical sovereignty of European thought' (1973: 376–7) – which permeates the idea of an independent logic of western progress, or civilization, or more recently modernization and development. The sense of the superiority of the west was clearly reflected in the modernization theory of the 1950s

and 1960s where, for instance, particular definitions of industrialization, agrarian change, and social and political development were rooted in a certain interpretation of Euro-American experience and then modelled for export to the Third World. The ethos of superiority fuels the drive to expand, and this aspect of westocentrism has been carefully and critically evaluated by Blaut (1993) in his discussion of colonialism and diffusionist models of western development.

Third, the development of the west within a Euro-Americanist frame is held to constitute a *universal* step forward for humanity as a whole (for a critical discussion, see Shohat and Stam, 1994). Such a view has been reflected in both traditional Marxist views of a progressive succession of modes of production, and in the well-known Rostowian notion of the 'stages of economic growth' (Rostow, 1960) with the west offering a mirror of development for the future of non-western societies. At the end of the 1950s, in a symptomatic expression of this orientation, the prominent American social scientist Daniel Lerner, in his well-known text on *The Passing of Traditional Society*, asserted that Middle East modernizers would do well to study the historical sequence of western growth since 'what the West is … the Middle East seeks to become' (1958: 47). Such a standpoint finds a historical precedent in the nineteenth-century writing of the influential philosopher Hegel (1967: 212), who defined the principle of the modern world, i.e. Europe, as thought and the universal. And subsequently in the twentieth century Husserl stated that 'philosophy has constantly to exercise through European man its role of leadership for the whole of mankind' (1965: 178).

Taking these three elements as a whole, the primary, the internally independent and the universal, we also find that in the representation of the history of relations between the west and the non-west, the other, i.e. the non-west, is portrayed in ways which are pervasively *negative*, thus constituting a positivity for the west. The complexities and pluralities of both the west and the non-west are reduced to a driving vector of meaning that assigns to the west the key historical and geopolitical significance of being the essential and superior motor of progress, civilization, science, development and modernity.

On the negative essentialization of the non-western other, Mbembe (2001), in his thought-provoking text on the 'postcolony', reminds us that in the context of western representations of Africa, the negativity is always present and that the African human experience continually appears as an experience that can only be understood through a negative interpretation. Africa, he argues, is never seen as exhibiting attributes that are properly part of humanity or, when it is, these attributes are generally seen as of lesser value. Mbembe suggests that in western social theory African politics and economics only appear as the sign of a lack, while the discourses of political science and development economics have become characterized by the search for the origins of that posited lack. In this frame, war is seen as all-pervasive, and the African continent is seen as powerless, engaged in rampant self-destruction. These negative images and one-dimensional representations, present in the post-colonial period of separation from western coloniality, are traced back to the violence of colonialism which Mbembe treats as having three forms: (1) the founding violence of colonial sovereignty which created the space over which conquest was exercised and also the space within which colonial power could introduce its own laws, where its supreme right was also the supreme denial of the right of the colonized; (2) a violence of legitimation where the language and models of colonial rule were introduced as part of a universalizing mission to cement into place a new institutionalizing authority; and (3) a banal and everyday violence of cultural rule, expressed in a 'gradual accumulation of numerous acts and rituals' so that there was a cultural imaginary that the state shared with society as a way of reproducing colonizing power through the intricate web of social relations (2001: 25).

I have referred to these passages from Mbembe because they point to a rather important component of the discussion of Euro-Americanism: the place of violence and conquest in the history of relations, representations and positionalities when thinking about west/non-west encounters. This point can be illustrated in the context of the conquest of the Americas and its close relation to western thought.

For example, the Argentinian philosopher Enrique Dussel (1998) strongly argues that the Spanish *conquisto* (or *vinco*), i.e. I conquer, must be given historical and ontological priority over the Cartesian *cogito, ergo sum* (I am thinking, therefore I exist) as the first determination of the subject of modernity. In this sense *conquisto* means to take possession of the land and the people of the conquered territory so that any subsequent formulation of thought and truth must already presume, implicitly or explicitly, a territorialization based on a self/other split which can only be properly understood in the frame of conquest. For example, whilst Hernán Cortés (1521), preceded the *Discours de la méthode* (1636) by more than a century, it is important to recall that Descartes studied at La Flèche, a

Jesuit college with a religious order that, at that time, had deep roots in America, Africa and Asia, and in its teaching the 'barbarian' was the obligatory context for all reflection on subjectivity, reason and the *cogito*. For Dussel, modernity was born when Europe constituted itself as the centre of world history in 1492. Only with the invasion of the 'New World' did Europe enjoy a true springboard that allowed it to supersede other regional social systems. The idea that 'I conquer' constitutes the practical foundation of 'I think' has been taken up by both Moreiras (2000) and Spanos (2000) in their discussion of the crucial relation between geopolitical power and the territorialization of thought, and this linkage, as I shall show, is highly relevant for subsequent sections of my argument. In sum then Dussel makes three interesting points.

First, the Hispanic Conquest of 1492 provides an essential context for understanding the subsequent evolution of thinking on subjectivity, reason and culture. Second, a great part of the achievements of modernity were not exclusively European but grew out of the continuous dialectic of impact and counter-impact between modern Europe and its periphery, including the constitution of modern subjectivity. And third, for Dussel, Eurocentrism consisted precisely in confusing or identifying aspects of human abstract universality in general with moments of European particularity, which was in fact the first 'global particularity' (1998: 132).

This last suggestion, Dussel's definition of Eurocentrism, leads me to the final element of this section of the chapter. What difference does 'America' make to our perspective on western universalism, or how might 'Euro-Americanism' be different from 'Eurocentrism'?

These questions can potentially lead us into a long discussion of the historical, socio-economic and political differences between the United States and Europe. Along this pathway, one can mention Lipset (1991: 40) who, in attempting to define 'American exceptionalism', notes that the United States is the least statist western nation in terms of public effort, benefits and employment, and, *inter alia*, more religious, more patriotic, more populist and less law-abiding than other developed countries. Daniel Bell (1991), for his part, draws our attention to the lack of a feudal history, the relative absence of socialism, and the abiding sense of a moral purpose, expressed perhaps most emblematically in the notion of 'manifest destiny'. Bell writes, for instance, that 'from the start there had been the self-consciousness of a destiny that marked this country as being different from others ... that the greatness was laid out like a magnetic field which would shape the contours of the nation from one ocean to the other, and finally, when it confronted the rest of the world from that magnetic core, this would become "the American Century"' (1991: 48). Fundamentally, however, for Bell, it has been the strength of its civil society in relation to state power that has given the United States its exceptional nature. American civil society is couched in terms of an emphasis on the voluntary association, on the church and community, on the self-management of resources on a local scale, outside the bureaucracy and the state, and above all in terms of a transformation that has gone from a version of 'republican virtue' through to the contrasting but complementary practices of rugged individualism and radical populism.

These lines provide us with a flavour of one reading of the 'exceptionalism' of America, which contrasts with an earlier article by the political geographer John Agnew (1983) who outlined a more critical vision of 'American exceptionalism' in relation to US foreign policy. More recently, other writers have sought to place the 'American Century' in a relatedly critical context (see, for example, Guyatt, 2000; Slater and Taylor, 1999). However, for the purposes of my argument here I simply want to signal the existence of three specificities of the United States which are particularly pertinent to a consideration of the relations between the US and the non-western world, distinctions which are important for our understanding of 'Euro-Americanism'.

The first specificity is that the United States, in contrast to west European nations, as well as Japan, has a history of spreading imperial power that is also rooted in post-coloniality. Today, the United States is not only the lone superpower, but also the only post-colonial imperial power, whereby a project of empire emerged out of an original anti-colonial struggle for independence from British rule. In the proliferating literature on post-colonialism the United States is customarily listed as a post-colonial country together with a whole range of Third World societies, but the 'exceptional' juxtaposition of post-coloniality and imperial power is often ignored. There are two facets to this juxtaposition.

First, in looking at the geopolitics of US interventionism in the countries of the global south, the coalescence of these two realities, of a belief in the rightness of the self-determination of peoples, together with a belief in the global destiny of 'America', constitutes a salient and frequently contradictory specificity. Historically, the contradiction between a belief in the rights of peoples to decide their own fate and a belief in the geopolitical predestination of America has been ostensibly transcended through an invocation of democracy that is valid at home and abroad. In the 1960s, for example, in a context formed by military intervention in the Dominican Republic (1965), the war in Vietnam and a social crisis in the cities of the United States, President Lyndon B. Johnson made it clear that the domestic and the foreign were two sides of the same coin, that promoting democracy at home meant securing it abroad, that the United States was a great, liberal and progressive democracy up to its frontiers, and 'we are the same beyond'. Let us never imagine, he continued, 'that Americans can wear the same face in Denver and Des Moines and Seattle and Brooklyn and another in Paris and Mexico City and Karachi and Saigon' (1969: 7). This sense of indivisibility, of global preeminence in the proclamation of liberal democracy, carries a universality that is based in the particularity of the United States, and by providing a horizon for other peoples this kind of enunciation also attempts to encapsulate the struggles and destinies of non-American peoples within an American vision.

Second, the primacy of self-determination is important in explaining the dichotomy frequently present in American interventions where a split is made between the governed (the people) and the governors (the rulers). Given the historic differentiation of the New World from the Old, and the support for anti-colonial struggles, perceived threats to US security have not infrequently been accompanied by this kind of separation between an oppressed people and tyrannical rulers. For example, in the context of past revolutionary breaks that were associated by the US government with 'communist subversion', it was the people who needed to be rescued from their undemocratic rulers, as clearly represented in the case of US military intervention in Grenada in 1983. In the long-standing case of US hostility towards the Cuban Revolution, a strong distinction has been made between the Cuban people, who are portrayed as being oppressed by their communist rulers, and the Castro regime. For example, in the earlier sections of the Helms–Burton Act of 1996, one reads that 'the consistent policy of the United States towards Cuba since the beginning of the Castro regime … has sought to keep faith with the people of Cuba', whilst 'sanctioning the totalitarian Castro regime'. Further on, the document continues, 'the Cuban people deserve to be assisted in a decisive manner to end the tyranny that has oppressed them for 36 years, and the continued failure to do so constitutes ethically improper conduct by the international community'.[1] The Act specifically argues that measures are needed to restore the values of 'freedom and democracy' and, above all, the sovereign and national right of self-determination to the Cuban people. The mode of representation at work here can be explained in terms of the presumed right to be able to designate the political future for a people whose sovereignty is envisaged as being usurped by a posited unrepresentative and tyrannical regime.

These two elements lend specificity to a certain post-colonial nature of imperial power that distinguishes the United States from other western societies, and gives a significant vector of meaning to the term 'Euro-Americanism' .

The second specificity is that, in the territorial formation of the United States, a formation that was intrinsically tied to war and the expansion of a 'civilizing' frontier, there were encounters with three significant others: the indigenous peoples of North America (the 'Indian'), the Hispanic and Indian population of Mexico in the US–Mexico War of 1846–8, and the African American in the initial context of slavery and the Civil War of the 1860s. In the constitution of mission, destiny and an Anglo-Saxon Americanization of the continent, the identification of internal enemies and shifting frontiers came to play a key role in the formation of a new nation. In the example of the decimation of the native peoples of the continent, white America's violent encounter with its Indian other came to form a deeply significant element of the nation's collective memory. It not only figured in the production of films about how 'the West was won' but also found expression in twentieth century warfare and foreign policy. In the 1960s, for example, during the Vietnam War, American troops described Vietnam as 'Indian country' and President Kennedy's ambassador to Vietnam justified military escalation by citing the necessity of moving the 'Indians' (the North Vietnamese) away from the 'fort' so that the 'settlers' could plant 'corn' (Slotkin, 1998: 3). More recently too, as Campbell (1999: 237) indicates in his analysis of the contradictions of a lone superpower, an American diplomat referred to Bosnian Serb territory as 'Indian country', whilst US units named their bases and areas using frontier references (for example, Fort Apache).

Since there is no space here to develop this argument in further detail, I simply want to state that my overall contention is that, notwithstanding the historical and cultural differences between them, these *founding* three encounters with internal others generated forms of subordinating representation and mechanisms of power that prefigured subsequent relations of power over Third World societies (Slater, 1999). Before the United States became a global power, these encounters provided an original reservoir of imperial experience that was not irrelevant to many of the interventions pursued by the United States in the twentieth century. In comparison to the colonizing nations of western Europe,[2] in the case of the United States the internal territorial constitution of the nation-state comprised a series of violent encounters with other peoples that took place on its own soil and intimately moulded its evolving sense of empire and mission, which in many ways has been most acutely reflected in the continuing significance given to notions of 'the frontier'.

The third specificity is that, historically, the United States has been portrayed by its leading political figures as the original haven of a New World. From the Monroe Doctrine of 1823 and Thomas Jefferson's twin notion of 'America' having a 'hemisphere to itself', and being an 'Empire for Liberty' , through to the Roosevelt Corollary of 1904 and the Rio Pact of 1947, the United States has staked out for itself an original heartland that was clearly delineated as a separate domain from the Old World of Europe. This demarcation of geopolitical domains, or the establishment in the western hemisphere of a 'grand area' of geo-strategy, constituted what I consider to be the first phase of a US strategy of containment. This first phase, which dated from the Monroe Doctrine, was characterized by a strategy for the establishment of US hegemony in the Americas, and the setting of limits for European influence. The second phase of containment, which was initiated with the Cold War and the rivalry between the superpowers, saw the United States as a global power developing a strategy of containment for what was perceived to be the communist threat to the 'free world'. This classic phase of containment was played out on the global stage from the late 1940s to 1989, with a short intermezzo in the late 1970s under the presidency of Jimmy Carter. The final phase of containment, in which we are living in the current era, relates to the specific targeting of what are portrayed as 'rogue states' such as Iraq, North Korea and Libya – states which are considered to be the instigators and/or protectors of terrorist groups and organizations. The containment and isolation of these specific states is part of an American strategy aimed at 'global pre-eminence' (Klare, 2000), a strategy which has important implications for the way we frame our discussions of the global and the democratic.

The evolution of these interconnected phases in the development of 'global America' (Valladão, 1998) provides a further specificity to the treatment of the United States within a western context. In sum these three points, schematically presented above, capture the existence of important constitutive differences between the United States and the rest of the west, and I have introduced them as one way of underlining the need to give more analytical oxygen to the specificity or 'exceptionalism' of the United States, and especially in relation to the projection of US power. Furthermore, in any discussion of the difference between Eurocentrism and Euro-Americanism these three factors of delineation provide one relevant basis for understanding the contrasts as well as commonalities within the west.

There are two further elements here which because of the limits of space can only be swiftly 'flagged up'. One concerns the need to further delineate the differences within the west, that is not only the need to give greater thought and attention to the particularity of the United States within the universe of the west, but also to realize that within western Europe itself there are other differences which relate to the historical specificities of colonialism – for example, the earlier historical cases of Spain and Portugal (see Coronil, 1996; Mignolo, 2000). A second element concerns what is meant by 'America' and how the multiple meanings and histories of this signifier can take us into a discussion of the place of the indigenous peoples of the Americas and their specification of what 'America' signifies in their histories and cultural foundations (see, for example, Brysk, 2000). The term 'America' also raises the issue of how Latin Americans are both 'americanos' and citizens of different nation-states in the Americas. In other words, as a general observation, it is important to be aware of the need to avoid conflating the United States of America with 'America'.

Having outlined certain important aspects of Euro-Americanism, including a short section on the difference that 'America' makes, it is now necessary to take our discussion into the area of democratic theory, envisaged globally.

FOR A POST-COLONIAL PERSPECTIVE ON DEMOCRATIC THEORY

It can be suggested that democracy has become one of the most pervasive signifiers of political thought. Perhaps like Coca-Cola, democracy needs no translation to be understood virtually everywhere. But the ostensible universality of democracy can act as a screen behind which lies the complexity of its multiple meanings. Democracy is of course a classic example of a floating signifier, open to a variety of discursive frames. In this context, we have a veritable plethora of descriptors that vie for our attention: to take one selection we might think of 'direct democracy', 'social democracy', 'liberal democracy', 'associational democracy', 'representative democracy', 'participatory democracy', 'popular democracy', 'radical democracy', 'cosmopolitan democracy', 'market-led democracy' and so on. These adjectival markers testify to the continuing struggles over the meaning, definition, content and political direction of what is to be meant by democracy. How then do we connect with the limitations of Euro-Americanism discussed above? And how might we use the insights of an enabling post-colonial perspective to go beyond these constraints so that our thinking about the democratic and the global might be able to avoid the customary pitfalls of the ethnocentric universalism so characteristic of western thought?

In an important statement on the cultural particularity of liberal democracy, the political theorist, Parekh (1993: 167) made the point that western liberal democracy has often imposed on other countries systems of government that were not relevant to the skills and talents of non-western countries, and that this kind of imposition has tended to destroy the coherence and integrity of their ways of life, reducing them to mimics, unable or unwilling to be true to either their own traditions or those of the alien norms imported from outside. As Parekh appropriately concludes, 'the cultural havoc caused by colonialism should alert us to the dangers of an over-zealous imposition of liberal democracy' (1993: 167). Not only Parekh but other non-western writers and analysts (for example, Dhaliwal, 1996; Rivera, 1990; Sheth, 1995) have pointedly observed that western representations of democracy and liberalism frequently presume a universal relevance for institutional arrangements and cultural values that may not be equally applicable in other regions of the world. Moreover, the historical and contemporary context of the exclusionary nature of democratic societies in relation, for instance, to questions of race and ethnicity, as well as the geopolitical association of democracy with imperialism, define a rather salient but often neglected thematic focus. It is exactly these kinds of observations and critical interventions, emanating from the work of non-western social scientists, that can help us develop a series of questions concerning the geographies of our analytical reference as well as the prevalence of western ethnocentrism in the conceptualization of democracy.

As a way of structuring my commentary, I want to discuss five interrelated problems which are symptomatic of much analysis of democracy as a form of rule or political system and the process of democratization in its social and political dimensions. All five points relate to the way we think about democracy in a global context and they all impinge on the presence within any global frame of the significance of west/non-west encounters.

In the first place, it is worthwhile recalling that many significant historical and geopolitical events occurring away from the heartlands of the capitalist west, events which have had a profound impact on the course of social struggles, have not infrequently been excluded from western writings on global history. As one example, which connects back to the prioritization of the French Revolution, mentioned earlier, we might refer to the work of Trouillot (1995). Trouillot, in his examination of what he denotes as 'the silencing of the past', shows how, in much western scholarship, both Anglo-American and French, the Haitian Revolution, with its crucial connection to the struggle against racism, slavery and colonialism, has largely been either erased from theoretical treatments of democracy or trivialized in terms of its wider import. A long process of social rebellion from an initial slave uprising in 1791 through to the proclamation of independence in 1804 represented an indigenous struggle for freedom, dignity and autonomy. This rebellion played a central role in the collapse of slavery and it also placed on the political agenda the crucial connections between democratic struggles and opposition to racism and coloniality.

However, such events have been customarily overshadowed by a concentration on the founding importance of the French Revolution for the future of democracy, writ universally. But a global, post-colonial perspective can be deployed to connect the two revolutions, thereby questioning the sedimented centrality of the 'European moment'. As Dubois (2000) argues, the slave insurgents claiming republican citizenship and racial equality expanded the idea of rights so that developments in the Antilles actually went beyond the political imagination of the metropole, transforming the content of citizenship and challenging the ethnocentric limits of western political thought in general. But the idea that a region of the periphery, where the agents of change are non-western and non-white, might actually have been more politically advanced than the centres of western civilization has been given, perhaps not entirely unexpectedly, little oxygen.

Although there are many other experiences from the periphery – insurgent movements such as the Zapatistas in Mexico, or experiments in participatory democracy as in Porto Alegre in Brazil – that hold out lessons for a global context, the example mentioned above is quite emblematic. This is so since the French Revolution is always taken as a key origin of the theoretization of democracy and human rights, especially in treatments of radical democracy, and yet that other Haitian Revolution of the same era tends to be shrouded in silence. One can be reminded of the Nietzschean point that at all 'origins' there is diversity, and in this example that diversity can be fruitfully used to disrupt and to displace one influential current of western political theory. Furthermore, this point applies not only to the French Revolution and the European experience but also to discussions of the origins of democracy in the United States, where of course the abolition of slavery followed on some time after the Haitian Revolution.

Second, when considering the established view that the west has diffused and continues to diffuse democracy to other parts of the globe, it is necessary to remember that the West, and in particular the United States, has intervened geopolitically in societies of the periphery to replace one government by another. Transgressions of national sovereignty have been well documented. Niess (1990: 208–9), for instance, records as many as 33 major armed interventions by the United States in Latin America alone from 1853 through to Grenada in 1983. It is important here to distinguish two elements: (1) interventions which have led to the replacement of one regime for another where the overthrown regime may not have been democratic, as in the case of the Manuel Noriega regime in Panama in 1989; and (2) interventions which have led to the replacement of democratically elected governments which were developing policies independent of the United States. This is not to implicitly condone the transgression of sovereignty when an undemocratic regime is involved, or to forget the enduring nature of US and western support for a variety of dictatorial regimes in the Third World (e.g. Chile, Indonesia, Zaire), but rather to underline the fact that US interventions have on a number of occasions led not to the creation of democracy but to its *termination*.

With reference to Latin America, that region of the Third World with the longest history of independent governments, the post-Second World War period was witness to a number of key US interventions that ended particular democratic experiments. Guatemala in 1954, the Dominican Republic in 1965, Chile in 1973 and Nicaragua in the 1980s are all examples where interventions effectively terminated democratic processes. With respect to Guatemala, a particularly tragic example, in 1954 a democratically elected government was overthrown by a CIA-organized military coup (Cullather, 1999) that changed that country's political landscape for ever. By the end of the 1990s an estimated 200,000 people had died in the civil war that followed the military *coup d'état*. This figure represented approximately 2 per cent of the total population, the rough equivalent of around 5 million deaths in a war within the United States. In all these examples, the democratically elected governments that were overthrown (Guatemala, the Dominican Republic and Chile) or undermined and destabilized through the financing of contra-guerrillas as in the Nicaraguan case, were developing policies that sought to redistribute wealth and income, introduce land reform, and construct a nationalist programme of development. They were not one-party states as was Cuba, but they were regarded as a threat to the United States because they represented a democratic alternative and genuine 'third way' between capitalist underdevelopment under US hegemony and socialist revolution within the sphere of influence of the Soviet bloc.[3]

In our times of geopolitical amnesia, it is not only important to continually recall these events, but also to think through their meaning, linked into the politics of memory. It is surely neither justifiable nor wise, especially writing after 11 September 2001, to indulge in what Trouillot (2000) calls the 'abortive rituals' of governmental apology when the conditions and ruling ideas that make possible western interventions have not been transformed. The rule of hegemonic western representation seeks to convince us that interventions from the centres of modern 'civilization' have always been marked by the pursuit of justice and democracy.[4] The geopolitical record shows otherwise, but its reality must be continually reactivated and remarked.

Third, a particularly western notion of 'democracy' and the desire to defend it have provided a justification for a variety of geopolitical interventions, as was so clear in the Central America and Caribbean of the 1980s. Falk (1995) has referred to this phenomenon as the geopolitical appropriation of 'democracy', pointing to the pivotal significance of the continuing struggles over the meanings of democracy. In the 1980s, the Reagan administration launched 'Project Democracy' and the 'Democracy Program' to promote, as Huntington put it, 'democratic institutions in other societies' (1984: 193). Conversely, and during the same years, the Sandinista government in Nicaragua was attempting to develop its own combination of representative democracy with popular democracy, having won

a resounding vote of confidence in the 1984 elections. Two visions of democracy and the unequal powers behind them came into open confrontation. The first, which was a market-led and US-friendly version, based on the Schumpeterian notion of the 'democratic method', whereby individuals are supposed to acquire the power to decide by means of a competitive struggle for the people's vote, eventually triumphed owing to the greater geopolitical power of the United States. The second, which was a revolutionary model that combined electoral competition between political parties with the encouragement of popular organizations and an anti-imperialist strategy, lost out in the 1990 elections where a US-supported opposition acquired governmental power.

What needs to be stressed here is that a particular vision of western liberal democracy is used as a gauge or model for judging the success or failure of non-western societies to develop democratically. It is always worthwhile remembering that there are many definitions of 'democracy', a classic example of a polysemic term. A post-colonial perspective would underscore the particularity and limits of western visions which purport to have universal relevance, and which are frequently employed in discussions on aid and development as a kind of gold standard for what are seen as aspiring democracies in the global south.

Fourth, in any critique of Euro-Americanism one of the salient elements concerns the complex interweaving of cultural representation and geopolitical power. In the historical annals of western democratic theory, one can encounter defining examples of a strong universalist ambition which has prioritized certain cultural practices and invested them with spatial power.

Taking an example which is relevant for its general theoretical influence as well as for its roots in a western perception of the experience of the United States, it is instructive to refer to aspects of Tocqueville's work on democracy in America. Towards the middle of the nineteenth century Tocqueville argued that the historical consolidation of what he referred to as the civic-territorial complex required the elimination of the Indian, the original American other who had the right of neither soil nor sovereignty and had to be cleansed from the founding of American democracy. For Tocqueville (1990) civilization had to be seen as the result of a long social process, which takes place in the same spot and is handed down from one generation to another. Consequently, peoples who are nomadic can never attain the status of being civilized, or as Tocqueville expressed it, 'civilization began in the cabin, but soon retired to expire in the woods' (1990: 342–3). Thus, as Connolly has helpfully suggested, Tocqueville's depreciation of nomadic life forfeits insights into how the American state might modify its own tendencies to centralization and 'fend off its cultural drive to sustain the purity of civilization through the extermination of the other' (1994: 31). But such a forfeiting of insight is sharply conditioned by a racially prejudiced vision.

This is clearly shown in Tocqueville's overall discussion of the 'three races in the United States', in which one reads that ' the most formidable of all the ills that threaten the future of the Union arises from the presence of a black population upon its territory' (1990: 356). This is seen as the case since the civilized European can scarcely acknowledge the common features of humanity in this stranger whom slavery has brought among us, and we Europeans are 'almost inclined to look upon him as a being intermediate between man and the brutes' (1990: 358). Such a viewpoint finds an echo in other writings of Tocqueville, including his interpretation of French colonialism in Algeria. His support for colonialism is expressed in clear and stark terms. He wrote, for example, that burning harvests, emptying granaries and seizing women and children are 'unfortunate necessities that any people wishing to make war on the Arabs must accept'. He went on to argue that in Algeria there should be two quite distinct legislations for there are two very separate communities, the colonizers and the colonized, and it was only the colonizers who would have their rights legally protected. War, he added, was a science that had been well developed by the French in Africa. These short quotations come from a recent article by the French political scientist Le Cour Grandmaison, who refers the reader to Tocqueville's complete works republished in 1991. Le Cour Grandmaison (2001: 13–14) asks the question: why is it that Tocqueville's open support for French colonialism and his justification of violence against the Algerian population are very rarely if ever mentioned by French political theorists in their consideration of Tocqueville's contribution to political thought? Whilst Trouillot refers to the 'silencing of the past' in the case of the Haitian Revolution, we might suggest that the racist and colonialist prejudice present in the writings of founding figures such as Tocqueville is also passed over in silence. Is it because contemporary writers think that such prejudices are no longer significant, simply being residues from a more racist age? Have such prejudices disappeared? Or must we preserve the honour and integrity of our founding western theoreticians?[5]

In a similar vein to Tocqueville, and writing a little later, the liberal theorist J.S. Mill (1989) drew a connection between the right to social justice and liberty and the existence of a 'civilized community'. For Mill, the principles of justice only applied to human beings in the maturity of their faculties, so that one could leave out of consideration 'those backward states of society in which the race itself may be considered as in its nonage' (1989: 13). The ethnocentric ground on which Mill and also Tocqueville built their arguments was not unique for the nineteenth century, or for following periods, and the binary splits between civilized and barbarian, or peoples with history and those without, received later elaborations in the twentieth-century context of modern versus traditional and developed versus developing.

Overall, my point in this section is to re-emphasize that the way the temporal and geopolitical configurations of the democracy problematic have been and continue to be interpreted is centrally affected by an ethnocentric universalism that is profoundly rooted in the formation of occidental thought. And that anchorage is frequently avoided in the contemporary western literature. But if we are to construct more equal forms of cultural dialogue in times of acute political instability, it is crucial to be continually aware and critical of these historical roots if we are to ever go beyond them.

Finally, a post-colonial perspective ought to encourage us to cast a more critical eye on the evolving nature of western democracy from within. This is not a task only relevant to such a perspective since increasingly critical questions are being asked from a wide range of positions. For example, it has been noted that with the continual extension of the powers of surveillance (Boyne, 2000), coupled with the extension and deepening of the bureaucratization of social life, a notion of 'totalitarian democracy' might not seem too inappropriate (Fiske, 1998). Fiske exemplifies his argument by referring to the extension of electronic surveillance, particularly noticeable with the growth in coverage of CCTV, reflected for instance by the fact that the whole of the downtown in cities like Minneapolis, Newark and Detroit are now covered by cameras that can zoom in to read a credit card. But also relevant, for Fiske, are appeals to moral totalism, intensified policing and the appearance of charismatic leaders, a point perhaps that should not be over-emphasized given the banality of much current political discourse. Along a related analytical pathway, Lacoue-Labarthe and Nancy (1997) make the point that the distinction between totalitarianism and democracy can sometimes be made too simple. It is true, they suggest, that 'we do not have camps and our police ... are not omnipresent political police', but nothing guarantees that our democracy 'is not in the process of secreting something else, a new form of totalitarianism' (1997: 128).

We can suggest here that the visibility of the structures of democracy can conceal totalitarian undercurrents and at the same time this visibility offers a legitimation to those who want to defend the idea of the existence of an open and democratic polity as a continuing reality. In the 1950s, for example, the House UnAmerican Activities Committee launched a witch-hunt against a wide range of citizens who were accused of not being faithful to 'American values'. This was a totalitarian initiative that unfolded within the structures of a democracy, being intimately interwoven with the international situation of superpower rivalry. The inside and outside were intertwined and a culture of containment permeated the arenas of both domestic politics and foreign policy. Interventions and invasiveness abroad were paralleled by the policing of difference and dissent at home. Here, a post-colonial perspective would encourage us to foreground not only the 'totalitarian' within the 'democratic' but also the imbrication of the internal and the external and the impossibility of understanding any western power as a self-contained entity.

Another example of the significant imbrication of the inside and the outside as well as the blindness that exists concerning the growth of counter-democratic trends can be illustrated from one of Fiske's (1998) examples – that of 'non-racist racism'. This, he contends, is a racism that has been developed by white-powered nations that avow themselves to be non-racist. In the United States this racism is recoded into ostensibly race-neutral discourses, such as those of economics, law, education and housing. Each of the social domains within which these discourses operate has racially differentiated effects for which the causes can always be made to appear non-racial. Indeed racism is illegal in most domains of US public life and many whites, Fiske goes on, profess to believe that racism is now a non-problem. Conversely, in black America there is a widespread knowledge that racism is waxing not waning, and clearly economic and educational indicators show that the gaps between white and black Americans are increasing not narrowing. This view finds an echo in Rae's (1999) recent study, where he points to the development of a 'segmental democracy' with decisions made by and for relatively homogeneous populations in specific areas, and the emergence of sophisticated forms of discrimination

in real-estate markets supplanting more vulgar practices made unlawful by the civil rights legislation of the 1960s.

The trends towards a diminishing of the openness required by democracy if it is to flourish are reflected in a variety of ways. Western democracies are far from being tyrannies, but the notion of a 'totalitarian democracy' does capture a disturbing trend whereby the institutions and formal arrangements of democratic rule may well remain in place but the participatory and empowering substance of those arrangements and mechanisms is being insidiously degraded. A recrudescence of racism, manifested in increased attacks on asylum seekers and ethnic minorities, and the attempts by democratically elected leaders to nurture a monoculture of political debate, supported by sections of the print media that consistently express xenophobic sentiments, both undermine key values on which the democratic process rests. A post-colonial view might seek to draw out the connections between the western undermining of democratic experiments abroad and the chilling of plurality and critical thought at home. There is too, especially visible in the aftermath of 11 September 2001, a resurfacing within the west of imperial sentiment and a reassertion of occidental supremacy. Both positions need to be continually confronted.

ON THE POLITICS OF 'THE BEYOND'

The title of this chapter invokes the need to go beyond Euro-Americanism, and suggests the potential relevance of a post-colonial perspective. The analytical scope clearly outstrips the confines of any chapter. All I have been able to do in these few pages is to alert the reader to some of the problems and challenges that we are all going to have to deal with in the future, and with far more rigour and attention.

One of the advantages of a crisis can be that the contours of debate are starkly clarified. Within the realm of the social sciences, Francis Fukuyama (2001), in a short intervention entitled 'The West has Won', gives expression to a virulent form of arrogance and ignorance that not only places the west at the heart of world civilizations, past, present and future, but at the same time denigrates the non-west and specifically Islamic civilization in a way that replicates the Hegel of the mid nineteenth century. In the domain of politics, the Italian prime minister Silvio Berlusconi asserts the supposed superiority of western civilization over all others and especially the

Islamic world (see *The Guardian*, London, 3 October 2001, p. 18). There are of course many other examples of such views and my point here is to remind ourselves that a detailed and continuing critique of positions which proclaim the supremacy of the west remains vital, for scientific and political reasons.

In some ways it might be seen as natural, in confronting the above types of view, to proclaim the moral superiority of the non-west and the corruption and decadence of the west itself. We then have an unproductive exchange of different essentializations which can never lead to the kind of critical cross-cultural exchange and understanding that is so urgently needed. What is needed is the continual search for hybrid forms of knowledge and interpretation so that we avoid unhelpful essentializations and the economy of stereotype. In rethinking our views on the geographies of the cultural and the political, and in the specific context of the global and the democratic, the following three interconnected points strike me as being particularly relevant.

First, as one example, if we consider the Zapatista uprising in Mexico, we will find that the way the Zapatistas have conceptualized democracy does not flow out of the terminology of Euro-American political philsopy but rather emanates from Maya social organization, in which reciprocity, communal values and the validity of wisdom are seen as central. This does not mean that the Zapatistas have the 'correct interpretation', but rather that employing the term 'democracy' does not necessarily hold the Zapatistas to any one-dimensional interpretation. Instead when the term 'democracy' is deployed by the Zapatistas, it can become a connector through which liberal concepts of democracy and indigenous concepts of reciprocity and community can be related to each other in a process of respectful and critical dialogue. In this sense then the meanings of 'democracy' become hybridized, plural and culturally mixed, enriching our political vocabularies and enhancing our cross-cultural understanding.

Second, the significance of a broad cross-cultural understanding has implications within nation-states as well as across them. Thus, in the Bolivian case, where ethnic and cultural difference is such a crucial factor, the Bolivian anthropologist Rivera (1990) reminds us of the realities of 'internal colonialism'. She argues that the ideal of equality has continued to be based on a western model of citizenship, where notions of being 'modern', 'rational' and 'proprietary' have prolonged a process of exclusion which is anchored in the colonial experience. Consequently, a genuinely democratic reform will have to contain some form of articulation between the direct democracy of the indigenous *ayllu* communities and the representative democracy of the nation-state as a whole. A key issue here is the conceptualization of citizenship in relation to a multicultural reality where the indigenous has been historically subjugated. Institutional reforms in this context also require profound changes in outlook and a 'radical decolonization' of Bolivia's social and political structures. The question of decolonization relates not only to 'structures' but also to imaginations and, as Rivera indicates for Bolivia, profound changes of outlook are required if new forms of hybrid democracy can emerge – forms which will have the support of the majority of the population in a society increasingly riven by political turmoil.

Finally, and centrally, radical changes in outlook are also needed to think through the relevance of non-western forms of democratic practice. Notions of reciprocity in the sharing of resources, communal values and redistribution, cooperative forms of labour, the preservation of fragile environments and the prioritization of ethical values which elevate wisdom over epistemology, exemplify some aspects of indigenous approaches to democratic organization.

Reciprocity also has to be rooted in recognition and respect, and one of the key problems we face in the west is to find ways of expanding our geographies of reference and learning so we do not reproduce the arrogance and ignorance of self-contained visions of superiority. This does not mean that the meaning of the west begins and ends with such visions. Differences and struggles within are a key part of the continuing dynamic within both west and non-west and in their inescapable encounters. Going beyond the veil of Euro-Americanism is a continuing struggle itself and one that assumes increasing urgency in our contemporary world.

NOTES

1 See the Cuban Liberty and Democratic Solidarity (LIBERTAD) Act of 1996, Public Law 104–114, 12 March 1996, 110 Stat. 785–824, Washington DC, United States General Printing Office, pp. 786–8.

2 In the case of Britain, one obviously thinks of the example of the colonization of Ireland where subordinating modes of representation also accompanied imperial rule (see, for example, Perry Curtis Jr, 1997), so I am not arguing for a total distinction between the United States and all the countries of the west, but the formative and exceptional encounters with three internal others – Indian, Hispanic and African – does seem to be particularly relevant to US–Third World interactions in the twentieth century.

3 For a critical and detailed analysis of the historical aspects of US–Latin American relations, see Schoultz (1999) and Smith (2000), who both give many highly germane examples of the continuing projection of US power into one of the other Americas – 'the Latin'.
4 For example, the military intervention in Panama in December 1989 was defined as 'Operation Just Cause', and in late September 2001 we heard from Washington of the cause of 'infinite justice', later changed to 'enduring freedom'.
5 I am reminded here of a recent and interesting article on globalization as democratic theory, where a concluding and positive reference is made to Tocqueville's notions of freedom and democracy: see Rosow (2000).

REFERENCES

Adas, M. (1990) *Machines as the Measure of Men*. Ithaca: Cornell University Press.

Agnew, J. (1983) 'An excess of "national exceptionalism": towards a new political geography of American foreign policy', *Political Geography Quarterly* 2 (2): 151–66.

Bell, D. (1991) 'The "Hegelian secret": civil society and American exceptionalism', in B.E. Shafer (ed.) *Is America Different?*. Oxford: Clarendon. pp. 46–70.

Blaut, J. (1993) *The Colonizer's Model of the World*. London: Guilford.

Boyne, R. (2000) 'Post-panopticism', *Economy and Society* 29 (2): 285–307.

Brysk, A. (2000) *From Tribal Village to Global Village*. Stanford: Stanford University Press.

Campbell, D. (1999) 'Contradictions of a lone superpower', in D. Slater and P.J. Taylor (eds) *The American Century: Consensus and Coercion in the Projection of American Power*. Oxford: Blackwell. 222–40.

Castoriadis, C. (1998) *El Ascenso de la Insignificancia*, Madrid: Ediciones Cátedra,

Connolly, W.E. (1994) 'Tocqueville, territory and violence', *Theory, Culture and Society* 11 (1): 19–41.

Coronil, F. (1996) 'Beyond occidentalism: toward nonimperial geohistorical categories', *Cultural Anthropology* 11 (1): 51–87.

Cullather, N. (1999) *Secret History*. Stanford: Stanford University Press.

Dhaliwal, A. (1996) 'Can the subaltern vote? Radical democracy, discourses of representation and rights and the question of race', in D. Trend (ed.) *Radical Democracy*. London: Routledge. pp. 42–61.

Dubois, L. (2000) '*La République Métissée*: citizenship, colonialism and the borders of French history', *Cultural Studies* 14 (1): 15–34.

Dussel, E. (1998) *The Underside of Modernity* New York: Humanity Books.

Falk, R. (1995) *On Humane Governance*. Cambridge: Polity.

Fiske, J. (1998) 'Surveilling the city: whiteness, the black man and democratic totalitarianism', *Theory, Culture and Society* 15 (2): 67–88.

Foucault, M. (1973) *The Order of Things*. New York: Vintage.

Fukuyama, F. (2001) 'We remain at the end of history', *The Independent* 11 October: 5.

Gramsci, A. (1971) *Selections from the Prison Notebooks*. London: Lawrence and Wishart.

Gregory, D. (1995) 'Imaginative geographies', *Progress in Human Geography* 19 (4): 447–85.

Guyatt, N. (2000) *Another American Century?* London: Zed.

Hegel, G.W.F. (1967) *Philosophy of Right (1821)*, (ed.) T.M. Knox. London: Oxford University Press.

Huntington, S. (1984) 'Will more countries become democratic?', *Political Science Quarterly* 99: 193–218.

Husserl, E. (1965) *Phenomenology and the Crisis of Philosophy*. New York: Harper and Row.

Johnson, L.B. (1969) 'Our foreign policy must always be an extension of this nation's domestic policy (April 1966)', in W. LaFeber (ed.) *America in the Cold War: Twenty Years of Revolution and Response, 1947–1967*. London: Wiley.

Klare, M.T. (2000) 'Permanent preeminence: US strategic policy for the 21st century', *NACLA Report on the Americas* XXXIV (3): 8–15.

Lacoue-Labarthe, P. and Nancy, J.-L. (1997) *Retreating the Political*. London: Routledge.

Le Cour Grandmaison, O. (2001) 'Liberty, equality and colony', *Le Monde Diplomatique* June; *The Guardian Weekly* 12–13.

Lerner, D. (1958) *The Passing of Traditional Society*. Glencoe: Free Press.

Lipset, S.M. (1991) 'American exceptionalism reaffirmed', in B.E. Shafer (ed.) *Is America Different?* Oxford: Clarendon. pp. 1–45.

Massey, D. (1995) 'Thinking radical democracy spatially', *Environment and Planning D: Society and Space* 13: 283–8.

Mbembe, A. (2001) *On the Postcolony*. Berkeley: University of California Press.

Mignolo, W.D. (2000) *Local Histories/Global Designs*. Princeton: Princeton University Press.

Mill, J.S. (1989) *On Liberty and Other Writings*, (ed.) Stefan Collini. Cambridge: Cambridge University Press.

Moreiras, A. (2000) 'Ten notes on primitive imperial accumulation', *Interventions* 2 (3): 343–63.

Niess, F. (1990) *A Hemisphere to Itself*. London: Zed.

Parekh, B. (1993) 'The cultural particularity of liberal democracy', in D. Held (ed.) *Prospects for Democracy*. Cambridge: Polity. pp. 93–111.

Perry Curtis, L. Jr (1997) *Apes and Angels: The Irishman in Victorian Caricature*. Washington: Smithsonian Institution Press.

Rae, D. (1999) 'Democratic liberty and the tyrannies of place', in I. Shapiro and C. Hacker-Cordón (eds) *Democracy's Edges*. Cambridge: Cambridge University Press. pp. 165–92.

Rivera, S. (1990) 'Liberal democracy and *ayllu* democracy in Bolivia: the case of northern Potosí', *Journal of Development Studies* 26 (4): 97–121.

Robinson, J. (1998) 'Spaces of democracy: remapping the apartheid city', *Environment and Planning D: Society and Space* 16: 533–48.

Rorty, R. (1999) *Philosophy and Social Hope*. Harmondsworth: Penguin.

Rosow, S.J. (2000) 'Globalisation as democratic theory', *Millennium* 29 (1): 27–45.

Rostow, W.W. (1960) *The Stages of Economic Growth*. Cambridge: Cambridge University Press.

Sheth, D.L. (1995) 'Democracy and globalization in India: post-Cold War discourse', *Annals of the American Political Science Association* 540: 24–39.

Shohat, E. and Stam, R. (1994) *Unthinking Eurocentrism*. London: Routledge.

Schoultz, L. (1999) *Beneath the United States*. Cambridge, MA: Harvard University Press.

Sidaway, J.D. (2000) 'Post-colonial geographies: an exploratory essay', *Progress in Human Geography* 24 (4): 591–612.

Slater, D. (1998) 'Spatialities of democratization in global times', *Development* 41 (2): 22–9.

Slater, D. (1999) 'Situating geopolitical representations: inside/outside and the power of imperial interventions', in D. Massey, J. Allen and P. Sarre (eds) *Human Geography Today*. Cambridge: Polity. pp. 62–84.

Slater, D. and Taylor, P.J. (1999) (eds) *The American Century: Consensus and Coercion in the Projection of American Power*. Oxford: Blackwell.

Slotkin, R. (1998) *The Gunfighter Nation: The Myth of the Frontier in Twentieth Century America*. Norman: University of Oklahoma Press.

Smith, P.H. (2000) *Talons of the Eagle*. Oxford: Oxford University Press.

Spanos, W.V. (2000) *America's Shadow: An Anatomy of Empire*. Minneapolis: University of Minnesota Press.

Tocqueville, A. de (1990) *Democracy in America* (1840), vol. 1. New York: Vintage.

Trouillot, M.-R. (1995) *Silencing the Past*. Boston: Beacon.

Trouillot, M.-R. (2000) 'Abortive rituals: historical apologies in the global era', *Interventions* 2 (2): 171–86.

Valladão, A.G.A. (1998) *The Twenty-First Century Will Be American*. London: Verso.

Weber, M. (1978) *Economy and Society*, vol. 1. Berkeley: University of California Press.

Weber, M. (1992) *The Protestant Ethic and the Spirit of Capitalism* (1930). New York: Routledge.

Žižek, S. (1998) 'A leftist plea for "eurocentrism"', *Critical Inquiry* Summer: 988–1009.

23

Alternative Modern – Development as Cultural Geography

Michael Watts

'Development' in other words is Orientalism transformed into a science for action in the contemporary world. (Akhil Gupta, 1998)

There is never atomistic and neutral self-understanding; there is only a constellation (ours) which tends to throw up the myth of this self-understanding as part of its imaginary. This is the essence of a cultural theory of modernity. (Charles Taylor, 2001)

BAD LATITUDE, RESURGENT GEOGRAPHY

Economist Jeffrey Sachs is indisputably part of the development establishment. Currently the Director of the Center for International Development at Harvard University, he is an international figure in the debates over economic reform in the post-socialist bloc (as a fearless advocate of 'shock therapy'), a sometime critic of the World Bank and the International Monetary Fund austerity programs, and latterly a tireless promoter of the idea that geography causes poverty. Virtually all of the tropics remain poor, says Sachs, because climate 'accounts for a quite significant proportion of cross-national and cross-regional disparities of world income' (2001: 9). Global production is 'highly concentrated in the coastal regions of the temperate zones', the 'proximate countries' (Sachs et al., 2000: 73). As his Harvard colleague (and former Marxist) Ricardo Haussman puts it, the problem is a case of 'bad latitude'. The 'new'

geography of development – bestsellers *Guns, Germs and Steel* (1997) by Jared Steel and *The Wealth and Poverty of Nations* (1999) by David Landes both extol the virtues of advantageous geography in the long march of economic development – endorses what in a June 2000 address to the then US Treasury Secretary (and now President of Harvard University) Lawrence Summers dubbed 'the tyranny of geography'. Isolation, poor soil, erratic climate, inaccessibility, low agricultural productivity and infectious disease mutually reinforce a vicious cycle of destitution and underdevelopment. Geography need not be destiny, they say, but for much of the world it seemingly is. The solution for the geographically challenged is a good dose of globalization (Hausmann, 2001: 53).

For the geographer, the 'return' to the role of geography in human affairs – or more properly to nature, location and topography as determinants of growth and welfare – simply recapitulates ideas of great genealogical complexity and historical depth (Glacken, 1967). In the Sachs/ Harvard model, however, ecology, disease and isolation carry a pungent, late-Victorian imperial odor. Is this not the Dark Continent all over again? Well, yes and no. In the Sachs account nature and geography take pride of place, but unlike the discourses of a century ago, the role of culture is for the most part invisible. Hausmann, Sachs and friends do indeed refer to 'social institutions' and markets as necessary building blocks for growth, but the tropes of sloth, fecundity, racial inferiority and an irredentist anti-market mentality – the hallmarks of the Victorian imperium – are largely absent. Geography is, rather, a material impediment to growth, its biophysical properties capable of wreaking unprecedented social havoc: 'It is no accident', says Sachs in an astonishing *aperçu,* 'that genocide took place in Rwanda' (2000: 15)! Central African ethnic bloodletting, produced by a lethal combination of organic nationalism and fascist politics (HRC, 1999), is here read as the product of 'geographical disadvantage'. Running through the 'new economic geography of development' is a gesture to Elsworth Huntingdon, a cavalier ratification of economics and economic growth as the *sine qua non* of development, and an unquestioned hymnal to the Olympian powers of the market and 'modern science and technology'.

If geography has a new cachet in economics, political science cannot of course be far behind. Princeton political scientist Jeffrey Herbst (2000: 156–69), for example, sees in the weakness of African postcolonial states more or less 'favorable geography'. It is the size and shape of inherited national territories, coupled with population density, that determine the 'broadcasting' of political authority as a precondition for political stability. In its 'national designs,' he argues, Africa has much to be desired.

All of this work, which strikes the contemporary geographer as a very bad case of Victorian recidivism, does pose sharply the question of what a critical geography, and for our purposes a critical *cultural* geography, of development might look like. That is to say, a geography capable of understanding actual development practices (institutions, knowledges, professions, accumulation of wealth, forms of state intervention and so on) in cultural (semiotic, representational, discursive) and spatial (regional, territorial, global) terms. In one sense, culture, understood broadly as practices which produce meanings constitutive of what Raymond Williams (1961) called 'a whole way of life', has *always* been present in much of what passes as post-war development theory, and indeed in colonial development. Colonial bureaucrats were obsessed with the place of 'tradition' in relation to imperial power and colonial stability. In the case of British indirect rule, African 'custom' was deliberately retained (but of course necessarily transformed) in the creation of what Mamdani (1995) calls 'decentralized despotism'. Ethnicity, whether construed in terms of the tribe or the indigene, was frequently the modality of colonial governance, though the 'tribals' in India or 'minorities in Nigeria were partly colonial inventions. Colonial governmentality, then, was cultural in disposition: indeed, it worked through local culture to gradually construct, or attempt to construct, a new sort of (colonial) subject (Scott, 1995). The destruction of the local cultural community by an unfettered colonial capitalism

was, in this sense, radically destabilizing for the colonial state, and politically threatening to metropolitan authority. Culture – embodied in the Gambian lineage or Rhodesian chieftainship – was to be the mechanism for stable, self-reproducing colonialism, to help manufacture docile colonial subjects ('happy natives'). At the same time, tradition was an obstacle to the Schumpeterian energies and technological innovation boiling in the incubus of colonial capitalism (Cooper and Stoler, 1995).

The same might be said of 1950s growth theory or the import-substitution industrialization of the 1960s. They were variants of cultural theory too, peddling certain sorts of cultural artefacts and regimes of consumption in the name of an *acultural* theory of modernity; as Taylor (2001) says, this sort of theory is culture-neutral because *any* traditional culture could experience the benefits of reason. The economists who invoked the backward-sloping supply curve rested their claim on purported peasant economic 'irrationality' or a *culture* of poverty: in short, a cultural disposition to produce more as prices collapsed because of risk aversion, least effort or a lack of need achievement. Cultural beliefs and ideologies were oppositional to capitalist ethics and to the idea of *Homo economicus.* Modernization theories in fact could never see *beyond* culture – what they saw as the dead weight of tradition – and specifically beyond the panoply of culture forms antithetical to self-sustaining capitalist growth. It was excess of social ascription (caste) in India, or a radical lack of 'need achievement' ('n-ach') in Ghana, or the absence of indices of modernity (postboxes, telephones, cars) in Burma, that signaled economic underdevelopment – never mind that such indices and proxies reached their historical zenith in Nazi Germany. Modernization theorists, of course, conspicuously read culture *out* of their own theory of the modern!

The cultural content of development theory understood in this non-reflexive way – culture is what others have – has never really gone away. The weight afforded to values in the newly industrializing 'miracles' of North East Asia – Taiwanese familialism or South Korean Confucianism (Hefner, 1998) – is one example of such cultural developmentalism. The purportedly parasitic tendencies of the African moral economy (Hyden, 1980) – the 'captured' state becomes, as Nigerian novelist Chinua Achebe (1988) put it, 'one big crummy family' – is another. But such analyses do little more than identify the ways in which cultural difference makes a difference for growth: how it facilitates, or more typically how it hinders, capitalist accumulation at the periphery of the world system. Is this all there is, or might be, to a cultural geography of development?

The answer to this question is, not unexpectedly, neither simple nor straightforward, in part because geographers (and others) have explored a variety of intellectual avenues – anti-modernist populism, postcolonial alternatives, hybrid development to name but a few – in their search for such a cultural geography. I want to organize my remarks around three broad arguments. The first is that critical cultural work on development deliberately locates development on the larger landscape of the modern but in doing so marks a shift from one theory of modernity (acultural) to another (cultural). That is to say, there is a shift from a theory in which development describes a transition from tradition to modernity as 'a form of life toward which all cultures converge' to one which sees modernity as unfolding within a cultural or civilizational context, always sensitive to the different starting points of the transition and its different outcomes (Taylor, 2001). Development is, in other words, culture and site specific; it is irreducibly *cultural geographic*. That said, depositing development on a culture-sensitive landscape confers no unity of purpose or theoretical coherence. Some see development as failed or something to be abandoned; others seek to specify in detail the cultural and historical grounding of the west's self-representation.

My second argument examines work that explores the cultural grounding of development – of projects, of development institutions and so on – *ethnographically*: that is to say, using the tools of the anthropologist to explore the cultural content of development processes. There is no assumption that this work has a common agenda but it typically offers a sort of institutional ethnography of development, seeking to understand how particular development categories are constituted, and attentive to how forms of development authority are produced and represented. For some, such a project has a reformist purpose – to reform the World Bank; others want to create a space for indigenous development (whatever that may mean). The third argument raises the question of alterity: that is to say, what difference cultural difference makes for the purposes of imagining and reimagining development, and specifically for thinking about alternatives. Here I try and engage with the related ideas of hybridity and cosmopolitanism.

In stitching these arguments together I want to suggest that a cultural geography of development is extremely powerful but that the search for alternatives is often politically myopic, misreads the insights to be gained from radical or

unconventional development theory (Marxism quite specifically), has surprisingly little to say about economics and economic alternatives, and in some cases is simply politically reactionary and crudely anti-modern in ways that do no justice to the very idea of modernity itself. Indeed, some of the work on alternative modernities – that is to say, a postcolonial account of development that sees modernity 'as not one but many', as 'old and familiar' and 'as necessarily incomplete' (Gaonkar, 2001: 23) – acknowledges the uncontested hegemony of capitalist modernization. There is no escaping it, says Dipesh Chakrabarty, anywhere in the world (2001: 123)! In this sense, I pose the question of whether the critical cultural geography of development has adequately come to terms with the overwhelming powers of capitalist modernity and the old grand theories of the modern that Marx among others long ago invoked and that are now all too out of fashion. Are the claims for hybrid development and 'creative adaptation' simply the desperate attempts to read some small victories against the grain of a Whiggish theory of modernization?

MODERNITY

> The concept of progress is to be grounded in the Idea of catastrophe. That things 'just go on' is the catastrophe. (Walter Benjamin, 1969)

Critical and culturally informed work on development during the 1980s and thereafter turned in large measure on seeing the development project itself as a form of modernity (Escobar, 1992; Ferguson, 1999; Mills, 1999; Parajuli, 1991; Rofel, 1999; Sachs, 1992). For what has now become the 'post-development' school (Rahnema, 1997), development was moreover a *failed* modernity of catastrophic proportions.[1] Development as an immanent process and as a set of intentions accordingly became, to return to Raymond Williams (1973), a keyword in relation to the modern, its meanings unstable, and always wrapped up with the problems it was being used to discuss. These problems are distinctively western and modern. Development 'rehearses, in a virtually unchanged form,' says Gupta, 'the chief premises of the self-representation of modernity' (1998: 36): progress, science, reason, universal history. This self-representation of modernity by the west via the other travels in a variety of colonial and postcolonial modalities, and in so doing becomes 'an inescapable feature of everyday life' (1998: 37) in the Third World.

An archaeology of development demands, then, a full grasp of location, situating it *historically* (in tracing its complex genealogy and meanings, particularly to the eighteenth century), *geographically* (in relation to sites of productions, routes of movement and patterns of reception) and *culturally* (in relation to the west's self-representation, of reason and the Enlightenment). In sum, development commands a complex historical-semantic field. Of course this critical modernist turn had a pre-history in the sense that a few economic historians like Alexander Gerschenkron and the development economist Albert Hirschmann were deeply sensitive to the *longue durée* within which any critical examination of development must be located. Hirschmann's (1986) brilliant account of the relation of market-based development to the Enlightenment idea of a perfect social order is simply one early contribution to what became a veritable landslide of scholarship, heeding Fredric Jameson's injunction to 'always historicize'.

The cultural and historical turn in the study of development was not all of a theoretical or political piece. To simplify a complex intellectual landscape I want to highlight four broad genealogical threads.

New historicism

Fred Cooper and Randall Packard acutely observe, in their foundational text on development and the social sciences, that the concept of development must be located in: 'historical conjunctures … understood in relation to intellectual trends, shifts in global economic structures, political exigencies and institutional dynamics' (1997: 29). I have chosen to characterize such an approach, borrowing from the work of Stephen Greenblatt and others, as 'new historicism'. An important early effort in this genre was the work of Arndt (1987) who, in a text mercifully unencumbered by discourse or high theory, charted the shifting intellectual trends and the circulation of ideas and people from the 1950s to the neoliberal counter-revolution three decades later. To simplify, one might say that the immediate postwar period was dominated on the one hand by variants of growth theory, in which the state was seen to have a legitimate and active role in human capital formation and infrastructure, and on the other by trade pessimism coupled with a proclivity for import-substitution industrial (ISI) strategies and a recognition of international protectionism.

By the mid 1960s what Arndt calls 'social objectives' were very much in evidence. A shift from interstate inequality and economic growth to a concern with employment, welfare and poverty was marked by the growing role of UNRISD and ILO in the international sphere and by the central role of figures such as H.W. Singer, Mohammed ul Haq and Richard Jolly, who ushered in a development discourse of redistribution with growth (RWG), the priority of basic needs and the powers of the informal sector. RWG contained a strong populist tone – 'small is beautiful', 'informal business' – rather than an affirmation of a 'strong' redistributive platform. Nonetheless, the fact that Robert McNamara, the new president of the World Bank (elected in 1968), was talking the language of equity and the poorest 40 per cent, highlights the slide in thinking away from both the glorification of economic growth *per se* and GNP as its lodestar. Dudley Seers' important 1969 address questioned whether the presumption that an economic growth rate of 6 per cent (the UN target) was always beneficial, and suggested that under some circumstances it might actually be politically dangerous. Seers had questioned, in his assault on GNP, the 'meaning of development'.[2]

The quickfire succession of development fashions, from social development to employment to equity to poverty eradication to basic needs, can be subject to all manner of cynical rebuttal, but Arndt and other new historicist work revealed something of its institutional and political dynamics and specifically the fundamental role of the University of Sussex, International Labor Organization and World Bank triumvirate who shifted the terms of the development debate. By the mid 1970s however the tenor had changed once more. Oil price increases, inflation, a recession and a grain shortage (coupled with worldwide drought) alerted McNamara and others to international inequity and the need for economic growth at all costs. Calls for a new international economic order in 1974 reflected the gravity of the global crisis, but as a redistributive (as opposed to a growth) program it was pretty much still born. As the 1970s wore on, all talk of basic needs and redistribution evaporated, overtaken by the neoliberal counter-revolution. By 1980 and the appearance of the famous Berg Report on Africa, laissez-faire capitalism was back in vogue (Toye, 1985).

The story of how and why this counter-revolution happened, and why it became hegemonic in the way it did, remains largely untold (Gore, 2000). To invoke the rise of Kohl and Thatcher and Reagan is part of the story, but understanding the circulation and institutionalization process – the key figures, the key institutions, the dispersion and reception of ideas, and so on – also requires a discipline which currently does not exist, namely the cultural study, or perhaps the ethnography, of development economics (Thrift, 2000). Rosen's (1985) book on the generation of US economists, including George Ackerloff and Stephen Hymer, who played a key role in India is one model of what is required to grasp the rise of a global neoliberalism; Hall's (1989) book on the circulation of Keyenesian ideas between the wars is another. I would add only that the so-called Washington consensus that marked the hegemony of neoliberalism in development thinking in the 1980s and 1990s is itself unstable.

The sort of struggle documented by Wade (1996) between Japanese policy makers and the US Treasury and the World Bank over the so-called 'East Asian miracle' is a vivid case in point; the political struggles within the World Bank associated with the departures of two senior bank officials, Joseph Stiglitz and Ravi Kanbur, is another (Wade, 2001). None of this should alter the fact that between 1965 and 1975 there was a curious sort of *intifadah* in conventional development discourse. One can argue about its longevity and how much was pure cant. But it marked a watershed in which the technocratic and growth-driven paradigm of the 1950s was put on the back burner. It is hard to imagine, for example, that 30 years later the UNDP could publish its *Human Development Report* – a document devoted to new measures of development and to the promotion of capabilities (Sen, 2000) over income or consumption – had there not been this 1960s *abertura*.

One of Arndt's insights, subsequently elaborated (see the contributions in Cooper and Packard, 1997; Crush, 1995), was that development, understood as a shifting set of knowledges and practices focused on the developing other, spoke as much to the realities of the advanced capitalist states and their internal problems as to the realities of the poorest 40 per cent. The link between the events of 1968 and the shift from development as growth to development as basic need provision can and should be read in precisely this way. None other than Richard Jolly, one of its key architects, explicitly attributed this shift to the 'growing questioning of Western consumer-urban-industrial models' (cited in Arndt, 1987: 108). Questioning the Holy Grail of GNP was derivative of the social crisis of the North Atlantic economies. Indian economist Deepak Lal (cited in Arndt, 1987: 67) is not far from the mark when he notes that the concern with the social, with equity and with redistribution reflected the failures of the American dream and of the inability to solve the race and inequality problem in the US in particular. It is no

accident that Robert McNamara's first speech which identified the poorest 40 per cent began with a discussion of poverty in the US (Watts, 2000)! The meanings of development and its political semiotics are seen to be profoundly dialectical; ideas and people travel between north and south and among and between sites of practice and knowledge production. The detection of some originary point in the genesis of development ideas proves to be particularly tricky.

Governmentality

A second line of work engages with institutional politics, self-reflexivity and the poststructural insight that development is a set of discursive practices and representations. Development practices can be construed as forms of what Michel Foucault called governmentality, the 'disabling of old forms of life by systematically breaking down their conditions and constructing in their place new conditions so as to enable ... new forms of life to come into being' (Scott, 1995: 193). To understand development is to grasp how 'the possible field of action of others' (Foucault, 1982: 221) is structured, how the triad of sovereignty, discipline and government through a variety of technics and micro-politics of power (from the map, to the national statistics, to forms of surveillance) accomplish stable rule through governable subjects (Li, 1999). Development came to be seen as an 'everyday form of state formation' (Joseph and Nugent, 1994). Some of the earliest work by Escobar (1992; 1995) and Mitchell (1995) saw the business and apparatuses of development as a power–knowledge nexus, but it was weak empirically and shallow in its grasp of the development institutions themselves.

A raft of new scholarship – prompted by Ferguson's (1990) excellent ethnographic account of a development project in Botswana – has begun to explore development in a much more grounded institutional and textual way, posing hard questions about how development ideas are institutionalized and how particular development interventions may generate conflict as much as consent; and it has begun to examine the internal dynamics and complexities of large, internally differentiated multilateral development organizations. This tack takes the *social construction* of knowledge (by whom, which what materials, with what authority, with what effects) and the relations between knowledge and institutional practice very seriously, and in so doing can identify struggles and spaces in which important changes can be and are made. Greene's (1999) account of how Malthusian ideas travel over time and space, attaching themselves to neo liberal, feminist and environmental agendas in specific places and times, is simply one illustration of what in the earlier work was passed over as the unproblematic 'transmission' of ideas.

Gardner and Lewis' (2000) careful analysis of the Blair government's White Paper on international development can show, from the inside so to speak, how changes of personnel and the shifting balance of powers among interest groups within the Department of International Development in the UK have consequences for both the design and the practices of actual development work. Many of the contributions to *International Development and the Social Sciences* (Cooper and Packard, 1997) focus on the relations between the academic disciplines and development practice. Sharpless (1997) reveals how demography as a practical discipline required the conjunction of intellectual and political processes in which the large US foundations such as Rockefeller and Ford played a pivotal mediating role. A similar global, institutional and political story has been told by Perkins (1997) in his exemplary account of the intersection of wheat, genetics and the Cold War in the genesis and diffusion of the green revolution technologies. Some of the work described by McEwan in Chapter 21 in this volume on feminisms and their relationship to development has a close affinity to such scholarship.[3]

Some of the most compelling work in this vein examines environment and national/multilateral development institutions. Candace Slater's (1994) excellent book on Amazonia, for example, focuses on the ways in which particular regions or ecologies are construed for developmental purposes. She discloses a transnational popular imagery of the region and analyses the literary, media and other cultural machinery contributing to what I have elsewhere called a 'discursive ecological formation' (Peet and Watts, 1996; see also Guthman, 1997). In her account, the Edenic or naturalized narratives of the Amazon always silence (the Indians have no voice or no voice of their own), exclude or distort. Slater ends with the provocation that there is an absence of competing images of Amazonia: but under what conditions can competing images *really* compete? Do these images and constructions of landscape really have the power and effect implicit in these accounts of narratives? Are they 'just' images and irrelevant to the hard edges of political economy and environmental destruction? Mitchell's (1991) account of the developer's imagery of the

Nile Valley – quaint, overpopulated, without history – attempts to deconstruct conventional development models and presumptions in a similar way, affirming what Alfred Hirschmann noted almost four decades ago, namely that development 'depend[s] upon a set of more or less naive, unproven and simplifying assumptions' (1968: 23).

Environmental and other forms of development knowledge production can also be approached through the 'epistemic community', or communities of developmental expertise. Here the knowledge is western science, and more properly the cosmopolitan scientist, expert and policy maker. Peter Haas (1990) has argued in the context of understanding regional and global (multilateral) conventions that the process of consensus building and collective action more generally is *knowledge-based* and *interpretive*. That is to say, international regulatory co-operation is fueled by fundamental scientific uncertainty about the environment which ensures that governments seek out authoritative advisers (experts) who, to the extent they are part of epistemic communities, are more important to the political solution than the content of the ideas *per se*. Cross-national differences in state behavior are determined by the variation in the penetration and institutionalization of experts (epistemic communities). Biodiversity and stratospheric ozone cooperation are seen in this way as instances of the cognitive and bureaucratic power of scientific experts. This is an argument that has also been made for NAFTA by Benton (1996) who argues that the trade and environmental constituencies brought together around tariff reduction actually created a dialogue – a transnational community of experts – that had not hitherto existed. Of course these epistemic communities are porous and in local settings knowledge carries between scientific and indigenous knowledge communities; at these 'knowledge interfaces' (Long and Long, 1993) the questions of how interpretation, authority and practice operate – for example at the farmer level – are rather more complicated, suggesting complex forms of hybrid knowledge in circulation (Gupta, 1998).

The epistemic community has an affinity to political ecology's notion of 'conventional models' (Leach and Mearns, 1996) developed in and around particular institutions and practices that assume a hegemonic, and often uncontested, status. Some of the most interesting geographical research has examined the politics of colonial and postcolonial conservation. Fiona Mackenzie (1997) has traced discourses of soil erosion and land conservation in the 1930s to the complex political struggles among and between the colonial state, white settlers and the native reserves. Neumann's excellent book *Imposing Wilderness* (1999), on the creation of the Arusha national park in Tanzania and the ideas of landscape and nature which lay behind state appropriations of land from local peasant communities, is an especially compelling illustration of how cultural and historical representations of nature intersect with colonial and postcolonial rule.

Leach and Mearns' (1996) reinterpretation of the West African forest–savanna mosaic is a careful deconstruction deploying historical studies coupled with detailed local analysis of agro-ecology to confirm what the new 'non-equilibrium' ecology posits, namely that climax models of ecological stasis are unhelpful. These static models however do enter into administrative practice (colonial and postcolonial) to reinforce the idea of Guinea's forest cover as 'relic' (which Mearns and Fairchild see as the basis for driving 'repressive policies designed to reform local land use practice') rather than as the outcome of intentional local management practice. Swift (1997) has shown how the assumptions about desertification in the West African Sahel rest not only on remarkably sparse evidence but on questionable models of the dynamics of semi-arid rangelands calibrated through linear, cybernetic models of ecological structure and temporalities, and neo-Malthusian models of social change. The source of this authoritative knowledge is exceedingly complex and needs to be tracked to colonial forestry and planning agencies, to the ecological sciences and the rigidity of particular models of the semi-arid tropics, and to the thorny question of why some ideas and some reports circulate faster and more effectively than others, and why some individuals have a legitimacy far in excess of their scientific abilities (Baker, 2000; Davis, 2000; Taylor, 1996). Political ecology sees the 'new' development-environment practices as necessarily plural (at the level of truth claims), democratic (to open up the practices of policy making to other voices) and complex and flexible (sensitive to local conditions and historical dynamics).

Post-development

The third thread of critical development work is broadly Third Worldist, and much more radical and self-consciously normative in a way that marks it off from the new historicists. I shall refer to it as postmodern/poststructural, though it traces its lineage to the work in the 1960s of Ivan Illich (1971), and earlier still to some of the

populist and civic theory associated with Prudhon, the Owenite socialists and others. Associated with a number of public intellectuals and activists largely but not wholly from the south, it is a variegated community that has marched under the banner of 'post-development'. The intellectual field which constitutes these radical critiques of development – one thinks of the work of Arturo Escobar, Gustavo Esteva and Wolfgang Sachs and the new *Post Development Reader* (Rahnema, 1997) as its compendium – is replete with the language of crisis, failure, apocalypse and renewal, and most especially of subaltern insurgencies which are purportedly the markers of new histories, social structures and political subjectivities (Pieterse, 1996).

The Delhi Centre for Developing Societies – to invoke one such important and visible cluster of erstwhile anti-development Jacobins, latterly referred to by Fred Dallmayr (1996) as a Third World Frankfurt School – includes among its pantheon the likes of Ashis Nandy, Rajni Kothari and Shiv Visvanathan who in their own way represent a veritable heteroglossia of alternative voices from the south encompassing a massive swath of intellectual and political territory on which there is often precious little agreement. I have chosen, however, to provide a unity to these critiques – drawn variously from post-Marxism, ecofeminism, narrative analysis, poststructuralism, postcolonial theory and postmodernism – by emphasizing their confluences around development as a flawed, in some quarters a catastrophically failed, modernist project (see Corbridge, 1998). Much but by no means all of this critique draws sustenance from the idea of the third leg of modernity – the dark side of modernity and the Enlightenment which produced the new human sciences and the disciplines – as much as from the Marxian leg of capitalist exploitation and the Weberian (and Habermasian) leg of the colonization of the lifeworld by monetization, rationalization, calculation and bureaucratization.

This tale of disenchantment carries much of the tenor and timbre of earlier critiques of development – most vividly of the 1960s but also of the 1890s and earlier, as Michael Cowen and Robert Shenton have admirably demonstrated in *Doctrines of Development* (1996) – readily apportioning blame to the multinational behemoths (corporate and multilateral) of global capitalism. Running across this body of work is the notion of development as an essentially western doctrine whose normalizing assumptions must be rejected: 'it [development] is the problem not the solution' (Rist, 1997). The sacred cows – for Esteva and Prakash (1998) they are 'the myth of global thinking', 'the myth of the universality of human rights' and 'the myth of the individual self' – must be substituted by what two of the post-development field's key voices have called 'grassroots postmodernism',

Arturo Escobar's book *Encountering Development* (1995) is the most developed account of thinking about the development industry in grand poststructural terms, offering a vision of sub-altern and indigenous social movements as vehicles for other ways of doing politics (non-party, non-mass, autopoietic and self-organizing) and doing 'post-development' (decentralized, community-based, participatory, indigenous and autonomous). Interestingly, this post-development movement met up with and cross-fertilized with a largely western academic development community energized by what was dubbed the 'impasse in development' debate of the 1980s and 1990s (Booth, 1994; Schurman, 1993). In effect this was a debate within the walls of Marxist development theory between its 'neo' and 'structural' schools over the extent to which Third World socialism suffered from many of the trappings of industrial capitalism (and many others unique to it!), and a theory captured by economic essentialism, class reductionism and teleological thinking.

One can argue whether this characterization of Marxist development theory is plausible or indeed an adequate account of Marxism itself in its panoply of guises (Watts, 1989). But the impasse debate spawned important new inter-sections between postcolonial and post-Marxist thinking, providing a fertile ground on which development could be refigured by a careful reading of Ranajit Guha or Gyatri Spivak or Edward Said (see Gupta, 1998). There is little theoretical coherence in the 'impasse work' – actor network approaches, a focus on identity politics and the cultural construction of class, a shift to 'responsible politics' (Booth, 1994) – but Corbridge is nonetheless right to emphasize that it, like the post-development work, reinforced the need to see 'the ways in which the West represents its non-western others' and forces us to ask 'What is development? Who says that is what it is? Who aims to direct it and for whom?' (1993: 95).

Diversity became the new watchword (Booth, 1994: Chapter 1; see also Gibson-Graham and Ruccio, 2001), but it carried its own burdens (who wants a few more neofascist development movements in the name of letting a thousand flowers bloom?). At the same time the postcolonialists' proper emphasis on writing history differently – signaling, as Stuart Hall says, the 'proliferation of histories and temporalities, the intrusion of difference and specificity into generalizing

Eurocentric post-Enlightenment grand narratives' (1996: 248) – in turn often mistook the word for the world, and populist incantation for 'new politics', and opted for a heavy dose of wishful thinking ('in the heartlands of the West,' said David Slater, 'modernity is in question [a]nd the fixed horizons for development and progress [are melting away]' (1993: 106)!

Is this new deconstruction and reimagining of development a distinctively original vision? What sort of vantage point does it provide for a post-development imaginary? To employ Arturo Escobar's own language in representing 1950s development economics, what sort of 'world as a picture' is contained within the scopic regime of alternatives to development? On the one hand there is a certain sense of 1960s *déjà vu* (Lehmann, 1997). A number of accounts of globalized political economy in this work – in spite of its aversion to metanarratives and totalizing history – rest clumsily on a blunt, undifferentiated account of world capitalism, in which institutions like the World Bank have untrammeled hegemonic power, and the Third World appears as a monolithic, caricatured and often essentialized realm of at worst normalized subjects and at best hybridized, subaltern emancipatory potential. Has Ernest Gellner's (1979) Big Ditch simply been replaced by the Big Panopticon (see Watts and McCarthy, 1995; Watts, 2000)? There is in any case an unfortunate hyperbole in some of this work. Rist for example argues that:

> Development (as a program for collective happiness) no longer exists except as a virtual reality, as synthetic image in the full-length film of globalization. It is like a dead star whose light can still be seen, even though it went out for ever long ago. (1997: 230)

As an account of the presence of development practices in some contemporary global world, this strikes me as implausible, unhelpful and rather ignorant.

To say there is much hazy utopianism and populist sentimentalizing here is something of an understatement but, as David Lehmann (1997) has noted, the critics reserve their most virulent prose for the development establishment – the development business – itself. Much of this work rightly takes on the professions of development, proposing ethnographies of those development institutions which in the name of building freedoms (from hunger, from oppression, from arbitrary rule) create forms of classification, exclusion, individuation, normalization and discipline: in short, the conversion of 'a dream into a nightmare'.[4] Globalization – the dialectic of indigenization and cosmopolitanism – now projects Foucault's idea of the birth of the clinic onto an unsuspecting Third World. In a world in which discourse seems to carry implausibly robust, powerful and hegemonic efficacy, post-development and alternatives reside in the hybrid, in critical traditionalism, in strategic essentialism – in the 'discursive insurrection' of the Third World, as Escobar put it.

What *is* different from the 1960s crisis of development is the degree to which the state as a necessary and appropriate vehicle for national aspirations, and the universalistic (and anti-imperialistic) claims for liberation, are no longer axiomatic and taken for granted. Locality, culture, authenticity are the forms of identification which stand in opposition to states, and the very fictions of the nation-state and nationalism are supplanted by what Lehmann calls 'multinational populist subcultures' in search of cultural difference: 'cultural difference is at the root of postdevelopment', as Escobar (1995: 225) says.[5] One might say that the practical and strategic content of this vision is rooted firmly in the soil of civil society rather than in the state or market. But it is civil society of a particular sort: of grassroots movements, of subaltern knowledge, of cultural economics, of hybrid politics, of the defense of the local, of cybercultural posthumanism. Much less is said about the civil society capable of engendering violence, genocide and fragmentation.

This post development corpus has opened up, initially through Esobar's provocation, important new avenues for understanding development practices. But it has left its own problematic legacy. First, there is the curious, and perhaps appropriately ironic, way in which a postmodern or poststructural sensibility is attached to claims of extraordinary totalizing power, certainty and rectitude. Development, as Escobar has it, is 'a historically *singular* experience' (1995: 10). Second, the unalloyed celebration of popular energies of grassroots movements is not subject to the sort of hypercritical discourse analysis which might permit an understanding of their achievements, their political strategies, the limits of their horizons and vision. Third, there is a curious confluence between elements of the neoliberal counter-revolution (the World Bank's account, for example, of Africa's post-colonial modernization failure, its anti-statism and the need to harness the energies of 'the people') and the uncritical celebration, and often naive acceptance, of post-development's new social movements. And fourth, the important critique of economic reduction and class determinism (the Marxian master narrative) – and, it should be added, the deconstruction of the free-market myopia (the Smithian master narrative) – has produced, to

quote Stuart Hall, not alternative ways of thinking about basic economic questions but instead 'a massive, gigantic and eloquent disavowal' (1996: 258). Where are the post-development studies of macro-economic management, of cooperatives, of participatory budgeting?

Some of this work has curiously not engaged *sufficiently* with the idea of development as modernity. The creative destruction of capitalist development has, as Marshall Berman notes in *All That Is Solid Melts into Air*, typically produced the experience of, and the reactions to, the solid melting into air:

> There is a mode of vital experience – experience of space and time, of the self and others, of life's possibilities and perils – that is shared by men and women all over the world today. I will call this body of experience 'modernity'. To be modern is to find ourselves in an environment that promises us adventure, power, joy, growth, transformation of ourselves and the world – and, at the same time, that threatens to destroy everything we have, everything we know, everything we are. Modern environments and experiences cut across all boundaries of geography and ethnicity, of class and nationality, of religion and ideology: in this sense, modernity can be said to unite all mankind. But it is a paradoxical unity, a unity of disunity: it pours us all into a maelstrom of perpetual disintegration and renewal, of struggle and contradiction, of ambiguity and anguish. To be modern is to be part of a universe in which, as Marx said, 'all that is solid melts into air'. (1982: 15)

Modernity *contains* the tragedy of underdevelopment: development and its alternatives are dialectically organized oppositions within the history of modernity. This is not to simply fold the current antipathy to development into the master narrative of modernity. Rather it is to observe that there is a danger of not learning from history, of losing touch with the roots of our own modernity, of not recognizing that modernity cannot be unproblematically located in the west, and of not seeing development and its alternatives as oppositions that contain the other (Chakrabarty, 2001; Harvey, 1996).

Ur-history

A final line of critical development theory straddles the new historicism and the poststructural/postcolonial. It addresses what one might call the *ur*-history of development – the historical semantics that Williams (1973) invoked – filling out the enormously labyrinthine genealogies of development itself. Much of the post-development work does this by starting with a rather caricatured sense of the Enlightenment or modernist self-representation and then charting one of its dimensions through the nineteenth- and twentieth-century trajectory of development (see Schech and Haggis, 2000: 5). For Esteva this is individualism, for Alvarez science, for Shanin progress, for Rist reason, for Kothari state rationality (see their contributions in Rahnema, 1997). For some, like Escobar (1995), this post-Enlightenment legacy congeals in a foundational moment to produce the 'invention' of development in 1945 in President Truman's famous inaugural address in 1949. The development industry then commences, through its national accounts data and statistical inventories and forms of governmentality, to produce a Third World subject (Mitchell, 2001).

There is something very clunky about all of these narratives. Even Gupta's (1998) sophisticated account is wide of the mark: are, as he says, the tropes of modernity in development 'rehearsed unchanged'? Is the Third World past always historically 'depoliticized' in development discourse? Several observations are in order. First, the Enlightenment itself has all of its contradictions, complexities and tensions *read out* of the script of post-development. Second, modernity as a contradictory experience (Berman, 1982) is not *read into* the shifting meanings of development. And third, the key difference between development as an immanent process and as a series of intentions is typically occluded. It is for this reason that *Doctrines of Development* (1996) is such an important intervention: Cowen and Shenton precisely show how in the course of the late eighteenth and nineteenth centuries an immanent conception becomes intentional development, something to be ordered to 'ameliorate the disordered fruits of progress' (1996: 7).

Development as a state project in Europe was a response to the creative destruction of industrial capitalism. How it travels – as trusteeship in the colonial period, or as structural adjustment in our own epoch – should not obscure this *ur*-history which is not readily captured in a claim that developmentalism is essentialist, homogenizing and evolutionary (Gupta, 1998: 33). Who would make this alloyed claim of the UNDP *Human Development Report*? Perhaps more than anything this UN agency has, through the work of Sen (2000) on capabilities and freedom, begun to radically and philosophically rethink the idea of development as economy and in a way which surely belies Gupta's homilies.

One of the strengths of all of this work nevertheless has been to chart for example the recycling of foundational ideas – for example populism (Brass, 2000; Watts, 2000) or

Malthusianism (Greene, 1999) – rooted in specific geographical and cultural-historical circumstances: say villagization in Tanzania, Maoism in China, or micro-credit in Bangladesh. Another is that it scrutinizes the complexities, ruptures and disjunctions associated with the circulation and reception of ideas, representations and institutional practices, always attentive to the 'traffic' in development which produces hybrid ideas (not simple mimicry or adoption), and processes of 'indigenization' (the 'domestication' of state planning and Keynesianism in post-independence India). Culture and geography are self-evidently in play throughout. This is nowhere more the case than in the relations between postcolonial development and the nation. Gupta's (1998) excellent book *Post-colonial Development* reveals that agricultural development was rooted firmly in Cold War Indian nationalist discourses but that the developmental packages were nonetheless struggled and fought over and radically reshaped ('indigenized') by the unanticipated consequences of populist politics in the 1960s and especially by Mrs Gandhi's political ploy to decentralize to the masses. In this sense the critical development field has contributed in some small way to what Hall calls the renarrativization which 'displaces the story of capitalist modernity from its European centering to its dispersed global peripheries' (1996: 250). It is an interruption of the grand *historiographical* narrative. But has it made the overwhelming material powers of capitalist expansion and modernity any less overwhelming or salvageable?

Curiously all of these critical perspectives – each sensitive in different ways to place and culture – emerged at a moment when conventional development practice seemed to have hardened, in the wake of the neoliberal counter-revolution, into an overwhelming free-market Washington consensus. The fall of the Berlin Wall added fuel to the capitalist fire as the likes of the World Bank crowed at the disappearance of alternatives, while radical critics like Perry Anderson (2000) could solemnly note that the arrival of the new millennium marked a massive, premeditated defeat. It was against this backdrop – one of political defeatism on the left and free-market triumphalism on the right – that the call for alternatives to development, for the radical deconstruction of development, and later for the anti-globalization movements came to fruition. This can be read in a number of ways: as a death throe, as myopia, as a reaction to the creative destruction of a global capitalism and so on. But it is incontestable that the growing clamor to 'provincialize Europe', to see outside and beyond the hegemony of the west, and to abandon the myth of development were forged – and perhaps reached their apogee – in the crucible of what Perry Anderson calls 'the neo-liberal grand slam' (2000: 15) and the disappearance of the entire political horizon of the 1960s generation (2000: 17). His call for 'renewal' in terms of 'accommodation' and 'consolidation' must cast a long shadow over the prospects for doing development differently.

ETHNOGRAPHY

At the end of October 2000 I attended a small seminar organized by the World Bank, the ground zero of conventional development practice, on social capital (Woolcock and Narayan, 2000). The meteoric rise and institutionalization of social capital within the Bank, one of the most important multilateral regulatory institutions, is an intriguing issue in itself – and indeed was one of the topics under discussion. What was it about this nebulous and rather elastic term, or perhaps about Robert Putnam (2000) and his social capitalist boosters, that provided a language, within a powerful and complex institution dominated by economists and engineers, capable of addressing social relations, poverty and the role of civics? How, in other words, do we grasp social capital's institutionalization – a discursive embedding – in the World Bank in terms of knowledge, power and practice? And what on earth might it represent, if anything, for the way the World Bank does business? Is it something that the anti-Bank NGO community might put its collective shoulder behind? Might the fact that a group of left-of-center sociologists and economists singing the praises of social networks, peasant confederations and decentralized social funds under the flag of the International Bank for Reconstruction and Development give the street protestors of Seattle and Washington DC any reason to rethink the nature of the beast?

Most of those present felt marginal to the Bank: economists who had fled in moral apoplexy from the harsh realities of implementing austerity measures, and anthropologists soured by the Bank's resettlement and indigenous peoples policies. Among the Bank's disenfranchised, social capital had, it turns out, a strategic appeal: it was a useful way to stimulate debates within and among the Bank's divisions and to provide a ground on which the hegemony of the neoliberal doctrine might be engaged. It was a discourse, moreover, with which some economists felt a degree of intellectual comfort. The workshop

revealed, on the other hand, that in terms of staffing, resources and legitimacy within the Bank structure, the social capital group remains wholly insignificant. The center of gravity within the institution resides elsewhere, quite specifically with the unassailable power of the 'structural adjustment people'.

Yet the overwhelming sense from the assembled Bank operatives – project managers in Indonesia, members of the operations and evaluation division, researchers, and one high-ranking administrator – was that the World Bank was in some sort of 'crisis'. On its face this is not exactly plausible. But at stake in this invocation of crisis was a series of internally and externally generated struggles and conflicts best understood in terms of Gramsci's notion of 'moral and intellectual leadership'. It had, after all, been a rough year for the world's major development institution, the global purveyor of conventional development wisdom. Not only could large numbers of low-income countries not afford to borrow from the World Bank – and a raft of middle-income states were not interested in borrowing – but the total Bank lending ($24 billion) was in any case trivial in relation to private capital flows.

There were various 'rebellions' – a term deployed by a high-ranking Bank official – over indigenous peoples and biodiversity policies; the Bank, along with the International Monetary Fund (IMF) and World Trade Organization (WTO), has borne the brunt of worldwide popular protest as the emblem of all that is wrong with globalization; and to round off the year, there were two high-visibility resignations. Joseph Stiglitz, the Vice-President of Research – indisputably the voice of theoretical and ideological leadership in the conventional development community – resigned, and was then fired at the behest of US Treasury Secretary Lawrence Summers. Stiglitz himself referred to the prospect of being 'muzzled', a product of his frontal assault on US Treasury/IMF adjustment policies and their handling of the Asian financial crisis. Several months after Stiglitz's exposé of the Bank and the IMF in the *New Republic* (2000), Ravi Kanbur, a long-standing Bank insider who at the time was overseeing the drafting of the World Bank's flagship text, the *World Development Report 2000* (WDR 2000), on 'attacking poverty', also resigned in another bout of internal warfare (Wade, 2001).

In so far as poverty reduction and its relation to growth have been a source of substantial debate and conflict within and outside the Bank since the 1980s, WDR 2000 – a bellwether and reference point for future development policy – was naturally going to be passionately debated.

The 'red cover' version accepted economic growth as the engine of poverty reduction but attached enormous weight and significance to 'empowerment, security and opportunity'. A ferocious debate ensued. Kanbur conceded to certain changes, but in the course of a May 2000 Bank meeting, in which US Treasury pressure was coupled with strong lobbying from powerful Bank economists (including sympathy for the 'growth first' position expressed by Bank President James Wolfenson), the rift between those defending the red cover version and the pro-growth faction deepened. Kanbur, realizing that the entire report was deeply vulnerable, promptly resigned. The final published version actually retained much of the thrust of the original draft, suggesting in some quarters a sort of victory for the Kanbur group.

What is one to make of this tale of intrigue, knowledge and 'discursive contestation' at 1818 H Street in Washington, DC? Any answer to this question would rest on some sort of institutional ethnography, that is to say applying the ethnographer's lens to the development institutions themselves (see Herbert, 2000). Ferguson's (1990) was the first attempt at an ethnography of the ways in which development problems are construed and represented – typically, he discovered, by rendering social conflicts as technical, 'depoliticized' problems; development was an 'anti-politics machine'. Yet much of this early work in attacking the development business as a sort of failed modernity was empirically weakest where it seemed theoretically strongest: it rested not on careful institutional ethnographies, or on discursive analysis of economic development as a profession or practice, or indeed on the careful reconstruction of how ideas develop and travel, but rather on poorly substantiated claims about purportedly all-powerful development institutions. All of this had the perhaps unintended effect of endorsing the supposedly unassailable powers of the development project itself. Ferguson's work also seem to run against the intuitive grain by suggesting that development projects were implemented without politics, when evidence abounded that development was *generative* of struggle (Schroeder, 1999) – a hyperpolitics machine!

New work has turned the cultural lens on development institutions and practices, and on resistance movements, by providing a deeper, thicker sort of ethnography (see Burawoy et al., 1995; 2000). Moore (2001) has examined multilateral agencies in semi-arid West Africa with an eye to the production of authoritative development and knowledge, and she does this as a sort of participant observer in a wholly understudied part of the development world, namely the global development consultant. Moore raises a panoply of important issues that shuttle between the politics of Burkina Faso and the upper echelons of CILSS and the Paris Club. Her starting point is the 'high modernism' (Scott, 1998) of the Sankara regime, which attempts to construct a new order on the backs of African 'decentralized despotism'. The result is a vast Ptemkin village in which nearly all the claims of animation and revolutionary practice are fraudulent; its legacy in Burkina is one part surrealism and one part high comedy, or perhaps farce (the governor's TV monitor), what Mbembe (1992) has called the 'banality of the postcolony'. Moore is concerned to show, however, that the Burkina government is especially adept at responding to the latest global development discourse by cooking up projects to acquire Official Development Assistance (ODA) monies: it might be community development, it might be decentralization, but either way the state is skilled at reworking discourses and texts to gain access to global development resources through the World Bank, USAID or CILSS. As Moore says, 'the reproduction of a donor way of thinking about development occurs very quickly' (2001: 171). In the Burkina case it is to be recalled that 20 per cent of GDP is derived from ODA (this is larger than all state current and capital expenditures combined, and more than all government consumption!) and hence there is a larger political dynamic at work – what Castells (1998) trenchantly calls the 'political economy of begging' – on which the discursive creativity (and fictions) of working over the foreign assistance establishment are predicated.

As the product of a consultant around one set of development interventions – the national resource management program – Moore's work raises some key questions. Neither the state operatives nor multilateral agents believe *any* of the public claims about compliance and accountability. In other words, a part of her concern with conditionality is actually a sort of 'public secret': everyone knows this is nonsense, but they all pretend otherwise. How does such a public secret operate, in what arenas can it take hold, what are its limits, and how can it be challenged precisely by some of the conditionalities that Moore seeks to document? Goldman's (2001) analysis of the World Bank explores these conundrums further by posing the question: how is a development institution 'greened'? He focuses on dams, one of the most controversial of Bank activities. Dams and resettlement have of course projected the Bank into a highly contested area, in which transnational activists and NGO networks have played a central role both in attempting to hold the Bank accountable and in preventing precisely the type of 'silencing' that he refers to in his Lao case study.

Goldman, like Moore, also raises a panoply of questions about authoritative knowledge. He rightly emphasizes how in project design, the sorts of appraisals conducted are dictated by speed – all development agencies are in the business of moving money quickly – and how rapid rural appraisals often generate information without local input and on the basis of ignorance of local conditions. In a curious way Goldman has less to say on the institutional and discursive greening of the Bank as such. Wade (1997) has documented the stages through which the environmental division was established, from frontier economics before 1987, to 'environmental protection' in the early 1990s, to 'comprehensive environmental management' thereafter, and how this was driven by NGO pressure. In turn McAfee (1999) has shown that this greening was flawed in at least two respects: analytically, it produced 'ecological economics' and what she calls 'green developmentalism', which is to say a discourse entirely compatible with neoclassical economics; and empirically, it created a division in which limited resources and a lack of project accountability rendered the environmental monitoring of projects a relatively weak reed.

The enthnographic turn in development studies, then, has opened up new avenues for cultural and geographical analysis. But the work of Pigg (1992) in Nepal, and Schroeder (1999) in West Africa, to take two examples, reveals that the project implementation stage of development poses knotty questions for how we think about the transmission and reception of ideas, and how and whether something like hegemony adequately captures the relation between peasant client and project manager. Goldman's study identified a community of experts whose authority is totally unquestioned, whose openness to other voices is limited, and whose ideas and models of development are entirely self-referential, but it is still not entirely clear why this community does possess such immunity and authorized power.

ALTERITY

[P]ostcolonial settings provide the rationale for the idea of alternative modernities ... where incommensurable conceptions and ways of life implode into one another, scattering rather than fusing, into strangely contradictory yet eminently 'sensible' hybridities'. (Akhil Gupta, 1998)

It is perhaps a sign of our times that any discussion of *reimagining* development or development alternatives in the 1990s begins with the word, with language and with discourse. And from there it is a very short step to the 'idea' of poverty, to the 'invention' and social construction of development. Alternatives, like everything else, can be imagined at will, but alternatives must be built *with*, not on, the ruins of capitalist modernization. One of the productive ways in which alternatives have been explored is through local 'development' knowledge, often in realms seemingly dominated by scientific agronomy or metallurgy or public health (Baker, 2000; Broch-Due and Schroeder, 2000). Indigenous technical knowledge (ITK) has been widely explored (and there are a number of international organizations devoted to its generation, propagation and use) and now widely understood within academic and activist circles (Richards, 1985). Perhaps the best political ecological study to address the question of peasant experimentation and practice (and the threats which this world confronts) is Zimmerer's book *Changing Fortunes* (1996), which examines biodiversity and peasant livelihoods in the Peruvian Andes.

In problematizing environmental knowledges, political ecology has identified a number of core issues: first, it must be recognized that environmental knowledge is unevenly distributed *within* local societies; second, it is not necessarily right or best just because it exists (i.e. it can be often wrong or inappropriate); and, third, traditional or indigenous knowledge may often be of relatively recent invention (which is to say these knowledges are not static or stable but, as Paul Richards, 1985, suggests, may be predicated on forms of ongoing experimentation). ITK may not be indigenous as such but rather is *hybrid* (Agrawal, 1999). Most knowledges are not simply local but cosmopolitan, complex 'compounds' drawing upon all manner of circulating knowledges: farmers in India may simultaneously employ concepts from humoral theories and modern green revolution technologies (Gupta, 1998). Gupta makes the important point that the hybrid qualities only are made visible by an emphasis on practice rather than on a narrow construal of knowledge. At the same time, ITK can also take on mystical and ideological forms, as in Vandana Shiva's (1989) account of Indian women as 'natural' peasant scientists (see Jackson, 1997). In so far as local actors know a great deal about local ecology and this knowing is typically culturally 'institutionalized' and 'embedded' in a variety of persons, offices, rituals and customary practices, the unanswered questions are: why has this knowledge been so difficult to legitimate, under what circumstances can such knowledge/practice be institutionalized without cooption or subversion, and how might it be systematized in some way?

The community, and the cultural community in particular, looms large in the discussion of development alternatives: communities that resist or stand at the edge of the hegemonic system, or communities defined by their purported autonomy or form of knowledge and practice, or communities defined increasingly by political articulations and identifications that are at odds with the developmental nation (see Corbridge and Harriss, 2000). But the community turns out to be – along with its lexical affines, namely tradition, custom and indigeneity – a keyword whose meaning is not easy to divine (Clarke, 2000). The community is important because it is typically seen as: a locus of *knowledge*; a site of *regulation* and management; a source of *identity* and a repository of 'tradition'; the embodiment of various *institutions* (say property rights) which necessarily turn on questions of representation, power, authority, governance and accountability; an object of *state control*; and a theater of *resistance* and struggle (of social movement, and potentially of *alternative visions of development*). It contains, in other words, different sorts of imaginaries. It is often invoked as a unity, as an undifferentiated entity with intrinsic powers, which speaks with a single voice to the state, to transnational NGOs or the World Court. Communities, of course, are nothing of the sort (Zerner, 2000).

One of the problems is that the community expresses quite different sorts of social relations and forms: from a nomadic band to a sedentary village to a confederation of Indians to a transnational virtual network ('netwar' as the Rand Corporation dubs it). It is usually assumed to be the natural embodiment of 'the local' – configurations of households, lineages, longhouses – which has some territorial control over resources which are historically and culturally constructed in distinctively local ways. A community, then, typically involves a territorialization of history ('this is our land and resources which can be traced in relation to these founding events') and a naturalized history ('history becomes the history of my people and not of our relations to others'). Communities fabricate, and refabricate through their unique histories, the claims which they take to be naturally and self-evidently their own. This is why communities have to be

understood in terms of hegemonies: not everyone participates or benefits equally in the construction and reproduction of communities, or from the claims made in the name of community interest. And this is exactly what is at stake in the current work on indigenous peoples and other movements such as the infamous tree-hugging or Chipko movement in north India (Rangan, 2000; Sinha et al., 1997) or the Ogoni movement in Nigeria (Watts, 2000).

Far from the mythic community of tree-hugging, unified, undifferentiated women articulating alternative subaltern knowledges for an alternative development – forest protection and conservation by women in defense of customary rights against timber extraction – we have three or four Chipkos each standing in quite different relationship to development, modernity, alterity, the state and local management. It was a movement with a long history of market involvement, of links to other political organizing in Garawhal, and with aspirations for regional autonomy. Watts (2000) shows how Saro-Wiwa created a particular Ogoni identity against the Shell oil company and the Nigerian state to create a strikingly modernist movement for whom a goal was more fruits of modernity, not fewer. Tsing (1999) and others (see Zerner, 2000) have shown how indigenous green development movements may create customary knowledges and practice as the currency with which development monies may be parlayed. Tradition or custom hardly captures what is at stake in the definition of the community or indigenous alternatives.

There are a number of implications that stem from an examination of the role of community and indigeneity in development alternatives. First, and most obviously, the forms of community regulations and access to resources are invariably wrapped up with questions of identity (Watts, 2000). Second, these forms of identity (articulated in the name of custom and tradition) are not stable (their histories are often quite shallow), and may be put to use (they are interpreted and contested) by particular constituencies with particular interests. Third, images of the community, whether articulated locally or nationally, can be put into service as a way of talking about, debating and contesting various alternatives to development or indeed for development. Fourth, to the extent that communities can be understood as differing fields of power – communities are internally differentiated in complex political, social and economic ways – then to that same extent we need to be sensitive to the internal political forms of resource use or conservation (there may be three or four different Chipkos or Ogoni movements within this purportedly coherent community struggle). Fifth, communities are rarely corporate or isolated, which means that the fields of power are typically non-local in some way (ecotourism working through local chiefs, local elites in the pay of the state or local logging companies, and so on). And not least, the community – as an object of social scientific analysis or of practical politics – has to be rendered politically; it needs to be understood in ethnographic terms as consisting of multiple and contradictory constituencies and alliances.

Here I might refer to the excellent work of Brosius (1997) and Li (1996; 1999) in Indonesia, who conducted comparative community work in two seemingly similar local communities to show how the type and fact of resistance varied dramatically between the two communities which were in many respects identical 'cultural' communities, and how these differences turn on a combination of contingent but nonetheless important historical events. Brosius found that the radical differences in resistance to logging companies between two communities turned on their histories with respect to colonial forces, their internal social structure, their autonomy and closed, corporate structure, and the role of transnational forces (environmentalists in particular). The point is that some communities do not resist (which disappoints the foreign or local academic) and may not have, or have any interest in, local knowledge. By the same token local 'traditions' can be discovered (not necessarily by the community and often driven by academic work on local traditions drawn from elsewhere) which can be put to the service of the new political circumstances in which villages and states find themselves (Zerner, 1994). Indeed, we know that some groups within communities are happy to take on board essentialism and wrong-headed 'local traditions' peddled by foreign activists or investors, in order to further local struggles. What is striking in the critical work on development is the great scarcity of work on alternative movements, projects and communities which brings the same ferocious sensitivity to questions of power, transparency, accountability, rights and so on that is the hallmark of its scrutiny of formal development institutions.

Indigeneity – whether indigenous development or indigenous alternatives – is a particularly complicated word in development and anti-development discourse. It obviously carries the baggage of authenticity and tradition, yet Gupta has reminded us that it is 'the desire for the indigenous that enables the West to construct its own identity through alterity' (1998: 239).[6] He argues that deployment of the term can only be redeemed if it is recognized that indigeneity is 'an invented space of authenticity'. But does the indigenous always resist the hegemonic, and how does it stand in relation to the modern? The now famous case of the Zapatista in Mexico reveals something on both counts: Subcommandante Marcos may talk the language of indigeneity (in a strikingly postmodern guise!) but it is well to recall that Chiapas was unthinkable outside the democratic processes unfurled by the slaughter of Mexican students in Tlateloco Square.

In his massive biography of Mexico, Enrique Krause notes that 1968 was 'both the high point of authoritarian power and the beginning of its collapse' (1999: 736). Second, the genesis of the Chiapas rebellion must be traced to the maelstrom of the 1960s, throwing together the church, Indian movements and left activism. The long fuse of the Zapatista Front was ignited by Bishop Ruiz and the Catechist 'Apostles' movement (liberated by the Medellin episcopal assembly of 1968), by Maoist insurgents in Monterrey and Chihuahua (established in the late 1960s) who helped form SLOP (with Ruiz) and Unión de Uniones/Asociación Rural de Interés Colectivo (ARIC), and of course by the burgeoning of Indian movements brought together in the 1974 Indian Conference. The trail from the Armed Forces of National Liberation to the Zapatista Army of National Liberation can, and must be, traced to the late 1960s even if, as Krause (1999) rightly shows, it was the period between 1983 and 1989, when the Diocese, the Zapatistas (EZLN), SLOP and ARIC worked together, that proved to be the revolutionary crucible in which the events of 1994 matured and ultimately combusted. Chiapas appears, in other words, as a case of stunning hybridity and of complex interscalar processes and territorialities that, to return to Gupta (1998), allows for the play of difference. This story could moreover be replicated in a number of circumstances but they are all of special significance to the cultural geographer because they highlight the cultural politics of territoriality, and the conflicts between different spatialities and representational politics (see Bryan, 2001; Brysk, 2000).

Indigeneity, identity and community are, of course, the very stuff of nation-building, and it is one of the strengths of the cultural geographical work on development that it has compelled us to rethink the relations between nation and development. The development project was, and has always been, as much about the task of putting reason to the service of the nation as the economy. Agricultural policies in India, for example, spoke powerfully to the need to construct an imagined community in the aftermath of a multicultural colonial creation called independent India (Anderson, 1998; Chatterjee, 1993; Gupta, 1998). Nation-building has of course proven to be a violent and contradictory process, no more so than in the context of global forces and flows that have created and crated territories (and territorial identities) at odds with the very idea of a national territory and identity (Ignatieff, 1994; 1999; Watts, 2000). For this reason some of the most important cultural geography work on development precisely occurs in the borderlands between development and nation, development and citizen, development and rights (Bernal, 1997; Mohan, 2000; Radcliffe, 1999; Schech and Haggis, 2000; Watts, 1998).

PLURAL MODERN? HYBRIDS AND COSMOPOLITANS

A cultural theory directs one to examine how 'the pull of sameness and the forces of making difference' interact in specific ways under the exigencies of history and politics to produce alternative modernities at different national and cultural sites. (Dilip Gaonkar, 2001)

If development is a resolutely dialectical process – a self-representation of modernity that refers to the non-developed other, and in turn travels to and is indigenized by the local other, in a way that may come to shape, indeed destabilize, that very self-representation (!) – then it is no surprise that cultural geographers have come to see development on the ground as a sort of mixing, syncretism and cross-fertilization rather than a crude mimicry or replication. Hybridity is the *nom de rigueur* (Bhabha, 1989; Canclini, 1996; Ferguson, 1999; Gilroy, 1993). The hybrid annexes a particular intellectual territory which sees postcolonial settings as borderlands and spaces of marginality, replete with instabilities for the West that emerge in Said's (2000) view from the exile and displacements of a global cultural and political economy. Gupta's (1998) account of agronomic knowledges blending humoral and scientific practices – a lived unity of incommensurability – becomes constitutive in fact of what it means to be postcolonial. As he says:

It is this unobtrusive intermingling and coexistence of incommensurable beliefs that makes it impossible to position ... peasants as occupying a space of pure difference ... As hybridized, syncretic, inappropriate postcolonials they enter a disturbing presence that continuously interrupts the redemptive narratives of the West. (1998: 232)

The ceaseless traffic in translation and mimicry (Bhabha, 1989) not only unsettles spatialized notions of culture, nation and so on, but posits the existence of what one might call non-Kantian forms of cosmpolitanism. Cosmopolitanism, with its obvious reference to the European experience, is now deployed to identify both precolonial or premodern cosmopolitanism outside the west – for example the Asia-wide circulation of Sanskrit poetry in the first millennium – and contemporary non-western cosmopolitan capitalism – for example Diouf's (2000) account of the global religio-economic network of the rural Mourides. Cosmopolitanism here refers both to the sense of an enthusiasm for customary difference (against a unified polychromatic culture) and to some sense of global citizenship (Brennan, 2001). This is the cosmopolitanism not of universality, rationality and progress but of the *victims* of modernity, what Pollock et al. call 'minoritarian modernity' (2000: 582). With good reason one might think that this is a terrifying case of turning adversity into advantage.

The role of the nation-state in these articulations is often ambiguous but they obviously speak simultaneously both to globalized forms of governance outside what Brennan calls 'manageable nations' and to a recognition of local development that is, as Diouf notes, inexplicable outside globalization. Here some of the most intriguing work focuses not so much on the authoritative knowledges of global institutions, as on the discursive power of particular ideas that travel across space through transnational networks, be they international NGOs, donors or global advocacy networks. They assert the staying power of the local in the face of global development flows that are not immutable and reified, but may be worked around and against. Thayer (2001) provides an extraordinary account of a rural women's landless movement in the arid Brazilian north-east, illustrating the double movement of global forces which can simultaneously exclude economically and yet empower symbolically or discursively. The Brazilian *sertão* could easily be yet another sad case of Castells' (1998) 'black holes' within the global network society. But Thayer shows how rural women workers are firmly embedded in global flows and are able to use transnational movements and resources – the work of Joan Scott in particular! – to facilitate new kinds of mobilization that are at once hybrid and networked. Rather than erase the local or marginalize a 'marginal' movement, the transnational flows of ideas and resources helped detonate a 'subaltern public'.[7]

Does all of this work add up to, or confirm, the notion of alternative or plural modernities? And what might it imply for development? Doubtless there is a sort of ineluctable logic that has led from the posing of development as a form of modernity to the recognition that in a world of globalization and a global development industry, what is at stake is the making of alternative modernities. But to make such a claim is on its face far removed from some of the post-development work. To think of alternative modernities is to admit modernity is inescapable and, as Gaonkar (2001: 1) says, to desist from speculations of its end. In the same way, when Diouf (2000) says that the cosmopolitanism of the rural Senegalese Mourides traders who now operate sophisticated global economic networks in Turin, Paris and New York is not simply informed by the western trajectory of modernity alone, he has done nothing but affirm the powers of non-western *capitalist* accumulation (indeed his article reads like African nationalism meeting the chamber of commerce!). All of which is to say that the powers of capitalist modernity are undiminished. Could not Diouf be read as a variant of Weber's Protestant ethic ('the Muslim ethic and the rise of capitalism')?

Chakrabarty at the end of his 'postcolonial' *Provincializing Europe* (2000) says his task has been to create conjoined and disjunctive genealogies for European categories of political modernity, to keep in tension the necessary dialogue between the universal history of capital ('History I') and the diverse ways of being human ('History II'). But what exactly has this distinction yielded as regards the prospects for the on-the-ground hegemony of development practice? Not much, I fear. In the same way Gibson-Graham and Ruccio (2001), in a critical 'poststructural' rereading of my own work, desperately search for some spheres in which non-capitalism resides untouched by the overwhelming powers of capitalism (as they interpret my analysis). Whether their reading is right or wrong – I would say for the record that certain threads of Marxism have always been concerned with diversity and the spaces opened within capitalism for non-capitalist production, and their interpretation of *my* Marxism is totally wrong-headed – matters less than the fact that they do not to seek to abandon progress and development but rather try to understand the possibilities for non-exploitative and fair forms of produced wealth, which sounds like most of left-of-center conventional development theory.

It is also striking that the cultural creativity and the 'creative adaptation' (Gaonkar, 2001) that are emphasized in this work bear the hallmark of great familiarity. In Gibson-Graham and Ruccio's reading of my Gambia work, the women's work gangs and the struggles over the conjugal contract are (cultural specificity notwithstanding) very *familiar* stories. After reading Chakrabarty's account of Calcutta, it is not the sense of hybridity or the difference that remains, but the extraordinary resonances with Marshall Berman's account of (western) modernity. One wonders whether the renarrativizing and the hybridities and the cosmopolitan capitalisms can really substantiate a claim that the 'not-quite-modern' disrupts the complacent march of progress (Gupta, 1998: 233) or unsettles 'the representational efficacy of the relations of global inequality' (1998: 231).

In some quarters, then, the cultural geography of development has, along with much post-colonial and poststructural theory, come to reluctantly admit the universality of capitalist modernity. There is, it must be said, an obvious tension between those who stand at a critical angle to western Enlightenment and who trumpet *grassroots postmodernism*, and those who in acknowledging the inescapability of the modern invoke a multiplicity of *other modernities*. In the former there is a danger of the worst of populist myopia. In the latter there is the vain hope that in the renarrativization of western hegemony (the discovery of alternative modernities) resides a sort of civilizational parity. Both tend to occlude the terrible realities of unprecedented global economic inequality and the crude violence of twenty-first-century empire. A 1950s modernization theorist might well offer a wry smile.

NOTES

1   The point being that 50 years of the development business has witnessed very little in terms of the radical improvement of poverty and material circumstances of the poor, and that development has little to offer in the face of growing economic inequality and ecological misery. Obviously a long line of radical Marxist development theory preceded this critical turn in the 1980s, stretching back to Marx himself, to the early theorists of imperialism (Lenin, Luxemburg, Trotsky) and their twentieth-century followers (Bill Warren) and subsequently to Third World theorists of underdevelopment (the Latin American dependistas, the French Marxists such as Claude Meillassoux and Pierre-Philippe Rey, the Indian mode of production debates, and the grand theory of Samir Amin). For a review see Richard Peet (2000).

2   'It looks as if economic growth may not merely fail to solve social and political difficulties; certain types of economic growth may actually cause them' (Seers, 1969).

3   In this chapter I have not referred to the relevant work on feminism that is covered by McEwan in Chapter 21.

4   In his new book *Seeing Like a State* (1998), James Scott provides a complementary critique, seeing in 'high modernism' the desire to link the high ideals of

developmentalism with the needs of states to be legible. The result is the crushing of all sense of popular creativity and the loss of metis, the sorts of subaltern practice that might hold out alternative models of human advancement.

5   There is strong continuity here between Escobar and the cultural emphasis in the work of Kothari (1988) (and his stress on alternative modes of thought) and in Ashis Nandy's (1987) work on 'cultural frames' and critical traditionalism.

6   Kingsbury (2000) has shown how the contested nature of indigeneity and community has a counterpoint in international law. The UN, the ILO and the World Bank have, as he shows, differing approaches to the definition of indigenous peoples. The complexity of legal debate raised around the category is reflected in the vast panoply of national, international and inter state institutional mechanisms deployed, and the on-going debates over the three key criteria of non-dominance, special connections with land/territory, and continuity based on historical priority. These criteria obviously strike to the heart of the community debate, and carry the additional problems of the normative claims which stem from them (rights of indigenous peoples, rights of individual members of such groups, and the duties and obligations of states). Whatever the current institutional problems of dealing with the claims of non-state groups at the international level (and there are knotty legal problems, as Kingsbury demonstrates), the very fact of the complexity of issues surrounding 'the indigenous community' makes for at the very least what Kingsbury calls 'a flexible approach to definition', and at worst a litigious nightmare.

7   This is a particularly interesting case because we know that in other settings and around other issues – for example transnational green networks (Keck, 1995) – the outcome was quite different. Brosius (1997) shows how activists can be guided by self-centered interests in program building that rest on mis-leading stereotypes of the community, just as Tsing (1999) decuments the ways in which Meratu commu-nity leaders play to a 'fantasy' of tribal green wisdom to mobilize international attention (see Zerner, 1994; 2000). A number of the large transnational NGOs (TNGOs) have themselves been shaped by the chang-ing political and market-driven winds in the west, pro-ducing a sort of in-house corporate environmentalism ('green corporatism') within the larger TNGO com-munity (Bailey, 2000). This itself raises the question of how large TNGOs as major donors change the domestic politics and structure of the local NGO com-munities in the south, and how foreign and local NGOs actually build political strategy and alliances. In these cases, transnational NGO networks produced something much more akin to Ferguson's 'anti-politics machine', but worse. Here the likes of the World Wildlife Fund and their 'green corporatism' were able to impose an agenda on weak local green social green social movements and NGOs, and indeed they weak-ened such movements by splitting and dividing a fragile coalition of activists.

REFERENCES

Achebe, C. (1988) *Anthills of the Savannas*. New York: Vintage.

Agrawal, A. (1999) *Greener Pastures*. Durham: Duke University Press.

Anderson, B. (1998) 'Nationalism, identity and the world-in-motion', in P. Cheah and B. Robbins (eds) *Cosmopolitics*. Minneapolis: University of Minnesota Press.

Anderson, P. (2000) 'Renewals', *New Left Review* 1: 1–24.

Arndt, W. (1987) *Economic Development*. Chicago: University of Chicago Press.

Bailey, J. (2000) 'Green corporatism'. PhD dissertation, University of California, Berkeley.

Baker, K. (2000) *Indigenous Land Management in West Africa*. London: Oxford University Press.

Benjamin, W. (1969) *Illuminations*. New York: Schoken.

Benton, L. (1996) 'The greening of free trade?', *Environment and Planning A* 28: 2155–77.

Berman, M. (1982) *All That Is Solid Melts into Air: The Experience of Modernity*. New York: Penguin.

Bernal, V. (1997) 'Islam, transnational culture and modernity in rural Sudan', in M. Grosz-Ngate and O. Kokole (eds) *Gendered Encounters*. New York: Routledge. pp. 131–51.

Bhabha, H. (1989) *The Location of Culture*. London: Routledge.

Booth, David (ed.) (1994) *Rethinking Social Development*. London: Methuen.

Brass, T. (2000) *Peasants and Postmodernism*. London: Cass.

Brennan, T. (2001) 'Cosmopolitan versus international', *New Left Review* 7: 75–84.

Broch-Due, V. and Schroeder, R. (2000) *Producing Nature and Poverty in Africa*. New Brunswick: Transactions.

Brosius, P. (1997) 'Prior transcripts, divergent paths', *Comparative Studies in Society and History* 39 (3): 468–510.

Bryan, J. (2001) 'By reason or force: territory, state power and Mapuche land rights in southern Chile'. MA thesis, Department of Geography, University of California, Berkeley.

Brysk, A. (2000) *From Tribal Village to Global Village*. Palo Alto: Stanford University Press.

Burawoy, M. et al. (1995) *Ethnography Unbound*. Berkeley: University of California Berkeley.

Burawoy, M. et al. (2000) *Global Ethnography*. Berkeley: University of California Press.

Canclini, N. (1996) *Hybrid Cultures*. Minneapolis: University of Minnesota Press.

Castells, M. (1998) *The Network Society*. Oxford: Blackwell.

Chakrabarty, D. (2000) *Provincializing Europe*. Princeton: Princeton University Press.

Chakrabarty, D. (2001) 'Adda, Calcutta: dwelling in modernity', in D. Daonkar (ed.) *Alternative Modernities*. Durham: Duke University Press. pp. 123–64.

Chaterjee, P. (1993) *The Nation and its Fragments*. Princeton: Princeton University Press.

Clarke, C. (2000) *Class, Ethnicity and Community in Southern Mexico*. London: Oxford University Press.

Cooper, F. and Packard, R. (eds) 1997 *International Development and the Social Sciences*. Berkeley: University of California Press.

Cooper, F. and Stoler, A.L. (eds) (1995*) Tensions of Empire: Colonial Cultures in a Bourgeois World*. Berkeley: University of California Press.

Corbridge, S. (1993) 'Ethics in development studies', in F. Schurmann (ed.) *Beyond the Impasse*. London: Zed. pp. 123–39.

Corbridge, S. (1998) 'Beneath the pavement only soil', *Journal of Development Studies*, 24: 138–48.

Corbridge, S. and Harris, J. (2000) *Reinventing India*. London: Polity.

Cowen, Michael and Shenton, R. (1996) *Doctrines of Development*. London: Routledge.

Crush, J. (ed.) (1995) *Power of Development*. London: Routledge.

Dallmayr, F. (1996) 'Global development?', *Alternatives* 21: 259–82.

Davis, D. (2000) 'Overgrazing the range'. PhD dissertation, University of California, Berkeley.

Diouf, M. (2000) 'The Senegalese Murid trade diaspora and the making of a vernacular cosmopolitanism', *Public Culture* 12 (3): 679–702.

Escobar, A. (1992) 'Imagining a post-development era? Critical thought, development and social movements', *Social Text* 31 (32): 20–56.

Escobar, A. (1995) *Encountering Development*. Princeton: Princeton University Press.

Esteva, G. and Prakhash, M. (1998) *Grassroots Postmodernism*. London: Zed.

Ferguson, J. (1990) *The Anti-Politics Machine*. Cambridge: Cambridge University Press.

Ferguson, J. (1999) *Expectations of Modernity: Myths and Meanings of Urban Life in the Zambian Copperbelt*. Berkeley: University of California Press.

Foucault, M. (1982) 'The subject and power', in H. Dreyfus (ed.) *Beyond Structuralism and Hermeneutics*. Chicago: University of Chicago Press.

Gaonkar, D. (ed.) (2001) *Alternative Modernities*. Durham, NC: Duke University Press.

Gardner, K. and Lewis, D. (2000) 'Dominant paradigms or "business as usual"?', *Critique of Anthropology* 20: 15–29.

Gellner, E. (1979) *Spectacles and Predicaments*. Cambridge: Cambridge University Press.

Gibson-Graham, K. and Ruccio, D. (2001) 'After Development', in J. Gibson-Graham, S. Resnick and R. Wolff (eds) *Re/Presenting Class*. Durham, NC: Duke University Press. pp. 158–81.

Gilroy, Paul (1993) *The Black Atlantic: Modernity and Double-Consciousness*. Cambridge, MA: Harvard University Press.

Glacken, C. (1967) *Traces on the Rhodesian Shore*. Berkeley: University of California Press.

Goldman, M. (2001) 'The birth of a discipline', *Ethnography* 2 (2): 191–219.

Gore, C. (2000) 'The rise and fall of Washington consensus as a paradigm for the developing countries', *World Development* 28 (5): 789–804.

Greene, R. (1999) *Malthusian Worlds*. Boulder: Westview.

Gupta, A. (1998) *Postcolonial Development*. Durham, NC: Duke University Press.

Guthman, J. (1997) 'Representing the crisis', *Development and Change* 28: 45–69.

Haas, P. (1990) *Saving the Mediterranean*. New York: Columbia University Press.

Hall, P. (1989) *The Political Power of Economic Ideas*. Princeton: Princeton University Press.

Hall, S. (1996) 'When was the post-colonial?', in I. Chambers and L. Curti (eds) *The Post-Colonial Question*. London: Routledge. pp. 242–60.

Harvey, D. (1996) *Justice, Nature and the Geography of Difference*. Oxford: Blackwell.

Hausmann, R. (2001) 'Prisoners of geography', *Foreign Policy* January: 44–56.

Hefner, R. (ed.) (1998) *Market Cultures*. London: Allen and Unwin.

Herbert, S. (2000) 'For ethnography', *Progress in Human Geography* 24 (4): 550–68.

Herbst, J. (2000) *States and Power in Africa*. Princeton: Princeton University Press.

Hirschmann, A. (1968) *Development Projects Observed*. Washington: Brookings Institution.

Hirschmann, A. (1986) *Rival Views of Market Society*. London: Harvard University Press.

HRC (1999) *Leave None To Tell the Story*. Washington: Human Rights Watch.

Hyden, G. (1980) *Beyond Ujamaa*. Berkeley: University of California Press,

Ignatieff, M. (1994) *Blood and Belonging*. New York: Norton.

Ignatieff, M. (1999) 'Nationalism and the narcissism of minor difference', in R. Beiner (ed.) *Theorizing Nationalism*. Albany: State University of New York Press. pp. 91–102.

Illich, I. (1971) *Celebration of Awareness*. London: Boyars.

Jackson, S. (1997) 'Postpoverty, gender and development', *IDS Bulletin* 28 (3): 145–53.

Joseph, G. and Nugent, D. (eds) (1994) *Everyday Forms of State Formation*. Durham: Duke University Press.

Keck, M. (1995) 'Social equity and environmental politics in Brazil', *Comparative Politics* 27 (4): 409–24.

Kingsbury, B. (2000) 'The international concept of indigenous peoples in Asia', in D. Bell and J. Bauer (eds) *Human Rights and Economic Development in East Asia*. Oxford: Oxford University Press.

Kothari, R. (1988) *Rethinking Development: In Search of Humane Alternatives*. Delhi: Ajanta.

Krause, E. (1999) 'Chiapas: the Indians' prophet', *New York Review of Books*, 16 December, 46 (20): 65–73.

Landes, D. (1999) *The Wealth and Poverty of Nations*. New York: Knopf.

Leach, M. and Mearns, R. (1996) 'Challenging the received wisdom in Africa', in J. Fairhead and M. Leach (eds) *The Lie of the Land*. London: Curry.

Lehmann, D. (1997) 'An opportunity lost', *Journal of Development Studies* 33: 568–78.

Li, T. (1996) 'Images of community', *Development and Change* 27: 501–27.

Li, T. (1999) *Transforming the Indonesian Uplands.* London: Harwood.

Long, N. and Long, A.A. (eds) (1993) *Battlefields of Knowledge.* London: Routledge.

Mackenzie, F. (1997) *Land, Ecology and Resistance In Kenya.* London: International of African Institute Press.

Mamdani, M. (1995) *Citizen and Subject.* Princeton: Princeton University Press.

Mbembe, A. (1992) 'Notes on the post colony', *Africa* 62 (1): 3–37.

McAfee, K. (1999) 'Selling nature to save it?', *Society and Space* 7 (2): 155–74.

Mills, M.B. (1999) *Thai Women in the Global Labor Force: Consuming Desires, Contested Selves.* New Brunswick: Rutgers University Press.

Mitchell, T. (1995) 'The object of development', in J. Crush (ed.) *Power of Development.* London: Routledge. pp. 129–57.

Mitchell, T. (2001) *Out of Egypt.* Berkeley: University of California Press.

Mohan, G. (2000) 'Postcolonialism, development geography and African politics'. Paper presented to the AAG, New York.

Moore, S. (2001) 'The international production of authoritative knowledge', *Ethnography* 2 (2): 161–90.

Nandy, A. (1987) *Traditions, Tyranny and Utopia: Essays in the Politics of Awareness.* Oxford: Oxford University Press.

Neumann, R. (1999) *Imposing Wilderness.* Berkeley: University of California Press.

Parajuli, P. (1991) 'Power and knowledge in development discourse', *International Social Science Journal* 127: 173–90.

Peet, R. (2000) *Theories of Development.* New York: Guilford.

Peet, R. and Watts, M. (eds) (1996) *Liberation Ecologies.* London: Routledge.

Perkins, D. (1997) *Geopolitics and the Green Revolution.* London: Oxford University Press.

Pieterse, J. (1996) 'My paradigm or yours? Alternative development, post development, reflexive development'. The Hague, Institute of Social Studies, Working Paper 229.

Pigg, Stacy Leigh (1992) 'Inventing social categories through place: social representations and development in Nepal', *Comparative Studies in Society and History* 34 (3): 491–513.

Pollock, S., Bhabha, H., Breckenridge, C. and Chakrabarty, D. (2000) 'Cosmopolitanisms', *Public Culture* 12 (3): 577–90.

Putnam, R. (2000) *Bowling Alone.* New York: Simon and Schuster.

Radcliffe, S. (1999) 'Popular and state discourses of power', in D. Massey, J. Allen and P. Sarre (eds) *Human Geography Today.* London: Polity. pp. 219–42.

Rahnema, M. (ed.) (1997) *The Postdevelopment Reader.* London: Zed.

Rangan, P. (2000) *Of Myths and Movements.* London: Verso.

Richards, P. (1985) *Indigenous Agricultural Revolution.* London: Hutchinson.

Rist, G. (1997) *The History of Development.* London: Zed.

Rofel, L. (1999) *Other Modernities: Gendered Yearnings in China after Socialism.* Berkeley: University of California Press.

Rosen, G. (1985) *Western Economists and Eastern Societies.* Baltimore: Johns Hopkins University Press.

Sachs, J. (2001) 'The geography of economic development'. Occasional Paper no 1, United States Naval War College, Washington, DC.

Sachs, J. et al. (2000) 'The geography of poverty', *Scientific American* March: 70–5.

Sachs, W. (ed.) (1992) *The Development Dictionary.* London: Zed.

Said, E. (2000) *Reflections on Exile.* Cambridge, MA: Harvard University Press.

Schech, S. and Haggis, J. (2000) *Culture and Development.* London: Blackwell.

Schroeder, R. (1999) *Shady Acts.* Berkeley: University of California Press.

Schurman, F. (ed.) (1993) *Beyond the Impasse.* London: Zed.

Scott, D. (1995) 'Colonial governmentality', *Social Text* 43: 191–220.

Scott, J. (1998) *Seeing Like a State.* New Haven: Yale University Press.

Seers, D. (1969) *The Meaning of Development.* IDS Working paper. University of Sussex.

Sen, A. (2000) *Freedom as Development.* New York: Knopf.

Sharpless, J. (1997) 'Population science, private foundation and development aid', in F. Cooper and R. Packard (eds) *International Development and the Social Sciences.* Berkeley: University of California Press. pp. 176–202.

Shiva, V. (1989) *Staying Alive.* London: Zed.

Sinha, S. et al. (1997) 'The new traditionalist discourse of Indian environmentalism', *Journal of Peasant Studies* 24 (3): 65–99.

Slater, C. (1994) *The Dance of the Dolphin.* Chicago: University of Chicago Press.

Slater, D. (1993) 'The political meanings of development', in F. Schurmann (ed.) *Beyond the Impasse.* London: Zed. pp. 93–112.

Steel, J. (1997) *Guns, Germs and Steel.* London: Harper.

Stiglitz, J. (2000) 'What I learned at the world economic crisis', *New Republic*, 17 April: 12–17.

Swift, J. (1997) 'Narratives, winners and losers', in M. Leach and R. Mearns (eds) *The Lie of the Land.* London: Currey. pp. 73–90.

Taylor, C. (2001) 'Two theories of modernity', in D. Gaonkar (ed.) *Alternative Modernities.* Durham: Duke University Press. pp. 172–96.

Taylor, P. (1996) 'Re/constructing socio-ecologies: system dynamics modelling of nomadic pastoralists in sub-Saharan Africa', in A. Clarke and J. Fujimura (eds)

The Right Tools for the Job. Princeton: Princeton University Press.

Thayer, M. (2001) 'Transnational feminism', *Ethnography* 2 (2): 243–72.

Thrift, N. (2000) 'Pandora's box? Cultural geographies of economies', in G. Clark, M. Feldman and M. Getler (eds) *The Oxford Handbook of Economic Geography.* London: Oxford University Press. pp. 698–704.

Toye, J. (1985) *The Dilemmas of Development.* Oxford: Blackwell.

Tsing, A. (1999) 'Becoming a tribal elder and other green development fantasies', in T. Li (ed.) *Transformation of the Indonesian Uplands.* London: Harwood. pp. 159–202.

Wade, R. (1996) 'Japan, the World Bank and the art of paradigm maintenance', *New Left Review* 217: 3–36.

Wade, R. (1997) 'Greening the Bank', in R. Kanbur, J. Lewis and R. Webb (eds) *The World Bank.* Washington: Brookings Institution. pp. 611–734.

Wade, R. (2001) 'The US and the World Bank', *New Left Review* 7: 124–37.

Watts, M. (1989) 'Deconstructing determinism', *Antipode* 20 (2): 142–68.

Watts, M. (1998) 'Islamic modernities?', in J. Holston (ed.) *Cities and Citizenship.* Durham: Duke University Press. pp. 67–102.

Watts, M. (2000) *Struggles over Geography.* Heidelberg: University of Heidelberg, Hettner lectures.

Watts, M. and McCarthy, J. (1995) 'Nature as artifice, nature as artifact', in R. Lee and J. Wills (eds) *Geographies of Economies.* London: Arnold. pp. 71–86.

Williams, R. (1961) *The Long Revolution.* London: Chatto.

Williams, R. (1973) *Keywords.* Oxford: Oxford University Press.

Woolcock, M. and Narayan, D. (2000) 'Social capital', *World Bank Research Observer* 15: 225–49.

World Bank (2000) *World Development Report.* New York: Oxford University Press.

Zerner, C. (1994) 'Telling stories about biological diversity', in S. Brush (ed.) *Valuing Local Knowledge.* New York: Island. pp. 68–101.

Zerner, C. (ed.) (2000) *People, Plants and Justice.* New York: Columbia University Press.

Zimmerer, K. (1996) *Changing Fortunes.* Berkeley: University of California Press.

Mexico, *Tijuana* 1995 © Alex Webb/Magnum Photos

GEOPOLITICAL CULTURES Edited by Gerard Toal
and John Agnew

Introduction: Political Geographies, Geopolitics and Culture

Gerard Toal and John Agnew

The common-sense meaning of political geography is the study of how politics is informed by geography. For a long time this meant trying to show how the physical features of the earth – the distribution of the continents and oceans, mountain ranges and rivers – affected the ways in which humanity divided the world up into political units such as states and empires and how these units competed with one another for global power and influence. Today the dominant meaning has changed considerably (Agnew et al., 2002). On one side, geography is now understood as including social and economic differences between places without necessarily ascribing these to physical differences. On the other, politics has been broadened to include questions of political identity (how social groups define themselves and their political objectives) and political movements (why did this movement or political party start here and why does it have this or that geographical pattern of support?). Even more fundamentally, 'geography' is itself now thought of as the selection and ranking of certain themes and issues – from the naming of the continents and the division of the world into regions to the identification of certain regions as more or less 'strategically' important – rather than a

set of objective facts beyond dispute (Lewis and Wigen, 1997). In this understanding, knowledge cannot be readily separated from power. Those with power, the ability to command others, are able to define what counts as geography. From this point of view, the meaning of political geography is completely reversed: it now becomes the study of how geography is informed by politics.

POLITICAL GEOGRAPHY AND CULTURE

The famous American cultural geographer Carl Sauer once described political geography as the 'wayward child' of geography. Political geography in the interwar years of the twentieth century was certainly controversial, dominated in many countries like Germany, Italy and Japan by militaristic and nationalistic geopolitical thought devoted to each country's imperial ambitions (Parker, 1998). Whereas cultural geography achieved a certain degree of respectability and intellectual status in the United States from the scholarly output of Sauer and his 'Berkeley School', political geography suffered from the stigma of its association with, on the one hand, militarism and nationalism and, on the other, clear political commitments that ruled it 'unscientific' in the eyes of Sauer and his peers. Indeed, the ascendancy of cultural geography was facilitated by active dissociation from political issues. Cultural geographers showed as little interest in politics as political geographers did in culture. After World War II, if the political geography that developed was studiously disinterested in offering explicit political judgement and aspired to a form of technical objectivity, concentrating for the most part on the description and classification of states and their borders, or on studies of electoral geography, then cultural geography was totally blind to both the political contexts of the cultural issues it studied (the diffusion of technologies, ethnic geographies, and the distribution of place names) and the political implications of what was studied (preferring the world before 1800 so as to avoid contemporary politics altogether) (Smith, 1989). As in so many fields, the Cold War of 1947–89 set narrow limits to the study of politics and to politically informed critique in the United States and elsewhere.

Intellectually, part of the problem lay with how culture was defined (Duncan, 1980). It was seen very much as the effects of 'tradition' or past symbolically powerful strategies for organizing life that lived on into the present, either through intergenerational transfer or as a result of living in places where the 'culture' was everywhere 'in evidence'. Of course, this meant giving little or no attention to relations of power within local groups or seeing how cultures change as a result of conquest, invasion and suppression. The emphasis on the diffusion of traits as the source of cultural change led to a systematic neglect of political relations of domination and subordination within a given culture and in relation to others. Culture existed almost, and ironically, as a state of nature, without the divisions and conflicts associated with politics.

Only with the reinvention of the concept of culture, understood now as a system of signs that give meaning to other activities, has the concept come to have much relevance for political analysis (for example, Jackson, 1989). It does so by drawing attention to the dependence of politics (and political geography) on

prefigured cultural identities and geographical taxonomies. From this point of view, political action, communication and representation are only possible when based on commonly understood cultural signs and symbols. A culturally informed political geography is about identifying and showing the mutually constitutive effects of cultural signs and symbols, on the one hand, and political acts, on the other, in so far as they involve geographical sites, presuppositions and taxonomies.

CULTURAL GEOPOLITICS

The social and intellectual waves that helped transform geography as a whole from the 1970s onwards also washed over the subfield of political geography, renewing its intellectual importance and revitalizing its intellectual core. The renewed political geography that has defined the discipline for the last 20 years is characterized by a variety of perspectives: spatial-analytical approaches that continue the subfield's long-established concern with spatial patterns and forms of political organization (O Loughlin, 1986), political-economic perspectives that address the structural processes of state power and the world economy, and culturalist trajectories of research that explore the multiple entanglements of geopolitics and states with questions of cultural identity, political discourse and the cultural reproduction of everyday life (Agnew, 1997).

As the field of cultural geography was transformed by the cultural turn across the social sciences in the 1980s, the intimate connections of the cultural to the political were made visible and increasingly acknowledged. Cultural geographies are now seen as entwined within political geographies and political geographies are now seen as sustained and transformed by cultural discourses and practices (Godlewska and Smith, 1984). Studies of national boundaries provide a particularly vivid example of the effect of the emphasis on cultural interpretation. If at one time political boundaries were viewed predominantly in a naturalistic light, as produced by natural features and conditions, they have been increasingly seen as the result of processes of cultural definition and negotiation. Anssi Paasi (1996), for example, has been a persuasive advocate

INTRODUCTION

of the position that national boundaries are the outcome of changes in political consciousness and are thus culturally contingent and not fixed features of the global political landscape. Perhaps the most visible element in the intellectual cross-fertilization between new understandings of culture and political geography has been the expropriation of precisely the most militaristic and nationalistic strain in political geography: geopolitics (Ó Tuathail, 1996; Ó Tuathail and Agnew, 1992).

The term *geopolitics* was invented by the Swedish political scientist Rudolf Kjelln in 1899 to refer to the harnessing of geographical knowledge to further the aims of specific national states. If Kjelln was concerned to dispute the claim of Norwegian nationalists (Norway was part of Sweden until 1905) that the mountain spine down Scandinavia constituted a natural boundary between two distinctive peoples by arguing that seas and rivers were much more significant, the term 'geopolitics' came to be applied by German thinkers in the 1920s and 1930s, most notoriously Karl Haushofer, to formal models of great power enmities based on their relative global location and need to establish territorial spheres of influence to feed their urge to expand. Through this formalization, Friedrich Ratzel's idea of states as organic entities came to inform German foreign policy after the Nazi accession to governmental power in Germany in 1933. After World War II the association with the Nazis gave the word 'geopolitics' a negative connotation. Though used informally to refer to the geographical structure of international relations in the 1950s and 1960s, it has only been since the 1970s that the word has re-entered political geography as a key concept.

Classically concerned with the direct impact of physical geography and the relative location of states on international conflict, imperial expansion and interstate war, geopolitics can be seen, in a radically alternative conception, as the deeply social and cultural process by which leaders and ordinary citizens in some of the world's most powerful states make geopolitical sense of the world. Instead of the leadingly deterministic question 'How does the soil or geographic location of a state affect its foreign policy?', a more culturalist question emerged: 'How does the geopolitical culture of a state spatialize world politics and fill it with certain defining dramas and dangers, friends and enemies?' In this latter question the notion of 'geopolitics' as a fixed and given fact of nature is challenged by an understanding of it as a tradition of discourse and debate about territorial nation-state identity and its relationship to the wider world. 'Geopolitics' becomes 'geopolitical culture'. This latter understanding of geopolitics reinterprets seemingly naturalist geopolitical arguments – that, for example, the United States acts the way it does because it is surrounded by two large oceans and separated from the other power centers of the world – as geopolitical traditions of argumentation and discourse (on Russian geopolitical culture see Smith, 1999). From within its perspective, the geopolitical culture of the United States is distinguished by competing traditions of universalism (the United States represents the aspirations of humankind) and isolationism (the United States is a distinct and virtuous sphere and should minimize its dealing with the corruption beyond its borders), both of which cast themselves in abstract philosophic not geographic terms, and are also mediated by a functional geographic illiteracy about the world beyond America's shores (Agnew, 1984). Such a critical geopolitics draws attention to how geographical claims and arguments are used to direct and justify various foreign policy actions and inform environmental and economic development practices.

This approach is not without considerable internal contention. If one tendency involves the relative commitment to postmodernist understandings of geopolitics as a type of discourse, another concerns the degree to which states should retain the centrality of focus. Some scholars remain relatively more committed to modernist theoretical approaches, involving claims about technological and economic processes as 'behind' discursive shifts, whereas others reject the possibility of such ontologies, preferring to focus largely on the discursive constitution of ontological claims (for example, Agnew and Corbridge, 1995, versus Ó Tuathail, 1996). Again, whereas some inhabit a world still largely made up of and by states, others are sceptical of such a singular division of the world, preferring to think in terms of the relative 'balance' between processes at different geographical scales or across actor networks (for example, Sharp, 2000, versus Thrift, 2000).

The chapters in this section showcase cultural geopolitics in all of its richness, featuring scholars who are today making signal contributions to cultural geopolitics beyond the confines of this *Handbook* (Dalby, 2002; Paasi, 1996; Sharp, 2000; Sparke, forthcoming). All of the essays challenge, transgress and transcend the antinomies and oppositions that characterize classic modernist thought: the inside/outside of states, the private/public of gendered political life, the local/national/global of modern scalar relations, and nature/culture, the last being one of the most foundational oppositions of modernism. Events like the 11 September attacks on the Pentagon and the World Trade Center reveal the importance of moving beyond these dichotomies in order to understand our world and the challenges it faces (Luke, 2002).

THE CHAPTERS

The first chapter, 'Boundaries in a Globalizing World' by Anssi Paasi, reviews the traditional political geography concern with borders and boundaries and demonstrates how these notions have been reconceptualized in the wake of the 'cultural turn' across the social sciences. Boundaries, Paasi argues, are dynamic cultural processes that require consideration and investigation of the cultural politics of state formation, national identity construction and everyday processes of spatial socialization. Banal state sponsored practices like waving a flag or studying borders in a classroom serve to support cultural systems of meaning which spatialize the world into a familiar 'here' and a potentially dangerous 'there'. Today, however, the spatial socialization promoted by the state has competition from more diffuse and transnational forms of spatial socialization like MTV, online chat rooms and the latest Hollywood movie. Supposed examples of globalization like wireless telecommunication, global television and the development of the internet – media that touch the lives of only a small minority of the world's population – have inspired considerable hyperbole about the 'deterritorialization' and the 'disappearance of boundaries',

glib rhetoric that Paasi dismisses while acknowledging that the meaning of boundaries is constantly under negotiation and revision. Huge divisions of wealth and affluence still characterize our world, yet these divides are now connected by pervasive concern with international terrorism as the affluent United States confronts war-ravaged and famine-torn states like Afghanistan, Somalia and Iraq.

The chapter 'Gender in a Political and Patriarchal World' by Joanne Sharp explores how the cultural turn has affected the study of geopolitics by giving rise to a critical geopolitics. This approach deconstructs the normalizing and naturalizing practices of geopolitical discourses, the ways in which the world is textually and rhetorically divided and labeled, treating these practices as active geo-graphings of world politics rather than objective descriptions of the realm of necessity in world affairs. Geopolitics is the geographical politics of the pursuit of identity and security by state elites, a geopolitics delimited and circumscribed by everyday gendered distinctions between inside and outside, 'us' and 'them', the private and the public. Sharp's own work highlights the importance of the mundane operation of a 'popular geopolitics' that is embedded in popular culture and mass media, in the pages of *Reader's Digest* and the films of John Wayne. She also discusses the gendered supports of many central conceptions in political life from images of the nation to the motivations of nationalism, conceptions of responsibility and visions of the military, security and violence. For Sharp, the personal is not only political but geopolitical.

The third chapter, 'The Cultural Geography of Scale', explores the ways in which scale is negotiated and reproduced in the political and cultural arena. The particular focus of the chapter is on 'scale-jumping', conceptualized from a Marxist-inspired perspective as the temporary transcendence of tensions and contradictions thrown up by dominant spatial and scalar order. The relationships of the local to the regional, national, international and global, according to Newstead, Reid and Sparke, are generally fixed within a relatively permanent order of scalar relations determined by geographies of political governance and capital accumulation. Breaking free of this order of scalar relations through 'scale-jumping' creates opportunities for particular interests to transcend certain problems and political stalemates. Businesses, for example, can disinvest in certain communities where they are experiencing labor strife and relocate to communities where the power of local labor is very weak. Local social movements, in contrast, can nationalize or internationalize a particular local struggle in order to break stalemated struggles of power in particular places. Scale-jumping creates opportunities for transforming power relations. The chapter considers two different ideological attempts to imagine and promote transnational regions: a Cascadia across the Pacific north-western US–Canada border, and the Caribbean Common Market and Community (CARICOM). In both instances, these efforts are suffused with contradictions and traces of alternative scale-jumping 'imaginations'. The chapter also considers the subaltern scale-jumping of Hispanic migrants traversing the US–Mexican border and the cultural geographic landscape they have created to recognize and memorialize their struggles to transcend the dominant order of scalar relations.

The final chapter in the section develops a critical geopolitical critique of the dichotomies that inform not only orthodox geopolitical thinking but modernity in general and contemporary urban life in particular. The persistent separation and denial of nature at the expense of 'culture', Dalby argues, has created a planetary ecological crisis that requires new ways of thinking and living. Humans do not live in cities but in complex ecological networks that have planetary ramifications and impacts. They are not on planet earth but of planet earth. The history of geopolitics is a history not only of great power imperialism and struggles for primacy but also of ecological imperialism and the triumph of an environmentally unsustainable urban form of life that is polluting the planet and proliferating risks of unknowable consequence. It is this deeper ecological form of imperialist geopolitics that requires exposure and challenge by political and cultural geographers. The crucial geopolitics of the twenty-first century lies not in the world of states but rather in the world that technoscientific modernity has wrought. Already we see evidence of this with a coalition of major states declaring war against global terrorist networks that are exposing the vulnerabilities of technoscientific modernity and seeking to acquire the weapons of mass destruction once monopolized only by states. Dramatic struggles with terrorists are likely to capture media interest but it is the slower structural struggles to address the global energy demands and pollution of technoscientific modernity that will condition the interplay of political geographies, geopolitics and culture in the twenty-first century.

FUTURE ISSUES

Future research in a culturally informed critical geopolitics will inevitably be shaped by the developing culture of the geopolitical world order. The 11 September 2001 terrorist attacks on the World Trade Center and Pentagon and the subsequent delivery of anthrax to the US Capitol appear to have provoked a significant change in that order. The 11 September was a global media event that implicated a worldwide television audience in America's trauma, generating a significant outpouring of sympathy for the United States. The geopolitical culture of the United States appears to have been significantly altered, with its sense of distance and invulnerability from the routine violence of world politics shattered. The attacks were compared to 'Pearl Harbor' and widely accepted as a 'wake-up call' for the United States. In their wake US Secretary of State Colin Powell declared the 'post-Cold War era' over, anticipating a new epoch defined by the US as the world hegemon leading an international coalition of states against terrorist networks with global reach. Events associated with the 11 September have justified passage of 'counter-terrorist' legislation in many states, stimulated new levels of spending on 'homeland defense', and forged a general consensus for aggressive military operations against those deemed to be supporting or aiding 'terrorists'. They have also led to a surprising alliance between the United States and Russia against radical Islamic terrorist networks, with both states actively cooperating to topple the Taliban regime in Afghanistan.

These developments pose intellectual challenges to those working in cultural and critical geopolitics. First, there is a need to grasp the contours and contradictions of the postmodern geopolitical condition, a condition characterized by 'deterritorialized threats' embedded within a 'world risk society' (Beck, 1999; Ó Tuathail, 2000). Contemporary life in advanced technoscientific states is dependent upon deep and pervasive technological systems – financial flows, energy grids, nuclear power stations, transportation networks, water and sewer treatment complexes, commodity markets, medical centers, information technology routers and webs – that are often unstable and vulnerable to predictable yet potentially catastrophic accidents. As is now evident, they are also exceedingly vulnerable to the malign intent of determined non-state actors who can, in utilizing the airline transportation system or the postal service as weapons, pose a general threat to the state and society they target. Powerful states are now justifiably concerned about 'asymmetrical threats' from weapons of mass destruction. Not only can the few threaten the many but relatively marginal groups can inflict catastrophic harm upon the populations of the mightiest and most affluent states. Protecting against such asymmetrical threats is virtually impossible, for technoscientific systems are pervasive and integral to the routine functioning of everyday life in advanced industrial states. Nevertheless, the public effort to securitize these infrastructures is likely to be a permanent feature of life in advanced industrial states.

Second, there is the need to understand how the new hegemonic global discourse of 'counter-terrorism' will work itself out in regional geopolitical contexts where state territorial conflicts persist. How are 'terrorism' and 'counter-terrorism' to be defined in specific geopolitical circumstances? Will the discourse of 'counter-terrorism' function in the same way as anti-communism and containment did during the Cold War, allowing authoritarian leaders and pivotal states to proclaim their enemies 'terrorists' and justify wars as necessary 'counter-terrorism'? The US–Russian alliance against 'Islamic terrorism' is intriguing in this respect, for the Putin administration has represented its war against

the breakaway republic of Chechnya as its own national variation on a global theme. Counter-terrorism becomes an alibi for human rights abuses. Conflicts in the Balkans, Israel–Palestine, Sri Lanka and India–Pakistan have also been characterized by tactical localizations of the global discourse of counter-terrorism. Whether the United States will gradually lose enthusiasm for its global crusade against terrorism as the costs mount is also an open question.

Finally, there is the question of identity and otherness. Identities, as the chapters in this section underscore, are performative cultural understandings about self and other, friends and enemies. One of the consequences of 11 September and the subsequent war against al-Qa'ida and the Taliban in Afghanistan is that it has renewed civilizational and religious storylines about identity and difference (Huntington, 1997). Radical Islam is the outside by which a new community of 'civilized states' is being reimagined and reconstituted. In Europe this storyline has resulted in a diminution of the discourse of Europe versus Russia and a revitalization of the discourse of Europe versus Islam. Russia, the United States and Europe are now ostensibly part of the international coalition of 'civilized states' against radicalized Islam. It is even proposed that Russia might at some future date join NATO, an alliance established to contain its power. In the months following 11 September Italian Prime Minister Silvio Berlusconi suggested that the west was inherently superior to the Islamic world and 'bound to conquer and occidentalize new peoples. It has done it with the Communist world and part of the Islamic world, but unfortunately a part of the Islamic world is 1,400 years behind' (quoted in Ash, 2001). While he subsequently distanced himself from these remarks, Berlusconi nevertheless owes part of his electoral success to his deft use of widespread fears of a 'secret invasion' of Italy by immigrants, particularly Muslims. Other politicians are sounding similar themes, shifting the emphasis of Europe's civilizational border from the east to the south.

These are still early tendencies and only a few of the issues thrown up by the cultural geopolitics of 11 September. Dramatic struggles against spectacular acts of terrorism are always likely to dominate the media, but it is the slower structural tendencies tied to technoscientific modernity – relentless technoscientific change, turbulent capitalist globalization, unstable energy markets, failing states and systems, emerging diseases, proliferating weapons of mass destruction and environmental degradation – and the persistence of territorial struggles within and between nation-states in the interstate system that will continue to delimit the struggles over cultural geopolitics in the twenty-first century.

REFERENCES

Agnew, J. (1984) 'An excess of national exceptionalism: towards a political geography of American foreign policy', *Political Geography Quarterly* 2: 151–66.

Agnew, J. (1997) *Political Geography. A Reader.* London: Arnold.

Agnew, J. and Corbridge, S. (1995) *Mastering Space: Hegemony, Territory and International Political Economy.* London: Routledge.

Agnew, J., Mitchell, K. and Toal, G. (eds) (2002) *A Companion to Political Geography.* Oxford: Blackwell.

Ash, T.G. (2001) 'Europe at war', *The New York Review of Books* XLVIII (20 December): 66–8.

Beck, U. (1999) *World Risk Society.* Cambridge: Polity.

Dalby, S. (2002) *Environmental Security.* Minneapolis: University of Minnesota Press.

Duncan, J. (1980) 'The superorganic in American cultural geography', *Annals of the Association of American Geographers* 70: 181–98.

Godlewska, A. and Smith, N. (eds) (1984) *Geography and Empire.* Oxford: Blackwell.

Huntington, S. (1997) *The Clash of Civilizations and the Remaking of World Order.* New York: Simon and Schuster.

Jackson, P. (1989) *Maps of Meaning: An Introduction to Cultural Geography.* London: Routledge.

Lewis, M. and Wigen, K. (1997) *The Myth of Continents: A Critique of Metageography.* Berkeley: University of California Press.

Luke, T. (2002) 'Postmodern geopolitics: the case of the 9/11 attacks', in J. Agnew, K. Mitchell and G. Toal (eds) *A Companion to Political Geography.* Oxford: Blackwell.

O Loughlin, J. (1986) 'Spatial models of international conflicts: extending current theories of war behavior', *Annals of the Association of American Geographers* 76: 63–80.

Ó Tuathail, G. (1996) *Critical Geopolitics.* Minneapolis: University of Minnesota Press.

Ó Tuathail, G. (2000) 'The postmodern geopolitical condition: states, statecraft, and security at the millennium', *Annals of the Association of American Geographers* 90 (1): 166–78.

Ó Tuathail, G. and Agnew, J. (1992) 'Geopolitics and discourse: practical geopolitical reasoning and American foreign policy', *Political Geography* 11: 155–75.

Paasi, A. (1996) *Territories, Boundaries and Consciousness.* Chichester: Wiley.

Parker, G. (1998) *Geopolitics: Past, Present and Future.* London: Pinter.

Sharp, J.P. (2000) *Condensing the Cold War: Reader's Digest and American Identity.* Minneapolis: University of Minnesota Press.

Smith, G. (1999) 'The masks of Proteus: Russia, geopolitical shift and the new Eurasianism'. *Transactions of the Institute of British Geographers* 24 (NS): 481–500.

Smith, N. (1989) 'Geography as museum: private history and conservative idealism in *The Nature of Geography*', in J.N. Entrikin and S. Brunn (eds) *Reflections on Richard Hartshorne's The Nature of Geography.* Washington: Association of American Geographers. pp. 89–120.

Sparke, M. (forthcoming) *Hyphenated Nation-States.* Minneapolis: University of Minnesota Press.

Thrift, N. (2000) 'It's the little things', in K. Dodds and D. Atkinson (eds) *Geopolitical Traditions.* London: Routledge.

24

Boundaries in a Globalizing World

Anssi Paasi

> Contemporary frontiers are not simply
> lines on maps, the unproblematic givens
> of political life, where one jurisdiction
> or political authority ends and another
> begins; they are central to understanding
> political life. Examining the justifica-
> tions of frontiers raises crucial, often
> dramatic, questions concerning citizen-
> ship, identity, political loyalty, exclu-
> sion, inclusion and the ends of the state.
> (M. Anderson, 1996: 1)

International boundaries have long been among
the most important objects of research in political
and cultural geography. They have been understood as material elements in cultural landscape
and seen as lines that distinguish 'power structures' and sovereign states, and which provide
opportunities for cooperation and discord. Yet
boundaries are much more than marks of 'the
limits of sovereignty' (Prescott, 1987: 80).
During the 1990s, borders and boundaries have
become keywords in social science and cultural
studies as the world around us has changed.
Researchers have challenged the ideas of fixed
boundaries, identities, truths and power and
instead have put stress on the fragmentary and
the impermanent nature of boundaries. In this
situation, borders and boundaries are increasingly understood as 'zones of mixing, blending,
blurring and hybridizations' where both material
and symbolic dimensions and power relations
come together (Bhabha, 1994; Thrift, 1996).

There are many reasons for the current interest
in boundaries. Firstly, the fall of the Berlin Wall
and the fading of superpower conflict meant that
both 'east' and 'west' lost the 'others' that were
used in constructing geopolitical practices and
images of threat (Figure 24.1). Secondly, violent
redefinition of territorial and ethnic identities
around the world has displayed the persistent
violence and contested role that boundaries still
play in the lives of many of the world's peoples.
Thirdly, the process of globalization and the
unprecedented flows of capital, goods and ideas
across the world today have problematized our
governing notions of borders, boundaries and
sovereignty. Finally, environmental risks have
revealed the porousness of state boundaries.

These dynamics are unfolding in different
ways in different places. In my own city of Oulu
in northern Finland, the cultural geographies of
boundaries are very dynamic. Located on the
western coast of Finland, Oulu is about 130 kilometres from the Finnish–Swedish boundary, and
more than 250 km from the Finnish–Russian
border. The former has been open and practically
insignificant for decades, and became even more
so when both Finland and Sweden joined the
European Union. The Finnish border with Russia
has been crucial for the narratives of Finnish
national identity since the nineteenth century and
for the practices of foreign policy since Finland
gained its independence in 1917. It was strictly
controlled during the years of Cold War and the
Soviet Union: basically all forms of cross-border
interaction were decided at the level of the two
states (Paasi, 1996). A several kilometres wide
frontier zone was established on both sides of the
borderline after World War II. On the Finnish
side people needed permission to move on the
zone, issued to Finnish citizens by the Finnish
Border Guard Detachment and to foreigners by
the Security Police. The Soviet zone was kept
almost empty. After the collapse of the west–east
divide the modes of interaction became more
versatile. Other actors, like municipalities and
firms, now shape the everyday cultural geography

Figure 24.1 *The Berlin Wall 2000*

of the boundary, supplementing the power of the state. The number of crossing points has increased dramatically, being now more than 20. The border, however, has not disappeared and is still controlled, but now in less exclusive ways, displaying the changing meanings of boundaries and territoriality. Further, it is not only a state-to-state border but also the sole border between the EU and Russia and one of the deepest divides in the world as far as the standards of living in the respective states are concerned.

Most people outside Finland may not know that Oulu is one of the central bases of operation of Nokia, a traditional Finnish manufacturing firm that has rapidly become the most significant name in the mobile phone marketplace. Its slogan 'Nokia – connecting people' celebrates the opposite of boundaries and borders. With R&D centres in 15 countries on four continents, products sold in 130 states, and more than 90 per cent of its stock owned by foreigners, Nokia is a global company that has apparently transcended the borders of the nation-state. Along with the rise of Nokia and other high-tech firms, Oulu has become increasingly international, partly because of the links between these firms and the university, both of which draw educated people to the city, many of them crossing the borders between states, cultures and social roles. The city is the centre of the province of Northern Ostrobothnia, which means that much of the regional decision making within the context of the EU is located

there, again passing national boundaries. This relativitization of the location of Oulu during the 1990s seems to illustrate the changing functions of political boundaries in a globalizing networking world. Superficially at least, it seems to support arguments about the disappearance of borders and nation-states in a world characterized by networking people, transnational corporations and suprastate levels of governance (Ohmae, 1995).

However, the disappearance of boundaries has been more celebrated in the catchy logos of transnational corporations than realized in practice. Boundaries are dense and multilayered processes whose meanings derive not merely from economic forces (or slogans!) but also from the accumulated histories and cultures of political entities that are still very much alive: states. The world political map is made up of almost 200 different states. Between them are hundreds of boundaries, some open and peaceful, some closed and full of political dynamite. All of these borders are human creations and a good number of them have contested stories of identity, struggle, desire and history embedded in their existence (Kirby, 1996). As the Finnish–Swedish and Finnish–Russian borders demonstrate, even different boundaries of the same state may have diverging, historically contingent narratives of identity associated with them. These narratives and their everyday meanings may also vary in the socio-spatial consciousness of various generations – those conditioned by the Cold War or those conditioned by Nokia's utopian world of happy communication – living in the same territorial context (Paasi, 1996).

In this chapter we will consider the changing and contested roles of political boundaries in the everyday making of identities and territories. Historical perspective helps us to understand how boundaries have become part of the material practices, ideologies and narratives through which territorial groups and their identities are constituted. The central contention of the chapter is that boundaries are dynamic cultural processes. They are more than 'lines on a map': they have crucial links with identity, action, mobility and power that we need to grasp if we are to understand the changing spatialities of our globalizing world. Boundaries are not, as traditional political geography once took them to be, timeless, neutral lines and absolute limits of sovereignty. They are much more complex and interesting.

We will begin by briefly considering the construction of state territoriality and boundaries. Since geographical discourses have been crucial in this process, the invention of a specific boundary language will be traced to unpack the spatialities of this language. Identity and boundaries are typically seen as two sides of the same coin, but often so that physical and symbolic boundaries are regarded as exclusive constituents of identity (Conversi, 1995; Hall, 1996). Thus the links between state, nation and identity need to be discussed to show how state boundaries – as specific spatialized symbols and institutions – become part of the practices and discourses of daily life. The contested interpretations of the effects of globalization on boundaries will then be discussed, and the challenges facing boundary studies as cultural geographies today will be briefly considered in the conclusion.

STATES, BOUNDARIES AND THE 'TERRITORIAL TRAP'

Because boundaries are manufactured by human cultures, they are political entities. Their creation involves choices between contesting visions of how to divide up space. They give expression to power relations since they inevitably order and shape the social relations of the peoples affected by them. They involve the politics of delimitation, the politics of representation, and the politics of identity, i.e. they keep things apart one from another, their meanings are represented in specific ways, and they enable certain expressions of identity while blocking others.

Although we do not always think about it, we live our daily life in a complex network of socio-spatial practices conditioned by the fact that the world political map is divided into states. In spite of centuries of human movements across the planet and cultural mixing and hybridization, the link between states, territoriality and sovereignty has been so dominating that it has been almost impossible to avoid what Agnew (1998) calls the 'territorial trap'. This notion refers to a way of thinking and acting that has three distinguishing features. Firstly, modern state sovereignty, security and political life require clearly bounded territorial spaces. Secondly, a fundamental opposition between domestic and foreign affairs exists. Thirdly, the territorial state acts as the geographical 'container' of modern society, i.e. boundaries of the state are the boundaries of political and social processes. Hence, the world is understood as consisting of bounded, exclusive territories that have their own identity.

These assumptions look relatively simple, but they are extremely powerful since they are sedimented both in material practice and in ideologies. Through legislation, media and the national education system they usually become forces through which the territoriality of the state and its limits become taken-for-granted constituents of social order and everyday experience (Paasi, 1996; Radcliffe and Westwood, 1996). The territorial trap effectively hides the fact that collective identities are not naturally generated but are partly produced through the construction of various forms of exclusion and inclusion, that is in defining who 'we' are, who belongs with 'us' and who are to be excluded because they are 'different'.

One significant element in the naturalization of the territorial trap is the broadly accepted narrative of the development of the international system of states. This narrative assumes that the modern state system began in sixteenth-century Europe with the codification of a world of nominally sovereign states in the Treaty of Westphalia in 1648. Triumphing over alternative organizations and visions of world political space, such as a unitary world of Christendom or a world empire dominated by an all-powerful dynasty, the modern state system slowly evolved to become organized around the principle of popular 'national sovereignty'. According to the creed of nationalism which triumphed in nineteenth-century Europe, the world's varied peoples were assumed to be made up of different nations and the most powerful and organized of these nations controlled their own states, which they called 'nation-states'. This nation-state model of the interstate system was exported by Europe to the rest of the world through colonization and then decolonization (see M. Anderson, 1996; Giddens, 1987; MacLaughlin, 2001; Smith, 1991). A well-known analyst of nationalism, Ernest Gellner, has depicted the power of the modern state as follows:

> [C]onsider the history of the national principle; or consider two ethnographic maps, one drawn up before the age of nationalism, and the other after the principle of nationalism had done much of its work. The first map resembles a painting by Kokoschka. The riot of diverse points of colour is such that no clear pattern can be discerned in any detail, though the picture as a whole does have one. A great diversity and plurality and complexity characterizes all distinct parts of the whole … Look now instead at the ethnographic and political map of an area of the modern world. It resembles not Kokoschka, but, say, Modigliani. There is very little shading; neat flat surfaces are clearly separated from each other, it is generally plain where one begins and another ends, and there is little if any ambiguity or overlap. Shifting from a map to the reality mapped, we see that an overwhelming part of political authority has been concentrated in the hands of one kind of institution, a reasonably large and well-centralized state. In general, each such state presides over, maintains, and is identified with, one kind of culture, one style of communication, which prevails within its borders. (1983: 139–140)

The pure world of Westphalian exclusive nation-state actors has never existed, since states have always shared their power with other states and organizations (Agnew and Corbridge, 1995). The territorial system created by the interstate system is always in perpetual transformation. Hence the number of states and boundaries has increased continually following the processes of secession, unification and decolonization. While some 50 states existed in 1900, currently almost 200 states are linked together with more than 300 land boundaries that have a unique, often violent history. Similarly the number of international non-governmental organizations has increased enormously since World War II. Also the roles and relative power of states in the global system of states and the ideas of sovereignty have changed perpetually (Murphy, 1996). Some 'micro-states', such as Singapore, Monaco or the Bahamas, are significant nodes in the globalizing network economy even if they may lack the institutional and ideological infrastructure of the modern state (Nairn, 1998). Simultaneously only a few states are capable of monitoring the territories of all other states with their satellite systems. States are, in sum, not equal in their sovereignty and their boundaries have very different meanings and functions.

Geo-power: national socialization and the attribution of meaning to boundaries

The modern state has a monopoly over power and violence, which it uses in the construction and maintenance of the territorial trap. The state has not only the negative power to repress those that challenge its power within its borders, but also the 'positive power' to 'civilize' and socialize its inhabitants through its educational structures and communication systems. The 'territorial trap' gets reproduced as a deep cultural geography of everyday life, for it is hidden in numerous state-based institutions and practices that produce and reproduce the symbolic boundaries of a territory. Through education, the media and the cultural institutions of civil society, national socialization takes place, the process by which people internalize collective cultural geographies of identity and division. Recent geographic research has focused on the process through which reified and naturalized national representations are constructed and reproduced by elites across the institutions of states. In Finland, for instance, the border with Russia has been used in school geography textbooks since the nineteenth century as a medium to represent Russia as the national other, a threat from the east that must be resisted. Changes of these representations have followed the transformations of broader geopolitical spaces and conditions (Paasi, 1996). The importance of boundaries in building national identity becomes clear also in a study on the Ecuadorian nation-building process (Radcliffe and Westwood, 1996).

The production of knowledge has been particularly important in the governance of the state system and in the creation of the significance of territory. Many academic disciplines have their origin in the practical interests of the state to inscribe territories with a content, a history and a meaning (Krishna, 1994; Walker, 1993). The institutionalization of geography at the end of the nineteenth century was a manifestation of the rise of the modernist ideology and all-encompassing nation-state. Major motives behind its institutionalization were nationalism and colonialism, showing that geography is a form of power/knowledge itself (Ó Tuathail, 1996). Murphy (1996) labels the view of sovereignty that dominated the end of the nineteenth century *anarchic*. This view emerged from a convergence of rising positivism, nationalism and Darwinian political geographic thought and these modes of thought raised the territorial ideal of sovereignty to new heights. Building strong, competitive nation-states resulted in the control over bounded territories.

Political boundaries are always expressions of *geo-power*, the use of geographical knowledge in the governance and management of territories (Ó'Tuathail, 1996). Geographers and their ideas of boundaries have been exploited in many states by power holding elites, both in the demarcation of concrete boundaries and in more abstract ways in the creation of geopolitical visions and identities. Geographical (and historical) knowledge that is used in giving names and boundaries to regions and territories is instrumental in the demarcation and control of territories. No wonder then that 'the boundary' became immediately one of the key categories in political geography. The so-called 'father of political geography', Friedrich Ratzel, greatly influenced thinking about boundaries with his notion that states are organisms and that the boundaries of a state are its peripheral organs that can and should be expanded as stronger states grow at the expense of weaker ones. More than acknowledging that borders are never static, Ratzel's ideas were justifications for imperialism and state adventurism. In his organismic thinking all dynamic states try to expand their spatial extent and extension.

Many of the ideas Ratzel championed were already significant. Particularly influential was the French concept of *les limites naturelles*, according to which every state has 'natural boundaries' which that state should pursue until it has obtained. As might be expected, adjacent states argued about where their 'natural boundaries' existed – the very disagreement undermining the notion that there really were 'natural boundaries'! Nevertheless, this idea was particularly influential in the eighteenth and nineteenth centuries. In most cases, historical arguments were used to justify the 'natural boundary of the state', though arguments frequently evoked a higher authority like God or natural law in order to justify particular claims (Pounds, 1954). Natural boundaries were seen as the only real borders because they emerged from nature – whether being God-given or not. The boundaries drawn by people were regarded as arbitrary. Yet the arbitrariness of even 'obvious' natural boundaries was not acknowledged. Mountains, after all, can be seen as 'natural communities', rivers as organically connected regions and seas as transportation highways rather than as 'natural limits'.

Political geographers have paid increasing attention to the *meanings* attached to border landscapes since the 1990s (Rumley and Minghi, 1991). They also became interested in the processes of inclusion and exclusion: how boundaries are used in the construction of communities, territorial identities and representations of 'us' and the Other in the 'purification of space', i.e. in the construction of images of culturally homogeneous territorial groups (Sibley, 1995). The social and discursive construction of boundaries has been typically studied on a national or subnational scale (Falah and Newman, 1995; Paasi, 1996; Radcliffe and Westwood, 1996), while geopolitical oppositions have been analysed on larger spatial scales (Newman, 1999; Ó'Tuathail, 1996). These studies have shown that the purification of space, the rejection of difference, the securing of boundaries and symbols to maintain solidarity in social communities are employed in the exclusion and inclusion of social groupings at all spatial scales, varying from the territories of local gangs in large cities to nations and global geopolitical spaces (Newman and Paasi, 1998).

BOUNDARIES AND THE NATION: NATION-STATES OR MULTINATIONAL STATES?

New interest in the cultural significance of political boundaries emerges from the observation that boundaries are not merely technical instruments exploited by the state but also constituents of (national) identity and the media of power that are used in the 'naturalization' of nations as individuals, i.e. they are major elements in the making of the territorial trap. Most theorists of nationalism note the significance of territory and boundaries in the construction of national communities and the images of their past, present and future. For some authors nationalism is a process of border maintenance and creation (Conversi, 1995) while others argue more generally that human consciousness and social organization are conditioned by territory and boundaries (M. Anderson, 1996). Nationalism promoted by the state exploits both history and territoriality to make this boundedness appear natural. The history of boundaries is a major component in the geopolitical imagination of most states (Radcliffe and Westwood, 1996: 57). History and territoriality are used in both the construction and the reproduction of the territory and citizens, the latter understood – or state institutions tending to represent them – as constituting 'the nation'. The nation is not only a political unit but also a cultural system of signification. A state that successfully reproduces itself as a nation must have specific symbolic and institutional practices for narrating, signifying and legitimating the nation and the bounded territory that it occupies. In this context, boundaries become part of the ways by which people try to make sense of the world at all spatial scales. Boundaries not only divide but also define and regulate social action. They are, in a way, 'forces at work' (Rée, 1998).

One significant ideological manifestation of the territorial trap has been the understanding of 'national cultures' as homogeneous coherent phenomena so that territory and exclusiveness become 'natural' constituents of cultures. Boundaries become, thus, an integral part of the very understanding of a 'national culture', and a particular problem for those who inhabit borderlands or cross them as immigrants, refugees or exiles. Yet cultures have always been based on the exchange of ideas and material innovations, and it is cultural flows that have drawn nations together (Featherstone, 1995). National cultures also change perpetually: new generations produce new national sharings, amalgamate them with 'tradition' and may generate perspectives on territory, boundaries and identity that differ considerably from those of previous generations (Paasi, 1996). This dynamic underscores once again how group identities based on bounded cultures are not natural but are created in specific contexts and in response to certain forces. 'Cultural identity' is, therefore, both the scene and the object of political struggles (Cohen, 1998: Jackson and Penrose, 1993). To take but one illustration, in Bosnia, for instance, Muslim, Croat and Serb children learn their own 'truths' of the national history. Hence, Gavrilo Princip, the Bosnian Serb who shot Archduke Franz Ferdinand in Sarajevo and set World War I in motion, is a 'hero and poet' for Serbs, an 'assassin' for Croats and a 'nationalist' for Bosnian Muslims (Hedges, 1997).

The contemporary world harbours hundreds of ethno-national groups; by some estimates there are as many as 5000 of these 'nations'. They coexist with some 200 bounded states that represent a vast array of internal differences in 'national' cultures, economies, identities and backgrounds of inhabitants. In fewer than 20 states the minorities account for less than 5 per cent of the population. This suggests that even if the attempt to fuse national identity and national state has been the major motive in European and world history, most modern states are plural multinational and multicultural states (Schaeffer, 1997; Smith, 1995). As a result numerous social groups in Europe and elsewhere regard themselves as 'nations' and struggle for self-determination or a state of their own. In some places this process occurs peacefully (Scotland); in other contexts violence has been part of the process (Basque country, Northern Ireland, Kurdistan, Israel); in the extreme case the result has been a terrible civil war (Sierra Leone, East Timor). This suggests that territory and boundaries are still vital in national imagination, symbolism and rhetoric.

While the world presented in maps is often ahistoric and stable, it is the postcolonial struggles that give the current geopolitical map much of its dynamic, especially inside the existing boundaries. While the international border disputes between India and Pakistan and between Ethiopia and Eritrea have recently been very visible in the media, in fact most border conflicts occur nowadays inside states. Since 1995 only one or two conflicts per year have been between states, whereas some 25–30 have been within states. In Africa, for instance, most conflicts occur inside the existing colonial boundaries (Biger, 1995; Shapiro, 1999). While also interstate conflicts occur now mainly in poor Third World countries, nationhood, forms of identity and even national iconographies are increasingly contested in most contemporary states. Also the 'first nations' all around the world are good examples of internal challenges to boundaries. Often supporting environmental values, traditional community life and identities, and people's rights to land and old territories, they struggle to transform the legislation and territorial governance that have been created by dominant national groups. Their interests may also cross the existing state borders – often by using modern information technology.

Displaced people and immigrants also question the significance of boundaries and the state-centred territorial and 'cultural' order. The intensifying interaction between states has created a 'perpetual motion machine' where refugees, migrants and tourists cross borders and create networks of translocalities, places where processes occurring from local to global scale come together (Appadurai, 1996). The separation of people from their 'native culture' through either physical dislocation or displacement – the colonizing imposition of 'foreign culture' – has been one of the most formative experiences of this century, and the estimated number of refugees alone has been 60–100 million people since World War II (Bammer, 1994). Whereas the refugees at the end of World War II were mainly Europeans, this group is increasingly heterogeneous today, the main groups now being refugees from Afghanistan, Iran, Bosnia and some African countries. UNHCR's statistics show that more than 22 million asylum seekers, refugees and displaced people existed in 1998.

While the pressures for transformation challenge the existing territorial order, states struggle to maintain their bounded national territories and identities, and to control, marginalize or destroy opposition (Shapiro and Alker, 1996). Various strategies exist to promote social integration, such as assimilation encouraged by the state, and various forms of socio-cultural autonomy and language accommodation for minorities (Knight, 1985). The statistics of Amnesty International show that not only symbolic but also physical violence is a much used instrument in territorial control: in 1998 human rights were violated in 142 states, political murders occurred in 47 states and people were arrested in 78 states because of their opinions. For many in the power structures of the state that are dominated by the majority nation, the perspectives of other nations and peoples within the borders of the state are ignored, denied and repressed. Many states have difficulty acknowledging that they are not the nation-states of their images and myths but multinational and multicultural states with many different kinds of peoples and perspectives.

BOUNDARIES IN A TRANSFORMING WORLD

No contemporary discussion of state borders can avoid addressing globalization since this process implies 'border crossings' and blurring of the spatial categorizations between 'us' and 'them' (Anderson and O'Dowd, 1999). Authors do not always make it clear whether they talk about the globalization of institutions (economy, culture), consciousness or communication networks. Recent comments suggest that a satisfactory definition must capture such elements as extensity (stretching of social, political and economic affairs), intensity, velocity ('speeding up') and impact (Held et al., 1999: 15). All these elements imply that boundaries between domestic and 'global' affairs are increasingly blurred, challenging one of the key assumptions of the territorial trap (see Rosenau, 1997).

It is useful, however, to make a distinction between 'strong' and 'weak' versions of the globalization thesis that respectively imply different views on the future of the state and boundaries (Anderson and O'Dowd, 1999). The proponents of the strong version put primary stress on economics and technology, with a secondary emphasis on culture. States are perceived as less important than transnational corporations or social movements and communities, which do not respect boundaries. This view seriously underestimates the cultural element that serves as an important 'glue' in the territorial trap. In Finland, for instance, the success of Nokia in the expanding mobile phone markets is viewed not only as an international but also as a national success story, inspiring Finnish pride. In this way the factor that is hollowing the territorial trap in an economic sense is strengthening it in a symbolic or cultural sense. This indicates that globalization has perhaps changed but not inevitably diminished the significance of cultural or jurisdictional barriers (see Cohen, 1998). Most strong thesis supporters see boundaries as fixed entities and territoriality or sovereignty in essentialist terms, i.e. they lean on the modernist language of traditional political geography. Hence boundaries are understood as lines dividing social entities, not as discursive formations and processes that are sedimented in the social and cultural practices of these very entities (Paasi, 1998). Authors like Ohmae (1995) use the ideas of a borderless world as *metaphors* depicting the condition of economic liberalism and expanding capitalism, rather than discussing concrete state borders. As such they are 'big metaphors' (Barnes and Duncan, 1992) since they change the rhetoric that is used in research.

'Weak versionists' (or 'sceptics') see internationalization as more significant than globalization (Anderson and O'Dowd, 1999; Held et al., 1999). For them the state is still the major context in which people organize their daily life (Hirst and Thompson, 1996). Accordingly, territorial states are not withering away but simply operate in a dynamic, global context. Moreover, globalization represents not the end of territorial distinctions and distinctiveness, but rather new influences on local (economic) identities and development capacities (Amin and Thrift, 1995). The ultimate question is what will be the balance between markets and state in the emerging systems of global governance?

As far as the changing meanings of boundaries are concerned, one important part of the globalization discourse has been the rise of specific rhetoric that in principle calls into question the image of a fixed territorial order, existing territorial traps and boundaries. Castells has been extremely influential in providing spatialized metaphors and arguments for current debates. His idea of the *space of flows* implies the decreasing power of sovereignty and a challenge to national identities and national boundaries (Castells, 1989). Other much used notions are *deterritorialization* and *reterritorialization*, adopted from Deleuze and Guattari (1984), who used them to describe the effects of capitalism on previous fixed orders of class, kinship and space. These notions have been significant for critical geopoliticians and IR scholars who have analysed the representational practices that have been hidden in the construction of the territorial trap. They have noted the importance of the boundary-drawing practices that are used in the spatialization of identity, nationhood and threats (Ó Tuathail and Dalby, 1998).

Nowhere has the metaphor of 'border crossing' been more evident than in discourses on *cyberspace* and its effects on current spatializations. It has been argued that virtual spaces will give rise to new global geographies that will change the ways we think, the nature of sexuality, the forms of communities and our very identities (Turkle, 1996: 9). Similarly electronic space has been seen potentially as a major theatre for capital accumulation (Sassen, 1999). Cyberspace also enables the creation of identity communities, which are not territorially bounded and may challenge the territorial trap of existing nation-states. The internet has been an important tool for social movements (for example, first nations) that operate both inside states and across borders (Routledge, 1998).

Leaders in many states (for example, Japan and Singapore) feel that cyberspace may threaten their cultural identities and have therefore been hesitant about enabling their citizens to connect to the internet at all (Stratton, 1997). This shows that the role of traditional/new institutions that control and reproduce territoriality/boundaries and national socialization will remain strong. In Finland, for instance, the number of mobile phones is about 75 per 100 inhabitants, which is among the highest rates in the world, and also internet links are part of everyday life for almost one-third of the population. These links have not, however, led to the disappearance of national boundaries. While new devices are 'connecting people' – also across the national boundaries – they do not inevitably change the content of national socialization and identity discourses, neither do they remove totally the territorial elements from journalism, media space or legislation. Also military and foreign policy practice and the border guarding systems have remained the same, even if the territorial context is now not only the Finnish state but also the EU (Paasi, 1998).

New technology may also create new boundaries. Cyberspace may be used in the reterritorialization and construction of images of 'natural boundaries', of 'us' and enemies, as is evident in the conflicts in the former Yugoslavia and Rwanda (Campbell, 1998). Like globalization as a whole, new technology also produces inequality: not everybody can afford new technological devices. In many countries in Africa and Asia, for instance, both the low level of literacy and almost non-existing telephone links among ordinary people make a mockery of most utopian visions of the power of global cyberspace. According to the statistics of the UN only 2.4 per cent of the world population has used the internet. As Ó Tuathail and Dalby (1998: 13) have noted, the digital nation may transgress state boundaries but it will remain the virtual 'home' of a small elite fraction of the world's overall population.

Are boundaries disappearing?

Previous analysis shows that very different views of the current roles of boundaries exist. These views are based, firstly, on diverging theoretical and conceptual frameworks that are used in interpreting the meanings of globalization and sovereignty. Secondly, they indicate that boundaries have many functions: they are elements in the international governance, instruments of state policy and territorial control, but also constituents and challengers of existing social identities. Thirdly, varying interpretations display that knowledge and understanding are situated categories. Thus changing economic, political and cultural contexts affect how researchers shape the categories that they use to interpret current territorial transformations. Boundaries, their disappearance, globalization or sovereignty mean different things not only for researchers coming from various states and 'academic territories' but also for politicians, international capitalists, business gurus, military leaders, refugees and displaced people or ordinary people.

The ideas and 'truths' of boundaries or the nation-state are themselves products of contested discourses. The meanings attached to boundaries are often expressions of various ideologies and rhetorics of power. The members of the elite often have conflicting aims with regard to boundaries. While economic actors may struggle to promote cross-border activities, military and political actors try usually to keep the state, its instruments of violence and narratives of 'the nation under threat' in operation (Paasi, 1998).

Previous contested visions raise the serious question of whether boundaries are really disappearing, or whether our conceptualizations are inadequate for understanding their current, complicated roles. It is clear that we are moving towards a situation where exclusive state boundaries are, at least in some areas, becoming porous. Increasing cross-border interaction and new sub- and suprastate regionalizations are leading to a change of formerly closed 'alienated border-lands' to 'interdependent' and perhaps finally to 'integrated borderlands' (Martinez, 1994). The assumed isomorphism between territorial and national integrity is increasingly problematic, and territory as the source for loyalty and identity is increasingly divorced from territory as the site of sovereignty and state control.

On the other hand, the continually increasing number of states and boundaries suggests quite opposite tendencies than ideas of the disappearance of borders. The emergence of new states does not mean that their boundaries should be closed. Most contemporary boundaries do not mark the limits where 'politics ends because the community ends' (Balibar, 1998: 220). Cross-border cooperation has increased during the whole post-war period, particularly in the current EU area. Still, some borders are more closed than others. New emerging spatializations, such as economic and institutional 'fortress Europe' (!) or, on a broader spatial scale, geopolitically laden ideas like the 'clash of civilizations' may concomitantly become instruments of exclusion and a basis for new images of threat against outsiders (Paasi, 2001).

The debates on globalization and boundaries force us to reflect on the connections between territory, political community and democracy. Democratic societies in the future are likely to involve increasing openness and porosity of borders. For example, the relationships between citizenship and sovereignty, national identity and political community, and cultural order and global flows are likely to be negotiated and renegotiated. If nations are 'imagined communities', it is the same imagination, Appadurai (1996) argues, that will have to carry us beyond the nation, even if the contemporary 'national imaginary' has not yet given in to the rise of non-national, transnational or postnational claims on loyalty and identity. There is no reason to assume that moral boundaries or even notions of 'home' and 'the nation' should overlap with the boundaries of our territorial traps in the current world (Agnew and Corbridge, 1995). Peace activists, feminists, environmentalists and anti-capitalist movements, for instance, increasingly 'cross' the state borders in their argumentation and activities. Similarly environmental problems show the porousness of boundaries: risk society knows no national boundaries (Kuehls, 1996; Lash and Urry, 1994).

FUTURE CHALLENGES FOR BOUNDARY STUDIES

A major challenge for research is to develop new conceptualizations of the changing meanings of boundaries. Blossoming studies on the roles of boundaries, cross-border cooperation and trans-border regionalization in Europe and elsewhere show that boundaries, identity and citizenship still matter and will provide important subject material for border scholars. In this context the political inevitably bleeds into economic, cultural and regional. Boundaries should not be regarded as static territorial lines but should be understood from a broader socio-cultural perspective. Researchers need to emphasize the social production and everyday reproduction of territories and boundaries, and their symbolic meanings in discourses and institutional practices that occur at all spatial scales.

Boundaries are cultural processes. They are both symbols and institutions that produce distinctions between social groups and are produced by them. As institutions and discourses, boundaries are 'spread' everywhere into the society, not only in the border areas. Therefore, boundaries are one part of the discursive landscape of social power that exists in numerous social practices and discourses in the fields of economy, administration, legislation and culture. This is why boundaries do not inevitably vanish when some structures and practices – for example, in the field of economy – change as a consequence of globalization. To the extent that we have social power networks we will have boundaries.

The boundaries that states create are no longer the only boundaries that matter, as the state is re-scaled from above by the processes of globalization and from below by the everyday operation of flows of peoples, ideas and commodities. We also have to be sensitive to boundary claims and representations that social groupings make at supra- and substate scales. These changes pose challenges for the politics of boundaries, i.e. the production and reproduction of physical and symbolic boundaries in the life of a society, in which nation and state, for instance, no longer fit neatly into the territorial trap.

All these processes are expressions of the fact that states and their identities are never finished as entities but are ongoing processes and projects (Campbell, 1992). They force scholars to reflect on how the meanings and interpretations attached to boundaries in the processes of state- and nation-building express state ideologies, reactions to broader international geopolitical and economic landscapes and localized, popular interpretations.

Boundaries are present in national iconographies (flags, coats of arms, statues), commemorations, military parades, literature, songs and folklore, graffiti, sites of battles and landscapes that all signify and symbolize national identity. Border scholars should analyse how these ideas have gained their importance in the constitution of territorial entities and the spatial identities of the people. Therefore researchers need to use critically many kinds of materials: media discourses (TV, movies, newspapers), manifestations of 'high' and popular culture, educational materials, etc. One important theme for boundary studies is the analysis of the contextual forms of national socialization (Paasi, 1996; Radcliffe and Westwood, 1996). This theme includes a critical political geography of boundary maintaining institutions: the military-industrial complex, religion, education, racism, foreign and security policies, for instance. All are crucial elements in the construction of geographies of inclusion and exclusion.

Security has been closely linked in national narratives with national political identities that are represented as depending on the construction of boundaries with the other. In a rapidly transforming world of foreign policy, the elites of states try to redefine territoriality, foreign/security policy and identity – often in the form of new images of threat. Therefore, researchers have to pay attention to analysing how the legitimation of boundaries and the production of their social and cultural meanings take place.

Boundaries are with us not least because they are part of social identities and the making of territory and place (Massey et al., 1999). Their contextual functions and meanings imply that we cannot *write* them away, as some globalization theorists seem to suggest. Boundaries, identities and difference construct the 'space of agency', the mode of participation in which we act as citizens in the complicated polities to which we belong (Yuval-Davis, 1997). This does not mean that the identities of people or places are pure and fixed. Identity is the result of myriad interactions that occur 'inside' (and with the outside) of a space whose boundaries are not clearly defined (Mouffe, 1994). This means that boundaries do not have any universal essence and do not represent permanent 'truths', but are social constructs, results of struggles and power relations. The boundary is, therefore, a space for struggle and a zone of negotiation and reflection (Massey, 1995). Researchers are therefore forced to reflect on the paradoxes, contradictions and contrasts that are hidden in the practices and narratives of boundaries, identities and exclusions. It is important to analyse who constructs the hegemonic practices and narratives on boundaries, on 'us' and others, and why, how and in which institutional practices these narratives are (re-)produced. Border scholars need to reflect upon the structural and ideological constituents of boundary formation while they also have to be sensitive to the ethnographies of daily life, where boundaries are ultimately reproduced.

REFERENCES

Agnew, J. (1998) *Geopolitics*. London: Routledge.

Agnew, J. and Corbridge, S. (1995) *Mastering Space*. London: Routledge.

Amin, A. and Thrift, N. (1995) 'Territoriality in the global political economy', *Nordisk Samhällsgeografisk Tidskrift* 20: 3–16.

Anderson, J. and O'Dowd, L. (1999) 'Border, border regions and territoriality: contradictory meanings, changing significance', *Regional Studies* 33: 593–604.

Anderson, M. (1996) *Frontiers*. Cambridge; Polity.

Appadurai, A. (1996) 'Sovereignty without territoriality: notes for a postnational geography', in P. Yager (ed.) *The Geography of Identity*. Ann Arbor: University of Michigan Press.

Balibar, E. (1998) 'The borders of Europe', in P. Cheah and B. Robbins (eds) *Cosmopolitics*. Minneapolis: University of Minnesota Press.

Bammer, A. (ed.) (1994) *Displacements: Cultural Identities in Question*. Bloomington: Indiana University Press.

Barnes, T. and Duncan, J. (eds) (1992) *Writing Worlds*. London: Routledge.

Bhabha, H. (1994) *The Location of Culture*. London: Routledge.

Biger, G. (1995) *The Encyclopedia of International Boundaries*. New York: Facts on File.

Campbell, D. (1992) *Writing Security*. Manchester: Manchester University Press.

Campbell, D. (1998) *National Deconstruction*. Minneapolis: University of Minnesota Press.

Castells, M. (1989) *The Informational City*. Oxford: Blackwell.

Cohen, A. (1998) 'Boundaries and boundary-consciousness: politicizing cultural identity', in M. Anderson and E. Bort (eds) *The Frontiers of Europe*. London: Pinter.

Conversi, D. (1995) 'Reassessing current theories of nationalism: nationalism as boundary maintenance and creation', *Nationalism and Ethnic Politics* 1: 73–85.

Deleuze, G. and Guattari, F. (1984) *Anti-Oedipus*. London: Athlone.

Falah, G. and Newman, D. (1995) 'The spatial manifestation of threats: Israelis and Palestinians seek a "good" border', *Political Geography* 14: 189–206.

Featherstone, M. (1995) *Undoing Culture.* London: Sage.

Gellner, E. (1983) *Nations and Nationalism.* London: Blackwell.

Giddens, A. (1987) *The Nation-State and Violence.* Berkeley: University of California Press.

Hall, S. (1996) 'Introduction: who needs identity?', in S. Hall and P. du Fay (eds) *Questions of Cultural Identity.* London: Sage.

Hedges, C. (1997) 'Kolme erilaista historiaa Bosniasta (Three different histories on Bosnia)', *Helsingin Sanomat* 30 November.

Held, D., McGrew, A., Goldblatt, D. and Perraton, J. (1999) *Global Transformations.* Cambridge: Polity.

Hirst, P. and Thompson, G. (1996) *Globalization in Question.* Cambridge: Polity.

Jackson, P. and Penrose, J. (eds) (1993) *Constructions of Race, Place and Nation.* London: UCL Press.

Kirby, K.M. (1996) *Indifferent Boundaries.* New York: Guilford.

Knight, D.B. (1985) 'Territory and people or people and territory: thoughts on postcolonial self-determination', *International Political Science Review* 6: 248–72.

Krishna, S. (1994) 'Cartographic anxiety: mapping the body politic of India', *Alternatives* 19: 507–21.

Kuehls, T. (1996) *Beyond Sovereign Territory.* Minneapolis: University of Minnesota Press.

Lash, S. and Urry, J. (1994) *Economies of Sign and Space.* London: Sage.

MacLaughlin, J. (2001) *Reimagining the Nation-State.* London: Verso.

Martinez, O.J. (1994) 'The dynamics of border interaction: new approaches to border analysis', in C.H. Schofield (ed.), *Global Boundaries: World Boundaries,* vol.1. London: Routledge.

Massey, D. (1995) 'The conceptualization of place', in D. Massey and P. Jess (eds) *A Place in the World.* Oxford: Open University.

Massey, D., Allen, J. and Sarre, P. (eds) (1999) *Human Geography Today.* Cambridge: Polity.

Mouffe, C. (1994) 'For a politics of nomadic identity', in G. Robertson, M. Mash, L. Tickues, J. Bird, B. Cuetis and T. Putnam (eds) *Travellers' Tales.* London: Routledge.

Murphy, A. (1996) 'The sovereign state system as political-territorial ideal: historical and contemporary considerations', in T.J. Bierstaker and C. Weber (eds) *State Sovereignty as Social Construct.* Cambridge: Cambridge University Press.

Nairn, T. (1998) 'After Brobdingnag: micro-states and their future', in M. Anderson and E. Bort (eds) *The Frontiers of Europe.* London: Pinter.

Newman, D. (ed.) (1999) *Boundaries, Territory and Postmodernity.* London: Cass.

Newman, D. and Paasi, A. (1998) 'Fences and neighbours in the postmodern world: boundary narratives in political geography', *Progress in Human Geography* 22: 186–207.

Ohmae, K. (1995) *The End of the Nation State.* New York: Free Press.

Ó Tuathail, G. (1996) *Critical Geopolitics.* London: Routledge.

Ó Tuathail, G. and Dalby, S. (eds) (1998) *Rethinking Geopolitics.* London: Routledge.

Paasi, A. (1996) *Territories, Boundaries and Consciousness: The Changing Geographies of the Finnish-Russian Border.* Chichester: Wiley.

Paasi, A. (1998) 'Boundaries as social processes: territoriality in the world of flows', *Geopolitics* 3: 69–88.

Paasi, A. (2001) 'Europe as a social process and discourse: considerations of place, boundaries and identity', *European Urban and Regional Studies* 8: 7–28.

Pounds, N. (1954) 'France and "les limites naturelles" from the seventeenth to the twentieth centuries', *Annals of the Association of American Geographers* 44: 51–62.

Prescott, J.V.R. (1987) *Political Frontiers and Boundaries.* London: Unwin Hyman.

Radcliffe, S. and Westwood, S. (1996) *Remaking the Nation.* London: Routledge.

Rée, J. (1998) 'Cosmopolitanism and the experience of nationality', in P. Cheah and B. Robbins (eds) *Cosmopolitics.* University of Minnesota Press: Minneapolis.

Rosenau, J.N. (1997) *Along the Domestic–Foreign Frontier.* Cambridge: Cambridge University Press.

Routledge, P. (1998) 'Going globile: spatiality, embodiment, and mediation in the Zapatista insurgency', in G. Ó Tuathail and S. Dalby (eds) *Rethinking Geopolitics.* London: Routledge.

Rumley, D. and Minghi, J. (eds) (1991) *The Geography of Border Landscapes,* London: Routledge.

Sassen, S. (1999) 'Digital networks and power', in M. Featherstone and S. Lash (eds) *Spaces of Culture.* London: Sage.

Schaeffer, R. (1997) *Understanding Globalization.* Boston: Rowman and Littlefield.

Shapiro, M. (1999) 'Triumphalist geographies', in M. Featherstone and S. Lash (eds) *Spaces of Culture.* London: Sage.

Shapiro, M.J. and Alker, H.R. (eds) (1996) *Challenging Boundaries.* Minneapolis: University of Minnesota Press.

Sibley, D. (1995) *Geographies of Exclusion.* London: Routledge.

Smith, A. (1991) *National Identity.* Reno: University of Nevada Press.

Smith, A. (1995) *Nations and Nationalism in a Global Era.* Cambridge: Polity.

Stratton, J. (1997) 'Cyberspace and the globalization of culture', in D. Porter (ed.) *Internet Culture.* London: Routledge.

Thrift, N. (1996) *Spatial Formations.* London: Sage.

Turkle, S. (1996) *Life on the Screen.* London: Weidenfeld and Nicolson,

Walker, R.B.J. (1993) *Inside/Outside: International Relations as Political Theory.* Cambridge: Cambridge University Press.

Yuval-Davis, N. (1997) *National Spaces and Collective Identities: Borders, Boundaries, Citizenship and Gender Relations.* London: Greenwich University Press.

25

Gender in a Political and Patriarchal World

Joanne P. Sharp

The last decade has witnessed a convergence between political and cultural geography. The effect of the 'cultural turn' on political geography has generated a turn toward non-traditional geographical knowledges and a concern with the everyday as a valid space of political analysis. The development of 'critical geopolitics' has been of particular significance in these changes. This has facilitated the breaking down of boundaries within the discipline of political geography – 'a geopolitical perspective on the field of geopolitics' (Ashley, 1987: 407), as one commentator put it – to examine those relationships that were previously taken for granted.

Less territorial but no less spatialized divisions have also been examined so that the political geographies of gender relations have begun to emerge as important areas of study. Both the themes and the values of feminist geography can be seen in critical geopolitics, although not necessarily with reference to their feminist heritage. This chapter will consider the direction and strengths of the cultural turn on political geography. It will examine the extent to which feminist issues have emerged strongly as a result of these changes and look to the future research directions that might emerge.

A CRITICAL GEOPOLITICS OF POLITICAL GEOGRAPHY

One of the most significant impacts of the convergence between cultural and political geography has come in 'critical geopolitics'. A conventional geopolitics is an approach to the practice and analysis of statecraft and international relations more generally which considers geography and

spatial relations to play a significant role in the constitution of international politics (Smith, 1994: 228). The spaces analysed by geopoliticians were those of the state, region and globe and provided the backdrop to the playing out of global politics. Various laws could be created, whether of the relationship of distance to political threat or the effect of particular environments on the construction of particular political cultures. Those who made geopolitics were the statesmen and leaders of powerful countries and their advisers.

More recent engagements with the political world have a close link to some of the main characteristics of the cultural turn. One aspect in particular which dominates is the recognition of the power of language and discourse in the construction of the world around us. 'All language,' states Eagleton, 'is ineradicably metaphorical, working by tropes and figures; it is a mistake to believe that any language is literally literal' (1983: 145). For critical geopolitics, this attention to language is key and explains the ways in which global space and geography have been constructed and interpreted as being of importance to political process.

Critical geopolitics seeks to broaden our understanding of the relationships between geography and politics via a number of engagements. The first is a rethinking of the meaning of space itself. Rather than the common-sense understanding of space as a simple container or backdrop to international politics, a more culturally and historically specific understanding is offered. This critique of the dominant modernist Cartesian conceptualization of space as an empty framework sees space as power. Rather than an unchanging backdrop, space has a history and has changed as it is written and rewritten by various powers. Ó Tuathail (1996b) insists on considering the writing of geopolitics as based upon 'geo-graphing' – earth-writing – to emphasize the creativity inherent in the process of using geographical reasoning in the practical service of power (Ó Tuathail and Agnew, 1992).

Linked to this then is a questioning of the language of geopolitics, or 'geopolitical discourse'. Cultural geography has taught that language is not a transparent form of communication somehow *simply* describing what is there. 'Geography' is not an order of facts and relationships 'out there' in the world awaiting description, but is instead created by key individuals and institutions and then imposed upon the world. There is always a choice of words and metaphors. The types of term used – the conceptual links made – affect the meaning of what is being described. There is, as a consequence, a politics of language.

Finally critical geopolitics has uncovered the previous overemphasis on the state as the main, or only, actor in international politics. Clearly other powers are involved both at the substate level, such as ethnic, regional and place-based groups, and at the suprastate level, such as transnational corporations and international organizations including the UN and NATO. The state-centred realist approach makes the state seem uncomplicated, unified and ordered – and safe (a point to which I will return). A more critical political geography offers an engagement with the practices of geo-graphing at a number of scales.

Critical geopolitical approaches seek to examine how it is that international politics is imagined spatially or geographically and in so doing to uncover the politics involved in writing the geography of global space. It therefore challenges the links between spaces and identities that are commonly accepted, and which act as the basis of much theoretical and practical work requiring an examination of the practices that maintain the boundary of inside and outside. This means that in the current world order, most often critical geopolitics offers 'counter narratives of the nation' (Ó Tuathail and Dalby, 1998). The nation is perhaps the most prevalent form of identity despite, perhaps because of, the internationalization of global politics and society.

Nationalist rhetoric is of an *a priori* identity, a people with long and natural links to territory achieving the rule that is naturally theirs. Most theorists are of the opinion that this is not the case but instead that the imagining of nations is orchestrated as an ongoing process of boundary-making which separates the members of the nation from those outside. For Benedict Anderson (1991), the community is imagined into existence. He argues that although any member of a nation cannot know all other members of the nation, he or she feels part of a distinct community sharing history and certain characteristics. A clear sense of boundedness is part of this imaginary in that those who lie outside the nation are different. Nations are written into the world not only as independent entities but in relation to the international. There is a clear sense of 'our space' and 'their space', us and them, inside and outside. Rather than being the product of the expression of some 'natural' identity tied to the territory of nation, then, it is possible to see nations imagined against what the nation is not, the outside. By defining the space of outside (other nations, the international) as 'other than us', a coherent sense of identity is created (see Campbell, 1992; Dalby, 1990b).

The clearest example of this process can be seen in the political culture of national identity in the USA during the Cold War. It has been noted that if all nations are imagined, then the USA is the imagined community *par excellence* (Campbell, 1992). More than in any other nation, American national identity has been organized around the impetus to articulate danger, the specification of difference and the figuration of otherness (1992: 251). This led to the production of a Cold War moral geopolitical model of good and evil in which, for example, the territory reduced to the signification of 'communist' could be variously depicted as threatening, perverse and inhuman in distinction to the 'democratic' or 'free' space of safety, progress and civilization (see Campbell, 1992; Dalby, 1990b; Sharp, 2000b). Here the 'imagined community' of American citizens had a common goal, one required of them because of their historical role and manifest destiny. America was to triumph over communism just as it had triumphed over the wilderness in its original imaginings of the frontier: a clearly inscribed battleground over which American national citizens could triumph, as could the values seen as identifying the American national character. The USSR offered a mirroring conceptual space to that occupied by America: into this space was projected negative characteristics against which a positive image of American character could be reflected. This is central to the stories Americans tell themselves about themselves from the scripts of westerns and war films, through the speeches of elite political figures, to the narratives of school textbooks. Hollywood films constructed the USSR as a threat – these scripts were often quoted by political figures (Reagan quoted Rambo, Quail looked to Tom Clancy for inspiration) – and textbooks reproduced the notion of America's manifest destiny as 'land of the free'. These different social locations converge in the everyday constructions of national identity.

Here then was the creation of a particular geopolitical model. A particular image of the USSR was created which had direct consequences for the nature of US identity and what was expected of its citizens. Like the orient to Said's (1978) occident, then, this international representation told us more about how those who created the representations imagined America than it did the material realities of the Soviet Union. Into this alternate space was projected all that America refused to see in itself. The boundaries between inside and outside were 'the result of domesticating the self through the transfer of differences within society to the inscription of differences between society' (Campbell, 1990: 273). As a result, then, the creation of a clearly defined space of difference is also 'about stifling domestic dissent; the presence of external threats provides the justification for limiting political activity within the bounds of the state' (Dalby, 1991: 172). The practices of security are inherent in the production of a coherent sense of national identity (Dalby, 1990a). There is an inherent link then between the scales of international and national in that the images of threats and dangers outside the nation offer a reason for national unity to resist them and, in their negative image, project a positive alternative to which the national citizens should aspire.

This suggests then that fairly mundane, everyday actions and identities might be significant to the construction and reconstruction of the nation. Feminist theorist Judith Butler (1990) offers inspiration here. She considers the ways in which gendered identities are reproduced through the repetition of mundane activities rather than there being any essentialist biological definition of gender, or any stable identity established through social construction. It is the deed and not the doer that is of significance. The notion of a coherent and independent identity – the subject – is the effect of constant performance. On the whole, she argues, repetition works to reinforce the norm of heterosexuality. It is only through the constant repetition of heterosexualized actions that the illusion of a heterosexual norm can emerge. Minor practices – advertising images, soap opera storylines, pictures of families on office desks – unselfconsciously reproduce heterosexuality as the norm, which queer politics resists. From a mass of possible sexual performances emerges a conceptual map on which clear and distinct lines can be drawn dividing 'straight' from 'gay', 'normal' from 'deviant'.[1]

This has direct implications for the notion of border formation and political identity in political geography and related fields, in that it would appear that we are interested in 'the constitution of political community, not something that takes place within it' (Mouffe, quoted in Yuval-Davis, 1997: 73). Following Butler, the boundary of one nation-state and the next is not the innocent marker of the spatial extents of different cultural groupings but is instead integral to the construction of the identities it nominally illustrates. International relations, then, is not so much about protecting an identity which already exists, but about constantly creating and recreating identities. Through the repeated insistence that those outside the national boundary – those who occupy other spaces – are different or other, national identity can be reproduced as a coherent and universal form. National identity, rather than something that is retrieved from the past or protected from modernity, is in fact the effect of the modern practice of national rituals of reading national newspapers, singing national songs, waving flags at sports events and so on. It is the unthinking reproduction of these ideas that ensures the maintenance of distinct national identities (Billig, 1995).

This more cultural approach to national identity illustrates that the incredible power of national identity stems from its mundaneness, or banality. Shotter and Billig suggest that it is not the spectacular or conscious acts which underpin national identity but the minor events. They argue that the enunciation of the definite article in certain terms and phrases heard each day assumes the nation's boundaries:

> It points to the homeland: but while we, the readers or listeners, understand the pointing, we do not follow it with our consciousness – it is a 'seen but unnoticed' feature of our everyday discourses. (Shotter and Billig, quoted in Thrift, 2000: 384)

This challenge to the public–private divide suggests a broadening of the possibilities of what politics is and where it occurs. This has been used to define the political away from the other events of daily life which, in contrast, are assumed to be apolitical. Much of the political map is thus hidden from analysis that focuses on the formal acts of political citizenship or on the pronouncements of political leaders. Theorists influenced by Gramsci (1971) highlight the power of culture and have stressed the importance of the civic sphere in the maintenance of state power. For Gramsci (1971: 245), it is here that a complex set of cultural dominances forms the norms regulating social behaviour, the creation of hegemonic values. Hegemony is constructed not only through political ideologies but more immediately through the detailed scripting of some of the most ordinary and mundane aspects of everyday life (Holub, 1992: 104). The political significance of these cultural characteristics challenges the binary of public/political, private/apolitical.

Once we accept the importance of everyday geographical imaginations for the construction of political identity, and the operation of politics more generally, the cultural context of elite discourses must be understood. It is here that 'metacultural' values are reproduced. These inherited values stand as 'common sense' in relationship to the rest of the world, and point to the importance of the cultural norms established through which both political elites and their constituents are socialized. In the introduction to his influential cultural analysis of the figure of 'John Wayne', Gary Wills says he was often asked:

> Why *him*? When I began this project 3 years ago, that was the question most often asked when anyone learned of it. I had received no such queries when I said I was writing about Richard Nixon or Ronald Reagan. They, after all, held political office, formed political policy, and depended upon a political electorate. People cast *votes* for them. They just bought *tickets* for John Wayne's movies. Yet it is a very narrow definition of politics that would deny John Wayne political importance. The proof of that is Richard Nixon's appeal to Wayne's movie *Chisum* when he wanted to explain his own views on law and order. Nixon had *policies*, but beneath those positions were the *values* Wayne exemplified. (1997: 29)

Nixon could claim to regaining law and order in American society by reference to Wayne's performance in *Chisum* without having to explain the importance of this cultural reference. Instead he could assume the existence of a set of stories about America which his audience had learned through popular culture. Various media exemplify these foundational myths and stories. Wills' study refers to the genre of Western films and stories, most prominently the figure of Wayne. The values that the films extol and which Wayne personifies are so powerful and work so perniciously because dominant forms of political theorizing, in the academy but also in wider American political culture, assume that these are 'just' movies and work only in the realm of entertainment. As stories, they are seen as apolitical entertainment. Yet, the values that Wayne epitomized have celebrated, reinforced, strengthened and in many ways made possible the decisions and actions of statesmen and women. So, Nixon can refer to *Chisum* and the values of the Western, and a majority of the populace will know his reference (however subconsciously or tenuously), and understand its origin in relation to an imagined geography of America. The real values underlying the most significant political pronouncement then are found circulating not in the realm of public discourse but in the sphere of the private of leisure and recreation.

This turning toward the everyday has significant implications for attempts to retheorize the nation and international from a feminist perspective.

THE GEOPOLITICS OF THE PUBLIC–PRIVATE DIVIDE

Taking seriously the mundane acts of national identification suggests that they have great importance for the politics of the everyday, of the private, in addition to the more obvious politics of the public space of state and international politics. The public–private dichotomy produces the sense of a political and non-political sphere but also assigns gendered characteristics to each. The historic separation of spheres of public and private have had profound implications for the creation of gender roles and identities.

As feminist theorists have pointed out, from the traditions of Greek and Roman political thought onwards in the west there has been a conceptual and actual (spatial) division between public and private realms. The public is the conceptual and actual space of transcendence, of action, of production, of wages, of politics, of men; whereas on the contrary, the private is the location of the family, of recreation, of unpaid labour, of women. The construction of public sphere, space and subjects in political thought is dependent upon a private other. Therefore, rather than being separate spheres, this suggests that the identities of the public and private spheres might be intertwined.

With this challenge to the public–private binary in mind, we can return to the nation. The rhetoric of national identity has suggested that all members of a nation are equal. Thus the populist appeal of this form of identity: no-one can be more or less national. This rhetoric of a horizontal national bond produces a sense of the nation moving through time, encompassing the entire citizenship. But, just as the national community is imagined, so are the citizens who occupy it. The politics of public–private ensure that there are in fact differences in the access people have to the nation. Anderson argues that 'No more arresting emblems of the modern culture of nationalism exist than cenotaphs and tombs of Unknown Soldiers' (1991: 9). The power of this particular image is that it could be any member of the fraternity of the imagined community lying in the tomb. However, until recently in many nations, this was not the case and even today in many nations it is unimaginable that women are in the tomb (Sharp, 1996).

For Anderson, the power of nationalism lies in the horizontal fraternity of national citizens. This fraternity emerges from the fact that any man could be the unknown soldier who has laid down his life for his country, providing the ultimate sacrifice. In the national imaginary, women are mothers of the nation or vulnerable citizens to be protected. Women are not equal to the nation – each standing for it in the way that men can as warrior citizens. Women are symbolic of the nation (McClintock, 1993). Many nations are figuratively female (for example, Britannia, Marianne, Mother Russia).[2] In the national imaginary women are mothers of the nation (its biological origin) or vulnerable citizens to be protected by the bond of male citizens.

While men are heroic in their defence of the nation, women are heroic in their reproduction of it and nurturing of its future. The public–private divide ensures different roles for male and female citizens (Yuval-Davis, 1997). But this division is also played out at a larger scale. It is the fear of *foreign* violation of women as symbols of nation that is key.[3] The division of the international (the sphere of unregulated anarchy) and the state (the protected and safe space of home) parallels the division between public and private. The private is supposedly the safe sphere. The public is dangerous: it is here that women are out of place and even perceived as partially culpable if attacked. As feminist geographers like Pain (1991) and Valentine (1989) have suggested, if, despite this image, it is in the private space of the home that women face the greatest danger, then it is possible that this is also the case for the nation. Dalby suggests that:

> Just as some feminists challenge the ideology of the family in suggesting that private spaces are 'safe' because of the presence of a male protector, whereas public spaces are dangerous to women … it is a simple extension of these arguments to argue that states do not really protect all their (domestic) citizens while providing protection from the perils of the anarchy beyond the bounds of the state. (1994: 531)

Certainly women seeking to adopt the role of protector in military service is an issue that seems central to many debates on national strength. Liberal feminism argues that keeping women from their right and responsibility to risk their lives for their country reduces them to second-class citizens. This is integral to the liberal feminist argument 'that sees women's differences used to keep women out of public power; seeks their equal admission to the state and an end to the male monopoly on legitimate violence' (Pettman, 1996: 147).

As theorists like Anderson see this as linked to ultimate citizenship, liberal feminists might be right in arguing that its attainment by women would move towards equality. However this would be to accept a culture of masculinism. Second- and third-wave feminists have argued that it is not sufficient merely to succeed within masculinist culture – to succeed only in some ways by adopting masculinist traits themselves. Instead they argue for a more thorough challenge to the nature of the values underlying society. Enloe (1989: 44) has suggested that nationalism is an inherently masculinist form of identity, based around 'anger at being emasculated' at some point in history. Similarly, Newsinger (1993: 126) asks whether violence – 'the ability both to inflict and to take it – is portrayed as an essential part of what being a man involves' in certain national cultures. This offers the possibility that this form of political identification has its roots in masculinist culture so that it cannot accommodate a feminist challenge. Research has suggested that women deployed in traditionally masculinized jobs have generated hostility and concern about the morale of men 'who had been recruited partly with the promise that joining the military would confirm their manliness' (Enloe, 1993: 214).

Women in the military are often seen to challenge masculinist pride and identity. Dowler's (1998) research into the role of women in the IRA has highlighted the difficulties that this militarized form of national identity has with the inclusion of women's violence and struggle. As a result women and their actions and sacrifices are often not recognized in the reproduction of national images. Dowler's interviews with women from the IRA demonstrate attempts to rein in women to the public sphere to perform their role in the struggle. One woman told her:

> The people here definitely view us differently than they do the men. For instance there is always a big party for a man when he gets out of prison. A hero's return. I didn't have such a welcome home. It was as if, 'thank Jesus that's over, now I can get my dinner on the table'. They also look at us differently than other women. First off if we were in prison we weren't having wee ones [children] which is what we were supposed to be doing. I think if my husband could have got me pregnant in jail he would have because that was what he was supposed to be doing. A lot of men think I'm wild because I did my whack [time in prison], because women aren't supposed to be doing the same thing that the men are in this war. (1998: 167–8)

There is here a clear sense of what women and men can and should do, and these are separate. When women transgress these boundaries the response is an attempt to get things back to normal as quickly as possible. This is not an unusual response. Various commentators discussing diverse national movements including India and countries of formerly Soviet eastern Europe highlight the perceived importance of preserving or recovering 'traditional' (meaning patriarchal) gender relations in an attempt to regenerate national character (see Chatterjee, 1993; Molyneux, 1991; Todorova, 1993). Dowler (1998) also uncovered alternative forms of resistance and conflict enacted by women in the IRA but again this was not recognized as a legitimate form of political action in the nationalist struggle. For example, although not actually jailed themselves, some interviewees explained that their long years visiting first husbands then sons in prison, and coping alone with raising a family, might as well have been a prison sentence for women like themselves who could never escape these responsibilities (1998: 165).

In other cases it could be argued that the integration of women and women's concerns into the military might start to work changes into this institution and the nature of masculinity. Pettman wonders whether, with changes to the nature of modern war, different forms of masculinity might be required: in addition to the 'brute force of the footsoldier' the military might now also require 'the rational planning of the military strategists and commanders, the intellectual and scientific masculinism of defence researchers' (1996: 95). This points to the existence of more than one version of masculinity: on the one hand the phallic, erect and strong embodiment of the son, on the other the objective and calculating brains of the father (see Nast, 1998: 193). It is interesting to note however that although it is increasingly acceptable that women play an active role in the military, the reverse process is not so clear: there is little evidence of men taking on the feminized roles of nurturers and carers of national populations.

On the other hand, many feminists have drawn power from a pacifist stance which rejects militarized and divisive national borders. This approach would avoid national boundaries, loyalties and practices altogether and seek alternative forms of community relations and governance.

CHALLENGING BOUNDARIES

Many postmodern theorists have argued that the world is becoming ever smaller, that the globe is deterritorializing, becoming ever more connected and more fluid. Borders and boundaries are dissolving, according to this narrative. At first glance this might seem to be conducive to a feminist political geography that resists the inherited patriarchal images of international and national politics. Continental feminists, most prominently Luce Irigaray, have argued that femininity is constructed as 'that which disrupts the security of the boundaries separating spaces and must therefore be controlled by a masculine force' (Deutsche, 1996: 301). A major project of which masculinity is the effect is the ability to re-establish and reinforce such boundaries. In his influential work, Theweleit (1987) suggests that masculinity has a great deal to do with purification of both borders and peoples so that control and bounding are inherently entwined with hegemonic projects of masculinity. Nast argues that this has a sexed dimension:

> In the context of *trans*nationalism and the emergence of megastates and/or supranational organizations, we have all kinds of contemporary, 'unconsciously' registered anxieties over the heterosexualized pure and solidly bordered body of the nation being penetrated, threatened, overcome, and/or dissolved by a plethora of frightening foreign microbes and dangers. National hysterias have emerged over various kinds of 'transgressive' movements: from illegal immigrants in the USA, to (carriers of) AIDS globally, to legal but now economically redundant foreign workers in western Europe; all transgressors are denigratorily racialized through constructions of associational (for example, metonymic and metaphoric) links with disease, death, floods and filth. (1998: 195)

This can be seen in recent anti-immigration legislation in the USA and 'fortress Europe', in anxiety about the breakdown of family values in Britain and the US, and in the rise of reactionary forms of national identity in eastern Europe, amongst other examples. In the UK this can be seen particularly clearly in debates over the scrapping of Section 28, the bill that prevented the 'promotion' of homosexual values in schools. This led, in Scotland, to a referendum funded privately by millionaire businessman Brian Souter about whether or not homosexuality could be discussed in schools as a normal form of human relationship. This provoked highly charged debates in the tabloid press about the threat of certain groups who, they claimed, sought to destroy family values and social cohesion.

If it is indeed a masculinist urge to contain and bound, it might be argued that a feminist political geography would necessarily embrace fluidity and borderlessness. Some feminists have abandoned territorial identity and all that it entails and instead looked to global communities of oppressed. Robin Morgan's 'global sisterhood' was one such attempt to look at the commonality which was the 'result of a *common condition* which, despite variations in degree, is experienced by all human beings who are born female' (1984: 4).

However, the global sisterhood has been critiqued by Third World feminists who argued that this image ignores all of the differences, inconsistencies and histories which make up the notion of womanhood in different places. For Mohanty this automatic alliance erases the agency of women in particular historical struggles, and requires that 'the categories of race and class have to become invisible for gender to become visible' (1997: 83). For Third World feminists like Mohanty, the global sisterhood image silences the histories of colonialism, imperialism and racism from which western feminists still benefit. Second World – perhaps now more appropriately termed 'post-communist' – feminists have similarly critiqued western feminism for its liberal, middle-class assumptions.

Responding to Virginia Woolf's claim that 'as a woman I have no country. As a woman I want no country. As a woman my country is the whole world', Adrienne Rich explains how this feminist dream of universal sisterhood is unobtainable. She insists that:

> As a woman I have a country; as a woman I cannot divest myself of that country merely by condemning its government or by saying three times 'As a woman my country is the whole world.' Tribal loyalties aside, and even if nation-states are now just pretexts used by multinational conglomerates to serve their interests, I need to understand how a place on the map is also a place in history within which as a woman, a Jew, a lesbian, a feminist I am created and trying to create. (1986: 212)

In these 'notes towards a politics of location', Rich insists that womanhood is constructed specifically in different locations, as a result of many geographies – and historical geographies – playing out local and global relationships, of colonialism, trade, exploration, struggle and so on. Rich's opportunities, experiences, expectations and actions are both constrained and made possible by her multiple positionings within different power 'containers', perhaps most significantly the nation-state within which she is a citizen.

So an innocent view of movement and fluidity is problematic. For a select few there is a dissolving of boundaries and a shortening of distance as the world apparently shrinks. For many others, however, daily life becomes more fixed and travel more of a challenge. For the poor and the marginal, borders become more difficult to traverse, not less. In her work on the progressive sense of place and power geometries, Massey (1991) critiques the easy image of global shrinkage offered by some theorists. Although clearly technology is facilitating easy information connections and movement for some, for many others places are actually becoming more remote as the world 'shrinks'. People without resources – money and knowledge – find it less easy to link and move. Rather than embrace an unbordered and accessible global space, Massey (1993) sees the existence of 'power geometries' which act to constrain and facilitate the movement of different groups of people. This is not to suggest that the world remains bounded by all-powerful divisions. Massey (1991) calls this an 'extroverted sense of place' which can be imagined to form around networks of relations and connections rather than being enacted by boundaries and through exclusions. This requires us to think of gender as something formulated through localized networks, which are nevertheless inherently linked into global processes.

Much place-based politics is organized around demonstrating and protecting the borders of places, defending them from infringements to their 'authentic' tradition, landscape and identity. Place is seen as bounded. Place- or territory-based identity is usually organized around constructing a sense of otherness or difference against which the place can be defined. Definitions of who belongs in a country revolve around establishing heritage or genealogy, drawing a border to keep out those who do not belong.

Massey (1991) suggests that this is not necessarily how place should be understood. It can be shown that in actuality the clear difference between inside and outside, self and other, does not exist in real life: reality does not possess the hard lines of a map. Massey is encouraging us to acknowledge the extra-local similarities and linkages that make up a place, in addition to the differences. However, given the existence and increasing prevalence of the 'feminization of poverty', women are more likely to be trapped by globalizing processes. Borders cannot simply be wished away.

And not all feminists would want them to be. Anzaldúa (1987) argues for the importance of place and identity in resistance to dominant global powers. She offers a problematized celebration of identity based on impurity, mixing and diversity rather than singularity – recognition of the importance of the historically specific border between Mexico and the US. Anzaldúa's is an ambivalent geography which recognizes the constructedness of identity but also the reality of historically constructed divisions such as national borders. The boundary is a lived reality: not just a mark on the map but inscribed over and over again on her own body. What is of particular interest in the context of the cultural influences on understandings of national and international politics is Anzaldúa's style of writing. She does not write in standard academic prose but offers a series of interventions on questions of identity, territory and borders. She alternates her writing between English, Spanish and Chicana, sometimes translating, sometimes not. There is then a political geography to Anzaldúa's texts, sometimes allowing a monolingual reader into her community, at other times excluding the reader:[4]

> To live in the Borderlands means you
>   are neither *hispana india negra espanola*
>   *ni gabacha, eres mestiza, mulata,* half-breed
>   caught in the crossfire between camps
>   while carrying all five races on your back
>   not knowing which side to turn to, run from …
>
> In the borderlands
>   you are the battleground
>   where enemies are kin to each other;
>   you are at home, a stranger,
>   the border disputes have been settled
>   the volley of shots have shattered the truce
>   you are wounded, lost in action
>   dead, fighting back. (1987: 194)

EMBODYING THE INTERNATIONAL

The work of Anzaldúa and other feminists implicitly warns of the danger of going too far down the line of ironic postmodern distanced analysis. Just as with the work it seeks to challenge, the textual critiques of critical geopolitics can privilege the words and texts of elites, so perpetuating the silencing of those who are seen as outside of this.

On a practical level, this makes it all too easy for those outside these critical circles to ignore the critique and carry on regardless. As Demeritt (1996) observes for human geographers' critiques of scientific and objective forms of knowledge more generally, concentrating on this level of complex discourse makes it too easy for critiques to be ignored. Similarly, Peck (1999) has illustrated the difficulties of getting 'deep' or critical academic work taken seriously by policy- and decision-makers who are looking for more straightforward answers. The cultural turn has led many geographers to think carefully about

the voices of those being represented in their texts. Perhaps it is now time to think more carefully about the ears of our audiences.

Of equal importance, there is a danger in going too far along the poststructural path as it can erase agency. Perhaps by moving too far into the realm of the cultural, the political, has been undervalued: by labelling all as political, perhaps a sense of what might comprise more important issues has become lost. Similarly by shifting attention from the actor to the action some of the important historical contexts for struggle have been hidden. Although critical geopolitics might offer very eloquent deconstructions of dominant political discourse, there is often little sense of alternative possibilities. As Paasi (2000: 284) suggests, there is a need to move from text and metaphor which had dominated radical political geography.

Just as the formal actors of international politics have been disembodied, offering a 'spectator' theory of knowledge (George, 1996: 42), undifferentiated by the marks of gender, race, class, sexuality or physical ability, in some ways so do their critics. Women and others omitted from this tradition have not generally been included on the pages of the international texts. Thus they remain invisible to critical geopoliticians for whom resistance is a textual intervention, a subversion of a sign or displacement of meaning. Ó Tuathail's (1996b) influential *Critical Geopolitics* offers an intellectual history of geopolitical practitioners and critical geopoliticians as a history of 'big men' (in order): Mackinder, Ratzel, Mahan, Kjellen, Hausehoffer, Spykman, Wittfogel, Bowman, Lacoste, Ashley and Dalby. A few women are allowed into the footnotes, but the central narrative is one of the exploits and thoughts of men. The history of struggles for space and representation are reduced to a male genealogy when discussing not just the masculinist history of geopolitical strategies of elite practitioners, but also the interventions of 'critical geopoliticians' (Sharp, 2000a).

This illustrates a need to move to what Ó Tuathail (1996a) in another guise has called an 'anti-geopolitical eye', an embodied and situated geographical vision which avoids the God trick that is both everywhere and nowhere. This position takes responsibility for its representation from somewhere. The knowledge produced by an anti-geopolitical eye emphasizes moral proximity and anger: it is not distanced and dispassionate, even-handed or ironic. Ó Tuathail (1996b) offers Maggie O'Kane's impassioned reports of the war in Bosnia as a situated, moral and subjective alternative to the distanced all-seeing eye of the traditional geopolitician. Her reports emphasized the agency and acts of people, and the materiality of violence, rather than the active or abstract lessons of history and rules of geopolitics.

Like that of Thrift (2000), then, this position argues for the need to think of bodies as sites of performance in their own right rather than simply simple surfaces for discursive inscription. Discourses do not simply write themselves directly onto the surface of bodies as if those bodies offered blank surfaces of equal topography. Instead these concepts and ways of being are taken up and used by people who make meaning of them in the different global contexts in which they operate. This will bring women and other marginalized figures back into the sight of critical geopolitics. Although women's bodies are inherently caught up in international relations, this is often at mundane or everyday levels, and so they are not written into the texts of political discourse. Women's places in international politics tend to be not as decision-makers but as international labourers and migrants, as images in international advertising, and as 'victims' to be protected by international peacekeepers. However, this does not mean that women have no role in the recreation of international orders, simply that their agency is hidden from the traditional gaze of geopolitics. As Enloe argues in her attempt to make feminist sense of international politics, 'if we employ only the conventional, ungendered compass to chart international politics, we are likely to end up mapping a landscape peopled only by men, mostly elite men' (1989: 1).

This is not to suggest that to understand the nation and international it is necessary to abandon discourse; instead, it is necessary to see it in a broader way that is less dominated by representation alone and more attuned to actual practices. Political geographies can be regarded as emerging from the textualized practices and discourses that actually draw people in as subjects. Women, caught up in different forms of international traffic, are especially vulnerable to racialization and eroticization of their bodies and labour. National security defines women's bodies as requiring protection, but this is often defined from a masculinist position. Women's bodies become quite literally a part of making 'the international' (Pettman, 1991; 1996); for example, in the recent conflict in Kosovo, NATO went to war to protect some of the most patriarchal kinship structures in Europe.

A culturally informed but nevertheless still resolutely political take on nation and international relations could consider the practices and institutional locales of international relations and nations. This would involve continued engagement with the discourses and narratives which structure these geographies (it is important to deconstruct common sense, to stop it from working without thought), but would also go further to see how these discourses actually work in everyday life and how they make subjects of people.

Thus, for example, an understanding of how the media work to incorporate people as subjects would require examination of the representational content of the media texts but also the 'content of the form', the ways in which people are drawn into the media's representation of the world to become complicit with it – how they become active political citizens. Certain media institutions, such as the *Reader's Digest* (Sharp, 2000b), actively constitute the reader's identity as an American citizen through the form of address of its articles in addition to the actual topics being discussed. This structures a sense of what all 'good' American citizens must know: 'What is being planned for you?', 'Are we worthy of our destiny?' (Sharp, 2000b). This ties the individual reader into the narratives and identities of international and national politics in exactly the ways theorized by Billig and Shotter mentioned earlier. In addition to the representations of the world, here there is also a sense of how the readers are supposed to make sense of these global geographies: how they are to incorporate them into their daily lives in the constant reproduction of self as a national citizen. Of course there is still the question of the extent to which people will follow these 'instructions'.

Sparke (1998) works the relationship between representation and practices through the layers of his analysis of the construction of the political subject, Timothy McVeigh, the Oklahoma bomber. McVeigh, Sparke argues, was the subject of American discourses of inside–outside, of the safety and value of American culture and identity, and of the threats from those others beyond the boundary who sought the destruction of America. McVeigh was interpolated into these discursive practices as a consumer of American culture throughout his life but also most extremely in his experience in the military in the Gulf War. He was in fact awarded a medal for his actions in the conflict and so regarded himself as a patriot. On his return to the US, McVeigh apparently developed a sense that America had lost its way. He was a loner, feeling marginalized by the dominant society which, he thought, was unable to see the rot that had set in. Reconstructed through the narratives of warrior masculinity articulated in Rambo films, this subjectivity merely reinforced his sense of patriotism. With the representations of American global geography that he had experienced there was little difference between turning the people working for the federal government into minions of an evil state apparatus and turning the people of Iraq into minions of an evil state apparatus (1998: 202).

Sparke shows how the Gulf representations have played out – in a very specific way – in this person's biography. Analysis of the discourses significant to this story might suggest that the danger would always lie outside the boundaries of the US. However Sparke's almost ethnographic account of the production of geopolitical images and their actual impact on people's daily lives shows how these have been remade in this case to provide rather different results. A broadening of methodology from textual analysis to what might be considered an anthropology of international relations offers exciting possibilities for future understandings of the complex local embodied geographies that reconstruct the nation and the geography of international relations.

CONCLUSION

In her 'notes towards a politics of location', Rich (1986) beautifully articulates a problem that faces feminist theorists approaching global geographies: how to engage with the various exploitations and oppressions of women around the globe and at various regional and local scales, without producing an insipid image of global sisterhood which ignores all of the differences, inconsistencies and histories which make up the notion of womanhood in different places.

A feminist take on the patriarchal world cannot be a simple and naive abandonment of borders. These are accepted as social constructs but this in no way reduces their power over the individuals and communities that need to negotiate them on a daily basis. Lives are constructed and reconstructed around political and patriarchal boundaries through discourses which apparently operate at the global and national scales. Attempts to understand the complex relations between the international and the everyday demonstrate the importance of ensuring that the smallest, mundane daily practices of everyday life are not silenced from reconstructions of the international. At the same time the impacts of the movement of global geopolitical discourses on individual bodies need to be examined. For instance, the recent protests against the impacts on communities around the world of World Trade Organization decisions are testament to an

emerging politics which recognizes and challenges the complex processes linking bodies and nations, communities and globe.

Methodologically there needs to be a movement beyond the text to draw in other actions and practices, to look at the relations between discourse and practice, to see how the discourses work in a material sense and how they become embodied in expected and unexpected ways when actually used by different people and different communities in the pursuit of their lives around the globe. It is this drawing together of the global in the local, and the complex embodiment of geopolitical discourses, that offer one possibility for the production of new political geographical imaginations.

NOTES

1 Butler's is not a monolithic theory. There is always the possibility of resistance and transgression in this model which is so dependent on correct repetition. Alternative practices – whether consciously performed or not – can destabilize and ultimately undermine these fragile assemblages.

2 See Nast (1998) on the differences between nations gendered as male and female.

3 Although there is evidence that women in the military might fear more physical and verbal violence from their own comrades (see Enloe, 1993: 223).

4 Other feminists writing political geographies have similarly developed different writing styles. Both Enloe (1989; 1993) and Seager (1993) write in an informal style, rarely referencing the 'great men' of political geography or international relations. The authorities to which they refer are embedded in real-life political struggles rather than the discursive exchanges of the academy.

REFERENCES

Anderson, B. (1991) *Imagined Communities*, 2nd edn. London: Verso.

Anzaldúa, G. (1987) *Borderlands/La Frontera: The New Mestiza*. London: Routledge.

Ashley, R. (1987) 'The geopolitics of geopolitical space: towards a critical social theory of international politics', *Alternatives* XIV: 403–34.

Billig, M. (1995) *Banal Nationalism*. London: Sage.

Butler, J. (1990) *Gender Trouble: Feminism and the Subversion of Identity*. London: Routledge.

Campbell, D. (1990) 'Global inscription: how foreign policy constitutes the United States', *Alternatives* XV: 263–86.

Campbell, D. (1992) *Writing Security: United States Foreign Policy and the Politics of Identity*. Minneapolis: University of Minnesota Press.

Chatterjee, P. (1993) *The Nation and Its Fragments*. Princeton: Princeton University Press.

Dalby, S. (1990a) 'American security discourse: the persistence of geopolitics', *Political Geography Quarterly* 9 (2): 171–88.

Dalby, S. (1990b) *Creating the Second Cold War: The Discourse of Politics*. New York: Guilford.

Dalby, S. (1991) 'Critical geopolitics: difference, discourse and dissent', *Environment and Planning D: Society and Space* 9 (3): 261–83.

Dalby, S. (1994) 'Gender and critical geopolitics: reading security discourse in the new world order', *Environment and Planning D: Society and Space* 12: 525–42.

Demeritt, D. (1996) 'Social theory and the reconstruction of science and geography', *Transactions of the Institute of British Geographers* 21 (NS): 484–503.

Deutsche, R. (1996) *Evictions*. Cambridge, MA: MIT Press.

Dowler, L. (1998) '"And they think I'm just a nice old lady": women and war in Belfast, Northern Ireland', *Gender, Place and Culture* 5 (2): 159–76.

Eagleton, T. (1983) *Literary Theory*. Minneapolis: University of Minnesota Press.

Enloe, C. (1989) *Bananas, Beaches and Bases: Making Feminist Sense of International Relations*. Berkeley: University of California Press.

Enloe, C. (1993) *The Morning After: Sexual Politics at the End of the Cold War*. Berkeley: University of California Press.

George, J. (1996) 'Understanding international relations after the Cold War', in D. Shapiro and R. Alker (eds) *Challenging Boundaries*. Minneapolis: University of Minnesota Press. pp. 33–79.

Gramsci, A. (1971) *Selections from the Prison Notebooks*, (ed.) and trans. Q. Hoare and G. Nowell Smith. London: Lawrence and Wishart.

Holub, R. (1992) *Antonio Gramsci: Beyond Marxism and Postmodernism*. London: Routledge.

Massey, D. (1991) 'A global sense of place', *Marxism Today* June: 24–9.

Massey, D. (1993) 'Politics and space/time', in M. Keith and S. Pile (eds) *Place and the Politics of Identity*. London: Routledge. pp. 141–61.

McClintock, A. (1993) 'Family feuds: gender, nationalism and the family', *Feminist Review* 44: 61–80.

Mohanty, C.T. (1997) 'Feminist encounters: locating the politics of experience', in L. McDowell and J. Sharp (eds) *Space, Gender, Knowledge: Feminist Readings*. London: Arnold. pp. 82–97.

Molyneux, M. (1991) 'Interview with Anastasya Posadskaya (25 September 1990)', *Feminist Review* 39: 133–40.

Morgan, R. (1984) *Sisterhood Is Global: The International Women's Movement Anthology*. New York: Anchor/Doubleday.

Nast, H. (1998) 'Unsexy geographies', *Gender, Place and Culture* 5 (2): 191–206.

Newsinger, J. (1993) '"Do you walk the walk?": aspects of masculinity in some Vietnam War films', in P. Kirkham and J. Thumin (eds) *You Tarzan: Masculinity, Movies and Men*. London: Lawrence and Wishart.

Ó Tuathail, G. (1996a) 'An anti-geopolitical eye: Maggie O'Kane in Bosnia, 1992–93', *Gender, Place and Culture* 3 (2): 171–85.

Ó Tuathail, G. (1996b) *Critical Geopolitics*. Minneapolis: University of Minnesota Press.

Ó Tuathail, G. and Agnew, J. (1992) 'Geopolitics and discourse: practical geopolitical reasoning in American foreign policy', *Political Geography* 11 (2): 190–204.

Ó Tuathail, G. and Dalby, S. (1998) 'Introduction: rethinking geopolitics: towards a critical geopolitics', in G. Ó Tuathail and S. Dalby (eds) *Rethinking Geopolitics*. London: Routledge. pp. 16–38.

Paasi, A. (2000) 'Review of *Rethinking Geopolitics*', *Environment and Planning D: Society and Space* 18 (2): 282–4.

Pain, R. (1991) 'Space, sexual violence and social control: integrating geographical and feminist analyses of women's fear of crime', *Progress in Human Geography* 15 (4): 415–31.

Peck, J. (1999) 'Editorial: grey geography?', *Transactions of the Institute of British Geographers* 24 (NS): 131–5.

Pettman, J.J. (1991) *Worlding Women*. London: Routledge.

Pettman, J.J. (1996) 'Border crossings/shifting identities: minorities, gender and the state in international perspective', in D. Shapiro and R. Alker (eds) *Challenging Boundaries*. Minneapolis: University of Minnesota Press. pp. 261–83.

Rich, A. (1986) *Blood, Bread and Poetry: Selected Prose 1979–1985*. New York: Norton.

Said, E. (1978) *Orientalism*. New York: Vintage.

Seager, J. (1993) *Earth Follies*. New York: Routledge.

Sharp, J. (1996) 'Gendering nationhood: a feminist engagement with national identity', in N. Duncan (ed.) *BodySpace: Destabilizing Geographies of Gender and Sexuality*. London: Routledge. pp. 97–108.

Sharp, J. (2000a) 'Remasculinising geo(-)politics? Comments on Gearóid Ó Tuathail's *Critical Geopolitics*', *Political Geography* 19 (3): 361–4.

Sharp, J. (2000b) *Condensing the Cold War: Reader's Digest and American Identity, 1922–1994*. Minneapolis: University of Minnesota Press.

Smith, G. (1994) 'Geopolitics', in R. Johnston, D. Gregory and D. Smith (eds) *The Dictionary of Human Geography* (3rd edition). Oxford: Blackwell. pp. 228–30.

Sparke, M. (1998) 'Outside inside patriotism: the Oklahoma bombing and the displacement of heartland geopolitics', in G. Ó Tuathail and S. Dalby (eds) *Rethinking Geopolitics*. London: Routledge. pp. 198–239.

Theweleit, K. (1987) *Male Fantasies. Vol. 1: Women, Floods, Bodies, Histories*. Cambridge: Polity.

Thrift, N. (2000) 'It's the little things', in K. Dodds and D. Atkinson (eds) *Geopolitical Traditions: A Century of Geopolitical Thought*. London: Routledge. pp. 380–7.

Todorova, M. (1993) 'The Bulgarian case: women's issues or feminist issues?', in N. Funk and M. Mueller (eds) *Gender Politics and Post-Communism:*

Reflections from Eastern Europe and the Former Soviet Union. New York: Routledge. pp. 30–8.

Valentine, G. (1989) 'The geography of women's fear', *Area* 21: 385–90.

Wills, G. (1997) *John Wayne: The Politics of Celebrity*. London: Faber.

Yuval-Davis, N. (1997) *Gender and Nation*. London: Sage.

26

The Cultural Geography of Scale

Clare Newstead Carolina K. Reid, Matthew Sparke

Far from neutral and fixed, therefore, geographical scales are the products of economic, political and social activities and relationships; as such, they are as changeable as those relationships themselves. At the very least, different kinds of society produce different kinds of geographical scale for containing and enabling particular forms of social interaction. (Neil Smith, 1995: 60–1)

Geographical theories of scale have come a long way since the days when geographers used to invoke fixed notions of local, regional and national scales as if they were universally understood and unchanging analytical categories. Recent research has highlighted the *social construction of scale* and the ways in which scale is negotiated and reproduced. Such insights invite scholars to move beyond seeing scale as a politically neutral container of social processes or a methodological abstraction, and to consider instead how it is produced through socio-economic struggles and transformations. These approaches have largely developed out of Marxist theories of the production of space, most notably Neil Smith's (1984) arguments about the creative destruction of scale wrought by capitalist processes of uneven development. Smith argued that particular consolidations of capitalist territoriality – the formation of regional clusters or cities or even nation-states, for example – need to be seen as transient scalar fixes which, however concretized they may seem, are always vulnerable to the transformations brought about by new rounds of capitalist investment and disinvestment. In this chapter we begin from these basic Marxist insights into what has come to be known as 'scale-jumping'. However, in introducing them into an

arena of examination addressed by cultural geography – the production and contestation of cultural landscapes – we also argue that the Marxian focus on capitalist economic determinations needs to be radically supplemented by attention to cultural-political forces of ideology, resistance and the construction and negotiation of cultural identity. In this way we seek to explore the production of scale as the overdetermined effect of diverse cultural, political and economic power relations. Reciprocally, we also argue that, because scale-jumping represents the reconfiguration of the territorial *scope* of power relations, it provides a particularly powerful entry point into empirical research on the ideological overdetermination of particular cultural geographies. We will elaborate on our understanding of ideology and overdetermination shortly, but first we need to clarify the basic argument about scale-jumping from which we are beginning.

The Marxian formulation of scale-jumping develops directly out of Smith's radical interpretation of uneven development as a product of the tension between capitalist tendencies towards territorial equalization and differentiation, tendencies which themselves relate to the tension between competition and cooperation in capitalism (see also Smith, 2000). Smith argues that scalar fixes emerge as partial and temporary resolutions of the capitalist tensions between equalization and differentiation and it follows that they are frequently superseded by new spatial resolutions in response to capitalist reorganization. In this way, capitalist territorial organization 'jumps' scale in the context of overaccumulation or other moments of capitalist crisis and crisis management (see also Harvey, 1999). For example, Smith (1995) points to the development of the

European Union to illustrate how the scale at which the capitalist flux of cooperation and competition is mediated can 'jump', at a tremendously general and systemic level, from the scale of the nation-state to that of the continental state (see also Swyngedouw, 1992; 1997a). Through such examples we can gain a vivid sense of how scale is reproduced and transformed through dynamics of political-economic change. Indeed, one of the epistemological advantages of an analytical focus on scale-jumping is that it actually helps to make more clear what scale in fact is: the temporary fixing of the territorial scope of particular modalities of power. In the same way, scale-jumping enables us to theorize the framing effects of particular scales without ignoring the general fluidity of scale and resorting to the old fixed assumptions. It enables us to describe the moments at which boundaries are reconfigured and struggles rearticulated. Because such moments of scale-jumping are also often moments when cultural landscapes are redrawn or reimagined, the resulting cultural geographies register the jumping of scales. This is the basic insight that guides our approach to the case studies presented in this chapter.

All of the above is not to say that the notion of scale-jumping in the abstract makes the *substance* of scalar configurations transparent. Such ontological questions of substance relate to the particular objects of research under examination, and these can range from economic concerns to other dynamics as diverse as the sexual, the ecological and the racial. In Smith's analyses of the EU's development and of gentrification (Smith, 1995) it is economic (including class) relationships that are foregrounded, and thus scale comes to name the territorial scope of particular political-economic power relations. However, the notion of using scale-jumping remains useful as an epistemological entry point into investigating the territorial scope of other power relations too, others that are interarticulated with those of capitalism but in such a way as to remain relatively autonomous. In this chapter we seek to widen our analysis to encompass such variant power relations through a focus on the cultural geographies of ideology and resistance. This does not move us all that far from the basic Marxian attention to capitalism and its discontents, but it does bring into focus the ways in which economic transformations are mediated and sometimes contested by the production of new visions, new ideas, new feelings and new ways of being in the world, all of which remain just as profoundly geographical, in their shifting territorial scales, as the brute economic geographies of capitalist creative destruction.

Linked with the territorial reconfiguration of economic coordination is the representational question of how such reconfigurations are framed ideologically. Corporate elites, for example, often shift the scales at which production is represented in order to avoid regulation and accountability. Sometimes they stake their claim to a national scale of operations, claiming the protection of national sovereignty against threatening international laws. However, at other times they present their operations as footloose and global in order to discipline national governments and labor movements by threatening the loss of local jobs. Such neoliberal ideological maneuverings are one of the concerns raised by the case studies in this chapter, which we examine through the lens of cultural politics.

Following work in cultural studies that has reinterpreted ideology through the conceptual apparatuses of Gramsci's concept of 'hegemony', Althusser's concept of 'ideological interpellation' and Foucault's concept of 'discourse', we invoke ideology to describe hegemonic discursive formations in which people's subjectivities are formed and through and against which counter-hegemonic resistance is enabled (see, in particular, Hall, 1988; Laclau and Mouffe, 1985; Smith, 1988; Spivak, 1988). In a similar way, we follow the post-Freudian/post-Althusserian use of 'overdetermination' to signal our understanding of ideological representation (including landscape representations) as being constitutively founded on moments of ideological condensation and displacement (see Silverman, 1983: 62, 90; Sparke, forthcoming). In this register, any particular cultural geography needs therefore to be understood not only as the geographic condensation of diverse cultural, political and economic determinations, but also as a certain form of displacement which, in geographic terms, takes the form of a reterritorialized *place*ment, a particular if still transient understanding and experience of place that at first sight hides its ideological underpinnings.

Much of the best recent work in cultural geography has already drawn on such expanded and culturally nuanced understandings of ideology and overdetermination (Anderson and Gale, 1991; Brown, 2000; Gregory, 1996; Henderson, 1999; D. Mitchell, 2000; K. Mitchell, 1996; Moore, 1997; Sharp et al., 2000; Wright, 2001). Our main goal here is to explore how such widened analysis of cultural geographies can both inform and be informed by the study of scale-jumping. As well as seeking to make a contribution to cultural geography in this way, our reciprocal aim is to expand the conceptual relevance of scale-jumping beyond its traditional focus on the economic geographies of capitalist transformation.

The notion of supplementing the Marxian focus on the economic production of scale is by no means original to this chapter. Other geographers have already sought to bring such concerns into communication with the Marxian literature on scale. Most notably, Sallie Marston (2000) has argued for coupling Marxian attention to the sphere of economic production with analysis of the sphere of social reproduction and its coactive impact on the creative destruction of scale. Marston's particular concern is with the changing relevance of the home and the public realm as scalar fixes for the expression of feminist agency. In addition to foregrounding the way in which scales are fixed and undone through processes of cultural conflict and negotiation, Marston's work emphasizes the need to come to terms with scale-making as an arena in which domination and resistance are interrelated (see also Brown, 1997). It is this double concern with how ideological domination and counter-hegemonic resistance are together worked out in an uneven field of circulating power relations that has guided our approach here. Like Marston, we do not view scale-jumping as being neatly dichotomized between ideologically dominant and resistant forms. Instead, we see the resulting cultural geographies as reflecting a spectrum of combinations, sometimes instantiating more the reconfiguration of the scope of dominative power relations, sometimes embodying more the reconfiguration of the scope of resistant power relations, but always emerging as an entangled and hybrid product of negotiation and contestation (see also Sharp et al., 2000).

In order to illustrate such varied combinations of ideological dominance and resistance we have chosen three examples, each of which illustrates scale-jumping from the national to the transnational scales. The first, the construction of a cross-border region called Cascadia, illustrates a dominant neoliberal elite's cultural geography of scale-jumping: a region invented to expand and entrench entrepreneurial governance across the border between Canada and the US on the Pacific Coast. Instantiating neoliberal hegemony as it does, though, this elite cross-border vision of Cascadia does not completely obliterate more counter-hegemonic scale-jumping visions of the same region as a landscape of ecocentric governance. Our second case, the landscape vision attending the development of the Caribbean trading community (CARICOM), also illustrates a neoliberal cultural geography of scale-jumping. But in contrast to the dominant Cascadian vision, the imagined cultural geography of a common

Caribbean landscape underpinning a united CARICOM is also closely articulated with the postcolonial reimagination of the Caribbean as coherent and united *despite* the legacies of interimperial division and rivalry. The depiction of a CARICOM landscape, then, represents a more ambivalent scale-jump, one that is simultaneously neoliberal *and* postcolonial. Our third example represents a more subaltern act of scale-jumping. It consists of the counter-hegemonic landscape visions of Mexican–US transnationals whose border-crossing ways of being in, seeing and depicting space actively contest the hegemonically divided landscape of policing and violence at the US–Mexican border. We do not seek to romanticize this resistant landscape vision, and only aim to underline how it exemplifies counter-hegemonic scale-jumping in a time and space that has been predominantly shaped by the hegemonic cross-border scale-jumping of production and finance under North American free trade.

Clearly, the overarching context of all three of our examples remains that of neoliberalism, today's dominant ideology of laissez-faire capitalist deregulation and global market-based governance. In this sense, therefore, we build on the work of economic geographers concerned with the renegotiation of scale in the context of globalization (for example, Roberts, 1998; Swyngedouw, 1992; 1997a; 1997b). However, in acknowledging this important economic context we by no means take neoliberal ideology as some sort of disembodied logic emanating out of the economic ether. Indeed, one of the signal ideological features of neoliberalism is precisely the way it is so often presented as a *dis*placed, non-political, post-ideological global economic imperative. As the cultural anthropologists Jean and John Comaroff underline, 'there is a strong argument to be made that neoliberal capitalism, in its millennial moment, portends the death of politics by hiding its own ideological underpinnings in the dictates of economic efficiency: in the fetishism of the free market, in the inexorable expanding needs of business, in the imperatives of science and technology' (2000: 322). Against this pattern of (post-) ideological dissembling, it is especially important to examine the grounded embodiment of both neoliberalism and its discontents in particular cultural geographies of scale-jumping. Thus while our first reason for including three case studies here is to explore the varied combinations of ideological domination and resistance in scale-jumping, our second reason is to flesh out empirically the varied cultural geographies of neoliberalism on the ground. *Placing* the problem in this way allows us

Figure 26.1 *Poster for Cascadia*

to problematize a dualistic depiction of domination and resistance. But more than this, it helps us to underline how the cultural geographies of neoliberalism are overdetermined by processes of ideological negotiation and contestation. Thus, while each of our cases represents a reflection of the expansion and entrenchment of neoliberal agreements and policy-making across the Americas, they each also show how the reterritorialization of economic relations is complexly displaced and replaced in new, sometimes radically resistant, cultural geographies of scale-jumping.

SELLING SCALE-JUMPING: THE TWO-NATION VACATION AND THE BRANDING OF CASCADIA

Cascadia, gateway to the Pacific North-west and the Two-Nation Vacation, consists of the American states

of Washington and Oregon and the Canadian province of British Columbia. It's an advantageous location of international tourism and trade … There's something magnetic here for a certain kind of soul … one who appreciates natural beauty, limitless recreational opportunities, and the vibrant blend of international influences that have produced Cascadia's diverse culture and thriving economy. Many people have decided to call this region home which is a decision you'll understand once you see Cascadia for yourself … Washington, Oregon and British Columbia. That's where Cascadia is. But once you've experienced this magical place, its going to be somewhere else as well. It'll be in your heart and on your mind … forever. (*The Cascadian Traveler pamphlet,* undated)

Here is a landscape conjured up to appeal directly to international tourists. Packaged as a transnational space for a so-called 'Two-Nation Vacation', this vision has been developed over the last decade as one of the main promotional gambits of Cascadia's business-oriented boosters.

More than just words, the landscape has also been depicted in images and maps, and at a conference of tourism firms in June 1996 in Seattle, the resulting landscaping of Cascadia was presented in the form of a glossy poster (see Figure 26.1). The assembled images in the poster serve at once to evoke an ancient history and a sublime naturalness for the rescaled cross-border region, representing it as rooted as deeply in the soil as the actual forests on the slopes of the Cascade mountains. The result is a graphic that involves the whole panoply of iconic commodification. From the trees themselves to the treeless golf courses, every object and activity is effectively marked as open for the new cross-border business of tourism. Meanwhile, the images are put together with a map that lends a sense of objectivity and historicity to the resulting Cascadian landscape.

This glossy attempt to project the region's binational diversity means that little escapes the image's instrumentalizing embrace. Native artifacts, waterfalls, bears, eagles, salmon, trees and orcas are all packaged together into the advertisement. In this superficially aestheticizing way, they are all also reduced to serving as objects for the long-distance touristic gaze (for more on the effects of such a gaze on landscape imagery see Duncan and Gregory, 2000). The objects in the image thus form a fantastic landscape to be viewed from afar, a culturally coded rescaled place that links the entrepreneurial vision of the promoters with would-be vacationers' visions generated on the other side of the world. While serving thus as touristic objects, they also function for the promoters as a means of fashioning a natural Cascadian future out of the region's objectified natural history. Indeed, while the 'Two-Nation Vacation' advertisements reflect an attempt to market Cascadia's novelty and diversity as a cross-border region, it is equally notable that they also reflect an attempt to suggest that this is the way things naturally always should have been. It is this effort to naturalize the cross-border region's status as a consolidated region that reflects in turn the politics and economics of scale-jumping at play. Moreover, the work of the poster itself serves as a form of epistemological framing device which, targeted at tourism agencies as much as at tourists themselves, aims to reframe at a new scale the previously disconnected destinations of British Columbia and the US Pacific Northwest, representing them as a unified regional unit. The very semiotic lengths to which the poster goes in order to entrench a sense of the region's naturalness and ancient history are therefore themselves part as well as parcel of the processes of scale-jumping.

Although the promoters never use the language of scale-jumping themselves, they are often quite explicit about the political-economic context they see as necessitating their cross-border rescaling schemes. 'We are competing for tourists in a global market' Alan Artibise, a Canadian academic and promoter, explained in 1995: 'To maintain our market share, and indeed increase it, we can do very well by marketing a region that crosses international borders' (quoted in Webb, 1995: A5). More than just creating a novel niche region with which to attract international consumption spending, though, there is another still more profound political-economic imperative at work behind the development of the cross-border landscape. This basic imperative is interpreted by the local elites as a need to 'cooperate regionally in order to compete globally' (Chapman, 1996), and, just as with a number of other cooperative ventures aimed at marketing Cascadia as a site for foreign direct investment, the binational tourism projects are conceived, if not practiced, as another key area for cross-border cooperation. Launched primarily at the instigation of the Port of Seattle, the 'Two-Nation Vacation' has also been supported by BC tourism interests as a way of appealing to long-distance tourists from the UK, Germany and Australia. As a marketing concept it simply illustrates an attempt to twin the post-NAFTA notion of a borderless region with the economizing notion that Cascadian tourists can explore two nations and all of their collective recreational diversity for the price of just one long-distance plane ride. However, as part of the larger, scale-jumping dynamics from the national to the continental associated with NAFTA, the resulting Cascadian landscape is also envisioned very much as a sign of the free-trade times (Sparke, 2000). Here, for example, is a typical epochal invocation of the region's raison d'être:

> The lines imposed over 100 years ago have simply been transcended by contemporary cultural and economic realities ... Cascadia is organizing itself around what will be the new realities of the next century – open borders, free trade, regional cooperation, and the instant transfer of information, money and technology. The nineteenth- and twentieth-century realities of the nation-state, with guarded borders and nationalistic traditions, are giving way. (Schell and Hamer, 1995: 141)

Other visionaries of the Cascadian landscape have argued that its special future as some sort of neoliberal utopia is already underwritten by a vast cross-section of cross-border economies of scale (Goldberg and Levi, 1993). In the context of free trade, they argue, these economies are only going to grow, and the result will be a rescaled cross-border region unfettered by old

regimes of national governance. This leads them
to claim boldly that Cascadia is 'as meaningful
an economic entity as California' (1993: 29).
However, as they go about repeatedly presenting
the region in such exaggerated ways, it starts to
become clear that, just as in the poster for the
'Two-Nation Vacation', the boosters call upon
and depend upon the more general contours of
the Cascadian landscape to do the work of
ideologically legitimating the scaled-up cross-
border development plans. It is this landscape
and the diversity of peoples and opportunities
placed upon it, then, that help naturalize and dis-
seminate the concept of a scaled-up cross-border
regional identity to outside visitors. In this way,
the geographical representation of a rescaled
Cascadian region serves to justify and facilitate
the very processes it presents as a *fait accompli*.

As a basic rationale for Cascadian development,
the rescaled binational scope of the landscape
vision is often called upon to explain the opportunities that lie in store. In this way it is said to
be natural for the separate parts of Cascadia to
cooperate locally and build a regional alliance
in the context of global interdependencies (for
example, Artibise, 1994: 4). Moreover, in terms
of explaining why the cross-border scope and
scale of Cascadia put the region on a trajectory
towards high-tech growth, the promoters also
often invoke the landscape's more aesthetic qualities, arguing that its natural beauty and diversity
create the basis for attracting and nurturing a
highly educated professional workforce. The
lifestyle appeal of the landscape is what is used in
turn in the naturalizing accounts of Cascadians'
special destiny as citizens of a rescaled free-trading
node of neoliberal opportunity (for example,
Sutherland, 1997: 42). There is an unhappy irony
in all this in so far as the promoters of Cascadia
coopted the concept from its original ecotopian
roots in the work of local bioregionalists (see
Henkel, 1993). Much cross-border environmentalism persists in the region, and, despite the
boosters' attempts to harness it to Cascadian sustainable development discourse, it is sometimes
connected to radically alternative views of ecocentric governance (for example, Schoonmaker
et al., 1997). Such visions of governance are
oriented by mappings of the region's ecological
diversity and vulnerability. But these more
counter-hegemonic representations of the cross-
border landscape remain fundamentally displaced
and replaced by the Cascadia of recreational
diversity and high-tech business parks envisioned
by the neoliberal promoters.

Ultimately, as a rescaled landscape vision put
to work in widespread rhetorics and plans
for regional development, Cascadia is also an
ideological coproducer of the very changes it is
supposedly meant just to reflect. It actually helps
to frame and naturalize a terrain that can then be
said to have magnetism, soul and magic. Perhaps
the most impressive, even magical, aspect of this
ideological rescaling process is that it also conjures up idealized citizen-subjects for the landscape too. Such citizens, or rather post-citizens
as they are imagined in the various promotional
projects, are basically just potential tourists and
investors. But as independent agents bringing
money and desires that respond to price signals,
they would seem to represent the practically perfect inhabitants of a rescaled post-national region
that is imagined as a neoliberal utopia.

STRATEGIC SCALE-JUMPING: SUN, SEA AND CARIBBEAN UNITY

The logo of the Caribbean Common Market and
Community (CARICOM) was introduced in
1983, 10 years after the organization's initial
inception (Figure 26.2). As a flag and an emblem
on official documents, the logo symbolizes and
names an organization empowered to promote
supranational regional integration, trade and
economic cooperation among the small economies
of the Caribbean archipelago. In this sense the
logo registers a process of scale-jumping, of
extending and integrating economies beyond the
borders of the nation. The logo invokes a discourse
of islandness in which the region is naturalized
through reference to ecological similarity and, in
the absence of contiguous borders, a locational
proximity in the Caribbean Sea. Much more than
this though, through its reference to a collective
history of domination and subordination the logo
also invokes memories of struggle, and as such
points to the complex and ambivalent tensions
between ideology and resistance in the production of scale. Highlighting a particular sense of
Caribbean specificity, the logo represents a
complex hybrid of neoliberal and postcolonial
remapping.

Like the landscapes used to promote Cascadia,
the logo of CARICOM naturalizes the supranational region through recurrent reference to local
images of the physical and ecological landscape.
Reflecting CARICOM's origins among ex-British
colonies, the logo situates the Caribbean Sea as a
backdrop upon which islandness, natural vegetation and sunshine are scripted as characteristic of
the region (CARICOM, 1999). The iconification
of the entire Caribbean, as geographically
unanchored islands in the sun, erases history and
geography in favor of a homogeneous and

Figure 26.2 *Logo for CARICOM*

harmonious ecological connectivity that, despite differences in language, ethnicity and historical experience, gives the region a materiality that might otherwise be difficult to imagine. The two large black Cs at the foreground of the logo double as both the initials of the organization, and two open links in a chain. As links in a chain they symbolize unity and interdependency among members, yet as *broken* links they mark freedom from the chains of colonial bondage. The dual symbolism presented by the chains/links testifies to a fundamental ambivalence underlying regionalism in the Caribbean.

In the service of a neoliberal agenda, the logo conveys an image of a region open for business: an interconnected economic space ready to service the needs of multinational capital. Not dissimilar to Cascadia's glossy tourist brochure, CARICOM's logo erases geographic difference and opens the whole region as a marketable tourist destination. The ever expanding cruise ship industry is attracted to the harmonious and indifferent 'Caribbean' as a vacation destination distant from the messy histories that define and distinguish places. Indeed, when 'real' places threaten to shatter this image, companies such as Disney and Royal Caribbean have simply purchased their own private islands and reconstructed the 'Caribbean Island' where the real cannot puncture the ideal (Orenstein, 1997; Weinbaum, 1997). In addition to place marketing, the image of the island Caribbean also serves to rework local identities in favor of the service industry. Writing about the Bahamas,

Alexander suggests that the island discourse works to rescript racialized bodies and identities as serviceable, compliant and available to satisfy a 'white European longing for what is "rare and intangible"' (1997: 96). Thus, like the boosters of Cascadia, states in the Caribbean find a strategic way to redefine their role in the global economy through the effective mobilization of regionalized images and identities.

While redefining the role of the region on the global political and economic stage, regionalism also performs an ideological function in managing and controlling internal disruptions. Obscuring difference and providing a unitary bond supposedly creates a sense of a larger goal, an ambition that, if rightly orchestrated, could serve to circumvent the local reactions to intensifying globalization. Economic restructuring designed to increase international competitiveness and openness to trade exerts downward pressures on wages, limits market opportunities for small producers, and undermines the power of unions to collectively organize in the region. Accompanied by state deregulation and the privatization of welfare services, the consequences of trade-led development have resulted in accelerated declines in the living standards of poor populations in the Caribbean (Safa and Antrobus, 1992). As the effects of globalization are increasingly felt at the local level, state political actors and corporate elites can sidestep the political pressures arising at the scales where the hardship of poverty and structural adjustment are most felt. In this context state leaders can be repeatedly heard blaming CARICOM and the regional movement for its failure to address their national economic, political and social problems, problems for which they have no effective resolution at the national scale.

Embedded within this discourse of economic competitiveness and global participation, however, is also a discourse of anti-imperialism and a struggle for independence. For small dependent states, regionalism has been a means to protect and extend economic and political independence in an increasingly neoliberal global economy. In the 1980s, in the context of mounting debts, growing internationalization of production, and US military intervention, independence in the region was increasingly defined in terms of strengthening links with global powers other than Europe. The signing of the Caribbean Basin Initiative (CBI) with the US in 1982 promised a preferential package of trade, investment and aid that tied the region much more closely to the American political economy. Introduced in 1983, the CARICOM logo celebrates independence, at the same time announcing the arrival on the global stage of the Caribbean as a region made newly independent and unconstrained by the forces of British colonialism. In the context of growing fears about being left out of an increasingly competitive and integrated global economy, regionalism continues to be invoked as the solution to the region's marginalization and as a source of independence in an increasingly interdependent global economy (Conway, 1998; Elbow, 1997).

Parallel to concerns about economic self-reliance is also a desire to right historic wrongs, which in an exercise of domination have divided the peoples of the region. Writing about the history of regionalism in the Caribbean, Manuel Zapata Olivella suggests:

> the Americas and the Caribbean have been arbitrarily divided up in accordance with the whims of Popes, empires, geographers and politicians ... seen as separate economic units, split up in accordance with the interests of those who were unaware of the existence of ecological ties between our ancient regions, ethnic groups and civilizations. (1999: 165)

Here Olivella challenges the perception that nations in the Caribbean are the natural geographic containers of Caribbean culture, politics and economics. With other proponents of integration, he considers regionalism to be a strategy of post-colonial resistance, a strategic scale-jump, necessary for the fulfillment of a historical destiny denied to Caribbean peoples by the violent interruption of colonialism. CARICOM retrieves this shared cultural history to convey a common ground upon which regionalism can be promoted as the right and fitting conclusion to centuries of displacement, domination and separation. The double play on chains, as both links between islands and graphic reminders of enslavement and bondage, plainly marks the significance of the anti-colonial struggle underlying Caribbean regionalism. Jumping up to the scale of the region can therefore be interpreted as an attempt to find a spatial resolution to the twin forces of colonial domination and global competition. Again, however, it is an ambivalent process that revives memories of economic exclusion and discourses of subordination at the same time as promising a neoliberal future based on global economic participation. On the one hand the appeal to this collective history is an appropriation and depoliticization of a radical historical memory. On the other hand, however, the articulation of the anti-colonial discourse into ideologies of neoliberal economic growth revives the memory and inserts it into the contemporary period where struggles against marginalization and domination are constantly made and remade.

While the logo of Caribbean unity culturally constructs a sense of a unified economic region, it contains within it traces of alternative and often conflicting readings of regional space and belonging. Unraveling the cultural, political and economic processes producing scales draws attention to these alternative scriptings, and highlights potential fissures or moments of political resistance within the play on unity. The emphasis on a history of colonialism in the Caribbean keeps alive a collective spirit of resistance that many non-governmental organizations are now utilizing to hold CARICOM accountable for its neoliberal development policies. As scale-jumping attempts to transform the spaces in which economies are organized, it simultaneously transforms the spaces of everyday life and creates new opportunities for the articulation of collective demands at a scale other than the nation-state. In the next section, we examine in detail how non-state, non-corporate actors are also able to jump scales. Like many groups within the Caribbean, Mexican migrants are themselves reworking the transborder space of the US–Mexico border on their own terms.

FINDING AGENCY IN SCALE-JUMPING: CROSSING LA FRONTERA

The caption on the cultural artifact in Figure 26.3 may be translated as follows:

Figure 26.3 *The retablo of Braulio Barrientos* (from Durand and Massey, 1995 © The Arizona Board of Regents 1995)

Rancho Palencia, San Diego de la Unión, Guanajuato. January 11, 1986. On this date I dedicate the present retablo to the Virgin of San Juan for the clear miracle she granted on the date of June 5, 1985. Re-emigrating to the United States with three friends, the water we were carrying ran out. Traveling in such great heat and with such thirst, and without hope of drinking even a little water, we invoked the Virgin of San Juan and were able to arrive at our destination and return to our homeland in health. In eternal gratitude to the Virgin of San Juan de los Lagos from the place where you find Braulio Barrientos.

Striking in their color and stark beauty, retablos hold an important place in Mexican culture and art. Retablos are small tin paintings left at religious shrines to offer public thanks to a divine image for a miracle or favor received. In *Miracles on the Border* (1995), Durand and Massey bring together a vivid collection of retablos commissioned by Mexican migrants. In doing so, they argue that the retablos 'provide a spiritual and cultural anchor for Mexicans in the northern diaspora, giving them a familiar cultural lens through which they can interpret and assimilate the fragmented and often disorienting experiences of life in an alien land' (1995: 4).

In the retablo of Braulio Barrientos, the author thanks the Virgin of San Juan for bringing him and his friends safely to the United States. The image brilliantly evokes the danger of the border. The sun bears down on the men as one sits dejectedly with his empty water jug among the cacti and scrub brushes. The retablo is a graphic reminder of the fact that a 'borderless' world is only really borderless for a few, that the scale-jumping that seems so easy for neoliberal boosters and planners reflects an altogether much more embodied and difficult movement for the subaltern migrant. The result is a landscape vision of the area around the border as a transnational space fraught by the violence of border policing and the perils of border crossing. Yet, the retablo also evokes the migrants' agency in negotiating that space. It is clear from the retablo's words that this is not the first time the men have made the journey to *el norte*. The narrative is one of re-emigration, and the supplications to the Virgin are thanks not only for surviving the journey to the United States, but also for returning them safely to their homeland, Guanajuato, Mexico.

The retablo thus tells one migrant's story of the emergence of a transnational region between the United States and Mexico. The retablo captures the everyday practices of migrants operating in the US–Mexico transnational space, and thus registers their own scale-jumping reterritorialization of the two nations. The landscape vision also works to facilitate further transmigration, as stories of successful crossings and passages home become woven into the Mexican cultural fabric.

The transnational migration networks that have developed across the US–Mexico border are embedded in historical geopolitical and economic relations, and have arguably existed for many centuries. As with the Canada–US relationship, however, NAFTA has extended economic and political links between the United States and Mexico, consolidating a transnational economy within which commodities and investments move freely across the border. Paradoxically, the porosity of borders to the flows of capital has led to a contradictory increase in the militarization of the US–Mexico border that attempts to limit the flow of (some) people across it (Andreas, 1998–9; Connolly, 1996; Nevins, 2001; ÒTuathail et al., 1998). This concomitant opening of the borders to trade and closing of the borders to immigrants from the south is legitimized through a neoliberal agenda that not only promotes freer markets but also argues for the scaling back of government expenditures (such as education, healthcare and welfare benefits) for 'undeserving' populations.

The retablo shown, however, suggests that migrants are not passive agents but are in fact also 'jumping scales', and in the process actively transforming the spaces of the US–Mexico border. While the US government continues to construct political franchise and access to social services within its own borders, migrants are engaging in transnational practices that delink social relations from a territorially defined nation-state. It is worthwhile to be careful about construing transmigration as an essentially oppositional or subversive practice (see Guarnizo and Smith, 1998; Mitchell, 1997; Ong, 1999). It is often as much a strategy for enabling capitalist flexibility as it is for contesting hegemonic narratives. However, to miss the way in which Braulio Barrientos is reframing and thus rescaling territory from his perspective is to ignore the way in which human agency and subjectivity formation are just as much part of the construction of scale as are the systemic transformations of a capitalist political economy.

One way we can see transmigrants jumping scales pertains to the ways in which they are responding to the erosion of the welfare state and the denial of public benefits for immigrant families. In interviews undertaken by one of the authors in southern California,[1] migrants discussed how economic instability and uncertainty are forcing them to devise economic strategies that rely on flexibility and distributing resources across multiple locales. Paula, a Mexican migrant who works as a maid in a hotel in San Diego, recognizes her tenuous position in the US economy and sees maintaining ties to Mexico as an important escape route: 'I listen carefully and when things here start to get too bad I can go back to Mexico … [with my] savings I can survive there.' Paula knows she will not be eligible for government welfare assistance if she loses her job. But because 'a dollar counts more' in Mexico, she is expecting to use her savings to settle there rather than in San Diego. Ironically, Paula is capitalizing on the same boundary of 'difference' constructed by the scale of the nation-state as are multinational corporations when they move south of the border to escape labor or environmental regulations. Other interviews illuminated the ways in which migrants remained politically active in their home villages (see also Basch et al., 1994; Smith, 2001). Such transnational practices problematize the idea that politics are always local, and force us to focus instead on how migrants construct their own cultural geographies across multiple scales simultaneously (Grewal and Kaplan, 1994).

Whether or not all migrants are conscious of the political ramifications of their transnational practices, they are nevertheless an important force rescaling political and economic relations between Mexico and the United States. The Mexican state has begun to extend citizenship rights – including health and welfare benefits, property rights and voting rights – to nationals living in the United States, thus reconceptualizing the historic links between citizenship and the nation-state to accommodate them. Although this can also be seen as an attempt to coopt the wealth and political support of those abroad, it nevertheless shows the way in which migrants themselves play a role in creating a new 'scalar fix' for the Mexican state. Like the retablo, then, the embodied and clearly voiced cultural geographies of migrants' daily practices underline the need to consider the role that transnational practices 'from below' play in the counter-hegemonic construction of scale (see also Silvern, 1999).

CONCLUSION

The three case studies discussed in this chapter give empirical specificity to the spatial transformations resulting from and contributing to the scale-jumping associated with the recent neoliberal restructuring of the Americas. They highlight not only how scale reflects the changing territorial scope of capitalist economic organization, but also the complex ways in which this is ideologically refracted, culturally coded and resisted. The boosterish visions of a neoliberal utopia in Cascadia, the struggle for postcolonial independence in an interdependent global economy in the Caribbean, and the lives of migrants operating at the US–Mexico border all highlight how scale is reproduced through the overdetermination of particular cultural geographies. Scale in these contexts is not simply rewritten by the juggernaut of global capital, but in each landscape it is produced through negotiation with individual and collective subjects, local histories and the environment.

We have presented the social construction of scale – at the moment of scale-jumping – as a complex and contradictory process engaged by multiple actors in political struggles that span continental geographies and the spaces of everyday life. On the one hand, scale-jumping provides an abstract framework through which it becomes theoretically possible to witness the re-placing and remapping of the scope of power relations. On the other, cultural geographers' attention to landscape, text and identity illuminates how socially and culturally inscribed agents struggle over ideology and meaning systems, and in turn interact with patterns of governance to form a scalar fix (Marston, 2000). This approach both deepens our understanding of cultural geography and opens a conceptual space to broaden our definition of what constitutes relevant subject matter for understanding the rescaling of the territorial scope of power relations. Paintings by migrants, organizational logos and tourist brochures suddenly become critical sites for understanding patterns of economic and political restructuring and the ideological (dis)placements through which power is mediated, organized and struggled over.

Our objective in this chapter has been to open geographers' research and praxis to a greater critical sensitivity to how power operates through the overdetermination of scale. Placing the problem of scale-jumping, as we have in our three examples, works to undermine dualistic notions of domination and resistance and suggest more creative ways to approach the (re)production of cultural landscapes of power, domination and resistance under neoliberal restructuring. Our case studies also show how cultural landscapes of scale-jumping are multiply determined sites of contestation and struggle. Just as they are the sites through which power is exerted, they are also constantly being rewritten and reworked through the production of collective subjectivities and sometimes radically resistant and profoundly human geographies too.

NOTE

1 The interviews cited in this chapter were conducted as part of Carolina Katz's (2000) master's thesis research. For additional information about methodology and content, please contact this author directly.

REFERENCES

Alexander, J.M. (1997) 'Erotic autonomy as political decolonization: an anatomy of feminist state practice in the Bahamas tourist economy', in J.M. Alexander and C.T. Mohanty (eds) *Feminist Genealogies, Colonial Legacies, Democratic Futures*. New York: Routledge.

Anderson, K. and Gale, F. (1991) *Inventing Places: Studies in Cultural Geography*. New York: Wiley.

Andreas, P. (1998–9) 'The escalation of U.S. immigration control in the post-NAFTA era', *Political Science Quarterly* 3 (4): 591–615.

Artibise, A. (1994) *Opportunities of Achieving Sustainability in Cascadia*. Vancouver: International Center for Sustainable Cities.

Basch, L., Schiller, N.G. and Szanton Blanc, C. (1994) *Nations Unbound: Transnational Projects, Post-colonial Predicaments and Deterritorialized Nation-States*. New York: Gordon and Breach.

Brown, M. (1997) *RePlacing Citizenship: AIDS Activism and Radical Democracy*. New York: Guilford.

Brown, M. (2000) *Closet Space: Geographies of Metaphor from the Body to the Globe*. London: Routledge.

CARICOM (1999) 'The CARICOM standard'. Accessed via the internet on 10 March 2001 at http://www.caricom.org/standard.htm.

Chapman, B. (1996) 'Cooperation not competition, key to Cascadia Region success', *The Seattle Post Intelligences*, 14 June: A16.

Comaroff, J. and Comaroff, J. (2000) 'Millennial capitalism: first thoughts on a second coming', *Public Culture* 12 (2): 291–344.

Connolly, W.E. (1996) 'Tocqueville, territory, and violence', in M.J. Shapiro and H.R. Alker (eds) *Challenging Boundaries: Global Flows, Territorial Identities*. Minneapolis: University of Minnesota Press. pp. 141–64.

Conway, D. (1998) 'Microstates in a macroworld', in T. Klak (ed.) *Globalization and Neo-Liberalism: The Caribbean Context*. Oxford: Rowman and Littlefield.

Duncan, J. and Gregory, D. (2000) *Writes of Passage: Reading Travel Writing*. New York: Routledge.

Durand, J. and Massey, D.S. (1995) *Miracles on the Border: Retablos of Mexican Migrants to the United States*. Tucson: University of Arizona Press.

Elbow, G.S. (1997) 'Regional cooperation in the Caribbean: the Association of Caribbean States', *Journal of Geography* 96 (1): 13–22.

Goldberg, M.A. and Levi, M.D. (1993) 'The evolving experience along the Pacific Northwest corridor called Cascadia', *New Pacific* 3 (Winter): 29–32.

Gregory, D. (1996) *Geographical Imaginations*. Oxford: Blackwell.

Grewal, I. and Kaplan, C. (1994) 'Introduction: transnational feminist practices and questions of postmodernity', in I. Grewal and C. Kaplan (eds) *Scattered Hegemonies: Postmodernity and Transnational Feminist Practices*. Minneapolis: University of Minnesota Press. pp. 1–33.

Guarnizo, L.E. and Smith, M.P. (1998) 'The locations of transnationalism', in M.P. Smith and L.E. Guarnizo (eds) *Transnationalism from Below*. New Brunswick: Transaction. pp. 3–34.

Hall, S. (1988) *The Hard Road to Renewal: Thatcherism and the Crisis of the Left*. London: Verso.

Harvey, D. (1999) *The Limits to Capital*. New York: Verso.

Henderson, G. (1999) *California and the Fictions of Capital*. Oxford: Oxford University Press.

Henkel W. (1993) 'Cascadia: a state of (various) mind(s)', *Chicago Review* 39: 110–18.

Katz, C. (2000) 'Remapping rights and responsibilities: a legal geography of the 1996 welfare and immigration reforms'. MA thesis, University of Washington, Seattle, Washington.

Laclau, E. and Mouffe, C. (1985) *Hegemony and Socialist Strategy: Towards a Radical Democratic Politics*. London: Verso.

Marston, S. (2000) 'The social construction of scale', *Progress in Human Geography* 24 (2): 219–42.

Mitchell, D. (2000) *Cultural Geography: A Critical Introduction*. Oxford: Blackwell.

Mitchell, K. (1996) 'Visions of Vancouver: ideology, democracy and the future of urban development', *Urban Geography* 17 (6): 478–501.

Mitchell, K. (1997) 'Different diasporas and the hype of hybridity', *Environment and Planning D: Society and Space* 15: 533–53.

Moore, D. (1997) 'Remapping resistance: "ground for struggle" and the politics of place', in S. Pile and M. Keith (eds) *Geographies of Resistance*. London: Routledge.

Nevins, J. (2001) *Operation Gatekeeper: The Rise of the 'Illegal Alien' and the Remaking of the US-Mexico Boundary*. New York: Routledge.

Olivella, Z. (1999) 'The Caribbean Basin: a global overview', *Association of Caribbean States: Integrating the Caribbean*. London: International Systems and Communications Limited in conjunction with the Association of Caribbean States. pp. 165–70.

Ong, A. (1999) *Flexible Citizenship: The Cultural Logics of Transnationality*. Durham: Duke University Press.

Orenstein, C. (1997) 'Fantasy island: Royal Caribbean parcels off a piece of Haiti', *The Progressive* 61 (8): 28–31.

Ò Tuathail, G., Herod, A. and Roberts, S. (1998) 'Negotiating unruly problematics', in A. Herod, G. Ò Tuathail and S. Roberts (eds) *An Unruly World? Globalization, Governance and Geography*. London: Routledge. pp. 1–24.

Roberts, S. (1998) 'Geo-governance in trade and finance and political geographies of dissent', in A. Herod, G. Ó Tuathail and S. Roberts (eds) *Unruly World? Globalization, Governance and Geography*. New York: Routledge. pp. 116–34.

Safa, H.I. and Antrobus, P. (1992) 'Women and the economic crisis in the Caribbean', in L. Beneria and S. Feldman (eds) *Unequal Burden: Economic Crisis, Persistent Poverty, and Women's Work*. Boulder: Westview.

Schell, P. and Hamer, J. (1995) 'Cascadia: the new binationalism of western Canada and the U.S. Pacific Northwest', in R. Earle and J. Wirth (eds) *Identities in North America: The Search for Community*. Palo Alto: Stanford University Press. pp. 140–56.

Schoonmaker, P., von Hagen, B. and Wolf, E. (1997) *The Rain Forests of Home: Profile of a North American Bioregion*. Washington: Island.

Sharp, J., Routledge, P., Philo, C. and Paddison, R. (eds) (2000) *Entanglements of Power: Geographies of Domination/ Resistance*. New York: Routledge.

Silverman, K. (1983) *The Subject of Semiotics*. New York: Oxford University Press.

Silvern, S.E. (1999) 'Scales of justice: law, American Indian treaty rights and the political construction of scale', *Political Geography* 18: 639–68.

Smith, M.P. (2001) *Transnational Urbanism: Locating Globalization*. Oxford: Blackwell.

Smith, N. (1984) *Uneven Development: Nature, Capital, and the Production of Space*. New York: Blackwell.

Smith, N. (1995) 'Remaking scale: competition and cooperation in prenational and postnational Europe', in H. Eskelinen and F. Snickars (eds) *Competitive European Peripheries*. Berlin: Springer. pp. 59–74.

Smith, N. (2000) 'Author's response', *Progress in Human Geography*, 24 (2): 271–7.

Smith, P. (1988) *Discerning the Subject*. Minnesota: University of Minnesota Press.

Sparke, M. (2000) 'Excavating the future in Cascadia: geoeconomics and the imagined geographies of a cross-border region', *BC Studies* 127 (Autumn): 5–44.

Sparke, M. (forthcoming) *Hyphen-Nation-States: Critical Geographies of Displacement and Disjuncture*. Minneapolis: University of Minnesota Press.

Spivak, G.C. (1988) *In Other Worlds: Essays in Cultural Politics*. New York: Routledge.

Sutherland, J. (1997) 'Natural selection', in M. Beebe (ed.) *Cascadia: A Tale of Two Cities, Seattle and Vancouver, B.C.* New York: Abrams. pp. 40–3.

Swyngedouw, E. (1992) 'The mammon quest: glocalization, interspatial competition and the monetary order: the construction of new scales', in M. Dunford and G. Kafkalas (eds) *Cities and Regions in the New Europe: The Global–Local Interplay and Spatial Development Strategies*. New York: Wiley. pp. 39–67.

Swyngedouw, E. (1997a) 'Neither global nor local: "glocalization" and the politics of scale', in K. Cox (ed.) *Spaces of Globalization*. New York: Guilford. pp. 137–66.

Swyngedouw, E. (1997b) 'Excluding the other: the production of scale and scaled politics', in R. Lee and J. Wills (eds) *Geographies of Economies*. London: Arnold. pp. 167–76.

Webb, A. (1995) 'Promoting the two nation vacation', *Puget Sound Business Journal* November: A5.

Weinbaum, B. (1997) 'Disney-mediated images emerging in cross-cultural expression on Isla Mujures, Mexico', *Journal of American Culture: Studies of Civilization* 20 (2): 19–29.

Wright, M. (2001) 'A manifesto against femicide', *Antipode* 33 (3): 550–66.

27

Environmental Geopolitics – Nature, Culture, Urbanity

Simon Dalby

GEOPOLITICS AND THE PROBLEM WITH NATURE

Geopolitics is about the largest scale considerations of power and space in human affairs. From the beginning of the twentieth century in the writings of Halford Mackinder, Rudoph Kjellen and Friedrich Ratzel through to the end of the century volumes by Samuel Huntington (1996) and Zbigniew Brzezinski (1997), empires, civilizations and states have been understood as the primary territorial entities in competition for power, space and influence (Dodds and Atkinson, 2000). But this tradition, understood as one concerned with great power rivalries and struggles for influence and control across a variegated and contested global political space, has also been one in which large scale assumptions about nature and the natural environment have been unavoidably present. They are so both because they imply assumptions about the philosophical questions of humanity, its place and purpose in a larger cosmos, and hence the appropriate organization of human affairs at the largest scales; and also because matters of power politics and superpower rivalry have always been about the expropriation of resources and the destruction of environments in the search for military and economic supremacy. As critical geopolitics writers made clear in the 1990s, geopolitical views are about the construction of the planet as an object of knowledge: practices of knowing the world as a whole that facilitate its division and administration (Agnew, 1998; Ó Tuathail, 1996).

Such imperial vision was in the past frequently the prerogative of emperors, generals, political elites and map makers in capital cities. Whether it was environmental determinism or organicist models of states in competition, the links between traditional geopolitics and questions of nature are far more direct than many formulations of either geography or politics often suggest. But now, as the twenty-first century opens, humanity is becoming an urban species and the geopolitical view of a planet to be divided and ruled is interconnected with urban thinking; nature has been turned into a global environment which has to be managed in the interests of western consumers (Luke, 1999). Now geopolitics is also about the administration of the natural world through management practices of science, ecology and, yes, geography. Where Mackinder (1904) argued that political space was closed at the end of the nineteenth century, geopolitics now also implicitly assumes that nature is now known, explored, enclosed, divided up, a matter for management by the globalized urban culture that dominates human affairs (Dalby, 2002). The implications for this shift require some fundamental rethinking of the premises for cultural geography, and some careful self-reflection on the cultural identities of the geographers who write the planet in such ways.

The argument in this chapter emphasizes that the urban culture within which most geographers live, and which usually specifies itself as separate from wild untamed nature, is also one that has a long colonial history of drawing boundaries and dividing nature into spaces which can be administered and altered to make them orderly (Driver, 2001). But land use change is inevitably also a matter of environmental change, and the consequences of large scale disruptions of 'nature' are now fundamentally challenging the modern assumptions of urban humanity as separate or somehow in control of that nature. What is defined as nature and environment is changing as a result of climate change, ozone holes and such phenomena as radioactive fallout from Chernobyl. So too, once again, is the understanding of humanity's place in relation to 'nature'. In the process the culture that defines itself as apart from nature is being challenged to rethink the assumption of this separation (Beck, 1992; Latour, 1993).

Whether we understand ourselves as modern, urban and apart from a nature that is of no concern to our lives, or as humans who live in an active nature that we change by our everyday actions, has profound consequences for our identities and for how we do geography. City 'planning' and colonial reorganization of nature to conserve 'resources' are both modern projects that share the assumptions of a nature out there to be controlled by being spatially reordered to human design. Globalization means that these themes now matter at a planetary scale and our thinking has to reflect this change, not least by incorporating them into a critique of the taken for granted premises of the modern geopolitical vision that surveys the whole world as an object of knowledge (Luke, 1997).

GEOGRAPHY AS URBAN SCHOLARSHIP

At the end of the 1980s Margaret FitzSimmons (1989) addressed these issues in the discipline, arguing that many of the more radical and critical approaches to geography had problems thinking about nature because of the preoccupation with spatial themes. She drew on Richard Peet's (1977) earlier argument to suggest why space rather than nature was the focus of concern. Space was an important theme in the 'new' geography of the 1950s and 1960s which applied statistical modeling and mapping to examining economies and cities. When the radical approaches, and the Marxist inspired literature in particular, criticized the liberal assumptions in the new geography, it also focused on spatial matters and related questions of justice in urban settings. FitzSimmons went on to suggest that the difficulty of dealing with nature is in part also

because intellectuals live and work mainly in the artificial spaces of large cities. Coupled to this was the growing role of science in shaping the intellectual practice of these urban intellectuals, most obviously in geography in terms of biophysical research and resources management. All this, she suggests, adds up to a substantial blindness to the complex abstractions of 'nature.'

This is not, of course, the whole disciplinary story. David Harvey (1974) wrote a powerful critique of Malthusian arguments in the early 1970s which had implications for the analysis of questions of natural resources. In the 1980s the concern with the politics of nature became part of discussions in geography and more critical perspectives in particular as the themes of political ecology emerged. These used analyses of underdevelopment to investigate poverty and linked them to practical understandings of environmental change, literally 'on the ground' in terms of soil erosion (Blaikie and Brookfield, 1987). This concern with environmental change was connected directly to the daily struggles of rural people, and farmers specifically, in the face of the disruptive encroachments by capitalist agriculture. Extended to rethink the 'environmental' assumptions in hazards research, such considerations have made clear how the uneven geography of social systems is a key factor causing vulnerability (Blaikie et al., 1994).

FitzSimmons (1989) argued that urbanization is about the reconstitution of social life and about the geographical differentiation between city and countryside. It also encourages a distinction between the urban and a pristine nature with humans constructed as external to nature. These ontologies reflect some of the most powerful dichotomies in the Enlightenment culture that has shaped knowledge production and science in particular (Latour, 1993). Distinctions between nature and culture are some of the most persistent cultural dualisms of our time. The efficacy of numerous moves of ideological 'naturalization' to render matters 'true' because 'natural', and hence beyond political debate, suggests its practical importance. FitzSimmons finally suggests that the sociological fissures in the discipline of geography accentuated these conceptual difficulties with urban and economic geographers emphasizing 'scientific' epistemologies and methods and rural geographers frequently making the anthropological emphasis on cultural analysis.

In the 1990s, partly influenced by the debates in social theory, geographers produced scholarship that bore out some of FitzSimmons' hopes that critical geography would overcome these difficulties. The further elaboration of work on political ecology has strengthened the links between the rural and the urban (Peet and Watts, 1996). Cultural geography has begun to think about nature in innovative ways as recent work on youth culture in particular suggests. Work by Cindy Katz (1998) on teenagers in rural Sudan and urban New York makes links between the processes of globalization in very different circumstances. The confrontation between rural English residents and the itinerant caravans of 'new age travelers' has emphasized the importance of conceptions of nature and landscape in contemporary politics (Hetherington, 1998). But, with the obvious exception of such work as Joanne Sharp's (2000) consideration of the construction of American identities in the Cold War discourses of popular geopolitics, the new cultural geography has yet to be extended in any great detail to the largest of scales and linked to the spatial concerns with geopolitics. Likewise this is only beginning in the theoretical discussions of political ecology (Braun and Castree, 1998).

Extending the themes of both cultural geography and political ecology to the largest scale emphasizes the importance not only of understanding capitalism as simultaneously constructing space and nature, as Neil Smith (1990) has argued, but of conceptualizing human urban culture as an increasingly important 'natural' force changing the biosphere. In Bruno Latour's (1993) terms we have to think in terms of hybrids now, or as the argument in this chapter suggests following Latour's inspiration, in terms of a single planet sized hybrid where artificial factors increasingly shape ecosystems and change the composition of the biosphere overall. The implications of such an ontological shift, this chapter suggests, are not just deconstruction of the nature–culture and urban–rural dichotomies, but a recognition of the geopolitical situatedness of social entities in a single changing planetary ecosystem. The emergence of such a planetary understanding of interconnections has significant implications for geographical thinking in the new millennium. Mackinder (1904) suggested that the end of the nineteenth century marked the end of the 'Columbian' age, in that most of the globe had been explored and drawn into the European dominated global economy. Now the concern with ozone holes, climate change, ocean fish stock depletion, declining biodiversity and other 'environmental' hazards extends this concern with various limits and suggests the end of the age of assumptions of infinite economic opportunity based on natural resource exploitation.

Put bluntly in the vernacular: the widespread cultural assumptions that 'we' are 'here' in a specific place 'on' earth has to be exposed for the powerful political illusion that it has so long been. We are not on earth. We are earth. We are not 'here', safe in our cities. We are interconnected through a complex web of economic activities and communications, which are also ecological actions, to remote supplies of resources, as well as to distant factories, which have complex consequences in a way that ensures that our actions 'here' are inevitably also actions 'there'. The implications of such understandings inevitably mean challenging the geopolitical categories that so inaccurately specify the political identity that legitimates 'us' in our various places while simultaneously obscuring the consequences of the interconnections that make 'our' urban based political economy possible. The converse argument is likewise crucial: invocations of the global danger of environmental threats can also be used to once again invoke the colonial necessity for 'us' to intervene to 'manage' change in particular places that may or may not meet the expectations and approval of whoever lives 'there'. This is where geopolitics connects directly to matters of 'global' environmental management (Dalby, 2002).

GEOPOLITICS

The complex politics of the cultural specifications of our identity as geographers unavoidably link relatively new concerns with cultural studies analyses in the discipline to the older themes of political geography. Earlier geographical analyses of the struggles for world power understood that geography is an important factor. Mackinder's famous heartland argument is formulated in part in terms of the drainage patterns of Eurasia. Moodie noted that while human geographers focus on geographical regions as their unit of analysis, 'the areal unit of the political geographer is the State' (1949: 13), which rarely coincides with natural regions. The mismatch between states and environments has long been a matter of concern, although the particular preoccupation with attempting to derive comprehensive regional designations for the earth's surface, that informed Moodie's writing, has declined in importance as the new geography shifted focus away from regional themes.

Nonetheless the important implication in these formulations is the frequency with which states import resources precisely because they do not coincide with the geography of the resources that are used within their boundaries. The traditional geopolitical focus on the spatial circumstances of disputes over access and boundaries or control over key geostrategic features emphasized military matters and particular spaces. In so far as resources are understood as natural phenomena with a complicated geography, access to them in the Cold War linked the spatial control themes to discussions of nature (Lipschutz, 1989). Although frequently used as rationalizations for other political actions the geopolitical arguments about resources remain especially powerful when petroleum supplies are in question, especially in the Middle East. This argument was usually subsumed in a larger formulation of the importance of maintaining western control over the oceans to ensure trade routes and military supply lines around the world. Extended to a discussion of the interactions of nature and humanity at the largest of scales these trade patterns now show the global reach of the resource use by consumers in the largest of cities (Redclift, 1996). Prawns from fish farms, fuel in taxicabs, vegetables flown across the world, the flowers at Kensington Palace when Princess Diana died, are all taken for granted items of London life which come from distant locations and depend on a particular geopolitical order for their provision.

This extension of the scale of analysis is one of the themes that distinguishes some earlier geopolitics from current arguments:

> The perspective of the old geopolitics focused on one particular segment of territory, the state, and on the single-minded pursuit of what were regarded as being its best interests even if these led, as all too often they did, to confrontation and war. Its recurring themes were space and power and the relationship between the two. The perspective of the new geopolitics on the other hand, is global, and its fundamental proposition is that the world as a whole is the proper unit for the addressing of those issues which have global repercussions. There is no such thing as the solution of a 'local' issue seen in isolation and out of its wider context. (Parker, 1998: 55)

While this is a broad generalization, the important point that many of the analyses of critical geopolitics have been making in the last decade is that the common sense and taken for granted geographical specifications of political identity that structure discussions of these things are part of what needs to be analyzed (Ó Tuathail, 1996). The cultural categories through which environmental and geopolitical matters are discussed are important to understanding how politics works.

Cultural change is about political change, not least because the language and categories of what are acceptable behaviors, and who is empowered to act and make decisions, are part of the cultural repertoire available for debate. Many of these themes are especially clear in current discussions about globalization which update and extend the assumptions built into earlier geopolitical thinking. Parker's comment about the change of scale needs careful explication if the old geopolitics of statist conceit is not to once again constrain the political imaginary.

The collapse of the obvious distinctions between local and distant, large and small, us and them, is much of what the discussion about globalization is about. Although it is understood sometimes narrowly as a matter of the ideological logic of neoliberalism, with the repeated ideological theme of its inevitability (Harvey, 2000), or alternatively as a matter of the cultural homogenization brought about through market integration and the export of American movies and television programs, globalization is also about the cultural impacts of the change in spatial sensibilities. This is often resisted by calls for the restriction of immigration and the protection of geographically distinct cultures in tropes that emphasize orderliness with everything fixed in its place, and assumptions of stable organic communities.

The urban aesthetic of orderliness and the necessity to civilize wilderness have had other powerful manifestations in recent history. The fascination with English style lawns in North America has produced a cultural naturalization of this constructed artifact of status and political stability which effectively aestheticizes lawns at the cost of environmental health (Feagan and Ripmeester, 1999). The European colonial project in Africa was frequently about the desire to bring orderliness to what were understood to be chaotic landscapes. But this was a European sense of orderliness which suggested that wilderness be tamed and organized to be productive in terms of specifically European models of agriculture and property (Doty, 1996). Native understandings of landscape and ecology did not fit into these productivist specifications of land as resource and of monoculture as the highest form of the agricultural art. The dispossession of those who did not fit the colonial managers' landscape sensibilities was simply part of the process of 'improvement'. Postcolonial states often continue such policies, using urban based expert knowledge to render rural ecologies into resource producing regions (Scott, 1998).

The design of Central Park in New York is usually taken to be paradigmatic of the nineteenth-century city garden model of the park, but the point is that Olmstead's design is part of a larger aesthetic of nature that was to be incorporated into the later movement for national parks in the United States and elsewhere. Parks and moral spaces are directly linked here in the urban geopolitics of policing and controlling specific areas in a city and in the assumptions about hygiene that underpin both the rationales for 'green spaces' and police discourses of security (Herbert, 1997). In both cases the assumption of nature as effectively devoid of humans was important. In the garden parks people visited but did not live; likewise in nature 'reserves', game parks and national parks. Human habitation is precisely what parks do not have. This argument can be extended to colonial practices also where native inhabitants were excluded from traditional lands to set them aside as game reserves for affluent tourists. Moving homeless people out of urban parks follows the same general pattern (Smith, 1997).

In this sense, might we read the 'problem' of dispossessed peoples destroying marginal environments in the 'south' as being analogous to the problems of homeless peoples in American cities? In both cases marginal populations violate the spatial codes of empty parks and orderly landscapes. But as conservation practitioners have learned to their cost over many decades, the strategy of drawing distinct boundaries around areas slated for preservation in hopes of preserving 'nature' intact is usually doomed to failure because ecological processes are not limited to such bounded spaces (Botkin, 1990). But nature and culture are not so easily separated; their interconnections are unavoidable, no matter what conventions cartographers use to render these landscapes. Environmental matters are about interaction and connection, not about discrete places and dividing lines, whether these are property lines, park boundaries or state borders.

The aesthetics of planning and urban parks are ones of order and symmetry, nature controlled and tamed prior to its reorganization and presentation in a fashion conducive to providing the benefits of 'nature' to urban populations. Gardens and interpretation centers for parks provide orderly landscapes in which the wild chaos of nature is controlled and presented in a manner that acts to distance the park visitor from that nature. The 'experience' of nature is one of separation and the production of 'experience' rather than an unmediated encounter. In Derek Gregory's (1994) terms, nature is produced as an exhibition for modern consumers. Ironically of course such manufactured experiences have value not only to the customer of the park, but to the preservation of the environment where ignorant urban dwellers are kept off the most easily disrupted ecological spaces.

The importance of Neil Smith's (1990) analysis of capitalism as simultaneously producing nature and space becomes especially clear in the context of thinking about environmental matters at the largest scales. Historically the largest changes humanity caused in the biosphere were matters related to agriculture and 'clearing' land (Diamond, 1997). The ecological changes of habitat were obviously also enhanced by hunting practices. Capitalism has furthered these processes of habitat change through the accelerated commodification of numerous facets of nature and the expansion of the system through colonialism and subsequently globalization. State boundaries at the largest scale have enclosed 'nature' and facilitated its transformation into 'property'. Property has in turn been cleared to 'improve' the land and make it ready to produce commodities for the cities. Rural and urban are profoundly interconnected even as the powerful rhetoric of such places obscures these connections.

Henri Lefevre (1991) notes this cartography of urban and rural at the scale of the French state which links directly once again to matters of geopolitics. Writing in the 1970s he noted that there were two maps of France: 'From Berre-l'Étang to Le Havre via the valleys of the Rhône (the great Delta), the Saône and the Seine, this stripe represents a narrow over-industrialized and over-urbanized zone which relegates the rest of dear old France to the realm of underdevelopment and "touristic" potential' (1991: 84). But Lefevre notes that these areas are at once places where historic authenticity is coded for the visitors' edification and also places where the military has appropriated large territories for its purposes. This division of landscape is not unique to France. Whether it is the remote lands of Labrador where NATO planes disrupt the caribou herds, the glens of the highlands of Scotland where RAF Tornados practice low level flying over hiking trails, or the military destruction of the supposedly empty desert of 'Marlboro country' in the United States (Davis, 1993), such hinterlands are at once nature 'reserves' and the training grounds for the sophisticated technologies of contemporary geopolitical power exercised in capital cities.

GOOD PLANETS ARE HARD TO FIND

The rise in concern about environmental issues was especially prominent in the 1960s in the aftermath of the publication of Rachael Carson's critique of pesticide technology, a spinoff of military research too, in *Silent Spring* (1962). High profile oil tanker disasters alerted publics to the hazards of petroleum transport. Poisoning by mercury effluents in Japan gave the world the phrase 'Minimata disease'. Pollution became a cause that mobilized political constituencies to demand control over corporate behavior. In the 1960s, too, widespread fascination with the NASA moon exploration program gave humanity powerful photographic images of the earth as a single entity. The theme of an endangered and unique planet shaped the background report prepared for the 1972 United Nations conference on the human environment, titled *Only One Earth* (Ward and Dubos, 1972). The cover of the paperback edition of this volume is, not surprisingly, a NASA photograph of a small blue earth set against a large, dark and empty sky.

Through the 1970s these cultural themes played their part in controversies over the potential impact of planned supersonic air travel and the first substantial scare concerning ozone depletion which also focused on CFC propellants in aerosol cans. The OPEC oil boycott of states supplying arms to Israel during the October war of 1973 was followed by large price increases for petroleum products. The link to matters of geopolitics was also made when concern about ozone depletion in the event of a nuclear war was expressed in the 1970s. This theme was dramatically extended in the early 1980s when the likely ecological consequences of a nuclear war between the superpowers gave additional support to the arguments for nuclear disarmament and brought climate change research directly into high politics in the 'nuclear winter' debate (Turco et al., 1983).

Given that the atmosphere we breathe is measurably different from that which our grandparents inhaled, the division between culture and a nature understood as something separate from culture is breaking down. Just as technology is increasingly producing cyborgs and genetically modified organisms, nature is increasingly constructed by cultural activity. Politics is now about what Beck (1992) calls fabricated uncertainties and risk society, where hazards and threats escape both the constraints of national spaces and the containment of conventional categories of risk assessment. What then is natural in these circumstances?

The speed of the urban transformation has obviously been greatest in the last few decades in the sprawling megacities of the 'south', but the overall tendency has been for growing populations to be increasingly urbanized. As such the demands for resources from afar have accelerated and the impact on distant environments has increased (Gadgil and Guha, 1995). The interconnections that are key to the processes of globalization are obviously about the processes of urbanization. Given the accelerating links between the major urban centers of the world economy, the so-called global cities, it may now be more helpful to consider matters in terms of one global city (Magnusson, 1996). While we obviously do not live in a system that has the whole planet as one continuous built-up landscape, the degree of interconnection of global markets, the ubiquity of the cleverly named VISA cards, and the worldwide interconnection of airline schedules suggests at least an embryonic single system. Might globalization be better understood as accelerated urbanization, with cross-boundary flows of resources and communications simply reflecting the growth in scale of the urban appropriation of resources from various hinterlands? The apparently declining importance of state boundaries might then be understood as an update of Moodie's (1949) observation of half a century ago that the state was a political entity that did not fit into classifications of natural regions. Clearly cities' hinterlands overlap and stretch around the globe. Understood as the global city, the whole planet becomes an interconnected hinterland.

Linked to an understanding of globalization as a process of accelerated interconnections, and with a sense of common destiny in a single biosphere, the possibilities of breaking down the powerful dualism of nature and culture combine with the need to overcome the constraints of urban and rural in political thinking. Thought about at the largest geopolitical scale, this is also about rethinking the patterns of resource use that date from colonial periods. The growth of European dominance on the world scene is mirrored in the continuing patterns of resource extractions from distant places under the rubric of globalization. The scale of these mining, farming and fishing activities has expanded through the twentieth century and integrated world markets link the urban to the rural much more quickly in the global city (Redclift, 1996).

In combination these connections are now beginning to change the biosphere as a whole. Humanity has become a major geomorphic and climate changing force in the planetary eco-system (McNeill, 2000). We are literally remaking some important planetary systems, a process that alarms many people who look to the future of humanity and our habitat.

TERRAFORMING

Scenarios for the future are stock in trade for the science fiction genre, one which is all too easily dismissed with some disparaging comments about either the patent absurdities of 'Star Trek' and the cult of trekkies, or the transparency of its allusions to Cold War geopolitics and political stereotypes. The *Star Wars* movie series' marriage of American revolutionary warfare themes to imperial Rome also captures the geopolitical themes in a pale imitation of Isaac Asimov's (1951) much earlier musings on themes of imperialism, technological innovation and cultural change. The theme of the single completely urbanized planet remains a fantasy of science fiction, the latest rendition, in the '*Phantom Menace*' episode of the *Star Wars* series, being the planet 'Coruscant', easily read as an update of Fritz Lang's 1920s movie images of the '*Metropolis*', as well as another reference to Roman themes. But to lightly dismiss science fiction is to ignore an important cultural genre that is worthy of careful scrutiny by the geography discipline.

More important, it is to ignore some of the most imaginative attempts to think through the likely outcomes of current political and technical difficulties. Of importance to the argument here is the rise in concern with ecological themes and in particular the questions of 'terraforming' or literally planetary engineering necessary to make other celestial bodies inhabitable by humans. Discussed in great detail in Kim Stanley Robinson's (1993; 1994; 1996) *Mars* trilogy, a fictionalization of 'Mars Project' plans for the future (Zubrin, 1996), the impact of space exploration on ecological thinking is unavoidable. It is worth remembering that James Lovelock's (1979) innovative formulation of the Gaia hypothesis was initially spurred by the crucial question posed by NASA about how to decide whether there was life on other planets, and Mars in particular. Arthur C. Clarke first posited the use of geosynchronous communication satellites long before he collaborated with Stanley Kubrick on *2001* (Clarke, 1968). Carl Sagan's popularizations of planetary exploration were not unconnected with either his subsequent work on nuclear winter or his science fiction novels (Turco et al., 1983).

Science fiction novels are frequently driven by some variation of post-apocalyptic struggles or by attempts to deal with the consequences of the twentieth century. Many deal imaginatively with the politics of space travel in the context of struggles on earth over resources and the politics of a polluted and unjustly divided planet (Barnes, 1994). Some explicitly deal with questions of environmental change and speculative discussions that pick up on some of the themes in the contemporary scholarly literature on environmental security (Suliman, 1999). They do so in ways that offer both a window into the contemporary angst about the future and also ways of conceptualizing matters that challenge conventional academic analyses. Some of them explicitly deal with the remaking of suburban landscapes using green technologies and the possibilities of the internet (Clee, 1998).

The best thinking about these themes manages to avoid the twin concerns that worry David Harvey (2000): the utopias of the spatial fix, the environmental engineering that assumes utopia is a geographical arrangement; and the utopias of social process, which ignore the practical geographical realities of lived experience. While the temptations of a technical fix are always present in the genre, the questions that they often explore are the questions of subjectivities and the possibilities inherent in technological arrangements that fundamentally alter the assumptions about the relationships between humans and ecosystems. This science fiction offers a potentially useful pedagogic tool for cultural geographers as well as a useful foil for critical reflection on the cultural assumptions about nature that modern geography has taken for granted for so long. If the current biophysical changes at the planetary scale are to be dealt with seriously, assumptions about consequenceless consumption will need to be fundamentally challenged. Science fiction offers ways of reflecting on such possibilities precisely because it so effectively facilitates a critique of the ontological categories of modern culture and in the process raises questions of how to rethink environmental geopolitics.

ECOPOLITICS

There are some grounds for optimism about the potential for new thinking as ecological economics and planning increasingly connect in innovative ways that are starting to have practical implications precisely because they are starting to ask questions about the distant consequences of local actions (Corbridge, 1998). The links between environmental degradation and human rights and the complex cultural contextualizations of resistance to these forms of development suggest the future directions for a cultural geography sensitive to these matters (Johnston, 1994; Watts, 1998). The Wuppertal Institute's imaginative program to 'green' the north goes further to explicitly think about the implications of changing 'northern' consumption and the possible impacts of such 'greening' on the patterns of southern development (Sachs et al.,1998).

One of the key themes in this thinking, and in the larger literature on post-development that links to such green thinking (Rahnema and Bawtree, 1997), is the old geographical concern with the importance of land use considerations. In particular the use of automobiles in suburbs and edge cities causes both direct pollution problems and indirect damage through the vast amounts of space that roads and car parks require. Higher density living reduces both, and when coupled to attempts to provide food supplies from local organic farms, the total environmental impact of urban living can be substantially reduced. Culture, space and environment are inseparable. In regard to the concern with 'footprints', 'environmental rucksacks' or the environmental space beyond the urbanized areas needed to support these geographies of consumption (Sachs et al., 1998), geopolitics links directly to the geography of 'lifestyle'.

The larger cultural politics of these matters can often also lead to pessimism when the magnitude of instigating such social changes is contemplated, not least because of the speed with which corporations have moved to invoke environmental themes in advertising numerous ecologically dubious products. Representations of nature as a challenge to be overcome by the technical acumen of the sports utility vehicle driver once again reproduce the dichotomies of culture and nature in a way that reinforces the idea of nature as a challenge to be subdued. The Nissan Pathfinder advertisements of 1999 in North America, in which 'nature is more civilized in a Pathfinder', is emblematic of the packaging of environment in the sales of dangerous and fuel inefficient luxury vehicles. But these images are cultural constructions and as such can be contested, as the road protests in Britain in the 1990s have made evident (Routledge, 1996).

Nissan commercials are a long way from Peter Taylor's (1996) speculations about the possibilities of the cultural changes needed for a 'deep green' hegemony of the future. His suggestion of the need for a culture of 'conspicuous asceticism' as a necessary part of a green future illustrates the distance that at least North American life has to move to become in any meaningful sense sustainable. Nonetheless tackling themes such as the advertising representations of nature and how these are used in constructing identity has considerable pedagogic potential for a critical geography (McHaffie, 1997). What such an integration of themes does offer is a way of linking culture and environment, representations of nature and technology and the politics of subjectivity in ways that also engage with the practical politics of how representation, knowledge and power are unavoidable in geographical studies.

Pushing such ways of thinking about cultural geography a little further also suggests both that matters of spatial scale are unavoidable, and that big scale issues are often not best tackled by large scale organizations. The most innovative changes to American automobile culture are likely to come as a matter of urban politics. Los Angeles has apparently finally reached the limit of tolerance for the disutility of urban automobiles. The attempts to respond there by mandating emission limits for vehicles are likely to have impacts on urban planning and automobile technologies well beyond southern California. European cities have also been experimenting with numerous devices to reduce the impact of the automobile just as the elites of the cities of the south embrace the technology as a symbol of their social success (Paterson, 2000).

But while the sense of shared vulnerability symbolized by the icon of the spinning blue marble has considerable political resonance, the assumption that the vulnerability is shared frequently obscures powerful discrepancies in who is currently in danger and how matters might be rearranged to reduce the hazards to the most vulnerable (Athanasiou, 1996). This raises profound questions of justice on a planet with an unsustainable appropriation of natural resources (Harvey, 1996). It also forces scholars to think hard about the categories and definitions they use and matters of the appropriate scales for considering politics and justice.

CULTURE, NATURE, GEOPOLITICS

Rethinking scale alone is not enough without a recognition of the persistent dangers of thinking in what Michael Curry (1992) calls, following Kant, architectonic terms. The impulse to universal claims to correct knowledge is always in danger of linking up to authoritarian aspirations to manage the globe (Gill, 1995). The corporate agenda at the UNCED meetings in Rio in 1992 was obviously partly about linking concerns with global danger to management by the rich and powerful (Chatterjee and Finger, 1994). The temptation to survey and control environments from afar is a colonial temptation both for environmentalists who advocate enhanced policies of resource management (Luke, 1997), and for military technologists who want to use remote sensing and satellite technology to 'monitor' environmental change (Deibert, 1999). Linked to financial panopticons (Ó Tuathail, 1997), the project of global mastery by the urban colonial system that has caused so much damage in the first place moves a few steps further.

Nonetheless there is also considerable intellectual activity that points in the direction of innovative thinking about the big questions in geography. The political ecology literature has begun to link up with the critical literature in cultural studies and the debates about the social construction of science in potentially very productive ways (Braun and Castree, 1998). McHaffie (1997) has suggested the importance of understanding the politics of representation around themes of advertising and corporate specifications of globality. Harvey (1996) makes the question of justice as an integral part of environmental thinking unavoidable, but does not work the geopolitical scale explicitly into his thinking. Local ecological knowledge challenges assumptions of jurisdiction and control from the other end of the geographical scale (Doubleday, 1993). Much remains to be done in problematizing questions of property and sovereignty, to mention only the most obvious spatial categories that are in question at both the macro scale of globalization, and the micro scale of genetically modified organisms and the debates about patenting genes (Miller, 2001). What local community means in a global marketplace is becoming ever more dubious despite its ritual invocation by politicians and activists of all political stripes.

Relating this point to the discussion of the production of nature and the fabrication of uncertainties brings us back to Glacken's (1967) theme of the earth as the divinely designed home of humanity. Clearly the expansion of the scale of economic activity through the twentieth century has changed the terms of this discussion. Humanity, or perhaps more precisely that part of humanity that is making most of the consequential political and economic decisions, is effectively now redesigning the planet. Discourses of architectural deities have to be replaced by discourses of artificial habitat. The implications of such a shift are profound, but they follow from the contemporary scale of human activity and the recognition of the theme of the single lonely planet. Culture and nature are now unavoidably matters of geopolitics.

What is clear from this attempt to think about geopolitics at the largest scale is the irony of the current situation where urban assumptions about both culture and nature are so inappropriate as guides to what Barbara Ward and René Dubos (1972) called 'the care and maintenance of a small planet'. The required cultural shift to recognition of the biosphere as an entity to be lived in, rather than as a planet to be lived on, challenges the boundaries in the geographical discipline just as much as it challenges larger ontological schemes of modern politics (Dalby, 2002). Such a line of argument suggests that crucial questions for the discipline are confronted here at the very largest scale, that of the planet as a whole. Traditions of the discipline focusing on the earth as the home of humanity need to be rethought as a result of such ontological refocusing. Questions of the practicalities of dwelling in a finite but changing planet and the possibility of harmonious or sustainable living are within the ambit of traditional disciplinary concerns. But the assumptions of modernization as the answer have long since lost connection with the pressing matters of what the big geographical questions ought to be on an endangered planet.

The geopolitical view has always been a modern one, an architectonic impulse to view the whole from a distance with the intention of managing it for various purposes, which has shaped practical knowledge construction in many disciplines (Agnew, 1998). The consequences of modern attempts to manage a complex unruly world by simplifying, conquering and rendering it an orderly landscape are precisely the impulses that are often aggravating current difficulties (Scott, 1998). Recent critiques of colonialism provide ample warning against the appropriation of crisis narratives by the world's political elites to justify various political strategies to reorder the landscape by urban fiat (Shiva, 1994). In contrast ecological thinking, coupled to sensitivities for justice and the importance of connection and context, offer some useful intellectual tools for engagement with the more innovative of current social movements. Shifting the focus from development to sustainability might also shake loose some of the remaining vestiges of the categorical structures derived from the colonial discourses of the past, although the political questions of who and what are to be sustained remain paramount.

An ontologically self-aware geography has the advantage of the influence of recent critical theory and postmodernism and their sensitivity to how such discourses play in particular contexts. Global dangers there may undoubtedly be, but the impacts of specific dangers vary widely. Ontological critique of the sort that this chapter offers might usefully link up with the innovative thinking of ecological and development activists to make geographical inquiry more sensitive to the cultural premises of a living biosphere. But it is unquestionably clear that neither urban architectonic conceit, nor the celebration of modern urban and virtual identities while obscuring the ecological consequences of their production, are appropriate modes for a critical cultural geography in the new millennium.

REFERENCES

Agnew, J. (1998) *Geopolitics: Revisioning World Politics* London: Routledge.

Asimov, I. (1951) *The Foundation Trilogy*. New York: Doubleday.

Athanasiou, T. (1996) *Divided Planet: The Ecology of Rich and Poor*. Boston: Little Brown.

Barnes, J. (1994) *Mother of Storms*. New York: Tor.

Beck, U. (1992) *Risk Society: Towards a New Modernity*. London: Sage.

Blaikie, P. and Brookfield, H. (1987) *Land Degradation and Society*. London: Methuen.

Blaikie, P., Cannon, T., Davis, I. and Wisner, B. (1994) *At Risk: Natural Hazards, People's Vulnerability and Disasters*. London: Routledge.

Botkin, D. (1990) *Discordant Harmonies: A New Ecology of the Twenty First Century*. New York: Oxford University Press.

Braun, B. and Castree, N. (eds) (1998) *Remaking Reality: Nature at the Millennium*. London: Routledge.

Brzezinski, Z. (1997) *The Grand Chessboard: American Primacy and its Geostrategic Imperatives*. New York: Basic.

Carson, R. (1962) *Silent Spring*. New York: Houghton Mifflin.

Chatterjee, P. and Finger, M. (1994) *The Earth Brokers: Power, Politics and World Development*. New York: Routledge.

Clarke, A.C. (1968) *2001: A Space Odyssey*. London: Arrow.

Clee, M. (1998) *Overshoot*. New York: Ace.

Corbridge, S. (1998) 'Development ethics: distance, difference, plausibility', *Ethics, Place and Environment* 1 (1): 35–53.

Curry, M.R. (1992) 'The architectonic impulse and reconceptualization of the concrete in contemporary geography', in T.J. Barnes and J.S. Duncan (eds) *Writing Worlds: Discourse, Text and Metaphor in the Representation of Landscape*. London: Routledge. pp. 97–117.

Dalby, S. (2002) *Environmental Security*. Minneapolis: University of Minnesota Press.

Davis, M. (1993) 'Dead west: ecocide in Marlboro country', *New Left Review* 200: 49–73.

Deibert, R. (1999) 'Out of focus: U.S. military satellites and environmental rescue', in D. Deudney and R. Matthew (eds) *Contested Grounds: Security and Conflict in the New Environmental Politics*. Albany: State University of New York Press. pp. 267–88.

Diamond, J. (1997) *Guns, Germs and Steel: The Fates of Human Societies*. New York: Norton.

Dodds, K. and Atkinson, D. (eds) (2000) *Geopolitical Traditions: Critical Histories of a Century of Geopolitical Thought*. London: Routledge.

Doty, R.L. (1996) *Imperial Encounters: The Politics of Representation in North–South Relations*. Minneapolis: University of Minnesota Press.

Doubleday, N.C. (1993) 'Finding common ground: natural law and collective wisdom', in J.T. Inglis (ed.) *Traditional Ecological Knowledge: Concepts and Cases*. Ottawa: International Development Research Council.

Driver, F. (2001) *Geography Militant: Cultures of Exploration and Empire*. Oxford: Blackwell.

Feagan, R. and Ripmeester, M. (1999) 'Contesting naturalized lawns: a geography of private green space in Niagara region' *Urban Geography* 20 (7): 617–34.

FitzSimmons, M. (1989) 'The matter of nature', *Antipode* 21 (1): 106–20.

Gadgil, M. and Guha, R. (1995) *Ecology and Equity: The Use and Abuse of Nature in Contemporary India*. London: Routledge.

Gill, S. (1995) 'The global panopticon? The neo-liberal state, economic life and democratic surveillance', *Alternatives* 20 (1): 1–50.

Glacken, C. (1967) *Traces on the Rhodian Shore*. Berkeley: University of California Press.

Gregory, D. (1994) *Geographical Imaginations*. Oxford: Blackwell.

Harvey, D. (1974) 'Population, resources and the ideology of science', *Economic Geography* 50 (3): 256–77.

Harvey, D. (1996) *Justice, Nature and the Geography of Difference*. Oxford: Blackwell.

Harvey, D. (2000) *Spaces of Hope*. Berkeley: University of California Press.

Herbert, S. (1997) *Policing Space: Territoriality and the Los Angeles Police Department*. Minneapolis: University of Minnesota Press.

Hetherington, K. (1998) 'Vanloads of uproarious humanity: new age travellers and the utopics of the countryside', in T. Skelton and G. Valentine (eds) *Cool Places: Geographies of Youth Cultures*. London: Routledge. pp. 328–42.

Huntingdon, S.P. (1996) *The Clash of Civilizations and the Remaking of World Order*. New York: Simon and Schuster.

Johnston, B.R. (ed.) (1994) *Who Pays the Price? The Sociocultural Context of Environmental Crisis*. Washington: Island.

Katz, C. (1998) 'Disintegrating developments: global economic restructuring and the eroding ecologies of youth', in T. Skelton and G. Valentine (eds) *Cool Places: Geographies of Youth Cultures*. London: Routledge. pp. 130–44.

Latour, B. (1993) *We Have Never Been Modern*. Cambridge, MA: Harvard University Press.

Lefevre, H. (1991) *The Production of Space*. Oxford: Blackwell.

Lipschutz, R.D. (1989) *When Nations Clash: Raw Materials, Ideology and Foreign Policy*. Cambridge, MA: Ballinger.

Lovelock, J.E. (1979) *Gaia: A New Look at Life on Earth*. Oxford: Oxford University Press.

Luke, T. (1997) *Ecocritique: Contesting the Politics of Nature, Economy and Culture*. Minneapolis: University of Minnesota Press.

Luke, T. (1999) *Capitalism, Democracy, Culture: Departing from Marx*. Urbana: University of Illinois Press.

Mackinder, H. (1904) 'The geographical pivot of history', *Geographic Journal* 23: 421–37.

Magnusson, W. (1996) *The Search for Political Space*. Toronto: University of Toronto Press.

McHaffie, P. (1997) 'Decoding the globe: globalism, advertising and corporate practice', *Environment and Planning D: Society and Space* 15 (1): 73–86.

McNeill, J.R. (2000) *Something New Under the Sun: An Environmental History of the Twentieth Century*. New York: Norton.

Miller, M.A.L. (2001) 'Tragedy of the commons: the enclosure and commodification of knowledge', in Dimitris Stevis and Valerie Assetto (eds) *The International Political Economy of the Environment*. Boulder: Lynne Rienner. pp. 111–34.

Moodie, A.E. (1949) *Geography behind Politics*. London: Hutchinson.

Ó Tuathail, G. (1996) *Critical Geopolitics: The Politics of Writing Global Space*. Minneapolis: University of Minnesota Press.

Ó Tuathail, G. (1997) 'Emerging markets and other simulations: Mexico, the Chiapas revolt and the geofinancial panopticon', *Ecumene* 4 (3): 300–17.

Parker, G. (1998) *Geopolitics: Past Present and Future*. London: Pinter.

Paterson, M. (2000) *Understanding Global Environmental Politics*. London: Macmillan.

Peet, R. (1977) 'The development of radical geography in the United States', in R. Peet (ed.) *Radical Geography*. Chicago: Maaroufa. pp. 6–30.

Peet, R. and Watts, M. (eds) (1996) *Liberation Ecologies: Environment, Development, Social Movements*. New York: Routledge.

Rahnema, M. and Bawtree, V. (eds) (1997) *The Post-Development Reader*. London: Zed.

Redclift, M. (1996) *Wasted: Counting the Costs of Global Consumption*. London: Earthscan.

Robinson, K.S. (1993) *Red Mars*. New York: Bantam.

Robinson, K.S. (1994) *Green Mars*. New York: Bantam.

Robinson, K.S. (1996) *Blue Mars*. New York: Bantam.

Routledge, P. (1996) 'The imagineering of resistance: Pollok Free State and the practice of postmodern politics', *Transactions of the Institute of British Geographers* 22: 359–76.

Sachs, W., Loske, R. and Linz, M. (1998) *Greening the North: A Post-Industrial Blueprint for Ecology and Equity*. London: Zed.

Scott, J. (1998) *Seeing Like a State: How Certain Schemes to Improve the Human Condition Have Failed*. New Haven: Yale University Press.

Sharp, J. (2000) *Condensing the Cold War: Reader's Digest and American Identity*. Minneapolis: University of Minnesota Press.

Shiva, V. (1994) 'Conflicts of global ecology: environmental activism in a period of global reach' *Alternatives* 19 (2): pp. 195–207.

Smith, N. (1990) *Uneven Development: Nature, Capital and the Production of Space*. Oxford: Blackwell.

Smith, N. (1997) 'Social justice and the new American urbanism: the revanchist city', in A. Merrifield and E. Swyngedouw (eds) *The Urbanization of Injustice*. New York: New York University Press. pp. 117–36.

Suliman, M. (1999) *Ecology, Politics and Violent Conflict*. London: Zed.

Taylor, P. (1996) *The Way the Modern World Works: World Hegemony to World Impasse*. London: Wiley.

Turco, R.P., Toon, O.B., Ackerman, T.P., Pollack, J.B. and Sagan, C. (1983) 'Nuclear winter: global consequences of multiple nuclear explosions', *Science* 222: 1283–92.

Ward, B. and Dubos, R. (1972) *Only One Earth: The Care and Maintenance of a Small Planet*. Harmondsworth: Penguin.

Watts, M. (1998) 'Nature as artifice and artifact', in B. Braun and N. Castree (eds) *Remaking Reality: Nature at the Millennium*. London: Routledge. pp. 243–68.

Zubrin, R. (1996) *The Case for Mars: The Plan to Settle the Red Planet and Why We Must*. New York: Free Press.

Fish-eye view of the Space Shuttle Atlantis as seen from the Russian Mir space station (Source: NASA)

Section 9

SPACES OF KNOWLEDGE   Edited by John Paul Jones III

Introduction: Reading Geography through Binary Oppositions

John Paul Jones III

In this section of the *Handbook* four chapters take up the question of knowledge in cultural geography. Ulf Strohmayer explores the imbrication of knowledge and culture in the western tradition, examining key themes from the Renaissance to contemporary poststructuralist thought. Francis Harvey surveys different critical perspectives on technology, each of which stands in opposition to the naive empiricist view that technology is little more than a tool, one that is exogenous to social relations. Audrey Kobayashi offers a historical analysis of the concept of race in geography. She argues that our disciplinary knowledge is racialized, and she suggests ways to subvert the dominant whiteness that infuses Anglo-American geographic thought and practice. The section ends with a chapter by Richard Howitt and Sandra Suchet-Pearson. They criticize disciplinary knowledge from the perspective of postcolonial theory, showing how cultural geography's understanding of key concepts such as nature and landscape are limited by the western bias of much geographic research.

In this introduction to the section, I offer a general context for these essays by exploring the evolution of Anglo-American geographic thought as seen through the lenses of its most prominent binary oppositions. These infuse both what and how we know, the study of which is, respectively, ontology and epistemology. My survey is grounded in the various 'paradigmatic' shifts enveloping geography over the past century or so. I conclude the introduction with a brief review of the chapters in this section.

WHAT IS KNOWLEDGE?

Defining knowledge is a slippery endeavor, for the concept skates back and forth between ontology and epistemology. The latter term refers to theories concerned with how we understand or know the world. It encompasses the theoretical study of science and interpretation, as well as many aspects of the metatheoretical perspectives associated with these endeavors. Of those that have been historically influential in geography, we can point to empiricism, positivism, critical realism, humanism, structuralism, and poststructuralism. In the west, our understanding of epistemology is founded upon – one might even say congealed around – a number of key binary oppositions. These include: objectivity and subjectivity; determination and uncertainty; rigor and play; explanation and description;

Ontology

Epistemology		Orderly	Chaotic
	Objective	Spatial science	Critical realism
	Subjective	Humanism	Poststructuralism

Important binaries in geography and their associated meta-theoretical perspectives (adapted for geography from Burrell and Morgan, 1979)

and generality and particularity. Each of these aspects of knowing the world structures different theoretical approaches: scientific understandings typically assign positive valences to the former terms in the above binaries, while interpretive approaches typically privilege the latter terms. Different thinkers have different stances with respect to these terms, including not only their utility in helping us conduct research, but also what they mean and, even, their very possibility. Furthermore, how we know the world at the everyday, non-reflective, level is in part determined (with more or less certainty) by how we negotiate these oppositions through our experiences, language and, of course, culture (see Strohmayer, Chapter 28 in this section).

Ontology refers to the theoretical study of what the world is like. It is similarly structured, in the western imagination, along a number of key binaries. Studies of the ontological character of the world within geography have tended to focus on the distinctions between: nature and culture; individual and society; order and chaos; space and time; space and society; and the world of ideas and discourses versus the world of material objects and concrete social processes. Even our disciplinary distinctions – between say, geography and sociology (that is, space and society) – are structured along ontological assumptions.

That knowledge is bound up in both epistemology and ontology implies that these terms are themselves interconnected. For instance, if we assume that the world comprises discrete objects and events located in time–space, and that these have measurable characteristics and effects on others, then it seems plausible to adopt a scientific epistemology that privileges

objectivity and explanation. On the other hand, if we assume that we are immersed in a world of meanings, and that we can only at best describe the world through culturally and temporally specific languages, then an interpretive approach based on a different epistemology seems more ready to the task. In general terms, then, our knowledge of what the world is like suggests how we should study it. But as we shall see below, it is not necessary to put ontology before epistemology, for it is altogether possible to focus on how we know and to use the knowledge gained to assess what we think the world is like.

A synoptic perspective on contemporary theoretical perspectives in geography and two of the pairs of binary oppositions that underwrite them is found in the figure shown (above). In this exercise, I have chosen one aspect of epistemology and ontology. The former is divided into objective and subjective approaches; the latter into whether the world is conceived as orderly or chaotic. These pairings contextualize the four most significant theoretical perspectives in contemporary geography, from the objective and orderly assumptions that authorize scientific geography to the subjective and chaotic worldview of poststructuralism. The figure is, however, simply a heuristic device: in both theory and practice things are more complicated. For one, these are not the only oppositions that distinguish contemporary theoretical perspectives; one could make equally plausible accounts of contemporary metatheories based on the idealism/materialism and generality/particularity oppositions. Second, each perspective tends to rest on different definitions of the same binary terms. For example, the definition of objectivity is not the same in spatial science and critical realism, nor do critical realists and poststructuralists have the same understanding of what is meant by a chaotic or disorderly world. Third, each term in the binaries has developed in tandem – or relationally – with its opposition; this implies that the metatheoretical perspectives are not so easily separated from one another (Dixon and Jones, 1996). Further complicating matters is the fact that the binary oppositions are subject to redefinition – both across disciplines and over time. We can, within the field of geography, discern some of this historical contingency by examining key programmatic statements in the field and asking to what extent – and what versions of – epistemology and ontology reign at different paradigmatic junctures. One such analysis, albeit greatly abbreviated, follows.

EPISTEMOLOGY AND ONTOLOGY IN MODERN GEOGRAPHY

Historically speaking, geographers have tended to weight their programmatic injunctions about the character of the discipline on the shoulders of ontology. In part, this was because, for much of the discipline's history, geographers assumed, along with most other social and natural scientists, that their field was a science: with few exceptions, they tended not to question the value of objectivity; they sought determinations; and they aimed to rigorously explain the presumed (ontological) order of the world by providing general accounts that could be tested in different parts of the world. During the period in which an empiricist scientific epistemology held sway in geography, paradigmatic discussions and geographic practice tended to be structured around the ontological distinction between nature and culture (and with the latter, race often reigned; see Kobayashi, Chapter 30 in this section). The disciplinary effects of this binary are notable in the writings characterizing the age of scientific exploration in the nineteenth century. Objects existing in the world tended to be classified into one or another category, with field mapping and travel accounts organized according to distinctions between physical aspects of the earth and the character and activities of its

inhabitants. For example, George Dawson's accounts of physical and human phenomena in the Pacific Northwoods during the 1880s and 1890s were penned on separate pages of his travel log (Willems-Braun, 1997). The nature–culture binary was further codified, as geography's raison d'être, in the form of environmental determinism, the programmatic charge of which was to determine cultural responses to environmental conditions. Key proponents of determinism included Halford Mackinder (1887), Ellsworth Huntington (1924), Griffith Taylor (1914), Ellen Churchill Semple (1911), and William Morris Davis (1909, first published 1906). For Davis:

> any statement is of geographical quality if it contains a reasonable relation between some inorganic element of the earth on which we live, acting as a control, and some element of the existence or growth or behavior or distribution of the earth's organic inhabitants, serving as a response. (1909: 8)

Geography's release from the methodological straitjacket implied by Davis' definition of geography is largely attributed to two schools of thought: the cultural landscape approach advocated by Carl Sauer and other geographers with linkages to Berkeley geography, and the regional approach, which was popular in both the United States and Europe. Importantly from the standpoint of ontology, neither of these schools challenged the nature–culture binary *per se*. They did, however, complicate matters by overlaying it with another ontological opposition: the distinction between idealism and materialism. For the Berkeley School, culture was splayed across both: it was on the one hand a mental construct, a template for interpreting social life through common values, mores, worldviews, and languages. This aspect of culture was largely left to anthropologists, while the Berkeley geographers focused their attention on the other side of the binary – on landscapes and other aspects of material culture, such as housing types and agricultural practices. In a famous phrase that incorporates both the nature–culture and idealist–materialist binaries, Sauer proclaimed: 'culture is the agent, the natural area is the medium, the cultural landscape is the result' (1925: 46).

The regional school offered its own elaboration. In the beginning, the region had a concrete status: it was there that natural and cultural phenomena interacted, giving a distinctiveness to different parts of the world and legitimizing geography as the integrative science (Fenneman, 1919). Yet the certainty of the region *qua* ontological object was difficult to sustain, and later commentators had to acknowledge that the region was not so much a thing unto itself but a mental construct – a heuristic device for organizing the study of places (Hartshorne, 1939; James, 1952). With this admission, the distinction between idealist and materialist approaches was further embedded in the discipline. Hartshorne (1939: 193–201) used the opposition to criticize Sauer and others for their views that, for something to be geographic, it had to be visible (that is, materially concrete). Against this, Hartshorne argued that non-material aspects of social life varied spatially, and that these too were part of geography (see Jones, 1995, for a discussion). Thus, the idealism and materialism binary found an early portal into geography – how often we tend to assume that it is a relatively recent concern – but in no sense did it displace nature versus culture. Regional geographers were comfortable with that opposition, using it to organize both their research efforts and many an undergraduate textbook (see, for example, James, 1942).

In the three post-World War II decades, geography underwent a so-called scientific-theoretical or quantitative 'revolution' (Gould, 1979, gives the most amusing account; Barnes, 1995, the most thoughtful). As Gregory (1994) points out, there was little offered by the spatial scientific school that was not consistent with the regional school that went before it, yet it also seems gracious enough to give its practitioners credit for explicitly theorizing both ontology and epistemology, and for standing their ground on the relative merits of the binary terms that underwrite them. Onto-logically speaking, they maintained a strict division between space and time and between space and society. Schaefer (1953), for example, argued that geographers needed to discover *spatial* laws: all other laws could be left to other fields. His paper, which was largely an attack on Hartshorne's book *The Nature of Geography* (1939), can also be interpreted as an effort to insert the order versus chaos ontological binary into geography; the emphasis

on laws betrays his allegiances. And epistemo-logically, his paper was unique for, alongside the presumption of an orderly world given over to laws, we can read a call for determi-nation over indeterminacy and generality over particularity. Though Hartshorne had negoti-ated these divisions in *Nature*, he did so halt-ingly, without the confidence that laws would ever be found. In his view, things were simply more chaotic than that, a result of the inter-actions of phenomena in regional contexts.

Nystuen (1963), in a still under-appreciated ontological paper, put forth several geographic primitives: distance, direction, and connectivity were the most important. His empirical exam-ple, the spatial layout of students listening to a teacher in a mosque, is a case study in the separation of space and social relations. Nystuen did, however, offer a brief commen-tary on time (an 'accumulated … legacy of the past' that continues to have effects, presaging Massey's 1984 geological metaphor). But more commonly, space–time in scientific geography tended to be conceived through 'slices' in the geographic data matrix (Berry, 1964), an onto-logical conceptualization of the world that rested on the strict division between space, time, and systematic characteristics. All of this theorizing was buttressed by epistemological certitude: the paradigm was characterized by a largely unquestioned faith in objectivity, the search for generalities, the determinative and rigorous discovery of orderly causal processes, and a realist approach to representation. The nature–culture opposition did not recede over the horizon, but it held less weight under spatial science, for both could be studied with the same methodology. And it wasn't until the advent of behavioral geography – an offshoot of spatial science (Golledge et al., 1972) – that those operating within this paradigmatic framework directly considered the individual versus society opposition.

Wright (1947) and Lowenthal (1961) offered the first serious challenges to scientific epistemology. The subsequent rise of humanis-tic geography – most of whose practitioners would not refuse the label 'cultural geogra-pher' – deepened such reflection. They explic-itly questioned the hegemony of scientific ways of knowing, and substituted in its place a hermeneutic concern for understanding and interpretation. During the 1970s, humanistic geographers made significant inroads into ontology, deploying concepts such as 'being' (á la Heidegger), the lifeworld, hearth and cosmos, authenticity, intentionality, and sense of place (Buttimer, 1976; Relph, 1970; Tuan, 1975; 1976; see Entrikin, 1976, for a cogent review and Pickles, 1985, for a critical evaluation). Nonetheless, much of the writing of this period continued to adhere to the idealist-materialist binary. Perhaps the humanists' most lasting contribution is the extent to which they destabilized the field's traditional adherence to objectivist epistemology.

The rise of Marxist approaches – at roughly the same time as humanistic critiques – further shook both the ontological and epistemological moorings of spatial science. Geography first saw an extended critique of the objective–subjective binary that had previously secured the foundation of scientific epistemology. The Marxist argument that all knowledge was social, and hence political, redrew the grounds upon which objectivity and subjectivity were conceived, but it did not jettison the opposition (see, for example, Harvey, Chapter 29 in this section). The former was still paramount, especially when compared to what was viewed as overly subjectivist formulations in humanistic geography. Objectivity was now conceived as a practiced and, importantly, achieved stance, one that relied upon the appropriate application of dialectical materialism. Marxism also offered a relational ontology that shifted the focus from external relations to internal ones (Cox, 1981). Through dialectics geographers were able to theorize society and space as intricately conjoined in a ceaselessly recursive and inseparable relation (Soja, 1980; 1989). The addition of a temporal dimension (Harvey, 1984) gave rise to a socio-spatial-historical 'trialectic' of sorts – which, with Marx, explicitly incorporated a dialectical approach to nature and culture (Smith, 1984).

Yet in geography as in other fields, Marxists were criticized for their neglect of still another binary: the individual versus the social. This distinction formed the foundation for most critical assessments of the differences between humanistic and Marxist geography (Gregory, 1981), with the former group being accused of volunteerism and the latter tainted with charges of structuralism (Duncan

and Ley, 1982; but see Peet, 1998). An influential reconceptualization of the binary was undertaken in the 1980s under the rubric of structuration theory (Giddens, 1979; 1986). Gregory (1981), Thrift (1983), and Pred (1984) each offered dialectical accounts of the individual and society, while at the same time explicitly integrating space–time – thought together – into their formulations.

In the past decade geography has witnessed considerable ontological and epistemological debate. A few movements are especially noteworthy. Once the dust had settled on structuration theory, most political economy researchers in geography turned sympathetically – if not always explicitly – to critical realism (Sayer, 1992). This perspective is significant in that it offers something of a middle ground between epistemology and ontology. Realists recognize the hermeneutic circle, but maintain nonetheless that there is an objective world of socio-spatial relations that can be understood through interrogations of actors' practical knowledge of causal mechanisms and structures. The realist notion of contingency (Jones and Hanham, 1995), which describes the deflection of a structure's mechanisms by relations embedded in local contexts, likewise represents an attempt to straddle the chaotic–orderly binary. And Sayer (1991) has offered an especially convincing negotiation of the general versus particular binary that had befuddled both regional and scientific geographers since the 1950s, and that continued on in a conflation with local–global and progressive–regressive oppositions (see Massey, 1991).

A second prominent theoretical movement was found in feminism, and here one can point to the influence of Gillian Rose's *Feminism and Geography* (1993). Her epistemological analysis of twentieth-century geographic research identifies a pervasive masculinism, a condition in which an omniscient, detached, and self-identified 'scientific' researcher is separated from his 'subjects'. Rose also contrasts modern geography's ontology of discreteness with that of a relationally constituted – and paradoxically juxtaposed – field of socio-spatial relations. In the process of rethinking methodology in light of such critiques, feminist geographers also redrew disciplinary understandings of difference, methodology, and representation (see the essays in Jones et al., 1997).

Not lastly, the 1990s saw the rise of post-structuralist influences in geography. Above all, poststructuralism was interpreted as an epistemological critique, with advocates arguing that previously sacrosanct ontological categories lacked foundational status (Dixon and Jones, 1998; Doel, 1999). Derrida's (1988) 'constitutive outside' has been influential in this regard. He proposes an anti-foundational theory of concept construction that rejects the structuralist positivity essential to grounding philosophical approaches to God, man and self (Derrida, 1970). Instead, knowledge is only possible through an exclusionary, negating process, one that leaves a (deconstructive) 'trace' of the other within the inscribed boundaries of the categories of knowing. This formulation has been used to destabilize notions of a fixed subject, replacing an essential identity with a socially constructed category defined by the constitutive outside – the raw material for the formation of identity (Laclau and Mouffe, 1985; Natter and Jones, 1997). Geographers have also employed Foucault's (1970; 1972) archaeological method to uncover how clusters of power/knowledge produce, fill, and maintain categories, such as nature (Willems-Braun, 1997; also see Philo, 1992). Both sorts of analysis typically dissolve ontological certainties: the point is no longer what we know, but how we came to know what we know in the first place. Such work is best done from a cross-cultural and historical perspective, one that admits that how we describe the world is constrained by the place- and time-bound languages that we have at our disposal (see Howitt and Suchet, Chapter 31 in this section).

In addition to nature–culture and identity, poststructuralist geographers have also called into question the concept of scale, which, rather than being viewed as an ontological category derived from a foundational spatiality (á la Nystuen, 1963), can be understood as both a discourse and the spatial counterpart to the general–particular epistemological opposition (Jones, 1998). Nor did the concept of culture escape the broom-sweep of discourse: Mitchell (1995) argues that we should pay attention not to culture's attributes – which he finds chaotically conceived – but to the work done in its name. Gibson-Graham (1996) makes parallel arguments for capitalism and other tropes related to the economy. Other key concepts subject to deconstruction include representation (Deutsche, 1991; Harley, 1989; Jones and Natter, 1999) and space/place (Hooper, 2001). There have, finally, been a few attempts to theorize a poststructuralist spatial ontology (Massey, 1994; Rose, 1993; Soja, 1996). The most radical account thus far appears in Doel's *Poststructuralist Geographies* (1999). Methodologically, he rejects dialectics in favor of deconstruction, and though he claims ontological agnosticism, he is also supportive of Deleuze and Guattari's (1987) folded, rhizomatic, and flowlike spatiality – what he terms 'scrumpled geography'.

In summary, geography's own knowledge production has relied upon historically contingent deployments of a handful of key binary oppositions. Over time, the assumption of an orderly world has given way to an assumption of disorder – even within physical geography (Phillips, 1999). Nature and culture, two concepts – or facts of life, depending on your perspective – once set the parameters within which geography was practiced and organized; after a period of relative neglect, this opposition is now one of the most intense areas of theoretical reflection in the field (see Braun and Castree, 1998; Castree and Braun, 2001). The problem of integrating the individual and society (or agency–structure), long overlooked and then seemingly solved through structuration theory during the 1980s, has re-emerged with the erasure of the self under poststructuralism's assault on identity, and with the arrival of psychoanalytic approaches (Nast, 2000; Pile, 1996). And though most geographers still tend to work with an ontological division between space and place (Entrikin, 1991) – a legacy of the separation between scientific and humanistic approaches – Hooper's (2001) analysis indicates that this too might be an epistemological division. Not lastly, geography continues to be haunted, along with most other disciplines, by the discursive versus 'real world' division (see Peet, 1998). This offshoot of the idealism–materialism opposition has created a considerable barrier between cultural and economic geographers, despite numerous attempts at integration (for example, Harvey, 1996; see Barnes, introduction to Section 2 in this volume, for a discussion).

THE CONTRIBUTIONS TO THIS SECTION

This section presents an exploration of the spaces of knowledge, from spaces apprehended through knowledge to those produced by knowledge, but with an emphasis on a critical engagement with epistemological developments in western theory. Ulf Strohmayer offers just such an account. His chapter touches upon some of the points I have addressed above, but with a wider historical sweep and less concern for geography than for the human sciences more generally. He begins his chapter by identifying a problematic relationship between knowledge and its objects. He identifies a circularity that, through the western tradition, is complicated by the recognition that knowledge is constructed through experience, discourse, and social practices (including those of science). Nor can knowledge be grounded in the certainty of identity. These complications undercut the traditional model of representational mimesis, yet Strohmayer is reluctant to give up on the analytic value of knowledge. He proposes instead a contextual and situated – and therefore ultimately spatial – form of cultural knowledge.

The editors also felt that the time was right for a contribution on technology. It was clear to us that this is an area that, like the internet and GIS, will continue to grow in importance. In his contribution, Francis Harvey argues on behalf of a thoroughly social and political stance with respect to technology: he takes as his point of departure the naive view that technology is merely a tool, and that it is, on its own terms, neutral. In his criticism of this view he reviews various theoretical perspectives on technology, from Marxism through the Frankfurt School to poststructuralist and science studies accounts. He then surveys some of the more important geographic works on mapping, technology and communication, GIS, and the internet. Clearly technology is an area that structures epistemology and ontology – Harvey's example of the light bulb as the quintessential human–machine binary relation is but one example – and should be a rich area for critical cultural geographies of the future.

We also sought a chapter on the relationships between knowledge production and 'race' – another area of increased attention on the part of critical cultural geographers. Audrey Kobayashi's historical survey identifies three phases of racialized thinking in the discipline. The early period explicitly deployed race as perhaps the most visible dimension of the nature–culture binary (see above). She shows how, from Kant to the early-twentieth-century environmental determinists, geography had something of an obsession for racial thinking. The second period is characterized by a kind of benign neglect of race and racialization; even when it was considered, as for example in early Marxist geography, it was as supplementary to other social processes. The third, and most recent, poststructuralist phase puts long-neglected attention on the construction of difference in the name of racial categories, but it, like the rest of geography, suffers from a whiteness that continues to infuse both geographic research and disciplinary institutions. Kobayashi concludes with some theoretical and practical directions for an activist and anti-racist geography.

Finally, we sought to undo the western preoccupations of geography by including a chapter on the challenges to geographic knowledge brought from non-western, postcolonial perspectives. Richard Howitt and Sandra Suchet-Pearson offer such a chapter using the metaphor of a hall of mirrors, a condition in which Eurocentric scholars – trapped in limited ontological and epistemological frameworks – see only their own reflections. They offer a detailed critique of these biases, showing how the hall of mirrors marginalizes and exoticizes indigenous, non-western knowledge. Their critique is extended into a consideration of alternative modes of theorizing space and time; writing cultural landscapes; deploying concepts of nature and culture; and conceptualizing identity. Each of these is developed through non-western examples. They conclude by calling for an ethic of openness to alternative ontologies – replacing the hall of mirrors with a set of windows offering new perspectives.

REFERENCES

Barnes, T. (1995) *Logics of Dislocation*. New York: Guilford.
Berry, B.J.L. (1964) 'Approaches to regional analysis: a synthesis', *Annals of the Association of American Geographers* 54: 2–11.

Braun, B. and Castree, N. (eds) (1998) *Remaking Reality: Nature at the Millennium*. London: Routledge.

Burrell, G. and Morgan, G. (1979) *Sociological Paradigms and Organizational Analysis: Elements of the Sociology of Contemporary Life*. London: Heinemann.

Buttimer, A. (1976) 'Grasping the dynamism of the life-world', *Annals of the Association of American Geographers* 66 (2): 277–92.

Castree, N. and Braun, B. (eds) (2001) *Social Nature: Theory, Practice, and Politics*. Oxford: Blackwell.

Cox, K.R. (1981) 'Bourgeois thought and the behavioral geography debate', in K.R. Cox and R.G. Golledge (eds) *Behavioral Geography Revisited*. New York: Methuen. pp. 256–79.

Davis, W.M. (1909) 'An inductive study of the content of geography' (1906), in D.W. Johnson (ed.) *Geographical Essays*. Boston: Ginn. pp. 3–22.

Deleuze, G. and Guattari, F. (1987) *A Thousand Plateaus: Capitalism and Schizophrenia*. Minneapolis: University of Minnesota Press.

Derrida, J. (1970) 'Structure, sign, and play in the discourses of the human sciences', in R. Macksey and E. Donato (eds) *The Structuralist Controversy*. Baltimore: Johns Hopkins University Press. pp. 247–65.

Derrida, J. (1988) *Limited Inc*. Evanston: Northwestern University Press.

Deutsche, R. (1991) 'Boy's town', *Environment and Planning D: Society and Space* 9 (1): 5–31.

Dixon, D. and Jones, J.P. III (1996) 'For a *Supercalifragilisticexpialidocious* scientific geography', *Annals of the Association of American Geographers* 86 (4): 767–79.

Dixon, D. and Jones, J.P. III (1998) 'My dinner with Derrida, or Spatial analysis and poststructuralism do lunch', *Environment and Planning A* 30 (2): 247–60.

Doel, M. (1999) *Poststructuralist Geographies*. Lanham: Rowman and Littlefield.

Duncan, J.D. and Ley, D. (1982) 'Structural Marxism and human geography: a critical assessment', *Annals of the Association of American Geographers* 72 (1): 30–58.

Entrikin, J.N. (1976) 'Contemporary humanism in geography', *Annals of the Association of American Geographers* 66 (4): 615–32.

Entrikin, J.N. (1991) *The Betweenness of Place: Towards a Geography of Modernity*. Baltimore: Johns Hopkins University Press.

Fenneman, N.M. (1919) 'The circumference of geography', *Annals of the Association of American Geographers* 9: 3–11.

Foucault, M. (1970) *The Order of Things: An Archaeology of the Human Sciences*. New York: Pantheon.

Foucault, M. (1972) *The Archaeology of Knowledge*. New York: Pantheon.

Gibson-Graham, J.K. (1996) *The End of Capitalism (As We Knew It): A Feminist Critique of Political Economy*. Oxford: Blackwell.

Giddens, A. (1979) *Central Problems in Social Theory*. Berkeley: University of California Press.

Giddens, A. (1986) *The Constitution of Society*. Berkeley: University of California Press.

Golledge, R.G., Brown, L.A. and Williamson, F. (1972) 'Behavioral approaches in geography: an overview', *Australian Geographer* 12 (1): 59–79.

Gould, P. (1979) 'Geography 1957–1977: the Augean period', *Annals of the Association of American Geographers* 69: 139–50.

Gregory, D. (1981) 'Human agency and human geography', *Transactions of the Institute of British Geographers* 6 (NS) (1): 1–18.

Gregory, D. (1994) *Geographical Imaginations*. Oxford: Blackwell.

Harley, J.B. (1989) 'Deconstructing the map', *Cartographica* 26 (2): 1–20.

Hartshorne, R. (1939) *The Nature of Geography*. Lancaster, PA: Association of American Geographers.

Harvey, D. (1984) 'On the history and present condition of geography: an historical materialist manifesto', *Professional Geographer* 36 (1): 1–11.

Harvey, D. (1996) *Justice, Nature and the Geography of Difference*. Oxford: Blackwell.

Hooper, B. (2001) 'Desiring presence, romancing the real', *Annals of the Association of American Geographers* 91 (4): 703–15.

Huntington, E. (1924) *The Character of Races, as Influenced by Physical Environment, Natural Selection and Historical Development*. New York: Scribners.

James, P. (1942) *Latin America*. Indianapolis: Bobbs-Merrill.

James, P. (1952) 'Toward a fuller understanding of the regional concept', *Annals of the Association of American Geographers* 42 (3): 195–222.

Jones, J.P. III (1995) 'Making geography objectively: ocularity, representation, and *The Nature of Geography*', in W. Natter, T.R. Schatzki and J.P. Jones III (eds) *Objectivity and its Other*. New York: Guilford. pp. 67–92.

Jones, J.P. III and Hanham, R. (1995) 'Contingency, realism, and the expansion method', *Geographical Analysis* 27 (3): 185–207.

Jones, J.P. III and Natter, W. (1999) 'Space "and" representation', in A. Buttimer, S.D. Brunn and U. Wardenga (eds) *Text and Image: Social Construction of Regional Knowledges*. Leipzig: Selbstverlag Institut für Länderkunde. pp. 239–47.

Jones, J.P. III, Nast, H. and Roberts, S. (eds) (1997) *Thresholds in Feminist Geography: Difference, Methodology, and Representation*. Lanham: Rowman and Littlefield.

Jones, K. (1998) 'Scale as epistemology', *Political Geography* 17 (1): 25–8.

Laclau, E. and Mouffe, C. (1985) *Hegemony and Socialist Strategy: Towards a Radical Democratic Politics*. London: Verso.

Lowenthal, D. (1961) 'Geography, experience, and imagination: towards a geographical epistemology', *Annals of the Association of American Geographers* 51 (3): 241–60.

Mackinder, H. (1887) 'On the scope and methods of geography', *Proceedings of the Royal Geographical Society* 9: 141–60.

Massey, D. (1984) *Spatial Divisions of Labour: Social Structures and the Geography of Production*. London: Macmillan.

Massey, D. (1991) 'The political place of locality studies', *Environment and Planning A* 23 (2): 267–82.

Massey, D. (1994) *Space, Place and Gender*. Minneapolis: University of Minnesota Press.

Mitchell, D. (1995) 'There's no such thing as culture: towards a reconceptualization of the idea of culture in geography', *Transactions of the Institute of British Geographers* 20 (1): 102–16.

Nast, H. (2000) 'Mapping the "unconscious": race and the Oedipal family', *Annals of the Association of American Geographers* 90 (2): 215–55.

Natter, W. and Jones, J.P. III (1997) 'Identity, space and other uncertainties', in G. Benko and U. Strohmayer (eds) *Space and Social Theory: Geographical Interpretations of Postmodernity.* Oxford: Blackwell. pp. 141–61.

Nystuen, J.D. (1963) 'Identification of some fundamental spatial concepts', *Papers of the Michigan Academy of Science, Arts and Letters* 48: 373–84.

Peet, R. (1998) *Modern Geographical Thought.* Oxford: Blackwell.

Phillips, J. (1999) *Earth Surface Systems: Complexity, Order, and Scale.* Oxford: Blackwell.

Philo, Chris (1992) 'Foucault's geography', *Environment and Planning D: Society and Space* 10 (2): 137–61.

Pickles, J. (1985) *Phenomenology, Science, and Geography: Spatiality and the Human Sciences.* Cambridge: Cambridge University Press.

Pile, S. (1996) *The Body and the City: Psychoanalysis, Space, and Subjectivity.* London: Routledge.

Pred, A. (1984) 'Place as historically contingent process: structurationism and the time-geography of becoming places', *Annals of the Association of American Geographers* 74 (2): 279–97.

Relph, E. (1970) 'An inquiry into the relations between phenomenology and geography', *Canadian Geographer* 14 (3): 193–201.

Rose, G. (1993) *Feminism and Geography.* London: Polity.

Sauer, C.O. (1925) 'The morphology of landscape', *University of California Publications in Geography* 2 (2): 19–53.

Sayer, A. (1991) 'Behind the locality debate: deconstructing geography's dualisms', *Environment and Planning A* 23 (2): 283–308.

Sayer, A. (1992) *Method in Social Science: A Realist Approach,* 2nd edn. London: Routledge.

Schaefer, F.K. (1953) 'Exceptionalism in geography: a methodological examination', *Annals of the Association of American Geographers* 43 (3): 226–49.

Semple, E.C. (1911) *Influences of Geographic Environment.* New York: Holt.

Smith, N. (1984) *Uneven Development.* Oxford: Blackwell.

Soja, E. (1980) 'The socio-spatial dialectic', *Annals of the Association of American Geographers* 70 (2): 207–25.

Soja, E. (1989) *Postmodern Geographies: The Reassertion of Space in Critical Social Theory.* London: Verso.

Soja, E. (1996) *Thirdspace: Journeys to Los Angeles and Beyond.* Oxford: Blackwell.

Taylor, G. (1914) 'The control of settlement by humidity and temperature'. Commonwealth Bureau of Meteorology, Bulletin no. 14, Melbourne.

Thrift, N.J. (1983) 'On the determination of social action in space and time', *Environment and Planning D: Society and Space* 1 (1): 23–58.

Tuan, Y.-F. (1975) 'Place: an experiential perspective', *Geographical Review* 65 (2): 151–65.

Tuan, Y.-F. (1976) 'Humanistic geography', *Annals of the Association of American Geographers* 66 (2): 266–76.

Willems-Braun, B. (1997) 'Buried epistemologies: the politics of nature in (post)colonial British Columbia', *Annals of the Association of American Geographers* 87 (3): 3–31.

Wright, J.K. (1947) '*Terrae incognitae*: the place of the imagination in geography', *Annals of the Association of American Geographers* 37 (1): 1–15.

28

The Culture of Epistemology

Ulf Strohmayer

THE CIRCULARITY OF KNOWLEDGE

Culture and knowledge do not sit together easily. Hence the need for a chapter on the ways they relate to one another. The problem can be stated simply: while every notion of 'culture' – especially in a book like the present one – represents a form of knowledge, this latter can never wholly be separated from cultural influences. To define 'knowledge' is thus to enter a twilight zone of sorts where success presupposes what it seeks to explain. In this respect, the scientific analysis of 'culture' is no different from other forms of scientific inquiry: it, too, is caught in a circle where knowledge and its objects are mutually constitutive elements. This chapter explores the ramifications of the initial arrangement set out in the preceding two sentences. It contends that both the concept of 'culture' and its academic pursuit can profitably be understood as attempts to address the involvement of knowledge in the workings of societies. In other words, 'culture' is one key epistemological answer to the mutual constitution of science and its objects and thus to the problem of circularity that has troubled and continues to trouble the (human) sciences.

To develop this argument, the chapter proceeds from a historical reconstruction of knowledge and the problems that have traditionally been associated with its constructed character to the role of 'culture' in the attempts to create novel and illuminating forms of insight. Already, this may strike some readers as a highly contrived setup for a chapter addressing the cultural side of epistemology. And I would agree that we do not normally perceive of knowledge in this manner – as an object itself, rather than a condition. Still, it is appropriate for a chapter addressing 'the nexus between culture and knowledge' to be clear about its own limitations. In fact, the impulse itself that makes us seek clarity concerning the boundaries to knowledge has been one of the guiding principles of epistemology, the branch of philosophy concerned with knowledge. Here scepticism has arguably been central, at least within the western canon of philosophical texts. Such scepticism has variously attached itself to the relationship between 'understanding' and 'knowledge' and that between 'experience' and 'knowledge', and arguably reached its pinnacle in the work of the seventeenth-century author René Descartes (Hookway, 1990).

We shall have numerous occasions to revisit the emerging and quintessentially modern battleground between the optimistic and pessimistic views as to how best to break free of the circularity at the root of epistemology – a circularity that binds 'knowledge' to its objects without allowing the latter ever to be independent from the former. Before we do so, we need to make clear that 'circularity' does not in and of itself pose a limit to any kind of knowledge. Rather, it imposes a no less important structural characteristic. Recognizing this trait, we can say that any attempt to delineate spaces of knowledge has to address what is often merely implicit in many other definitions of knowledge: the self-referential nature of its constitution. To give examples here, the uncritical invocation of 'knowledge' is rather akin to the explanation of an apple through reference to an apple tree or the definition of a 'thief' through reference to an activity called 'stealing': one has to be familiar with the one to make any sense of the other.

A still more fitting analogy would perhaps liken the limits to 'knowledge' to the closed reference system of a thesaurus or an encyclopedia, where in order to understand a definition a reader has to be able to comprehend the terms of the definition. We shall return to this analogy shortly. Before we do so, it is fitting to acknowledge that the circularity invoked here is not at all unusual or itself problematic. Since 'knowledge' is part of the human condition, it will come as little surprise to find the structures of knowledge intimately linked with everyday experience, thought processes and other forms of human practice. The emerging intimate relationship between experience, knowledge and practice is thus not inherently problematic; it is *made* awkward only because in the pursuit of 'cultural geography' or 'cultural studies' one of the elements – knowledge – is privileged over the others. Then, the reflexive impulse that seeks insights by asking 'why', 'how' or 'why not differently' – whether it leads to something called 'knowledge' or not – is often led into cul-de-sac of sorts: intellectually, starting anew would be the rigorous course of action, one that is pursued by only a handful of individuals. In short, the circularity of thinking hinders neither action nor the continuation of the reflexive impulse *per se*. 'Culture' is a way both of understanding this situation and of epistemologically addressing the issues it raises. Cultural geography and cultural studies represent some of the more intricate ways of addressing both the circularity of knowledge and the everyday context of experience that remains largely unhampered by epistemological concerns.

THE ROLE OF CULTURE

To recognize the involvement of 'culture' in the spaces of modern knowledge, we need to pry open the association between knowledge and everyday experience that we fashioned a moment ago. We need to acknowledge that even the reflexive variant of everyday experience hardly ever confronts problems of prioritization. We *know* that things exist because they play a role in our lives. Even the time-honoured distinction between subject and object that has long plagued epistemology is of little consequence for the continuation of everyday practice, as is testified by the hybrid etymology of the word 'existence'. Since the word 'culture' in a direct way captures this entanglement of existence with and in the world, rather than postulating some unattainable 'outside', it is tempting to rush to conclusions and proclaim that every form of knowledge is somehow 'cultural' and to render the 'cultural construction of reality' the *sine qua non* of each and every attempt towards understanding. Yet

however tempting, such a move would merely bypass, rather than address, the epistemological issues raised by 'the cultural problem'. Since avoided issues have a tendency to resurface at inopportune moments, this chapter centres its argument on the relationship between 'culture' and 'knowledge' without according either term an *a priori*, accepted status. In order for the 'spaces of knowledge' that are central to this *Handbook* – the spaces of body, region, the local and the nation, to mention but a few – to be knowable entities, rather than circular postulations of 'a will to culture', we have to justify the explanatory power of 'culture' as an analytical concept.

In other words, it is crucial to acknowledge that the entanglement of knowledge and its object is a fundamental dilemma for anyone interested in delimiting knowledge and in assuring the status of 'cultural' forms of learning as valid forms of research. This dilemma is caused by the standards adopted in the now dominant western and modernist manner of conceptualizing knowledge as that which provides a secure basis from which to understand the world. If premodern and protomodern sceptics from Pyrrho to Montaigne could afford freely to speculate about the grounding of respective claims to knowledge, such luxury was increasingly denied to their modern heirs apparent. Indeed, it was Descartes again who famously decreed the defeat of scepticism as the primary undertaking of modern philosophy. More than other words perhaps, it is the word 'fact' that reflects the novelty of this understanding of 'knowledge': the currently dominant mode of thinking in western societies dictates that once uncovered and determined, 'facts' become trustworthy points of departure for the pursuit of future, as yet unknown forms of knowledge. In Ludwig Wittgenstein's (1961) early dictum, it is the totality of facts that constitutes 'the world' for us. From its origin in the world of commerce, the 'modern fact' and the knowledges to which it led have created the sense of confidence necessary for the sciences to be conceptualized as progressive and mutually supportive enterprises (Poovey, 1998). At bottom, such trustworthiness has resulted in an implied optimism that has been characteristic of modernity in general and of modern science in particular throughout most of the last two centuries.

Yet 'facts' are facts not because scientists have recognized some innate quality residing within an object but because some knowledge of 'facticity' in general, or the 'factness' of facts, has been applied to an object. In short, we recognize something as a fact, because we were looking for facts – and not because something revealed itself as a 'fact'. Upon reflection, circularity yet again rears its untimely head. Rather than recognizing a 'fact', we accept that certain statements are true while others are not. In most cases, this recognition takes place subconsciously: it becomes part of 'common sense' in the form of a largely taken-for-granted context that we inherit. Who, for instance, would doubt that water flows downstream? And what point would there be in doubting that it does?

The circular notion at the heart of conceptualizations of knowledge is thus as unavoidable as it is real. If 'knowledge' cannot create the spaces it requires by its own, modern standards to be separated from other forms of reflexive behaviour, perhaps it is time to reconfigure 'knowledge' differently: perhaps as a particular form of 'practice' or 'culture' that operates in accordance with its own set of rules and regulations and according to the spaces that are produced by these 'practised knowledges'. Rather than insisting on the idea of knowledge as being 'external' to and detached from experience, we would then be free to speak of 'knowledge' as one form of culture, which resides alongside other forms of existence. Alternatively, such recognition could lead to a levelling of the distance between culture and knowledge, effectively elevating any kind of practice to the status of knowledge (and vice versa). In the first case, witchcraft and modern medicine would no longer be categorically distinct; in the second, the laboratory and the kitchen were both to emerge as spaces of knowledge. Perhaps. But what would initially have been gained by such a move if, to invoke Ian Hacking (1999), the emerging cultural construction of reality would leave both knowledge and culture ill defined and lacking in analytical rigour? In other words, would we not merely bypass what we need to understand? Faced with the difficulty of *defining* knowledge, we might arguably have been well advised to have begun by *placing* 'knowledge' in relationship to other concepts. After all, socially and culturally specific practices such as 'knowledge' acquire their meaning by being placed in relationship to other practices and concepts. To give an example: we can identify the meaning of 'employment' because we can place it within a spectrum of related activities that range from 'slavery' to non-profit forms of work such as say gardening. Alternatively, we employ a range of practices alluded to as 'leisure' activities to create a context that is different in kind from the one we designate as 'employment'. Together, these concepts and the semantic fields they characterize bestow meaning on the world we inhabit.

'Knowledge' is no different from other concepts: it, too, forms part of strings of concepts that allow us to differentiate practices from one another. But again there is one crucial difference: more than most practices, 'knowledge' is set aside from other concepts not merely with the help of dualisms that together form part of a semantic web. Rather, 'knowledge' becomes what it is perceived to be through the stabilization of dualisms into polar opposites of the 'either/or' kind (Doel, 1999; Hannah, 1999). The bipolarity of individualized phenomena, or the fact that propositions are either true or false, is translated into a model for knowledge as such (Wittgenstein, 1961). In western societies in particular, 'knowledge' is thus characterized as much by reference to what it is as by allusions to what it is not. This essentially Hegelian insight, the recognition that each and every 'positive' identification creates some form of 'negative' context, usually takes the form of an axis between what is identified as 'knowledge' and its opposite other, variously defined as ignorance, simplicity, error or other signs of absence. Arguably its most fundamental axis, however, is that between 'knowledge' and the form of 'non-knowledge' usually referred to as 'belief'. The solidification of the polarity between various forms of religious practice and scientific knowledge is finally one of the key defining elements of modernity.

In the rest of this chapter, I explore some of the key ramifications of this fundamental setup for the study of culture. I suggested earlier that perhaps it would be advisable not to approach 'knowledge' by the commonly accepted standards, which present it as an endeavour external to everyday routine and thus capable of avoiding the circularity that attaches to other forms of human practice. For such a critical approach to work, we will have to clarify where and why the adoption of this standard occurred – and why it continues to matter. Only then will this chapter be in a position to analyse the contribution of 'culture' to ongoing epistemological debates and to clarify any possible contribution that 'cultural approaches' could make to the human sciences.

THE RISE OF KNOWLEDGE

How, then, did 'knowledge' come to be what it is today – the most common of yardsticks for the evaluation of claims about various aspects of reality? The first and admittedly obvious point to note is the historical novelty of both the status and the current conceptualization of knowledge. The break most commonly associated with the twin occurrence of the Renaissance and the

Reformation in western Europe is of key significance precisely because it shattered the certainties of old and replaced them with the impersonal and accountable system most people in the modern world recognize as 'knowledge' (Jardine, 1996). Since much has been written about these key epochs in the history of the western world, I take the liberty of assuming a shared horizon with readers of this chapter – which allows me to be succinct. What remains of crucial importance from a cultural point of view is the close association between knowledge and the rise of the 'subject' as an identifiable entity (Cosgrove, 1989) autonomously capable of interpreting the world (Lambropoulos, 1993). The epistemological subject capable of conquering doubt that emerged in Descartes' infamous *cogito, ergo sum* thus has a practical complement in the legal subject. What unites both the epistemological and the legal subject is arguably the ability of both to empower across geographic distances: a reader need not guess but is invited to share the presence of something beyond doubt; and what could be less the subject of doubt than the affirmed presence of some*one*? Historically, the most significant example of this 'witnessing ego' can arguably be found in the figure of the nineteenth-century heroic explorer using all the financial muscle provided by western forms of capital and modern technological means of documentation to construct detached 'knowledge' about non-western societies (Driver, 2001; see also Mattelart, 2000).

As a result, the new importance attached to 'knowledge' some five centuries ago is still recognizably part of our contemporary intellectual landscape, western style, with its overt emphasis on individual freedom and responsibility, authorship and legal subjects – symbolically encapsulated in the importance attributed to individual signatures. But the link between the modern subject cum scientist and 'knowledge' is far from a straightforward one: while on the one hand the advances in knowledge have been closely associated with the names of key individuals (the 'canonical' figures of modern science), 'knowledge' as such is thought to exist independently of individual achievements. The word 'evidence' only highlights this seeming paradox (Ginzburg, 1989) while at the same time leading us to the key structural element uniting subject and knowledge: the long-underestimated importance of *representation* in the production of knowledge. Simply put, it is the role of the modern subject to represent truths about the world that subsequently – once they have been rendered communicatively available – take on a life of their own. The validity of this 'life' is thought

to rest entirely on extrinsic criteria and is constructed around the notion, so aptly deployed by Richard Rorty (1979), of the 'mirroring' capacity embodied by particular forms of representation.

What we witness here is an important step: although unthinkable without the originating presence of modern subjects, 'knowledge' requires these subjects – read: scientific authors – to recede into the background, to blend into the representations that empower knowledge. The book, the scientific article, the slide-show, the anatomical chart or the geological map all embody this key facet of knowledge: their 'production' virtually disappears behind their claim to validity (Nelson, 2000; Rose, 2000). This step gives birth to another and by no means less important distinction: the difference between 'subjective' and 'objective' forms of knowing the world. Although born of the reflexive mode which this chapter initially examined under the rubric of 'scepticism', 'knowledge' requires the 'tain' or silvering behind the mirror of nature not to be tainted by its construction or (social) practice but to embody an aspatial form of existence (Gasché, 1986).

The problem is that such a mirror does not exist: irrespective of their usefulness or the status they have acquired as 'objective' forms of representations, none of the scientific means of representation mentioned above can ever be disassociated from the context in which they arose. In fact, the very notion of 'objectivity', by virtue of being 'useful', can profitably be understood to characterize a particular form of 'practice' (or 'language game') itself which is circular no less for being called 'knowledge': its 'success' is ascertained purely through internal consistency. In this, 'knowledge' is no different from other practices.

THE SPATIALIZATION OF KNOWLEDGE

It is this train of thought that allows us to reveal the relevance of the above historical detour for the construction of 'cultural' knowledges. For me, the key consequence lies in the epistemological revaluation of the concept of 'culture' within the human sciences during the latter half of the nineteenth and well into the twentieth century. This first significance of 'the cultural' and 'culture' is by now well documented, as is its intellectual lineage from Matthew Arnold through Max Weber and the Frankfurt School to Raymond Williams and the birth and development of 'cultural studies' and 'cultural geography' in the 1970s (Bennett et al., 1981; Storey, 1993). From our current perspective it is vital to note that the importance of culture as an explanatory concept was fostered by the acute disappointment felt by many when applying the standard model of knowledge to the realm of human activities (Strohmayer, 1997). In other words, it was precisely the paradoxical relationship between reflexive origin and objectified status of knowledge unearthed in the preceding pages that has brought forth the most violent of critiques – spaces of knowledge of sorts – and was instrumental in the rise of culturally sensitive knowledges.

We can summarize these critiques as falling into two broad groups: the phenomenological and the social constructionst reappraisal of knowledge. Both share a view of classical 'representationism' (as developed above) as being idealist; they part company in the alternatives they offer for the scientifically interested practitioner. The former, phenomenology, radicalized an insight that was already familiar to empiricists during the eighteenth century: it reinstated the importance of experience by insisting that knowledge was always knowledge of something that someone had intentionally made the centre of her or his attention. Rather than establishing a neutral relationship between subject and object, the very existence of intentionality was thus indicative of an irreducibly personal aspect to the construction of knowledge.

The 'tain' of the mirror of representation – to use the earlier metaphor – could not but be shaped by likes and dislikes, biographies and individual constraints (Pickles, 1985; Spiegelberg, 1994). My own interest in epistemological questions, to give but one example, has been shaped by an earlier series of communication breakdowns in the process of 'doing research', by a host of personal moves between different countries and by the odd chance encounter. 'Objectively', this should not matter: whatever research I would produce would be evaluated independently from its context. Realistically, however, traces remain and idiosyncrasies proliferate: being white and male, any possible contribution to knowledge about, for example, slavery would undoubtedly remain tainted by this personal context. A similar conclusion was advanced by social constructionism. Broadly conceived, this includes any claim that the construction of knowledge needs to be understood within its proper historical, social and cultural context. Rather than offering timeless insights produced from a single, 'original' vantage point into the workings of societies, a relativized, scaled-down and locally sensitized form of knowledge was seen to be more appropriate.

Although vastly dissimilar in scale, if not in ambition, both critiques – and the notion of culture they sought to resurrect – had one point of convergence in common: both acknowledged the 'situated' nature of knowledge (Haraway, 1988). In other words, they derived a logically compelling place for the role of 'culture' in the production of knowledge. Methodologically, this acknowledgement often implied a turn to 'local' forms of knowledge. The introduction of this term into the realm of cultural and geographic knowledges was in essence a spatialization of what had become known as tacit knowledge amongst epistemologists (Geertz, 1983). 'Tacit' here circumscribes the implicit character of the many forms of knowledge that allow us as human beings to exist: from the biologically determined and largely instinctive (breathing) to the selectively acquired and trained (driving), we rely on internalized forms of knowing to live our lives.

Circumscribing the boundedness of knowledge within local traditions, 'local knowledge' furthermore highlighted the fact that knowledge is not just the outcome of academic or scientific ways of analysing the world, it also constituted a social practice (Thrift, 1983). In practice, this effectively enlarged the spectrum of possible knowledges through the designation of a spectrum stretching from 'knowing' one's way about within the confines of a familiar spatial context (say a kitchen) to the artificially created spaces of a laboratory. In the realm of science studies, this enlargement has by now, largely through the input made by feminist scholars, produced the most important results (Duden, 1993; Harding, 1992; Keller, 1985; Latour, 1987). Of particular interest for our given topic is the term 'epistemic cultures' (Knorr-Cetina, 1999; but see Hacking, 1999), which critiques the notion of a unified, progressive science that was so central to modernity as a whole.

Such a reappraisal of the cultures of science is all the more pertinent in the context of the present rush into as yet unstructured knowledge and expert societies. At the same time, however, the scale of individual objects of analysis was often drastically diminished: the centrality accorded to 'the body' (Butler, 1990; Nast and Pile, 1998; Shilling, 1993; Stratton, 2001) and/or to 'performance' (Crang, 1994; Hetherington, 1998; Lewis and Pile, 1996; McDowell, 1995) in the human sciences recently is indicative, amongst other things, of the reduction in scale that has marked the most recent developments in many human sciences. Knowledge *through* the body and *in* performance strives to be mobile and mimetically approximates the object of its curiosity:

what it lacks in transferability, it makes up in rigour and precision.

Crucially, 'culture' played a key role in the construction of such knowledges, as well as explaining the variations found in practices around the globe. It did so through the invocation of 'everyday' practices that have shaped and continue to shape the various histories and geographies that form the body of our knowledges (de Certeau, 1984). The fact that these latter now often appear with a plural 's' is indicative of this response: where 'knowledge' becomes spatially constructed – as it does once we allow for 'the local' or 'tacit knowledge' to become legitimate grounds for the construction of knowledge – a researcher will more likely than not find himself or herself confronting a plurality of 'knowledges' or spaces of knowledge. What is more, the fact that these 'knowledges' often compete with one another is no longer seen as a threat to the unity of the sciences but becomes part of their dialogical construction. In other words, where the traditional model of sciences, western style, acknowledged at best (in say the works of Karl Popper) the existence of temporal variations – often read as 'progressions' – in the truth content of scientific explanations, the 'cultural' sciences today accept the existence of spatial and hence cultural differences: scientific paradigms no longer merely succeed one another, they also coexist (Kuhn, 1962).

CULTURAL KNOWLEDGE

The acknowledgement of 'the cultural' over the last two centuries would have been unthinkable, however, without a realization that had its roots in both of the developments mentioned above: (1) the novel relevance being attributed to phenomena of everyday significance, and (2) the acknowledgement of 'representation' being a key epistemological issue for any production of knowledge. For me, both converge in what has become known as 'the linguistic turn' (Rorty, 1967) – the realization that 'language' is (at) the heart of knowledge and of cultural expressions, as well as cultural change. Within the field of epistemology, this insight was nothing new: language had always been central to the many and various theories of knowledge. What changed during the reappraisal of 'culture' in the 1970s was a new centrality that was accorded to discourse, a particular subcategory of linguistic enquiry that makes sense of language as practice (Benveniste, 1971). It was in this form that language entered into the cultural sciences and

into cultural geography (Barnett, 1998; Curry, 1996; Mills, 1997). Owing a great deal to Wittgenstein's (1953) designation of 'language games', the notion of 'discourse' underlined the need to understand knowledge as a particular, highly contextualized form of communication. In other words, 'discourse' signaled a turn to 'practices' in the human sciences and thus, albeit implicitly, legitimized 'culture' as a key concept within and beyond the human sciences.

The consequences of this development were far-reaching indeed. Not only was a rather amorphous interest in matters cultural concretized through the centrality accorded to language, but the interest in 'discourse' furthermore opened up whole new worlds to the epistemologically interested. These worlds included the novel interpretations of modern history, science and culture unearthed in the work of Michel Foucault (1979; 1989), the 'deconstruction' of the western philosophical tradition advanced by Jacques Derrida (1982; 1989), and the displacement of the subject from the throne rightly or wrongly assumed in the early modern period in the books of Gilles Deleuze and Félix Guattari (1987), Jacques Lacan (1977) and Bruno Latour (1993).

This is emphatically not the place for an in-depth review of all the trends mentioned above. Some signposts will thus have to suffice to clarify the connection between this kind of discourse-driven research and the study of culture. What is initially clear is that we are confronted with a novel impulse: rather than turn its attention to an allegedly 'progressing' march of scientific achievements, science now increasingly turned its attention onto itself and analysed itself – to use Foucault's term – as an archive. In this already, we can manifest a 'cultural' impulse in the sense that the differences between 'knowledge' and other forms of human practice were gradually erased. In fact, 'knowledge' often became a byword for human practice in general, thus losing the privileged position it had acquired since its inception some 500 years ago. We shall return to this theme later in this chapter.

Among the concepts that have proven to be fruitful within the culturally inspired human sciences, Derrida's notion of différance needs to be singled out. Radicalizing the earlier insistence upon difference that had already led to the acknowledgement of the 'local' characteristics of knowledge (as discussed above), différance effectively localized 'the local'. In denying stability to the concepts employed to ground epistemic differences in space and time, Derrida refocused the question of power within the human sciences. A similar argument emerged in the writings of Foucault, who read the accepted history of western progress against the grain and in this way uncovered the 'blind' spots of scientific discourses: the 'normalization' of western cultures here goes hand-in-hand with the creation of abnormal or non-normal 'others' – the sick, the homeless, those without work, the sexually perverted and so on.

Both themes have arguably been rendered most susceptible to cultural analyses in the work of Gayatri Spivak (1988) and Homi Bhabha (1994). What emerges as a key, if often overlooked, theme is a new centrality accorded to the construction and maintenance of *communities*. 'Knowledge' and the scientific communities it created are but one example amongst many of how the notion of a 'community' came to be constructed in the wake of modernity. The notion of 'common sense' developed earlier is again crucial in this respect in that it represents a shared if mostly unexamined communality at the heart of many communities. A 'cultural' trait itself, the discursive construction of communities around such shared – and culturally highly relevant – notions as 'fact' (the scientific community), 'representation' (the political community), 'the public' (the civic community) or 'wants' (economic discourses), became a legitimate area of research (Mouffe, 1993). This deconstruction of the discursive unity underlying various communities across the scales radicalized the very notion of culture as a recognizably unifying set of practices. In this, the study of culture followed a 'lead originally initiated by the Frankfurt School – whose pessimistic interpretation of modern knowledge was increasingly reconciled with the notion of 'culture' as lacking in clear direction (Adorno and Horkheimer, 1979). The link between discursive construction of communities and history in the work of Walter Benjamin in particular was to prove of singular importance within the field of cultural studies, broadly conceived (Benjamin, 1999).

Just as important, however, was the price incurred by a particular radicalization of 'the cultural' in its linguistic manifestations during the last two centuries: the (analytically rigorous) acknowledgement that 'knowledge' might vary geographically was often but a first step up the ladder of relativism. Unable or unwilling – in the words of Ludwig Wittgenstein (1961) – to 'forget' the ladder once it had served its purpose, cultural and other scientists suddenly found themselves facing the contingency of not merely the phenomena under investigation but also the claims to knowledge that they presented to the wider world. In short, cultural scientists of many persuasions were surrounded by a phenomenon

many saw fit to label 'postmodernism' (Bertens, 1998; Dear, 2001). 'Postmodern' cultural geography, although not a term commonly used by its practitioners, placed a particular emphasis on the plurality of cultural knowledges, rather than on a unity of knowledge created through the invocation of the term 'culture' – an impulse that was very much alive in cultural geography until the late 1970s. Others still saw the acknowledgement of relative knowledge less in terms of a departure from 'modernity' as such. Interpreting the stability traditionally provided by univocal forms of representation as a structural component of knowledge in general, this second group interpreted the move towards localized or otherwise situated types of knowledge as a departure from 'structural' forms of knowledge production, advocating 'poststructural' approaches instead (Doel, 1999).

For readers interested in cultural 'spaces of knowledge', the difference between the two positions outlined above is important and needs to be spelled out in greater detail. While culturally sensitive approaches to the construction of science are strictly speaking compatible with a postmodern point of view, they do not square easily with a denial of structures *per se*. At bottom, the difference attaches to a different interpretation of 'materiality' and thus of the status of 'culture' amongst the sciences. While a 'postmodern' view interprets the recognition of 'culture' as a change in the real makeup of societies – a view easily compatible with the development of 'multicultural' societies and the increasing obsolescence of other non-local forms of interpretation such as colonialism, Marxism or Fordism – the denial of structures refocused attention on the purely epistemological realm.

The resulting wavering between ontological and epistemological claims has left a clear mark on the study of 'culture'. Expressed perhaps most succinctly in the doubts raised by Don Mitchell (1995), the resulting tension is very much present today. On the one hand, the acknowledgement of the key role of 'culture' has left an imprint on all those intellectual developments that seek to localize powerful discourses, from postcolonialism (Barnett, 1998; Said, 1993; Sidaway, 1997; 2000) to post-Marxism (Gibson-Graham, 1996; Laclau and Mouffe, 1985) and post-Fordism (Amin, 1994). On the other hand, 'culture' continues to serve discursive strategies that unearth the relativity of claims to understanding. According to some observers, both strands of inquiry can be seen to converge upon the role of 'resistance' as a cultural expression of both the 'positive' existence of something and the 'negative' disruption to 'common sense' it creates

(Cresswell, 1996; Pile and Keith, 1997; Scott, 1985). However, the epistemological status of such claims remains unclear and continues to be vulnerable to sustained critique (Sharp, 2000).

It may be pertinent to attempt a first and tentative summary at this point in our discussion. Such an endeavour would posit 'culture' as a key concept in the attempts to address epistemological issues that have emerged in the nineteenth century. The recognition of 'representation' and 'discourse' in particular within theories of knowledge has been instrumental in the move towards experience, everyday life and other, local forms of knowledge. These latter in turn fostered an environment conducive to an interest in matters cultural. What remains unresolved, however, is the status of 'culture' as a concept: does it function according to the logic of established scientific discourse or does it operate in a different manner? Perhaps the avoidance of the term 'culture' by those interested in genuinely novel ways of creating insights is indicative in this respect. One could also argue with some degree of justification that the conceptual history of the term 'culture', with its associated uses especially during colonialism and the construction of nation-states, renders it ill suited to advance knowledge theoretically. Be that as it may, this author is quite at ease in allowing the perceptions created by the entanglements of culture and the spaces of knowledge to speak for themselves.

CULTURE AND REPRESENTATION

This chapter would be intellectually dishonest if it were to deny that (epistemologically speaking) the headlong flight into matters cultural was but one of many escape routes from the 'crisis of representation' (Dear, 1988) diagnosed during the last two decades of the twentieth century. As often, the roads not travelled are every bit as interesting as the paradigms actually developed. The resilience, for instance, of many in the human sciences towards the ideas initially proposed by Karl Popper (1962) – accepting 'falsification' as a means of establishing the temporary status of competing forms of representation – still surprises. Equally astonishing is perhaps the lack of any culturally inspired responses to the challenges born of 'the crisis of representation' that were developed in a pragmatist mode (Habermas, 1972; 1988). In particular, the lack of any sustained engagement between cultural theory and structurationism comes as a surprise given their mutual interest in the

avoidance of dualist patterns of thinking (Giddens, 1984; but see Shilling, 1993). A similarly unfortunate neglect characterizes the possibility of exchange between critical realism and cultural theory (Bhaskar, 1986; Hannah, 1999; Sayer, 1992).

There is, however, another form of critique that has proven to be equally important in the rise of culture to the status it currently enjoys. Here I am thinking of the critique of the partiality of many accepted forms of knowledge. The most pertinent of these critiques was and continues to be that launched under the heading of 'feminism'. At its best, this appraisal has avoided replacing one set of partial viewpoints with another and established genuinely novel forms of cultural analyses (Bordo, 1993; 1998). Take, for instance, the reappraisal of the importance of ocular metaphors in the realm of science (Levin, 1993; also Jay, 1993). Once thought to be innocent expressions of a universally shared desire to know, the very notion of *enlightening* cultural and other phenomena is now increasingly seen as part and parcel of an objectified worldview (Rose, 1993; but see Gould, 1999) and thus – crucially – of an embodied way of knowing (Pile and Thrift, 1995). This is also as good an example as any of the taken-for-granted nature of the construction of knowledge in western societies: the privileging of the 'eye' over and against other forms of connecting to the world not only comes naturally to most, it clearly is also an expression of a certain kind of culture that helps to ferment that culture through the delegitimation of alternative 'ways of seeing' (Rose, 2000; Ryan, 1997).

Not coincidentally, it is the eye with its propensity to declare things to be either present or absent that has become the Leitmotif of modern science: it nicely complements other strategies that modern science has utilized to its advantage. The sheer diversity of cultural expressions and the rather obvious deficiencies offered by an exclusively visual approach have made this critique appear perhaps to be less radical than it is, for the target of this and many other culturally minded approaches to knowledge has often been a concept at the heart of the modern scientific enterprise: the notion of *identity* (Cohen, 1999; Friese, 2001). Implied in much of what has been discussed in this chapter, it is perhaps fitting finally to arrive at some semblance of a centre. 'Identity' is of course one of the key concepts of philosophy writ large. It is also the term customarily deployed to signify those irreducible, non-circular elements that form the basis for structures in general. Amongst a developing set of concepts denoting 'identity' – including political concepts such as 'the nation', aesthetic concepts such as 'landscape' and economic concepts like weights and measurements – the (modern) subject occupied a privileged position. Linked with visual metaphors through the designation of clear and unequivocal positions to the human gaze, identity is constantly reaffirmed in an everyday context by just about everyone.

The acknowledgement of culture, situatedness and contextuality challenged the beautiful, if restrictive, geometry of the associated production of knowledge. Folded back upon its origin and the cultural context that surrounded it, 'identity' revealed itself to have been a social construction that masked often surprising forms of difference. Once more, a key initial impulse came from feminist cultural scholars who recognized the importance of gender differences in the construction of identities. Other differences soon followed in the wake of this recognition, quickly turning 'difference' into one of the key concepts of cultural analyses. But the critique of identity did not stop until it had coined a new category to designate the evaporation of identity into its opposite other, *hybridity* (Bhabha, 1994; Whatmore, 1997). Used both as an epistemological concept and as a strategic tool, the recognition of hybridity can thus been seen as the logical answer to a fundamental paradox of identity already acknowledged by Ludwig Wittgenstein: 'Incidentally, to say of *two* things that they are identical is nonsense, and to say of *one* thing that it is identical with itself is to say nothing at all' (1961: 5.5303). Verging between a highly suspicious 'as if' and meaningless tautology, identity as the touchstone of all knowledge becomes infested with power; as a consequence, in the fitting words of Gunnar Olsson, 'knowledge can be defined as the ability and the opportunity of saying that $a = b$ and be believed when one does it' (1998: 147–8, see also Olsson, 2000). The analysis of 'culture' is equally affected by this insight but suffers less from it given its overall structure and the reality of its existence as practice.

Of the many issues raised in conjunction with the concept of 'identity', none is perhaps more pertinent to the title of this section than the sheer diversity of abstract and practical approaches that different philosophical cultures the world over have developed to focus on its existence (Mbiti, 1990). Lacking the space and the expertise properly to explore non-western traditions, the present pages can merely allude to this most fundamental of differences in the hope for a more inclusive form of cultural knowledge. Identity may well be central to all of them; addressing it, however, does not always follow the same route, nor does it yield structurally comparable results.

The communitarian tradition that characterized much of African thinking and culture and which has left a mark on politics in Leopold Senghor's notion of *négritude*, for instance, does not in general favour the individualized notions of science and knowledge familiar to anyone in the west (Birt, 2001; Gottlieb, 1992; Kwame, 1997; Mudimbe, 1988; Senghor, 1964). The resulting social role of cultural practices such as animism, for instance, can arguably not be understood within a framework derived from and ultimately aimed at western forms of identity and subjects (Rooney, 2001). A similar disposition to seek knowledge at a scale larger than the individual has been characteristic of the Latin American experience (Schutte, 1993; Wolf, 2001), while the prevalence of holistic, non-dualistic traditions has seen Asian culture and thinking develop yet different approaches to the understanding of humankind and its cultures (Ames et al., 1994; Loy, 1988). The resulting set of questions has become central to the emergence of 'postcolonial' forms of geographic knowledges (Clayton, 2001; Sidaway, 2000) that are explored elsewhere in this volume.

CONCLUSIONS

All of this allows us at long last to approach the role of theory in the construction of knowledge. A vast terrain itself, the realm of 'theory' is often thought to be the antipode to empiricist notions of knowledge. Historically speaking this is not quite accurate given the existence of developed theoretical systems supporting the primacy of empirical approaches over and against other forms of knowledge production. Otherwise, the fortunes of 'theory' within the knowledge of culture have changed substantially since the days of Max Weber. Once thought to provide a context for the pursuit of knowledge, the invocation of 'theory' has a much looser meaning nowadays. We no longer speak of observations being 'in line with theoretical assumptions', opting instead for watered-down 'theoretically informed' ways of approaching cultural phenomena. All of which is perfectly in line with our discussion so far: the very idea of 'localized' knowledge, of different cultures or of a lack of stable identities implies a different, a substantively reduced status and applicability for the notion of transferable, generalizable and abstract claims to knowledge. Knowledge about culture will tend to be idiosyncratic knowledge. It will tend to focus on the concrete workings of particular cultural configurations and leave claims about the bigger picture to others. Instead of theory, what has emerged is 'metatheory' of the kind attempted in this chapter: the examination of possibilities, limitations and contradictions within, as well as between, various epistemological propositions (Bordo, 1998).

This chapter has sought to present the plurality of 'spaces of knowledge' as a logical outcome of epistemological debates within the human sciences. The concept of 'culture' attaches to these spaces – in fact, it becomes largely synonymous with these spaces – because unlike many other general concepts, 'culture' does not resist the reductions in claims to knowledge that is one of the main characteristics of contemporary knowledge in the human sciences. To my mind, this accounts for some of the attraction of 'cultural' forms of knowledge. As such, 'cultural' geography (or other 'culturally' sensitive approaches to the production of knowledge) offers a practical solution to the problem of circularity with which this chapter began; as solutions go, this one does not make the original problem disappear, but it has the advantage of dissolving into a form of practice what otherwise would remain hidden from view. The resulting 'denaturalization' of, amongst others, commonsense, taken-for-granted customs and methodological assumptions clearly is a benefit of the cultural reconsiderations of past decades.

There is, however, a danger that needs to be spelled out just as urgently. We speak of 'culture' in a global manner, thus subsuming what could – and often should – just as sensibly be analysed under the rubric of 'the economy', 'politics' or individual psychological categories. If experience and epistemological rigour invite us to these too, as tied into the workings of 'culture', we arguably stand to lose as much as we can expect to gain. At the very least, anyone interested in the cultural ways of knowledge should acknowledge that 'culture', too, can and must be subjected to the same localization that has previously produced so many unexpected insights. This last step would represent a genuine and much welcome addition to the 'spaces of knowledge' we create, nourish and inhabit.

REFERENCES

Adorno, T.W. and Horkheimer, M. (1979) *Dialectic of Enlightenment*. London: Verso.

Ames, R. Dissanayake, W. and Kasulis, T. (1994) *Self as Person in Asian Theory and Practice*. Albany: SUNY Press.

Amin, A. (ed.) (1994) *Post-Fordism: A Reader*. Oxford: Blackwell.

Barnett, C. (1998) 'Impure and worldly geography: the Africanist discourse of the Royal Geographical Society, 1831–1871', *Transactions of the Institute of British Geographers* 23: 239–51.

Beneviste, E. (1971) *Problems in General Linguistics*. Coral Gables: University of Miami Press.

Benjamin, W. (1999) *The Arcades Project*. Cambridge, MA: Belknap.

Bennett, T. et al. (eds) (1981) *Culture, Ideology and Social Process*. London: Batsford.

Bertens, H. (1998) *The Idea of the Postmodern*. London: Routledge.

Bhabha, H. (1994) *The Location of Culture*. London: Routledge.

Bhaskar, R. (1986) *Scientific Realism and Human Emancipation*. London: Verso.

Birt, R.E. (2001) *The Quest for Community and Identity: An Africana Philosophical Anthology*. Lanham: Rowman and Littlefield.

Bordo, S. (1993) *Unbearable Weight: Feminism, Western Culture, and the Body*. Berkeley: University of California Press.

Bordo, S. (1998) 'Bringing body to theory', in D. Welton (ed.) *Body and Flesh: A Philosophical Reader*. Oxford: Blackwell. pp. 84–97.

Butler, J. (1990) *Gender Trouble: Feminism and the Subversion of Identity*. London: Routledge.

Clayton, D. (2001) 'Questions of postcolonial geography', *Antipode* 33 (4): 749–51.

Cohen, A. (ed.) (1999) *Signifying Identities*. London: Routledge.

Cosgrove, D. (1989) 'Historical considerations on humanism, historical materialism and geography', in A. Kobayashi and S. Mackenzie (eds) *Rethinking Human Geography*. London: Unwin Hyman.

Crang, P. (1994) 'It's showtime: on the workplace geographies of display in a restaurant in south east England', *Environment and Planning D: Society and Space* 12: 675–704.

Cresswell, T. (1996) *In Place/Out of Place: Geography, Ideology and Transgression*. Minneapolis: University of Minnesota Press.

Curry, M. (1996) *The Work in the World: Geographical Practice and the Written Word*. Minneapolis: University of Minnesota Press.

Dear, M. (1988) 'The postmodern challenge: reconstructing human geography', *Transactions of the Institute of British Geographers* 13(NS): 262–74.

Dear, M. (2001) 'The postmodern turn', in C. Minca (ed.) *Postmodern Geography*. Oxford: Blackwell.

de Cereau, M. (1984) *The Practice of Everyday Life*. Berkeley: University of California Press.

Deleuze, G. and Guattari, F. (1987) *A Thousand Plateaus: Capitalism and Schizophrenia*. Minneapolis: University of Minnesota Press.

Derrida, J. (1982) 'Sending: on representation', *Social Research* 49 (2).

Derrida, J. (1989) *Edmund Husserl's Origin of Geometry: An Introduction*. Lincoln: University of Nebraska Press.

Doel, M. (1999) *Poststructuralist Geographies*. Edinburgh: Edinburgh University Press.

Driver, F. (2001) *Geography Militant: Cultures of Exploration and Empire*. Oxford: Blackwell.

Duden, B. (1993) *Disembodying Women: Perspectives on Pregnancy and the Unborn*. Cambridge, MA Harvard: University Press.

Foucault, M. (1979) *Discipline and Punish: The Birth of the Prison*. New York: Vintage.

Foucault, M. (1989) *The Birth of the Clinic*. London: Routledge.

Friese, H. (ed.) (2001) *Identities: Time, Difference, and Boundaries*. New York: Berghahn.

Gasché, R. (1986) *The Tain of the Mirror*. Cambridge, MA: Harvard University Press.

Geertz, G. (1983) *Local Knowledge: Further Essays in Interpretative Anthropology*. New York: Basic.

Gibson-Graham, J.K. (1996) *The End of Capitalism (As We Knew It)*. Oxford: Blackwell.

Giddens, A. (1984) *The Constitution of Society*. Berkeley: University of California Press.

Ginzburg, C. (1989) *Clues, Myths, and the Historical Method*. Baltimore: Johns Hopkins University Press.

Gottlieb, A. (1992) *Under the Kapok Tree: Identity and Difference in Beng Thought*. Bloomington: Indiana University Press.

Gould, P. (1999) 'Sharing a tradition', in *Becoming a Geographer*. Syracuse: Syracuse University Press.

Habermas, J. (1972) *Knowledge and Human Interests*. London: Heinemann.

Habermas, J. (1988) *Theory and Practice*. Cambridge: Polity.

Hacking, I. (1999) *The Social Construction of What?* Cambridge, MA: Harvard University Press.

Hannah, M. (1999) 'Skeptical realism: from either/or to both-and', *Environment and Planning D: Society and Space* 9: 309–27.

Haraway, D. (1988) 'Situated knowledges: the science question in feminism as a site of discourse on the privilege of partial perspective', *Feminist Studies* 14 (3): 575–99.

Harding, S. (1992) *Whose Science? Whose Knowledge? Thinking from Women's Lives*. Ithaca: Cornell University Press.

Hetherington, K. (1998) *Expressions of Identity: Space, Performance, Politics*. London: Sage.

Hookway, C.J. (1990) *Scepticism*. London: Routledge.

Jardine, L. (1996) *Worldly Goods: A New History of the Renaissance*. London: Macmillan.

Jay, M. (1993) *Downcast Eyes: The Denigration of Vision in Twentieth-Century French Thought*. Berkeley: University of California Press.

Keller, E.F. (1985) *Reflections on Gender and Science*. New Haven: Yale University Press.

Knorr-Cetina, K. (1999) *Epistemic Cultures: How the Sciences Make Knowledge*. Cambridge, MA: Harvard University Press.

Kuhn, T.S. (1962) *The Structure of Scientific Knowledge*. Chicago: University of Chicago Press.

Kwame, G. (1997) *Tradition and Modernity: Philosophical Reflections on the African Experience*. Oxford: Oxford University Press.

Lacan, J. (1977) *Écrits*. London: Routledge.

Laclau, E. and Mouffe, C. (1985) *Hegemony and Socialist Strategy*. London: Verso.

Lambropoulos, V. (1993) *The Rise of Eurocentrism: Anatomy of Interpretation*. Princeton: Princeton University Press.

Latour, B. (1987) *Science in Action: How to Follow Scientists and Engineers through Society*. Cambridge, MA: Harvard University Press.

Latour, B. (1993) *We Have Never Been Modern*. Cambridge, MA: Harvard University Press.

Levin, D.M. (1993) 'Introduction', in D.M. Levin (ed.) *Modernity and the Hegemony of Vision*. Berkeley: University of California Press. pp. 1–29.

Lewis, C. and Pile, S. (1996) 'Woman, body, space: Rio carnival and the politics of performance', *Gender, Space and Culture* 3 (1): 23.

Loy, D. (1988) *Nonduality: A Study in Comparative Philosophy*. New Haven: Yale University Press.

Mattelart, A. (2000) *Networking the World 1794–2000*. Minneapolis: University of Minnesota Press.

Mbiti, J. (1990) *African Religions and Philosophies*, 2nd edn. London: Heinemann.

McDowell, L. (1995) 'Body work: heterosexual gender performances in city workplaces', in D. Bell and G. Valentine (eds) *Mapping Desire: Geographies of Sexualities*. London: Routledge. pp. 75–95.

Mills, S. (1997) *Discourse*. London: Routledge.

Mitchell, D. (1995) 'There's no such thing as culture: towards a reconceptualization of the idea of culture in geography', *Transactions of the Institute of British Geographers* 20 (NS): 102–16.

Mouffe, C. (1993) *The Return of the Political*. London: Verso.

Mudimbe, V.Y. (1988) *The Invention of Africa: Gnosis, Philosophy and the Order of Knowledge*. Bloomington: Indiana University Press.

Nast, H. and Pile, S. (eds) (1998) *Places through the Body*. London: Routledge.

Nelson, R.S. (2000) 'The slide lecture, or the work of art history in the age of mechanical reproduction', *Critical Inquiry* 26 (3): 414–35.

Olsson, G. (1998) 'Towards a critique of cartographic reason', *Ethics, Place and Environment* 1 (2): 145–55.

Olsson, G. (2000) 'From $a = b$ to $a = a$', *Environment and Planning A* 32: 1235–44.

Pickles, J. (1985) *Phenomenology, Science, and Geography: Spatiality and the Human Sciences*. Cambridge: Cambridge University Press.

Pile, S. and Keith, M. (1997) *Geographies of Resistance*. London: Routledge.

Pile, S. and Thrift, N. (1995) *Mapping the Subject: Geographies of Cultural Transformation*. London: Routledge.

Poovey, M. (1998) *A History of the Modern Fact: Problems of Knowledge in the Sciences of Wealth and Society*. Chicago: University of Chicago Press.

Popper, K. (1962) *Conjectures and Refutations*. New York: Basic.

Rooney, C. (2001) *Literature, Animism and Politics*. London: Routledge.

Rorty, R. (ed.) (1967) *The Linguistic Turn: Recent Essays in Philosophical Method*. Chicago: University of Chicago Press.

Rorty, R. (1979) *Philosophy and the Mirror of Nature*. Princeton: Princeton University Press.

Rose, G. (1993) *Feminism and Geography*. Minneapolis: University of Minnesota Press.

Rose, G. (2000) 'Practising photography: an archive, a study, some photographs and a researcher', *Journal of Historical Geography* 26 (4): 555–71.

Ryan, J. (1997) *Picturing Empire: Photography and the Visualization of the British Empire*. London: Reaktion.

Said, E. (1993) *Culture and Imperialism*. New York: Knopf.

Sayer, A. (1992) *Method in Social Science: A Realist Approach*. London: Routledge.

Schutte, O. (1993) *Cultural Identity and Social Liberation in Latin American Thought*. Albany: SUNY Press.

Scott, J. (1985) *Weapons of the Weak: Everyday Forms of Peasant Resistance*. New Haven: Yale University Press.

Senghor, L.S. (1964) *On African Socialism*. New York: Praeger.

Sharp, J. (ed.) (2000) *Entanglements of Power: Geographies of Domination/Resistance*. London: Routledge.

Shilling, C. (1993) *The Body and Social Theory*. London: Sage.

Sidaway, J.D. (1997) 'The production of British geography', *Transctions of the Institute of British Geographers* 22: 488–504.

Sidaway, J.D. (2000) 'Postcolonial geographies: an exploratory essay', *Progress in Human Geography* 24 (4): 591–612.

Spiegelberg, H. (1994) *The Phenomenological Movement: A Historical Introduction*. Dordrecht: Kluwer.

Spivak, G. (1988) *In Other Worlds: Essays in Cultural Politics*. London: Routledge.

Storey, J. (1993) *An Introductory Guide to Cultural Theory and Popular Culture*. London: Harvester Wheatsheaf.

Stratton, J. (2001) *The Desirable Body: Cultural Fetishism and the Erotics of Consumption*. Urbana: University of Illinois.

Strohmayer, U. (1997) 'The displaced, deferred or was it abandoned middle: another look at the idiographic–nomothetic distinction in the German social sciences', *Revue of the Fernand Braudel Center* 10 (3/4): 279–344.

Thrift, N. (1983) 'On the determination of social action in space and time', *Environment and Planning D: Society and Space* 1: 23–58.

Whatmore, S. (1997) 'Dissecting the autonomous subject: hybrid cartographies for a relational ethics', *Environment and Planning D: Society and Space* 15: 37–53.

Wittgenstein, L. (1953) *Philosophical Investigations*. Oxford: Blackwell.

Wittgenstein, L. (1961) *Tractatus Logico-Philosophicus*. London: Routledge.

Wolf, E.R. (2001) *Pathways of Power: Building an Anthropology for the Modern World*. Berkeley: University of California Press.

29

Knowledge and Geography's Technology – Politics, Ontologies, Representations in the Changing Ways We Know

Francis Harvey

GEOGRAPHY AND TECHNOLOGY

This chapter examines how technology changes geography and, in the process, changes in the most fundamental ways what and how we know. Knowledge is imbricated in the cultural use of technology (Latour, 1999), yet geography has only recently begun to probe how technology influences the production and representation of geographic knowledge. This lacuna is widespread. Western civilization tends to represent technology as a neutral tool. Geographers often unquestioningly adopt this view. Technology has usually been thought of as a means to improve analyses of space and place (the study of geography) through better observation and analysis, or as the means through which humans intervene in the environment.

A good example of this is Simpson's (1966) article which appeared in the *Annals of the American Association of Geographers* with the apposite title 'Radar, geographic tool'. In it Simpson demonstrates the capabilities and utility of radar, a tool for geographers. Yet radar, even considering its uses in air traffic control and weather forecasting, is not just a neutral tool. Originally developed by the military, it allowed for the distant recognition of objects. Because it could detect objects far beyond the range of human sight, it became integral to a new form of command and control, requiring the organization of a new type of technocratic bureaucracy to direct and coordinate soldiers, ships, and planes. Radar changed the space of war, and the allied victory in World War II was seen as a victory for technology,

technological rational organization, and the creation of new spaces of control at a distance.

In this chapter, I present ways to understand the imbrication of technology in the production and representation of geographic knowledge. Following the introduction, the second section provides an overview of the theoretical corpus that addresses the manifold relationships between geography and technology. Shifts in the scholarly discourses surrounding these issues point to increasing awareness that, in the most fundamental ways, technology has reframed our thinking (Poster, 1990). This scholarship rests on Marxian and neo-Marxian readings of knowledge production and engagement with the substantial bodies of literature that since Heidegger and Ellul have examined the instabilities and complications of technology.

The third section situates current and recent engagements with the implications of technology's role in the production and representation of geographic thought in terms of three streams of thought. The first of these, a historical perspective on the development and use of technology in geography, points to key themes in human geography that have always relied on technology to overcome human limitations. The second, simulation and communication, is perhaps the most pervasive aspect of geographers' use of technology on behalf of state or corporate interests. GIS is included here as a technology (of simulation) that underpins both disembodied engagements with the world and reifications of the geographic imagination. The intermingling of simulation and communication points to the hybrid role technology plays in cultural discourse. Finally, this section points to cultural studies of the materiality of technological practices in human geography and the linkages to capitalist and modernist ideological knowledge production.

Radar is one example I will use to provide examples of how technology changes the ways geographers produce knowledge as well as the implications of that knowledge. In contrast to 'geographical engineering,' which included the grandiose plan to build a vastly larger Panama Canal using 300 atomic bombs to remove the mountain chains at the isthmus of Panama and conveniently vaporize most of the material, radar seems to be a far humbler technology. However, its political consequences are wider since it enables modern, instrumentalist command and control of space and time. The tragic social consequences of instrumentalist command and control were made ignobly clear during the Vietnam War debacle, when under the influence of Robert McNamara, Secretary of Defense from 1961 to 1968, radar technology was used to coordinate US attacks on the Ho Chi Minh trail and across southern Laos along what was called the McNamara line (Edwards, 1997). Managed by a command and control center called Igloo White, a tripped ground sensor would be registered by the observation center in Thailand and within five minutes an air strike would blast whatever happened to be at the sensor coordinates. To say, with hindsight, that this 'tool' or system failed to work, but could be improved, ignores the human tragedy caused by this indiscriminate use of violence and masks the political intent of these acts. Living in a techno-culture whose enamoring of cyberspace seems to climb rung after rung on the same rationalist's ladder, tool-makers rarely step back to see that technologies such as radar are never neutral: they embody and reinforce complex power relationships that alter society and culture. Reciprocally, the resulting cultural changes codetermine the next technological development. The tool metaphor fails to capture these complex relationships evident in geography's relationship to technology.

Just as the army staff of project Igloo White had no need to leave their Bangkok military bases to direct bombers over Vietnam to their targets, geography tends to stay more in the back offices of government and industry, and persist in its peculiar, Faustian, relationship with technology. Parts of the discipline have always depended on it and driven its development, other parts have used it, some have studied its use, and a few people have criticized it. Only a few have examined the political dimensions of geography's technology (Chrisman, 1987; Openshaw, 1991;

Smith, 1992). The imbricated relationships necessitate a broad review. The examination of literatures on culture, technology, and geography engages the political stakes involved in technology. Exploring the different theoretical dimensions helps to understand how we can theorize the relationship between technology and society, and by implication culture. This allows us to assess what technology is from an ontological or material standpoint and probe its political and social dimensions (Curry, 1998).

PERSPECTIVES ON TECHNOLOGY

Every human culture has the technological means to enhance and ensure its continued existence. From a normative perspective in western civilization, technology defines civilization (Mumford, 1967). Beginning with the most basic manual implements of varied practices, including planting, gathering, hunting, and sheltering, to organizing scribes to duplicate books or guaranteeing society's survival when faced with harsh environments, technology is inseparable from every culture's foundation. Along with the many mundane uses, technology also provides ways of producing, communicating, and archiving knowledge. Clearly, technology is integral to culture, but then what are its politics? What are its social and cultural impacts?

A great number of liberal and conservative scholars in the modern tradition have repeatedly asserted that technology is the great empowerment of current western civilization and is the means to uplift other civilizations. These positions are so ingrained that many people have difficulty recognizing the problematic assumptions inherent in the representation of human society's evolution from Stone Age tools to farming tractors to digital communication. The hype that accompanies releases of hardware and software often makes use of this millenarist and progressive imagery to invoke powerful suggestions of a better world that awaits the user of the newest invention. By returning to Marx's work and tracing the development of related political-economic perspectives on technology, this section presents theoretical dimensions to engage the politics, economics, and cultures of technological knowledge production and representation.

Marx and technology

The Marxian corpus provides much of the foundation for understanding the role of technology in geography. Although dated in many ways, Marx's classical analysis of production itself provides a useful opening for grasping technology's impact on cultural production. Specifically, the industrial revolution involved a change in the mode of production. Technology contributed to society's wealth by increasing the means of control, enhancing the exploitation of workers, and contributing to the growth of fixed capital held by a small class of capitalists. It is often overlooked that Marx also saw technology as a means to decrease working hours and to create time for proletarian educational and cultural activities. Neither dismissive nor condoning of technologies, Marx wrote about technological potentials and risks from a political-economic emphasis on production and class struggle.

Marx and Engel's work on labor theory of value and the mode of production is the foundation for some of cultural geography's engagement with technology. The theorized mode of production distinguishes itself from liberal analyses that emphasize technology as a tool in enhancing extractive capabilities. In Marx's detailed analysis of the role of technology in the production process, published in the *Grundrisse* (1857–8), the introduction of machinery requires an initial capital input, but overall requires less capital outlay because the production of surplus value is equal or greater to the surplus value produced by the workers whose labor has been replaced by technology. The capital input is made up for by a reduction in the costs of wage labor. In the process of substitution, the total sum of capital laid out diminishes and the surplus value of the retained workers increases. Politically, this is an appropriation of labor by capital, an act that Marx thoroughly understood as a Faustian bargain: 'Capital absorbs [machine] labor into itself – as though its body were by love possessed.' The exchange of living labor for objectified labor changes the situation of workers, as 'the creation of real wealth comes to depend less on labor time … but rather on the general state of science and on the progress of technology'. In consequence, 'He [the worker] steps to the side of the production process instead of being its chief actor.' In contrast to the miserable nineteenth-century 'theft of labor', Marx downplayed the problems of substitution and saw technology positively in the development of a social individual who acquires free time for artistic and scientific development. In the mode of production, Marx conceived of technology as the means of liberating workers from capitalist exploitation and an integral part of social reproduction: 'They [machines] are *organs of the human brain, created by the human hand;* the

power of knowledge objectified' (italics in the original).

Foreshadowing subsequent analysts (Habermas, 1984; Horkheimer and Adorno, 1944), Marx recognized that if the whole of society did not reduce labor time, then labor time would become the measure of value and the promise of technological liberation would encounter its dialectical antithesis when, in Marx's memorable words, the machine forces 'the worker to work longer than the savage does'. Having acknowledged western literature in the form of Goethe's Faustian bargain, Marx and the neo-Marxists rely on the 'selling of the soul versus modest existence' binary to critique the development and adoption of technology in capitalism, which emphasizes the short-term impacts on work time, papering over broader issues such as the impact of technological advances in production on society at large.

Frankfurt School influences

Addressing the limitations of Marx's analysis, along with Hannah Arendt (1958), a number of writers coming from or influenced by the Frankfurt School indicate that technology in the twentieth century has become a cultural intermediary for our engagement with the world. Breaking with Marx's Enlightenment assumptions about scientific liberation, Horkheimer and Adorno analyzed science and technology as ideology and a set of discourses that extends the power of dominant social forces (Horkheimer and Adorno, 1944). These analyses are characterized by a stark either/or binary: the unreflective adoption of new technologies, or critical reflection on their role in hegemonic cultural politics (Postman, 1993). Unintentionally keeping the binary between technology and society, the Frankfurt School writers aptly addressed the negative attributes and consequences of technology.

Technological determinism characterizes many of the other writings on technology in the twentieth century (Sejerstad, 1997). Jacques Ellul's *The Technological Society*, first published in English in 1964, is a ground-breaking analysis of the relationship between technology and society in the technological determinist framework. In these analyses, technology is culturally and politically problematic.

Structuralist analysis provides a robust framework for probing implications and effects. Kroker and Weinstein's *Data Trash: The Theory of the Virtual Class* (1994) stands out as a recent publication in this area. They skillfully analyze the cultural processes through which the development and fetishization of technology lead to a closure of cultural opportunities. An earlier publication, one emblematic for humanists, is Hannah Arendt's *The Human Condition* (1958). She assesses cultural changes in work in western civilization alongside the increasing fetishization of technology. Arendt's work provides a thoughtful examination of western civilization's presumption that technology is a mere tool. Her compatriot Martin Heidegger (1977) also engaged these issues, albeit from a more conservative posture in line with cultural and political attitudes in 1930s Germany.

Heidegger on technology

Renowned for his work in phenomenology and reviled for his support of the Nazi regime, Martin Heidegger's work on technology bears special attention. Similar in some ways to the Frankfurt School's focus on ideology, but more conservative, Heidegger's (1977) reflections on technology point to the significant cultural consequences of the modernist ideology and the rise of instrumental thought. While Heidegger is highly critical of technology, he does not suggest discarding, disregarding, or destroying technology (Dreyfus, 1995). Technologies for Heidegger intervene between human activities and the world, distancing and possibly separating people from nature. Ideas are not simply imposed on reality. Being calls forth thought. As William Lovitt writes in the preface to his translation of Heidegger's work on technology, 'in the modern "Cartesian" scientific age man does not merely impose his own construction upon reality. He does indeed represent reality to himself, refusing to let things emerge as they are. He does forever catch reality up in a conceptual system and find that he must fix it thus before he can see it at all. But man does this *both* as his own work *and* because the revealing now holding sway at once in all that is in himself bring it about that he should do so' (1977: xxviii, emphasis in the original).

From the ontological issue of being, Heidegger articulates the perspective that humans have always used technological devices, but must resist their domination. Technologies gather and focus human practices in line with technological uses, subverting pre-technology practices. Culturally technology supports new ways of being through shared practices. Modern technology, for example, emphasizes efficiency and effectiveness, ways of being that spill over into social relations between humans, and between humans and machines. By assuring that technology does not lead to the domination of

humans, we can assure that instrumentalism does not destroy our nature. The gathering and focusing of human practices through things present us with opportunities to develop new practices and ways of knowing.

The ubiquity of information technology, its mythological importance, and consumerism in late capitalism effectively veil these opportunities. At the same time, ontological and epistemological limits to modernist attempts to seek to develop logical science, rational reasoning, and flawless engineering point to the underlying cultural instability of technology. Heidegger's work theorizes a relationship between work and technology, but does not engage the implications that are intricately caught up in the day-to-day use of technology. This oversight is commonplace in progressive notions regarding the role of technology in society and has a broad influence on liberal agendas. Addressing this issue, Hubert Dreyfus (1992) presents a detailed critique of the inherent failings of artificial reasoning. Arising out of work on cybernetics, artificial reasoning has been a field characterized by the broadest assertions about the ability of scientists to develop information technology whose reasoning could not be differentiated from humans. Dreyfus' main argument is that their attempts fail because human thinking is not a 'closed system', but is open and characterized by a rich faculty for making associations that lie outside a rational 'problem area'. His well-thought-out analogy of a restaurant waiter using non-verbal clues to negotiate a customer's order point clearly to the marked limitations of so-called expert systems, which can only account for narrowly specified categories that have been defined *a priori*. While eggs can be readily stored by a system as a menu item for breakfast, a customer ordering 'eggs Benedict' may crash the system.

Poststructuralist turns

As technology became more ubiquitous in western civilization, analyses shifted from Marx's political-economic orientation to consider political and cultural dimensions. These works attempted to specify highly interconnected mind/machine relationships. Recently new and very diverse scholarship has turned to the imbricated relationships between technology and society which are no longer just matters of impacts, of structure and agency, but involve mutual constitution in webs of relationships.

Most directly related to the Franfurt School, the work of Dreyfus (1995), Winner (1986), Suchman (1987), and Agre (1992) point to new poststructuralist understandings of technology that emphasize its situated and embedded character. Situatedness exhibits several characteristics. Foremost of concern here is the insight that we develop (construct) technologies in relationship to a particular use and 'problem area'. The modernist understands technology as a problem of determining the proper tools and the correct steps to follow. While Marx's analysis went beyond liberal notions of technology as 'tools', it constrained itself to functions in the labor process. The poststructuralist concept of situatedness helps problematize this limited model, providing for a more culturally cogent understanding of the complexity of technology. Although authors in this area rarely directly address geographical dimensions of embeddedness, there is a clear awareness of spatial aspects of technology.

Second, poststructuralist recognition of the embeddedness of technology in cultural production points to the importance of reassessing cultural geography's engagement with technology. Technology, beginning with the humble binary on/off light switch, is at the heart of western culture's analytic approaches to the world and is decisive in arbitrating numerous categorizations and classifications. Poststructural theories provide the necessary purchase for engaging the hegemonic cyberepisteme of western civilization. They also provide a means of overcoming technology's modernist hegemony. Extending Marxian analysis of production, a key thought underlying poststructuralist reflections on technology is the political awareness that all technology is culturally mediated: inextricable from culture, technology used in the material production of knowledge is also imbricated in the cultural instability that occupies a key moment in poststructuralist thought. Authors draw on examples from geography to underscore this point (Latour, 1993; 1995; 1999). The now yellowed spaces on turn-of-the-century maps showing the extent of the British empire point to the role of geographic technologies in facilitating colonial hegemony. The changing technological production of culture continues, reflected in popular cyberepistemes that rely on epistemologies of virtual representation. For example, the disembodied view that, with technology, 'geography is dead' points to new instabilities in a modernist knowledge production that is intricately interwoven with technology (Castells, 1996).

Science and technology studies

Although it is impossible to demarcate science and technology studies (STS) from post-structuralist work, the distinction seems relevant given the pervasiveness of STS. A number of French scholars and a growing number of Anglo-American ones have been strongly influenced by the growing interdisciplinary corpus of STS. Although known largely in the Anglo-American world through the prolific writings of Bruno Latour, this area is far broader and is distinguished in a number of ways. Of particular importance is a distinction between the strong program of Edinburgh and the approaches taken by Michael Serres, Bruno Latour, and Madeline Akrich of Paris and John Law in the United Kingdom. The strong program was dedicated to studying science (physics, chemistry, etc.) by the same techniques that empirical science used in producing knowledge. Others challenged what they saw as the limitations of this approach and brought in anthropological, sociological, and critical philosophical concepts and methods to study science. Latour's aphorism 'Follow the actors' suggests the broader orientation of these approaches, but substantial distinctions remain, above all between actor network theories and symbolic interaction or activity theories. Whereas the former focus on a few key individuals and institutions (for example Louis Pasteur and the laboratory), the latter seek more to unravel the details of the mundane practices and unsung people struggling with conventions, rules, and guidelines (Clarke, 1990).

Symbolic interactionism and activity theory originated largely in the work of Anselm Strauss and are influenced by a variety of mid-twentieth-century scholarship, ranging across Ludwig Wittgenstein, Margaret Mead, John Dewey, Howard Becker, and the Frankfurt School, among others. Several key concepts underpin work in this area. First, they are anti-determinist, that is, there are no predetermined explanations: *'the meaning of knowledge is given in its consequences'* (Star, 1996: 303, emphasis in the original). Second, they are non-rational. Work in this vein has always emphasized collective ways in which we constitute others' ontologies and epistemologies through pragmatic and dialectical representations and technologies (Star, 1996). Knowledge is the representation of ordered local knowledges, which are always partial and flawed. Much work in this vein has focused on the role of technology in the production and ordering of knowledges.

Geographers have made significant contributions to this literature. The work of Trevor Barnes (1997) stands out for his historical analysis and engagement with the politics of the discipline. This resonates with feminist criticisms of actor network theory which have problematized the tacit assumptions that reproduce masculine value systems in much of this work (Star, 1991). Numerous other geographers have drawn on this work (including Alderman, 1998; Chrisman, 1999; Harvey, 2001; Murdoch, 1998; Thrift, 1996). These works are highly relevant to the growing body of research that probes the constitutive relationships between geography and technology, and they provide insights into the roles of geography's technology in the production of knowledge.

CONSIDERING KNOWLEDGE AND TECHNOLOGY

The previous section drew largely on work outside human geography to set the stage for reviewing recent work in cultural geography or technology. The work I discuss in this section predominantly relies on poststructuralist literature. This is distinct from 'classical' human geography work in the mid twentieth century. Human geographies have regularly engaged the consequences and implications of technology in a liberal, structuralist framework. Studies by Sauer (1925) make technological production an important part of cultural change, but he confined his consideration to the functions of the tool. More recent work in cultural ecology demonstrates the intricate interconnections between cultural practices and the socioeconomic situation. The cultural geography engagement with technology focuses particularly on cyberspace (Adams and Warf, 1997; Kitchin and Dodge, 2001). In the process, cultural geography is opening other forms of geographic information technology to critical reflection.

Although the most recent cultural geography of technology focuses on cyberspace, this narrow thematic focus hides a more profound shift in cultural geographers' interests (Mikesell, 1978). The radar example that started this chapter pointed to the political and social dimensions of technology in producing knowledge. In the remainder of this section, I focus on four exemplary areas to show the political and social dimensions of geographic technology, and to show the ontological shifts that technology involves.

Mapping

The discipline of geography, in the service of the state, military, royalty, wealthy individuals, or companies, has always made ample use of technology. Felix Driver (1992) discusses the imbrication of technology in the practice of geography, as do many other geographers exploring the historical development of the discipline. Whether for colonization or for local government, geographers have used these technologies to serve the state. Beyond the powerful bandwagon effect that technology has had in the twentieth century, which goes hand-in-hand with a fetishization of technology, the ideological dimensions of technology have frequently been studied in the context of particular activities. For example, Brian Harley's (1989; 2001) work is important in assessing the historic role of maps in constructing the landscape.

Harley's analysis of mapping draws on Foucault, among others, to articulate the nature of maps. He is especially keen to analyze what maps leave off – the silences, one could say. His essays evoke the political dimensions of cartography, particularly for colonial projects, and in this sense his work also suggests the influence of orientalism and its mapping practices (Said, 1978). Harley rests on a critique of the modern map-maker's assumptions of objectivity, detachment, neutrality, transparency, exactness, and accuracy. These attributes correspond to those characteristics usually ascribed to technology in western culture. This interesting link points to the substantial relationships between geography, technology, and society. Denis Wood (1992) draws on Harley's work and develops a semiotic analysis of mapping symbology and referentiality. Pat McHaffie (1995; 2000) has written about the production process and the organizational and individual dimensions of producing maps, thereby extending Harley's arguments about political and cultural influences.

Perhaps David Turnbull (1989) has elaborated the most thorough analysis based on Harley's ideas as he eloquently describes the erasure of non-western thinking in the colonization of Australia. Robert Rundstrom (1993; 1995) also takes up this charge in his assessment of the consequences of hegemonic practices on landscapes in North America and the Native American cultures who struggle under the dominance of European colonization. In both these cases, the interest is in the consequences of technologies, not the practices. Rundstrom's work bears poignant witness to the historic assimilationist politics of European Americans who fail to consider the cultures they represent in maps. They both create a colonial landscape that removes other cultures while representing them as attractions for Europeanized tourists seeking 'authentic' native relics.

Recently, Denis Cosgrove (2001) has published a book on the ideological history of geography that stands out for a number of reasons. First of all, it is a history of cartographic thought and the central role of authority and deities in crafting geography's Apollonian view of the world. Second, it exquisitely details the development of western civilization's geographic thought over the millennia. While not explicitly engaging the development of geographic technologies, Cosgrove clarifies the myriad connections between technology and the dominant, state- or monarch-oriented epistemologies in geography. This work builds on earlier research on the role of maps in western civilization, one example being Norman Thrower's accessible contribution *Maps and Man: An Examination of Cartography in Relation to Culture and Civilization* (1972).

Finally, authors as varied as Bruno Latour, Michael Goodchild, and Denis Cosgrove write about the map as a means of communication. Latour (1990; see also Latour, 1995) describes maps as 'centers of calculation' that permit colonial knowledge to be not only archived, but made transportable. In this way, they become a means for the political body that created the map to calculate and act upon the area represented. Michael Goodchild (2000) discusses the role of geographic information in a post-mapping, digital earth setting where access to information becomes the banner for new engagements with place. And Denis Cosgrove and Luciana Martins (2000) discuss the role of creative engagements with geographic information as a means to produce different geographies. They refer to this as performative mapping.

Transportation and communication

The ability of information technologies to rapidly increase the speed at which people communicate has attracted substantial interest among economic and cultural geographers (Harvey, 1996). The so-called death of geographic distance (Cairncross, 1997) has motivated many geographers to revisit core assumptions in geography and study the processes of globalization and economic change. A key finding of these studies is that while distance matters less, accessibility to information technology matters even more. Even cyberspace becomes difficult to navigate without spatial metaphors.

Accessibility is in and of itself a multifaceted concept. While for some authors, information is replacing energy as the driving force of social organization, other writers are more concerned about the possible potential information inundation. Dan Sui (2000) writes about the limits of the geographic concept of access to capture the

important dimensions of transportation and communication in the information age. He argues that the mechanical metaphor of the web is ill-suited to assess the actual processes of individuals, organizations, and societies and calls for geographers to rely on the biological metaphor of adaptability that underscores the ability of people to make sense in a constantly changing world by adjusting their tactics and strategies. A number of geographers have engaged these issues to study the networks that are becoming increasingly dominant in linking geographically distant places to coordinated and codependent economic units. The complex dynamics of the new internet economy call for non-linear studies that overcome the limitations of applying Newtonian gravity models to transportation and communication in the information age (Brunn and Leinbach, 1991).

New approaches to studying space–time dynamics have helped shed some light on individual tactics in day-to-day living in the information age. Mei-Po Kwan (2000) describes and analyzes the daily routines of a number of people who interact in various ways at home and work with people from all around the world. Instead of physical distance, the people she describes engage with many people in information spaces that are not physically connected. This leads to complex relationships that require the negotiation of disparate time–place constraints, beginning with time zone differences. This shows well the mutual influence that technology has on society and vice versa. Paul Adams' (1996) work on community and television also provides useful insights into changing place identities.

GIS

In *Ground Truth* (1995), John Pickles articulates the basis for understanding the uses and consequences of geographic information technologies in a neo-Marxian framework that trenchantly analyzes the problems of representation. This work problematizes the hegemonic politics of representation and the objectification of biased observations, going beyond Harley's critique of cartography to engage the multifaceted worlds of geographic information. Following Benjamin's work, Pickles extends the theoretization of simulation to consider the many ways in which geographic information is used. Of particular concern, and related to the neo-Marxian framework, are the implications of geographic information systems. Eric Sheppard (1993; 1995) has also written on these themes, but with more stress on the practices of producing geographic knowledge. Both of these geographers have been widely read and criticized by practitioners of GIS, many of whom retain the tool metaphor as their *modus operandi* (Wright et al., 1997). While a number of individuals have divined these struggles to be a revival of qualitative/ quantitative quarrels, recent work that has synthesized and pointed to substantial common ground indicates the coming together of these disparate approaches (Schuurman, 2000).

Nicholas Chrisman's work is instructive in this regard. He explores the linkages between institutions, society, and technological practices. For example, Chrisman (1991) points out that GIS data models are based on a system of axioms that limits them to the data for which they were developed and ignores the complexity of human geographic experience. Chrisman's work is notable among GIS practititioners because it indicates a rare interest in the social and cultural settings where GIS is developed. His analysis, although often structuralist in orientation, is nevertheless instructive in understanding the forces shaping GIS (Chrisman, 1987).

GIS represents a shift in various dimensions. Whereas previously, cartographers readily understood themselves as specialists in the cartographic representation of geographic knowledge, this activity has become so decentralized and widespread that geographers now see themselves as embedded in a large-scale re-engineering of cartographic products and information. While these activities are undoubtedly important, a focus on data and cartographic products impairs our ability to discern an even larger shift to alternative forms of representations, some of which have been brought about by artists and professionals to enhance the communicative capabilities of these technologies. Appropriating GIS and other spatial technologies for cultural discourses involves a direct engagement with the politics of representation. Clearly, geographic technology is tied up in our geographic imaginations (Gregory, 1994). Today, the cinematic landscape of cyberspace, which none of us empirically 'experiences', is as tied up in our concepts of geography as a weekly walk in the park. To enable alternative geographic information technologies we need to develop local technologies for local needs. This is more than just being involved in design; it implies participation, an engagement that involves 'simultaneously understanding and activating spatialities' (Cosgrove and Martins, 2000: 97). This is a politics that displaces hegemonic perspectives so as to facilitate pluralistic sets of meanings and communication – the creation of multiple ontologies (see Howitt and Suchet, Chapter 31 in this section).

For example, Cosgrove and Martins' (2000) performative mappings open ways for artistic engagement by using geographic technologies for the discursive production of culture. The map 'becomes a discursive expression of an active and participatory geographical place-making' (2000: 107). While they discuss this technique as a form of artistic expression, there is certainly great political and practical potential in using the discursive character of imagining and drawing a map. Registered, or transformed to a coordinate grid, these maps could easily be combined with other mappings and official maps as part of a communicative discourse.

A similar but more visual approach was developed by an international team studying the perceptions of housing projects in Holland (Gaver et al., 1999). Cultural probes are media given to residents of the project to document their experiences of the project. For instance, a postcard was given out at a large meeting inviting residents to indicate on a map the areas they felt threatened by, areas with too much traffic, etc. These images were used to start discussion in meetings, and the team recorded varying perspectives. This use of GIS is evidenced in earlier work on feathered layers, an idea presented in India during the 1950s (Khan, 1954). Sliced maps, as Khan also called them, were transparent overlays bound to a base map on each edge of the map. The four transparent sheets could be folded down individually or together to examine relationships between different themes visually. India continues to be the focus of participatory work (Hoeschele, 2000).

Public participation in GIS has become a vibrant area of research in its own right. A large body of work has been published in this area on topics ranging from citizen participation in North American inner cities to aiding non-profit conservation group activities (McMaster et al., 1997; Mugerauer, 2000; Schiffer, 1998; Sheppard, 1995; Sieber, 2000). While much of this work has focused on bettering existing institutional planning activities, there is a growing interest in grassroots activities (Craig et al., 2002; Harris and Weiner, 1998). The world wide web has also stimulated work by a number of researchers working to redress the lack of web-supported public participation GIS and the constraints posed by existing power structures and planning practices (Carver, 2001). This work all shares a motivation to improve the inclusivity of planning practices through communication.

Robert Mugerauer (2000) directly engages these issues, but with a focus on more fundamental questions about the role of geographic information technology. He points to the current limits of GIS and asks the fundamental question: what matters, in their own terms and value systems, to people who do not use GIS? This basic research is surely called for as a variety of authors point out (Chrisman, 1987; Pickles, 1995; Rundstrom, 1993; 1995; Turnbull, 1989). Mugerauer also calls for the development of a pluralistic-democratic GIS for mediation, drawing on Paulo Freire's emancipatory pedagogy. Learning local ways of knowing and helping individuals develop their own way of describing them in GIS is a fundamental form of empowerment that awaits.

Responding to human geographic critiques of geographic technologies and opening new ontological perspectives, these three approaches also help make GIS accessible and tangible for many people who lack exposure to information technologies. Instead of creating an immense hurdle by forcing people to learn a certain technology before they can share their ideas, these approaches and modest technologies provide readily accessible ways for people to interact and be involved in participatory development.

Cyberspace

Rob Kitchin and Martin Dodge's *Mapping Cyberspace* (2001), along with Ken Hillis' *Digital Sensations* (1999), are ideal starting points for picking up on some of the key questions geographers are asking about cyberspace. Clearly, cyberspace presents geographers with new challenges for how we think about nature, space, and places. Hillis asks why, at this cultural moment, do cyberspace and virtual reality emerge? He sees these developments in the context of embodiment. Whereas in geographic space the socio-cultural body provides us with a means for developing strategies and tactics to cope, cyberspace has yet to develop a strong culture. In this sense, Hillis notes that with the popularization and commercialization of cyberspace there has been a deepening sense of segmentation, in terms of both the body and space. This accompanies a Cartesian desire for disembodied, alienated subjectivity that finds transcendence in the physical limits of cyberspace. Cyberpunk literature highlights these dimensions of transcending physical space and the separation of the Cartesian mind from its physical surroundings. At the same time, Don Janelle's recognition that we cannot describe and navigate cyberspace without geographic metaphors points to the continued importance of geography (Janelle and Hodge, 2000).

Children growing up in the information age bear witness to this geographic complexity. Gill Valentine and Sarah Holloway (2001) point out that there has never been a sharp demarcation between, in their terms, 'online' and 'offline' spaces. Drawing on actor network theory, they argue that children's activities, such as playing, 'are shaped by shifting associations (and disassociations) between humans and nonhuman entities in which the properties of the technology are not inherent but emerge in practice' (2001: 75). The process of developing stable, culturally meaningful practices goes hand-in-hand with the stabilization of technology.

Finally, the importance of practice in the development of technology has been stressed by Nigel Thrift (1996). *Virtual Geographies* (Crang et al., 1999) extends Thrift's analysis, drawing substantially on science and technology studies to develop powerful insights about the cultural geography of the information age. In thinking geography's engagement with technology through the lens of structuralism, they make the evocative point that the virtual world created on the internet, cyberspace, or whatever it may be called, should be assessed first and foremost as an alternative, not as a copy or representation. This insight helps to move geographers beyond the limits of understanding technology solely in terms of results and consequences. It contributes to deepening geography's engagement with the politics of the practices that produce geographic knowledge.

The potential of cyberspace to substantially alter our experience of the world has profound implications. Through the increasing embedding of technology in our quotidian lives, more active and pervasive digital representations are replacing the human relationships that for millennia have defined society and humanity itself. Many people already regularly communicate with institutions such as banks, local government, power companies, etc., solely through automatic response systems. While not yet cyberspace, this is the beginning of a future experience in which it becomes increasingly difficult to distinguish human from computer.

POLITICALLY ENGAGING GEOGRAPHY'S TECHNOLOGY

Bonnie Nardi and Vicki O'Day (1999) write clearly about the political implications of confining technology to the tool metaphor. In their words, 'Using the tool metaphor to describe technology suggests several tactics to users. Before starting to work, it is important to choose the right tool for the job' (1999: 29). The tool metaphor is useful for talking about usability, training, and learning, but it offers few insights into how the tool is embedded in the larger context. As they put it: 'People who see technology as a tool see themselves controlling it. People who see technology as a system see themselves caught up inside it' (1999: 27). What is more, tools have politics. Construing the cultural implications of technology to be those of the tool maintains or reinforces the hegemony of those who develop and control the 'tools'. Embedded as infrastructure, geographic technologies are bound up in the production of geographic knowledge, and this, in turn, impacts culture more broadly.

Poststructuralist thought, informing recent cultural geography's probing of the destabilization and complication of technology, provides a multidimensional theoretical framework for understanding the manifold linkages between capitalist production, technology, culture, and knowledge. As technological artefacts become more pervasive they are changing the ways we understand and represent the world. The capitalist emphasis on technology to enhance the extraction of surplus value is not a means unto itself, but a political and social relationship that is fundamental to changing terms about humanity. Technology is a political way of producing knowledge that embeds values in artefacts (Callon, 1991). As technological objects become locationally aware through the global positioning system and artificial intelligence techniques, our ontological understanding of geography will alter to include extra-experiential dimensions of non-human geography. The consequences of this opening can be more profound insights into the dynamic relationships that constitute place and our embeddedness in various scales of relationships and flows that go beyond what is physically connected (Massey, 1993).

Many cultural geographers have extended our insights into the political dimensions that are opened up by technologically influenced ontologies. Oliver Froehling's (1999) work on the use of the internet by the Zapatista movement offers a lucid example of the processes of de- and reterritorialization that can occur by means of what may be loosely called cyberspace. The politics of using technology have been frequently studied and, as much of the work examined in this chapter points out, more stimulating work exploring the emergence of geography through cultural practices can be expected. What we should not overlook is the political tension within geography over the use of technologies.

What we do with technology is always political. It also always has real consequences, individually, disciplinarily, socially, and economically. These domains highlight the complexity of the political relationships and point to the inseparable relationship between geographic knowledge and technology. Technology is integral to our geographic understanding, our disciplinary identity, and our ontologies. The question asked at the beginning of this chapter about the connection between geographic knowledge and technology is also a question about the politics of geography's changing ontologies and epistemologies. Regardless of how we answer this question, the imbrication is ubiquitous and points to the inevitability of the answer that they are inseparable: technology always changes what we know and how we know.

NOTE

I would like to thank John Paul Jones III, Kay Anderson, Mona Domosh, Steve Pile, and Nigel Thrift for their helpful assistance in preparing this chapter.

REFERENCES

Adams, P. (1996) 'Protest and the scale politics of telecommunications', *Political Geography* 15 (5): 419–41.

Adams, P.C. and Warf, B. (1997) 'Introduction: cyberspace and geographical space', *The Geographical Review* 87 (2): 139ff.

Agre, P.E. (1992) 'Formalization as a social project', *Quarterly Newsletter of the Laboratory of Comparative Human Cognition* 14: 25–7.

Alderman, D.H. (1998) 'A vine for postmodern times: an update on kudzu at the close of the twentieth century', *Southeastern Geographer* 38 (2): 167–79.

Arendt, H. (1958) *The Human Condition*. Chicago: University of Chicago Press.

Barnes, T.J. (1997) 'A history of regression: actors, networks, machines, and numbers', *Environment and Planning A* 30: 203–23.

Brunn, S. and Leinbach, T.R. (eds) (1991) *Collapsing Space and Time: Geographic Aspects of Communications and Information*. London: Harper Collins.

Cairncross, F. (1997) *The Death of Distance: How the Communciations Revolution will Change our Lives*. Cambridge, MA: Harvard Business School Press.

Callon, M. (1991) 'Techno-economic networks and irreversibility', in J. Law (ed.) *A Sociology of Monsters: Essays on Power, Technology and Domination*. London: Routledge. pp. 132–61.

Carver, S. (2001) 'Public participation using web-based GIS (guest editorial)', *Environment and Planning B* 28: 803–4.

Castells, M. (1996) *The Rise of the Network Society*. Cambridge, MA: Blackwell.

Chrisman, N.R. (1987) 'Design of geographic information systems based on social and cultural goals', *Photogrammetric Engineering and Remote Sensing* 53 (10): 1367–70.

Chrisman, N. R. (1991) 'Beyond spatio-temporal data models: a model of GIS as a technology embedded in historical context'. Paper presented at the AutoCarto 11, Baltimore.

Chrisman, N.R. (1999) 'Speaking truth to power: an agenda for change', in K. Lowell and A. Jaton (eds) *Spatial Accuracy Assessment: Land Information Uncertainty in Natural Resources*. Chelsea, MI: Ann Arbor Press. pp. 27–31.

Clarke, A. (1990) 'A social worlds research adventure: the case of reproductive science', in S.E. Cozzens and T.F. Gieryn (eds) *Theories of Science in Society*. Bloomington: Indiana University Press. pp. 15–42.

Cosgrove, D. (2001) *Apollo's Eye: A Cartographic Genealogy of the Earth in the Western Imagination*. Baltimore: Johns Hopkins University Press.

Cosgrove, D. and Martins, L. (2000) 'Millennial geographics', *Annals of the American Association of Geographers* 90 (1): 97–113.

Craig, W., Harris, T. and Weiner, D. (eds) (2002) *Community Participation and Geographic Information Systems*. London: Taylor and Francis.

Crang, M., Crang, P. and May, J. (eds) (1999) *Virtual Geographies: Bodies, Space, and Relations*. New York: Routledge.

Curry, M. (1998) *Digital Places: Living with Geographic Information Technologies*. New York: Routledge.

Dreyfus, H.L. (1992) *What Computers Still Can't Do: A Critique of Artificial Reason*. Cambridge: MIT Press.

Dreyfus, H.L. (1995) 'Heidegger on gaining a free relation to technology', in A. Feenberg and A. Hannay (eds) *Technology and the Politics of Knowledge*. Bloomington: Indiania University Press. pp. 7–109.

Driver, F. (1992) 'Geography's empire: histories of geographical knowledge', *Environment and Planning D* 10: 23–40.

Edwards, P.N. (1997) *The Closed World: Computers and the Politics of Discourse in Cold War America*. Cambridge: MIT Press.

Ellul, J. (1964) *The Technological Society*. New York: Knopf.

Froehling, O. (1999) 'Internauts and guerrilleros: the Zapatista rebellion in Chiapas, Mexico', in M. Crang, P. Crang and J. May (eds) *Virtual Geographies: Bodies, Space, and Relations*. New York: Routledge. pp. 164–77.

Gaver, B., Dunne, T. and Pacenti, E. (1999) 'Cultural probes', *Interactions* (January/February): 21–9.

Goodchild, M.F. (2000) 'Communicating geographic information in a digital age', *Annals of the American Association of Geographers* 90 (2): 344–55.

Gregory, D. (1994) *Geographical Imaginations*. Cambridge: Blackwell.

Habermas, J. (1984) *Theory of Communicative Action*, vol. 1. London: Heinemann.

Harley, B.J. (1989) 'Deconstructing the map', *Cartographica* 26 (2): 1–29.

Harley, B. (2001) *The New Nature of Maps: Essays in the History of Cartography*. Baltimore: Johns Hopkins University Press.

Harris, T. and Weiner, D. (1998) 'Empowerment, marginalization, and community-integrated GIS', *Cartography and Geographic Information Systems* 25 (2): 67–77.

Harvey, D. (1996) *Justice, Nature and the Geography of Difference*. Oxford: Blackwell.

Harvey, F. (2001) 'Constructing GIS: actor networks of collaboration', *URISA Journal* 13 (1): 29–37.

Heidegger, M. (1977) 'The question concerning technology', in W. Lovitt (ed.) *The Question Concerning Technology and Other Questions*. New York: Harper Torchbooks. pp. 3–35.

Hillis, K. (1999) *Digital Sensations: Space, Identity, and Embodiment in Virtual Reality*. Minneapolis: University of Minnesota Press.

Hoeschele, W. (2000) 'Geographic information engineering and social ground truth in Attappadi, Kerala State, India', *Annals of the American Association of Geographers* 90 (2): 293–321.

Horkheimer, M. and Adorno, T.W. (1944) *Dialectic of Enlightenment*. New York: The Continuum, 1995.

Janelle, D.G. and Hodge, D.C. (eds) (2000) *Information, Place, and Cyberspace: Issues in Accessibility*. Berlin: Springer.

Khan, M.A.W. (1954) 'Sliced maps', *Indian Forester* 80 (2): 103–16.

Kitchin, R. and Dodge, M. (2001) *Mapping Cyberspace*. London: Routledge.

Kroker, A. and Weinstein, M.A. (1994) *Data Trash: The Theory of the Virtual Class*. New York: St. Martin's Press.

Kwan, M.-P. (2000) 'Human extensibility and individual hybrid-accessibility in space–time: a multi-scale representation using GIS', in D.G. Janelle and D.C. Hodge (eds) *Information, Place, and Cyberspace: Issues in Accessibility*. Berlin: Springer. pp. 241–56.

Latour, B. (1990) 'Drawing things together', in M. Lynch and S. Woolgar (eds) *Representation in Scientific Practice*. Cambridge, MA: MIT Press. pp. 19–68.

Latour, B. (1993) *We Have Never Been Modern*. Cambridge, MA: Harvard University Press.

Latour, B. (1995) *Le Métier de chercheur: regard d'un anthropologue*. Paris: INRA.

Latour, B. (1999) *Pandora's Hope: Essays on the Reality of Science Studies*. Cambridge, MA: Harvard University Press.

Marx, K. (1857–8) *Grundrisse*. Available at (23 April 2001): http://www.marxists.org/archive/marx/works/1857-gru/.

Massey, D. (1993) 'Power-geometry and a progressive sense of place'. In J. Bird, B. Curtis, T. Putnam, G. Robertson and L. Tickner (eds) *Mapping the Futures: Local Cultures, Global Change*. New York: Routledge.

McHaffie, P. (1995) 'Manufacturing metaphors: public cartography, the market, and democracy', in J. Pickles (ed.) *Ground Truth: The Social Implications of Geographic Information Systems*. New York: Guilford. pp. 113–29.

McHaffie, P. (2000) 'Surfaces: tacit knowledge, formal language, and metaphor at the Harvard Lab for Computer Graphics and Spatial Analysis', *International Journal of Geographic Information Science* 14 (8): 755–73.

McMaster, R., Leitner, H. and Sheppard, E. (1997) 'GIS-based environmental equity and risk assessment: methodological problems and prospects', *Cartography and Geographic Information Systems* 24 (3): 172–89.

Mikesell, M. (1978) 'Tradition and innovation in cultural geography', *Annals of the Association of American Geographers* 68 (1): 1–16.

Mugerauer, R. (2000) 'Qualitative GIS: to mediate, not dominate', in D.G. Janelle and D.C. Hodge (eds) *Information, Place, and Cyberspace: Issues in Accessibility*. Berlin: Springer. pp. 317–38.

Mumford, L. (1967) *The Myth of the Machine*. New York: Harcourt, Brace and World.

Murdoch, J. (1998) 'The spaces of actor-network theory', *Geoforum* 29 (4): 357–74.

Nardi, B.A. and O'Day, V.L. (1999) *Information Ecologies: Using Technology with Heart*. Cambridge, MA: MIT Press.

Openshaw, S. (1991) 'A view on the GIS crisis in geography, or using GIS to put Humpty Dumpty back together again (commentary)', *Environment and Planning A* 23: 621–8.

Pickles, J. (ed.) (1995) *Ground Truth: The Social Implications of Geographic Information Systems*. New York: Guilford.

Poster, M. (1990) *The Mode of Information*. Chicago: University of Chicago Press.

Postman, N. (1993) *Technopoly*. New York: Vintage.

Rundstrom, R.A. (1993) 'The role of ethics, mapping, and the meaning of place in relations between Indians and Whites in the United States', *Cartographica* 30 (1): 21–8.

Rundstrom, R. (1995) 'GIS, indigenous peoples, and epistemological diversity', *Cartography and Geographic Information Systems* 22 (1): 45–57.

Said, E. (1978) *Orientalism*. New York: Pantheon.

Sauer, C.O. (1925) 'The morphology of landscape', *University of California Publications in Geography* II: 19–53.

Schiffer, M. (1998) 'Multimedia GIS for planning support and public discourse', *Cartography and Geographic Information Systems* 25 (2): 89–94.

Schuurman, N. (2000) 'Trouble in the heartland: GIS and its critics in the 1990s', *Progress in Human Geography* 24 (4): 569–90.

Sejerstad, F. (1997) 'Beyond technical determinism'. Paper presented at the Society for the Social Studies of Science Annual Meeting (4S), Tucson.

Sheppard, E. (1993) 'Automated geography: what kind of geography for what kind of society', *The Professional Geographer* 45 (4): 457–60.

Sheppard, E. (1995) 'GIS and society: towards a research agenda', *Cartography and Geographic Information Systems* 22 (1): 5–16.

Sieber, R.E. (2000) 'Conforming (to) the opposition: the social construction of geographical information systems in social movements', *International Journal of Geographic Information Science* 14 (8): 775–93.

Simpson, R.B. (1966) 'Radar, geographic tool', *Annals of the American Association of Geographers* 56: 80–96.

Smith, N. (1992) 'History and philosophy of geography: real wars, theory wars', *Progress in Human Geography* 16 (2): 257–71.

Star, S.L. (1991) 'Power, technology and the phenomenology of conventions: on being allergic to onions', in J. Law (ed.) *A Sociology of Monsters: Essays on Power, Technology and Domination*. London: Routledge. pp. 26–56.

Star, S.L. (1996) 'Working together: symbolic interactionism, activity theory, and information systems', in Y. Engestrom and D. Middleton (eds) *Cognition and Communication at Work*. Cambridge: Cambridge University Press.

Suchman, L. (1987) *Plans and Situated Actions: The Problem of Human–Machine Communication*. Cambridge: Cambridge University Press.

Sui, D.Z. (2000) 'The e-merging geography of the information society: from accessibility to adaptability', In D.G. Janelle and D.C. Hodge (eds) *Information, Place, and Cyberspace: Issues in Accessibility*. Berlin: Springer. pp. 107–29.

Thrift, N. (1996) *Spatial Formations*. London: Sage.

Thrower, N.J.W. (1972) *Maps and Man: An Examination of Cartography in Relation to Culture and Civilization*. Englewood Cliffs: Prentice Hall.

Turnbull, D. (1989) *Maps are Territories: Science Is an Atlas*. Chicago: University of Chicago Press.

Valentine, G. and Holloway, S. (2001) 'On-line dangers? Geographies of parents' fears for children's safety in cyberspace', *Professional Geographer* 53 (1): 71–83.

Winner, L. (1986) *The Whale and the Reactor: A Search for Limits in an Age of High Technology*. Chicago: University of Chicago Press.

Wood, D. (1992) *The Power of Maps*. New York: Guilford.

Wright, D., Goodchild, M. and Proctor, J. (1997) 'GIS: tool or science? Demystifying the persistent ambiguity of GIS as "tool" versus "Science"', *Annals of the American Association of Geographers* 87 (2): 346–62.

30

The Construction of Geographical Knowledge – Racialization, Spatialization

Audrey Kobayashi

The construction of the human body is a historical form of geographical knowledge that reflects geography's ocularcentric past. Although they have seldom been explicit about doing so, geographers have usually followed dominant social norms and intellectual trends, placing human bodies in particular landscapes, setting spatial limits upon the activities of those bodies, and linking the characteristics of those bodies (their gender, 'race' or ability, for example) to specific places. In other words, much of the history of cultural geography is about how bodies are – and should be – *seen*. This ocularcentric tendency is strongly evident in the ways in which geographers have constructed racially visible bodies, and fundamentally linked to the ways in which society as a whole has categorized human beings according to a racialized vision.

In this short chapter, I take up the concept of the racialized body as a form of geographical knowledge. I situate the racialized body at three general moments in the history of the western disciplinary gaze: during the late eighteenth century, when Enlightenment thinking provided a scientific justification for racialized colonial expansion that culminated in a deeply racialized modern landscape a century later; during the post-World War II period, when the racialized body was stripped of its particular characteristics but the subsequent undifferentiated gaze resulted in a deepening of categories of difference; and finally during the poststructuralist period, in which the racialized body is given explicit recognition as a social construction and some geographers have attempted to disrupt racialized vision through anti-racist analysis. For all three moments, I attempt to link the geography of 'racial' knowledge and vision and the larger intellectual and social context.

THE COLONIAL OTHER

Scholars who study the process of racialization generally agree that the concept of 'race', at least as we now know it, had little social meaning prior to the Enlightenment period. Over the two centuries between the eighteenth century and the turn of the twentieth century it became a fundamental part of western understanding of what is human, normalized in every aspect of common discourse, from the intellectual to the political, the economic and the social. Enlightenment thinkers, one of the most notable of whom was the geographer Immanuel Kant, had set the intellectual world on a trajectory that defined 'race' as an irrevocable marker of human value. Kant's lectures on geography advanced the definitive position that skin colour is a result of distance from the equator, and that those of darker skin colour are possessed of inferior moral, social and intellectual qualities. He thus racialized the two primary forms of geographical knowledge, that of 'space' by asserting the relative position of persons of various skin colours across the surface of the earth, and that of 'place' by establishing what kinds of human landscapes (people in particular places) are most civilized.

Kant's view – which was widely and more and more commonly held by scholars – depicts human beings as a single genetic species, derived from what he calls a 'stem genus', and constituted as follows: 'the elemental determinants for a certain development which are inherent in the nature of an organic body (plant or animal) may be called ... *germs [Keime]*; but if this development concerns only the size or the relationships between parts, I call these determinants *natural dispositions* [*Anlagen*]' (1997a: 42). The phenomenon of 'race', however, is polygenetic, arising in different parts of the world as a result of the modification of germs and natural dispositions by specific features of climate (temperature and humidity combined) as well as in response to a variety of landscape stimuli, such as the abundance (or lack) of flora and fauna, or even the visual distance to the horizon, which might have an effect on human sight. In particular, *blackness*, the most significant 'proof' of both moral lassitude and stupidity, occurs because:

> the drying up [by the hot sun] of the vessels that carry the blood and serum under the skin brings about the lack of a beard and the short curly hair. Likewise, because the sunlight that falls through the surface skin into the dried up vessels eats up the reticular membrane, there arises the appearance of black color. (1997c: 61–2)

Kant believed it 'curious', however, that people of other 'races', particularly the European race, seemed immune to such genetic transformation, for 'The Europeans who live in this hot belt of the world do not become Negroes after many generations but rather retain their European figure and color' (1997c: 60).

Overlying the effects of climate in Kant's schematic are what he called 'national characteristics', which determine a society's level of appreciation for the beautiful and the sublime, the two most elevated aspects of aesthetic sense. At the top of his national scale are the Germans, who are able to appreciate both, while those of French and English heritage are more limited in their sensibilities, the French tending towards a sense of the beautiful, the English towards the sublime (1997b: 50–1). In the same schematic, while the natives of North America are capable of a certain nobility that does not equal that of the Europeans but is valuable in its own right, the African has 'by nature no feeling that rises above the trifling' (1997b: 55).

Kant's unequivocal views amount to a comprehensive justification for a sense of European moral superiority that underlies Enlightenment thought. The world was controlled by the European eye/I, whose gaze represented the judgement of the civilized mind. Ultimately, for Kant:

> The inhabitant of the temperate parts of the world, above all the central part, has a more beautiful body, works harder, is more jocular, more controlled in his passions, more intelligent than any other race of people in the world. That is why at all points in time these peoples have educated the others and controlled them with weapons. (1997c: 64)

Kant's views on 'race' are not original; as Livingstone points out, 'in large measure the specifics of his teaching were culled from the conventional German geographical lore of the day, and from Büsching and Varenius in particular' (1992: 114). Kant's ideas are important, however, for three reasons. First, as the inaugural holder of a chair in geography at a European university, he influenced a large number of students, and his lectures continued to have widespread authority for many years, indeed centuries, after they were presented. The fact that he was primarily known as one of the foremost philosophers of his time, and that his writings on the possibility of human knowledge have affected western thought at its most profound level, further enhanced his credibility as a teacher of geography. Second, as Livingstone (1992, 1994) also points out, Kant was responsible for elevating much human knowledge of the world from the realm of religious faith to that of scientific knowledge. Over the course of the next two centuries, while many of the specific facts of Kant's scientific explanations were modified, expanded or discarded, a fundamental belief in the absolute and scientific verifiability of racial difference never wavered. In particular, the belief in the environmental determinism of human value was to dominate the discipline of geography well into the twentieth century. And third, Kant's work – again, in the company of most of the influential writers of his day – provides an imperative for linking 'race' and human moral value, thus justifying the most heinous acts of colonial violence, subjugation and oppression against people deemed by European eyes to be inferior beings incapable of appreciating a better life. Kant's work was thus part of a larger intellectual context in which the classification of human beings, according to a putative scale of civilization that made the white more civilized and distant from nature than the black, justified and played into the interests of the colonial project (Anderson, 1998a; 2000).

Recent work by cultural geographers shows the tenacity of Enlightenment racialized knowledge as intellectual fuel for geography's emergence during the late nineteenth century as a leading international science. Two overwhelming and intersecting characteristics – its role in supporting and advancing the European colonial project, and its intellectual fascination with what Livingstone (1992) has called 'climate's moral economy' – dominated nineteenth-century geography. A growing number of historical geographers has documented the role of geography in the advancement of colonialism (Driver, 1992; Godlewska and Smith, 1994; Mayhew, 2000: Chapter 12; Withers, 1997). Mayhew (2000: 227–8) claims that during the late nineteenth century, at the height of British colonial nationalism, the more humanistic aspects of Enlightenment thought were shed in favour of a racialism based on the 'superiority' of the white 'race'. Geographers came to the fore as influential authors of gazetteers that not only perpetrated racist notions, but also claimed dominance as purveyors of information about the world. Environmental determinism added scientific credibility to geographic nationalism, by providing a plausible explanation for putative white superiority. The geographer's map had become a significant visual symbol of both moral and scholarly authority.

As the twentieth century progressed, environmental determinism was not, of course universally accepted. Carl Sauer and other cultural geographers opposed determinism by advancing theories of culture to explain human differences across the surface of the earth. The work of latter-day environmentalists such as Griffith Taylor became increasingly controversial in the face of 'possibilist' arguments that supported human agency as an explanation for human ability to modify the earth's surface in order to overcome environmental challenges. Indeed it might be said that no issue was of more significance to the discipline – and certainly none was more vehemently debated – than that of environmental determinism versus human agency. But none of the geographers involved in the debate prior to World War II took on the construction of 'race' *per se*, let alone challenged the damage to human rights that resulted from racism and colonialism. That project was begun in the late 1940s, when international opposition to Adolph Hitler's Nazism led scholars and society in general to begin to challenge actions based on beliefs in racial difference. But it would be some time before the effects of the human rights movement would begin to infiltrate the discipline of geography.

ABANDONING 'RACE': THE NEW SCIENCE OF EQUALITY

In the several decades following World War II, geography was contested territory. Particularly in North American geography departments, there was a strong push for disciplinary legitimacy, as 'human geographers increasingly sought a clear identity of their own within the social sciences' (Johnston, 1991: 95). One result was a push away from what was increasingly seen as the descriptive but theoretically vacant approach of regionalism to a social science in which the rules of 'space'

would become justification for intellectual independence. While a significant number, especially of cultural and historical geographers, challenged the spatial approach (Harris, 1971), as Johnston (1991: 186) points out, even such challenges continued to be cast in a positive rather than a normative light. Such cultural geographers sought, therefore, to depict the richness of cultural processes without going so far as to challenge the role of the discipline in actually constituting culture.

By the post-war period, nonetheless, there were few lingering traces of environmental determinism, and geographers were no longer involved explicitly in the project of colonialism. The Cold War discouraged geographers from openly advocating views that might be seen as politically radical, while the general attitude towards those non-aligned former colonial nations was that their development could occur through modernization projects whose ambitions were a testament to possibilism.[1] A lack of direct involvement, however, does not guarantee a lack of intellectual complicity in the re-creation of racialized visions.

Within this context, the lack of interest in questions of 'race' and racism among geographers makes sense. The fact that they were overwhelmingly white (and male) and therefore less likely to recognize the ways in which their own lives and their surroundings were racialized went largely unnoticed. They were generally apolitical, at a time when (as they do today) discussions of racism invariably provoked political confrontation. But most importantly in understanding the trajectory of geographical knowledge, the quest for disciplinary independence required a clear delineation of the geographer's intellectual mandate, and investigation of human bodies was deemed to be outside that mandate, in the realm of the anthropologist. Finally, faith in the power of modernity to overcome poverty and oppression encouraged, among western thinkers at least, a complacent faith that human divisions were increasingly meaningless. As a result most academics, geographers included, wrote as though 'race' did not exist, in direct contradiction to the growing gap, both internationally and in the cities of most developed countries, between whites and people of colour. Geographers were thus complicit in perpetuating the continued effects of colonialism, albeit in a new form of putative neutrality.

For present purposes, the point is that positive (as distinct from positivist) approaches to geography either had no interest in the phenomenon of 'race', or took it as an essential given. If 'race' was considered at all, it was as a human variable that needed to be explained, to the extent that it was correlated to some geographical criterion such as spatial pattern.

Within the new positive scientific framework, the earliest attempts to document 'race' as a geographic phenomenon (for example, Morrill, 1965), viewed it as a problem that could be 'fixed' given a better spatial model. 'Race' was taken as an invariant category, and spatial inequality as the problem. Space, not 'race', was the object of geographical knowledge. The most significant pioneer in understanding spatial inequality was Harold Rose, one of the first African Americans to practise geography and the first to address explicitly the effects of spatial inequality from within the black community (Rose, 1970; 1972). While Rose's work is perhaps the most important historical statement by a geographer that spatial inequality is foundational to American life, his concept of race remained a static one, for the problem he identified (at least in his geographical publications) is one of spatial distribution rather than racism. This theoretical conservatism foreclosed the possibility of construing anti-racism as anything but another problem in planning.

The early work by Morrill, Rose and a small handful of others notwithstanding, there was simply very little interest in questions of racism from a human rights perspective. This approach seems ironic among professional geographers occupying positions in geography departments that had burgeoned just after World War II, when the defeat of Nazism was widely viewed as a defeat of racism. But the immediate post-war period in advanced western, especially English-speaking, countries where geography flourished was the period of great denial. Such countries participated actively in creating international human rights documents, such as the United Nations' *International Declaration of Human Rights*, thereby imposing a particularly western vision of equality in which 'race' was denied as a means of creating human difference, but geographers erased such concerns from within their professional ambit. Perhaps they thought that the struggles of World War II had solved the problems; certainly they viewed racialization as outside the geographer's responsibility.

If racial difference was denied, so was racism; it would be some time before critics began to recognize that the practice of denial was in effect one of allowing segregation and the impoverishment of racialized groups, especially in the United States, but increasingly throughout the western world and in former colonized nations, now referred to as the 'developing world'. We live with the result today, in a dominant set of international human rights instruments produced in the advanced western nations according to their norms, with little regard for the ways in which the knowledge and human circumstances that inform those instruments might be quite different in other parts of the world, particularly those parts occupied by racialized others. The power of the colonial vision continues.

The great denial was reinforced by geographers and other social scientists who defined themselves as neutral observers, neither reflexive nor agents of social change. They could act as observers of and commentators upon the political processes that produced certain types of landscapes and spatial patterns, but they could not engage in what Pile (2000) has recently called 'political thinking', or thinking that would result in shifting the geography of the world according to a moral vision. Claiming a responsibility to remain thus detached served to reinforce the denial of racism as a social issue. This perspective began to shift, however, as more and more geographers recognized the impossibility of a neutral stance and began to place more emphasis upon the concerns that beset the society around them. Issues of racism began to see some discussion in the 'relevance debate' that swept the discipline during the early 1970s (for an overview see Johnston, 1991: Chapter 7). The debate ranged from a focus on the role of geography in influencing public policy in general (Chisholm, 1971) to addressing social inequality (Eyles, 1973) to redressing global inequalities (Slater, 1973). For the first time, geographers as a group began to debate the relationship between ethical and positive science.

The issue of racism was explicitly taken up with the establishment of the journal *Antipode* at the annual meeting of the Association of American Geographers, held at Ann Arbor, Michigan in 1971. As someone who claims that the discipline has paid too little attention to questions of 'race', it is interesting for me to look back on a comment by Johnston (2000), who has gone so far as to state that *Antipode* had 'little Marxist material in its earliest issues, and ... more focus on race than class'. True, the early 'radical' geography expressed strong, even overwhelming concern for the conditions of life in increasingly segregated American black neighbourhoods, as well as in developing countries. And they did not fail to point out that the majority of the impoverished and underprivileged of the world were people of colour (Blaut, 1974; Bunge, 1971; Harvey, 1972; 1973; Smith, 1974). Blaut's definition of imperialism as 'white exploitation of the non-white world' could not have been more explicit (1970: 65).

But, in a remarkably perceptive contemporary critique, Leach (1973) suggested that neither a radical approach based on defining social justice, nor a conservative approach based on idealizing spatial justice, could provide a social geography of 'black people' that would resonate for the racialized people who would become the objects of research. For both approaches involve constituting people as problems, rather than seeking direct engagement with their lives; and neither approach involves enhancing the power to change people's lives. For radical geographers, the desire for social change was certainly apparent, but the theoretical focus of the early radical works was not on understanding 'race' or the process of racialization, but on situating the production of colonial and class relations in the operations of capital. As radical geography became more and more sophisticated in its analysis throughout the late 1970s and 1980s – the continued work of a few people such as Jim Blaut notwithstanding – its focus shifted dramatically away from questions of racism. The explanation for racial inequality as an effect of class relations presented by subsequent Marxist analysis cut short an understanding of the phenomenon of 'race'.

At this point in the history of the discipline, a very different and, as it turns out, more sustained approach to issues of 'race' was adopted within the rapidly developing subfield of humanistic geography. Humanistic geography was also established as a radical approach fuelled by the 'relevance debate', but over the course of the next decade it took a very different theoretical road that was, at least initially, much less engaged than the Marxist approach in advocating change in everyday conditions of living. I am concerned here only with those aspects of the humanism in geography that contributed to an understanding of the concept of 'race'.

In 1981, Ceri Peach was to argue that Marxist perspectives had been unable either to describe accurately or to explain the phenomenon of racial segregation, while positivist empirical methods based on 'facts and observation' that address 'social forces over and above those of the economic structure' can more effectively provide not only understanding but social solutions (1981: 31–32). Humanistic geography, he argued, is simply an extension of established positivist analysis, extending deeper into the realm of the personal lifeworld, to describe lived conditions.

Peach was clearly conflating 'positivist' with 'positive' analysis, and isolating Marxist theory and methodology from its normative epistemology. Moreover, he was implicitly buying into the notion that instrumental knowledge could be a powerful agent in providing ideal solutions to

what were widely seen as social problems. This position is untenable, both theoretically and ethically. Nonetheless, what is more important about Peach's work is the inspiration that it helped to engender in a new generation of social geographers, whose work began to document the everyday experiences of racialized communities (Peach, 1975). For example, David Ley's The Black Inner City as Frontier Outpost (1974), while as yet uncritical of the concept of 'race', represents another frontier between disembodied spatial pattern and the embodiment of the everyday lifeworld. In a later statement that marks the major debut of humanistic perspectives within the discipline, Ley and Samuels (1978) stress the need to shift towards a normative interpretation of human experience, and the need to understand human experience and the products of human action – such as, for example, the city – as social constructions. Although this early humanistic work does not problematize 'race' as a social construction, it does contain the implicit theoretical underpinnings for such an understanding.

But humanistic geography has not easily shaken its Enlightenment beginnings. In a collection of writings on Social Interaction and Ethnic Segregation, Peter Jackson and Susan Smith (1981: 2) draw a strong connection between Robert Park's (1926) deeply Kantian notion of the relationship between social and physical distance, and the possibility for geographers to understand human relations as patterns of integration and segregation. Most of the papers that comprise this edited collection adopt such a 'social physics' approach. Jackson and Smith, however, were part of a generation that wanted to cast spatial understanding in much more complex philosophical terms (see also Entrikin, 1980), and who recognized in the work of Park and other American pragmatists the basis for a more relational understanding of the concept of 'race' that incorporates notions of power and conflict, ideology as popular knowledge, and the negotiation of social meaning and identity. While these concepts had some way yet to go in incorporating a broader poststructuralist perspective and a fully critical analysis of racialization, they nonetheless provide the most fully developed critique at that time of the various formulations of the concept of 'race'.

Marxian and humanistic geography, which began with common roots in the relevance debates of the early 1970s, diverged for the next decade and a half (Kobayashi and Mackenzie, 1989) until the deliberate attempt to learn from both occurred in the development of critical geographies of the late 1980s and the 1990s. The 'critical turn' that resulted is perhaps the most significant development allowing geographers, along with other social scientists, to move beyond an understanding of 'race' as a taken-for-granted fact to recognizing its socially constructed status.

CRITICAL 'RACE' THEORY AND GEOGRAPHY

The most significant theoretical development of the poststructuralist era is the recognition that 'race', like other forms of physical manifestation such as gender or sexuality, is an idea. The idea of 'race' has allowed the construction of the raced body according to historically, culturally and place-based sets of meanings. Thus the term 'racialization' refers to the process by which somatic characteristics (which *may* be phenotypical or genotypical) have been made to go beyond themselves to designate the socially inscribed value and the attributes of racialized bodies. Those bodies are the results of normative vision, constituted by the eye of the most powerful viewer. Such values determine how those bodies will be used, as slaves, as racialized labour or, in the case of 'white' bodies, in positions of power. Historically, both the siting and the sighting of the body have thus reinforced racialized notions.

Poststructural approaches see 'race' as a *historically* constructed idea that first entered the English language in the early seventeenth century and became commonly used in scientific writings throughout the Enlightenment period. By the nineteenth century, it had become fully accepted in popular discourse and had taken on the ideological trappings that made racism, capitalism and colonialism the dominant characteristics of human systems. The world had become *racialized* (Miles, 1989; 1993; see also Banton, 2000; Barzun, 1938; Guillaumin, 1972; Jordan, 1974; Malik, 1996; Montagu, 1997; Stepan, 1982). While there occurred major debates throughout the last three centuries on the actual effects of 'race', it was not until the second half of the twentieth century that it came to be viewed by critical theorists as what Montagu (1997) calls 'man's most dangerous myth'.[2]

The impetus for the social constructivist position combines two intellectual initiatives that have recently informed geographical knowledge. First, in the intense reaction to Hitler's Nazism following World War II, repudiation of the concept of 'race' became a major project of the United Nations, and a part of the campaign to create an international framework for human rights. The United Nations Educational Scientific and Cultural Organization (UNESCO) undertook a major research project that resulted in the publication of three volumes (Kuper, 1975; UNESCO, 1956; 1980) that bring together some of the world's most respected scientists and social scientists to refute the concept of 'race'. This repudiation has been interpreted through a poststructuralist lens to examine the particular forms of modern, or postmodern, social relations that sustain the discursive power of 'race' to create human difference. Theorists such as Goldberg (1993), Malik (1996) and Omi and Winant (1994) have extended our understanding of how 'racism' works within dominant cultural contexts and in a range of political contexts, and in the production of labour (Phizacklea and Miles, 1980).

While the poststructuralist literature is now very large, two writers bear mention for their overwhelming influence and their relevance to geographers. Colette Guillaumin's (1972; 1980) early work as part of the UNESCO project was perhaps more influential than any other, both in explaining the historical constitution of 'race' ideas and in linking the concepts of 'race' and gender (see Guillaumin, 1995). She simultaneously provides a scientific basis for questioning our assumptions about 'race' as a social given, and asks that we recognize the ways in which different forms of historical oppression intersect.

Franz Fanon (1952) not only provided the first anti-racist reading of major French poststructuralists, but also inspired a whole generation of, in particular, American scholars to understand anti-black racism and its relationship to the history of colonialism and slavery. Recent rereadings of his work continue to enrich this perspective, and now refer to the intellectual phenomenon of 'Fanonism' (see especially Gibson, 1999; Gordon et al., 1996). Fanon's work has been especially influential among geographers concerned with the historical intersections of racism and colonialism (for example, Blaut, 1993). As Pile has recently pointed out:

> What particularly attracted people was Fanon's refusal to allow the 'normal' categories of colonial life – such as 'black' or 'white', 'native' or 'foreigner' – to be authentic or stable. (2000: 262)

The practice of destabilizing life's normative categories is one that is deeply unsettling and fundamentally geographical:

> For Fanon, the colonial regime's imposition of skin hierarchies not only defines the visibility of the body, and also territorialises the body, but ... is also woven by the white man 'out of a thousand details, anecdotes,

stories' … He shows that black male identity is forged out of a set of identifications that are inherently anxious – simultaneously fearful and desiring. These identifications smuggle senses of self – black and white – across a fictional, though foundational, black/white border. The black/white epidermal schema is not just imposed from the outside, but … also inscribed in the movements of people, in their actions, thoughts and feelings. But it is the black who moves under the constant scrutiny of the fearful/fear-full master's 'blue' eyes. (2000: 264)

Pile's interpretation pushes the geographical imagination to go beyond naive observations of the ways in which places are the result of political processes, to engage social change itself. As a result, nothing will ever be normal again. Geography's ocularcentric vision has been shattered in favour of one that is multifocal.

At the centre of the constructivist position is the repudiation of any essential status for what have been constructed as racialized bodies, that is, the rejection of the belief that 'race' determines essential human traits that define moral, intellectual or cultural values or abilities, or anything else. The constructivist approach to 'race' is consistent with poststructuralist theory and particularly with feminist theory, which recognizes that essentialized notions act normatively to provide the basic template for structuring human relations through their discursive inscription of every human action and, thus, act to contour relations of power.

One of the most serious issues faced by 'race' theorists is to get beyond the assumption that it is human difference itself that needs to be explained, rather than the human tendency to create difference. As Banton points out, the idea of 'race' is so firmly fixed in modern thinking, and so profoundly validated through scientific, social and cultural means, that 'physical differences catch people's attention so readily that they are less quick to appreciate that the validity of "race" as a concept depends upon its value as an aid in explanation' (2000: 51–2). The theorist operates in a world in which taken-for-granted, essentialist notions of 'race' continue to dominate, and are difficult to overcome even for those who hold constructionist views. To do so has required a shift from the study of the racialized to those who have perpetuated the idea of 'race'; in other words, to shift from 'race' to racism (Jackson, 1987a). This shift recasts the 'problem' not as people of colour but as those responsible for historical discrimination. In so doing, it is important to recognize not only historical forms of racism, which are self-evident to many contemporary observers, but also the subtle and often unobserved – even by the most critical of

observers – discursive forms that continue to script the process of racialization today through socially taken-for-granted means.

For many recent writers, this challenge means going beyond the study of racism to the study of 'whiteness' as a historical form of racialization. Whiteness involves not only depicting those who are non-white in prejudicial terms, but also reinforcing the centrality and superiority of white cultural, social or aesthetic forms. It can be as much about the absence as the presence of people of colour, and it works independently of their existence. This recognition has led some researchers to declare that nothing short of 'abolishing the white race' (Ignatiev and Garvey, 1996) will solve the problem of racism. Nonetheless, the concept of whiteness is contradictory. Studies of whiteness run the risk of themselves ignoring non-white people (Bonnett, 1993; 1996a; 1996b), maintaining white privilege even from a critical perspective. The project of whiteness can be translated into myriad forms; indeed, recognition of its adaptability, flexibility and variability is essential to understanding its power. Those forms need to be understood, however, not only in their own terms but also according to their impact, and their susceptibility to resistance, among non-white people.

Geographers have been actively engaged in both constructivist and 'whiteness' studies for the past decade and a half, particularly in Britain where the work of Peter Jackson (1987b; 1989; 1992; 1993) has strengthened geographers' grasp of the importance of understanding 'race' as a historical construction, and has extended the notion of 'race' as an ideational construction to analyses of how it is constructed spatially. Whiteness works at its most powerful level when it is hegemonic, creating landscapes in which people of colour do not even figure (Kobayashi and Peake, 2000). In Canada, there is an especially rich understanding of how the city of Vancouver has been racialized through a dominant discourse of whiteness (see Anderson, 1987; 1991) and of how the state creates a national ideology of whiteness (Peake and Ray, 2000).[3] According to Anderson:

the insight that race identities are constructed out of specific historical and political contexts prompted a radical revision of theorizing about racial segregation in cities. It required geographers to adopt a more rigorous approach that critically examined the discursive leap made in Western cultures from visible differences to something more fundamental which has been called 'race'. It was a move that went beyond describing the spatial forms produced by commonsense notions of difference to deconstructing the processes of exclusion and inclusion out of which segregated cities were produced, both symbolically and materially. (1998b: 204–5)

According to Bonnett (1996c), in a comprehensive review of the study of 'race' by mainly British geographers, the constructionist approach has become the dominant paradigm, by contributing both an interdisciplinary understanding of racialization, and a specifically geographical understanding of how 'space' is racialized. We need to understand urban areas as places of race-based conflict and as specific sites in which the construction of the meaning of place varies according to history, experience and cultural practice, making it highly mutable and subject to sometimes rapid redefinition and reformation (1996c: 876–7, in reference to Keith, 1993).

The constructionist approach in geography has come under criticism, however, both for its limited success in 'relating the cultural and material aspects of race and race-based inequality in such a way as to demonstrate their mutual structuring' (Anderson, 1998b: 204) and for its failure to go beyond the understanding of 'race' as a social construct to understand it as a political construct (Kobayashi and Peake, 1994). Bonnett attributes the problems with the constructionist approach to disciplinary fetishism, claiming that geographers have never been 'completely confident that space really does "matter"', and that the importance of geography 'may only be made fully visible when "the geographical perspective" is finally abandoned' in favour of one in which 'geography, history and sociology are woven together as equally necessary components of a fully interdisciplinary account' (1996c: 880–1).

Elsewhere, Bonnett supports the contention, increasingly popular especially among North American scholars, that a more effective approach must take a position of anti-racism, which 'refers to those forms of thought and/or practice that seek to confront, eradicate and/or ameliorate racism' (2000: 4). Anti-racism is different from non-racism because it involves an active commitment to the project of overcoming racism and its effects. While anti-racism is diverse, sometimes contradictory, always complicated, it always also involves a specifically political project. Not satisfied with a critical understanding of racism, or whiteness, as historical constructions, the anti-racist seeks to place her or his scholarship within a historical context, with the express purpose of intervening politically, and of shifting the academic discourse from a synthetic analysis of the process of racialization to a diachronic analysis of the relationship between historical constructions and ongoing risks to the physical and emotional wellbeing of racialized peoples (Kobayashi and Ray, 2000).

Anti-racism requires a high level of reflexivity (Kobayashi, 1994; 2001), perhaps going so far as to establish political action and social change as the goal to which adequate theories are a tool, rather than the other way around. Reflexivity involves not only new knowledge, but also a new mode of engaged knowledge, abandoning faith in positive science and in the ability, necessity or desirability of the researcher to remain detached from his or her investigation. Reflexivity is not simply an improved form of methodology, therefore, but a normative, moral stand. While constructivist theories may have contributed to the recent move towards reflexivity on the part of geographers, therefore, they are by no means sufficient to the development of political commitment or social action. To put it simply, anti-racist geography is about caring about the conditions of life for specific living individuals, more than it is about theorizing 'race'.

But reflexivity is also contradictory, especially in a situation where the field of anti-racism studies – in geography, in particular – is dominated by white scholars, for whom reflexivity means on the one hand moral introspection both about their role in reinscribing racist relations and about their effectiveness as anti-racist activists, but on the other the risk of appropriating the moral ground of anti-racism with concerns about their own position. Reflexivity is a necessary aspect of contemporary scholarship, therefore, but one that must be constantly on guard against smugness, or against turning the anti-racist project itself into a white project.

There is a growing list of recent examples of work undertaken, especially, by geographers of colour who bring personal experience, commitment and passion to their work. Ruth Wilson Gilmore (1998–9) weaves a complex linkage between racism and globalization, anti-terrorism and the growth of prison networks in the United States to provide not only a compelling scholarly analysis of the processes that have targeted immigrants and prisoners in racialized ways, but also a sense of her own commitment to the cause of deracializing the criminal system, and of working directly with the people whose lives are most affected. Laura Pulido's (2000) work on environmental racism brings not only a commitment to anti-racism, but also deep environmental concern, and one of the strongest justifications for the dual role of academic and activist in bringing about social change. Clyde Woods' (1998) erudite treatment of the role of the blues in the southern USA emphasizes the recursive relationship between culture and politics, but also rings with his own sense of the importance of the blues as a statement of culture empowerment in black America. Finally, I can speak with personal conviction of my own experience in melding my commitment to equity and human rights with my academic work devoted singularly to issues of racism (Kobayashi, 1994; 2001). I believe that my work would be diminished in credibility, understanding and impact were it not based firmly in an activist agenda.

But what of Bonnett's contention that geographers need to give up their sense of discipline to fulfil an anti-racist agenda? Significantly, Bonnett's recent book *Anti-Racism* (2000), while appealing to a broad, interdisciplinary audience, makes virtually no mention of geography. Similarly, in another piece where he emphasizes the importance of a 'historical and geographical contingency of whiteness' (1996b: 104) he makes no mention of the work of geographers. His contention is that the disciplinary fetish has resulted in a failure to incorporate the concepts of whiteness or anti-racism sufficiently thus disqualifying the discipline from an adequate understanding of racialization. But Bonnett's failure to *engage* the geographical literature, especially that produced within the last five years, means that he has foreclosed the possibility of revising his own understanding of geographical contingency. Nor, does he make clear the important, indeed historically fundamental, relationship between 'race' and 'space' from an interdisciplinary perspective. That project, if done effectively, could make it more clear that sociology, history, and other social sciences, also cannot be supported by fetishes of 'society', 'time', and the like. We await, then, the work that will allow a transformation of geographical knowledge such that we are made more reflexively aware of the ways in which geographers, through their almost casual acceptance of the concept of 'space', have been unable to deal as effectively as they might with the ways in which 'race' and 'space' are intertwined.

Bonnett and others fail to explain adequately the point that 'space', like 'race', is a historically constructed concept that has been mobilized in the major projects of modernity. The construction of 'space' too needs to be taken back to its Enlightenment roots, coming full circle back to the ways in which Kant structured and racialized the knowledge of future geographers. The idea of 'space' was for Kant an essential ontological category, necessary for engagement between human and world, and also for structuring the fabric of human action. I have argued elsewhere (Kobayashi and Peake, 1994: 239; see also Kant, 1970: 55) that modern spatial strategies, including the designation of particular places as accessible or inaccessible, the exclusion of particularly women and people of colour from certain sites, the development of a sense of spatial territoriality through both colonialism and the concept of private property, need to be understood in the light of efforts by Kant and other Enlightenment thinkers both to construct 'space' and to encode its particularly public use with ideological designs.

The scope of this chapter does not allow for a full discussion of the development of the concept of 'space' as an essentialized category, but it is important to note that 'space' and 'race' share a similar heritage. Both play a dual role as public discursive categories rooted in Enlightenment logic and essential to the aims of both colonialism and capitalism and, simultaneously, as objects of scientific investigation, albeit that 'race' came under scientific scrutiny a couple of centuries before 'space' became so elevated in the twentieth century. Both have been obscured by ideology from critical assessment. Indeed the ideological project of racialization is equally a project of spatialization. Both projects are a fundamental part of the construction of geographical knowledge. And that knowledge has been based upon a historically peculiar form of moral vision in which the geographer's eye has been trained to cast itself upon the world, organizing, mapping and constructing its human dimensions according to essentialized notions used to ideological ends as constructions that organize, constrain and categorize human experience.

Given its historical construction, then, geography *matters* just as 'race' *matters*, not because either infers essentialized characteristics of human beings or landscapes, but because both result from processes of human differentiation. It is not so much a disciplinary fetishism as a spatial fetishism that has hampered geographers from a fuller recognition that they have been held intellectual hostage, in many ways, to a faith in 'space' as an object of enquiry. At root of the spatial fetish, as with the racial fetish, is an essentialist notion of our discipline, which moved through a series of other essential ideas before it became fixated upon 'space' (see Mayhew, 2000).

Notwithstanding recent work that explicitly develops the relational, ideological and historically constituted quality of 'space' (see Lefebvre, 1991; Soja, 1996), geographers have been very slow to give up the study of 'space', or to recognize how deeply essentialized that concept has become. For evidence, simply pick up any geographical journal, even the radical or critical journals, to see the uncritical way in which the majority of geographical language is peppered with the word 'space', used in much the same way that 'race' is used in popular discourse. Bonnett's challenge to overcome spatial fetishism needs to be taken seriously, then, but not at the expense of fully exploring the ways in which geographers throughout the history of the discipline have been complicit in the construction of 'space' and thereby in the construction of 'race'. Contrary to Bonnett, however, I believe that geographers' preoccupation with space means that it needs to be taken more seriously, so that we might work through this relationship to focus on racialization and spatialization.

But there is yet another problem that requires the development of activism in geography. As long as geographical knowledge of 'race' remains privileged as white knowledge, even the most critical and reflexive of scholars will have a difficult time sustaining or defending anti-racist scholarship. As Delaney opines:

It's my impression that Geography as an institution is nearly as white an enterprise as country and western music or professional golf ... most of the teachers are white, most of the students are white, most of the discussion of race in these contexts is among white folk. (1998: 25, 22)

But despite some of the encouraging scholarship discussed above, there is little evidence that those discussions are taking place on a major scale. Both the continued relative neglect of issues of racism, and the continuing scarcity of geographers of colour, are a direct result of the discipline's racialized past, entwined within a legacy of deep assumptions about the natural differences that 'race' creates. As Gerry Thomas points out:

The categorical ignoring by the discipline of issues confronting American society (and Western society and their colonies in general) with regards to race relations, race identity and race politics ... speaks to a discipline that is impregnated with a particular racial identity and ideology. (1998: 134)

In Britain, where much of the critical 'race' theory in geography has been produced, the discipline is almost completely white, and changing very slowly. In the United States and Canada, the situation is somewhat – but only somewhat – better. Unfortunately, the dialogue between the critical race theorists and activist geographers of colour has been extremely limited, either at the theoretical or at the political level. Recent events such as the NSF-funded workshop on 'race' and geography held at the University of Kentucky in 1998 have begun to chip away at these solitudes. That conference inspired the creation of an 'anti-racist manifesto' aimed at the discipline, as well as a project, supported to date by the Association of American Geographers and the Canadian Association of Geographers, to take measures to increase the participation of students of colour in graduate training programmes. The scholarly proceedings from the conference show for the first time a well-developed integration of critical theory and activism.[4]

As much as it needs further epistemological development of anti-racist theory, geography needs an anti-racist transformation, to make geographers, not just geography, representative of its political aims. Partly, this is an issue of human rights and accessibility, to allow people of colour access to the 'spaces' of knowledge that have previously been denied them. But equal opportunity is not the only issue. While there is no essential reason that geographers of colour should have a better perspective on questions of racism, it remains a fact that the majority of empirical work that maps the racialized exclusion of communities of colour has been done by geographers who have a personal attachment to those communities, an attachment born of commitment, personal experience, and an ethic of caring that stimulates their work. In an ideal anti-racist world, colour would not matter, either to knowledge or to political commitment; there is no theoretical reason that colour should affect either. But the very fact that geography remains such a white discipline shows that, at least at this time in our history, colour *does matter*, and until we redress the balance, our knowledge will remain hypocritical dogma. Critical geographers have established well the concept that knowledge does not stand independently of the knower, but they have been relatively ineffective as yet in shifting the balance of knowers. To do so, we need also to take seriously the spatiality of the discipline, the question of who gets to be in which disciplinary place, and to wield power over its occupation.

CONCLUSION

I have addressed three moments in the history of geographical knowledge of the idea of 'race', to show that geographers have moved over time through at least three significant moments. While these stages do not follow a strict chronology or even a consistent intellectual trajectory, and there are many individual outliers to the progression described here, these three moments depict, I believe, major conceptual shifts in the ways that geographers have studied 'race'. The first moment involved attempts to legitimate 'race' as a scientific basis for understanding human difference, but also for justifying the practices of racialization involved in subsequent colonial and capitalist domination. The second stage involved a liberal approach, grounded in positive social theories, that eschewed belief in 'race' as a determining factor, yet e-raced the effects of human differentiation by reducing racial characteristics to spatial patterns and by failing to address the ways in which human beings construct difference. In the third moment, the most recent constructivist approaches have recognized the idea of 'race' as the basis for the historical project of racialization, and have advanced critical understanding of the discursive forms that sustain racism. Yet such understanding has done little to deracialize the discipline of geography and has even reinforced the power of white geographers to define the intellectual terms in which we understand 'race'. All three moments have been fundamentally structured by Enlightenment notions of human and spatial differentiation, notions that have been reinforced by the ocularcentrism of society in general and of the discipline of geography in particular. Both notions have deep, albeit far from exclusive, roots in the work of the geographer/philosopher Immanuel Kant.

Geographical knowledge is indeed powerful – more powerful, perhaps, than we have given it credit for. The challenge before us, if we are to use geographical knowledge to overcome racism, is twofold: to explore further the recursive knowledge relationship between 'race' and 'space', between historical processes of racialization and spatialization; and to bring to geographical knowledge more reflexivity and stronger political commitment. For, as the history of racism shows, geographical knowledge and activism are synonymous.

NOTES

1 According to David Livingstone, possibilism is 'a thesis about the relationship between human culture and the natural environment which claims that the human species has the capacity to choose between a range of possible responses to physical conditions' (2000: 608–9). Originally advanced as a counter to environmental determinism, possibilism also carries the implicit rejection of ideas of racialization, but that line of discussion was not followed in the geographic literature, partly because of the geographic emphasis on culture as an epiphenomenon.

2 Much of the poststructuralist theoretical literature, particularly that produced in France, was written in the years before and immediately after World War II, and thus predates the so-called quantitative revolution in geography. I am not arguing for a straightforward chronology here, therefore. But in terms of the major developments in the discipline of geography, poststructuralism did not become a major paradigm until the 1980s. Positive theories, which include positivism on the part of a much smaller group, became popular in the late 1960s and through the 1970s, fuelling much of the discussion in the 'relevance debate'.

3 In tracing the progress of geographical knowledge, it is significant that current work on the white construction of racism is being undertaken by students of David Ley at the University of British Columbia, thus establishing a new generation of scholars with direct lineage to Ceri Peach and others at Oxford, who influenced Ley as well as Jackson and Smith. As an aside, as a former student of David Ley, I count myself as part of that lineage.

4 The proceedings of this conference will be published in 2002 in special issues of *The Professional Geographer* and *Social and Cultural Geography*.

REFERENCES

Anderson, K. (1987) 'The idea of Chinatown: the power of place and institutional practice in the making of a racial category', *Annals of the Association of American Geographers* 77 (4): 580–98.
Anderson, K. (1991) *Vancouver's Chinatown: Racial Discourse in Canada, 1875–1980.* Montreal: McGill–Queen's University Press.
Anderson, K. (1998a) 'Science and the savage: the Linnean Society of New South Wales', *Ecumene* 5: 125–43.
Anderson, K. (1998b) 'Sites of difference: beyond a cultural politics of race polarity', in R. Fincher and J.M. Jacobs (eds) *Cities of Difference*. New York: Guilford. pp. 201–25.
Anderson, K. (2000) '"The beast within": race, humanity and animality', *Environment and Planning D: Society and Space* 18: 301–20.
Banton, M. (2000) 'The idiom of race', in L. Back and J. Solomos (eds) *Theories of Racism: A Reader*. New York: Routledge. pp. 51–63. From M. Banton, *Research and Race Relations, Vol. 2*. London: JAI Press. pp. 21–40.
Barzun, J. (1938) *Race: A Study in Modern Superstition.* London: Methuen.
Blaut, J.M. (1970) 'Geographic models of imperialism', *Antipode* 2 (1): 65–85.
Blaut, J.M. (1974) 'The ghetto as an internal neo-colony', *Antipode* 6 (1): 37–41.
Blaut, J.M. (1993) *The Colonizer's Model of the World.* New York: Guilford.
Bonnett, A. (1993) *Radicalism, Anti-Racism and Representation.* London: Routledge.
Bonnett, A. (1996a) 'Forever "white"?: challenges and alternatives to a "racial" monolith', *New Community* 13 (2): 173–80.

Bonnett, A. (1996b) 'Anti-racism and the critique of "white identities"', *New Community* 22 (1): 97–110.

Bonnett, A. (1996c) 'Construction of "race", place and discipline: geographies of "racial" identity and racism', *Ethnic and Racial Studies* 19 (4): 864–83.

Bonnett, A. (2000) *Anti-Racism*. London: Routledge.

Bunge, W. Jr (1971) *Fitzgerald: The Geography of an American Revolution*. Cambridge: Cambridge University Press.

Chisholm, M. (1971) 'Geography and the question of "relevance"', *Area* 3: 65–8.

Delaney, D. (1998) 'The space that race makes'. Unpublished papers, National Science Foundation Research Workshop on Race and Geography, Department of Geography, University of Kentucky, 29 October to 1 November 1998. pp. 15–26.

Driver, F. (1992) 'Geography's empire: histories of geographical knowledge', *Environment and Planning D: Society and Space* 10: 23–40.

Entrikin, J.N. (1980) 'Contemporary humanism in geography', *Annals of the Association of American Geographers* 70: 43–58.

Eyles, J. (1973) 'Geography and relevance', *Area* 5: 158–60.

Fanon, F. (1952) *Peau Noire, Masques Blancs*. Paris: Éditions de Seuil. Translation 1967 *Black Skin, White Masks*. New York: Grove.

Gibson, N.C. (ed.) (1999) *Rethinking Fanon: The Continuing Dialogue*. Amherst: Humanity.

Gilmore, R. (1998–9) 'Globalisation and US prison growth: from military Keynesianism to post-Keynesian militarism', *Race and Class* 40 (2–3): 177–88.

Godlewska, A. and Smith, N. (1994) (eds) *Geography and Empire*. Oxford: Blackwell.

Goldberg, D. (1993) *Racist Culture: Philosophy and the Politics of Meaning*. Oxford: Blackwell.

Gordon, L.R., Sharpley-Whiting, T.D. and White, R.T. (eds) (1996) *Fanon: A Critical Reader*. Oxford: Blackwell.

Guillaumin, C. (1972) 'Caractères spécifiques de l'idéologie raciste', *Cahiers internationaux de sociologie* LIII: 247–74.

Guillaumin, C. (1980) 'The idea of race and its elevation to autonomous scientific and legal status', in UNESCO *Sociological Theories: Race and Colonialism*. Paris: UNESCO.

Guillaumin, C. (1995) *Racism, Sexism, Power and Ideology*. London: Routledge.

Harris, R.C. (1971) 'Theory and synthesis in historical geography', *The Canadian Geographer* 15: 157–72.

Harvey, D. (1972) 'Revolutionary and counter-revolutionary theory in geography and the problem of ghetto-formation', *Antipode* 4 (1): 1–13.

Harvey, D. (1973) *Social Justice and the City*. London: Arnold.

Ignatiev, N. and Garvey, J. (eds) (1996) *Race Traitor*. New York: Routledge.

Jackson, P.J. (ed.) (1987a) *Race and Racism: Essays in Social Geography*. London: Allen and Unwin.

Jackson, P.J. (1987b) 'The idea of "race" and the geography of racism', in P.J. Jackson (ed.) *Race and Racism: Essays in Social Geography*. London: Allen and Unwin. pp. 1–22.

Jackson, P.J. (1989) 'Geography, "race" and racism', in N.J. Thrift and R. Peet (eds) *New Models in Geography*, vol. 2. London: Allen and Unwin. pp. 176–95.

Jackson, P.J. (1992) 'The racialization of labour in postwar Bradford', *Journal of Historical Geography* 18: 190–209.

Jackson, P.J. (1993) 'Changing ourselves: a geography of position', in R.J. Johnston (ed.) *The Challenge for Geography*. Oxford: Blackwell. pp. 198–214.

Jackson, P. and Smith, S.J. (eds) (1981) 'Introduction', in *Social Interaction and Ethnic Segregation*. Institute of British Geographers Special Publication 12. London: Academic.

Johnston, R.J. (1991) *Geography and Geographers: Anglo American Human Geography since 1945*, 4th edn. London: Arnold.

Johnston, R.J. (2000) 'Radical geography', in R.J. Johnston, D. Gregory, G. Pratt and M. Watts (eds) *The Dictionary of Human Geography*. Oxford: Blackwell. pp. 670–1.

Jordan, W.D. (1974) *White Man's Burden: Historical Origins of Racism in the United States*. London: Oxford University Press.

Kant, I. (1970) 'An answer to the question, "What is Enlightenment?"', in H. Reiss (ed.) *Kant's Political Writings*. Cambridge: Cambridge University Press.

Kant, I. (1997a) 'On the different races of man' (1775), in E.C. Eze (ed.) *Race and the Enlightenment: A Reader*. Cambridge, MA: Blackwell. pp. 38–48. Originally published in *This is Race* (ed). E.D. Count. New York: Henry Schuman, 1950.

Kant, I. (1997b) 'On national characteristics', in E.C. Eze (ed.) *Race and the Enlightenment: A Reader*. Cambridge, MA: Blackwell. pp. 49–57. Originally published in *Observations on the Feeling of the Beautiful and Sublime*. Berkeley: University of California Press, 1960.

Kant, I. (1997c) 'On countries that are known and unknown to Europeans', in E.C. Eze (ed.) *Race and the Enlightenment: A Reader*. Cambridge, MA: Blackwell. pp. 58–64. Originally published in *Physical Geography* [*Physische Geographie*], in *Gesammelte Schriften*. Berlin: Reimer, 1900–66.

Keith, M. (1993) *Race Riots and Policing: Lore and Disorder in a Multi-racist Society*. London: University College London Press.

Kobayashi, A. (1994) 'Coloring the field: gender, "race", and the politics of fieldwork', *The Professional Geographer* 45 (1): 73–80.

Kobayashi, A. (2001) 'Negotiating the personal and the political in critical qualitative research', in M. Limb and C. Dwyer (eds) *Qualitative Methodologies for Geographers*. London: Arnold.

Kobayashi, A. and Mackenzie, S. (1989) 'Introduction: humanism and historical materialism in contemporary geography', in A. Kobayashi and S. Mackenzie (eds) *Remaking Human Geography*. Boston: Unwin Hyman. pp. 1–16.

Kobayashi, A. and Peake, L. (1994) 'Unnatural discourse: "Race" and gender in geography', *Gender, Place and Culture* 1 (2): 225–43.

Kobayashi, A. and Peake, L. (2000) 'Racism out of place: thoughts on an anti-racist agenda for geography in the new millennium', *Annals of the Association of American Geographers* 90 (1): 391–403.

Kobayashi, A. and Ray, B. (2000) 'Civil risk and landscapes of marginality in Canada: a pluralist approach to social justice', *The Canadian Geographer* 44: 401–17.

Kuper, L. (ed.) (1975) *Race, Science and Society*. Paris: UNESCO. London: Allen and Unwin.

Leach, B. (1973) 'The social geographer and black people: can geography contribute to race relations', *Race* 15 (12): 230–41.

Lefebvre, H. (1991) *The Production of Space*. Oxford: Blackwell.

Ley, D. (1974) *The Black Inner City as Frontier Out-post: Images and Behavior of a Philadelphia Neighbor-hood*. Washington: Association of American Geographers.

Ley, D. and Samuels, M. (eds) (1978) *Humanistic Geography: Prospects and Problems*. Chicago: Maaroufa.

Livingstone, D.N. (1992) *The Geographical Tradition*. Oxford: Blackwell.

Livingstone, D.N. (1994) 'Climate's moral economy: science, race and place in post-Darwinian British and American geography', in A. Godlewska and N. Smith (eds) *Geography and Empire*: Oxford: Blackwell. pp. 132–54.

Livingstone, D.N. (2000) 'Possibilism', in R.J. Johnston, D. Gregory, G. Pratt, and M. Watts (eds) *The Dictionary of Human Geography* (4th edn). Oxford: Blackwell.

Malik, K. (1996) *The Meaning of Race: Race, History and Culture in Western Society*. Basingstoke: Macmillan.

Mayhew, R. (2000) *Enlightenment Geography: The Political Languages of British Geography, 1650–1850*. New York: St. Martin's Press.

Miles, R. (1989) *Racism*. London: Routledge.

Miles, R. (1993) *Racism after 'Race Relations'*. London: Routledge.

Montagu, A. (1997) *Man's Most Dangerous Myth*, 6th edn. Walnut Creek: Altamira and Sage.

Morrill, R.L. (1965) 'The Negro ghetto: problems and alternatives', *The Geographical Review* 55: 339–61.

Omi, M. and Winant, H. (1994) *Racial Formation in the United States*. New York: Routledge.

Park, R.E. (1926) 'The urban community as a spatial pattern and a moral order', in E. Burgess (ed.) *The Urban Community*. Chicago: University of Chicago Press. pp. 1–19.

Peach, C. (ed.) (1975) *Urban Social Segregation*. London: Longmans.

Peach, C. (1981) 'Conflicting interpretation of segregation', in P. Jackson and S.J. Smith (eds) *Social Interaction and Ethnic Segregation*. London: Academic. pp. 35–58.

Peake, L. and Ray, B. (2000) 'Racializing the Canadian landscape: whiteness, uneven geographies and social justice', in A. Kobayashi (ed.) *Fifty Years After: Geographical Interpretations of Canada, The Canadian Geographer* 45: (1): 180–6.

Phizacklea, A. and Miles, R. (1980) *Labour and Racism*. London: Routledge and Kegan Paul.

Pile, S. (2000) 'The troubled spaces of Frantz Fanon', in M. Crang and N. Thrift (eds) *Thinking Space*. London: Routledge. pp. 260–77.

Pulido, L. (2000) 'Rethinking environmental racism: white privilege and urban development in Southern California', *Annals of the Association of American Geographers* 90 (1): 12–40.

Rose, H.M. (1970) 'The development of an urban subsystem: the case of the Negro ghetto', *Annals of the Association of American Geographers* 60: 1–17.

Rose, H.M. (1972) 'The spatial development of black residential subsystems', *Economic Geography* 48: 43–65.

Rothenberg, T.Y. (1994) 'Voyeurs of imperialism: *The National Geographic Magazine* before World War II', in A. Godlewska and N. Smith (eds) *Geography and Empire*. Oxford: Blackwell. pp. 155–72.

Slater, D. (1973) 'Geography and underdevelopment: Part I', *Antipode* 5 (3): 21–33.

Smith, D. (1974) 'Geography, racial equality and affirmative action', *Antipode* 6 (2): 34–41.

Soja, E. (1996) *Thirdspace: Journeys to Los Angeles and Other Real-and-Imagined Places*. Malden: Blackwell.

Stepan, N. (1982) *The Idea of Race in Science: Great Britain, 1800–1945*. London: Macmillan.

Thomas, G. (1998) 'Race and geography: a position paper'. Unpublished paper, National Science Foundation Research Workshop on Race and Geography, Department of Geography, University of Kentucky, 29 October to 1 November 1998. pp. 133–47.

UNESCO (1956) *The Race Question in Modern Science*. Paris: UNESCO.

UNESCO (1980) *Sociological Theories: Race and Colonialism*. Paris: UNESCO.

Withers, C. (1997) 'Geography, royalty and empire: Scotland and the making of Great Britain, 1603–61', *Scottish Geographical Magazine* 113: 22–32.

Woods, C. (1998) *Development Arrested: Race Power and the Blues in the Mississippi Delta*. London: Verso.

31

Ontological Pluralism in Contested Cultural Landscapes

Richard Howitt and Sandra Suchet-Pearson

Academic discourse typically represents its knowledge as detached, objective and universal. Contemporary institutions of teaching, research, governance and disciplinary thinking are profoundly influenced by Enlightenment thinking. Cultural geography reflects this inheritance. Yet cultural geography's discursive spaces often try to displace and unsettle it. The ontological and epistemological legacies of European colonialism, however, are highly resistant. Even the most liberal universities operate in ways that place substantial domains of human experience, thought and insight outside the conventional bounds of legitimate knowledge. So, despite its inclusive intent, much postcolonial cultural geography privileges abstract theory rather than grounded practice. Consequently, like other academic disciplines, it produces a 'hall of mirrors' in which self-consciously postcolonial theory reflects its own views rather than engaging with alternative ontologies – diverse ways of knowing, being-in-place and relating to complex, often contested cultural landscapes at various scales.

This chapter argues cultural geography should explore practical, theoretical and methodological implications of ontological pluralism because landscapes of cultural conflict are often as much about different knowledge systems as about contested claims to land, identity, resources or livelihood. It considers the implications of multiple knowledges (ontological diversity) for cultural geography and seeks to unsettle and challenge the dominance of Eurocentrism, which affects even the new cultural geography, by taking seriously the philosophies and experiences of indigenous groups.

'WESTERN' PHILOSOPHY AS A HALL OF MIRRORS

People interpret, make meaning and relate to themselves, other people and environments in many different ways. 'Western' philosophies, which we broadly define as 'Eurocentric' knowledges in this chapter, generally assume external, objective realities exist. Definitional categories, boundaries and relations are set as part of those realities and are easily accepted as a static, natural truth. This approach sets limits on how legitimate knowledge is constructed in 'western' philosophical traditions. Christie uses Latour's imagery to describe how:

> the production of knowledge business in the modern world has been likened to a railroad industry in which knowledge can only run on tracks already laid down from the laboratory out. (1992: 1)

Eurocentric thinking, drawing on Enlightenment science, industrial revolution technologies, market economics and/or Judeo-Christian philosophies, run on the tracks of a naturalized and externalized truth founded on a belief in atomism, where the world is divided into distinguishable segments with essential differences:

> In atomistic views of the world, identity is marked by irreducible essences, and by discontinuities – by boundaries between what (and where) something is, and what (and where) it is not. (1992: 2)

Belief in an external world, an objective reality that exists 'out there', disguises the cultural construction of ontologies as external, unbiased and naturalized (1992: 2). The assumption that universal truths can be discovered, and that Eurocentric knowledges have revealed at least some of them, means the idea of knowledge itself is often not problematized in academic discourse. This renders invisible processes that construct knowledge, and many of their consequences. Eurocentric knowledges, boundaries and relationships are conventionally treated in the academy as the only possible knowledges and as universally relevant.

The assumption of Eurocentric knowledges' universal relevance parallels the political processes of imperialism and displacement. Other knowledges are rendered silent. They are ignored, devalued and/or undermined so that Eurocentric knowledges see only themselves, becoming self-legitimating rather than self-aware. D. Rose eloquently describes the circular argument formed by these assumptions as an all-knowing self, centring itself in a hall of mirrors:

> The self sets itself within a hall of mirrors; it mistakes its reflection for the world, sees its own reflections endlessly, talks endlessly to itself, and, not surprisingly, finds continual verification of itself and its world view. This is monologue masquerading as conversation, masturbation posing as productive interaction; it is a narcissism so profound that it purports to provide a universal knowledge when in fact its violent erasures are universalizing its own singular and powerful isolation. It promotes a nihilism that stifles the knowledge of connection, disabling dialogue, and maiming the possibilities whereby 'self' might be captured by 'other'. (1999: 177)

Irigaray (1985a) also draws on metaphorical mirrors. She discusses how the male imaginary duplicates and reflects itself to ensure 'coherence' and legitimacy. Cultural geographer G. Rose expands further upon this in her 'dialogue' with Irigaray:

> And the mirrors are frozen ... Solidified in their repetitive reflection of the same, a solidity of morphological tumescence and of death. And mirrors can be walls. They cluster together, overlap, build a 'palace of mirrors' (Irigaray, 1985b: 137), provide 'solid walls of principle' (Irigaray, 1985a: 106). They give form, they turn ideas into structures, edifices, they produce 'the absolute power of form' (Irigaray, 1985a: 110), the solidity of concepts, boundaries and order (Irigaray, 1985a: 107). (1996: 67)

We seek to engage cultural landscapes on the other side of the mirrors constructed by Eurocentric knowledges. We aim to open readers to possibilities unthought of, and spaces inaccessible from within, the hall of mirrors. Blunt and Rose (1994: 15–16) identify two strategies used to open up spaces of resistance. One strategy draws on 'imagined geographies' as an 'imaginative resource' to challenge colonization. The other engages with the inherent limits of Eurocentric knowledge itself. Nader identifies three research directions that challenge notions of western rationality as the benchmark for all other cultural knowledge: describing knowledges in traditional societies; ethnographic studies of the socio-cultural context of western science; and linking studies of science with studies of other knowledges to encourage 'mutual interrogation' (1996: 6).

In this chapter, the self-defining limits of the circular argument which characterizes Eurocentric knowledges are exposed by deconstructing key concepts in cultural geography and investigating how Eurocentric knowledges become colonizing knowledges. Concurrently, situated knowledges and practices, diverse systems of local knowledge and social organization arising from cultures being-in-place, are drawn on to challenge and unsettle the position of Eurocentric knowledges within the hall of mirrors. This is done not to romanticize other knowledges but to challenge, unsettle and reconfigure the knowledge–power nexus constructed in cultural geography, and elsewhere in the academy.

GEOGRAPHICAL KNOWLEDGES IN THE HALL OF MIRRORS

Academic knowledges positioned within this metaphorical hall of mirrors play vital roles in making invisible, writing over and blocking out other knowledge systems. For example, it is no coincidence that many of the words associated with research are drawn from and embedded in colonizing discourses: re-searching, exploring, collecting, journeying, examining, investigating, travelling, discovering. Clifford identifies colonizing legacies in his examination of how anthropology problematizes the concept of fieldwork – a concept similarly cherished by geographers:

> Fieldwork has become a problem because of its positivist and colonialist associations (the field as 'laboratory', the field as place of 'discovery') ... they [anthropologists and ethnographers, and we might add geographers] have navigated in the dominant society, often enjoying white skin privilege and a physical safety in the field guaranteed by a history of prior punitive expeditions and policing. (1997: 194, 196)

Geographical research is certainly not immune from these colonizing legacies, contexts and practices. However, geographers are turning their attention to colonizing discourses and 'postcolonial', 'decolonizing' or 'counter-colonial' projects. They are probing, questioning and challenging the role the discipline has played in disempowering people and making multiple realities and imaginaries invisible through exploring, charting, locating, mapping and writing (for example, see Blunt and Rose, 1994; Driver, 1992; 2000; Godlewska and Smith, 1994; Howitt and Jackson, 1998).

Beyond research and fieldwork, geographical writing has also been scrutinized for colonizing power relations. Academic protocol has sought to imbue academic knowledge and texts as 'truth' with a neutral, objective character that cannot be challenged. This automatically excludes some narratives, alienates some audiences and discredits certain texts. Keith argues that distancing the author from the text powerfully authorizes the text and 'exemplifies the need of the powerful to rationalise, comprehend and control the seductively anarchic world of the irrational "other"' (1992: 563). Geographers, anthropologists, literary theorists and others have been considering the power relations that imbue a text with authority by exploring relations between authoring a text and the editorial control with which authors make decisions (Crang, 1992: 542). Much academic writing ignores and makes invisible these relations and the discursive communities they create. By implication this asserts 'authority over and ownership of the work' (McDowell, 1992b: 62).

A text is not a neutral, passive presentation of an external truth (Christie, 1992). It is a partial, active re-presentation of complex worlds using particular strategies to persuade and influence readers for specific purposes (Katz, 1992: 496). The power involved in re-presenting people in geographical texts also reflects and produces colonizing relationships as people's knowledges and practices are excluded or devalued. Attempting to include people and their perspectives does not mean power relations have been addressed and a 'solution' has been found to 'the problem' of representing 'others'. Processes of inclusion are as saturated with power relations as those of exclusion. Being able to write, the appropriation of other people's experiences, choosing whom to include and how to include them, the choices other people have made in representing themselves to the author and other authors, the ways the readers interpret the words and the ulterior motive for the usage of the 'voices', all involve relationships of power (Crang, 1992; Katz, 1992).

Human geographers have challenged attempts to include 'voices' in projects that aim to speak (or write) on *behalf* of 'others' who have been excluded in Eurocentric representations. The colonizing arrogance and politics of appropriation encompassed in such a notion have been challenged (McDowell, 1994: 242). In response there has been some 'opening up [of] spaces within geography for alternative voices to be heard [and read]' (1994: 243). This chapter seeks to open up such spaces, engaging the reader with a polyphony that challenges Eurocentric assumptions of universalism which have silenced and devalued other knowledges. This is the option identified by Crang as:

> a form of polyphony grounded much more firmly in recounting the lives of particular individuals, each becoming what we might call a bearer of cultural otherness without collectively forming an 'Other'. (1992: 536)

UNSETTLING KEY IDEAS IN CULTURAL GEOGRAPHY

Eurocentric knowledges typically assume an essentialized, naturalized truth where boundaries are seen as external to categorically separate entities. For example, philosophical discourse and social science method often assume a profound, categorical distinction between and around key concepts used in cultural geography, for example:

- a separation between space, scale and time
- language and meaning form singular entities
- a binary opposition between culture and nature
- identities that are singular and static.

Each of these assumptions has been challenged by recent cultural geographical research, but the expectation that such concepts should be categorically distinct and independent persists. Subsequently, academic discourse continues to proceed in a piecemeal way. The further assumption of universal applicability of this approach obscures alternative approaches behind a mass of mirrors. Human existence, communication, identity (and otherness) is always, inescapably and inevitably, embodied, emplaced and (geographically, historically and culturally) contextualized. Thus, conventions and assumptions of Eurocentric knowledge and the practices of academic discourse directly marginalize diverse human experience and contribute to wider political consequences.

SPACE, SCALE AND TIME

Following the suggestion of Horvath, Howitt (2001b) argues that five foundational concepts underpin the discipline of geography: space–time, place, nature, culture and scale. Human geographers have played a leading role in reconceptualizing dominant concepts of space. For example, Soja seeks 'to spatialize the historical narrative, to attach to durèe an enduring critical human geography' (1989: 1). Blunt and Rose argue that space is not a neutral given but that 'space itself could … be interpreted in multiple ways but only after its construction in the minds of those perceiving it' (1994: 12). Massey (for example, 1984) also eloquently explains that neither the spatial, nor the social, nor the natural should or can be theorized independently. These domains co-construct each other. They are coequal components of any sophisticated social scientific analytical framework.

In challenging notions of space as a non-active element in social relations, cultural geographers have actively investigated how people construct spaces, places, boundaries and relationships. It is not that spatial boundaries and relationships are illusionary or do not exist. Rather, it is necessary to recognize that 'the material and ideological are coconstitutive' (Jacobs, 1996: 5). Imaginary and real boundaries and relationships around, within and between spaces and places are crucial in understanding power relationships and consequently are vital in imagining and realizing relevant and contextualized processes in specific circumstances. Moore argues for a vision which 'insists on joining the cultural politics of place to those of identity' rather than 'viewing geographically specific sites as the stage – already fully-formed constructions that serve as settings for action – for the performance of identities' (1998: 347).

Spatial, cultural and natural processes and relationships are always constituted in time across and between scales. Howitt (1998; also forthcoming) discusses the multidimensional and simultaneous interactions that define and are defined within and between spaces. Drawing upon a philosophy of internal relations, Howitt argues the 'rigidity of many categorical definitions is unsustainable … Boundaries which previously separated clearly independent, even mutually exclusive, conceptual categories have been transgressed' (1993: 34). He emphasizes that 'scale, like all spatial relationships, is embedded in the dynamics of social life rather than imposed externally' (1993: 39). Relationships are contextualized across space, between places, across and between times, within and between groups and territories. This points to a commitment to radical contextualization – a commitment to taking context seriously as an ontological element rather than treating it as a superficially contingent element. The practical impact of radical contextualization is that cultural research must both investigate and debate, rather than either assume or ignore, the implications of geographical, historical, cultural, political and environmental context.

Despite the wide acknowledgement of space–time as an integrated concept (for example, Massey, 1984), geographers' debate of ideas about space has rarely been matched by similar interest in concepts of time. Time continues to be widely seen as a separate category, often being left for historians to explore. However time, like space, has been constructed in Eurocentric epistomologies on the basis that the world can be seen as it really is. Judeo-Christian and scientific discourses represent time through a time-line on which arrows point to the future, constructing a linear notion of time. Ideas of evolution and social Darwinism reinforce this as a definable progress – an inevitable movement towards a singular future. Notions of progress and development, cause and effect contribute towards this taken-for-granted sensation:

> Linear time underlies our most cherished notions of 'progress' – our collective faith in the inexorable, incremental refinement of human society, technology, and thought. (Knudtson and Suzuki, 1992: 143)

The discipline of history makes it 'possible to recover an absolute truth of what happened in history. History is not a story told by the present, it is a *fait accompli*, which we need carefully to uncover' (Christie, 1992: 2). Carter (1987: xiv) critiques histories based on and contributing to a notion of linear bound time. He argues that this type of history replaces spatial events with a historical stage whereby it 'is not the historian who stages events, weaving them together to form a plot, but History itself'. This conceptually separates space and time, and:

> [T]he fact that where we stand and how we go is history is not recognised. In a theatre of its own design, history's drama unfolds; the historian is an impartial onlooker, simply *repeating* what happened … Such history is a fabric woven of self-reinforcing illusions [placed within the hall of mirrors]. (1987: xv)

LANGUAGE, LANDSCAPE AND MEANING

Language reflects, shapes and limits how humans understand the world around us. It provides the building blocks of ontology and it simultaneously constructs and limits our vision. Language reflects and constructs power. For example, concepts of time are embedded in language. English, for example, constructs tenses in ways that reflect and reinforce a view of time as categorically distinct, as either past, present or future. In other words, the linearity implicit in much Eurocentric epistemology is embedded in English language, making it difficult to convey non-linear concepts of time and temporal relations, and their spatial implications. Consider, for example, the use of the verb 'come' in the following passage of Aboriginal English:

> *Kakawuli* (bush yam) come up from Dreaming. No matter what come up, they come up from Dreaming. All tucker come out from Dreaming. Fish, turtle, all come from Dreaming. Crocodile, anything, all come from Dreaming. (Big Mick Kankinang in D. Rose, 1996: 35)

In this passage Kankinang uses the verb 'to come' without conventional tense markers. For many English speakers this shift from standard English is read as an inability to express the past tense properly because they construct the Dreaming as a time in an ancient past. Yet Kankinang's grammar here precisely represents an ever-present Dreaming (what the anthropologist Stanner, 1979: 24, referred to as the 'everywhen'), where things did come, do come and will always come from the continually renewing relationships between people, place and other species and entities that are called 'Dreaming'. In this reading, the statement offers a potent challenge to conventional temporal thinking in English. It unsettles English tense boundaries and a Eurocentric notion of time by presenting time as simultaneously past, present and future. It very carefully constructs a cultural landscape that Eurocentric philosophies and most English speakers cannot easily comprehend.

In providing a culturally mediated relationship between foreground and background, between the here-and-now of place and the horizon of space (Hirsch and O'Hanlon, 1995: 4), the idea of landscape offers a metaphor for cultural relationships and processes in space–time, place and scale. Hirsch and O'Hanlon's useful discussion considers the notion of landscape within the discursive space of anthropology. They consider Carter's (1987) account of the colonial encounter in Botany Bay when James Cook's expedition encountered or intruded into the cultural landscapes of Eora people. The different cultural readings engaged in that encounter used different narrative forms to contextualize people and landscape. For the European imperial narrative, exploration, discovery and settlement are the central tropes. The other might be unknown, but the human other was constructed as naturally and inherently inferior (capable of being known and dismissed) and the non-human other, however exotic and bizarre, capable of discovery, exploitation, conquest and acquisition (see also Blaut, 1993).

Although the task of engaging with the reading of this distant encounter by contemporary Eora people is difficult, perhaps impossible, the central trope of the Eora narrative was the Dreaming. Although indigenous peoples' sense of place is often glossed as exemplifying a localized world view, the Dreaming offers a scale metaphor which encompasses the infinite within the immediate. It mediates relationships across space and time at vast scales, while retaining an embodiment and emplacement that is concrete, local and specific. Cultural geography's 'local sense of place' gloss for non-European or non-academic ontologies just will not do in such situations. It reserves the only cultural logic of multiple scales for the imperial, acquisitive European gaze, reducing the question of scale in cultural relations to an underlying economic and political logic that is Eurocentric. In the Dreaming, there is an ethical narrative that establishes a very different relationship between the here-and-now of place and the wider narrative of distant horizons of space, time and social and environmental order. For D. Rose (1996), the Dreaming nurtures the landscape as a nourishing terrain – country. This term in Aboriginal English encompasses people (countrymen), place (homeland) and past, here-and-now and horizon.

NATURE AND CULTURE

In the process, the Dreaming reveals and challenges another Eurocentric construction – the assumed categorical separation of culture and nature. This construction places the human (culture, society) in a binary opposition with the natural (animals, plants, landscapes, seascapes, lightning, thunder, etc.). Within the Judeo-Christian tradition, the creation process itself constructs naturalized boundaries between 'culture' and 'nature'. The human, or to be more specific the hierarchically privileged man, is seen as separate from and more powerful than other 'living creatures' (including woman), in large part owing to his ability to name them. Science produces classificatory systems which distinguish animals from plants and other things (Anderson, 1995; Whatmore and Thorne, 1998). Nader (1996: 3) sees the power of science not only in its naming and categorizing of the world, but also in the way that the label 'science' dismisses other knowledges as inferior or primitive. Science judges itself on its own terms, proclaims itself superior and legitimates its behaviour from within the hall of mirrors.

The characteristics of being able to consciously reason, be rational, and have intent and purpose have been the most pervasive attributes used to externalize society from nature, human from animal and even man from woman (Passmore, 1995; Plumwood, 1995). In reviewing Eurocentric attitudes towards nature and animals, Passmore (1995: 136) identifies two leading traditions in 'modern Western thought':

- One is Cartesian in inspiration, where matter (including nature and animals) is inert and passive, and has no inherent powers of resistance or agency, and humans relate to it in order to reshape and reform it.
- The other is Hegelian, where the human's task is to actualize nature and animals through art, science, philosophy and technology so that nature can be converted into something with which humans can feel 'at home' and not as something from which they are alienated.

Although some contemporary philosophers challenge the hierarchical separation of humans and nature, they often fail to get beyond their own epistemological mirrors. For example, in contrast to the relationship encompassed in the Dreaming, Passmore writes:

No doubt, men, plants, animals, the biosphere form parts of a single community in the ecological sense of the word: each is dependent upon the others for its continued existence. *But this is not the sense of community which generates rights, duties, obligations; men and animals are not involved in a network of responsibilities or a network of mutual concessions.* (1995: 140, our emphasis)

Eurocentric discourses not only classify animals as a part of nature, but also distinguish between categories of wild and domestic or tame:

And the man gave names to the cattle and to the birds of the sky and to all the wild beasts: Genesis 2: 19, 20. (Plaut et al., 1981: 30)

Such categories are easily naturalized in Eurocentric discourse, but they are better understood as reflecting Eurocentric assumptions about evolution and progress (Usher, 1995: 203). Linear notions of progress and development see humans progressing from a state of hunter-gatherer through that of pastoralist to the pinnacle of achievement as agriculturalists. Taming and domesticating wild nature and animals – civilization – is seen as a progression towards more developed forms of society, away from a primitive, wild existence as hunter-gatherers. The opposition of civilized to primitive ascribes characteristics of wild, untamed nature and animals to societies and people. This version of Eurocentrism justified exhibition of humans in 'wild' animal exhibits in some nineteenth-century zoos (Anderson, 1995: 292). Even in the 1950s, a wildlife refuge was created in Botswana (the Central Kalahari Game Reserve) to contain not only wild animals, but also wild Bushmen (Wilmsen, 1995: 222)!

More recently, positive notions of wild and wilderness as an escape, spiritual space and true research domain (or of noble savage as original conservationist or keeper of solutions) romanticize an illusion of a wild based on originality and authenticity, prior to and external from human control and interference. Experiences of removals, evictions, interventions, control and management are silenced and ignored. Langton (1998: 9) discusses the way Aboriginal people and their land management traditions have been rendered invisible by the application of notions of wilderness to Australian landscapes. She refers to this as a 'science fiction' that, like the legal fiction of *terra nullius* (Australia seen as unoccupied land in Eurocentric Australian law) arises from 'the assumption of superiority of Western knowledge over indigenous knowledge systems' (1998: 18).

The nature/culture binary, however, is not universally acknowledged in human thinking. There are multiple, shifting ways of organizing human experience. Christie, for example, reflects on the task of learning Yolngu-matha, the Aboriginal language of north-east Arnhem Land in Australia:

I failed as I struggled mentally to arrange all Yolngu-matha names into a hierarchy. I assumed, for example, that the distinction between 'plant' and 'animal' is a 'natural' one, an ontological distinction, a reality quite independent of human attempts to make sense of the world. But there is no Yolngu-matha word for either 'plant' or 'animal'. (1992: 5)

Similarly, Scott contrasts the embedded and naturalized 'Cartesian myths of the dualities of mind–body, culture–nature' with Cree epistemologies:

In Cree, there is no word corresponding to our term 'nature'. There is a word *pimaatisiiwin* (life), which includes human as well as animal 'persons'. The word for 'person', *iiyiyuu*, can itself be glossed as 'he lives'.

Humans, animals, spirits, and several geophysical agents are perceived to have qualities of personhood. All persons engage in a reciprocally communicative reality. Human persons are not set over and against a material context of inert nature, but rather are one species of person in a network of reciprocating persons. These reciprocative interactions constitute the events of experience. (1996: 72–3)

In characterizing what a non-anthropocentric cosmos looks like, D. Rose discusses the land ethic of Ngarinman and Ngaliwurru people from the Northern Territory in Australia. In contrast to Passmore's (1995: 140) denial of reciprocity, Ngarinman people say 'human life exists within the broader context of a living and conscious cosmos' (D. Rose, 1988: 379). Williams (cited in Langton, 1998: 27) also argues that 'Aboriginal people regard the environment as sentient and as communicating with them.' These statements fly directly in the face of all those knowledges which are based on a belief in the universal application of classifications based on the fact that only humans are conscious beings.

Many Aboriginal people have relationships with specific species of animals at personal and tribal scales. These relationships are based on an underlying understanding that through creation animals and humans were, are and will be inter-related. Dreaming stories inform relationships between humans and animals whereby responsibility to country is based on a common heritage and kinship (D. Rose; Bennett, 1983): 'Animals, they're related to us ... Animals were human before' (Napranum elders in Suchet, 1996: 211). Many indigenous people in Canada also relate to 'animals' in mutually conscious and reciprocal relationships. In Cree notions of knowing, signing and making meaning, animals are an integral part of knowledge systems:

Animal actions, particular qualities and features in the bodies of animals, weather, dream images and events, visions, and religious symbols all fall within the Cree notion of 'sign', with signs constituting knowledge or guidance for actors. Not only humans, but animals and other nonhuman persons send, interpret and respond to signs pertinent to various domains of human action: hunting success or failure, birth and death, and, implicit to these, the circumstances of reciprocity between persons in the world. (Scott, 1996: 73)

In Africa, the Ju/'hoansi of Nyae Nyae in north-eastern Namibia engage in a dialogue with elephants in which elephants:

must participate in the planning for the harvesting of *Marula* [a tree fruit] ... 'We share the resource with elephants and we have decided together, that is the elephants and us, which tree is to be used by the

Ju/'hoansi and which by the elephant.' (/Ai!ae/Oma, cited in Powell, 1998: 47)

African discourses also unsettle the naturalization of the categories wild, domestic and tame:

Animals have always helped us, and they still do ... The wild animals and the tame ones are the same to us. (respected elder Baha Mhlanga from Mahenye, cited in Hove and Trojanow, 1996: 22–9)

Treating binaries such as society/nature, human/animal and domestic/wild as self-evident epistemological givens, naturalizes assertions and impositions of power and control. The 'epistemological transformation' (Esteva, 1987: 138) that occurs when nature, wildlife and 'wild humans' are constructed as *resources*, legitimates the assertion and imposition of Eurocentric practices of management – intervening, taming, domesticating, controlling, subduing and dominating the wild-wildlife-wilderness. Managerial intervention, however, is not how all people interact with worlds. D. Rose (1988) argues that Ngarinman and Ngaliwurru people are hesitant to intervene in ecological processes. She notes that for many people 'non-intervention is frequently a virtue'. A survey of perceptions of Aboriginal people in Central Australia similarly highlights knowledges that are not based on beliefs in intervention and overt control:

Many people expressed a sense of loss that the [locally extinct 'native'] animals were no longer around but there was also a pervading sense of passive acceptance about what had happened. Rather than question why the animals had gone and then attempt to act to bring them back, Aboriginal people accept what they perceive as a change in circumstances which is beyond their control. (B. Rose, 1995: 95)

In Canada, what scientists perceive as population declines are understood very differently by the Inuit:

Elders say that any kind of animal moves away for a while but, according to the government, animals are in decline. To the Inuit, they have moved, but not declined. (Peter Alogut, cited in Freeman, 1999: 8)

[M]any Inuit do not believe that 'wildlife' can be 'harvested', 'managed' or 'conserved' as 'stocks' or 'populations'. Many of these concepts have no basis in Inuit reality. (Stevenson, 1996: 8)

Aboriginal people at Napranum on Cape York Peninsula insist they have their own way of relating to country, and this may or may not fit into Eurocentric ideas about what land and resource management should be. They often express anger and frustration at not having their ways recognized and respected by other people and cultures who only judge on their own values and priorities: 'Use your commonsense, but usually there's different commonsense' (in Suchet, 1999: 238).

Concepts and practices of management play an integral role in colonizing processes. The development and conservation of resources have been asserted and imposed through management mechanisms such as sovereignty, ownership, laws, institutions, scientific research, etc. The tension between indigenous groups' nominal ownership of their territories, and its appropriation through colonization processes embedded deep in management regimes, is a hotly contested political process in many places.

In the same way that exercising rights and responsibilities to care for (and to be cared for by) country are reconstituted in Eurocentric discourses as 'environmental' or 'wildlife management', and the ontological primacy of the human domain at the top of the hierarchical chain of being is surreptitiously embedded in 'management systems', discourses of human management have also harnessed efforts to liberate the objects of injustice and oppression to regressive structures of discipline and power. This can see indigenous self-determination reconstituted as 'community management'. Rendered invisible are processes of dispossession, theft and genocide (see Tatz, 1998; 1999) that produced what the Aboriginal affairs industry reconstitutes as 'communities', as well as assertions of sovereignty and identity and aspirations of being-in-place on one's own terms.

Within this nature- and management-centred view of change, the persistence of indigenous rights is seen as simply another element to be managed, another tool in the manager's toolkit. The notion that it is not only residual rights that persist, but epistemological systems, value systems, cultural institutions, systems of customary law, and deeply entrenched ways of being-in-place is only dimly glimpsed in the nature and management speak of so-called postcolonial discourse. In many places, diverse elements of indigenous society, economy and ecology continue to shape everyday life for large groups of people. However, ideology disciplines social change to conform to existing patterns, forms and explanations.

IDENTITY AND SUBJECTIVITY

This leads us more or less directly to Levinas and his consideration of otherness, lived experience and ethics. For Levinas (for example, 1989), the ambiguity of such spaces reflects the self–other relationship, which he sees as foundational in human existence. Levinas represents relations between the self and the other in terms of an ethical imperative in which the face-to-face encounter between the self and the other develops terms for understanding one's place in society. 'Intersubjective space,' he writes – that space in which one relates to the other(s) – 'is not symmetrical' (1989: 48). For him, this intersubjective space is a moral space. We occupy moral landscapes in which ethics (responsibility, reciprocity, proximity, collectivity and coexistence) frame and temper interpersonal, structural and political relationships. Cultural landscapes are, therefore, to be understood as simultaneously material and metaphorical.

Levinas' writing disrupts common binaries that underpin western philosophy's hall of mirrors. Even the self–other binary is disrupted to establish relationships between the self, the 'other that is like me' and the 'entirely other'. Levinas grapples to establish terms for engaging with relationships between entities whose coexistence is not reducible to a larger unity. In 'Time and the other', he asserts that 'existence is pluralist'. A plurality, he writes, 'insinuates itself into the very existing of the existent' (1989: 43). In grappling with the self–other relation, which is always contextualized 'because we are always immersed in the empirical world' (1989: 43), he targets that which is, by convention, unscaleable and immeasurable – the infinite.

In cultural geography, the shift between one-to-one, one-to-many and many-to-many is always troubling. Once this shift is spatialized, it clearly implicates the notion of scales and scale shifting. The link between psychoanalysis, psychology and sociology, however, is not reducible to a measurement or a formula. While individuals, social formations and cultural groupings may be mutually influential, there is no predetermined causal link between them. One simply cannot predict individual behaviour from the knowledge of social behaviour or cultural values, or vice versa. Neither can one read off from large structures the details of small events and processes, whether past, present or future (see Storper, 1988). This conundrum leads to many studies that deal with different scales as autonomous spheres of social action – as bounded domains. Yet such domains do interact. We do shift between geographical scales and sociological levels. Cultural geographers have begun to engage with issues of scale, but have generally avoided the metaphysical scale (the infinite), limiting their scope to a local–global binary.[1]

Geographical endeavours must not only challenge the fundamental building blocks of thought and understanding, but be constantly vigilant to the fundamental Eurocentric assumptions underlying the researching and writing of geographical knowledge. For example, it is easy for cultural geography to be captured in the hall of mirrors by representing complex worlds and multiple knowledges within its own terms. Bhabha warns that:

> Western connoisseurship is the capacity to understand and locate cultures in a universal time-frame that acknowledges their various historical and social contexts only to eventually transcend them and render them transparent. (1990: 208)

Always bound by one's own epistemological understandings, it can be difficult to envision, let alone adequately portray, other ways of knowing. As Christie states:

> most of us who have been counting ever since we can remember can have little hope of imagining what a world could look like in which reality is unquantifiable (1992: 2).

It is therefore necessary to critically consider the spatial and temporal settings examined in and formed by geographical knowledge. This can be a challenging task. Soja describes the 'linguistic despair' felt when:

> What one sees when one looks at geographies is stubbornly simultaneous, but language dictates a sequential succession, a linear flow of sentential statements bounded by the most spatial of earthly constraints, the impossibility of two objects (or words) occupying the same place (as on a page). (1989: 1–2)

SITUATED ENGAGEMENT: JUSTICE, COEXISTENCE AND OTHERNESS

Rising to the challenge of social justice solely by harnessing the tools of Eurocentric knowledges risks reinforcing colonizing relationships. Cultural geography's postcolonial projects often draw on notions of moving towards something better. Yet such formulations subtly reinforce the almost invisible epistemology of developmentalism, orienting thinking towards a linear narrative. The implicit symbolism is about direction, progression and control – exactly what this chapter seeks to challenge and unsettle. Le Guin (1989) undertakes a similar exercise of unsettling. She suggests that 'through long practice I know how to tell a story, but I'm not sure I know what a story is' (1989: 37). Her discussion of writing science fiction unsettles the assumption

that Eurocentric discourses can simply make the world as we wish it to be. She asks her readers to begin to see that every remote place is simultaneously somebody else's homeland. In her image of 'dancing at the edge of the world', she offers an escape from the tyranny of the linear narratives of developmentalism, and glimpses the seasonal and cyclical patterns of time's circle embedded in people–people and people–environment relationships and processes, alongside time's arrow. In such images, there are opportunities to rethink the epistemological foundations that are conventionally used to shape and reshape geographical imaginations so that they may be woven in ways that acknowledge and include those knowledges that are so often rendered invisible.

The idea of 'situated engagement' (Suchet, 1999) offers a way out of the confines of Eurocentric discourses, which so often render even well-intentioned cultural geographers relatively tongueless and earless in dealing with ontological pluralism. Simply acknowledging the existence and possibility of multiple knowledges is not enough. The ontologies of other peoples need to be understood and engaged with in active partnerships in the construction of knowledge (and power). New interactions and relationships open new possibilities:

> At the margins, within the domain of the 'other', one knows that the world, life and people express themselves with rich and interactive presences that are invisible from the viewpoint of deformed power, except, perhaps, as disorder or blockage. The dismantling of this oppressive and damaging pole is a necessary step in moving toward dialogue. Dismantling will fail if it is confined to monologue; we must embrace noisy and unruly processes capable of finding dialogue with the peoples of the world and with the world itself. (D. Rose, 1999: 177)

In these noisy and unruly spaces it is necessary to reconsider the implications of ontological pluralism. Boundaries around concepts can no longer be concrete, impenetrable no person's lands. Rather, they become blurry, fluid, complex, interacting and multiple. As with the metaphor of edges in the constantly shifting and changing tidal zone, boundaries and relationships are conceived as constructive places which 'entwine and interpenetrate in a complex and fecund embrace of coexistence' (Howitt, 2001a).

The discursive spaces of ontological pluralism are similar to what Bhabha calls third space – 'that position of liminality, in the productive space of the construction of culture as difference, in the spirit of alterity or otherness' (1990: 209). Others have embraced this sort of space. Lavie and Swedenburg (1996: 154, 174), for example, move beyond examining texts to situate their exploration of the boundaries of culture in the everyday, a 'terrain of practice and theory'. In her exploration of the postcolonial in the modern city, Jacobs uses the space of the contemporary city to embrace the 'unstable negotiation of identity and power' (1996: xi). She calls this space the 'edge'; 'the "unsafe" margin which marks not only a space of openness but also the very negotiation of space itself'.

Having challenged the hall of mirrors with its own contradictions, and through glimpses into multiple knowledges, we will now further explore the notion of situated engagement. Drawing on D. Rose's (1999) Levinasian notion of 'situated availability' and Jacobs and Mulvihill's (1995: 9) concept of 'viable interdependence', situated engagement is introduced as an approach which encourages noisy and unruly engagement in situated, interacting material, discursive and conceptual places. Situated engagement opens up these places in an ethical sense so that everyone's ground is destabilized and everyone expects to be surprised, challenged and changed (D. Rose, 1999). In a practical sense, self-reliance and equitable sharing are celebrated (Jacobs and Mulvihill, 1995: 9). Engaging (conversing, interacting, thinking, doing) therefore moves into the realm of considering not only how knowledges form, but also how they interact and how this matters.

From within the hall of mirrors it is almost impossible to imagine talking, thinking, writing, doing, smelling, imagining and realizing worlds without 'law', 'spaces', 'places', 'time', 'scale', 'nature' and 'self'. However, local and indigenous communities are doing this as they construct processes, experiences, thoughts and actions. In this diversity of experiences, there can be no singular, correct, model, process, alternative or notion of resistance, empowerment and decolonization that can apply globally. As Escobar argues:

instead of searching for grand alternative models or strategies, what is needed is the investigation of alternative representations and practices in concrete local settings ... One must ... resist the desire to formulate alternatives at an abstract, macro level. (1995: 19, 222)

If nothing can be generalized or universalized, how does one deal with diversity and multiplicity? By focusing, contextualizing and positioning in terms of specific material, conceptual and discursive places, one can practically and actively engage with and recognize diversity. What is vital is recognizing that these places are crossed, permeated and saturated with knowledges and experiences that the hall of mirrors is constantly straddling, conflicting, parallelling and/or not touching. It is necessary to simultaneously reach in, reach out and reach across (Ellis, 1998), so that one can recognize and engage with these processes, experiences, discourses, systems and structures. This avoids reinforcing and recentring practices and concepts within the constraints of the hall of mirrors, within the globalizing essentialism of the colonial and postcolonial narrative.

In the context of the confrontation of contested epistemologies, local and indigenous epistemologies confront multiple permutations of knowledge–power relations constructed by the hall of mirrors. Concurrently, within diverse local and indigenous settings, multiple individual and collective identities are formed and reformed. In terms of the analysis proposed here, reconstructing this multiplicity and complexity into a set of experiences relevant to specific situations can be neither solely abstract nor solely empirical. Rather, this occurs in the interface and interplay of discursive, material and conceptual spaces that occur in contextualized interactions and dialogue – in situated engagement. Communication and interactions aimed at breaking down assumptions and recognizing diversity and multiple knowledges are vital because:

The difference of cultures cannot be something that can be accommodated within a universalist framework ... The assumption that at some level all forms of cultural diversity may be understood on the basis of a particular universal concept, whether it be 'human being', 'class' or 'race', can be both very dangerous and very limiting in trying to understand the ways in which cultural practices construct their own systems of meaning and social organisation. (Bhabha, 1990: 209)

Fothergill's analysis of *Heart of Darkness* argues proximity challenges stereotyping:

when Marlow specifies the African subject's historical or political context, the representation tends to be critical of typical European representations. When Marlow erases the specific context, the representation tends to endorse the stereotype. (1992: 50)

However, Fothergill also warns that an engagement with the specific that brings with it its own cultural assumptions and projects these onto the 'other' can reinforce stereotypes by denying difference. Strategies to empower the 'colonized victim' often simply invert relationships. These 'new' relationships are still based on the same beliefs, with the local, traditional, community or indigenous represented as separate and contained, yet this time superior and progressive. Dialogue remains closed with the other's silence further reinforced behind mirrors of romantic stereotyping. Romancing unproblematized categories is dangerous and reinforces colonizing relationships as assumptions of universality are not challenged. McDowell draws upon hooks to argue that:

> it is not possible to merely invert or reverse old categories, rather we have to decolonize our minds and construct new alternatives. She [hooks] suggests that women and people of colour cannot possibly be immune from hegemonic notions of knowledge. There is no position outside the social construction of knowledge where an unsullied 'other' might speak from. 'Others' too have internalized that set of Western philosophical dualist concepts that structure knowledge – internalized and, often, inverted the dualisms, reluctant to consider the possibility that work is not necessary oppositional because it is created by women. (1992a: 411)

Esteva talks about the need to challenge and transcend the assertion and imposition of universal values by solidly grounding values in the experiences of 'daily life', of situated places:

> I am now more than convinced that if one fully accepts cultural relativism … one must also accept its consequences, i.e. the dissolution of universal values. This does not mean, of course, having no guiding principles to live in community. It means exactly the opposite, having them fully rooted in the perception and attitudes of daily life, instead of supplanting them with artificial constructs which are hypothetically universal and more or less ahistorical. (1987: 138)

CONCLUSION: SHATTERED MIRRORS AND REFLECTION ON REFLECTIONS

This chapter has argued that cultural geography often relies on Eurocentric discourses. Even self-consciously postcolonial discourses reflect and reinforce pervasive channels of power, such as education, research and governance, to privilege and reprivilege Eurocentric ontologies against diverse local, indigenous and non-Eurocentric traditions. Weaving together field experience and secondary literature, it has been argued that these hegemonic discourses construct a hall of mirrors. Many of the most problematic human and environmental relationships of contemporary experience are constructed within this hall of mirrors.

We have argued that cultural geography must consider what is involved in moving beyond this hall of mirrors. Shattering the mirrors and stepping beyond the solipsistic monologue of Eurocentric discourses is necessary, but not sufficient. Cultural geographers face a difficult balancing act in simultaneously nurturing one's expertise, and minimizing its value *per se*. We have advocated situated engagement as an approach that allows the discipline to address this. In the first instance, this dislodges Eurocentric knowledges from the frame of universalized constructs. Without this frame (or with it constructed as an assumption subject to further consideration) the hall of mirrors is less secure and one *has* to begin differently – to approach empirical questions, value questions and methodological questions as a coinvestigator with non-technical experts within the relevant local or indigenous groups. To do this requires consideration of histories, geographies, languages and powers; it involves simultaneously reaching in, reaching out and reaching across from the hall of mirrors – and in the process it reveals the logical flaws within the looking glass. We have aimed to open a discursive space that reaches out, across and into a wider discursive community. This engagement opens up situated, interrelated conceptual and discursive places and allows ideas, knowledges and thoughts to be recognized and understood.

The image of shattering the mirrors signals the urgent need to recognize the limits of Eurocentric knowledges. On reflection, shattering is perhaps too violent a response to the violations involved in colonizing processes. Perhaps an image of transforming mirrors into windows is more suitable. This would allow knowledges to remain embodied and emplaced (rather than lacerated by splintered glass), but perspectives gained from looking out of windows and seeing multiple knowledges would decentre the assumption that any single knowledge system is superior or universal. But even this metaphorical transformation is too limited. Looking out of windows is grossly inadequate as a basis for reconciling ontological diversity in real social, political and intercultural relations. Windows need to be opened. To allow a breath of fresh air in, these windows must stay open. This will encourage people to actively and intimately reach in, reach out and reach across to engage with each other in embodied and emplaced ways. Opening windows not only allows an engagement with other knowledges, but also opens a window to the soul as one engages on a personal level with one's own knowledges and understandings. Such reflection on one's own position shows that it is impossible, and counterproductive, to aim for complete empathy with all knowledges:

> The means by which we come to know the unknown Other will always be determined by our own terms of

reference, our own horizon of understanding. (Fothergill, 1992: 38–9)

Instead of attempting to induct everyone within an all-knowing gaze, situated engagement offers a way not only to identify differences, but also to celebrate and revel in their limits, tensions and transformative energy.

NOTES

We wish to thank Venessa Kealy, Emma Ignjic, Leah Gibbs, Elisabeth Ellis – especially for her discussion of 'reaching in, reaching out and reaching across' – Debbie Rose, Bob Fagan and Robyn Dowling for their contributions of ideas, debate and criticism that have so shaped our own nurturing and challenging discursive community.

1 For an expanded discussion of the relevance of the work of Levinas to cultural geography, see Howitt (2002). More generally, see Howitt (forthcoming) for a discussion of recent debates about geographical scale.

REFERENCES

Anderson, K. (1995) 'Culture and nature at the Adelaide Zoo: at the frontiers of "human" geography', *Transactions of the Institute of British Geographers* 20 (NS): 275–94.

Bennett, D.H. (1983) 'Some aspects of Aboriginal and non-Aboriginal notions of responsibility to non-human animals', *Australian Aboriginal Studies* 2: 19–24.

Bhabha, H. (1990) 'Interview with Homi Bhabha: the third space', in J. Rutherford (ed.) *Identity: Community, Culture, Difference*. London: Lawrence and Wishart. pp. 207–21.

Blaut, J.M. (1993) *The Colonizer's Model of the World: Geographical Diffusionism and Eurocentric History*. New York: Guilford.

Blunt, A. and Rose, G. (1994) 'Introduction: women's colonial and postcolonial geographies', in *Writing Women and Space: Colonial and Postcolonial Geographies*. New York: Guilford. pp. 1–25.

Carter, P. (1987) *The Road to Botany Bay: An Essay in Spatial History*. London: Faber and Faber.

Christie, M. (1992) 'Grounded and ex-centric knowledges: exploring Aboriginal alternatives to western thinking'. Paper presented at the Conference on Thinking, Townsville, July 1992.

Clifford, J. (1997) 'Spatial practices: fieldwork, travel, and the disciplining of anthropology', in A. Gupta and J. Ferguson (eds) *Anthropological Locations: Boundaries and Grounds of a Field Science*. Berkeley: University of California Press. pp. 185–222.

Crang, P. (1992) 'The politics of polyphony: reconfigurations in geographical authority', *Environment and Planning D: Society and Space* 10: 527–49.

Driver, F. (1992) 'Geography's empire: histories of geographical knowledge', *Environment and Planning D: Society and Space* 10: 23–40.

Driver, F. (2000) *Geography Militant: Cultures of Exploration and Empire*. Oxford: Blackwell.

Ellis, E. (1998) 'Reaching out/reaching in/reaching across: identity and the spatial politics of disablement and enablement. Unpublished BA (Hons) Thesis, Macquarie University, Sydney.

Escobar, A. (1995) *Encountering Development: The Making and Unmaking of the Third World*. Princeton: Princeton University Press.

Esteva, G. (1987) 'Regenerating people's space', *Alternatives* 12 (1): 125–52.

Fothergill, A. (1992) 'Of Conrad, cannibals, and kin', in M. Gidley (ed.) *Representing Others: White Views of Indigenous Peoples*. Exeter: University of Exeter Press. pp. 37–59.

Freeman, M.M.R. (1999) '"They knew how much to take": respect and reciprocity in Arctic sustainable use strategies'. Paper presented at the 1999 International Symposium on Society and Resource Management, Brisbane, Australia, 7–10 July 1999.

Godlewska, A. and Smith, N. (1994) 'Introduction: critical histories of geography', in A. Godlewska and N. Smith (eds) *Geography and Empire*. Oxford: Blackwell. pp. 1–8.

Hirsch, E. and O'Hanlon, M. (eds) (1995) *The Anthropology of Landscape: Perspectives on Place and Space*. Oxford: Clarendon.

Hove, C. and Trojanow, I. (1996) *Guardians of the Soil: Meeting Zimbabwe's Elders*. Harare: Baobab.

Howitt, R. (1998) 'Scale as relation: musical metaphors of geographical scale', *Area* 30 (1): 49–58.

Howitt, R. (1993) '"A world in a grain of sand": towards a reconceptualisation of geographical scale', *Australian Geographer* 24 (1): 33–44.

Howitt, R. (2001a) 'Frontiers, borders, edges: liminal challenges to the hegemony of exclusion', *Australian Geographical Studies* 39 (2): 233–45.

Howitt, R. (2001b) *Rethinking Resource Management: Sustainability, Justice and Indigenous Peoples*. London: Routledge.

Howitt, R. (2002) 'Scale and the other: Levinas and geography', *Geoforum* 33 (3): 299–313.

Howitt, R. (forthcoming) 'Scale', in J. Agnew, K. Mitchell and G. Ó Tuathail (eds) *A Companion to Political Geography*. Oxford: Blackwell.

Howitt, R. and Jackson, S. (1998) '"Some things do change": indigenous rights, geographers and geography in Australia', *Australian Geographer* 29 (2): 155–73.

Irigaray, L. (1985a) *The Sex Which Is Not One*. Ithaca: Cornell University Press.

Irigaray, L. (1985b) *Speculum of the Other Woman*. Ithaca: Cornell University Press.

Jacobs, J.M. (1996) *Edge of Empire: Postcolonialism and the City*. London: Routledge.

Jacobs, P. and Mulvihill, P. (1995) 'Ancient lands: new perspectives. Towards multi-cultural literacy in landscape management', *Landscape and Urban Planning* 32: 7–17.

Katz, C. (1992) 'All the world is staged: intellectuals and the projects of ethnography', *Environment and Planning D: Society and Space* 10: 495–510.

Keith, M. (1992) 'Angry writing: (re)presenting the unethical world of the ethnographer', *Environment and Planning D: Society and Space* 10: 551–68.

Knudtson, P. and Suzuki, D. (1992) *Wisdom of the Elders.* Sydney: Allen & Unwin.

Langton, M. (1998) *Burning Questions: Emerging Environmental Issues for Indigenous Peoples in Northern Australia.* Centre for Indigenous Natural and Cultural Resource Management, Northern Territory University, Darwin.

Lavie, S. and Swedenburg, T. (1996) 'Between and among the boundaries of culture: bridging text and lived experience in the third timespace', *Cultural Studies* 10 (1): 154–79.

Le Guin, U. (1989) *Dancing at the Edge of the World: Thought on Words, Women, Places.* London: Gollancz.

Levinas, E. (1989) 'Time and the other' (1947), in D. Hand (ed.) *The Levinas Reader.* Oxford: Blackwell. pp. 37–58.

Massey, D. (1984) 'Introduction: geography matters', in D. Massey and J. Allen (eds) *Geography Matters! A Reader.* Cambridge: Cambridge University Press. pp. 1–11.

McDowell, L. (1992a) 'Doing gender: feminism, feminists and research methods in human geography', *Transactions of the Institute of British Geographers* 17: 399–416.

McDowell, L. (1992b) 'Multiple voices: speaking from inside and outside "The Project"' *Antipode* 24: 56–72.

McDowell, L. (1994) 'Polyphony and pedagogic authority', *Area* 26 (3): 241–8.

Moore, D. (1998) 'Subaltern struggles and the politics of place: remapping resistance in Zimbabwe's Eastern Highlands', *Cultural Anthropology* 13 (3): 344–81.

Nader, L. (1996) 'Anthropological inquiry into boundaries, power, and knowledge', in L. Nader (ed.) *Naked Science: Anthropological Inquiry into Boundaries, Power, Knowledge.* London: Routledge. pp. 1–23.

Passmore, J. (1995) 'Attitudes to nature', in R. Elliot (ed.) *Environmental Ethics.* New York: Oxford University Press, pp. 129–41.

Plaut, W.G., Bamberger, B.L. and Hallo, W.W. (1981) *The Torah: A Modern Commentary. New York:* Union of American Hebrew Congregations.

Plumwood, V. (1995) 'Nature, self, and gender: feminism, environmental philosophy, and the critique of rationalism', in R. Elliot (ed.) *Environmental Ethics.* New York: Oxford University Press. pp. 155–64.

Powell, N. (1998) 'Co-management in non-equilibrium systems: cases from Namibian rangelands'. Doctoral thesis, Swedish University of Agricultural Sciences, Uppsala.

Rose, B. (1995) *Land Management Issues: Attitudes and Perceptions amongst Aboriginal People of Central Australia.* Alice Springs: Central Land Council.

Rose, D.B. (1988) 'Exploring an Aboriginal land ethic', *Meanjin* 47 (3): 378–86.

Rose, D.B. (1996) *Nourishing Terrains: Australian Aboriginal Views of Landscape and Wilderness.* Canberra: Australian Heritage Commission.

Rose, D.B. (1999) 'Indigenous ecologies and an ethic of connection', in N. Low (ed.) *Global Ethics and Environment.* London: Routledge. pp. 175–87.

Rose, G. (1996) 'As if the mirrors had bled: masculine dwelling, masculinist theory and feminist masquerade', in N. Duncan (ed.) *Bodyspace: Destabilizing Geographies of Gender and Sexuality.* London: Routledge. pp. 56–74.

Scott, C. (1996) 'Science for the west, myth for the rest? The case of James Bay Cree knowledge construction', in L. Nader (ed.) *Naked Science: Anthropological Inquiry into Boundaries, Power, Knowledge.* London: Routledge. pp. 69–86.

Soja, E.W. (1989) *Postmodern Geographies: The Reassertion of Space in Critical Social Theory.* London: Verso.

Stanner, W.E.H. (1979) *White Man Got No Dreaming: Essays 1938–1973.* Canberra: Australian National University Press.

Stevenson, M. (1996) 'In search of Inuit ecological knowledge: a protocol for its collection, interpretation and use. A discussion paper'. On behalf of the Hunters and Trappers Associations of the Qikiqtaaluk Region of Nunavut. Prepared for The Department of Renewable Resources, GNWT, Qikiqtaaluk Wildlife Board and Parks Canada.

Storper, M. (1988) 'Big structures, small events, and large processes in economic geography', *Environment and Planning A* 20: 165–85.

Suchet, S. (1996) 'Nurturing culture through country: resource management strategies and aspirations of local landowning families at Napranum', *Australian Geographical Studies* 34 (2): 200–15.

Suchet, S. (1999) 'Situated engagement: a critique of wildlife management and post-colonial discourse'. PhD thesis, Department of Human Geography, Macquarie University, Sydney.

Tatz, C. (1998) 'The reconciliation "Bargain"', *Melbourne Journal of Politics*, special issue on reconciliation, 25: 1–8.

Tatz, C. (1999) *Genocide in Australia.* Canberra: Australian Institute of Aboriginal and Torres Strait Islander Studies.

Usher, P.J. (1995) 'Comanagement of natural resources: some aspects of the Canadian experience', in D.L. Peterson, and D.R. Johnson (eds) *Human Ecology and Climate Change: People and Resources in the Far North.* Washington: Taylor and Francis. pp. 197–206.

Whatmore, S. and Thorne, L. (1998) 'Wild(er)ness: reconfiguring the geographies of wildlife', *Transactions of the Institute of British Geographers* 23 (4): 435–54.

Wilmsen, E.N. (1995) 'Primitive politics in sanctified landscapes: the ethnographic fiction of Laurens van der Post', *Journal of Southern African Studies* 21 (2): 201–23.

Index